Climate Change 1995

The Science of Climate Change

Climate Change 1995

The Science of Climate Change

**Edited by J.T. Houghton,
L.G. Meira Filho, B.A. Callander, N. Harris,
A. Kattenberg and K. Maskell**

Production Editor: J.A. Lakeman

Contribution of WGI to the Second Assessment Report
of the Intergovernmental Panel on Climate Change

Published for the Intergovernmental Panel on Climate Change

CAMBRIDGE
UNIVERSITY PRESS

Published by the Press Syndicate of the University of Cambridge
The Pitt Building, Trumpington Street, Cambridge CB2 1RP
40 West 20th Street, New York, NY 10011–4211, USA
10 Stamford Road, Oakleigh, Melbourne 3166, Australia

First published 1996

Printed in Great Britain at the University Press, Cambridge

A catalogue record for this book is available from the British Library

Library of Congress cataloguing in publication data available

ISBN 0 521 56433 6 hardback
ISBN 0 521 56436 0 paperback

GE

Contents

Foreword

The Intergovernmental Panel on Climate Change (IPCC) was jointly established by the World Meteorological Organisation and the United Nations Environment Programme in 1988, in order to: (i) assess available scientific information on climate change, (ii) assess the environmental and socio-economic impacts of climate change, and (iii) formulate response strategies. The IPCC First Assessment Report was completed in August 1990 and served as the basis for negotiating the UN Framework Convention on Climate Change. The IPCC also completed its 1992 Supplement and "Climate Change 1994: Radiative Forcing of Climate Change and An Evaluation of the IPCC IS92 Emission Scenarios" to assist the Convention process further.

In 1992, the Panel reorganised its Working Groups II and III and committed itself to complete a Second Assessment in 1995, not only updating the information on the same range of topics as in the First Assessment, but also including the new subject area of technical issues related to the economic aspects of climate change. We applaud the IPCC for producing its Second Assessment Report (SAR) as scheduled. We are convinced that the SAR, like the earlier IPCC Reports, will become a standard work of reference, widely used by policymakers, scientists and other experts.

This volume, which forms part of the SAR, has been produced by Working Group I of the IPCC, and focuses on the science of climate change. It consists of 11 chapters covering the physical climate system, the factors that drive climate change, analyses of past climate and projections of future climate change, and detection and attribution of human influence on recent climate.

As usual in the IPCC, success in producing this report has depended upon the enthusiasm and co-operation of numerous busy scientists and other experts world-wide. We are exceedingly pleased to note here the very special efforts made by the IPCC in ensuring the participation of scientists and other experts in its activities, in particular in the writing, reviewing, and revising of its reports. The scientists and experts from the developed, developing and transitional economy countries have given of their time very generously, and governments have supported them, in the enormous intellectual and physical effort required, often going substantially beyond reasonable demands of duty. Without such conscientious and professional involvement, the IPCC would be greatly impoverished. We express to all these scientists and experts, and the governments who supported them, our sincere appreciation for their commitment.

We take this opportunity to express our gratitude to the following individuals for nurturing another IPCC report through to a successful completion:

Prof. Bolin, the Chairman of the IPCC, for his able leadership and skilful guidance of the IPCC; the Co-Chairmen of Working Group I, Sir John Houghton (United Kingdom) and Dr. L.G. Meira Filho (Brazil); the Vice-Chairmen of the Working Group, Prof. Ding Yihui (China), Mr. A.B. Diop (Senegal) and Prof. D. Ehhalt (Germany);

Dr. B.A. Callander, the Head of the Technical Support Unit of the Working Group and his staff, Ms. K. Maskell, Mrs. J. A. Lakeman and Mrs. F. Mills, with additional assistance from Dr. N. Harris (European Ozone Research Co-ordinating Unit, Cambridge), Dr. A. Kattenberg (Royal Netherlands Meteorological Institute); and Dr. N. Sundararaman, the Secretary of the IPCC and his staff including Mr. S. Tewungwa, Mrs. R. Bourgeois, Ms. C. Ettori and Ms. C. Tanikie.

G.O.P. Obasi
Secretary-General
World Meteorological Organisation

Ms. E. Dowdeswell
Executive Director
United Nations Environment Programme

Climate Change 1995

The Science of Climate Change

Prepared by Working Group 1

IPCC reports are formally described as "approved" or "accepted". An "approved" report has been subject to detailed, line-by-line discussion and agreement in a plenary meeting of the relevant IPCC Working Group. For practical reasons only short documents can be formally approved, and larger documents are "accepted" by the Working Group, signifying its view that a report presents a comprehensive, objective and balanced view of the subject matter. In this report, the Summary for Policymakers has been approved, and the Technical Summary and Chapters 1 to 11 have been accepted, by Working Group I.

MAKING REFERENCE TO THIS REPORT

When citing information from a specific chapter of this report (as distinct from the Policymakers' or Technical Summary) please give proper credit to the Lead Authors. Do not simply refer to IPCC (1995) or the editors.

Preface

This report is the most comprehensive assessment of the science of climate change since Working Group I (WGI) of the IPCC produced its first report *Climate Change: The IPCC Scientific Assessment* in 1990. It enlarges and updates information contained in that assessment and also in the interim reports produced by WGI in 1992 and 1994. The first IPCC Assessment Report of 1990 concluded that continued accumulation of anthropogenic greenhouse gases in the atmosphere would lead to climate change whose rate and magnitude were likely to have important impacts on natural and human systems. The IPCC Supplementary Report of 1992, timed to coincide with the final negotiations of the United Nations Framework Convention on Climate Change in Rio de Janeiro (June 1992), added new quantitative information on the climatic effects of aerosols but confirmed the essential conclusions of the 1990 assessment concerning our understanding of climate and the factors affecting it. The 1994 WGI report *Radiative Forcing of Climate Change* examined in depth the mechanisms that govern the relative importance of human and natural factors in giving rise to radiative forcing, the "driver" of climate change. The 1994 report incorporated further advances in the quantification of the climatic effects of aerosols, but it also found no reasons to alter in any fundamental way those conclusions of the 1990 report which it addressed.

We believe the essential message of this report continues to be that the basic understanding of climate change and the human role therein, as expressed in the 1990 report, still holds: carbon dioxide remains the most important contributor to anthropogenic forcing of climate change; projections of future global mean temperature change and sea level rise confirm the potential for human activities to alter the Earth's climate to an extent unprecedented in human history; and the long time-scales governing both the accumulation of greenhouse gases in the atmosphere and the response of the climate system to those accumulations, means that many important aspects of climate change are effectively irreversible. Further, that observations suggest "a discernible human influence on global climate", one of the key findings of this report, adds an important new dimension to the discussion of the climate change issue.

An important political development since 1990 has been the entry into force of the UN Framework Convention on Climate Change (FCCC). IPCC is recognised as a prime source of scientific and technical information to the FCCC, and the underlying aim of this report is to provide objective information on which to base global climate change policies that will meet the ultimate aim of the FCCC – expressed in Article 2 of the Convention – of stabilisation of greenhouse gases at some level that has yet to be quantified but which is defined as one that will "prevent dangerous anthropogenic interference with the climate system". Because the definition of "dangerous" will depend on value judgements as well as upon observable physical changes in the climate system, such policies will not rest on purely scientific grounds, and the companion IPCC reports by WGII on *Impacts, Adaptations and Mitigation of Climate Change*, and by WGIII on *Economic and Social Dimensions of Climate Change* provide some of the background information on which the wider debate will be based. Together the three WG reports establish a basis for an IPCC synthesis of information relevant to interpreting Article 2 of the FCCC. An important contribution of WGI to this synthesis has been an analysis of the emission pathways for carbon dioxide that would lead to a range of hypothetical stabilisation levels.

This report was compiled between October 1994 and November 1995 by 78 lead authors from 20 countries. First drafts of the chapters were circulated for informal review by experts in early 1995, before further revision in March. At that time, drafts of the Summary for Policymakers and the Technical Summary were also prepared by the lead authors assisted by a few additional experts with experience of the science-policy interface. Formal review of the chapters and the summaries by governments, non-governmental organisations (NGOs) and individual experts took place during May to July. Over 400 contributing authors from 26 countries submitted draft text and information to the lead authors and over 500 reviewers from 40 countries submitted valuable suggestions for improvement during the review process. The hundreds of comments received were carefully analysed and assimilated in a revised document that was distributed to countries and NGOs six weeks in advance of the fifth session of WGI in Madrid, 27-29 November 1995. There, the Summary for Policymakers was approved in detail and the rest of the report accepted. Participants included 177 delegates from 96 countries, representatives from 14 NGOs and 28 lead authors.

We wish first of all to express our sincere appreciation to the lead authors whose expertise, diligence and patience have underpinned the successful completion of this effort, and to the many contributors and reviewers for their valuable and painstaking work. We are grateful to the governments of Sweden, UK and USA which hosted drafting sessions in their countries, and to the government of Spain which hosted the final session of Working Group I in Madrid at which the documents were accepted and approved. The IPCC Trust Fund, contributed to by many countries, supported the participation of many developing country scientists in the completion of this report. The WGI Technical Support Unit was funded by the UK government with assistance from the Netherlands, and we echo the appreciation expressed in the Foreword to the members of the Technical Support Unit. The graphics section of the UK Meteorological Office provided skilful assistance in the preparation of the hundreds of diagrams of this report.

Bert Bolin
IPCC chairman

John Houghton
Co-chair (UK) IPCC WGI

L. Gylvan Meira Filho
Co-chair (Brazil) IPCC WGI

Summary for Policymakers

This summary, approved in detail at the fifth session of IPCC Working Group I, (Madrid, 27-29 November 1995), represents the formally agreed statement of the IPCC concerning current understanding of the science of climate change.

Summary for Policymakers

Considerable progress has been made in the understanding of climate change[1] science since 1990 and new data and analyses have become available.

Greenhouse gas concentrations have continued to increase

Increases in greenhouse gas concentrations since pre-industrial times (i.e., since about 1750) have led to a positive *radiative forcing*[2] of climate, tending to warm the surface and to produce other changes of climate.

- The atmospheric concentrations of greenhouse gases, *inter alia* carbon dioxide (CO_2), methane (CH_4) and nitrous oxide (N_2O) have grown significantly: by about 30%, 145% and 15% respectively (values for 1992). These trends can be attributed largely to human activities, mostly fossil fuel use, land-use change and agriculture.

- The growth rates of CO_2, CH_4 and N_2O concentrations were low during the early 1990s. While this apparently natural variation is not yet fully explained, recent data indicate that the growth rates are currently comparable to those averaged over the 1980s.

- The direct radiative forcing of the long-lived greenhouse gases (2.45 Wm^{-2}) is due primarily to increases in the concentrations of CO_2 (1.56 Wm^{-2}), CH_4 (0.47 Wm^{-2}) and N_2O (0.14 Wm^{-2}) (1992 values).

- Many greenhouse gases remain in the atmosphere for a long time (for CO_2 and N_2O, many decades to centuries), hence they affect radiative forcing on long time-scales.

- The direct radiative forcing due to the CFCs and HCFCs combined is 0.25 Wm^{-2}. However, their *net* radiative forcing is reduced by about 0.1 Wm^{-2} because they have caused stratospheric ozone depletion which gives rise to a negative radiative forcing.

- Growth in the concentration of CFCs, but not HCFCs, has slowed to about zero. The concentrations of both CFCs and HCFCs, and their consequent ozone depletion, are expected to decrease substantially by 2050 through implementation of the Montreal Protocol and its Adjustments and Amendments.

- At present some long-lived greenhouse gases (particularly HFCs (a CFC substitute), PFCs and SF6) contribute little to radiative forcing but their projected growth could contribute several per cent to radiative forcing during the 21st century.

- If carbon dioxide emissions were maintained at near current (1994) levels, they would lead to a nearly constant rate of increase in atmospheric concentrations for at least two centuries, reaching about 500 ppmv (approaching twice the pre-industrial concentration of 280 ppmv) by the end of the 21st century.

- A range of carbon cycle models indicates that stabilisation of atmospheric CO_2 concentrations at 450, 650 or 1000 ppmv could be achieved only if global anthropogenic CO_2 emissions drop to 1990 levels by, respectively, approximately 40, 140 or 240 years from now, and drop substantially below 1990 levels subsequently.

- Any eventual stabilised concentration is governed more by the accumulated anthropogenic CO_2 emissions from now until the time of stabilisation, than by the way those emissions change over the period. This means that, for a given stabilised concentration value, higher emissions in early decades require lower emissions later on. Among the range of stabilisation cases studied, for stabilisation at 450, 650 or 1000 ppmv accumulated anthropogenic emissions over the period 1991 to 2100 are 630 GtC[3], 1030 GtC, and 1410 GtC respectively ($\pm$ approximately 15% in each case).

---

[1] Climate change in IPCC Working Group I usage refers to any change in climate over time whether due to natural variability or as a result of human activity. This differs from the usage in the Framework Convention on Climate Change where climate change refers to a change of climate which is attributed directly or indirectly to human activity that alters the composition of the global atmosphere and which is in addition to natural climate variability observed over comparable time periods.

[2] A simple measure of the importance of a potential climate change mechanism. Radiative forcing is the perturbation to the energy balance of the Earth-atmosphere system (in watts per square metre [Wm^{-2}]).

[3] 1 GtC = 1 billion (10^9) tonnes of carbon.

For comparison the corresponding accumulated emissions for IPCC IS92 emission scenarios range from 770 to 2190 GtC.

- Stabilisation of CH_4 and N_2O concentrations at today's levels would involve reductions in anthropogenic emissions of 8% and more than 50% respectively.

- There is evidence that tropospheric ozone concentrations in the Northern Hemisphere have increased since pre-industrial times because of human activity and that this has resulted in a positive radiative forcing. This forcing is not yet well characterised, but it is estimated to be about 0.4 Wm^{-2} (15% of that from the long-lived greenhouse gases). However the observations of the most recent decade show that the upward trend has slowed significantly or stopped.

Anthropogenic aerosols tend to produce negative radiative forcing

- Tropospheric aerosols (microscopic airborne particles) resulting from combustion of fossil fuels, biomass burning and other sources have led to a negative direct forcing of about 0.5 Wm^{-2}, as a global average, and possibly also to a negative indirect forcing of a similar magnitude. While the negative forcing is focused in particular regions and subcontinental areas, it can have continental to hemispheric scale effects on climate patterns.

- Locally, the aerosol forcing can be large enough to more than offset the positive forcing due to greenhouse gases.

- In contrast to the long-lived greenhouse gases, anthropogenic aerosols are very short-lived in the atmosphere, hence their radiative forcing adjusts rapidly to increases or decreases in emissions.

Climate has changed over the past century

At any one location year-to-year variations in weather can be large, but analyses of meteorological and other data over large areas and over periods of decades or more have provided evidence for some important systematic changes.

- Global mean surface air temperature has increased by between about 0.3 and 0.6°C since the late 19th

century; the additional data available since 1990 and the re-analyses since then have not significantly changed this range of estimated increase.

- Recent years have been among the warmest since 1860, i.e., in the period of instrumental record, despite the cooling effect of the 1991 Mt. Pinatubo volcanic eruption.

- Night-time temperatures over land have generally increased more than daytime temperatures.

- Regional changes are also evident. For example, the recent warming has been greatest over the mid-latitude continents in winter and spring, with a few areas of cooling, such as the North Atlantic ocean. Precipitation has increased over land in high latitudes of the Northern Hemisphere, especially during the cold season.

- Global sea level has risen by between 10 and 25 cm over the past 100 years and much of the rise may be related to the increase in global mean temperature.

- There are inadequate data to determine whether consistent global changes in climate variability or weather extremes have occurred over the 20th century. On regional scales there is clear evidence of changes in some extremes and climate variability indicators (e.g., fewer frosts in several widespread areas; an increase in the proportion of rainfall from extreme events over the contiguous states of the USA). Some of these changes have been toward greater variability; some have been toward lower variability.

- The 1990 to mid-1995 persistent warm-phase of the El Niûo-Southern Oscillation (which causes droughts and floods in many areas) was unusual in the context of the last 120 years.

The balance of evidence suggests a discernible human influence on global climate

Any human-induced effect on climate will be superimposed on the background "noise" of natural climate variability, which results both from internal fluctuations and from external causes such as solar variability or volcanic eruptions. Detection and attribution studies attempt to distinguish between anthropogenic and natural influences. "Detection of change" is the process of demonstrating that an observed change in climate is highly unusual in a statistical sense, but does not provide a reason for the change. "Attribution" is the process of establishing cause and effect relations, including the testing of competing hypotheses.

Since the 1990 IPCC Report, considerable progress has been made in attempts to distinguish between natural and anthropogenic influences on climate. This progress has been achieved by including effects of sulphate aerosols in addition to greenhouse gases, thus leading to more realistic estimates of human-induced radiative forcing. These have then been used in climate models to provide more complete simulations of the human-induced climate-change "signal". In addition, new simulations with coupled atmosphere-ocean models have provided important information about decade to century time-scale natural internal climate variability. A further major area of progress is the shift of focus from studies of global-mean changes to comparisons of modelled and observed spatial and temporal patterns of climate change.

The most important results related to the issues of detection and attribution are:

- The limited available evidence from proxy climate indicators suggests that the 20th century global mean temperature is at least as warm as any other century since at least 1400 AD. Data prior to 1400 are too sparse to allow the reliable estimation of global mean temperature.

- Assessments of the statistical significance of the observed global mean surface air temperature trend over the last century have used a variety of new estimates of natural internal and externally forced variability. These are derived from instrumental data, palaeodata, simple and complex climate models, and statistical models fitted to observations. Most of these studies have detected a significant change and show that the observed warming trend is unlikely to be entirely natural in origin.

- More convincing recent evidence for the attribution of a human effect on climate is emerging from pattern-based studies, in which the modelled climate response to combined forcing by greenhouse gases and anthropogenic sulphate aerosols is compared with observed geographical, seasonal and vertical patterns of atmospheric temperature change. These studies show that such pattern correspondences increase with time, as one would expect as an anthropogenic signal increases in strength. Furthermore, the probability is very low that these

correspondences could occur by chance as a result of natural internal variability only. The vertical patterns of change are also inconsistent with those expected for solar and volcanic forcing.

- Our ability to quantify the human influence on global climate is currently limited because the expected signal is still emerging from the noise of natural variability, and because there are uncertainties in key factors. These include the magnitude and patterns of long term natural variability and the time-evolving pattern of forcing by, and response to, changes in concentrations of greenhouse gases and aerosols, and land surface changes. Nevertheless, the balance of evidence suggests that there is a discernible human influence on global climate.

Climate is expected to continue to change in the future

The IPCC has developed a range of scenarios, IS92a-f, of future greenhouse gas and aerosol precursor emissions based on assumptions concerning population and economic growth, land-use, technological changes, energy availability and fuel mix during the period 1990 to 2100. Through understanding of the global carbon cycle and of atmospheric chemistry, these emissions can be used to project atmospheric concentrations of greenhouse gases and aerosols and the perturbation of natural radiative forcing. Climate models can then be used to develop projections of future climate.

- The increasing realism of simulations of current and past climate by coupled atmosphere-ocean climate models has increased our confidence in their use for projection of future climate change. Important uncertainties remain, but these have been taken into account in the full range of projections of global mean temperature and sea level change.

- For the mid-range IPCC emission scenario, IS92a, assuming the "best estimate" value of climate sensitivity[1] and including the effects of future increases in aerosol, models project an increase in

---

[1] In IPCC reports, climate sensitivity usually refers to the long term (equilibrium) change in global mean surface temperature following a doubling of atmospheric equivalent CO_2 concentration. More generally, it refers to the equilibrium change in surface air temperature following a unit change in radiative forcing ($°C/Wm^2$).

global mean surface air temperature relative to 1990 of about 2°C by 2100. This estimate is approximately one third lower than the "best estimate" in 1990. This is due primarily to lower emission scenarios (particularly for CO_2 and the CFCs), the inclusion of the cooling effect of sulphate aerosols, and improvements in the treatment of the carbon cycle. Combining the lowest IPCC emission scenario (IS92c) with a "low" value of climate sensitivity and including the effects of future changes in aerosol concentrations leads to a projected increase of about 1°C by 2100. The corresponding projection for the highest IPCC scenario (IS92e) combined with a "high" value of climate sensitivity gives a warming of about 3.5°C. In all cases the average rate of warming would probably be greater than any seen in the last 10,000 years, but the actual annual to decadal changes would include considerable natural variability. Regional temperature changes could differ substantially from the global mean value. Because of the thermal inertia of the oceans, only 50-90% of the eventual equilibrium temperature change would have been realised by 2100 and temperature would continue to increase beyond 2100, even if concentrations of greenhouse gases were stabilised by that time.

• Average sea level is expected to rise as a result of thermal expansion of the oceans and melting of glaciers and ice-sheets. For the IS92a scenario, assuming the "best estimate" values of climate sensitivity and of ice melt sensitivity to warming, and including the effects of future changes in aerosol, models project an increase in sea level of about 50 cm from the present to 2100. This estimate is approximately 25% lower than the "best estimate" in 1990 due to the lower temperature projection, but also reflecting improvements in the climate and ice melt models. Combining the lowest emission scenario (IS92c) with the "low" climate and ice melt sensitivities and including aerosol effects gives a projected sea level rise of about 15 cm from the present to 2100. The corresponding projection for the highest emission scenario (IS92e) combined with "high" climate and ice-melt sensitivities gives a sea level rise of about 95 cm from the present to 2100. Sea level would continue to rise at a similar rate in future centuries beyond 2100, even if concentrations of greenhouse gases were stabilised by that time, and would continue to do so even beyond the time of

stabilisation of global mean temperature. Regional sea level changes may differ from the global mean value owing to land movement and ocean current changes.

• Confidence is higher in the hemispheric-to-continental scale projections of coupled atmosphere-ocean climate models than in the regional projections, where confidence remains low. There is more confidence in temperature projections than hydrological changes.

• All model simulations, whether they were forced with increased concentrations of greenhouse gases and aerosols or with increased concentrations of greenhouse gases alone, show the following features: greater surface warming of the land than of the sea in winter; a maximum surface warming in high northern latitudes in winter, little surface warming over the Arctic in summer; an enhanced global mean hydrological cycle, and increased precipitation and soil moisture in high latitudes in winter. All these changes are associated with identifiable physical mechanisms.

• In addition, most simulations show a reduction in the strength of the north Atlantic thermohaline circulation and a widespread reduction in diurnal range of temperature. These features too can be explained in terms of identifiable physical mechanisms.

• The direct and indirect effects of anthropogenic aerosols have an important effect on the projections. Generally, the magnitudes of the temperature and precipitation changes are smaller when aerosol effects are represented, especially in northern mid-latitudes. Note that the cooling effect of aerosols is not a simple offset to the warming effect of greenhouse gases, but significantly affects some of the continental scale patterns of climate change, most noticeably in the summer hemisphere. For example, models that consider only the effects of greenhouse gases generally project an increase in precipitation and soil moisture in the Asian summer monsoon region, whereas models that include, in addition, some of the effects of aerosols suggest that monsoon precipitation may decrease. The spatial and temporal distribution of aerosols greatly influence regional projections, which are therefore more uncertain.

- A general warming is expected to lead to an increase in the occurrence of extremely hot days and a decrease in the occurrence of extremely cold days.

- Warmer temperatures will lead to a more vigorous hydrological cycle; this translates into prospects for more severe droughts and/or floods in some places and less severe droughts and/or floods in other places. Several models indicate an increase in precipitation intensity, suggesting a possibility for more extreme rainfall events. Knowledge is currently insufficient to say whether there will be any changes in the occurrence or geographical distribution of severe storms, e.g., tropical cyclones.

- Sustained rapid climate change could shift the competitive balance among species and even lead to forest dieback, altering the terrestrial uptake and release of carbon. The magnitude is uncertain, but could be between zero and 200 GtC over the next one to two centuries, depending on the rate of climate change.

There are still many uncertainties

Many factors currently limit our ability to project and detect future climate change. In particular, to reduce uncertainties further work is needed on the following priority topics:

- estimation of future emissions and biogeochemical cycling (including sources and sinks) of greenhouse gases, aerosols and aerosol precursors and projections of future concentrations and radiative properties;

- representation of climate processes in models, especially feedbacks associated with clouds, oceans, sea ice and vegetation, in order to improve projections of rates and regional patterns of climate change;

- systematic collection of long-term instrumental and proxy observations of climate system variables (e.g., solar output, atmospheric energy balance components, hydrological cycles, ocean characteristics and ecosystem changes) for the purposes of model testing, assessment of temporal and regional variability and for detection and attribution studies.

Future unexpected, large and rapid climate system changes (as have occurred in the past) are, by their nature, difficult to predict. This implies that future climate changes may also involve "surprises". In particular these arise from the non-linear nature of the climate system. When rapidly forced, non-linear systems are especially subject to unexpected behaviour. Progress can be made by investigating non-linear processes and sub-components of the climatic system. Examples of such non-linear behaviour include rapid circulation changes in the North Atlantic and feedbacks associated with terrestrial ecosystem changes.

Technical Summary

Prepared by the lead authors taking into account comments arising from extensive review by individual experts, governments and non-governmental organisations

This summary, part of the background material on which the Summary for Policymakers is based, has been accepted by the IPCC but not approved in detail. Acceptance by the IPCC signifies the Panel's view that the summary presents a comprehensive, objective and balanced view of the subject matter.

CONTENTS

A Introduction

The IPCC Scientific Assessment Working Group (WGI) was established in 1988 to assess available information on the science of climate change, in particular that arising from human activities. In performing its assessments the Working Group is concerned with:

- developments in the scientific understanding of past and present climate, of climate variability, of climate predictability and of climate change including feedbacks from climate impacts;
- progress in the modelling and projection of global and regional climate and sea level change;
- observations of climate, including past climates, and assessment of trends and anomalies;
- gaps and uncertainties in current knowledge.

The first Scientific Assessment in 1990 concluded that the increase in atmospheric concentrations of greenhouse gases since the pre-industrial period[1] had altered the energy balance of the Earth/atmosphere and that global warming would result. Model simulations of global warming due to the observed increase of greenhouse gas concentrations over the past century tended towards a central estimate of about 1°C while analysis of the instrumental temperature record, on the other hand, revealed warming of around 0.5°C over the same period. The 1990 report concluded: "The size of this warming is broadly consistent with predictions of climate models, but it is also of the same magnitude as natural climate variability. Thus the observed increase could be largely due to this natural variability; alternatively this variability and other human factors could have offset a still larger human-induced greenhouse warming."

A primary concern identified by IPCC (1990) was the expected continued increase in greenhouse gas concentrations as a result of human activity, leading to significant climate change in the coming century. The projected changes in temperature, precipitation and soil moisture were not uniform over the globe. Anthropogenic aerosols were recognised as a possible source of regional cooling but no quantitative estimates of their effects were available.

The IPCC Supplementary Report in 1992 confirmed, or found no reason to alter, the major conclusions of IPCC (1990). It presented a new range of global mean temperature projections based on a new set of IPCC emission scenarios (IS92 a to f) and reported progress in quantifying the effects of anthropogenic aerosols. Ozone

depletion due to chlorofluorocarbons (CFCs) was recognised as a cause of negative radiative forcing, reducing the global importance of CFCs as greenhouse gases.

The 1994 WGI report on Radiative Forcing of Climate Change provided a detailed assessment of the global carbon cycle and of aspects of atmospheric chemistry governing the abundance of non-CO_2 greenhouse gases. Some pathways that would stabilise atmospheric greenhouse gas concentrations were examined, and new or revised calculations of Global Warming Potential for 38 species were presented. The growing literature on processes governing the abundance and radiative properties of aerosols was examined in considerable detail, including new information on the climatic impact of the 1991 eruption of Mt. Pinatubo.

The Second IPCC Assessment of the Science of Climate Change presents a comprehensive assessment of climate change science as of 1995, including updates of relevant material in all three preceding reports. Key issues examined in the Second Assessment concern the relative magnitude of human and natural factors in driving changes in climate, including the role of aerosols; whether a human influence on present-day climate can be detected; and the estimation of future climate and sea level change at both global and continental scales.

The United Nations Framework Convention on Climate Change (FCCC) uses the term "climate change" to refer exclusively to change brought about by human activities. A more generic usage is common in the scientific community where it is necessary to be able to refer to change arising from any source. In particular scientists refer to past climate change and address the complex issue of separating natural and human causes in currently observed changes. However, the climate projections covered in this document relate only to future climate changes resulting from human influences, since it is not yet possible to predict the fluctuations due to volcanoes and other natural influences. Consequently the use of the term "climate change" here, when referring to future change, is essentially the same as the usage adopted in the FCCC.

B Greenhouse Gases, Aerosols and their Radiative Forcing

Human activities are changing the atmospheric concentrations and distributions of greenhouse gases and aerosols. These changes can produce a radiative forcing by changing either the reflection or absorption of solar radiation, or the emission and absorption of terrestrial radiation (see Box 1).

---

[1] The pre-industrial period is defined as the several centuries preceding 1750.

Box 1: What drives changes in climate?

The Earth absorbs radiation from the Sun, mainly at the surface. This energy is then redistributed by the atmospheric and oceanic circulation and radiated to space at longer ("terrestrial" or "infrared") wavelengths. On average, for the Earth as a whole, the incoming solar energy is balanced by outgoing terrestrial radiation.

Any factor which alters the radiation received from the Sun or lost to space, or which alters the redistribution of energy within the atmosphere, and between the atmosphere, land and ocean, can affect climate. A *change* in the energy available to the global Earth/atmosphere system is termed here, and in previous IPCC reports, a *radiative forcing*.

Increases in the concentrations of greenhouse gases will reduce the efficiency with which the Earth cools to space. More of the outgoing terrestrial radiation from the surface is absorbed by the atmosphere and emitted at higher altitudes and colder temperatures. This results in a positive radiative forcing which tends to warm the lower atmosphere and surface. This is the *enhanced* greenhouse effect – an enhancement of an effect which has operated in the Earth's atmosphere for billions of years due to the naturally occurring greenhouse gases: water vapour, carbon dioxide, ozone, methane and nitrous oxide. The amount of warming depends on the size of the increase in concentration of each greenhouse gas, the radiative properties of the gases involved, and the concentrations of other greenhouse gases already present in the atmosphere.

Anthropogenic aerosols (small particles) in the troposphere, derived mainly from the emission of sulphur dioxide from fossil fuel burning, and derived from other sources such as biomass burning, can absorb and reflect solar radiation. In addition, changes in aerosol concentrations can alter cloud amount and cloud reflectivity through their effect on cloud properties. In most cases tropospheric aerosols tend to produce a negative radiative forcing and cool climate. They have a much shorter lifetime (days to weeks) than most greenhouse gases (decades to centuries) so their concentrations respond much more quickly to changes in emissions.

Volcanic activity can inject large amounts of sulphur-containing gases (primarily sulphur dioxide) into the stratosphere which are transformed into aerosols. This can produce a large, but transitory (i.e., a few years), negative radiative forcing, tending to cool the Earth's surface and lower atmosphere over periods of a few years.

The Sun's output of energy varies by small amounts (0.1%) over an 11-year cycle, and variations over longer periods occur. On time-scales of tens to thousands of years, slow variations in the Earth's orbit, which are well understood, have led to changes in the seasonal and latitudinal distribution of solar radiation; these changes have played an important part in controlling the variations of climate in the distant past, such as the glacial cycles.

Any changes in the radiative balance of the Earth, including those due to an increase in greenhouse gases or in aerosols, will tend to alter atmospheric and oceanic temperatures and the associated circulation and weather patterns. These will be accompanied by changes in the hydrological cycle (for example, altered cloud distributions or changes in rainfall and evaporation regimes).

Any human-induced changes in climate will be superimposed on a background of natural climatic variations which occur on a whole range of space- and time-scales. Natural climate variability can occur as a result of changes in the forcing of the climate system, for example due to aerosol derived from volcanic eruptions. Climate variations can also occur in the absence of a change in external forcing, as a result of complex interactions between components of the climate system such as the atmosphere and ocean. The El Niño-Southern Oscillation (ENSO) phenomenon is an example of such natural "internal" variability. To distinguish anthropogenic climate changes from natural variations, it is necessary to identify the anthropogenic "signal" against the background "noise" of natural climate variability.

Information on radiative forcing was extensively reviewed in IPCC (1994). Summaries of the information in that report and new results are presented here. The most significant advance since IPCC (1994) is improved understanding of the role of aerosols and their representation in climate models.

B.1 Carbon dioxide (CO$_2$)

CO$_2$ concentrations have increased from about 280 ppmv in pre-industrial times to 358 ppmv in 1994 (Table 1, and Figure 1a). There is no doubt that this increase is largely due to human activities, in particular fossil fuel combustion, but also land-use conversion and to a lesser

Table 1: *A sample of greenhouse gases affected by human activities.*

	CO_2	CH_4	N_2	CFC-11	HCFC-22 (a CFC substitute)	CF_4 (a perfluoro-carbon)
Pre-industrial concentration	~280 ppmv	~700 ppbv	~275 ppbv	zero	zero	zero
Concentration in 1994	358 ppmv	1720 ppbv	312[§] ppbv	268[§] pptv[±]	110 pptv	72[§] pptv
Rate of concentration change*	1.5 ppmv/yr	10 ppbv/yr	0.8 ppbv/yr	0 pptv/yr	5 pptv/yr	1.2 pptv/yr
	0.4%/yr	0.6%/yr	0.25%/yr	0%/yr	5%/yr	2%/yr
Atmospheric lifetime (years)	50–200[††]	12[†††]	120	50	12	50,000

[§]   Estimated from 1992–93 data.

[†]   1 pptv = 1 part per trillion (million million) by volume.

[††]   No single lifetime for CO_2 can be defined because of the different rates of uptake by different sink processes.

[†††]   This has been defined as an adjustment time which takes into account the indirect effect of methane on its own lifetime.

*   The growth rates of CO_2, CH_4 and N_2O are averaged over the decade beginning 1984 (see Figures 1 and 3); halocarbon

   growth rates are based on recent years (1990s).

extent cement production (Table 2). The increase has led to a radiative forcing of about +1.6 Wm^{-2} (Figure 2). Prior to this recent increase, CO_2 concentrations over the past 1000 years, a period when global climate was relatively stable, fluctuated by about ±10 ppmv around 280 ppmv.

The annual growth rate of atmospheric CO_2 concentration was low during the early 1990s (0.6 ppmv/yr in 1991/92). However, recent data indicate that the growth rate is currently comparable to that averaged over the 1980s, around 1.5 ppmv/yr (Figure 1b). Isotopic data suggest that the low growth rate resulted from fluctuations in the exchanges of CO_2 between the atmosphere and both the ocean and the terrestrial biosphere, possibly resulting from climatic and biospheric variations following the eruption of Mt. Pinatubo in June 1991. While understanding these short-term fluctuations is important, fluctuations of a few years' duration are not relevant to projections of future concentrations or emissions aimed at estimating longer time-scale changes to the climate system.

The estimate of the 1980s' carbon budget (Table 2) remains essentially unchanged from IPCC (1994). While recent data on anthropogenic emissions are available, there are insufficient analyses of the other fluxes to allow an update of this decadal budget to include the early years of the 1990s. The net release of carbon from tropical land-use change (mainly forest clearing minus regrowth) is roughly balanced by carbon accumulation in other land ecosystems due to forest regrowth outside the tropics, and by transfer to

other reservoirs stimulated by CO_2 and nitrogen fertilisation and by decadal time-scale climatic effects. Model results suggest that during the 1980s, CO_2 fertilisation resulted in a transfer of carbon from the atmosphere to the biosphere of 0.5 to 2.0 GtC/yr and nitrogen fertilisation resulted in a transfer of carbon from the atmosphere to the biosphere of between 0.2 and 1.0 GtC/yr.

CO_2 is removed from the atmosphere by a number of processes that operate on different time-scales, and is subsequently transferred to various reservoirs, some of which eventually return CO_2 to the atmosphere. Some simple analyses of CO_2 changes have used the concept of a single characteristic time-scale for this gas. Such analyses are of limited value because a single time-scale cannot capture the behaviour of CO_2 under different emission scenarios. This is in contrast to methane, for example, whose atmospheric lifetime is dominantly controlled by a single process: oxidation by OH in the atmosphere. For CO_2 the fastest process is uptake into vegetation and the surface layer of the oceans which occurs over a few years. Various other sinks operate on the century time-scale (e.g., transfer to soils and to the deep ocean) and so have a less immediate, but no less important, effect on the atmosphericconcentration. Within 30 years about 40–60% of the CO_2 currently released to the atmosphere is removed. However, if emissions were reduced, the CO_2 in the vegetation and ocean surface water would soon equilibrate with that in the atmosphere, and the rate of removal would

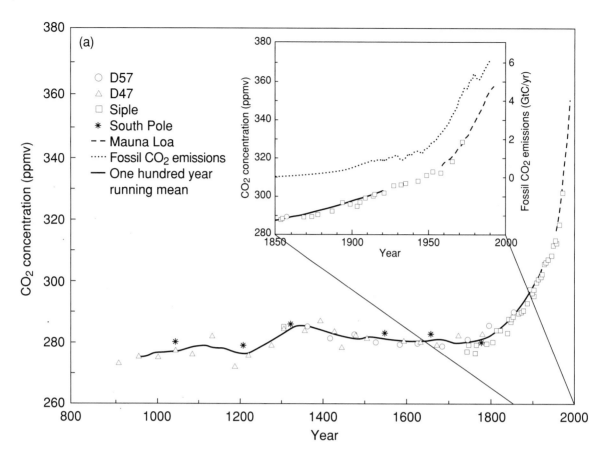

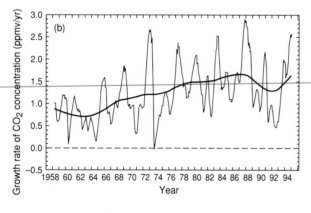

Figure 1: (a) CO_2 concentrations over the past 1000 years from ice core records (D47, D57, Siple and South Pole) and (since 1958) from Mauna Loa, Hawaii, measurement site. All ice core measurements were taken in Antarctica. The smooth curve is based on a hundred year running mean. The rapid increase in CO_2 concentration since the onset of industrialisation is evident and has followed closely the increase in CO_2 emissions from fossil fuels (see inset of period from 1850 onwards). (b) Growth rate of CO_2 concentration since 1958 in ppmv/yr at Mauna Loa. The smooth curve shows the same data but filtered to suppress variations on time-scales less than approximately 10 years.

then be determined by the slower response of woody vegetation, soils, and transfer into the deeper layers of the ocean. Consequently, most of the excess atmospheric CO_2 would be removed over about a century although a portion would remain airborne for thousands of years because transfer to the ultimate sink – ocean sediments – is very slow.

There is large uncertainty associated with the future role of the terrestrial biosphere in the global carbon budget for several reasons. First, future rates of deforestation and regrowth in the tropics and mid-latitudes are difficult to predict. Second, mechanisms such as CO2 fertilisation

remain poorly quantified at the ecosystem level. Over decades to centuries, anthropogenic changes in atmospheric CO2 content and climate may also alter the global distribution of ecosystem types. Carbon could be released rapidly from areas where forests die, although regrowth could eventually sequester much of this carbon. Estimates of this loss range from near zero to, at low probabilities, as much as 200 GtC over the next one-to-two centuries, depending on the rate of climate change.

The marine biota both respond to and can influence climate change. Marine biota play a critical role in

Table 2: *Annual average anthropogenic carbon budget for 1980 to 1989. CO_2 sources, sinks and storage in the atmosphere are expressed in GtC/yr.*

CO_2 sources	
(1) Emissions from fossil fuel combustion and cement production	5.5 ± 0.5*
(2) Net emissions from changes in tropical land-use	1.6 ± 1.0@
(3) Total anthropogenic emissions = (1) + (2)	7.1 ± 1.1
Partitioning amongst reservoirs	
(4) Storage in the atmosphere	3.3 ± 0.2
(5) Ocean uptake	2.0 ± 0.8
(6) Uptake by Northern Hemishere forest regrowth	0.5 ± 0.5#
(7) Inferred sink: 3–(4+5+6)	1.3 ± 1.5§

Notes:

* For comparison, emissions in 1994 were 6.1 GtC/yr.

@ Consistent with Chapter 24 of IPCC WGII (1995).

This number is consistent with the independent estimate, given in IPCC WGII (1995), of 0.7±0.2 GtC/yr for the mid- and high latitude forest sink.

§ This inferred sink is consistent with independent estimates, given in Chapter 9, of carbon uptake due to nitrogen fertilisation (0.5±1.0 GtC/yr), plus the range of other uptakes (0–2 GtC/yr) due to CO_2 fertilisation and climatic effects.

depressing the atmospheric CO_2 concentration significantly below its equilibrium state in the absence of biota. Changes in nutrient supply to the surface ocean resulting from changes in ocean circulation, coastal runoff and atmospheric deposition, and changes in the amount of sea ice and cloudiness, have the potential to affect marine biogeochemical processes. Such changes would be expected to have an impact (at present unquantifiable) on the cycling of CO_2 and the production of other climatically important trace gases. It has been suggested that a lack of iron limits phytoplankton growth in certain ocean areas. However, it is not likely that iron fertilisation of CO_2 uptake by phytoplankton can be used to draw down atmospheric CO_2: even massive continual seeding of 10–15% of the world oceans (the Southern Ocean) until 2100, if it worked with 100% efficiency and no opposing side-effects (e.g., increased N_2O production), would reduce the atmospheric CO_2 build-up projected by the IPCC (1990) "Business-as-usual" emission scenario by less than 10%.

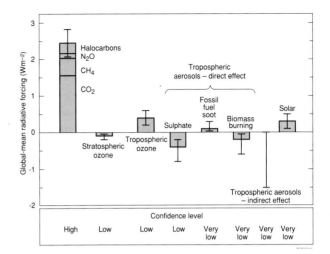

Figure 2: Estimates of the globally and annually averaged anthropogenic radiative forcing (in Wm⁻²) due to changes in concentrations of greenhouse gases and aerosols from pre-industrial times to the present (1992) and to natural changes in solar output from 1850 to the present. The height of the rectangular bar indicates a mid-range estimate of the forcing whilst the error bars show an estimate of the uncertainty range, based largely on the spread of published values; the "confidence level" indicates the author's confidence that the actual forcing lies within this error bar. The contributions of individual gases to the direct greenhouse forcing is indicated on the first bar. The indirect greenhouse forcings associated with the depletion of stratospheric ozone and the increased concentration of tropospheric ozone are shown in the second and third bar respectively. The direct contributions of individual tropospheric aerosol component are grouped into the next set of three bars. The indirect aerosol effect, arising from the induced change in cloud properties, is shown next; quantitative understanding of this process is very limited at present and hence no bar representing a mid-range estimate is shown. The final bar shows the estimate of the changes in radiative forcing due to variations in solar output. The forcing associated with stratospheric aerosols resulting from volcanic eruptions is not shown, as it is very variable over this time period. Note that there are substantial differences in the geographical distribution of the forcing due to the well-mixed greenhouse gases (mainly CO_2, N_2O, CH_4 and the halocarbons) and that due to ozone and aerosols, which could lead to significant differences in their respective global and regional climate responses. For this reason, the negative radiative forcing due to aerosols should not necessarily be regarded as an offset against the greenhouse gas forcing.

B.2 Methane (CH_4)

Methane is another naturally occurring greenhouse gas whose concentration in the atmosphere is growing as a result of human activities such as agriculture and waste disposal, and fossil fuel production and use (Table 3).

Table 3: *Estimated sources and sinks of methane for 1980 to 1990. All figures are in Tg[†](CH₄)/yr. The current global atmospheric burden of CH₄ is about 5000 Tg(CH₄).*

(a) Observed atmospheric increase, estimated sinks and sources derived to balance the budget.

		Individual estimates	**Total**
Atmospheric increase[*]			**37** (35–40)
Sinks of atmospheric CH_4:	tropospheric OH	**490** (405–575)	
	stratosphere	**40** (32–48)	
	soils	**30** (15–45)	
Total atmospheric sinks			**560** (460–660)
Implied sources (sinks + atmospheric increase)			**597** (495–700)

(b) Inventory of identified sources.

		Individual estimates	**Total**
Natural sources			**160** (110–210)
Anthropogenic sources:	Fossil fuel related	**100** (70–120)	
	Total biospheric	**275** (200–350)	
Total anthropogenic sources			**375** (300–450)
Total identified sources			**535** (410–660)

[†]   1Tg = 1 million million grams, which is equivalent to 1 million tonnes.

[*]   Applies to 1980–1990 average. Table 1 and the stabilisation discussion in Section B.9.2 use the average for 1984–1994.

Global average methane concentrations increased by 6% over the decade starting in 1984 (Figure 3). Its concentration in 1994 was about 1720 ppbv, 145% greater than the pre-industrial concentration of 700 ppbv (Table 1, Figure 3). Over the last 20 years, there has been a decline in the methane growth rate: in the late 1970s the concentration was increasing by about 20 ppbv/yr, during the 1980s the growth rate dropped to 9–13 ppbv/yr. Around the middle of 1992, methane concentrations briefly stopped growing, but since 1993 the global growth rate has returned to about 8 ppbv/yr.

Individual methane sources are not well quantified. Carbon isotope measurements indicate that about 20% of the total annual methane emissions are related to the production and use of fossil fuel. In total, anthropogenic activities are responsible for about 60–80% of current methane emissions (Table 3). Methane emissions from natural wetlands appear to contribute about 20% to the global methane emissions to the atmosphere. Such emissions will probably increase with global warming as a result of greater microbial activity. In 1992 the directradiative forcing due to the increase in methane concentration since pre-industrial times was about +0.47 Wm[-2] (Figure 2).

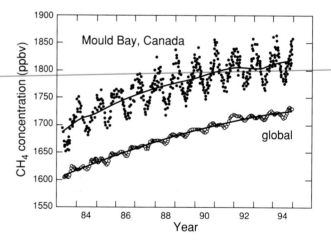

Figure 3: Global methane concentrations (ppbv) for 1983 to 1994. Concentrations observed at Mould Bay, Canada are also shown.

Changes in the concentration of methane have clearly identified chemical feedbacks. The main removal process for methane is reaction with the hydroxyl radical (OH). Addition of methane to the atmosphere reduces the concentration of tropospheric OH which can in turn feed back and reduce the rate of methane removal.

The adjustment time for a pulse of methane added to the atmosphere has been revised to 12 (±3) years (compared with 14.5 (±2.5) years in IPCC (1994)). Two factors are responsible for the change: (a) a new estimate for the chemical removal rate (11% faster); and (b) inclusion of the uptake of methane by soils. The revised global sink strength is 560 (±100) Tg(CH_4)/year, higher than the 1994 estimate, but still consistent with the previous range of global source strength.

B.3 Nitrous oxide (N_2O)

There are many small sources of nitrous oxide, both natural and anthropogenic, which are difficult to quantify. The main anthropogenic sources are from agriculture and a number of industrial processes (e.g., adipic acid and nitric acid production). A best estimate of the current (1980s) anthropogenic emission of nitrous oxide is 3 to 8 Tg(N)/yr. Natural sources are poorly quantified, but are probably twice as large as anthropogenic sources. Nitrous oxide is removed mainly by photolysis (breakdown by sunlight) in the stratosphere and consequently has a long lifetime (about 120 years).

Although sources cannot be well quantified, atmospheric measurements and evidence from ice cores show that the atmospheric abundance of nitrous oxide has increased since the pre-industrial era, most likely owing to human activities. In 1994 atmospheric levels of nitrous oxide were about 312 ppbv; pre-industrial levels were about 275 ppbv (Table 1). The 1993 growth rate (approximately 0.5 ppbv/yr) was lower than that observed in the late 1980s and early 1990s (approximately 0.8 ppbv/yr), but these short-term changes in growth rate are within the range of variability seen on decadal time-scales. The radiative forcing due to the change in nitrous oxide since pre-industrial times is about +0.14 Wm^{-2} (Figure 2).

B.4 Halocarbons and other halogenated compounds

Halocarbons are carbon compounds containing fluorine, chlorine, bromine or iodine. Many of these are effective greenhouse gases. For most of these compounds, human activities are the sole source.

Halocarbons that contain chlorine (CFCs and HCFCs) and bromine (halons) cause ozone depletion, and their emissions are controlled under the Montreal Protocol and its Adjustments and Amendments. As a result, growth rates in the concentrations of many of these compounds have already fallen (Figure 4) and the radiative impact of these compounds will slowly decline over the next century. The contribution to *direct* radiative forcing due to concentration increases of these CFCs and HCFCs since pre-industrial

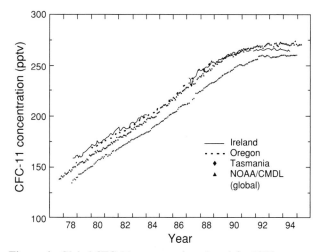

Figure 4: Global CFC-11 concentrations (pptv) for 1978 to 1994. As one of the ozone-depleting gases, the emissions of CFC-11 are controlled under the Montreal Protocol and its Adjustments and Amendments. Observations at some individual measurement sites are also shown.

times is about +0.25 Wm^{-2}. Halocarbons can also exert an *indirect* negative radiative forcing through their depletion of stratospheric ozone (see Section B.5.2).

Perfluorocarbons (PFCs, e.g., CF_4, C_2F_6) and sulphur hexafluoride (SF_6) are removed very slowly from the atmosphere with estimated lifetimes greater than 1000 years. As a result, effectively all emissions accumulate in the atmosphere and will continue to influence climate for thousands of years. Although the radiative forcing due to concentration increases of these compounds since pre-industrial times is small (about +0.01 Wm^{-2}), it may become significant in the future if concentrations continue to increase.

Hydrofluorocarbons (HFCs) are being used to replace ozone-depleting substances in some applications; their concentrations and radiative impacts are currently small. If emissions increase as envisaged in Scenario IS92a, they would contribute about 3% of the total radiative forcing from all greenhouse gases by the year 2100.

B.5 Ozone (O_3)

Ozone is an important greenhouse gas present in both the stratosphere and troposphere. Changes in ozone cause radiative forcing by influencing both solar and terrestrial radiation. The net radiative forcing is strongly dependent on the vertical distribution of ozone change and is particularly sensitive to changes around the tropopause level, where trends are difficult to estimate due to a lack of reliable observations and the very large natural variability. The patterns of both tropospheric and stratospheric ozone changes are spatially variable. Estimation of the radiative forcing due to changes in ozone is thus more complex than for the well-mixed greenhouse gases.

B.5.1 Tropospheric Ozone

In the troposphere, ozone is produced during the oxidation of methane and from various short-lived precursor gases (mainly carbon monoxide (CO), nitrogen oxides (NO_x) and non-methane hydrocarbons (NMHC)). Ozone is also transported into the troposphere from the stratosphere. Changes in tropospheric ozone concentration are spatially variable, both regionally and vertically, making assessment of global long-term trends difficult. In the Northern Hemisphere, there is some evidence that tropospheric ozone concentrations have increased since 1900, with strong evidence that this has occurred in many locations since the 1960s. However, the observations of the most recent decade show that the upward trend has slowed significantly or stopped. Model simulations and the limited observations together suggest that ozone concentrations throughout the troposphere may have doubled in the Northern Hemisphere since pre-industrial times, an increase of about 25 ppbv. In the Southern Hemisphere, there are insufficient data to determine if tropospheric ozone has changed, except at the South Pole where a decrease has been observed since the mid-1980s.

Changes in tropospheric ozone have potentially important consequences for radiative forcing. The calculated global average radiative forcing due to the increased concentration since pre-industrial times is +0.4 (±0.2) Wm^{-2}.

B.5.2 Stratospheric Ozone

Decreases in stratospheric ozone have occurred since the 1970s, principally in the lower stratosphere. The most obvious feature is the annual appearance of the Antarctic "ozone hole" in September and October. The October average total ozone values over Antarctica are 50–70% lower than those observed in the 1960s. Statistically significant losses in total ozone have also been observed in the mid-latitudes of both hemispheres. Little or no downward trend in ozone has been observed in the tropics (20°N–20°S). The weight of recent scientific evidence strengthens the previous conclusion that ozone loss is due largely to anthropogenic chlorine and bromine compounds. Since the stratospheric abundances of chlorine and bromine are expected to continue to grow for a few more years before they decline (see Section B.4), stratospheric ozone losses are expected to peak near the end of the century, with a gradual recovery throughout the first half of the 21st century.

The loss of ozone in the lower stratosphere over the past 15 to 20 years has led to a globally averaged radiative forcing of about –0.1 Wm^{-2}. This negative radiative forcing represents an indirect effect of anthropogenic chlorine and bromine compounds.

B.6 Tropospheric and stratospheric aerosols

Aerosol is a term used for particles and very small droplets of natural and human origin that occur in the atmosphere; they include dust and other particles which can be made up of many different chemicals. Aerosols are produced by a variety of processes, both natural (including dust storms and volcanic activity) and anthropogenic (including fossil fuel and biomass burning). Aerosols contribute to visible haze and can cause a diminution of the intensity of sunlight at the ground.

Aerosols in the atmosphere influence the radiation balance of the Earth in two ways: (i) by scattering and absorbing radiation – the *direct* effect, and (ii) by modifying the optical properties, amount and lifetime of clouds – the *indirect* effect. Although some aerosols, such as soot, tend to warm the surface, the net climatic effect of anthropogenic aerosols is believed to be a negative radiative forcing, tending to cool the surface (see Section B.7 and Figure 2).

Most aerosols with anthropogenic sources are found in the lower troposphere (below 2 km). Aerosols undergo chemical and physical transformations in the atmosphere, especially within clouds, and are removed largely by precipitation. Consequently aerosols in the lower troposphere typically have residence times of a few days. Because of their short lifetime, aerosols in the lower troposphere are distributed inhomogeneously with maxima close to the natural (especially desert) and anthropogenic (especially industrial and biomass combustion) source regions. Aerosol particles resulting from volcanic activity can reach the stratosphere where they are transported around the globe over many months or years.

The radiative forcing due to aerosols depends on the size, shape and chemical composition of the particles and the spatial distribution of the aerosol. While these factors are comparatively well-known for stratospheric aerosols, there remain many uncertainties concerning tropospheric aerosols.

Since IPCC (1994), there have been several advances in understanding the impact of tropospheric aerosols on climate. These include: (i) new calculations of the spatial distribution of sulphate aerosol largely resulting from fossil fuel combustion and (ii) the first calculation of the spatial distribution of soot aerosol. The impact of these developments on the calculation of aerosol radiative forcing is discussed in Section B.7.

B.7 Summary of radiative forcing

Globally averaged radiative forcing is a useful concept for giving a first-order estimate of the potential climatic importance of various forcing mechanisms. However, as was emphasised in IPCC (1994), there are limits to its utility. In particular, the spatial patterns of forcing differ between the globally well-mixed greenhouse gases, the regionally varying tropospheric ozone, and the even more regionally concentrated tropospheric aerosols, and so a comparison of the global mean radiative forcings does not give a complete picture of their possible climatic impact.

Estimates of the radiative forcings due to changes in greenhouse gas concentrations since pre-industrial times remain unchanged from IPCC (1994) (see Figure 2). These are +2.45 Wm^{-2} (range: +2.1 to +2.8 Wm^{-2}) for the direct effect of the main well-mixed greenhouse gases (CO_2, CH_4, N_2O and the halocarbons), +0.4 Wm^{-2} (range: 0.2 to 0.6 Wm^{-2}) for tropospheric ozone and –0.1 Wm^{-2} (range: –0.05 to –0.2 Wm^{-2}) for stratospheric ozone.

The total direct forcing due to anthropogenic aerosol (sulphates, fossil fuel soot and organic aerosols from biomass burning) is estimated to be –0.5 Wm^{-2} (range: –0.25 to –1.0 Wm^{-2}). This estimate is smaller than that given in IPCC (1994) owing to a reassessment of the model results used to derive the geographic distribution of aerosol particles and the inclusion of anthropogenic soot aerosol for the first time. The direct forcing due to sulphate aerosols resulting from fossil fuel emissions and smelting is estimated to be –0.4 Wm^{-2} (range: –0.2 to –0.8 Wm^{-2}). The first estimates of the impact of soot in aerosols from fossil fuel sources have been made: significant uncertainty remains but an estimate of +0.1 Wm^{-2} (range: 0.03 to 0.3 Wm^{-2}) is made. The direct radiative forcing since 1850 of particles associated with biomass burning is estimated to be –0.2 Wm^{-2} (range: –0.07 to –0.6 Wm^{-2}), unchanged from IPCC (1994). It has recently been suggested that a significant fraction of the tropospheric dust aerosol is influenced by human activities but the radiative forcing of this component has not yet been quantified.

The range of estimates for the radiative forcing due to changes in cloud properties caused by aerosols arising from human activity (the indirect effect) is unchanged from IPCC (1994) at between 0 and –1.5 Wm^{-2}. Several new studies confirm that the indirect effect of aerosol may have caused a substantial negative radiative forcing since pre-industrial times, but it remains very difficult to quantify, more so than the direct effect. While no best estimate of the indirect forcing can currently be made, the central value of –0.8 Wm^{-2} has been used in some of the scenario calculations described in Sections B.9.2 and F.2.

There are no significant alterations since IPCC (1994) in the assessment of radiative forcing caused by changes in solar radiative output or stratospheric aerosol loading resulting from volcanic eruptions. The estimate of radiative forcing due to changes in solar radiative output since 1850 is +0.3 Wm^{-2} (range: +0.1 to +0.5). Radiative forcing due to volcanic aerosols resulting from an individual eruption can be large (the maximum global mean effect from the eruption of Mt. Pinatubo was –3 to –4 Wm^{-2}), but lasts for only a few years. However, the transient variations in both these forcings may be important in explaining some of the observed climate variations on decadal time-scales.

B.8 Global Warming Potential (GWP)

The Global Warming Potential is an attempt to provide a simple measure of the relative radiative effects of the emissions of various greenhouse gases. The index is defined as the cumulative radiative forcing between the present and some chosen time horizon caused by a unit mass of gas emitted now, expressed relative to that for some reference gas (here CO_2 is used). The future global warming commitment of a greenhouse gas over a chosen time horizon can be estimated by multiplying the appropriate GWP by the amount of gas emitted. For example, GWPs could be used to compare the effects of reductions in CO_2 emissions relative to reductions in methane emissions, for a specified time horizon.

Derivation of GWPs requires knowledge of the fate of the emitted gas and the radiative forcing due to the amount remaining in the atmosphere. Although the GWPs are quoted as single values, the typical uncertainty is ±35%, not including the uncertainty in the carbon dioxide reference. Because GWPs are based on the radiative forcing concept, they are difficult to apply to radiatively important constituents that are unevenly distributed in the atmosphere. No attempt is made to define a GWP for aerosols. Additionally the choice of time horizon will depend on policy considerations.

GWPs need to take account of any indirect effects of the emitted greenhouse gas if they are to reflect correctly future warming potential. The net GWPs for the ozone-depleting gases, which include the direct "warming" and indirect "cooling" effects, have now been estimated. In IPCC (1994), only the direct GWPs were presented for these gases. The indirect effect reduces their GWPs, but each ozone-depleting gas must be considered individually. The net GWPs of the chlorofluorocarbons (CFCs) tend to be positive, while those of the halons tend to be negative. The calculation of indirect effects for a number of other

Table 4: *Global Warming Potential referenced to the updated decay response for the Bern carbon cycle model and future CO_2 atmospheric concentrations held constant at current levels.*

Species	Chemical Formula	Lifetime (years)	Global Warming Potential (Time Horizon)		
			20 years	100 years	500 years
CO_2	CO_2	variable§	1	1	1
Methane*	CH_4	12±3	56	21	6.5
Nitrous oxide	N_2O	120	280	310	170
HFC-23	CHF_3	264	9,100	11,700	9,800
HFC-32	CH_2F_2	5.6	2,100	650	200
HFC-41	CH_3F	3.7	490	150	45
HFC-43-10mee	$C_5H_2F_{10}$	17.1	3,000	1,300	400
HFC-125	C_2HF_5	32.6	4,600	2,800	920
HFC-134	$C_2H_2F_4$	10.6	2,900	1,000	310
HFC-134a	CH_2FCF_3	14.6	3,400	1,300	420
HFC-152a	C_2H4F_2	1.5	460	140	42
HFC-143	$C_2H_3F_3$	3.8	1,000	300	94
HFC-143a	$C_2H_3F_3$	48.3	5,000	3,800	1,400
HFC-227ea	C_3HF_7	36.5	4,300	2,900	950
HFC-236fa	$C_3H_2F_6$	209	5,100	6,300	4,700
HFC-245ca	$C_3H_3F_5$	6.6	1,800	560	170
Sulphur hexafluoride	SF_6	3,200	16,300	23,900	34,900
Perfluoromethane	CF_4	50,000	4,400	6,500	10,000
Perfluoroethane	C_2F_6	10,000	6,200	9,200	14,000
Perfluoropropane	C_3F_8	2,600	4,800	7,000	10,100
Perfluorobutane	C_4F_{10}	2,600	4,800	7,000	10,100
Perfluorocyclobutane	$c–C_4F_8$	3,200	6,000	8,700	12,700
Perfluoropentane	C_5F_{12}	4,100	5,100	7,500	11,000
Perfluorohexane	C_6F_{14}	3,200	5,000	7,400	10,700
Ozone-depleting substances†	e.g., CFCs and HCFCS				

§ Derived from the Bern carbon cycle model.

*   The GWP for methane includes indirect effects of tropospheric ozone production and stratospheric water vapour production, as in IPCC (1994). The updated adjustment time for methane is discussed in Section B.2.

† The Global Warming Potentials for ozone-depleting substances (including all CFCs, HCFCs and halons, whose direct GWPs have been given in previous reports) are a sum of a direct (positive) component and an indirect (negative) component which depends strongly upon the effectiveness of each substance for ozone destruction. Generally, the halons are likely to have negative net GWPs, while those of the CFCs are likely to be positive over both 20- and 100-year time horizons (see Chapter 2, Table 2.8).

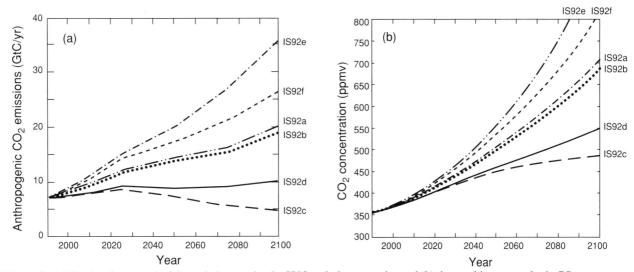

Figure 5: (a) Total anthropogenic CO_2 emissions under the IS92 emission scenarios and (b) the resulting atmospheric CO_2 concentrations calculated using the "Bern" carbon cycle model and the carbon budget for the 1980s shown in Table 2.

gases (e.g., NO_x, CO) is not currently possible because of inadequate characterisation of many of the atmospheric processes involved.

Updates or new GWPs are given for a number of key species (Table 4), based on improved or new estimates of atmospheric lifetimes, molecular radiative forcing factors, and improved representation of the carbon cycle. Revised lifetimes for gases destroyed by chemical reactions in the lower atmosphere (particularly methane, HCFCs and HFCs) have resulted in GWPs that are slightly lower (typically by 10–15%) than those cited in IPCC (1994). The IPCC definition of GWP is based on calculating the relative radiative impact of a release of a trace gas over a time horizon in a constant background atmosphere. In a future atmosphere with larger CO_2 concentrations, such as occur in all of the IPCC emission scenarios (see Figure 5b), we would calculate slightly larger GWP values than those given in Table 4.

B.9 Emissions and concentrations of greenhouse gases and aerosols in the future

B.9.1 The IS92 emission scenarios

The projection of future anthropogenic climate change depends, among other things, on assumptions made about future emissions of greenhouse gases and aerosol precursors and the proportion of emissions remaining in the atmosphere. Here we consider the IS92 emission scenarios (IS92a to f) which were first discussed in IPCC (1992).

The IS92 emission scenarios extend to the year 2100 and include emissions of CO_2, CH_4, N_2O, the halocarbons (CFCs and their substitute HCFCs and HFCs), precursors

of tropospheric ozone and sulphate aerosols and aerosols from biomass burning. A wide range of assumptions regarding future economic, demographic and policy factors are encompassed (IPCC, 1992). In this report, the emissions of chlorine- and bromine-containing halocarbons listed in IS92 are assumed to be phased out under the Montreal Protocol and its Adjustments and Amendments and so a single revised future emission scenario for these gases is incorporated in all of the IS92 scenarios.

Emissions of individual HFCs are based on the original IS92 scenarios, although they do not reflect current markets. CO_2 emissions for the six scenarios are shown in Figure 5a.

The calculation of future concentrations of greenhouse gases, given certain emissions, entails modelling the processes that transform and remove the different gases from the atmosphere. For example, future concentrations of CO_2 are calculated using models of the carbon cycle which model the exchanges of CO_2 between the atmosphere and the oceans and terrestrial biosphere (see Section B.1); atmospheric chemistry models are used to simulate the removal of chemically active gases such as methane.

All the IS92 emission scenarios, even IS92c, imply increases in greenhouse gas concentrations from 1990 to 2100 (e.g., CO_2 increases range from 35 to 170% (Figure 5b); CH_4 from 22 to 175%; and N_2O from 26 to 40%).

For greenhouse gases, radiative forcing is dependent on the concentration of the gas and the strength with which it absorbs and re-emits long-wave radiation. For sulphate aerosol, the direct and indirect radiative forcings were calculated on the basis of sulphur emissions contained in

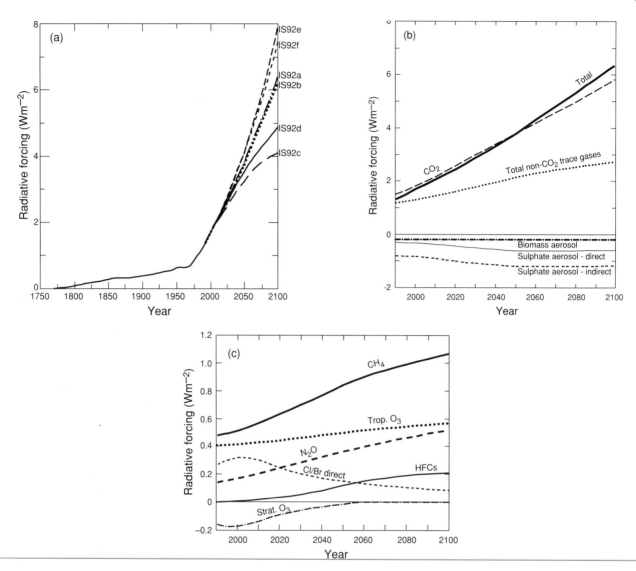

Figure 6: (a) Total globally and annually averaged historical radiative forcing from 1765 to 1990 due to changes in greenhouse gas concentrations and tropospheric aerosol emissions and projected radiative forcing values to 2100 derived from the IS92 emissions scenarios. (b) Radiative forcing components resulting from the IS92a emission scenario for 1990 to 2100. The "Total non-CO_2 trace gases" curve includes the radiative forcing from methane (including methane related increases in stratospheric water vapour), nitrous oxide, tropospheric ozone and the halocarbons (including the negative forcing effect of stratospheric ozone depletion). Halocarbon emissions have been modified to take account of the Montreal Protocol and its Adjustments and Amendments. The three aerosol components are: direct sulphate, indirect sulphate and direct biomass burning. (c) Non-CO_2 trace gas radiative forcing components. "Cl/Br direct" is the direct radiative forcing resulting from the chlorine and bromine containing halocarbons; emissions are assumed to be controlled under the Montreal Protocol and its Adjustments and Amendments. The indirect forcing from these compounds (through stratospheric ozone depletion) is shown separately (Strat. O_3). All other emissions follow the IS92a Scenario. The tropospheric ozone forcing (Trop. O_3) takes account of concentration changes due only to the indirect effect of methane.

the IS92 scenarios. The radiative forcing due to aerosol from biomass burning was assumed to remain constant at –0.2 Wm^{-2} after 1990. The contribution from aerosols is probably the most uncertain part of future radiative forcing.

Figure 6a shows a single "best estimate" of historical radiative forcing from 1765 to 1990 (including the effects

of aerosols), followed by radiative forcing for Scenarios IS92 a to f. Figures 6b and c show the contribution to future radiative forcing from various components of the IS92a Scenario; the largest contribution comes from CO_2, with a radiative forcing of almost +6 Wm^{-2} by 2100. The negative forcing due to tropospheric aerosols, in a globally averaged sense, offsets some of the greenhouse gas positive forcing. However, because tropospheric aerosols are highly variable regionally, their globally averaged radiative forcing will not adequately describe their possible climatic impact.

Future projections of temperature and sea level based on the IS92 emissions scenarios are discussed in Section F.

B.9.2 Stabilisation of greenhouse gas and aerosol concentrations

An important question to consider is: how might greenhouse gas concentrations be stabilised in the future?

If global CO_2 emissions were maintained at near current (1994) levels, they would lead to a nearly constant rate of increase in atmospheric concentrations for at least two centuries, reaching about 500 ppmv (approaching twice the pre-industrial concentration of 280 ppmv) by the end of the 21st century.

In IPCC (1994), carbon cycle models were used to calculate the emissions of CO_2 which would lead to stabilisation at a number of different concentration levels from 350 to 750 ppmv. The assumed concentration profiles leading to stabilisation are shown in Figure 7a (excluding 350 ppmv). Many different stabilisation levels, time-scales for achieving these levels, and routes to stabilisation could have been chosen. The choices made are not intended to have policy implications; the exercise is illustrative of the relationship between CO_2 emissions and concentrations. Those in Figure 7a assume a smooth transition from the current average rate of CO_2 concentration increase to stabilisation. To a first approximation, the stabilised concentration level depends more upon the accumulated amount of carbon emitted up to the time of stabilisation, than upon the exact concentration path followed *en route* to stabilisation.

New results have been produced to take account of the revised carbon budget for the 1980s (Table 2), but the main conclusion, that stabilisation of concentration requires emissions eventually to drop well below current levels, remains unchanged from IPCC (1994) (Figure 7b). Because the new budget implies a reduced terrestrial sink, the allowable emissions to achieve stabilisation are up to 10% lower than those in IPCC (1994). In addition, these calculations have been extended to include alternative pathways towards stabilisation (Figure 7a) and a higher stabilisation level (1000 ppmv). The alternative pathways assume higher emissions in the early years, but require steeper reductions in emissions in later years (Figure 7b). The 1000 ppmv stabilisation case allows higher maximum emissions, but still requires a decline to current levels by

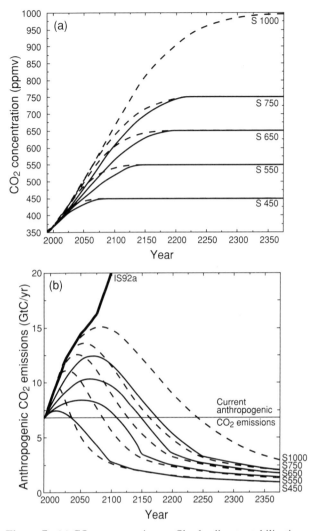

Figure 7: (a) CO_2 concentration profiles leading to stabilisation at 450, 550, 650 and 750 ppmv following the pathways defined in IPCC (1994) (solid curves) and for pathways that allow emissions to follow IS92a until at least 2000 (dashed curves). A single profile that stabilises at a CO_2 concentration of 1000 ppmv and follows IS92a emissions until at least 2000 has also been defined. (b) CO_2 emissions leading to stabilisation at concentrations of 450, 550, 650, 750 and 1000 ppmv following the profiles shown in (a). Current anthropogenic CO_2 emissions and those for IS92a are shown for comparison. The calculations use the "Bern" carbon cycle model and the carbon budget for the 1980s shown in Table 2.

about 240 years from now and further reductions thereafter (Figure 7b).

The accumulated anthropogenic CO_2 emissions from 1991 to 2100 inclusive are shown in Table 5 for the profiles leading to stabilisation at 450, 550, 650, 750 and 1000 ppmv via the profiles shown in Figure 7a and, for comparison, the IS92 emission scenarios. These values are calculated

Table 5: *Total anthropogenic CO_2 emissions accumulated from 1991 to 2100 inclusive (GtC). All values were calculated using the carbon budget for 1980s shown in Table 2 and the Bern carbon cycle model.*

Case		Accumulated CO_2 emissions 1991 to 2100 (GtC)
IS92 scenarios	c	770
	d	980
	b	1430
	a	1500
	f	1830
	e	2190

		Concentration profiles A[*]	Concentration profiles B[†]
Stabilisation at	450 ppmv	630	650
	550 ppmv	870	990
	650 ppmv	1030	1190
	750 ppmv	1200	1300
	1000 ppmv	-	1410

[*]   As in IPCC (1994) – see Figure 7a.

[†]   Profiles that allow emissions to follow IS92a until at least the year 2000 – see Figure 7a.

using the "Bern" carbon cycle model. Based on the results in IPCC (1994) it is estimated that values calculated with different carbon cycle models could be up to approximately 15% higher or lower than those presented here.

If methane emissions were to remain constant at 1984–1994 levels (i.e., those sustaining an atmospheric trend of +10 ppbv/yr), the methane concentration would rise to about 1850 ppbv over the next 40 years. If methane emissions were to remain constant at their current (1994) levels (i.e., those sustaining an atmosphere trend of 8ppbv/yr), the methane concentration would rise to about 1820 ppbv over the next 40 years. If emissions were cut by about 30 Tg(CH_4)/yr (about 8% of current anthropogenic emissions), CH_4 concentrations would remain at today's levels. These estimates are lower than those in IPCC (1994).

If emissions of N_2O were held constant at today's level, the concentration would climb from 312 ppbv to about 400 ppbv over several hundred years. In order for the concentration to be stabilised near current levels, anthropogenic sources would need to be reduced by more than 50%. Stabilisation of PFCs and SF_6 concentrations can only be achieved effectively by stopping emissions.

Because of their short lifetime, future tropospheric

aerosol concentrations would respond almost immediately to changes in emissions. For example, control of sulphur emissions would immediately reduce the amount of sulphate aerosol in the atmosphere.

C Observed Trends and Patterns in Climate and Sea Level

Section B demonstrated that human activities have changed the concentrations and distributions of greenhouse gases and aerosols over the 20th century; this section discusses the changes in temperature, precipitation (and related hydrological variables), climate variability and sea level that have been observed over the same period. Whether the observed changes are in part induced by human activities is considered in Section E.

C.1 Has the climate warmed?

Global average surface air temperature, excluding Antarctica, is about 15°C. Year-to-year temperature *changes* can be computed with much more confidence than the absolute global average temperature.

The mean global surface temperature has increased by about 0.3° to 0.6°C since the late 19th century, and by about 0.2° to 0.3°C over the last 40 years, the period with most credible data (see Figure 8 which shows data up to the end of 1994). The warming occurred largely during two periods, between 1910 and 1940 and since the mid-1970s. The estimate of warming has not significantly changed since the IPCC (1990) and IPCC (1992). Warming is evident in both sea surface and land-based surface air

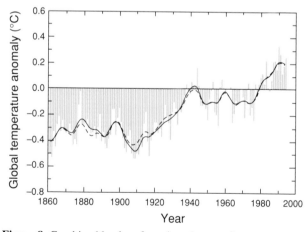

Figure 8: Combined land-surface air and sea surface temperatures (°C) 1861 to 1994, relative to 1961 to 1990. The solid curve represents smoothing of the annual values shown by the bars to suppress sub-decadal time-scale variations. The dashed smoothed curve is the corresponding result from IPCC (1992).

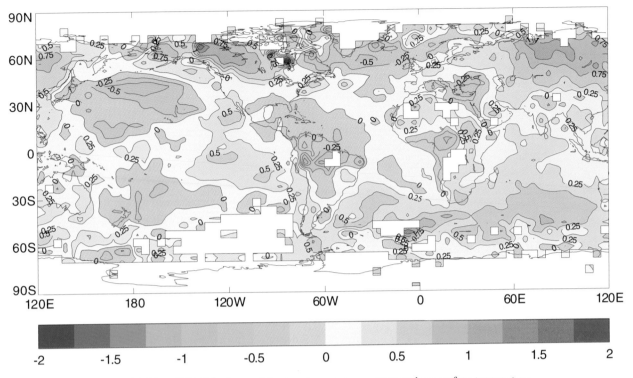

Figure 9: Change (from 1955–74 to 1975–94) of annual land-surface air temperature and sea surface temperature.

temperatures. Urbanisation in general and desertification could have contributed only a small part (a few hundredths of a degree) of the overall global warming, although urbanisation influences may have been important in some regions. Indirect indicators, such as borehole temperatures and glacier shrinkage, provide independent support for the observed warming. Recent years have been among the warmest since 1860, i.e., in the period of instrumental record.

The warming has not been globally uniform. The recent warmth has been greatest over the continents between 40°N and 70°N. A few areas, such as the North Atlantic Ocean north of 30°N, and some surrounding land areas, have cooled in recent decades (Figure 9).

As predicted in IPCC (1992) and discussed in IPCC (1994), relatively cooler global surface and tropospheric temperatures (by about 0.5°C) and a relatively warmer lower stratosphere (by about 1.5°C) were observed in 1992 and early 1993, following the 1991 eruption of Mt. Pinatubo. Warmer temperatures at the surface and in the lower troposphere, and a cooler lower stratosphere, reappeared in 1994 following the removal by natural processes of Mt. Pinatubo aerosols from the stratosphere.

The general tendency toward reduced daily temperature range over land, at least since the middle of the 20th

century, noted in IPCC (1992), has been confirmed with more data (which have now been analysed for more than 40% of the global land area). The range has decreased in many areas because nights have warmed more than days. Minimum temperatures have typically increased twice as much as maximum temperatures over the last 40 years. A likely explanation, in addition to the effects of enhanced greenhouse gases, is an increase in cloud cover which has been observed in many of the areas with reduced diurnal temperature range. An increase in cloud reduces diurnal temperature range both by obstructing daytime sunshine, and by preventing the escape of terrestrial radiation at night. Anthropogenic aerosols may also have an influence on daily temperature range.

Temperature trends in the free atmosphere are more difficult to determine than at the surface as there are fewer data and the records are much shorter. Radiosonde data which are available for the period since the 1950s show warming trends of around 0.1°C per decade, as at the surface, but since 1979 when satellite data of global tropospheric temperatures became available, there appears to have been a slight cooling (about –0.06°C per decade), whereas surface measurements still show a warming. These apparently contradictory trends can be reconciled if the diverse response of the troposphere and surface to short-term events such as volcanic eruptions and El Niño are taken into account. After adjustment for these transient effects, both tropospheric and surface data show slight warming (about 0.1°C per decade for the troposphere and nearly 0.2°C per decade at the surface) since 1979.

Cooling of the lower stratosphere has been observed since 1979 both by satellites and weather balloons, as noted in IPCC (1992) and IPCC (1994). Current global mean stratospheric temperatures are the coldest observed in the relatively short period of the record. Reduced stratospheric temperature has been projected to accompany both ozone losses in the lower stratosphere and atmospheric increases of carbon dioxide.

C.2 Is the 20th century warming unusual?

In order to establish whether the 20th century warming is part of the natural variability of the climate system or a response to anthropogenic forcing, information is needed on climate variability on relevant time-scales. As an average over the Northern Hemisphere for summer, recent decades appear to be the warmest since at least 1400 from the limited available evidence (Figure 10). The warming over the past century began during one of the colder periods of the last 600 years. Prior to 1400 data are insufficient to provide hemispheric temperature estimates. Ice core data from several sites suggest that the 20th century is at least as warm as any century in the past 600 years, although the recent warming is not exceptional everywhere.

Large and rapid climatic changes occurred during the last glacial period (around 20,000 to 100,000 years ago) and during the transition period towards the present interglacial (the last 10,000 years, known as the Holocene). Changes in annual mean temperature of about 5°C occurred over a few decades, at least in Greenland and the North Atlantic, and were probably linked to changes in oceanic circulation. These rapid changes suggest that climate may be quite sensitive to internal or external climate forcings and feedbacks. The possible relevance of these rapid climate changes to future climate is discussed in Section F.5.

Temperatures have been less variable during the last 10,000 years. Based on the incomplete evidence available, it is unlikely that global mean temperatures have varied by more than 1°C in a century during this period.

C.3 Has the climate become wetter?

As noted in IPCC (1992), precipitation has increased over land in high latitudes of the Northern Hemisphere, especially during the cold season. A step-like decrease of precipitation occurred after the 1960s over the subtropics and tropics from Africa to Indonesia. These changes are consistent with changes in streamflow, lake levels and soil moisture (where data analyses are available). Precipitation, averaged over global land areas, increased from the start of the century up to about 1960. Since about 1980 precipitation over land has decreased (Figure 11).

There is evidence to suggest increased precipitation over the central equatorial Pacific Ocean in recent decades, with decreases to the north and south. Lack of data prevents us from reaching firm conclusions about other precipitation changes over the ocean.

Estimates suggest that evaporation may have increased over the tropical oceans (although not everywhere) but decreased over large portions of Asia and North America. There has also been an observed increase in atmospheric water vapour in the tropics, at least since 1973.

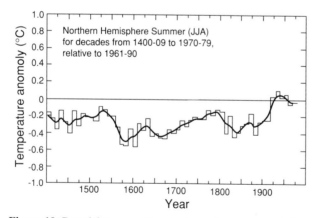

Figure 10: Decadal summer (June to August) temperature index for the Northern Hemisphere (to 1970–1979) based on 16 proxy records (tree-rings, ice cores, documentary records) from North America, Europe and East Asia. The thin line is a smoothing of the same data. Anomalies are relative to 1961 to 1990.

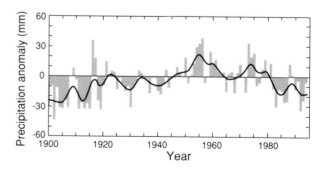

Figure 11: Changes in land-surface precipitation averaged over regions between 55°S and 85°N. Annual precipitation departures from the 1961–90 period are depicted by the hollow bars. The continuous curve is a smoothing of the same data.

(a) **Temperature indicators**

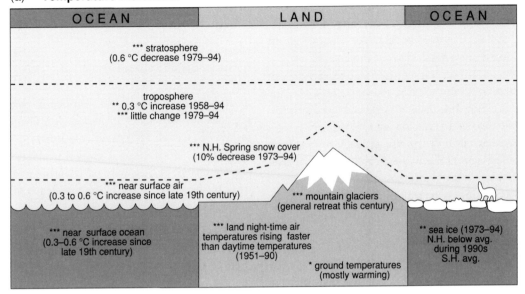

(b) **Hydrological indicators**

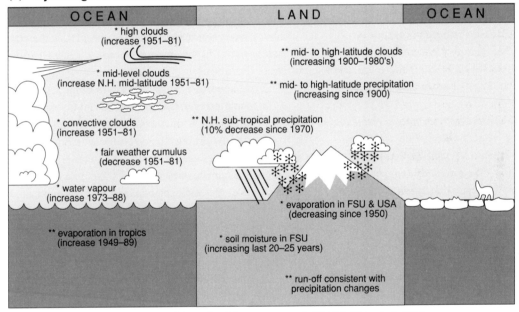

Asterisk indicates confidence level (i.e., assessment): * high, ** medium, * low**

Figure 12: Summary of observed climatic trends during the instrumental period of record.

Cloudiness appears to have increased since the 1950s over the oceans. In many land areas where the daily temperature range has decreased (see Section C.1), cloudiness increased from the 1950s to at least the 1970s.

Snow cover extent over the Northern Hemisphere land surface has been consistently below the 21-year average (1974 to 1994) since 1988. Snow-radiation feedback has amplified springtime warming over mid- to high latitude Northern Hemisphere land areas.

A summary of observed climate trends is shown in Figure 12.

C.4 Has sea level risen?

Over the last 100 years global sea level has risen by about 10 to 25 cm, based on analyses of tide gauge records. A major source of uncertainty in estimating the rate of rise is the influence of vertical land movements, which are included in sea level measurements made by tide gauges. Since IPCC (1990), improved methods for filtering out the effects of long-term vertical land movements, as well as greater reliance on the longest tide-gauge records for estimating trends, have provided greater confidence that the volume of ocean water has, in fact, been increasing and causing sea level to rise within the indicated range.

It is likely that much of the rise in sea level has been related to the concurrent rise in the global temperature over the last 100 years. On this time-scale, the warming and consequent expansion of the oceans may account for about 2 to 7 cm of the observed rise in sea level, while the observed retreat of glaciers and ice-caps may account for about 2 to 5 cm. Other factors are more difficult to quantify. Changes in surface and ground water storage may have caused a small change in sea level over the last 100 years. The rate of observed sea level rise suggests that there has been a net positive contribution from the huge ice sheets of Greenland and Antarctica, but observations of the ice sheets do not yet allow meaningful quantitative estimates of their separate contributions. The ice sheets remain a major source of uncertainty in accounting for past changes in sea level, because there are insufficient data about these ice sheets over the last 100 years.

C.5 *Has the climate become more variable and/or extreme?*

Many of the impacts of climate change may result from changes in climate variability or extreme weather events. Some reports have already suggested an increase in variability or extremes has taken place in recent decades. Do meteorological records support this?

There are inadequate data to determine whether consistent global changes in climate variability or extremes have occurred over the 20th century. On regional scales there is clear evidence of changes in some extremes and climate variability indicators (e.g., fewer frosts in several widespread areas; an increase in the proportion of rainfall from extreme events over the contiguous states of the USA). Some of these changes have been toward greater variability; some have been toward lower variability.

There have been relatively frequent El Niño-Southern Oscillation warm phase episodes, with only rare excursions into the other extreme of the phenomenon since 1977, as noted in IPCC (1990). This behaviour, and especially the persistent warm phase from 1990 to mid-1995, is unusual in the last 120 years (i.e., since instrumental records began). The relatively low rainfall over the subtropical land areas in the last two decades is related to this behaviour.

D Modelling Climate and Climate Change

Climate models which incorporate, in various degrees of complexity, mathematical descriptions of the atmosphere, ocean, land, biosphere and cryosphere, are important tools for understanding climate and climate change of the past, the present and the future. These models, which use primarily physical laws and physically based empirical relations, are very much more complete than, for example, models based on statistical relationships used in less quantitative disciplines. Detailed projections of future climate change rely heavily on coupled atmosphere-ocean models (see Box 2). How much confidence should we have in predictions from such models?

D.1 *The basis for confidence in climate models*

As discussed in Section B, changes in the radiatively active trace gases in the atmosphere produce radiative forcing. For equivalent CO_2 concentrations equal to twice the pre-industrial concentration, the positive radiative forcing is about +4 Wm^{-2}. To restore the radiative balance other changes in climate must occur. The initial reaction is for the lower atmosphere (the troposphere) and the Earth's surface to warm; in the absence of other changes, the warming would be about 1.2°C. However, heating not only changes temperatures, but also alters other aspects of the climate system and various feedbacks are invoked (see Section D.2). The key role of climate models is to quantify these feedbacks and determine the overall climate response. Further, the warming and other climate effects will not be uniform over the Earth's surface; an important role of models is to simulate possible continental and regional scale climate responses.

Climate models include, based on our current understanding, the most important large scale physical processes governing the climate system. Climate models have improved since IPCC (1990), but so too has our understanding of the complexity of the climate system and the recognition of the need to include additional processes.

In order to assess the value of a model for projections of future climate, its simulated climate can be compared with known features of the observed current climate and, to a less satisfactory degree, with the more limited information from significantly different past climate states. It is important to realise that even though a model may have deficiencies, it can still be of value in quantifying the climate response to anthropogenic climate forcing (see also Box 2). Several factors give us some confidence in the ability of climate models to simulate important aspects of anthropogenic climate change in response to anticipated changes in atmospheric composition:

Box 2: What tools are used to project future climate and how are they used?

Future climate is projected using climate models. The most highly developed climate models are atmospheric and oceanic general circulation models (GCMs). In many instances GCMs of the atmosphere and oceans, developed as separate models, are combined, to give a *coupled* GCM (termed here a coupled atmosphere-ocean model). These models also include representations of land-surface processes, sea ice related processes and many other complex processes involved in the climate system. GCMs are based upon physical laws that describe the atmospheric and oceanic dynamics and physics, and upon empirical relationships, and their depiction as mathematical equations. These equations are solved numerically with computers using a three-dimensional grid over the globe. For climate, typical resolutions are about 250 km in the horizontal and 1 km in the vertical in atmospheric GCMs, often with higher vertical resolution near the surface and lower resolution in the upper troposphere and stratosphere. Many physical processes, such as those related to clouds, take place on much smaller spatial scales and therefore cannot be properly resolved and modelled explicitly, but their average effects must be included in a simple way by taking advantage of physically based relationships with the larger scale variables (a technique known as parametrization).

Useful weather forecasts can be made using atmospheric GCMs for periods up to about ten days. Such forecasts simulate the evolution of weather systems and describe the associated weather. For simulation and projection of climate, on the other hand, it is the statistics of the system that are of interest rather than the day-to-day evolution of the weather. The statistics include measures of variability as well as mean conditions, and are taken over many weather systems and for several months or more.

When a model is employed for climate projection it is first run for many simulated decades without any changes in external forcing in the system. The quality of the simulation can then be assessed by comparing statistics of the mean climate, the annual cycle and the variability on different time-scales with observations of the current climate. The model is then run with changes in external forcing, for instance with changing greenhouse gas concentrations. The differences between the two climates provide an estimate of the consequent climate change due to changes in that forcing factor. This strategy is intended to simulate changes or perturbations to the system and partially overcomes some imperfections in the models.

Comprehensive coupled atmosphere-ocean models are very complex and take large computer resources to run. To explore all the possible scenarios and the effects of assumptions or approximations in parameters in the model more thoroughly, simpler models are also widely used and are constructed to give results similar to the GCMs when globally averaged. The simplifications may involve coarser resolution, and simplified dynamics and physical processes. An example is the upwelling diffusion-energy balance model. This represents the land and ocean areas in each hemisphere as individual "boxes", with vertical diffusion and upwelling to model heat transport within the ocean.

Early climate experiments, using atmospheric GCMs coupled to a simple representation of the ocean, were aimed at quantifying an equilibrium climate response to a doubling of the concentration of (equivalent) CO_2 in the atmosphere. Such a response portrays the final adjustment of the climate to the changed CO_2 concentration (see Glossary). The range of global warming results is typically between 1.5 and 4.5°C. The temporal evolution and the regional patterns of climate change may depend significantly on the time dependence of the change in forcing. It is important, therefore, to make future projections using plausible evolving scenarios of anthropogenic forcing and coupled atmosphere-ocean models so that the response of the climate to the forcing is properly simulated. These climate simulations are often called "transient experiments" (see Glossary) in contrast to an equilibrium response.

The main uncertainties in model simulations arise from the difficulties in adequately representing clouds and their radiative properties, the coupling between the atmosphere and the ocean, and detailed processes at the land surface.

(i)   The most successful climate models are able to simulate the important large-scale features of the components of the climate system well, including seasonal, geographical and vertical variations which are a consequence of the variation of forcing and dynamics in space and time. For example, Figure 13 shows the geographical distribution of December to February surface temperature and June to August precipitation simulated by comprehensive coupled atmosphere-ocean models of the type used for climate prediction, compared with observations. The large scale features are reasonably well captured by the

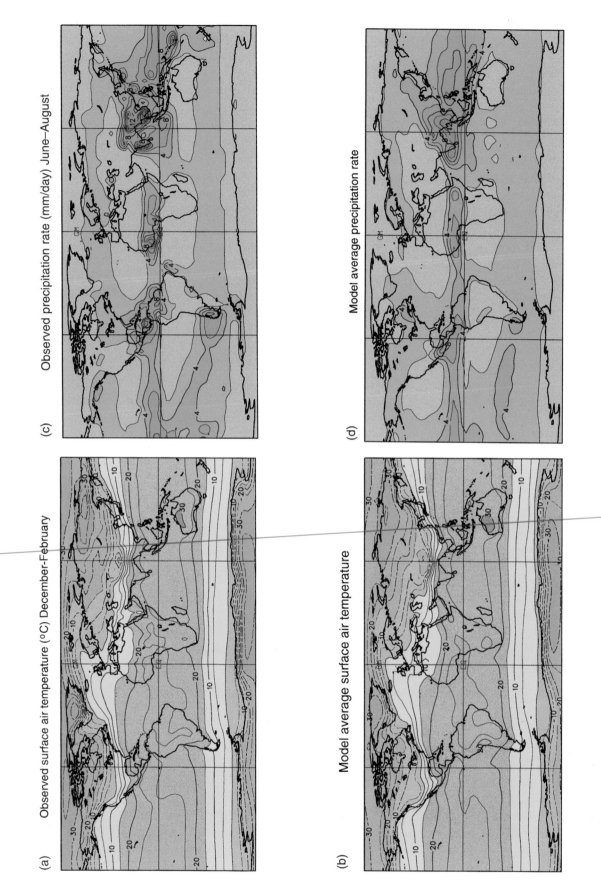

(a)   Observed surface air temperature (°C) December–February

(b)   Model average surface air temperature

(c)   Observed precipitation rate (mm/day) June–August

(d)   Model average precipitation rate

Figure 13: The geographical distribution of December to February surface temperature (a) and June to August precipitation (c) compared to that simulated by comprehensive coupled models of the type used for climate projection (b) and (d).

models, although at regional scales more discrepancies can be seen. Other seasons are similarly well simulated, indicating the ability of models to reproduce the seasonal cycle in response to changes in solar forcing. The improvement since IPCC (1990) is that this level of accuracy is achieved in models with a fully-interactive ocean as compared to the majority of models that employed simpler schemes used in 1990.

(ii) Many climate changes are consistently projected by different models in response to greenhouse gases and aerosols and can be explained in terms of physical processes which are known to be operating in the real world, for example, the maximum warming in high northern latitudes in winter (see Section F).

(iii) The models reproduce with reasonable fidelity other less obvious variations in climate due to changes in forcing:

- Some atmospheric models when forced with observed sea surface temperature variations can reproduce with moderate to good skill several regional climate variations, especially in parts of the tropics and sub-tropics. For example, aspects of the large scale interannual atmosphere fluctuations over the tropical Pacific relating to the El Niño-Southern Oscillation phenomenon are captured, as are interannual variations in rainfall in north-east Brazil and to some extent decadal variations in rainfall over the Sahel.

- As discussed in IPCC (1994), stratospheric aerosols resulting from the eruption of Mt. Pinatubo in June 1991 gave rise to a short-lived negative global mean radiative forcing of the troposphere which peaked at -3 to -4 Wm^{-2} a few months after the eruption and had virtually disappeared by about the end of 1994. A climate model was used to predict global temperature variations between the time of the eruption and the end of 1994 and the results agreed closely with observations (Figure 14). Such agreement increases confidence in the ability of climate models to respond in a realistic way to transient, planetary-scale radiative forcings of large magnitude.

- Previous IPCC reports demonstrated the ability of models to simulate some known features of palaeoclimate. Only modest progress has been made in this area, mainly because of the paucity of reliable data for comparison.

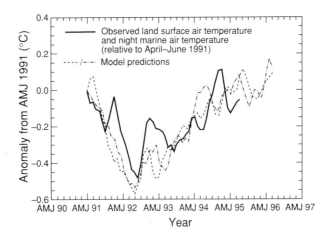

Figure 14: Predicted and observed changes in global land and ocean surface air temperature after the eruption of Mt. Pinatubo. Lines represent changes of three-month running mean temperature from April to June 1991 until March to May 1995. The two model lines represent predictions starting from different initial atmospheric conditions.

- Currently available model simulations of global mean surface temperature trend over the past half century show closer agreement with observations when the simulations include the likely effect of aerosol in addition to greenhouse gases (Figure 15).

(iv) The model results exhibit "natural" variability on a wide range of time- and space-scales which is broadly comparable to that observed. This "natural" variability arises from the internal processes at work in the climate system and not from changes in external forcing. Variability is a very important

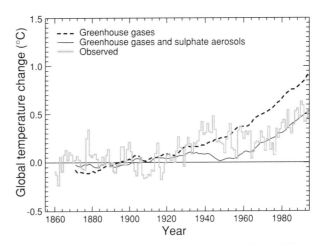

Figure 15: Simulated global annual mean warming from 1860 to 1990 allowing for increases in greenhouse gases only (dashed curve) and greenhouse gases and sulphate aerosols (solid curve), compared with observed changes over the same period.

aspect of the behaviour of the climate system and has important implications for the detection of climate change (see Section E). The year to year variations of surface air temperature for the current climate are moderately realistic in model simulations at the larger space-scales. For example the smaller variability over the oceans compared with continental interiors is captured. Too low interannual variability of the tropical east and central Pacific Ocean temperatures associated with the El Niño-Southern Oscillation (ENSO) phenomenon is one deficiency. No current coupled atmosphere-ocean model simulates all aspects of ENSO events, but some of the observed interannual variations in the atmosphere associated with these events are captured.

Climate models are calibrated, in part, by introducing adjustments which are empirically determined. The most striking example of these are the systematic adjustments (the so-called flux adjustments) that are used in some models at the atmosphere-ocean interface in order to bring the simulated climate closer to the observed state. These adjustments are used to compensate for model errors, for example inadequacies in the representation of clouds in atmospheric models. Flux adjustments, which can be quite large in some models, are used to ensure that the simulated present day climate is realistic and hence that climate feedback processes operate in the appropriate range in climate change simulations. Many features of the response of models with and without flux adjustments to increasing greenhouse gases are qualitatively similar. The most substantial differences in simulated climate change can generally be traced to deficiencies in the simulation of current climate in the unadjusted models, for example, systematic errors in sea ice. The main unknown regarding the use of adjustments in models is the extent to which they allow important non-linear processes to operate in the models. They have been tested with a good degree of success against known climate variations including the seasonal cycle and the perturbations mentioned above. This provides some confidence in their use for future climate perturbations caused by human activities.

In summary, confidence in climate models has increased since 1990. Primary factors that have served to raise our confidence are model improvements, e.g., the successful incorporation of additional physical processes (such as cloud microphysics and the radiative effects of sulphate aerosols) into global coupled models, and the improvement in such models' simulation of the observed changes in climate over recent decades. Further confidence will be gained as models continue to improve.

D.2 Climate model feedbacks and uncertainties

Warming from radiative forcing will be modified by climate feedbacks which may either amplify (a positive feedback) or reduce (a negative feedback) the initial response. The likely equilibrium response of global surface temperature to a doubling of equivalent carbon dioxide concentration (the "climate sensitivity") was estimated in 1990 to be in the range 1.5 to 4.5 °C, with a "best estimate" of 2.5°C. The range of the estimate arises from uncertainties in the climate models and in their internal feedbacks, particularly those concerning clouds and related processes. No strong reasons have emerged to change these estimates of the climate sensitivity. The present activities regarding incorporation of these feedback processes in models are described below.

Water vapour feedback

An increase in the temperature of the atmosphere increases its water holding capacity and is expected to be accompanied by an increase in the amount of water vapour. Since water vapour is a powerful greenhouse gas, the increased water vapour would in turn lead to a further enhancement of the greenhouse effect (a positive feedback).

About half of this feedback depends on water vapour in the upper troposphere, whose origin and response to surface temperature increase is not fully understood. Feedback by water vapour in the lower troposphere is unquestionably positive and the preponderance of evidence points to the same conclusion for upper tropospheric water vapour. Feedbacks resulting from changes in the decrease of temperature with height can partially compensate the water vapour feedback.

Cloud/radiative feedback

Several processes are involved in cloud/radiative feedback. Clouds can both absorb and reflect solar radiation (which cools the surface) and absorb and emit long-wave radiation (which warms the surface), depending on cloud height, thickness and cloud radiative properties. The radiative properties of clouds depend on the evolution of atmospheric water in its vapour, liquid and ice phases and upon atmospheric aerosols. The processes are complex and, although considerable progress has been made since IPCC (1990) in describing and modelling those cloud processes that are most important for determining radiative and hence temperature changes, their uncertainty represents a significant source of potential error in climate simulation.

This potential error can be estimated by first noting that if clouds and sea ice are kept fixed according to their observed distributions and properties, climate models would all report climate sensitivities in the range of 2 to 3°C. Modellers have shown for various assumptions that physically plausible changes in cloud distribution could either as much as double the warming expected for fixed clouds or, on the other hand, reduce it by up to 1°C. The range in estimated climate sensitivity of 1.5 to 4.5°C is largely dictated by this uncertainty.

Ocean circulation

Oceans play an important role in climate because they carry large amounts of heat from the tropics to the poles. They also store large amounts of heat, carbon and CO_2 and are a major source of water to the atmosphere (through evaporation). Coupling of atmospheric and oceanic GCMs (see Box 2) improves the physical realism of models used for projections of future climate change, particularly the timing and regional distribution of the changes.

Several models show a decrease or only marginal increase of sea surface temperatures in the northern North Atlantic in response to increasing greenhouse gases, related to a slowing down of the thermohaline circulation as the climate warms. This represents a local negative temperature feedback, although changes in cloud cover might be an important factor. The main influence of the oceans on simulations of climate change occurs because of their large heat capacity, which introduces a delay in warming that is not uniform spatially.

Ice and snow albedo feedback

An ice or snow covered surface strongly reflects solar radiation (i.e., it has a high "albedo"). As some ice melts at the warmer surface, less solar radiation is reflected leading to further warming (a positive feedback), but this is complicated by clouds, leads (areas of open water in sea ice) and snow cover.

The realism of simulated sea ice cover varies considerably between models, although sea ice models that include ice dynamics are showing increased accuracy.

Land-surface/atmosphere interactions

Anthropogenic climate changes, e.g., increased temperature, changes in precipitation, changes in net radiative heating and the direct effects of CO_2, will influence the state of the land surface (soil moisture, albedo, roughness, vegetation). In turn, the altered land surface can feed back and alter the overlying atmosphere (precipitation, water vapour, clouds). Changes in the composition and structure of ecosystems can alter not only physical climate, but also the biogeochemical cycles (see Section B). Although land-surface schemes used in current GCMs may be more sophisticated than in IPCC (1990), the disparity between models in their simulation of soil moisture and surface heat and moisture fluxes has not been reduced. Confidence in calculation of regional projections of soil moisture changes in response to greenhouse gas and aerosol forcing remains low.

Changes in vegetation can potentially further modify climate locally and regionally by altering the exchange of water and energy between the land surface and atmosphere. For example, forests spreading into tundra in a warmer world would absorb a greater proportion of solar energy and increase the warming. This feedback would be modified by land-use changes such as deforestation. Coupled atmosphere-ocean models used for climate change studies do not yet include such interactions between climate and vegetation, and such feedbacks may have important effects on regional climate change projection.

E Detection of Climate Change and Attribution of Causes

An important question is: does the instrumental record of temperature change show convincing evidence of a human effect on global climate? With respect to the increase in global mean temperature over the last 100 years, IPCC (1990) concluded that the observed warming was "broadly consistent with predictions of climate models, but it is also of the same magnitude as natural climate variability". The report went on to explain that "the observed increase could be largely due to this natural variability; alternatively this variability and other human factors could have offset a still larger human-induced greenhouse warming".

Since IPCC (1990), considerable progress has been made in the search for an identifiable human-induced effect on climate.

E.1 Better simulations for defining a human-induced climate change "signal"

Experiments with GCMs are now starting to incorporate some of the forcing due to human-induced changes in sulphate aerosols and stratospheric ozone. The inclusion of these additional factors has modified in important ways the picture of how climate might respond to human influences. Furthermore, we now have information on both the timing and spatial patterns of human-induced climate change from a large (>18) number of transient experiments in which coupled atmosphere-ocean models are driven by past and/or projected future time-dependent changes in CO_2 concentration (used as a surrogate to represent the combined effect of CO_2 concentration and other well-mixed greenhouse gases; see "equivalent CO_2" in the Glossary). Some of these experiments have been repeated with identical forcing but starting from a slightly different initial climate state. Such repetitions help to better define the expected climate response to increasing greenhouse gases and aerosols. However, important uncertainties remain; for example, no model has incorporated the full range of anthropogenic forcing effects.

E.2 Better simulations for estimating natural internal climate variability

In observed data, any "signal" of human effects on climate must be distinguished from the background "noise" of climate fluctuations that are entirely natural in origin. Such natural fluctuations occur on a variety of space- and time-scales, and can be purely internal (due to complex interactions between individual components of the climate system, such as the atmosphere and ocean) or externally driven by changes in solar variability or the volcanic aerosol loading of the atmosphere. It is difficult to separate a signal from the noise of natural variability in the observations. This is because there are large uncertainties in the evolution and magnitude of both human and natural forcings, and in the characteristics of natural internal variability, which translate to uncertainties in the relative magnitudes of signal and noise.

In the modelled world, however, it is possible to perform multi-century control experiments with no human-induced changes in greenhouse gases, sulphate aerosols or other anthropogenic forcings. Since 1990, a number of such control experiments have been performed with coupled atmosphere-ocean models. These yield important information on the patterns, time-scales, and magnitude of the "internally generated" component of natural climate variability. This information is crucial for assessing whether observed changes can be plausibly explained by internal climatic fluctuations, but constitutes only one part of the "total" natural variability of climate (since such control runs do not include changes in solar output or volcanic aerosols). Uncertainties still remain in estimates of both internal and total natural climate variability, particularly on the decadal-to-century time-scales.

E.3 Studies of global mean change

Most studies that have attempted to detect an anthropogenic effect on climate have used changes in global mean, annually averaged temperature only. These investigations compared observed·changes over the past 10-100 years with estimates of internal or total natural variability noise derived from palaeodata, climate models,

or statistical models fitted to observations. Most but not all of these studies show that the observed change in global mean, annually averaged temperature over the last century is unlikely to be due entirely to natural fluctuations of the climate system.

These global mean results cannot establish a clear cause and effect link between observed changes in atmospheric greenhouse gas concentrations and changes in the Earth's surface temperature. This is the attribution issue. Attribution is difficult using global mean changes only because of uncertainties in the histories and magnitudes of natural and human-induced forcings: there are many possible combinations of these forcings that could yield the same curve of observed global mean temperature change. Some combinations are more plausible than others, but relatively few data exist to constrain the range of possible solutions. Nevertheless, model-based estimates of global temperature increase over the last 130 years are more easily reconciled with observations when the likely cooling effect of sulphate aerosols is taken into account, and provide qualitative support for an estimated range of climate sensitivity consistent with that given in IPCC (1990) (Figure 16).

E.4 Studies of patterns of change

To better address the attribution problem, a number of recent studies have compared observations with model-predicted *patterns* of temperature-change in response to anthropogenic forcing. The argument underlying pattern-based approaches is that different forcing mechanisms ("causes") may have different patterns of response ("effects"), particularly if one considers the full three- or even four-dimensional structure of the response, e.g., temperature change as a function of latitude, longitude, height and time. Thus a good match between modelled and observed multi-dimensional patterns of climate change would be difficult to achieve for "causes" other than those actually used in the model experiment.

Several studies have compared observed patterns of temperature change with model patterns of change from simulations with changes in both greenhouse gases and anthropogenic sulphate aerosols. These comparisons have been made at the Earth's surface and in vertical sections through the atmosphere. While there are concerns regarding the relatively simple treatment of aerosol effects in model experiments, and the neglect of other potentially significant contributions to the radiative forcing, all such pattern comparison studies show significant correspondence between the observations and model predictions (an example is shown in Figure 17). Much of

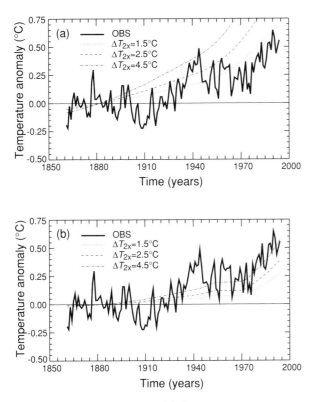

Figure 16: Observed changes in global mean temperature over 1861 to 1994 compared with those simulated using an upwelling diffusion-energy balance climate model. The model was run first with forcing due to greenhouse gases alone (a) and then with greenhouse gases and aerosols (b).

the model-observed correspondence in these experiments occurs at the largest spatial scales – for example, temperature differences between hemispheres, land and ocean, or the troposphere and stratosphere. Model predictions are more reliable at these spatial scales than at the regional scale. Increasing confidence in the identification of a human-induced effect on climate comes primarily from such pattern-based work. For those seasons during which aerosol effects should be most pronounced the pattern correspondence is generally higher than that achieved if model predictions are based on changes in greenhouse gases alone (Figure 17).

As in the global mean studies, pattern-oriented detection work relies on model estimates of internal natural variability as the primary yardstick for evaluating whether observed changes in temperature patterns could be due to natural causes. Concerns remain regarding the reliability of this yardstick.

E.5 Qualitative consistency
In addition to quantitative studies, there are broad areas of

qualitative agreement between observations and those model predictions that either include aerosol effects or do not depend critically on their inclusion. As in the quantitative studies, one must be cautious in assessing consistency because the expected climate change signal due to human activities is still uncertain, and has changed as our ability to model the climate system has improved. In addition to the surface warming, the model and observed commonalities in which we have most confidence include stratospheric cooling, reduction in diurnal temperature range, sea level rise, high latitude precipitation increases and water vapour and evaporation increase over tropical oceans.

E.6 Overall assessment of the detection and attribution issues
In summary, the most important results related to the issues of detection and attribution are:

- The limited available evidence from proxy climate indicators suggests that the 20th century global mean temperature is at least as warm as any other century since at least 1400 AD. Data prior to 1400 are too sparse to allow the reliable estimation of global mean temperature (see Section C.2).

- Assessments of the statistical significance of the observed global mean temperature trend over the last century have used a variety of new estimates of natural internal and externally forced variability. These are derived from instrumental data, palaeodata, simple and complex climate models, and statistical models fitted to observations. Most of these studies have detected a significant change and show that the observed warming trend is unlikely to be entirely natural in origin.

- More convincing recent evidence for the attribution of a human effect on climate is emerging from pattern-based studies, in which the modelled climate response to combined forcing by greenhouse gases and anthropogenic sulphate aerosols is compared with observed geographical, seasonal and vertical patterns of atmospheric temperature change. These studies show that such pattern correspondences increase with time, as one would expect as an anthropogenic signal increases in strength. Furthermore, the probability is very low that these correspondences could occur by chance as a result of natural internal variability only. The vertical patterns

(a)

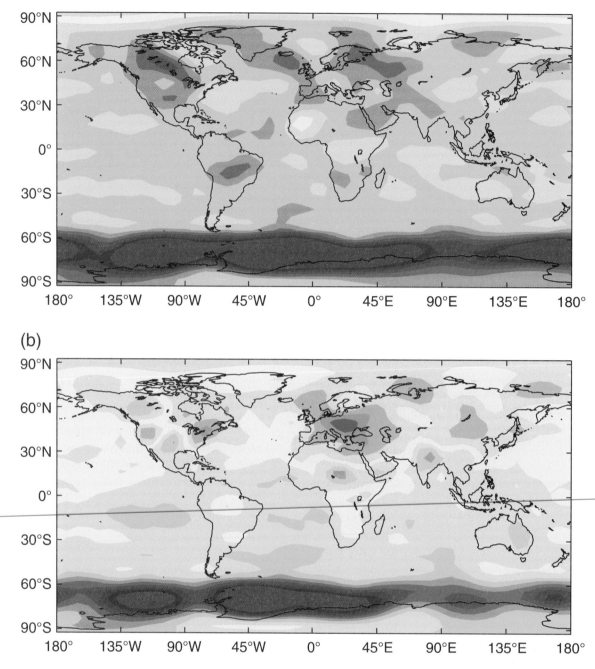

(b)

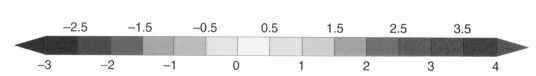

Figure 17: Annual mean near-surface air temperature changes (°C) from equilibrium response experiments with an atmospheric GCM with a mixed-layer ocean coupled to a tropospheric chemistry model, forced with present-day atmospheric concentrations of CO_2 (a) and by the combined effects of present-day CO_2 levels and sulphur emissions (b). Observed temperature changes from 1955–74 to 1975–94, shown in Figure 9, are qualitatively more similar to the changes in the combined forcing experiment than in the CO_2 only experiment.

of change are also inconsistent with those expected for solar and volcanic forcing.

- Our ability to quantify the human influence on global climate is currently limited because the expected signal is still emerging from the noise of natural variability, and because there are uncertainties in key factors. These include the magnitude and patterns of long-term natural variability and the time-evolving pattern of forcing by, and response to, changes in concentrations of greenhouse gases and aerosols, and land-surface changes. Nevertheless, the balance of evidence suggests that there is a discernible human influence on global climate.

F The Prospects for Future Climate Change

F.1 Forcing scenarios

Projections of future anthropogenic climate change depend, amongst other things, on the scenario used to force the climate model. The IS92 emission scenarios are used here for projections of changes in global mean temperature and sea level. The IS92 scenarios include emissions of both greenhouse gases and aerosol precursors (see Section B.9.1) and for the first time both factors have been taken into account in the global mean temperature and sea level projections (Section F.2).

In many coupled model experiments the forcing scenario is simplified by summing the radiative forcings of all the trace gases (CO_2, CH_4, O_3, etc.) and treating the total forcing as if it came from an "equivalent" concentration of CO_2. The rate of increase of "equivalent CO_2" in these experiments is often assumed to be a constant +1%/yr (compounded). For comparison the IS92a Scenario, neglecting the effect of aerosols, is equivalent to a compounded rate of increase varying from 0.77 to 0.84%/yr during the 21st century.

The projections of global mean temperature and sea level changes do not come directly from coupled atmosphere-ocean models. Though these are the most sophisticated tools available for making projections of future climate change they are computationally expensive, making it unfeasible to produce results based on a large number of emission scenarios. In order to assess global temperature and sea level projections for the full range of IS92 emission scenarios, simple upwelling diffusion-energy balance models (see Box 2) can be employed to interpolate and extrapolate the coupled model results. These models, used for similar tasks in IPCC (1990) and IPCC (1992), are calibrated to give the same globally

averaged temperature response as the coupled atmosphere-ocean models.

The climate simulations here are called *projections* instead of *predictions* to emphasise that they do not represent attempts to forecast the most likely (or "best estimate") evolution of climate in the future. The projections are aimed at estimating and understanding responses of the climate system to possible forcing scenarios.

F.2 Projections of climate change

F.2.1 Global mean temperature response to IS92 emission scenarios

Using the IS92 emission scenarios, which include emissions of both greenhouse gases and aerosol precursors (Section B.9.1) projected global mean temperature changes relative to 1990 were calculated for the 21st century. Temperature projections assuming the "best estimate" value of climate sensitivity, 2.5°C, (see Section D.2) are shown for the full set of IS92 scenarios in Figure 18. For IS92a the temperature increase by 2100 is about 2°C. Taking account of the range in the estimate of climate sensitivity (1.5 to 4.5°C) and the full set of IS92 emission scenarios, the models project an increase in global mean temperature of between 0.9 and 3.5°C (Figure 19). In all cases the average rate of warming would probably be greater than any seen in the last 10,000 years, but the actual annual to decadal changes would include considerable natural variability. Because of the thermal inertia of the oceans, global mean temperature would continue to

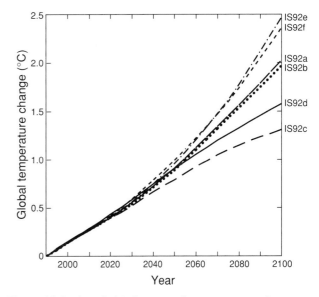

Figure 18: Projected global mean surface temperature changes from 1990 to 2100 for the full set of IS92 emission scenarios. A climate sensitivity of 2.5°C is assumed.

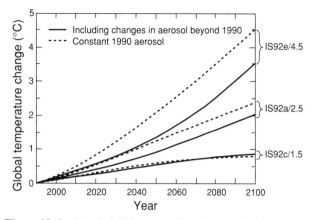

Figure 19: Projected global mean surface temperature change extremes from 1990 to 2100. The highest temperature changes assume a climate sensitivity of 4.5°C and the IS92e emission scenario; the lowest a climate sensitivity of 1.5°C and the IS92c emission scenario and the mid-range curve a climate sensitivity of 2.5°C and the IS92a Scenario. The solid curves include the effect of changing aerosol; the dashed curves assume aerosol emissions remain constant at their 1990 levels.

increase beyond 2100 even if concentrations of greenhouse gases were stabilised at that time. Only 50–90% of the eventual temperature changes are realised at the time of greenhouse gas stabilisation. All scenarios show substantial climate warming, even when the negative aerosol radiative forcing is accounted for. Although CO_2 is the most important anthropogenic greenhouse gas, the other greenhouse gases contribute significantly (about 30%) to the projected global warming.

To allow closer comparison with the projections presented in IPCC (1990) and IPCC (1992) and to illustrate the sensitivity of future global temperature to changes in aerosol concentrations, the same series of calculations were performed neglecting future aerosol changes, i.e. aerosol concentrations were held constant at 1990 levels. These lead to higher projections of temperature change. Taking account of the range in the estimate of climate sensitivity and the full set of IS92 emission scenarios, the models project an increase in global mean temperature of between 0.8 and 4.5°C. For IS92a, assuming the "best estimate" of climate sensitivity, the temperature increase by 2100 is about 2.4°C. For comparison, the corresponding temperature increase for IS92a presented in IPCC (1992) was 2.8°C. The projections in IPCC (1990) were based on an earlier set of emission scenarios, the "best estimate" for the increase in global temperature by 2100 (relative to 1990) was 3.3°C.

F.2.2 Global mean sea level response to IS92 emission scenarios

Using the IS92 emission scenarios, including greenhouse gas

and aerosol precursors, projected global mean sea level increases relative to 1990 were calculated for the 21st century. Sea level projections assuming the "best estimate" values for climate sensitivity and ice melt are shown for the full set of IS92 scenarios in Figure 20. For IS92a, the sea level rise by 2100 is 49 cm. For comparison, the "best estimate" of global sea level rise by 2100 given in IPCC (1990) was 66 cm. Also taking account of the ranges in the estimate of climate sensitivity and ice melt parameters, and the full set of IS92 emission scenarios, the models project an increase in global mean sea level of between 13 and 94 cm (Figure 21). During the first half of the next century, the choice of emission scenario has relatively little effect on the projected sea level rise due to the large thermal inertia of the ocean-ice-atmosphere climate system, but has increasingly larger effects in the latter part of the next century. In addition, because of the thermal inertia of the oceans, sea level would continue to rise for many centuries beyond 2100 even if concentrations of greenhouse gases were stabilised at that time. The projected rise in sea level is primarily due to thermal expansion as the ocean waters warm, but also due to increased melting of glaciers.

In these projections, the combined contributions of the Greenland and Antarctic ice sheets are projected to be relatively minor over the next century. However, the possibility of large changes in the volumes of these ice sheets (and, consequently, in sea level) cannot be ruled out, although the likelihood is considered to be low.

Changes in future sea level will not occur uniformly around the globe. Recent coupled atmosphere-ocean model

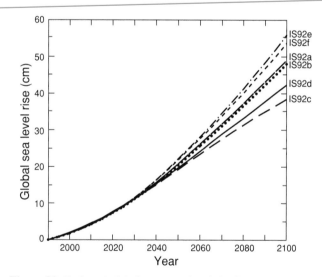

Figure 20: Projected global mean sea level rise from 1990 to 2100 for the full set of IS92 emission scenarios. A climate sensitivity of 2.5°C and mid-value ice melt parameters are assumed.

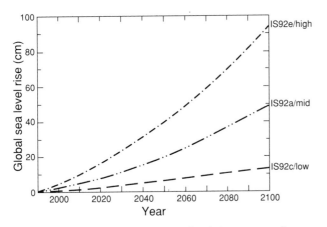

Figure 21: Projected global mean sea level rise extremes from 1990 to 2100. The highest sea level rise curve assumes a climate sensitivity of 4.5°C, high ice melt parameters and the IS92e emission scenario; the lowest a climate sensitivity of 1.5°C, low ice melt parameters and the IS92c emission scenario and the middle curve a climate sensitivity of 2.5°C, mid-value ice melt parameters and the IS92a Scenario.

experiments suggest that the regional responses could differ substantially, owing to regional differences in heating and circulation changes. In addition, geological and geophysical processes cause vertical land movements and thus affect relative sea levels on local and regional scales.

Tides, waves and storm surges could be affected by regional climate changes, but future projections are, at present, highly uncertain.

F.2.3 Temperature and sea level projections compared with IPCC (1990)

The global average temperature and sea level projections presented here for 1990 to 2100, both excluding and including changing aerosol emissions, are lower than the corresponding projections presented in IPCC (1990). Taking into account the negative radiative forcing of aerosols reduces projections of temperature and sea level rise. Those projections which *exclude* the effect of changing aerosol emissions are lower than IPCC (1990) for a number of reasons, mainly:

- The revised (IS92) emission scenarios have been used for all greenhouse gases. This is particularly important for CO_2 and CFCs.

- Revised treatment of the carbon cycle. The carbon cycle model used to calculate future temperature and sea level rise in IPCC (1990) and IPCC (1992) did

not incorporate the effect of carbon uptake through CO_2 fertilisation, resulting in higher future CO_2 concentrations for given emissions in IPCC (1990).

- The inclusion of aerosol effects in the pre-1990 radiative forcing history. The estimated historical changes of radiative forcing up to 1990, used in this report for global mean temperature and sea level projections, includes a component due to aerosols. This particularly affects projections of sea level rise, which are strongly influenced by the history of radiative forcing over the last century.

- Revised (and more realistic) parameters in the simple upwelling diffusion-energy balance climate model.

- The inclusion in the model of spatial variations in the climate sensitivity and the effect of changing strength of the thermohaline circulation, to accord with coupled atmosphere-ocean general circulation models.

- The use of improved models for the ice melt component of sea level rise.

F.3 Spatial patterns of projected climate change

Although in *global mean* terms, the effect of including aerosols is to reduce the projected warming (see Section F.2), it can be misleading to consider only the global mean surface temperature, which does not give an effective indication of climate change at smaller spatial scales.

Because aerosols are short-lived, they are unevenly distributed across the globe, being concentrated near regions where they are emitted. As a result, the spatial pattern of aerosol forcing is very different to that produced by the long-lived well-mixed greenhouse gases and, when considering patterns of climate change, their cooling effect is not a simple offset to the warming effect of greenhouse gases, as might be implied from the global mean results. Aerosols are likely to have a significant effect on future regional climate change.

Confidence is higher in hemispheric to continental scale projections of climate change (Section F.3.1) than at regional scales (Section F.3.2), where confidence remains low.

F.3.1 Continental scale patterns

In IPCC (1990), estimates of the patterns of future climate change were presented, the most robust of which related to continental and larger spatial scales. The results were based on GCM experiments which included the effect of greenhouse gases, but did not take into account the effects of aerosols.

The following provides some details of the changes on continental scales in experiments with greenhouse gases alone (generally represented by a 1%/yr increase in CO_2) and increases in greenhouse gas and aerosol concentrations (using aerosol concentration derived from the IS92a Scenario). It is important to realise that, in contrast to the many model results with CO_2 alone, there are only two recent coupled atmosphere-ocean model simulations that include the effects of both aerosols and CO_2, neither of which have yet been thoroughly analysed. We have concentrated on those changes which show most consistency between models, and for which plausible physical mechanisms have been identified.

Temperature and Precipitation

All model simulations, whether they are forced with increased concentrations of greenhouse gases and aerosols, or with increased greenhouse gas concentrations alone, show the following features:

- generally greater surface warming of the land than of the oceans in winter, as in equilibrium simulations (Figures 22 and 23);
- a minimum warming around Antarctica and in the northern North Atlantic which is associated with deep oceanic mixing in those areas;

- maximum warming in high northern latitudes in late autumn and winter associated with reduced sea ice and snow cover;
- little warming over the Arctic in summer;
- little seasonal variation of the warming in low latitudes or over the southern circumpolar ocean;
- a reduction in diurnal temperature range over land in most seasons and most regions;
- an enhanced global mean hydrological cycle;
- increased precipitation in high latitudes in winter.

Including the effects of aerosols in simulations of future climate leads to a somewhat reduced surface warming, mainly in middle latitudes of the Northern Hemisphere. The maximum winter warming in high northern latitudes is less extensive (compare Figures 22 and 23).

However, adding the cooling effect of aerosols is not a simple offset to the warming effect of greenhouse gases, but significantly affects some of the continental scale patterns of climate change. This is most noticeable in summer where the cooling due to aerosols tends to weaken monsoon circulations. For example, when the effects of both greenhouse gases and aerosols are included, Asian summer monsoon rainfall decreases, whereas in earlier simulations with only the effect of greenhouse gases represented, Asian summer monsoon rainfall increased.

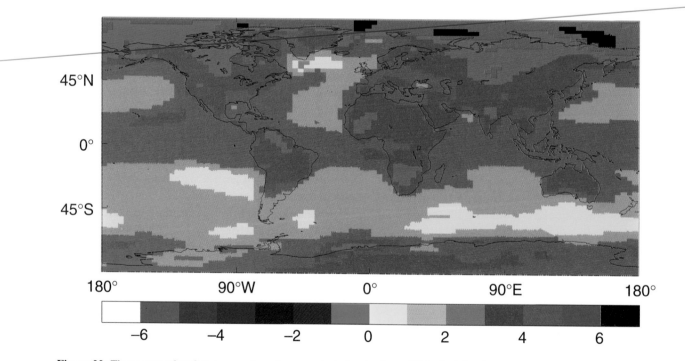

Figure 22: The pattern of surface temperature change projected at the time of CO_2 doubling from a transient coupled model experiment.

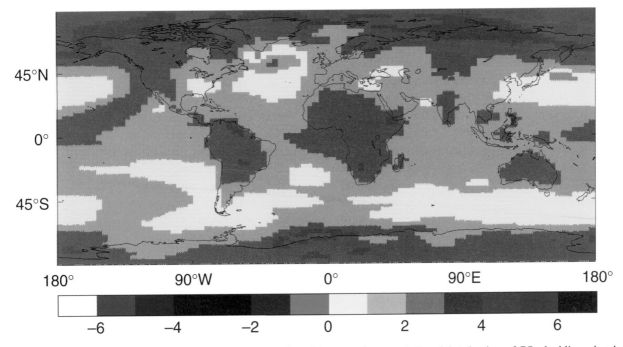

Figure 23: The pattern of surface temperature change projected by a transient coupled model at the time of CO_2 doubling when both CO_2 and aerosol concentration increases are taken into account.

Conversely, the addition of aerosol effects leads to an increase in precipitation over southern Europe, whereas decreases are found in simulations with greenhouse gases only. These changes will be sensitive to the aerosol scenario used, and the details of the parametrization of the radiative effects of aerosol. Other forcings, including that due to increases in tropospheric ozone, soot and the indirect effect of sulphate aerosols have been neglected and could influence these results. In general, regional projections are also sensitive to model resolution and are affected by large natural variability. Hence confidence in regional projections remains low.

With increases in CO_2 only, two coupled atmosphere-ocean models show a pattern of SST (sea surface temperature) change, precipitation change and anomalies in wind and ocean currents that resemble the warm phase of ENSO, as well as the observed decadal time-scale SST anomalies of the 1980s and early 1990s. This is characterised by a reduction of the east-west SST gradient in the tropical Pacific, though the magnitude of this effect varies among models.

Soil moisture
Although there is less confidence in simulated changes in soil moisture than in those of temperature, some of the results concerning soil moisture are dictated more by changes in precipitation and evaporation than by the detailed response of the surface scheme of the climate model. All model simulations, whether they are forced with increased concentrations of greenhouse gases and aerosols, or with increased greenhouse gas concentrations alone, produce predominantly increased soil moisture in high northern latitudes in winter. Over the northern continents in summer, the changes in soil moisture are sensitive to the inclusion of aerosol effects.

Ocean circulation
In response to increasing greenhouse gases, most models show a decrease in the strength of the northern North Atlantic oceanic circulation further reducing the strength of the warming around the North Atlantic. The increase in precipitation in high latitudes decreases the surface salinity, inhibiting the sinking of water at high latitude, which drives this circulation.

F.3.2 Regional scale patterns
Estimation of the potential impacts of climate change on human infrastructure and natural ecosystems requires projections of future climate changes at the regional scale, rather than as global or continental means.

Since IPCC (1990), a greater appreciation has been developed of the uncertainties in making projections at the regional scale. There are several difficulties:

- The global climate models used for future projections are run at fairly coarse resolution and do not adequately depict many geographic features (such as coastlines, lakes and mountains), surface vegetation, and the interactions between the atmosphere with the surface which become more important on regional scales. Considerable spread exists among model projections on the regional scale even when climate model experiments are driven by the same future radiative forcing scenario.

- There is much more natural variation in local climate than in climate averaged over continental or larger scales. This variation arises from locally generated variability from storms, interactions between the atmosphere and the oceans (such as ENSO), and from variations in soil moisture, sea ice, and other components of the climate system. Series or ensembles of model predictions started from different initial conditions allow both the mean climate and the superimposed variability to be determined.

- Because of their uneven spatial distribution, human induced tropospheric aerosols are likely to greatly influence future regional climate change. At present, however, there are very few projections of climate change with coupled atmosphere-ocean models (the type of model that gives more reliable information on the regional scale) which include the radiative effects of aerosols. Those that have been run include a very simplified representation of aerosol effects.

- Land-use changes are also believed to have a significant impact on temperature and precipitation changes, especially in the tropics and subtropics. Climate model experiments have shown the likelihood of substantial local climate change associated with deforestation in the Amazon, or desertification in the Sahel. Changes in land-use on small scales which cannot be foreseen are expected to continue to influence regional climate.

Because of these problems, no information on future regional climate change is presented here. However, this situation is expected to improve in the future as a result of:

- more coupled atmosphere-ocean model experiments with aerosol effects included;
- improvements in models, both from increased resolution and improved representation of small-scale processes;
- more refined scenarios for aerosols and other forcings.

F.3.3 Changes in variability and extremes

Small changes in the mean climate or climate variability can produce relatively large changes in the frequency of extreme events (defined as events where a certain threshold is surpassed); a small change in the variability has a stronger effect than a similar change in the mean.

Temperature

A general warming tends to lead to an increase in extremely high temperature events and a decrease in extremely low temperatures (e.g., frost days).

Hydrology

Many models suggest an increase in the probability of intense precipitation with increased greenhouse gas concentrations. In some areas a number of simulations show there is also an increase in the probability of dry days and the length of dry spells (consecutive days without precipitation). Where mean precipitation decreases, the likelihood of drought increases. New results reinforce the view that variability associated with the enhanced hydrological cycle translates into prospects for more severe droughts and/or floods in some places and less severe droughts and/or floods in other places.

Mid-latitude storms

In the few analyses available, there is little agreement between models on the changes in storminess that might occur in a warmer world. Conclusions regarding extreme storm events are obviously even more uncertain.

Hurricanes/Tropical cyclones

The formation of tropical cyclones depends not only on sea surface temperature (SST), but also on a number of atmospheric factors. Although some models now represent tropical storms with some realism for present day climate, the state of the science does not allow assessment of future changes.

El Niño-Southern Oscillation

Several global coupled models indicate that the ENSO-like SST variability they simulate continues with increased

CO_2. Associated with the mean increase of tropical SSTs as a result of increased greenhouse gas concentrations, there could be enhanced precipitation variability associated with ENSO events in the increased CO_2 climate, especially over the tropical continents.

F.4 Effects of stabilising greenhouse gas concentrations

Possible global temperature and sea level response to the scenarios for stabilising concentrations discussed in Section B.9.2 were calculated with the same upwelling diffusion-energy balance model used for the results in Sections F.2.1 and F.2.2.

For each of the pathways leading to stabilisation, the climate system shows considerable warming during the 21st century. Figure 24 shows temperature increases for the cases which stabilise at concentrations of 650 and 450 ppmv for different climate sensitivities. Stabilisation of the concentration does not lead to an immediate stabilisation of the global mean temperature. The global mean temperature is seen to continue rising for hundreds of years after the concentrations have stabilised in Figure 24 due to long time-scales in the ocean.

As shown in Figure 25, the long-term sea level rise "commitment" is even more pronounced. Sea level

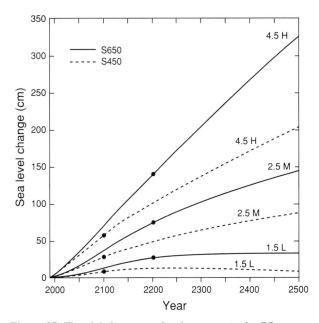

Figure 25: The global mean sea level response to the CO_2 concentration pathways leading to stabilisation at 450 (dashed curves) and 650 (solid curves) ppmv (see Figure 7a) for a climate sensitivity of 1.5, 2.5 and 4.5 °C. The changes shown are those arising from CO_2 increases alone. The date of concentration stabilisation is indicated by the dot. Calculations assume the "observed" history of forcing to 1990, including aerosol effects and then CO_2 concentration increases only beyond 1990.

continues to rise, at only a slowly declining rate, for many centuries after greenhouse gas concentrations and temperatures have stabilised.

F.5 The possibility of surprises

Unexpected external influences, such as volcanic eruptions, can lead to unexpected and relatively sudden shifts in the climatic state. Also, as the response of the climate system to various forcings can be non-linear, its response to gradual forcing changes may be quite irregular. Abrupt and significant changes in the atmospheric circulation involving the North Pacific which began about 1976 were described in IPCC (1990). A related example is the apparent fluctuation in the recent behaviour of ENSO, with warm conditions prevailing since 1989, a pattern which has been unusual compared to previous ENSO behaviour.

Another example is the possibility that the West Antarctic ice sheet might "surge", causing a rapid rise in sea level. The current lack of knowledge regarding the specific circumstances under which this might occur, either in total or in part, limits the ability to quantify this risk. Nonetheless, the likelihood of a major sea level rise by the year 2100 due to the collapse of the West Antarctic ice sheet is considered low.

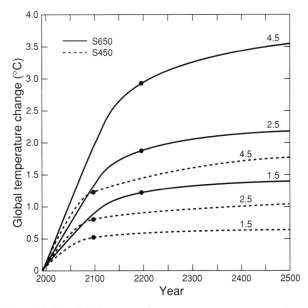

Figure 24: The global mean surface temperature response to the CO_2 concentration pathways leading to stabilisation at 450 (dashed curves) and 650 (solid curves) ppmv (see Figure 7a) for a climate sensitivity of 1.5, 2.5 and 4.5 °C. The changes shown are those arising from CO_2 increases alone. The date of concentration stabilisation is indicated by the dot. Calculations assume the "observed" history of forcing to 1990, including aerosol effects and then CO_2 concentration increases only beyond 1990.

In the oceans the meridional overturning might weaken in a future climate. This overturning (the thermohaline circulation) is driven in part by deep convection in the northern North Atlantic Ocean and keeps the northern North Atlantic Ocean several degrees warmer than it would otherwise be. Both the study of palaeoclimate from sediment records and ice cores and modelling studies with coupled climate models and ocean GCMs can be interpreted to suggest that the ocean circulation has been very different in the past. Both in these observations and in the ocean models, transitions between different types of circulation seem to occur on a time-scale of a few decades, so relatively sudden changes in the regional (North Atlantic, Western Europe) climate could occur, presumably mainly in response to precipitation and runoff changes which alter the salinity, and thus the density, of the upper layers of the North Atlantic. Whether or not such a sudden change can actually be realised in response to global warming and how strong a perturbation is required to cause a transition between types of circulation are still the subject of much debate.

In terrestrial ecological systems, there are thresholds in the sustained temperature and water availability at which one biological population is replaced by another. Some replacement, e.g., in tree species, is slow while some, e.g., in micro-organisms is rapid. Minimum temperatures exist for the survival of organisms in winter, and the populations of such organisms may move polewards as the climate and especially night-time temperatures warm. If the transitions are not orderly, sudden shifts in ecosystem functioning will occur. These may have impacts of direct human relevance (as discussed in IPCC WGII (1995)) but may also have surprising impacts on climate via effects on albedo, aerosol forcing, the hydrological cycle, evapotranspiration, CO_2 release and methane cycling, for example (see Sections B.1 and D.2).

G Advancing our Understanding

An important long-term goal is the accurate projection of regional climate change, so that potential impacts can be adequately assessed. Progress towards this objective depends on determining the likely global magnitude and rate of human-induced climate change, including sea level change, as well as the regional expressions of these quantities. The detection and attribution of human-induced climate change is also most important. To achieve these objectives requires systematic and sustained global observations of relevant variables, as well as requiring the

effective co-operation and participation of many nations. The most urgent scientific problems requiring attention concern:

(i)   the *rate* and *magnitude* of climate change and sea level rise:

- the factors controlling the distribution of clouds and their radiative characteristics;
- the hydrological cycle, including precipitation, evaporation and runoff;
- the distribution and time evolution of ozone and aerosols and their radiative characteristics;
- the response of terrestrial and marine systems to climate change and their positive and negative feedbacks;
- the response of ice sheets and glaciers to climate;
- the influence of human activities on emissions;
- the coupling between the atmosphere and ocean, and ocean circulation;
- the factors controlling the atmospheric concentrations of carbon dioxide and other greenhouse gases;

(ii)   *detection* and *attribution* of climate change:

- systematic observations of key variables, and development of model diagnostics relating to climate change;
- relevant proxy data to construct and test palaeoclimatic time-series to describe natural variability of the climate system;

(iii) *regional patterns* of climate change:

- land-surface processes and their link to atmospheric processes;
- coupling of scales between global climate models and regional and smaller scale models;
- simulations with higher resolution climate models.

The research activities for each objective are strongly interconnected. Such research is and needs to be conducted by individual investigators in a variety of institutions, as well as by co-ordinated international efforts which pool national resources and talents in order to more efficiently engage in large-scale integrated field and modelling programmes to advance our understanding.

References

IPCC, (Intergovernmental Panel on Climate Change) 1990: *Climate Change: The IPCC Scientific Assessment*, J.T. Houghton, G.J. Jenkins and J.J. Ephraums (eds.). Cambridge University Press, Cambridge, UK, 365 pp.

IPCC, 1992: *Climate Change 1992: The Supplementary Report to the IPCC Scientific Assessment*, J.T. Houghton, B.A. Callander and S.K. Varney (eds.). Cambridge University Press, Cambridge, UK, 198 pp.

IPCC, 1994: *Climate Change 1994: Radiative Forcing of Climate Change and an Evaluation of the IPCC 1S92 Emission Scenarios*, J.T. Houghton, L.G. Meira Filho, J. Bruce, Hoesung Lee, B.A. Callander, E. Haites, N. Harris and K. Maskell (eds.). Cambridge University Press, Cambridge, UK, 339 pp.

IPCC WGII, 1995: *Climate Change 1995-Impacts, Adaptations and Mitigations of Climate Change: Scientific-Technical Analyses: The Second Assessment Report of the Inter-Governmental Panel on Climate Change.* R.T. Watson, M.C. Zinyowera and R.H. Moss (eds.). Cambridge University Press, New York, USA.

Glossary

Aerosols	Airborne particles. The term has also come to be associated, erroneously, with the propellant used in "aerosol sprays".
Climate change (FCCC usage)	A change of climate which is attributed directly or indirectly to human activity that alters the composition of the global atmosphere and which is in addition to natural climate variability observed over comparable time periods.
Climate change (IPCC usage)	Climate change as referred to in the observational record of climate occurs because of internal changes within the climate system or in the interaction between its components, or because of changes in external forcing either for natural reasons or because of human activities. It is generally not possible clearly to make attribution between these causes. Projections of future climate change reported by IPCC generally consider only the influence on climate of anthropogenic increases in greenhouse gases and other human-related factors.
Climate sensitivity	In IPCC reports, climate sensitivity usually refers to the long-term (equilibrium) change in global mean surface temperature following a doubling of atmospheric CO_2 (or equivalent CO_2) concentration. More generally, it refers to the equilibrium change in surface air temperature following a unit change in radiative forcing (°C/Wm^{-2}).
Diurnal temperature range	The difference between maximum and minimum temperature over a period of 24 hours
Equilibrium climate experiment	An experiment where a step change is applied to the forcing of a climate model and the model is then allowed to reach a new equilibrium. Such experiments provide information on the difference between the initial and final states of the model, but not on the time-dependent response.
Equivalent CO_2	The concentration of CO_2 that would cause the same amount of radiative forcing as the given mixture of CO_2 and other greenhouse gases.
Evapotranspiration	The combined process of evaporation from the Earth's surface and transpiration from vegetation.
Greenhouse gas	A gas that absorbs radiation at specific wavelengths within the spectrum of radiation (infrared radiation) emitted by the Earth's surface and by clouds. The gas in turn emits infrared radiation from a level where the temperature is colder than the surface. The net effect is a local trapping of part of the absorbed energy and a tendency to warm the planetary surface. Water vapour (H_2O), carbon dioxide (CO_2), nitrous oxide (N_2O), methane (CH_4) and ozone (O_3) are the primary greenhouse gases in the Earth's atmosphere.
Ice-cap	A dome-shaped glacier usually covering a highland near a water divide.
Ice sheet	A glacier more than 50,000 km^2 in area forming a continuous cover over a land surface or resting on a continental shelf.

Radiative forcing	A simple measure of the importance of a potential climate change mechanism. Radiative forcing is the perturbation to the energy balance of the Earth-atmosphere system (in Wm^{-2}) following, for example, a change in the concentration of carbon dioxide or a change in the output of the Sun; the climate system responds to the radiative forcing so as to re-establish the energy balance. A positive radiative forcing tends to warm the surface and a negative radiative forcing tends to cool the surface. The radiative forcing is normally quoted as a global and annual mean value. A more precise definition of radiative forcing, as used in IPCC reports, is the perturbation of the energy balance of the surface-troposphere system, after allowing for the stratosphere to re-adjust to a state of global mean radiative equilibrium (see Chapter 4 of IPCC (1994)). Sometimes called "climate forcing".
Spatial scales	continental $10 - 100$ million square kilometres (km^2) regional 100 thousand – 10 million km^2 local less than 100 thousand km^2
Soil moisture	Water stored in or at the continental surface and available for evaporation. In IPCC (1990) a single store (or "bucket") was commonly used in climate models. Today's models which incorporate canopy and soil processes view soil moisture as the amount held in excess of plant "wilting point".
Stratosphere	The highly stratified and stable region of the atmosphere above the troposphere (*qv.*) extending from about 10 km to about 50 km.
Thermohaline circulation	Large scale density-driven circulation in the oceans, driven by differences in temperature and salinity.
Transient climate experiment	An analysis of the time-dependent response of a climate model to a time-varying change of forcing.
Troposphere	The lowest part of the atmosphere from the surface to about 10 km in altitude in mid-latitudes (ranging from 9 km in high latitudes to 16 km in the tropics on average) where clouds and "weather" phenomena occur. The troposphere is defined as the region where temperatures generally decrease with height.

1

The Climate System: an overview

K.E. TRENBERTH, J.T. HOUGHTON, L.G. MEIRA FILHO

CONTENTS

1.1 Climate and the Climate System

Changes in climate, whether from natural variability or anthropogenic causes and over a range of time-scales, can be identified and studied from different climate variables. Because humans live in and breathe the atmosphere it is natural to focus on the atmospheric changes where phenomena and events are loosely divided into the realms of "weather" and "climate". The large fluctuations in the atmosphere from hour-to-hour or day-to-day constitute the weather; they occur as weather systems move, develop, evolve, mature and decay as forms of atmospheric turbulence. These weather systems arise mainly from atmospheric instabilities and their evolution is governed by non-linear "chaotic" dynamics, so that they are not predictable in an individual deterministic sense beyond a week or two into the future.

Climate is usually defined to be average weather, described in terms of the mean and other statistical quantities that measure the variability over a period of time and possibly over a certain geographical region. Climate involves variations in which the atmosphere is influenced by and interacts with other parts of the climate system, and

"external" forcings. The internal interactive components in the climate system (Figure 1.1) include the atmosphere, the oceans, sea ice, the land and its features (including the vegetation, albedo, biomass, and ecosystems), snow cover, land ice (including the semi-permanent ice sheets of Antarctica and Greenland and glaciers), and hydrology (including rivers, lakes and surface and subsurface water). The greatest variations in the composition of the atmosphere involve water in various phases in the atmosphere, as water vapour, clouds containing liquid water and ice crystals, and hail. However, other constituents of the atmosphere and the oceans can also change thereby bringing in considerations of atmospheric chemistry, marine biogeochemistry, and land surface exchanges.

The components normally regarded as external to the system include the Sun and its output, the Earth's rotation, Sun-Earth geometry and the slowly changing orbit, the physical components of the Earth system such as the distribution of land and ocean, the geographic features on the land, the ocean bottom topography and basin configurations, and the mass and basic composition of the atmosphere and ocean. These components determine the

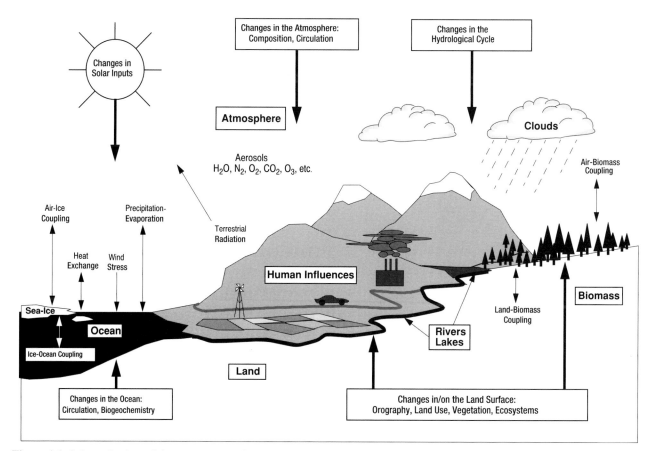

Figure 1.1: Schematic view of the components of the global climate system (bold), their processes and interactions (thin arrows) and some aspects that may change (bold arrows).

mean climate which may vary from natural causes. A change in the average net radiation at the top of the atmosphere due to perturbations in the incident solar radiation or the emergent infrared radiation leads to what is known as radiative forcing of the system. Changes in the incident radiation energy from the Sun or changes in atmospheric composition due to natural events like volcanoes are possible examples. Other external forcings may occur as a result of human activities, such as increases in greenhouse gases. In this report we are particularly concerned with the latter.

Changes in any of the climate system components, whether internal and thus a part of the system, or from the external forcings, cause the climate to vary or to change. Thus climate can vary because of alterations in the internal exchanges of energy or in the internal dynamics of the climate system. For example, El Niño-Southern Oscillation (ENSO) events arise from natural coupled interactions between the atmosphere and the ocean centred in the tropical Pacific. As such they are a part of climate and they lead to large and important systematic variations in weather patterns (events such as floods and droughts) throughout

the world from year to year. Often, however, climate is taken to refer to much longer time-scales – the average statistics over a 30-year period is a widespread and long-standing working definition. On these longer time-scales, ENSO events vanish from mean statistics but become strongly evident in the measures of variability: the variances and the extremes. However, the mean climate is also influenced by the variability. These considerations become very important as we develop models of the climate system as tools to predict climate change.

1.2 The Driving Forces of Climate

1.2.1 The Global Energy Balance
The source of energy which drives the climate is the radiation from the Sun. Much of this energy is in the visible part of the electromagnetic spectrum although some extends beyond the red into the infrared and some extends beyond the violet into the ultraviolet. The amount of energy per second falling on a surface one square metre in area facing the Sun outside the atmosphere is about 1370 W. Because of the spherical shape of the Earth, at any one

---

Climate Change Definitions

Although the common definition of climate refers to the average of weather, the definition of the climate system must include the relevant portions of the broader geophysical system which increasingly interacts with the atmosphere as the time period considered increases. For the time-scales of decades to centuries associated with the change of climate due to the effect of enhanced greenhouse warming, the United Nations Framework Convention on Climate Change defines the climate system to be "the totality of the atmosphere, hydrosphere, biosphere and geosphere and their interactions".

The concept of climate change has recently acquired a number of different meanings in the scientific literature and in relevant international fora. There is a simple view that climate change refers to any change of the classical 30-year climatology, regardless of its causes. Often "climate change" denotes those variations due to human interference while "climate variations" refers to the natural variations. Sometimes "climate change" designates variations longer than a certain period. Finally, "climate change" is sometimes taken in the literature to mean climate fluctuations of a global nature, which is an interpretation used in parts of this report and which includes the effects due to human actions, such as the enhanced greenhouse effect, and those due to natural causes such as stratospheric aerosols from volcanic eruptions. One complication with this definition is the anthropogenic changes of climate on a restricted space scale, a good example of which is the heat island phenomenon by which highly urbanised areas may have a mean temperature which is higher than it would otherwise have been.

Nevertheless, for the purposes of the United Nations Convention on Climate Change, the definition of climate change is: "a change of climate which is attributed directly or indirectly to human activity that alters the composition of the global atmosphere and which is in addition to natural climate variability observed over comparable time periods." Further, the Convention considers, for purposes of mitigation measures, only those greenhouse gases which are not controlled by the Montreal Protocol and its Amendments (i.e., ozone-depleting substances such as CFCs and the HCFCs), presumably on the grounds that the latter are covered by a separate international legal instrument. This definition thus introduces the concept of the difference between climate with the effect of human-induced increase in the concentration of greenhouse gases and that which would be realised without such human interference. This point is important scientifically for both detection and prediction because at least one of these climate states has to be modelled.

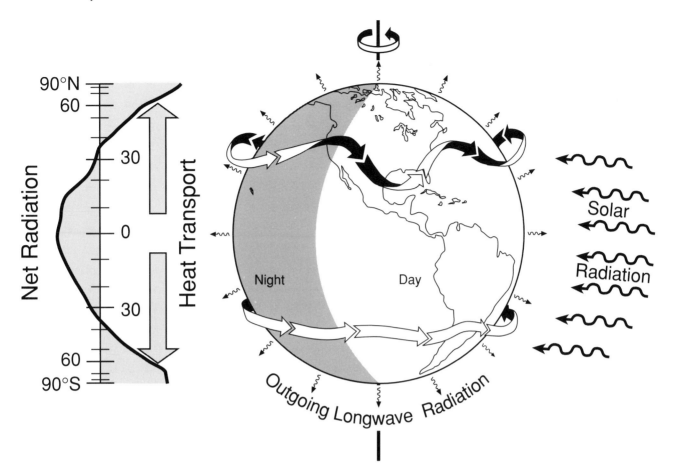

Figure 1.2: The incoming solar radiation (right) illuminates only part of the Earth while the outgoing long-wave radiation is distributed more evenly. On an annual mean basis, the result is an excess of absorbed solar radiation over the outgoing long-wave radiation in the tropics, while there is a deficit at middle to high latitudes (far left), so that there is a requirement for a poleward heat transport in each hemisphere (arrows) by the atmosphere and the oceans. This radiation distribution results in warm conditions in the tropics but cold at high latitudes, and the temperature contrast results in a broad band of westerlies in the extra-tropics of each hemisphere in which there is an embedded jet stream (shown by the "ribbon" arrows) at about 10 km above the Earth's surface. The flow of the jetstream over the different underlying surface (ocean, land, mountains) produces waves in the atmosphere and adds geographic spatial structure to climate. The excess of net radiation at the equator is 68 Wm^{-2} and the deficit peaks at -100 Wm^{-2} at the South Pole and -125 Wm^{-2} at the North Pole; from Trenberth and Solomon (1994).

time half the Earth is in night (Figure 1.2) and the average amount of energy incident on a level surface outside the atmosphere is one quarter of this or 342 Wm^{-2}. About 31% of this energy is scattered or reflected back to space by molecules, microscopic airborne particles (known as aerosols) and clouds in the atmosphere, or by the Earth's surface, which leaves about 235 Wm^{-2} on average to warm the Earth's surface and atmosphere (Figure 1.3).

To balance the incoming energy, the Earth itself must radiate on average the same amount of energy back to space (Figure 1.3). It does this by emitting thermal "long-wave" radiation in the infrared part of the spectrum. The amount of thermal radiation emitted by a warm surface depends on its temperature and on how absorbing it is. For

a completely absorbing surface to emit 235 Wm^{-2} of thermal radiation, it would have a temperature of about $-19°C$. This is much colder than the conditions that actually exist near the Earth's surface where the annual average global mean temperature is about 15°C. However, because the temperature in the troposphere – the lowest 10-15 km of the atmosphere – falls off quite rapidly with height, a temperature of $-19°C$ is reached typically at an altitude of 5 km above the surface in mid-latitudes.

1.2.2 The Greenhouse Effect
Some of the infrared radiation leaving the atmosphere originates near the Earth's surface and is transmitted relatively unimpeded through the atmosphere; this is the

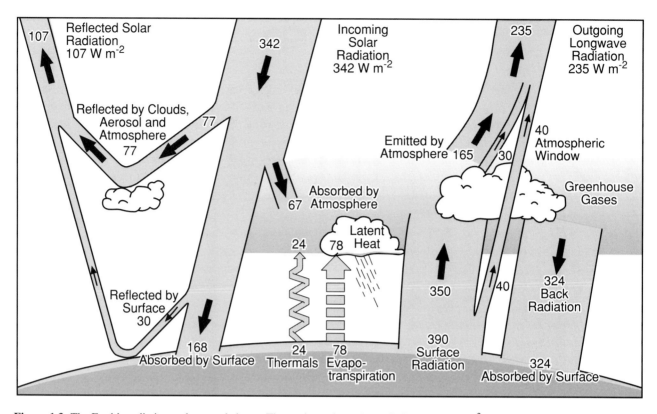

Figure 1.3: The Earth's radiation and energy balance. The net incoming solar radiation of 342 Wm⁻² is partially reflected by clouds and the atmosphere, or at the surface, but 49% is absorbed by the surface. Some of that heat is returned to the atmosphere as sensible heating and most as evapotranspiration that is realised as latent heat in precipitation. The rest is radiated as thermal infrared radiation and most of that is absorbed by the atmosphere which in turn emits radiation both up and down, producing a greenhouse effect, as the radiation lost to space comes from cloud tops and parts of the atmosphere much colder than the surface. The partitioning of the annual global mean energy budget and the accuracy of the values are given in Kiehl and Trenberth (1996).

radiation from areas where there is no cloud and which is present in the part of the spectrum known as the atmospheric "window" (Figure 1.3). The bulk of the radiation, however, is intercepted and absorbed by the atmosphere which in turn emits radiation both up and down. The emissions to space occur either from the tops of clouds at different atmospheric levels (which are almost always colder than the surface), or by gases present in the atmosphere which absorb and emit infrared radiation. Most of the atmosphere consists of nitrogen and oxygen (99% of dry air) which are transparent to infrared radiation. It is the water vapour, which varies in amount from 0 to about 2%, carbon dioxide and some other minor gases present in the atmosphere in much smaller quantities which absorb some of the thermal radiation leaving the surface and emit radiation from much higher and colder levels out to space. These radiatively active gases (see Chapter 2 for details) are known as greenhouse gases because they act as a partial blanket for the thermal radiation from the surface and enable it to be substantially warmer than it would

otherwise be, analogous to the effects of a greenhouse. This blanketing is known as the natural greenhouse effect.

Clouds also absorb and emit thermal radiation and have a blanketing effect similar to that of the greenhouse gases. But clouds are also bright reflectors of solar radiation and thus also act to cool the surface. While on average there is strong cancellation between the two opposing effects of short-wave and long-wave cloud radiative forcing (Chapter 4) the net global effect of clouds in our current climate, as determined by space-based measurements, is a small cooling of the surface.

1.2.3 Mars and Venus

Similar greenhouse effects also occur on our nearest planetary neighbours, Mars and Venus. Mars is smaller than the Earth and possesses, by Earth's standards, a very thin atmosphere (the pressure at the Martian surface is less than 1% of that on Earth) consisting almost entirely of carbon dioxide which contributes a small but significant greenhouse effect. The planet Venus, by contrast, has a much thicker atmosphere, largely composed of carbon dioxide, with a surface pressure nearly 100 times that on Earth. The resulting greenhouse effect on Venus is very large and leads to a surface temperature of about 500°C more than it would otherwise be.

1.2.4 Spatial Structure of Climate and Climate Change

For the Earth, on an annual mean basis, the excess of solar over outgoing long-wave radiation in the tropics and the deficit at mid- to high latitudes (Figure 1.2) sets up an equator-to-pole temperature gradient that results, with the Earth's rotation, in a broad band of westerlies in each hemisphere in the troposphere. Embedded within the mid-latitude westerlies are large-scale weather systems which, along with the ocean, act to transport heat polewards to achieve an overall energy balance. These weather systems are the familiar migrating cyclones and anticyclones (i.e., low and high pressure systems) and their associated cold and warm fronts.

Because of the land-ocean contrasts and obstacles such as mountain ranges, the mid-latitude westerlies and the embedded jet stream (Figure 1.2) in each hemisphere contain planetary-scale waves. These waves are usually geographically anchored but can change with time as

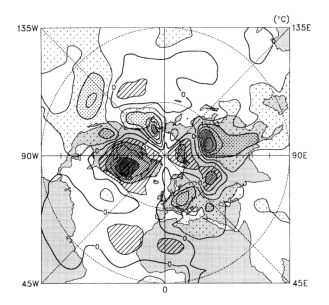

Figure 1.4: The anomalies in surface temperature over the Northern Hemisphere for the winter of December 1993 to February 1994 relative to the mean for 1951 to 1980. Temperature anomalies exceeding 1°C are stippled and those below −1°C are hatched. The spatial structure seen here is inherent in the atmospheric dynamics and regions of below-normal temperatures should be expected even in the presence of an overall mean that is 0.2°C above normal. Data courtesy of David Parker.

heating patterns change in the atmosphere. A consequence is that anomalies in climate on seasonal-to-annual time-scales typically occur over large geographic regions with surface temperatures both above and below normal in different places. An example for the Northern Hemisphere winter season, December 1993 to February 1994, is shown in Figure 1.4. The strong cold anomaly over north-eastern parts of North America was accompanied by many cold and snowy outbreaks, yet warmer-than-normal conditions prevailed over most of the rest of the hemisphere. Relative to 1951 to 1980, the result was a hemispheric anomaly in surface temperature of +0.2°C. Extensive regions of above and below normal temperatures are the rule, not the exception, as should clearly be expected from the wave motions in the atmosphere. A bout of below average temperatures regionally may not be inconsistent with global warming, just as an interval of above normal temperatures may not indicate global warming.

1.3 Anthropogenic Climate Change

Climate can vary for many reasons. In particular, human activities can lead to changes in atmospheric composition and hence radiative forcing through, for instance, the burning of fossil fuels or deforestation, or through processes which increase the number and distribution of aerosols. Altered properties of the surface because of changes in land-use can also give rise to changes in climate. It is especially these effects due to human activities with which we are concerned in this report.

1.3.1 The Enhanced Greenhouse Effect

The amount of carbon dioxide in the atmosphere has increased by more than 25% in the past century and since the beginning of the industrial revolution, an increase which is known to be in large part due to combustion of fossil fuels and the removal of forests (Chapter 2). In the absence of controls, projections are that the future rate of increase in carbon dioxide amount may accelerate and concentrations could double from pre-industrial values within the next 50 to 100 years (IPCC, 1994).

The increased amount of carbon dioxide is leading to climate change and will produce, on average, a global warming of the Earth's surface because of its enhanced greenhouse effect – although the magnitude and significance of the effects are not yet fully resolved. If, for instance, the amount of carbon dioxide in the atmosphere were suddenly doubled, but with other things remaining the same, the outgoing long-wave radiation would be reduced by about 4 Wm^{-2}. To restore the radiative balance, the atmosphere must warm up and, in the absence of other changes, the warming at the surface and throughout the troposphere would be about 1.2°C. However, many other factors will change, and various feedbacks come into play (see Section 1.4.1), so that the best estimate of the average global warming for doubled carbon dioxide is 2.5°C (IPCC, 1990). Such a change is very large by historical standards and would be associated with major climate changes around the world.

Note that if the carbon dioxide were removed from the atmosphere altogether, the change in outgoing radiation would be about 30 Wm^{-2} – 7 or 8 times as big as the change for doubling – and the magnitude of the temperature change would be similarly enhanced. The reason is that the carbon dioxide absorption is saturated over part of the spectral region where it absorbs, so that the amount of absorption changes at a much smaller rate than the concentration of the gas (Chapter 2). If concentrations of carbon dioxide are more than doubled, then the relationship between radiative forcing and concentration is such that each further doubling provides a further radiative forcing of about 4 Wm^{-2}.

Several other greenhouse gases are also observed to be increasing in concentration in the atmosphere because of human activities (especially biomass burning, landfills, rice paddies, agriculture, animal husbandry, fossil fuel use and industry). These include methane, nitrous oxide, and tropospheric ozone, and they tend to reinforce the changes in radiative forcing from increased carbon dioxide (Chapter 2). The human-introduced chlorofluorocarbons (CFCs) also produce a greenhouse effect although offset somewhat by the observed decreases in lower stratospheric ozone since the 1970s, caused principally by the CFCs and halons (IPCC, 1994).

1.3.2 Effects of Aerosols

Human activities also affect the amount of aerosol in the atmosphere which influences climate in other ways. The main direct effect of aerosols is the scattering of some solar radiation back to space, which tends to cool the Earth's surface. Some aerosols can also influence the radiation budget by directly absorbing solar radiation leading to local heating of the atmosphere and, to a lesser extent, by absorbing and emitting thermal radiation. A further influence of aerosols is that many of them act as nuclei on which cloud droplets condense. A changed concentration therefore tends to affect the number and size of droplets in a cloud and hence alters the reflection and the absorption of solar radiation by the cloud.

Aerosols occur in the atmosphere from natural causes; for instance, they are blown off the surface of deserts or

dry regions. The eruption of Mt. Pinatubo in the Philippines in June 1991 added considerable amounts of aerosol to the stratosphere which, for about two years, scattered solar radiation leading to a loss of radiation at the surface and a cooling there. Human activities contribute to aerosol particle formation mainly through injection of sulphur dioxide into the atmosphere (which contributes to acid rain) particularly from power stations, and through biomass burning.

Because human-made aerosols typically remain in the atmosphere for only a few days they tend to be concentrated near their sources such as industrial regions. The radiative forcing therefore possesses a very strong regional pattern (Chapters 2, 6 and 8), and the presence of aerosols adds further complexity to possible climate change as it can help mask, at least temporarily, any global warming arising from increased greenhouse gases. However, the aerosol effects do not cancel the global-scale effects of the much longer-lived greenhouse gases, and significant climate changes can still result.

1.4 Climatic Response

1.4.1 Feedbacks

The increases in greenhouse gases in the atmosphere and changes in aerosol content produce a change in the radiative forcing (Chapter 2). The determination of the climatic response to this change in forcing is complicated by feedbacks. Some of these can amplify the original warming (positive feedback) while others serve to reduce it (negative feedback) (Chapter 4). An example of the former is water vapour feedback in which the amount of water vapour in the atmosphere increases as the Earth warms and, because water vapour is an important greenhouse gas, it will amplify the warming. However, increases in cloud may act either to amplify the warming through the greenhouse effect of clouds or reduce it by the increase in albedo (which measures reflectivity); which effect dominates depends on the height and type of clouds and varies greatly with geographic location and time of year. Ice-albedo feedback is another potentially important process that may lead to amplification of temperature changes in high latitudes. It arises because decreases in sea ice, which has high albedo, decrease the radiation reflected back to space and thus produces warming which may further decrease the sea ice extent. However, increased open water may lead to more atmospheric water vapour and increased fog and low cloud amount, offsetting the change in surface albedo.

There are a number of feedbacks involving the biosphere which are especially important when considering details of the carbon cycle and the impacts of climate change. The behaviour of ecosystems on the surface of the Earth (Chapter 9) and biogeochemical processes within the oceans (Chapters 2 and 10) are greatly influenced by changes in atmospheric composition and climate. The availability of surface water and the use of the Sun's energy in photosynthesis and transpiration in plants influence the uptake of carbon dioxide from the atmosphere as plants transform the carbon and water into usable food. Changes in vegetation affect surface albedo, evapotranspiration and roughness. Much remains to be learned about these feedbacks and their possible influences on predictions of future carbon dioxide concentrations and climate, and models used for future projections have not yet incorporated them.

1.4.2 The Role of the Oceans

The oceans cover 70% of the Earth's surface and through their fluid motions, their high heat capacity, and their ecosystems they play a central role in shaping the Earth's climate and its variability. Wind stress at the sea surface drives the large-scale ocean circulation in its upper layers. Water vapour, evaporated from the ocean surface, provides latent heat energy to the atmosphere. The ocean circulation is an effective means of redistributing heat and fresh water around the globe. The oceans store heat, absorbed at the surface, for varying durations and release it in different places thereby ameliorating temperature changes over nearby land and contributing substantially to the variability of climate on many time-scales. Additionally, the ocean thermohaline[1] circulation allows water from the surface to be carried into the deep ocean where it is isolated from atmospheric influence and hence it may sequester heat for periods of a thousand years or more. The oceans absorb carbon dioxide and other gases and exchange them with the atmosphere in ways that alter with ocean circulation and climate variability. In addition, it is likely that marine biotic responses to climate change will result in feedbacks (Chapter 10).

Any study of the climate and how it might change must include an adequate description of processes in the ocean (Chapter 4) together with the coupling between the ocean and the atmosphere (Chapter 6).

1.4.3 The Role of Land

The heat penetration into land associated with the annual cycle of surface temperature is limited to about the uppermost 2 m and the heat capacity of land is much less

than that of a comparable depth of ocean. Accordingly, land plays a much smaller role in the storage of heat. A consequence is that the surface air temperature changes over land occur much faster and are much larger than over the oceans for the same heating and, because we live on land, this directly affects human activities. The land surface encompasses an enormous variety of topographical features and soils, differing slopes (which influence runoff and radiation received) and water capacity. The highly heterogeneous vegetative cover is a mixture of natural and managed ecosystems that vary on very small spatial scales. Changes in soil moisture affect the disposition of heat and whether it results in sensible heating or evapotranspiration (and subsequently latent heating) and changes in vegetation alter the albedo, roughness, and evapotranspiration. The land surface and its ecosystems play an important role in the carbon cycle (Chapter 9), the hydrological cycle and in surface exchanges of trace gases. Currently, many of these land surface processes are only crudely represented in global climate models.

1.5 Observed Climate Change

Given that climate change is expected from anthropogenic effects, what have the observed changes been? Because the high quality of much-needed long time-series of observations is often compromised, special care is required in interpretation. Most observations have been made for other purposes, such as weather forecasting, and therefore typically suffer from changes in instrumentation, exposure, measurement techniques, station location and observation times, and there have been major changes in the distribution and numbers of observations. Adjustments must be devised to take into account all these influences in estimating the real changes that have occurred. For the more distant past, proxy data from climate-sensitive phenomena, such as from tree rings, ice cores, coral cores, and pollen in marine sediments are used.

Questions of how the climate has varied in the past, whether there has been recent warming and the structure of climate change in three dimensions are addressed in Chapter 3. Analysis of observations of surface temperature show that there has been a global mean warming of 0.3 to 0.6°C over the past one hundred years. The observed trend of a larger increase in minimum than maximum temperatures is apparently linked to associated increases in low cloud amount and aerosol as well as to the enhanced greenhouse effect (Chapters 3 and 4). There is good evidence for decadal changes in the atmospheric circulation which contribute to regional effects, and some

---

1 The circulation driven by changes in sea water density arising from temperature (thermal) or salinity (haline) effects.

evidence for ocean changes. Changes in precipitation and other components of the hydrological cycle vary considerably geographically. Changes in climate variability and extremes are beginning to emerge, but global patterns are not yet apparent.

Changes in climate have occurred in the distant past as the distribution of continents and their landscapes have changed, as the so-called Milankovitch changes in the orbit of the Earth and the Earth's tilt relative to the ecliptic plane have varied the insolation received on Earth, and as the composition of the atmosphere has changed, all through natural processes. Recent new evidence from ice cores drilled through the Greenland ice sheet have indicated that changes in climate may often have been quite rapid and large, and not associated with any known external forcings. Understanding the spatial scales of this variability and the processes and mechanisms involved is very important as it seems quite possible that strong nonlinearities may be involved. These may result in large changes from relatively small perturbations by provoking positively reinforcing feedback processes in the internal climate system. Changes in the thermohaline circulation in the Atlantic Ocean are one way such abrupt changes might be realised (Chapters 4 and 6). An important question therefore is whether there might be prospects for major surprises as the climate changes.

Rates of change of radiative forcing induced by human activities are exceedingly rapid compared with the historical record. This raises questions about how, for instance, surface ecosystems might adapt to such change (Chapter 9).

1.6 Prediction and Modelling of Climate Change

To quantify the response of the climate system to changes in forcing it is essential to account for all the complex interactions and feedbacks among the climate system components (Figure 1.1). It is not possible to do this reliably using empirical or statistical models because of the complexity of the system, and because the possible outcomes may go well beyond any conditions ever experienced previously. Instead the response must be found using numerical models of the climate system based upon sound well-established physical principles.

1.6.1 Climate Models

Global climate models include as central components atmospheric and oceanic general circulation models (GCMs), as well as representations of land surface processes, sea ice and all other processes indicated in Figure 1.1. Models and their components are based upon

physical laws represented by mathematical equations that describe the atmospheric and oceanic dynamics and physics. These equations are solved numerically at a finite resolution using a three-dimensional grid over the globe. Typical resolutions used for climate simulations in 1995 are about 250 km in the horizontal and 1 km in the vertical in atmospheric GCMs. As a result, many physical processes cannot be properly resolved but their average effects must be included through a parametric representation (called parametrization) that is physically based (Chapter 4).

An essential component of climate models is the description of the interactions among the different components of the climate system. Of particular importance is the coupling between the two fluid components, the atmosphere and the ocean. Ensuring this is adequately simulated is one of the greatest challenges in climate modelling (Chapter 4). A frontier and future research challenge is to bring more complete chemistry, biology, and ecology into the climate system models (Chapters 9, 10, and 11) and to improve the representation of physical processes. Once validated, these models will become valuable tools for advancing our understanding and quantifying and reducing the uncertainty in future predictions.

Comprehensive climate models are very complex and take large computer resources to run. To explore all the possible scenarios and the effects of assumptions or approximations in parameters in the model more thoroughly, simpler models are also widely used and are constructed to give similar results to the GCMs when globally averaged (Chapter 6).

1.6.2 Climate Predictability

An important and fundamental question concerns the extent to which the climate is predictable; i.e., are there climate "signals" large enough to be distinguished from the "noise" of natural variability that may be potentially predictable. Reliable weather forecasts can be made using atmospheric GCMs for periods up to ten days (Chapter 5), beyond which time detailed predictability is lost because of the dominance of chaotic dynamics in weather systems. For some parts of the world, however, some predictability exists for statistical averages of weather (i.e., the climate) up to a year or so ahead. Such predictability is largely due to the influence of the patterns of sea surface temperatures on the atmosphere. El Niño events provide the dominant example. Prediction of changes in sea surface temperatures in turn requires that climate models be able to adequately simulate the coupling between ocean and atmosphere.

The predictability considered so far concerns the internal variability. Climate changes arising from changes in external forcing can also be predictable, as is evidenced from several sources. Firstly, there is the mean annual cycle which climate models simulate very well. Secondly, there is the existence of the regularities observed in past climates which were forced by changes in the distribution of solar radiation arising from the variations in the geometry of the Sun-Earth orbit. Models show some success in simulating these past climates. Thirdly, there is the success of climate models in simulating the changes due to the effects of stratospheric aerosols from the Mt. Pinatubo volcanic eruption. In addition, there is the evidence provided by the performance of the models themselves in simulating the effects of hypothetical situations, such as changes in solar radiation; the resulting climate changes are largely reproducible and thus potentially predictable.

Predictability is a function of spatial scales. Atmospheric variability arising from internal instabilities is huge on small scales; it is mainly the variability on larger scales influenced by the interactions of the atmosphere with other parts of the climate system that is predictable. Figure 1.5 shows the natural variability of the annual mean surface temperature on several different spatial scales from a climate model simulation for 200 years. The vertical scale is the same on all three plots, and the standard deviation goes from 0.1°C for the Southern Hemisphere to 0.5°C for Australia to 0.8°C for a grid square with sides about 500 km in south-east Australia. This example highlights the much greater natural variability that can be experienced on smaller scales which makes detection of the small systematic signal, such as might arise from enhanced global mean greenhouse forcing, much more difficult to achieve on regional scales.

1.6.3 Climate Projections

When a model is employed for climate prediction it is first run for many simulated decades without any changes in external forcing in the system. The quality of the simulation can then be assessed by comparing the mean, the annual cycle and the variability statistics on different time-scales with observations of the climate. In this way the model is evaluated (Chapter 5). The model is then run with changes in external forcing, such as with a possible future profile of greenhouse gas concentrations (Chapter 2). The differences between the climate statistics in the two simulations provide an estimate of the accompanying climate change (Chapter 6).

A long-term change in global mean surface air temperature arising from a doubling of carbon dioxide is

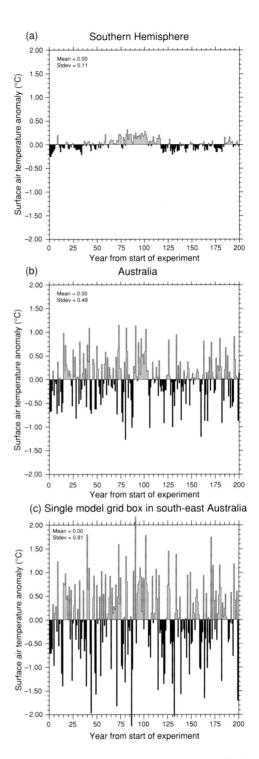

Figure 1.5: Annual-average surface air temperature (°C) from a 200-year integration of a coupled atmosphere-ocean model from the Geophysical Fluid Dynamics Laboratory run at low resolution of about 500 km spacing. The model is in statistical equilibrium and has no trends in climate forcing applied. The three panels show results for (a) the Southern Hemisphere, (b) Australia, and (c) a grid box in south-east Australia. (Courtesy J. Mahlman, S. Manabe, T. Delworth and R. Stouffer).

often used as a benchmark to compare models and as an indication of the climate sensitivity of the models. The range of results is typically an increase of 1.5° to 4.5°C. However, the concentrations of greenhouse gases will not level off at doubling and the regional patterns of climate change depend significantly on the time dependence of the change in forcing. It is important, therefore, to make future projections using plausible evolving scenarios of anthropogenic forcing so that the response of the climate to the forcing is properly simulated.

Accordingly, the focus of Chapter 6 is on projections of future climate using possible scenarios of greenhouse gas and aerosol emissions. Because of uncertainties in the scenarios, it is worth noting that these outlooks are not predictions so much as climate change estimates which can be used to assess possible impacts on the environment and society (IPCC Working Group II), such as the changes in sea level (Chapter 7), and for planning and policy purposes (IPCC Working Group III). However, definitive projections of possible local climate changes, which are most needed for assessing impacts, are the most challenging to do with any certainty. Further, it is desirable to examine and evaluate the past observational record by running models forced with realistic radiative forcing. It is in this way that it may be possible to attribute the observed changes to particular changes in forcing, such as from volcanic or solar origins, and to achieve detection of the effects of human activities and specifically the effects from increases in aerosols and greenhouse gases (Chapter 8).

The models used in climate projections are valuable tools for helping to quantify possible outcomes under various scenarios. They are used with the observations and all the other evidence to make the best assessments possible. The models are still undergoing development and their capabilities will improve in the future as past and new observations are analysed and improved understanding is obtained.

References

IPCC (Intergovernmental Panel on Climate Change), *1990: Climate Change: The IPCC Scientific Assessment*, J.T. Houghton, G.J. Jenkins and J.J. Ephraums (eds.). Cambridge University Press, Cambridge, UK, 365 pp.

IPCC, 1994: *Climate Change 1994. Radiative Forcing of Climate Change and an Evaluation of the IPCC IS92 Emission Scenarios*. J.T. Houghton, L.G. Meira Filho, J. Bruce, H. Lee, B.A. Callander, E. Haites, N. Harris and K. Maskell (eds). Cambridge University Press, Cambridge, UK, 339 pp.

Kiehl, J.T. and K.E. Trenberth, 1996: Earth's annual global mean energy budget. *Bull. Am. Met. Soc.* (submitted).

Trenberth, K.E. and A. Solomon, 1994: The Global Heat Balance: Heat transports in the atmosphere and ocean. *Clim. Dyn.*, **10**, 107–134.

2

Radiative Forcing of Climate Change

D. SCHIMEL, D. ALVES, I. ENTING, M. HEIMANN,
F. JOOS, D. RAYNAUD, T. WIGLEY (2.1)
M. PRATHER, R. DERWENT, D. EHHALT, P. FRASER,
E. SANHUEZA, X. ZHOU (2.2)
P. JONAS, R. CHARLSON, H. RODHE, S. SADASIVAN (2.3)
K.P. SHINE, Y. FOUQUART, V. RAMASWAMY,
S. SOLOMON, J. SRINIVASAN (2.4)
D. ALBRITTON, R. DERWENT, I. ISAKSEN, M. LAL, D. WUEBBLES (2.5)

Contributors:
*F. Alyea, T.L. Anderson, M. Andreae, D. Blake, O. Boucher, C. Brühl, J. Butler,
D. Cunnold, J. Dignon, E. Dlugokencky, J. Elkins, I. Fung, M. Geller, D. Hauglustaine,
J. Haywood, J. Heintzenberg, D. Jacob, A. Jain, C.D. Keeling, S. Khmelevtsov,
H. Le Treut, J. Lelieveld, I. Levin, M. Maiss, G. Marland, S.F. Marshall, P. Midgley,
B. Miller, J.F.B. Mitchell, S Montzka, H. Nakane, P. Novelli, B. O'Neill, D. Oram,
S. Penkett, J.E. Penner, S. Pinnock, R. Prinn, P. Quay, A. Robock, S.E. Schwartz,
P. Simmonds, A. Slingo, F. Stordal, E. Sulzman, P. Tans, A. Wahner, R. Weiss, T. Whorf*

CONTENTS

SUMMARY

Climate change can be driven by changes in the atmospheric concentrations of a number of radiatively active gases and aerosols. We have clear evidence that human activities have affected concentrations, distributions and life cycles of these gases. These matters, discussed in this chapter, were assessed at greater length in IPCC WGI report "Radiative Forcing of Climate Change" (IPCC 1994). The following summary contains some material more fully discussed in IPCC (1994): bullets containing significant new information are marked "***"; those containing information which has been updated since IPCC (1994) are marked "**"; and those which contain information which is essentially unchanged since IPCC (1994) are marked "*".

Carbon dioxide (CO_2)

* Carbon dioxide concentrations have increased by almost 30% from about 280 ppmv in the late 18th century to 358 ppmv in 1994. This increase is primarily due to combustion of fossil fuel and cement production, and to land-use change. During the last millennium, a period of relatively stable climate, concentrations varied by about ±10 ppmv around the pre-industrial value of 280 ppmv. On the century time-scale these fluctuations were far less rapid than the change observed over the 20th century.

*** The growth rate of atmospheric CO_2 concentrations over the last few years is comparable to, or slightly above, the average of the 1980s (~1.5 ppmv/yr). On shorter (interannual) time-scales, after a period of slow growth (0.6 ppmv/yr) spanning 1991 to 1992, the growth rate in 1994 was higher (~2 ppmv/yr). This change in growth rate is similar to earlier short time-scale fluctuations, which reflect large but transitory perturbations of the carbon system. Isotope data suggest that the 1991 to 1994 fluctuations resulted from natural variations in the exchange fluxes between the atmosphere and both the land biota and the ocean, possibly partly induced by interannual variations in climate.

*** As well as the issue of natural fluctuations discussed above, other issues raised since IPCC (1994) have been addressed. There are some unresolved concerns about the ^{14}C budget which may imply that previous estimates of the atmosphere-to-ocean flux were slightly too high. However, the carbon budget remains within our previously quoted uncertainties and the implications for future projections are minimal. Suggestions that the observed decay of bomb-^{14}C implies a very short atmospheric lifetime for CO_2 result from a mis-understanding of reservoir lifetimes. Current carbon cycle modelling is based on principles that have been well-understood since the 1950s and correctly accounts for the wide range of reservoir time-scales that affect atmospheric concentration changes.

** The major components of the anthropogenic perturbation to the atmospheric carbon budget, with estimates of their magnitudes over the 1980s, are: (a) emissions from fossil fuel combustion and cement production (5.5 ± 0.5 GtC/yr); (b) atmospheric increase (3.3 ± 0.2 GtC/yr); (c) ocean uptake (2.0 ± 0.8 GtC/yr); (d) tropical land-use changes (1.6 ± 1.0 GtC/yr); and (e) Northern Hemisphere forest regrowth (0.5 ± 0.5 GtC/yr). Other potential terrestrial sinks include enhanced terrestrial carbon storage due to CO_2 fertilisation (0.5–2.0 GtC/yr) and nitrogen deposition (0.2–1.0 GtC/yr), and possibly response to climatic anomalies. The latter is estimated to be a sink of 0–1.0 GtC/yr over the 1980s, but this term could be either a sink or a source over other periods. This budget is changed from IPCC (1994) by a small adjustment (from 3.2 to 3.3 GtC/yr) to the atmospheric rate of increase and a corresponding decrease in "other terrestrial sinks" from 1.4 to 1.3 GtC/yr.

* In IPCC (1994) calculations of future CO_2 concentrations and emissions from 18 different carbon cycle models were presented based on the IPCC (1992) carbon budget. Concentrations were derived for the IS92 emission scenarios. Future CO_2 emissions were derived leading to stable CO_2 concentration levels at 350, 450, 550, 650 and 750 ppmv. Inter-model differences varied with time and were up to ±15% about the median value. Biogeochemical (apart from CO_2 fertilisation) and climate feedbacks were not included in these calculations. The results showed that in order for atmospheric concentrations to stabilise at 750 ppmv or below, anthropogenic emissions must eventually fall well below today's levels. Stabilisation of emissions at 1990 levels is not sufficient to stabilise atmospheric CO_2: if anthropogenic emissions are held constant at 1990 levels, modelled atmospheric concentrations of CO_2 continue to increase throughout the next century and beyond.

*** These calculations have been re-run based on the revised IPCC (1994) carbon budget discussed above. These changes result in higher concentration projections by 15 ppmv (IS92c) to 40 ppmv (IS92e) in the year 2100. In terms of radiative forcing, the additional amounts in the year 2100 are 0.2–0.3 Wm^{-2}. Emissions requirements to achieve stabilisation are correspondingly reduced by up to about 10%. These changes are within the overall uncertainties in the calculations.

*** The implied future CO_2 emissions were calculated for additional cases to investigate the effect of lower reductions in CO_2 emissions in the early years by using CO_2 concentration profiles which closely followed the IS92a emission scenario for 10–30 years after 1990. Concentrations stabilised at the same date and levels as in the earlier scenarios, *viz* 350, 450, 550, 650 and 750 ppmv. Stabilisation at 1000 ppmv, above the range previously examined, was also investigated. These new calculations, while not necessarily spanning the full range of future emission options, still show the same characteristic long-term emissions behaviour as found previously; namely an eventual decline to emissions well below present levels.

Methane (CH₄)

** The atmospheric methane concentration increased from about 700 ppbv in pre-industrial times to 1721 ppbv in 1994.

** Over the last 20 years, there has been a decline in the methane growth rate: in the late 1970s the concentration was increasing at about 20 ppbv/yr; during the 1980s the growth rate dropped to 9–13 ppbv/yr. Around the middle of 1992, methane concentrations briefly stopped growing, but since 1993 the global growth rate has returned to about 8 ppbv/yr. Several causes for the anomaly in 1992/93 have been proposed, but none is individually able to explain all the observations.

*** The loss of methane through reaction with tropospheric hydroxyl radical, the main sink of CH_4, is 11% faster than recommended in IPCC (1994) owing to an improved estimate of the concentration of methyl chloroform in the atmosphere. The new recommendation for the CH_4 turnover time, 8.6yr (which includes tropospheric OH, stratospheric loss and uptake by soils), results in an inferred global loss of 560 ± 100 $Tg(CH_4)$/yr.

** Methane has clearly identified chemical feedbacks. Addition of CH_4 to the atmosphere reduces the concentration of tropospheric hydroxyl radical which can subsequently reduce the rate of CH_4 removal, thus lengthening the effective duration of the added CH_4 and any derived perturbations (e.g., in tropospheric ozone). Our estimate of the adjustment time for an added pulse of CH_4 is 12.2 ± 3 yr compared with the IPCC (1994) recommendation of 14.5 ± 2.5 yr that did not include soil sinks and whose estimated uncertainty only included the range in the feedback factors.

** If methane emissions were held constant at 1984–1994 levels (i.e., those sustaining an atmospheric trend of +10 ppbv/yr), CH_4 would rise to about 1850 ppbv over the next 40 years, an additional radiative forcing of around +0.05 Wm^{-2}. If emissions were cut by about 30 $Tg(CH_4)$/yr (about 8% of current anthropogenic emissions), CH_4 concentrations would remain at today's levels. These estimates are reduced from those in IPCC (1994) which were based on the average 1980s trend of 13 ppbv/yr and a longer adjustment time. Estimates of current CH_4 emissions are roughly 20–40% from natural sources, 20% from anthropogenic fossil fuel related sources and the remaining 40–60% from other anthropogenic sources.

Nitrous oxide (N₂O)

* Atmospheric N_2O concentrations have increased from about 275 ppbv in pre-industrial times to 311 ppbv in 1992. The trend during the 1980s was +0.25%/yr with substantial year-to-year variations. This growth rate, 0.8 ppbv/yr, corresponds to an imbalance between sources and sinks of about 3.9 $Tg(N_2O)$/yr. If these emissions were frozen then N_2O levels would rise slowly to about 400 ppbv over the next two centuries, an additional radiative forcing of about 0.3 Wm^{-2}. Natural sources are probably twice as large as anthropogenic ones.

Halogenated compounds

*** The atmospheric abundance of methyl chloroform (CH_3CCl_3) increased steadily by about 4–5%/yr from 1978 until mid-1990, when it levelled off. In 1994 it decreased by about 8%. This was the first reported global decrease in a halocarbon whose emissions are regulated by the Montreal Protocol and its Amendments. The rates of decrease are consistent with the reported phase-out of industrial emissions. Measurements show that the 1993 growth rate of CFC-12 has further declined, and that CFC-11 concentrations peaked at about 265 pptv.

*** Global HCFC-22 levels have continued to grow. The mean concentration in 1994 was approximately 110 pptv, with a growth rate (1992 to 1994) of about 5 pptv/yr. Global abundances of HCFC-142b and -141b are still small, approximately 6 and 2 pptv, respectively. However, they grew by 20% and 65%, respectively, in 1994, presumably in response to their increasing use as CFC substitutes.

*** Perfluorocarbons and sulphur hexafluoride (SF_6) are removed very slowly from the atmosphere with estimated lifetimes greater than 1000 years. As a result, effectively all emissions accumulate in the atmosphere and will continue to influence climate for thousands of years. Currently their concentrations and associated radiative forcings are low but they may become significant in the future if emissions continue.

** Hydrofluorocarbons (HFCs) are being used to replace ozone-depleting substances in some applications; their concentrations and radiative impacts are currently negligible. However, with increased use, their contribution to radiative forcing will increase.

Tropospheric ozone (O₃)
Tropospheric ozone (O_3)
* Model simulations and limited observations together suggest that tropospheric ozone has increased, perhaps doubled, in the Northern Hemisphere since pre-industrial times. In the 1980s, however, the trends were variable, being small or non-existent. At the South Pole, a decrease has been observed, and in the Southern Hemisphere as a whole, there are insufficient measurements to draw strong inferences about long-term changes.

* Uncertainties in the global budget of tropospheric ozone are associated primarily with our lack of knowledge of the distribution of ozone, its short-lived precursors (NO_x, hydrocarbons, CO) and atmospheric transport. These uncertainties severely limit our ability to model and predict tropospheric ozone on a global scale. We cannot determine with confidence the relative importance of anthropogenic

sources (principally surface combustion of fossil fuel and biomass, and aircraft) compared to natural sources (lightning, soils, stratospheric input) in controlling the global NO_x distribution.

* The possible increase in tropospheric ozone resulting from emissions of subsonic aircraft cannot currently be calculated with confidence. The resulting radiative effect may be significant, but it is unlikely to exceed the effect of the CO_2 from the combustion of aviation fuel, about 3% of current fossil fuel combustion.

*** The IS92 scenarios for emissions of greenhouse and related gases are used to calculate future concentrations of greenhouse gas and aerosol concentrations (and hence radiative forcing) up to 2100. These calculations use a simple model based on current lifetimes, except for CH_4 whose lifetime responds to its own increase. Changes in NO_x, hydrocarbon and CO emissions (IS92) would be expected to alter tropospheric OH and ozone, but there is no current consensus how to include growth in these emissions. The changes in ozone resulting from the estimated changes in CH_4 are calculated.

Stratospheric ozone (O_3)
* Trends in total column ozone at mid-latitudes in both hemispheres are significant in all seasons (averaging about −4 to −5%/decade since 1979) and are largely attributable to increases in halocarbons. Little or no change has occurred in the tropics.

* Unusually low values of total ozone (lower than would be expected from an extrapolation of the 1980s trend) were observed in the 1991 to 1993 period, especially at Northern mid- and high latitudes. Antarctic ozone "holes" in the 1990s have been the most severe on record; for instance, parts of the lower stratosphere contained extremely low amounts of ozone, corresponding to local depletions of more than 99%.

Aerosols
* Atmospheric aerosol in the troposphere influences climate in two ways, directly through the scattering and absorption of solar radiation, and indirectly through modifying the optical properties and lifetime of clouds.

* Estimation of tropospheric aerosol radiative forcing is more complex and uncertain than radiative forcing due to the well-mixed greenhouse gases for several reasons. First, both the direct and indirect radiative effect of aerosol

particles depend strongly on particle size and chemical composition and cannot be related to aerosol mass source strengths in a simple manner. Second, the indirect radiative effect of aerosols depends on complex processes involving aerosol particles and the nucleation and growth of cloud droplets. Last, aerosols in the lower troposphere have short lifetimes (around a week) and so their spatial distribution is highly inhomogeneous and strongly correlated with their sources.

* Many lines of evidence suggest that anthropogenic aerosol has increased the optical depth (a measure of the direct aerosol effect) over and downwind of industrial regions and that this increase is very large compared with the natural background in these regions. Major contributions to the anthropogenic component of the aerosol optical depth arise from sulphates (produced from sulphur dioxide released as a result of fossil fuel combustion) and from organics released by biomass burning.

*** Preliminary results suggest that a substantial fraction of the soil dust aerosol is subject to influence by human activity.

*** Improvements have been made to models of the distribution of anthropogenic aerosols and these calculations have now been extended to soot aerosols. As a result, more confidence can be placed on the distributions used for calculations of direct aerosol forcing.

* Future concentrations of anthropogenic sulphate aerosols will depend on both fossil fuel use and emission controls. Even if the total global emissions of SO_2 were stabilised, there are likely to be major changes in the geographical distribution of the SO_2 emissions and, hence, the aerosol concentration, in the 21st century.

Radiative forcing

* The use of global mean radiative forcing remains a valuable concept for giving a first-order estimate of the potential climatic importance of various forcing mechanisms. However, there are limits to its utility; in particular, the spatial patterns of forcing due to the well-mixed greenhouse gases and tropospheric aerosols are very different, and a comparison of the global mean radiative forcings does not give a complete picture of their possible climatic impact. For example, if the global mean radiative forcing were to be zero due to cancellation of positive and negative forcings from different mechanisms, this cannot

be taken to imply the absence of regional-scale or possibly even global climate change.

** Estimates of the radiative forcing due to changes in greenhouse gas concentrations since pre-industrial times remain unchanged from IPCC (1994). These are $+2.45$ Wm^{-2} (with an estimated uncertainty of 15%) for the direct effect of the well-mixed greenhouse gases (CO_2, CH_4, N_2O and the halocarbons), between $+0.2$ and $+0.6$ Wm^{-2} for tropospheric ozone and -0.1 Wm^{-2} (with a factor of two uncertainty) for stratospheric ozone. Chemical feedbacks from stratospheric ozone depletion (i.e., increased UV in the troposphere) have been proposed that could amplify its negative forcing but these are not yet well quantified.

*** The total direct forcing due to tropospheric aerosols (sulphate aerosols, fossil fuel soot and aerosols from biomass burning) is estimated to be -0.5 Wm^{-2} (with a factor of two uncertainty). The uncertainty in the direct forcing due to sulphate aerosols resulting from fossil fuel emissions and smelting has been slightly reduced since IPCC (1994) owing to a re-evaluation of the available modelled aerosol spatial distributions; the radiative forcing relative to pre-industrial times is estimated to be -0.4 Wm^{-2} (with a factor of two uncertainty). An estimate of $+0.1$ Wm^{-2} (with at least a factor of three uncertainty) is proposed for the radiative forcing due to soot from fossil fuel sources. The direct radiative forcing since 1850 from particles associated with biomass burning remains unchanged from IPCC (1994) at -0.2 Wm^{-2} (with a factor of three uncertainty).

** The radiative forcing due to changes in cloud droplet radii (and possible associated changes in cloud water content, amount and thickness) as a result of aerosols arising from human activity (the indirect effect) remains very difficult to quantify, more so than the direct effect. Several new studies support the view that it has caused a negative radiative forcing since pre-industrial times. The estimate from IPCC (1994) is unchanged at between 0 and -1.5 Wm^{-2}.

** Large volcanic eruptions can significantly increase the aerosol content in the stratosphere and cause a radiative forcing for a few years. For instance, the eruption of Mt. Pinatubo in 1991 caused a transient but large global mean radiative forcing that reached between -2 and -4 Wm^{-2} for about one year. The uneven distribution of volcanic eruptions during the past century or so means that the transient variations in volcanic forcing may have been important in some of the observed climate variations on the decadal scale.

* Extension of current understanding of the relationship between observed changes in solar output and other indicators of solar variability suggests that long-term increases in solar irradiance since the 17th century Maunder Minimum might have been climatically significant. A global mean radiative forcing of a few tenths of a Wm^{-2} since 1850 has been suggested, but uncertainties are large.

Global Warming Potential (GWP)

* Policymakers may need a means of estimating the relative radiative effects of the various greenhouse gases. The GWP is an attempt to provide such a measure. The index is defined as the cumulative radiative forcing between the present and some chosen later time "horizon" caused by a unit mass of gas emitted now, expressed relative to some reference gas (here CO_2 is used). The future global warming commitment of a greenhouse gas over the reference time horizon is the appropriate GWP multiplied by the amount of gas emitted. For example, GWPs could be used to estimate the effect of a given reduction in CO_2 emissions compared with a given reduction in CH_4 emissions, for a specified time horizon.

** The set of Global Warming Potentials (GWPs) for greenhouse gases that is presented in this chapter is an update and expansion of that presented in IPCC (1994). Three new gases have been added to the suite of GWPs

presented. The typical uncertainty is ±35% relative to the carbon dioxide reference. The time horizons of the GWPs are 20, 100, and 500 years.

* GWPs have a number of important limitations and underlying assumptions (see IPCC (1994)). For example, the GWP concept is currently inapplicable to gases and aerosols that are very unevenly distributed, as is the case for tropospheric ozone and aerosols and their precursors. Further, the indices and the estimated uncertainties are intended to reflect global averages only, and do not account for regional effects.

*** The net GWPs for the ozone-depleting gases, which include the direct "warming" and indirect "cooling" effects, have now been estimated. In IPCC (1994), only the direct GWPs were presented for these gases. The indirect effect reduces their net GWPs: those of the chloroflurocarbons (CFCs) tend to be positive, while those of the halons tend to be negative. The calculation of indirect effects for a number of other gases (e.g., NO_x) is not currently possible because of inadequate characterisation of many of the atmospheric processes involved.

*** Revised lifetimes for gases destroyed by chemical reactions in the lower atmosphere (particularly methane and the CFC substitutes) are based upon a recent reference gas re-calibration, resulting in GWPs that are slightly lower (typically by 10–15%) than those cited in IPCC (1994).

Introduction

In 1994, IPCC WGI produced a report entitled "Radiative Forcing of Climate Change". Two main topics were addressed:

(a) the relative climatic importance of anthropogenically induced changes in the atmospheric concentrations and distribution of different greenhouse gases and aerosols;

(b) possible routes to stabilisation of greenhouse gas concentrations in the atmosphere.

Each section of this chapter contains a summary and update of the material presented in one of the first 5 chapters of the 1994 report to which the reader is referred for a fuller discussion.

Clear evidence has been presented in past IPCC WGI reports (1990, 1992, 1994) that the atmospheric concentrations of a number of radiatively active gases have increased over the past century as a result of human activity. Most of these trace gases possess strong absorption bands in the infrared region of the spectrum (where energy is emitted and absorbed by the Earth's surface and atmosphere) and they thus act to increase the heat trapping ability of the atmosphere and so drive the climate change considered in this report. The other atmospheric constituents which are important in climate change are aerosols (suspensions of particles in the atmosphere) which tend to exert a cooling effect on the atmosphere.

The relative importance of the various constituents is assessed using the concept of radiative forcing. A change in the concentration of an atmospheric constituent can cause a radiative forcing by perturbing the balance between the net incoming radiation and the outgoing terrestrial radiation. A radiative forcing is defined to be a change in average net radiation (either solar or terrestrial in origin) at the top of the troposphere (the tropopause). As defined here, the incoming solar radiation is not considered a radiative forcing, although a change in the amount of incoming solar radiation would be a radiative forcing. Similarly changes in clouds and water vapour resulting from alterations to the general circulation of the atmosphere are considered to be climate feedbacks (see Chapter 4) rather than radiative forcings. However, changes caused, for example, in clouds through the indirect aerosol effect and in water vapour through the oxidation of methane in the stratosphere, are counted as indirect radiative forcings. Radiative forcing was discussed in detail in IPCC (1994): Section 2.4 contains a summary and update of that material.

The concentration of an atmospheric constituent depends on the size of its sources (emissions into and production within the atmosphere) and sinks (chemical loss in the atmosphere and removal at the Earth's surface). These processes act on different time-scales which, singly and on aggregate, define the various lifetimes of the atmospheric constituent (see box "Definition of Time-scales"). The past changes in the concentrations of the stable gases are relatively well known as they can be found by measuring the concentrations in air bubbles trapped in ice cores in Greenland and Antarctica. The past changes in the concentrations of the less stable gases and of aerosols are harder to quantify.

Estimation of possible future changes in the concentration of an atmospheric constituent requires a quantitative understanding of the processes that remove the constituent, and estimates of its future emissions (and/or atmospheric production). The former is discussed in this chapter for a wide range of atmospheric constituents. Estimating the emissions, however, is outside the scope of this report and the so-called IS92 scenarios (IPCC, 1992) are used here because they cover a wide range of possible future emissions and because both their strengths and weaknesses are relatively well-known (Alcamo *et al.*, 1995).

The first three sections of this chapter discuss the factors that affect the atmospheric abundance of carbon dioxide (Section 2.1), the other trace gases (Section 2.2) and aerosols (Section 2.3). Section 2.4 discusses the radiative forcing that arises from changes in the concentrations of these atmospheric constituents and from changes in solar output. IPCC (1990) introduced the global warming potential (GWP) as a measure of possible future commitments to global warming resulting from current anthropogenic emissions. The GWP is defined as the cumulative radiative forcing between the present and some chosen later time "horizon" caused by a unit mass of gas emitted now, expressed relative to that of some reference gas (CO_2 has typically been used). A GWP is thus not a simple measure as it involves a number of assumptions (e.g., the time-scale, the reference gas) and a number of scientific uncertainties (e.g., lifetime, radiative properties – discussed in IPCC (1994)). These issues and revised values of the GWPs of a wide range of gases are presented in Section 2.5.

An intriguing, but still unresolved issue is whether there is an underlying cause for the variations in the trends of a number of trace gases including carbon dioxide (CO_2), methane (CH_4) and nitrous oxide (N_2O) which are discussed individually later in this chapter. In the early 1990s, particularly between 1991 and 1993, the rates of increases of these gases became smaller and, at certain times and places, negative. A number of mechanisms have been proposed for each gas, but no quantitative resolution has yet been made. One common factor is the eruption of Mt. Pinatubo in June 1991 which may have affected the sources and sinks of CO_2, CH_4 and N_2O through changes in meteorology, atmospheric chemistry and/or biogeochemical exchange at the Earth's surface (see Sections 2.1.2, 2.2.2.1 and 2.2.2.2). However, no consensus has been reached as to whether the changes in the growth rates are linked or not.

2.1 CO_2 and the Carbon Cycle

2.1.1 Introduction

Two factors have increased attention on the carbon cycle: the observed increase in levels of atmospheric CO_2 (~280 ppmv in 1800; ~315 ppmv in 1957; ~358 ppmv in 1994); and the Framework Convention on Climate Change (FCCC) under which nations have to assess their contributions to sources and sinks of CO_2 and to evaluate the processes that control CO_2 accumulation in the atmosphere. Over the last several years, our understanding of the carbon cycle has improved, particularly in the

---

DEFINITION OF TIME-SCALES

Throughout this report different time-scales are used to characterise processes affecting trace gases and aerosols. The following terminology is used in this chapter.

Turnover time (T) is the ratio of the mass (M) of a reservoir – e.g., a gaseous compound in the atmosphere – and the total rate of removal (S) from the reservoir: $T = M/S$

In cases where there are several removal processes (S_i), separate turnover times (T_i) can be defined with respect to each removal process: $T_i = M/S_i$

Adjustment time or response time (T_a) is the time-scale characterising the decay of an instantaneous pulse input into the reservoir. Adjustment time is also used to characterise the adjustment of the mass of a reservoir following a step change in the source strength.

Lifetime is a more general term often used without a single definition. In the Policymakers Summary and elsewhere in this report, lifetime is sometimes used, for simplicity, as a surrogate for adjustment time. In atmospheric chemistry, however, lifetime is often used to denote the turnover time.

In simple cases, where the global removal of the compound in question is directly proportional to the global reservoir content ($S = kM$, with k, the removal frequency, being a constant), the adjustment time almost equals the turnover time ($T_a = T$). An example is CFC-11 in the atmosphere which is removed only by photochemical processes in the stratosphere. In this case $T = T_a = 50$ years.

In other situations, where the removal frequency is not constant or there are several reservoirs that exchange with each other, the equality between T and T_a no longer holds. An extreme example is that of CO_2. Because of the rapid exchange of CO_2 between the atmosphere and the oceans and the terrestrial biota, the turnover time of CO_2 in the atmosphere (T) is only about 4 years. However, a large part of the CO_2 that leaves the atmosphere each year is returned to the atmosphere from these reservoirs within a few years. Thus, the adjustment time of CO_2 in the atmosphere (T_a) is actually determined by the rate of removal of carbon from the surface layer of the oceans into the deeper layers of the oceans. Although an approximate value of about 100 years may be given for the adjustment time of CO_2 in the atmosphere, the actual adjustment is faster in the beginning and slower later on.

Methane is another gas for which the adjustment time is different from the turnover time. In the case of methane the difference arises because the removal (S) – which is mainly through chemical reaction with the hydroxyl radical in the troposphere – is related to the amount of methane (M) in a non-linear fashion, i.e., $S = kM$, with k decreasing as M increases.

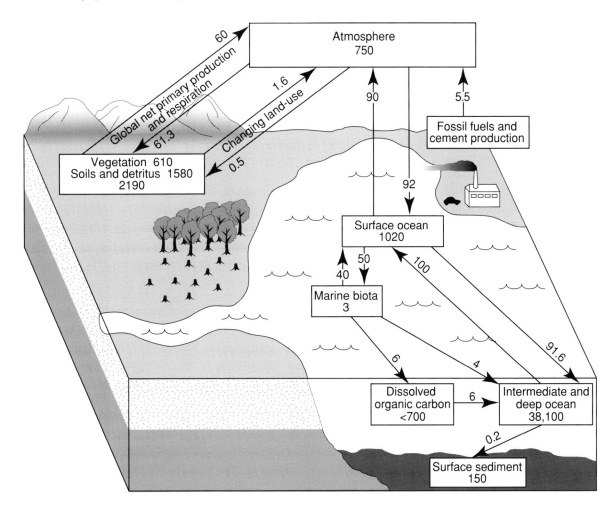

Figure 2.1: The global carbon cycle, showing the reservoirs (in GtC) and fluxes (GtC/yr) relevant to the anthropogenic perturbation as annual averages over the period 1980 to 1989 (Eswaran *et al.*, 1993; Potter *et al.*, 1993, Siegenthaler and Sarmiento, 1993). The component cycles are simplified and subject to considerable uncertainty. In addition, this figure presents average values. The riverine flux, particularly the anthropogenic portion, is currently very poorly quantified and so is not shown here. Evidence is accumulating that many of the key fluxes can fluctuate significantly from year to year (terrestrial sinks and sources: INPE, 1992; Ciais *et al.*, 1995a; export from the marine biota: Wong *et al.*, 1993). In contrast to the static view conveyed by figures such as this one, the carbon system is clearly dynamic and coupled to the climate system on seasonal, interannual and decadal time-scales (e.g., Schimel and Sulzman, 1995).

quantification and identification of mechanisms for terrestrial exchanges, and in the preliminary quantification of feedbacks. An overview of the carbon cycle is presented in Figure 2.1.

IPCC (1994) (Schimel *et al.*, 1995) specifically addressed four areas:

(a) the past and present atmospheric CO_2 levels;
(b) the atmospheric, oceanic, and terrestrial components of the global carbon budget;
(c) feedbacks on the carbon cycle;
(d) the results of a model-based examination of the relationship between future emissions and atmospheric concentrations (addressing, in particular,

the requirements for achieving stabilisation of atmospheric CO_2 concentrations).

In this introduction we summarise the earlier review, updating it where appropriate, before discussing four specific issues.

In Section 2.1.2 we re-address the issue of the early 1990s slow-down in the atmospheric growth rate of CO_2, as recent evidence shows that growth rates are once again rising. We also present modelled emissions and concentrations calculated on the basis of the carbon budget presented in 1994, and compare the results with those presented earlier based on the 1992 budget. In addition, we address some concerns of the reviewers of the 1994 report. For example, an expanded analysis of the sensitivity of the model calculations to the extremely uncertain estimates of global net deforestation is presented in Section 2.1.3.2. Further, we discuss the published suggestion that the carbon cycle calculations carried out to date could be in error (Starr, 1993), and we explain why these suggestions are wrong. Lastly, we present recent evidence that suggests a slightly smaller magnitude for the oceanic CO_2 sink, and discuss the implications of this in terms of concentration and emissions projections.

Atmospheric CO_2 levels

Precise, direct measurements of atmospheric CO_2 started in 1957 at the South Pole, and in 1958 at Mauna Loa, Hawaii. At this time the atmospheric concentration was about 315 ppmv and the rate of increase was ~0.6 ppmv/yr. The growth rate of atmospheric concentrations at Mauna Loa has generally been increasing since 1958. It averaged 0.83 ppmv/yr during the 1960s, 1.28 ppmv/yr during the 1970s, and 1.53 ppmv/yr during the 1980s. In 1994, the atmospheric level of CO_2 at Mauna Loa was 358 ppmv. Data from the Mauna Loa station are close to, but not the same as, the global mean.

Atmospheric concentrations of CO_2 have been monitored for shorter periods at a large number of atmospheric stations around the world (e.g., Boden *et al.*, 1991). Measurement sites are distributed globally and include sites in Antarctica, Australia, Asia, Europe, North America and several maritime islands, but, at present, nowhere on the continents of Africa or South America. The globally averaged CO_2 concentration, as determined through analysis of NOAA/CMDL data (Boden *et al.*, 1991; Conway *et al.*, 1994), increased by 1.53 ± 0.1 ppmv/yr over the period 1980 to 1989. This corresponds to an annual average rate of change in atmospheric carbon of 3.3 ± 0.2 GtC/yr. Other carbon-containing compounds like methane, carbon monoxide and larger hydrocarbons contain ~1% of the carbon stored in the atmosphere (with even smaller percentage changes) and can be neglected in the atmospheric carbon budget. There is no doubt that the increase shown by the atmospheric record since 1957 is due largely to anthropogenic emissions of CO_2. The record itself provides important insights that support anthropogenic emissions as a source of the observed increase. For example, when seasonal and short-term interannual variations in concentrations are neglected, the rise in atmospheric CO_2 is about 50% of anthropogenic emissions (Keeling *et al.*, 1989a, 1995) with the inter-hemispheric difference growing in parallel to the growth of

fossil emissions (Keeling *et al.*, 1989b; Siegenthaler and Sarmiento, 1993).

The atmospheric CO_2 concentration records prior to 1957 mainly come from air bubbles in ice cores, although some values have been inferred indirectly from isotopic data. Ice cores provide a direct record of past atmospheric composition back to well before the industrial revolution. Information between 1000 and 9000 years ago is less certain because of ice defects. Prior to that, the data show natural variations, the most noticeable of which is an increase in CO_2 level of about 80 ppmv that paralleled the last interglacial warming. Throughout this record, there is no direct evidence that past changes in CO_2 levels were as rapid as those of the 20th century; early indications that such rapid changes may have occurred during the last glacial period have not been confirmed (Neftel *et al.*, 1988). Over the last 1000 years, CO_2 concentrations in the atmosphere have fluctuated ±10 ppmv around 280 ppmv, until the recent increase to a concentration of ~358 ppmv, with a current rate of increase of ~1.6 ppmv/yr (1994).

The Anthropogenic Carbon Budget

The major components of the atmospheric carbon budget are anthropogenic emissions, the atmospheric increase, exchanges between the ocean and the atmosphere, and exchanges between the terrestrial biosphere and the atmosphere (Table 2.1). Emissions from fossil fuels and cement production averaged 5.5 ± 0.5 GtC/yr over the decade of the 1980s. In 1990 the emissions were 6.1 ± 0.6 GtC. The measured average annual rate of atmospheric increase in the 1980s was 3.3 ± 0.2 GtC/yr. Average ocean uptake during the decade has been estimated by a combination of modelling and measurements of carbon isotopes and atmospheric oxygen/nitrogen ratios to be 2.0 ± 0.8 GtC/yr.

Averaged over the 1980s, terrestrial exchanges include a tropical source of 1.6 ± 1.0 GtC/yr from ongoing changes in land-use, based on land clearing rates, biomass inventories and modelled forest regrowth. Recent satellite data have reduced uncertainties in the rate of deforestation for the Amazon, but rates for the rest of the tropics remain poorly quantified. For the tropics as a whole, there is incomplete information on initial biomass and rates of regrowth. There is currently no estimate available for the years 1990 to 1995. The tropical analyses do not account for a number of potential terrestrial sinks. These include the regrowth of mid- and high latitude Northern Hemisphere forests (1980s mean value of 0.5 ± 0.5 GtC/yr), enhanced forest growth due to CO_2 fertilisation (0.5–2.0 GtC/yr), nitrogen deposition (0.2–1.0 GtC/yr)

Table 2.1: *Average annual budget of CO_2 perturbations for 1980 to 1989. Fluxes and reservoir changes of carbon are expressed in GtC/yr, error limits correspond to an estimated 90% confidence interval.*

	IPCC 1992[†]	IPCC 1994[*]	IPCC 1995
	Estimates for 1980s budget		
CO_2 sources			
(1) Emissions from fossil fuel combustion and cement production	5.5 ± 0.5[Δ]	5.5 ± 0.5	5.5 ± 0.5[§]
(2) Net emissions from changes in tropical land-use	1.6 ± 1.0[Δ]	1.6 ± 1.0	1.6 ± 1.0[§]
(3) Total anthropogenic emissions = (1)+(2)	7.1 ± 1.1	7.1 ± 1.1	7.1 ± 1.1
Partitioning amongst reservoirs			
(4) Storage in the atmosphere	3.4 ± 0.2[Δ]	3.2 ± 0.2	3.3 ± 0.2[§]
(5) Ocean uptake	2.0 ± 0.8[Δ]	2.0 ± 0.8	2.0 ± 0.8[§]
(6) Uptake by Northern Hemisphere forest regrowth	not accounted for	0.5 ± 0.5	0.5 ± 0.5[§]
(7) Other terrestrial sinks = (3)–((4)+(5)+(6))			
(CO_2 fertilisation, nitrogen fertilisation, climatic effects)	1.7 ± 1.4	1.4 ± 1.5	1.3 ± 1.5

† Values given in IPCC (1990, 1992).

* Values given in IPCC (1994).

Δ Values used in the carbon cycle models for the calculations presented in IPCC (1994).

§ Values used in the carbon cycle models for the calculations presented here.

and, possibly, response to climatic anomalies (0–1.0 GtC/yr). The latter term, although thought to be positive over the 1980s, and in 1992 to 1993 in response to the Mt. Pinatubo eruption of 1991 (Keeling *et al.*, 1995), could be either positive or negative during other periods. Partitioning the sink among these processes is difficult, but it is likely that all components are significant. While the CO_2 fertilisation effect is the most commonly cited terrestrial uptake mechanism, existing model studies indicate that the magnitudes of the contributions from each process are comparable, within large ranges of uncertainty. For example, some model-based evidence suggests that the magnitude of the CO_2 fertilisation effect is limited by interactions with nutrients and other ecological processes. Experimental confirmation from ecosystem-level studies, however, is lacking. As a result, the role of the terrestrial biosphere in controlling past atmospheric CO_2 concentrations is uncertain, and its future role is difficult to predict.

The Influence of Climate and other Feedbacks on the Carbon Cycle

The responses of terrestrial carbon to climate are complex, with rates of biological activity generally increasing with warmer temperatures and increasing moisture. Storage of

carbon in soils generally increases along a gradient from low to high latitudes, reflecting slower decomposition of dead plant material in colder environments (Post *et al.*, 1985; Schimel *et al.*, 1994). Global ecosystem models based on an understanding of underlying mechanisms are designed to capture these patterns, and have been used to simulate the responses of terrestrial carbon storage to changing climate. Models used to assess the effects of warming on the carbon budget point to the possibility of large losses of terrestrial carbon (~200 GtC) over the next few hundred years, offset by enhanced uptake in response to elevated CO_2 (that could eventually amount to 100–300 GtC). Effects of changing land-use on carbon storage also may be quite large (Vloedbeld and Leemans, 1993). Experiments that incorporate the record of past changes of climate (e.g., the climate of 18,000 years ago) suggest major changes in terrestrial carbon storage with climate. See Chapter 9 for a more complete discussion of these issues.

Storage of carbon in the ocean may also be influenced by climate feedbacks through physical, chemical and biological processes. Initial model results suggest that the effects of predicted changes in circulation on the ocean carbon cycle are not large (10s rather than 100s of ppmv in the atmosphere). However, exploration of the long-term impacts of warming on ocean circulation patterns has just begun; hence, analyses of climate impacts on the oceanic carbon cycle must be viewed as preliminary. See Chapter 10 for a more complete discussion of these issues.

Modelling Future Concentrations of Atmospheric CO_2

For IPCC (1994), modelling groups from many countries were asked to use published carbon cycle models to evaluate the degree to which CO_2 concentrations in the atmosphere might be expected to change over the next several centuries, given a standard set of emission scenarios (including changes in land-use) and levels for stabilisation of CO_2 concentrations (350, 450, 550, 650 and 750 ppmv (S350–S750)). Models were constrained to balance the 1992 IPCC version of the 1980s mean carbon budget and to match the atmospheric record of past CO_2 variations using CO_2 fertilisation as the sole sink for the terrestrial biosphere. These analyses were re-done for this report, using the Bern (Siegenthaler and Joos, 1992; Joos *et al.*, 1996), Wigley (Wigley, 1993), and Jain (Jain *et al.*, 1995) models as representatives of the whole model set. The new calculations incorporated more recent information about the atmospheric CO_2 increase and a revised net land-use flux. As complete data are not yet available for a 1990s budget, the 1980s mean values were again used as a reference calibration period to account for the influence of natural variability. This choice does not affect the results in any significant way (see Chapter 9).

The stabilisation analyses explored the relationships between anthropogenic emissions and atmospheric concentrations. The analyses were based on a specific set of concentration profiles constrained to match present-day (1990) conditions and to achieve stabilisation at different levels and different future dates. The levels chosen (350–750 ppmv) spanned a realistic range (which has been extended in this report). Stabilisation dates were chosen so that the emissions changes were not unrealistically rapid. Precise pathways were somewhat arbitrary, loosely constrained by the need for smooth and not-too-rapid emissions changes (see Enting *et al.* (1994) for detailed documentation). Alternative pathways are considered in this report. The models were used to perform a series of inverse calculations to determine fossil emissions (land-use emissions were prescribed). These calculations:

(1)   determined the time course of carbon emissions from fossil fuel combustion required to arrive at the selected CO_2 concentration stabilisation profiles while matching the past atmospheric record and the 1980s mean budget; and

(2)   assessed (by integration) the total amount of fossil carbon released.

For stabilisation at 450 ppmv in the 1994 calculations, fossil emissions had to be reduced to about a third of today's levels (i.e., to about 2 GtC/yr) by the year 2200. For stabilisation at 650 ppmv, reductions by 2200 had to be about two-thirds of current levels (i.e., 4 GtC/yr). This clearly indicates that stabilisation of emissions at 1990 levels is not sufficient to stabilise atmospheric CO_2. These results are independent of the assumed pathway to concentration stabilisation (as shown later in this report), although the detailed changes in future anthropogenic emissions do depend on the pathway selected.

For the 18 models used for IPCC (1994), the implied fossil emissions differed. Initial differences were small, increasing with time to span a range up to ±15% about the median. In addition, the maximum range of uncertainty in fossil emissions associated with the parametrization of CO_2 fertilisation (evaluated with one of the models) varied ±10% about the median for low stabilisation values and ±15% for higher stabilisation values. For these 1994 calculations, the use of CO_2 fertilisation alone to control terrestrial carbon storage, when in fact other ecological mechanisms are likely to be involved, probably results in an underestimate of future concentrations (for given emissions) or an overestimate of emissions (for the stabilisation profiles). The revised calculations presented here use a lower fertilisation factor and so the associated bias is likely to be smaller. IPCC (1994) reported these results in terms of total anthropogenic emissions, directly addressing the requirements of the FCCC and coincidentally removing the loss of generality arising from choosing a specific land-use flux.

2.1.2 Atmospheric CO_2 Concentrations and the Status of the CO_2 Growth Rate Anomaly

The decline in the growth rate of CO_2 in 1992 is one of the most noticeable changes in the carbon cycle in the recent record of observation (Conway *et al.*, 1994). The decline in the short time-scale growth rate (based on interannual data filtered to remove annual cycle variations and considering the mean growth rates for overlapping 12-month periods) to 0.6 ppmv/yr in 1992 is considerable when contrasted with the mean rate over 1987 to 1988 of 2.5 ppmv/yr, until recently the highest one-year mean growth rate ever recorded. Examination of the CO_2 growth rate record reveals, in fact, considerable variability over time (Figure 2.2, which is the updated version of Figure 1.2 in Schimel *et al.*, 1995). The magnitude of individual anomalies in the

Attacks on IPCC Report Heat Controversy Over Global Warming

The claim that human activities significantly influence global climate has long been hotly debated. So when the influential Intergovernmental Panel on Climate Change (IPCC) gave its support to this claim for the first time in its recently published "Second Assessment Report" (SAR) on global climate change, the report came under immediate attack by a small but vocal community consisting of the Global Climate Coalition (GCC) (a Washington, DC–based association representing about 60 companies from the energy sector) and a small number of scientific skeptics. These critics claim that the IPCC broke its own procedural rules in preparing the report, and that the essence of a crucial chapter (see box on page 56) that deals with the detection and attribution of global warming was altered in the process. Scientists and officials of the IPCC, for their part, deny any wrongdoing.

The IPCC was established in 1988 by the United Nations Environment Programme and the World Meteorological Organization to provide periodic comprehensive assessments on climate change to guide policymakers internationally. The SAR, a three-volume report consisting of assessments of the science of climate change prepared by Working Group I, impact and response strategies by WG II and economic and social implications by WG III, involved about 2500 scientists worldwide and was reviewed extensively by scientific experts and by national governments. The report will be used to advise the United Nations Framework Convention on Climate Change (FCCC). "The most immediate influence will be on the Conference on the Parties [a UN body that is responsible for coordinating the implementation of the FCCC]," says Bronson Gardner, a consultant to the GCC. "If they are convinced that climate change is sufficiently threatening, they may add additional requirements to the FCCC, which could commit many countries to adopt new regulations."

"Chapter 8 ['Detection of Climate Change and Attribution of Causes'] was the most politically charged and hotly debated subject"

> A statement by scientists that the "balance of evidence suggests a discernible human influence on global climate" has stirred up a political controversy that seems to be about everything except the science.

at WG I's plenary meeting in Madrid (27–29 November 1995), says Gardner, a view echoed by many others. The main focus of the Madrid meeting, however, was to prepare the Summary for Policy Makers (SPM), based on the eleven chapters of the WG I report. The SPM concludes that "the balance of evidence suggests a discernible human influence on global climate," a statement that has potentially enormous implications for policy-making, for the energy industry and for the global economy.

"It took hours and hours of painstaking negotiation," Gardner says, adding that there was "some high-powered deal cutting" in agreeing to the SPM. But all 96 nations finally approved the SPM line by line, and no changes were made to it after the Madrid meeting.

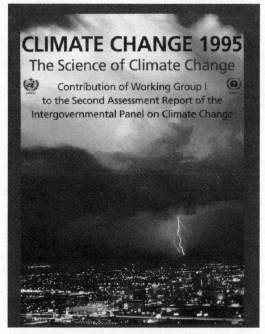

COVER OF THE WG I contribution to the 1995 IPCC report on climate change, published this May by Cambridge University Press.

The individual chapters, in contrast, were not approved line by line, but rather were "accepted," a technical term used by the IPCC to denote that a document "presents a comprehensive, objective and balanced view" of the subject matter. John Houghton, cochairman of WG I, stresses that acceptance of the chapters meant that changes would be made to reflect the discussions at the Madrid meeting.

The GCC and its allies are concerned—after all, policies aimed at curtailing fossil fuel emissions would have a major impact on the energy interests that the GCC represents. (Since 1991, representatives of the coal and oil industries have mounted a series of attacks on both the science of climate change and the scientists of the IPCC, according to an article in last December's *Harper's Magazine.*) The gist of the allegations is that changes made to chapter 8 after the Madrid meeting were "unauthorized" and in violation of the IPCC's procedural rules, and that the revised chapter suppresses scientific uncertainty to present more strongly the case for the influence of human activity on global climate.

But the WG I scientists and IPCC officials staunchly disagree with the allegations, which they say are unfounded. "These attacks mean that the energy industry is taking the science seriously, and that they acknowledge that the implications are serious for them," says Alan Robock, a climatologist at the University of Maryland at College Park and a contributor to chapter 8. "They would have found something to make trouble over, just to stall things," he adds. The debate has evolved into a fervid and ongoing brouhaha, and the heart of the controversy—the issue of human influence on global climate—seems sometimes to get lost in the process. Each side suspects political motives on the part of the other; the lead author of chapter 8, Benjamin Santer, an atmospheric scientist at Lawrence Livermore National Laboratory, feels his scientific integrity is being challenged; and the GCC and other detractors have questioned the credibility of the IPCC.

The allegations . . .

Articles attacking the writing of chapter 8 have appeared in the *Wall Street Journal*, *Financial Times Energy Economist* and *Energy Daily*, among other places, and have also been distributed to reporters and congressional representatives. A nine-page GCC analysis entitled "The IPCC: Institutionalized 'Scientific Cleansing' " compared the 9 October 1995 draft of chapter 8, sent out before the Madrid meeting, to the published version, and repeatedly called the latter "revisionist" and "scientifically cleansed." This analysis claimed that the changes made "change the fundamental character of the chapter, for they obscure, and in several important instances entirely delete, scientific analysis that casts serious doubts about current ability to attribute climate change to human activities." John Schlaes, executive director of the GCC, says, "The context of the report was changed. . . . Words *are* important." And Donald Rheem, a GCC spokesman, says, "We do not oppose the conclusions of the 1995 SAR. We support more scientific research, and object that the changes [to chapter 8] make the claim [of a human influence on climate] appear to be more certain than it is." Gardner says, "The changes look like political manipulation, and they shift the essence of chapter 8."

Similar charges have been made by others, such as Frederick Seitz, whose impressive résumé includes stints as president of the American Physical Society, president of Rockefeller University and president of the National Academy of Sciences. He is currently the chairman of the George C. Marshall Institute, a conservative Washington, DC–based think tank. "I have never witnessed a more disturbing corruption of the peer-review process than the event that led to this IPCC report," Seitz wrote in an op-ed published in the 12 June *Wall Street Journal*. "If the IPCC is incapable of following its most basic procedures, it would be best to abandon the entire IPCC process, or at least that part that is concerned with the scientific evidence on climate change, and look for more reliable sources of advice to governments on this important question." Santer says that Seitz never contacted him or any of the other lead authors or IPCC officials before writing his op-ed piece.

(The Marshall Institute puts out pamphlets and distributes them on Capitol Hill. The introduction to a 1996 pamphlet, "Are Human Activities Causing Global Warming?" reads, in part: "The most recent Marshall Institute review of scientific evidence on

Global Warming and Chapter 8

The global mean surface temperature of Earth has increased by 0.3–0.6 °C in the past 100 years. This rise in global temperature—which is accompanied by other climate effects such as a rise in sea level—is partly due to perturbations in the Earth–atmosphere energy balance that are associated with "radiative forcing," more popularly known as the greenhouse effect. "Greenhouse gases" such as carbon dioxide, methane and nitrous oxide accumulate in the atmosphere and warm it by absorbing heat that is radiated from Earth, which in turn was absorbed from sunlight. This process results in more longwave radiation being transferred from the atmosphere to Earth's surface, and hence in a warmer climate.

But atmospheric processes are complex and are not yet fully understood. They involve such regional mitigating effects as sulfate aerosols in the atmosphere, which block the Sun's radiation from reaching Earth, and also affect cloud reflectivity.

One of the things that chapter 8 of *Climate Change 1995: The Science of Climate Change*, volume 1 of the three-volume Second Assessment Report of the Intergovernmental Panel on Climate Change (IPCC), is concerned with is the possible impact on climate of human-induced enhancement of the naturally occurring greenhouse effect. The chapter evaluates the detection of changes in the global climate system, and considers the extent to which the observed changes can be attributed to human activities.

Detection of a "change in climate" requires that the observed change be proved statistically significant compared to natural background variability. (Natural variability results from a combination of internal factors—interactions within the coupled atmosphere-ocean-land-biosphere system—and external variables—primarily variability of solar energy input to the atmosphere and effects of volcanic eruptions.)

Attribution is made by carrying out numerical model simulations that consider variables such as the spatial distributions of temperature, trace gas concentrations, wind speed, rainfall and water vapor to determine the climate response signal for different hypothesized causes, and by then comparing these predictions to observed changes. Unique attribution of a detected "significant" climate change to human activities requires consideration and elimination of other plausible mechanisms.

Since the IPCC issued its first comprehensive assessment report in 1990, the observed measurements and simulation models have become more sophisticated: Estimates of natural background variability as well as statistical analysis applied to modeling results have improved; models can now incorporate the effects of sulfate aerosols (from human activity) as well as greenhouse gases; and pattern-based simulations consider spatial variability (rather than global mean temperature). Taken together, the advances increase the confidence level with which attribution can be made.

Chapter 8 discusses the recent advances and the scientific uncertainties. The chapter concludes cautiously: "The body of statistical evidence, when examined in the context of our physical understanding of the climate system, now points towards a discernible human influence on global climate. Our ability to quantify the magnitude of this effect is currently limited by uncertainties of key factors, including the magnitude and patterns of longer-term natural variability and the time-evolving patterns of forcing by (and response to) greenhouse gases and aerosols."

---

climate change confirms the earlier conclusion that predictions of an anthropogenic global warming have been greatly exaggerated. . . . Spread over a century, a temperature rise of this magnitude will be lost in the noise of natural climate fluctuations." Asked whether such publications are peer reviewed, Seitz said no, but explained that they "represent opinion." This proviso is not noted in the pamphlet, however.)

. . . are called unfounded

Santer takes full responsibility for all changes made to chapter 8 after the WG I meeting in Madrid, and he categorically denies all allegations that the tone or content of the chapter was altered, or that any IPCC rules were broken. Immediately after the Madrid meeting, Santer says, he "spent two days in a hotel room in England re-

viewing comments received between 9 October and up through the Madrid meeting, and revising the chapter." Although the main focus of the Madrid meeting was the SPM, the ambiguities that emerged during extensive discussions inevitably were relevant to the underlying chapters of the report as well, says Santer. "It seemed that some people were willfully misinterpreting things" in chapter 8. All changes, he adds, "were made—and delivered to WG I—before the full plenary session in Rome," held on 11–15 December and attended by all three working groups.

On 25 June the *Wall Street Journal* published a response to Seitz's op-ed. Written by Santer and cosigned by 40 lead authors and contributors to the WG I report, the letter emphasized that "IPCC procedures *required* changes in response to these comments [those re-

BENJAMIN SANTER, lead author of chapter 8.

JOHN SCHLAES, executive director of the Global Climate Coalition.

ceived on the 9 October draft] in order to produce the best possible and most clearly explained assessment of the science." A companion letter from Bert Bolin, the chair of the IPCC, showed clearly that the IPCC shares this view. "No one could have been more thorough and honest in undertaking [the] task [of the lead author of chapter 8]," wrote Bolin.

"The important message that the human signal was 'distinguishable beyond the noise' needed to be made more explicit" in chapter 8, says Michael MacCracken, director of the interagency Office of the US Global Change Research Program, and the US coordinator of the review process for the Working Group I report. "It was clear from comments [received both before and during the meeting] that the authors [of chapter 8] needed to more clearly express the level of confidence in their results." Also, in July 1995 (after completion of the full country review process) the authors of chapter 8 had decided to include some new and highly relevant information on vertical temperature change, recounts Mac-Cracken. (See this issue, page 9.) "The period before the Madrid plenary would give reviewers a chance to consider the additional findings," he says. "It is beyond me why they [the detractors] claim it [the 9 October draft] should be considered the final version," he adds.

But some—such as Donald Pearlman, an attorney and executive director of the Climate Council (a group of energy and transportation concerns), an ally of the GCC and one of those who helped write the IPCC procedural rules—maintain that the 9 October draft was "final" and should not have been altered, except for minor editorial changes. Pearlman, who was at the meetings in Madrid and Rome, con-

tends that "those changes were never proposed by the lead authors to the full working group, and they [the lead authors] had no unfettered right to make them."

"The changes to chapter 8 did not violate IPCC procedural rules," counters Kathy Maskell, a member of the WG I Technical Support Unit, which ensures that things proceed on schedule and according to the principles of the IPCC, and oversees coordination with the other two working groups. Maskell says that the draft of the entire WG I report was sent out on 9 October to IPCC governments and accredited nongovernmental organizations (of which the GCC is one) specifically so that they would have a chance to comment on it, and to check whether their comments from the earlier rounds of review had been taken into account. "That's why it was labeled 'DRAFT,'" she says.

One factor that has added to the confusion is that delegates to the full plenary meeting in Rome in December

received the pre-Madrid (9 October) version of the WG I report. That happened because of the tight timing of the plenary meetings, which were less than two weeks apart: Back in 1993, the IPCC had agreed on a concurrent review process by scientific experts and governments for WG I because of the group's other commitments, Maskell explains.

"Most important," Santer says, "the bottom-line conclusion of the chapter— that the balance of scientific evidence points towards a human influence on global climate—is the same in the 9 October and the published versions of chapter 8, and these conclusions were unanimously accepted by the IPCC governments at the Madrid meeting." Santer adds that key uncertainties, which are an integral part of the climate change and attribution problem, are discussed at length in the chapter, and that "claims that the chapter has been 'cleansed' of underlying uncertainties are just plain false."

"I am troubled that this controversy has surfaced. I had hoped that any controversy regarding the 1995 IPCC report would focus on the science itself, and not on the scientists," says Santer. A US government official who was a delegate to both the Madrid and Rome meetings says that all the procedural rules were followed legitimately, but that the rules themselves may be a bit sloppy. He adds that the scientists and policymakers should have—and still need to—adapt to each other to achieve better communication.

The hectoring is not yet over: Now the GCC is calling for an "independent review" of the changes made to chapter 8 of the WG I volume of the IPCC report. "They [the GCC and its allies] want to put a caveat on everything, rather than focus on what we have learned in the past five years," says Santer.

TONI FEDER

Bryn Mawr Physics Is Going Strong

A surprising thing about Bryn Mawr, a women's liberal arts college in a quiet suburb of Philadelphia, is the number of women who take bachelor's degrees in physics. The number has been climbing fairly steadily for the past 20 years, whereas the nationwide total for men and women combined has dropped 8% in the past 5 years. This spring, Bryn Mawr had 40 declared physics majors (sophomores, juniors and seniors) in a student body of only about 1200. In hard numbers, MIT and Harvard University are the only schools in the country that

graduated more women in physics during the 1990–94 period. "Simple projection shows that only the college's strict admissions limit precludes the day when all women physicists will have a degree from Bryn Mawr," jokes Neal Abraham, one of the department's four professors.

What is Bryn Mawr doing right? Being a women's college—which for some women can provide an environment in which pursuit of traditionally male-dominated fields is more comfortable than at coed schools—surely plays a role in the remarkably high number

DEAN E. EASTMAN

109
YEARS OF
WOMEN IN
CLASSICALLY
FORBIDDEN
REGIONS

MOTTO on the t-shirt designed by Bryn Mawr's physics department in 1994.

of women in physics and the other natural sciences. Bryn Mawr is also the only US women's college that has a PhD program in physics. It is a small coed program, but physics professor Elizabeth McCormack believes it has a significant impact: "Undergraduates get to take part in sophisticated research and can see early on if they like it," she says. Bryn Mawr physics majors also go on to pursue doctorates in physics at a rate (52%) significantly higher than the average from women's colleges (35%), or the average of women from PhD-granting departments (41%). (These figures, for 1990–94, are from the American Institute of Physics.)

"Students like the sociology of the department," says department chair Peter Beckmann. "It's like a big family," says junior Heather Fleming. Emily Peterson, a 1996 graduate, believes that "the department's biggest strength is that they [the faculty] really care about the students' opinions." Student representatives get to participate in twice-monthly departmental meetings, at which issues such as course curricula, computer software purchases and space allocation are discussed.

The department also offers career counseling. "We have built up a wide network of contacts," says physics professor Alfonso Albano, adding that the department also succeeds in finding nearly all juniors, as well as many sophomores, summer jobs in labs throughout the country. Albano goes on to describe how faculty members actively recruit students: "If we see a student who seems promising, we try to convince her to major in physics." One example is Jennifer Mosher, a 1996 graduate who is working this summer at the National Institute of

Standards and Technology in Gaithersburg, Maryland, and plans to go to graduate school in physics. She had her "heart set on poli sci," and took a course in conceptual (nonmath-based) physics to satisfy a college requirement. "Physics was a gap in my education, and my friends said Neal [Abraham] was a god," explains Mosher. "Physics is harder than poli sci, but I get more out of it." At a dinner for sophomores, "talking with Aurora [Vicens, a visiting faculty member who is no longer at Bryn Mawr] and Peter [Beckmann] convinced me to switch majors," she continues.

"The attitude at Bryn Mawr is, if someone is interested, then she should be able to do physics, even if she is not a genius," says Peterson. But she adds that this attitude might be a mistake: "It may convince some people to major in physics who otherwise would not have considered it, and since other people in the field may not be as supportive as they are at Bryn Mawr, it would always be an uphill battle." Abraham takes a different view: "Why sacrifice interested students on the altar of toughness?" He adds that "rigor is not being traded for a caring and supportive environment."

Beckmann stresses that teaching is a high priority at Bryn Mawr: "There is no bad teaching here." And all of the physics faculty members believe that physics is a good education, "no matter what one decides to do later," as one says. They also encourage each student to pursue her own interests—graduate school, industry, medicine, school teaching, finance or anything else.

In the end, though, it is impossible to pinpoint why physics is so popular at Bryn Mawr. After all, many schools offer similar resources. Ted Ducas, a physics professor at Wellesley College, another women's college that has a strong physics department, may be on the mark when he says, "Sometimes one or more charismatic individuals can make all the difference."

TONI FEDER

Eastman Succeeds Schriesheim as Argonne Director

On 15 July, Dean E. Eastman became director of Argonne National Laboratory. He succeeded Alan Schriesheim, who retired on 1 July, which also happened to mark the date of the lab's founding 50 years earlier.

Eastman, who holds a PhD in elec-

trical engineering from MIT, had worked at the IBM Corp since 1963. His rise through the ranks in the IBM research division included posts as manager of the photoemission and surface physics group, director of the advanced packaging laboratory and, most recently, vice president of systems technology and science. During the past several years, Eastman led IBM's development reengineering efforts to make its hardware business units more competitive. His research areas have included condensed matter physics, surface science and photoelectron spectroscopy using synchrotron radiation.

Located about 25 miles southwest of Chicago, Argonne is operated by the University of Chicago for the US Department of Energy. During Schriesheim's tenure, the lab's operating budget and staff nearly doubled. But like the rest of the DOE national lab complex, it has undergone close scrutiny in recent years and has also endured some funding cuts. This year, for example, the lab's operating budget dropped by 2%, to $485 million, and the work force was pared by 6%, to about 4500 people. Even so, Schriesheim says, Argonne's future appears more secure than that of some other DOE labs.

Schriesheim, who now holds the position of director emeritus of Argonne, says he plans to continue working on science policy and technology transfer issues. Prior to joining Argonne in 1983, he worked for many years at Exxon Research and Engineering C. Among the major projects to be und taken during his tenure at Argo was the construction of the Adv Photon Source, a hard-x-ray sy tron light source (see PHYSIC May 1995, page 59). Experir

CO_2 growth rate depends on what is defined as "normal" for the long-term trend. At Mauna Loa, for example, the 1987 to 1988 increase was similar to a variation which occurred in 1972 to 1973 (see Figure 2.2).

How are these changes explained, and do they matter? Analyses of the recent changes suggest that both the relative and absolute magnitudes of ocean and terrestrial processes vary substantially from year-to-year and that these changes cause marked annual time-scale changes in the CO_2 growth rate. Although the oceanic and biospheric sources of CO_2 cannot be distinguished by examining the concentration data alone, they can be distinguished by looking at observations of the stable carbon isotope ratio (^{13}C to ^{12}C) of atmospheric CO_2. Carbon dioxide release from the ocean has nearly the same ^{13}C to ^{12}C ratio as atmospheric CO_2, whereas carbon of biospheric origin is substantially depleted in ^{13}C. As such, it is clearly distinguishable from carbon originating from the ocean. The use of atmospheric transport modelling to analyse spatial distributions of CO_2 and $^{13}CO_2$ provides additional information (Ciais *et al.*, 1995a, b). To study interannual CO_2 fluctuations, Keeling *et al.* (1989a, 1995) removed the seasonal cycle and the long-term fossil fuel-induced trend from the direct atmospheric record to obtain anomalies in

CO_2 concentration at Mauna Loa and the South Pole. Anomalies of the order of 1–2 ppmv are found; a small signal compared to the long-term increase (about 80 ppmv). The interannual CO_2 variations must reflect imbalances in the exchange fluxes between the atmosphere and the terrestrial biosphere and/or the ocean.

The isotopic data and modelling studies suggest that earlier short-term CO_2 anomalies on the El Niño-Southern Oscillation (ENSO) time-scale reflected two opposing effects: reduced net primary production of the terrestrial biosphere possibly due to reduced precipitation in monsoon regions, and a concomitant, temporary increase in oceanic CO_2 uptake due to a reduction of the CO_2 outgassing in the Pacific equatorial ocean (Keeling *et al.*, 1989b, 1995; Volk, 1989; Siegenthaler, 1990; Winguth *et al.*, 1994). The most recent anomaly seems to be unusual as it cannot be directly related to an ENSO event. Keeling and co-workers suggest that it may have been induced by a global anomaly in air temperature. While the evidence for pronounced interannual variability in both marine and terrestrial exchange fluxes is strong (Bacastow, 1976; Francey *et al.*, 1995) the exact magnitude and timing of the variations in these exchange fluxes remains controversial (Keeling *et al.*, 1989a, 1995; Feely *et al.*, 1995; Francey *et al.*, 1995).

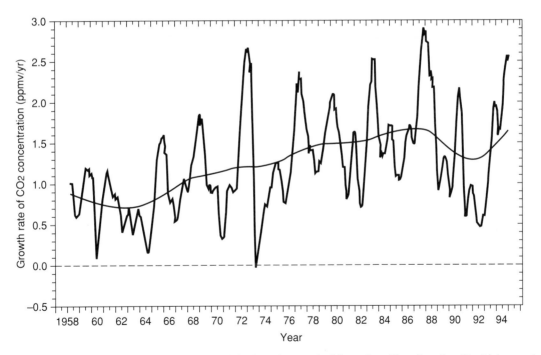

Figure 2.2: Growth rate of CO_2 concentrations since 1958 in ppmv/year at the Mauna Loa, Hawaii station. The high growth rates of the late 1980s, the low growth rates of the early 1990s, and the recent upturn in the growth rate are all apparent. The smoothed curve shows the same data but filtered to suppress variations on time-scales less than approximately 10 years. (Sources: C.D. Keeling and T.P. Worf, Scripps Institute of Oceanography, and P. Tans, NOAA CMDL. The Keeling and NOAA results are in close agreement. The Mauna Loa Observatory is operated by the NOAA.)

Most of the global carbon cycle models used in IPCC (1994) (Schimel *et al.*, 1995) address only the longer term (10 year time-scale) direct perturbation of the global carbon cycle due to anthropogenic emissions. These models assume that the physical, chemical, and biological processes that control the exchange fluxes of carbon between the different reservoirs change only as a function of the carbon contents. Hence these models ignore any perturbations due to fluctuations in climate and are thus not able to reproduce these shorter-term atmospheric CO_2 variations. Nevertheless, some of the more complex models are founded on physico-chemical and biological principles that also operate on shorter time-scales and first attempts to model these interannual variations have been attempted (Kaduk and Heimann, 1994; Winguth *et al.*, 1994; Sarmiento *et al.*, 1995). An improved understanding of the rapid fluctuations of the global carbon cycle would help to further develop and validate these models and eventually should help to quantify some of the potential feedbacks addressed in Chapters 9 and 10.

We now know that short-term growth rates increased markedly through 1993 and 1994, and are currently at levels above the long-term (decadal time-scale) mean (Figure 2.2). As best as can be established, the anomaly of the early 1990s represented a large but transient perturbation of the carbon system.

2.1.3 Concentration Projections and Stabilisation Calculations

Fossil fuel burning and cement manufacture, together with forest harvest and other changes of land-use, all transfer carbon (mainly as CO_2) to the atmosphere. This anthropogenic carbon then cycles between the atmosphere, oceans and the terrestrial biosphere. Because an important component of the cycling of carbon in the ocean and terrestrial biosphere occurs slowly, on time-scales of decades to millennia, the effect of additional fossil and biomass carbon injected into the atmosphere is a long-lasting disturbance of the carbon cycle. The record itself provides important insights that support anthropogenic emissions as a source of the observed increase. For example, when seasonal and short-term interannual variations in concentrations are neglected, the rise in atmospheric CO_2 is about 50% of anthropogenic emissions (Keeling *et al.*, 1989a) with the inter-hemispheric difference growing in parallel to the growth of fossil emissions (Keeling *et al.*, 1989a; Siegenthaler and Sarmiento, 1993). These aspects are in accord with our understanding of the carbon cycle, and agree with model simulations of it. An additional important indicator of

anthropogenically-induced atmospheric change is provided by the ^{14}C levels preserved in materials such as tree rings and corals. The ^{14}C concentration measured in tree rings decreased by about 2% during the period 1800 to 1950. This isotopic decrease, known as the Suess effect (Suess, 1955), provides one of the most clear demonstrations that the increase in atmospheric CO_2 is due largely to fossil fuel inputs.

In this section, the relationships between future concentration changes and emissions of CO_2 are examined through use of models that simulate the major processes of the carbon cycle. In the context of future climate change, carbon cycle model calculations play a central role because the bulk of projected radiative forcing changes comes from CO_2. In IPCC (1994), two types of carbon cycle calculations relating future CO_2 concentration and emissions were presented: concentration projections for the IPCC 1992 (IS92) emission scenarios; and emissions estimates for a range of concentration stabilisation profiles directly addressing the stabilisation goal in Article 2 of the FCCC.

As noted above, these emissions-concentrations relationships have been re-calculated because the previous results were based on a 1980s mean budget from IPCC (1992) that has been revised in three ways:

(1)   In the 1992 budget the net land-use flux for the decade of the 1980s was 1.6 GtC/yr and the atmospheric increase was 3.4 GtC/yr. The more recent estimate of the net land-use flux, however, is significantly lower (1.1 GtC/yr).

(2)   IPCC (1994) presented a change in atmospheric CO_2 over the 1980s of 3.2 GtC/yr, a value that has been further revised to 3.3 GtC/yr (Komhyr *et al.*, 1985; Conway *et al.*, 1994; Tans, pers. comm).

(3)   Minor changes have been made to the industrial emissions data for the 1980s, although these do not noticeably change the 1980s mean. These changes mean that the additional sinks required to balance the budget (which were and still are assigned to the CO_2 fertilisation effect in the model calculations) are smaller than assumed in calculations presented in IPCC (1994). Consequently, when the new budget is used, future concentration projections for any given emission scenario are larger relative to those previously presented, and the emissions consistent with achieving stabilisation of concentrations are lower.

Three different models were used to carry out these calculations: the Bern model (Siegenthaler and Joos, 1992; Joos *et al.*, 1996) and the models of Wigley (Wigley, 1993) and Jain (Jain *et al.*, 1995). Changes relative to the previous calculations are virtually the same for each model. The methodology used was the same as in IPCC (1994), as documented in Enting *et al.* (1994). New results were produced for concentrations associated with emission scenarios IS92a-f and emissions associated with stabilisation profiles S350–S750. Impulse response functions were also revised for input into GWP calculations.

In addition, stabilisation calculations were performed for a profile with a 1000 ppmv stabilisation level (S1000) and for variations on S350–S750 in which the pathway to stabilisation was changed markedly. These latter profiles (WRE350–WRE750), from Wigley *et al.* (1995), were designed specifically to follow closely the IS92a "existing policies" emission scenario for 10–30 years after 1990. Coupled with S350–S750 results, these new profiles give insights into the range of emissions options available for any given stabilisation level, although they still do not necessarily span the full option range. It should be noted that, even though concentration stabilisation is still possible if emissions initially follow an existing policies scenario, this cannot be interpreted as endorsing a policy to delay action to reduce emissions. Rather, the WRE scenarios were designed to account for inertia in the global energy system and the potential difficulty of departing rapidly from the present level of dependence on fossil fuels. Wigley *et al.* (1995) stress the need for a full economic and environmental assessment in the choice of pathway to stabilisation. Both the S350–S750 and the WRE350–WRE750 profile sets require substantial reductions in emissions at some future time and an eventual emissions level well below that of today.

The S1000 profile was designed to explore the emissions consequences of an even higher stabilisation level; the emissions implied by this case follow IS92a closely out to 2050. The results show that, even with such a high stabilisation level, the same characteristic emissions curve arises as found for S350–S750. This scenario is by no means derived to advocate such a high stabilisation level. The environmental consequences of such a level have not been assessed, but they are certain to be very large. It should be noted that recent emissions are low compared to IS92a-projected emissions (emissions were essentially the same in 1990 and in 1992, at 6.1 GtC/yr), although the slow-down may be temporary.

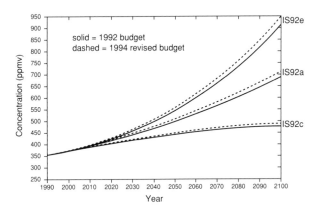

Figure 2.3: CO_2 concentrations resulting from emission scenarios IS92a, c, and e plotted using both the 1992 budget and the 1994 budget. Solid lines represent model results based on the 1992 budget calibration, dashed lines, calibration with the 1994 budget. The differences between results is small (of order 20 ppmv change in projected concentration by the year 2100). The changes from revising the budget result in changes in radiative forcing of approximately 0.2 Wm^{-2}.

2.1.3.1 Effects of carbon cycle model recalibration
The effects of using the 1992 versus the 1994 budget with IS92 Scenarios a, c, and e are shown in Figure 2.3. The results show that the impact of the new budget on projected future concentrations is noticeable but relatively minor. In the year 2100, the concentrations increase by about 15 ppmv (IS92c), 25 ppmv (IS92a) and 40 ppmv (IS92e). All three models gave similar results. In terms of radiative forcing increases, the 2100 values are 0.20 Wm^{-2} (IS92c), 0.25 Wm^{-2} (IS92a), and 0.26 Wm^{-2} (IS92e). These changes are all small compared to the overall forcing changes over 1990 to 2100 for these scenarios.

Figure 2.4 shows changes in the emissions requirements for the S450 and S650 profiles due to changes between the 1992 and 1994 budgets. The net effect of these changes is quite complex as it involves compensating effects due to different budget factors. Changing the 1980s mean net land-use flux from 1.6 GtC/yr to 1.1 GtC/yr (by reducing the fertilisation factor required to give a balanced budget) leads to emissions that are about 10% lower than given previously (Schimel *et al.*, 1995). This reduction is modified slightly by the changes in concentration history and profiles, and by an even smaller amount due to the change in the fossil fuel emissions history, leading to the results presented in Figure 2.4. The changes shown in Figure 2.4 arise from the complex interplay of a number of factors; but the overall conclusion (that emissions must eventually decline to substantially below current levels) remains unchanged.

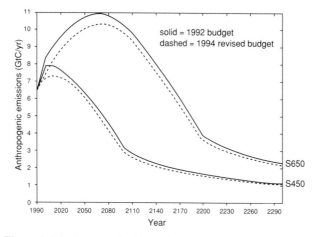

Figure 2.4 Anthropogenic CO_2 emissions for revised concentration stabilisation profiles S450 and S650 resulting from model projections initialised with the 1992 budget (solid curve: net land-use flux = 1.6 GtC/yr) and the 1994 budget (dashed curve: net land-use flux = 1.1 GtC/yr).

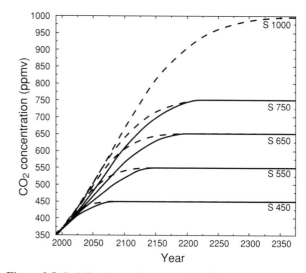

Figure 2.5: Stabilisation pathways used to illustrate sensitivity of allowed emissions to choice of pathway. The solid curves are based on the revised versions of the original S450 to S750 stabilisation profiles, incorporating minor changes to account for the revised concentration history. The dashed curves show slower approaches to stabilisation and are from Wigley *et al.* (1995). Stabilisation at 1000 ppmv is plotted for comparison (S1000).

We have carried out several new calculations to illustrate the effects of the choice of pathway leading to stabilisation of modelled emissions and uptake. In Figures 2.5 and 2.6 we show the effect of pathway in more detail by employing the profiles of Wigley *et al.* (1995) (WRE below). Figure 2.5 shows the different pathways, reaching stabilisation at the same time in each case, for target levels of 450 ppmv to 750 ppmv. The corresponding anthropogenic emissions are shown in Figure 2.6, which shows two things: first, that the emissions required to achieve any given stabilisation target are quite sensitive to the pathway taken to reach that target; and second, that it is possible to achieve stabilisation even if the initial emissions pathway follows the IS92a Scenario for some period of time. There is, of course, a penalty for this: having higher emissions initially requires a larger and possibly sharper drop in emissions later, to lower levels than otherwise (also see Enting, 1995). This is a direct consequence of the fact that cumulative emissions tend to become (at the stabilisation point or beyond) similar no matter what the pathway. If cumulative emissions are constrained to be nearly constant, then what is gained early must be lost later.

Figures 2.5 and 2.6 also show the case for stabilisation at 1000 ppmv (in the year 2375). The profile here was constructed to follow the IS92a Scenario out to 2050 (using the model of Wigley (1993)) and is just one of a number of possible pathways. Even with this high target, to achieve it requires an immediate and substantial reduction below IS92a after 2050, with emissions peaking around 2080 and then undergoing a long and steady decline to, eventually,

values well below the current level. Although offset substantially, the general shape of the emissions curve for S1000 is similar to all other curves.

2.1.3.2 *Effects of uncertainties in deforestation and CO_2 fertilisation*

The future uptake of CO_2 by the terrestrial biosphere is critical to the global carbon balance. In the above concentration projections and emissions results, the terrestrial biosphere uptake was determined by the CO_2 fertilisation factor, which in turn is determined mainly by the value assumed for the 1980s mean land-use emissions (if the ocean uptake is taken as a given). Thus, as in Wigley (1993) and Enting *et al.* (1994), we can assess the effects of fertilisation and future terrestrial uptake uncertainties by changing the 1980s mean land-use term used in initialising the carbon cycles models. Because the ocean uptake is held constant, this results in changing the modelled CO_2 effect. The sensitivity of the calculated emissions for CO_2 stabilisation at 450 and 650 ppmv to these uncertainties is shown in Figure 2.7 (cf. Figure 1.16 in Schimel *et al.*, 1995).

The results from all three models used in the present analysis show that uncertainty in the strength of the CO_2 fertilisation effect leads to significant differences in emissions deduced for any specified concentration profile. Reducing the uncertainty in emissions due to past land-use

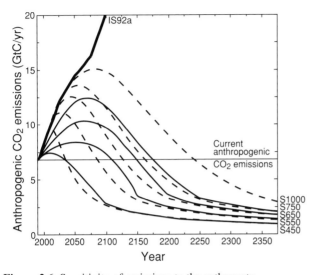

Figure 2.6: Sensitivity of emissions to the pathway to stabilisation. The pathways, which stabilise at 450 ppmv to 750 ppmv, are shown in Figure 2.5. Results for stabilisation at 1000 ppmv (also from Figure 2.5) are shown for comparison. (The results shown are calculated using the Bern model.)

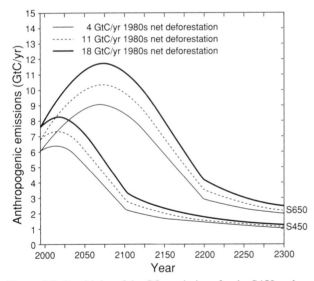

Figure 2.7: Sensitivity of the CO_2 emissions for the S450 and S650 concentration stabilisation profiles to the magnitude of the net land-use flux. The 1994 "best guess" for the magnitude of this flux is 1.1 GtC/yr (dashed curve). The range shown is for net land-use fluxes of 0.4–1.8 GtC/yr. The 1992 "best guess" was 1.6 GtC/yr. This sensitivity is equivalent to assessing the sensitivity to the magnitude of the CO_2 fertilisation effect. (The results shown are calculated using the Bern model.)

change is therefore very important in order to constrain the overall range of projections from carbon cycle models. At the same time, independent estimates of the global fertilisation effect may be used as a direct test of the value

used in model calculations and so can also help to reduce uncertainties in projections. Estimates of the CO_2 fertilisation effect from terrestrial ecosystem models are now becoming available (see Chapter 9).

2.1.4 Bomb Lifetime vs. Perturbation Lifetime

Some recently published work (Starr, 1993) has argued that the post-industrial increase of atmospheric CO_2 may be due largely to natural variation rather than to human causes. This argument hinges on a presumed low value for the "lifetime" of CO_2 in the atmosphere. The argument is as follows: if the atmospheric lifetime were short, then large net CO_2 fluxes into the atmosphere would have been required to produce the observed concentration increases; if it were long, smaller fluxes would yield the observed increases. Proponents of a natural cause assert that the CO_2 lifetime is (and has remained) only a decade or less. The "evidence" for such a short lifetime includes the rapid decline of bomb ^{14}C in the atmosphere following its 1964 peak. A lifetime this short would require fluxes much larger than those estimated from fossil fuel burning in order to produce the observed atmospheric increase; therefore, the reasoning goes, the build-up can only be partly human-induced.

Atmospheric CO_2 is taken up by the ocean and the biospheric carbon pools if a disequilibrium exists between atmosphere and the ocean (i.e., if the partial pressure of CO_2 in the atmosphere is higher than in the surface ocean) and if the uptake by plant growth is larger than the CO_2 released by the biosphere by respiration and by decay of organic carbon (heterotrophic respiration). For terrestrial ecosystems, the removal of radiocarbon and that of anthropogenic CO_2 from the atmosphere are different. Carbon storage corresponds to the difference in net primary production (plant growth, NPP) and decay of organic carbon plus plant respiration. Radiocarbon, however, is assimilated in proportion to the product of NPP and the atmospheric ^{14}C/^{12}C ratio, and is released in proportion to the decay rate of organic matter and its ^{14}C/^{12}C ratio.

For oceans, one needs to differentiate between the amount or concentration of radiocarbon in the atmosphere and the isotopic ratio ^{14}C/[12]C. The ratio, a non-linear function of the CO_2 and ^{14}C concentration, is often confused with concentration, giving rise to misunderstanding. The removal of bomb-radiocarbon atoms is governed by the same mechanisms as the removal of anthropogenic CO_2. The removal of small pulse inputs of CO_2 and ^{14}C into the atmosphere by air-sea exchange follows the same pathways for both model tracers. However, an atmospheric perturbation in the isotopic ratio disappears *much faster* than the perturbation in the number of ^{14}C atoms as shown in Figure 2.8 (Siegenthaler and Oeschger, 1987; Joos *et al.*, 1996).

Thus a ^{14}C perturbation, such as the bomb ^{14}C input into the atmosphere, induces a net transfer of ^{14}C into oceans and biosphere. Because the $^{14}CO_2$ molecules generated from the bomb tests represent a negligible CO_2 excess and, more importantly, because the atmospheric $^{14}C/^{12}C$ ratio is changed considerably by the ^{14}C input, this results in a large isotopic disequilibrium between the atmosphere and the oceanic and terrestrial biospheric reservoirs.

Because of the different dynamic behaviour, ^{14}C can not be taken as a simple analogue tracer for the excess anthropogenic CO_2, a fact that has long been known to the carbon cycle modelling community (e.g., Revelle and Suess, 1957; Oeschger *et al.*, 1975; Broecker *et al.*, 1980).

2.1.5 Recent Bomb Radiocarbon Results and their Implication for Oceanic CO₂ Uptake

Two recent assessments of bomb radiocarbon in the entire global carbon system (i.e., the troposphere, stratosphere, terrestrial biosphere and the ocean (Broecker and Peng, 1994; Hesshaimer *et al.*, 1994)) imply a surprising global imbalance during the decade after the bomb tests, when,

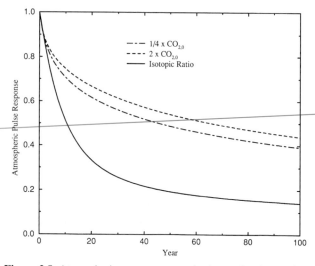

Figure 2.8: Atmospheric response to a pulse input of carbon at time t = 0 for CO_2 and for a perturbation of the isotopic ratio ($^{13}C/^{12}C$ or $^{14}C/^{12}C$), obtained by using the ocean-atmosphere compartments of the Bern model with no biosphere. The response of the atmospheric CO_2 concentration to a pulse input depends on the pulse size and the CO_2 background concentration. The dashed line is for a doubling of pre-industrial CO_2 concentration (280 ppmv) at t = 0; the dashed-dotted line is for an increase of pre-industrial CO_2 concentration by one quarter (70 ppmv) at t = 0; the solid line shows the decrease of an isotopic perturbation. Note the more rapid decline in the isotopic ratio, in accord with observations.

apart from the relatively very small natural cosmic production, no major ^{14}C source existed.

This imbalance may be explained either by a hidden high (above 30 km height) stratospheric bomb ^{14}C inventory not detected in the existing observations, which is unlikely, or by an overestimated bomb radiocarbon inventory in the terrestrial biosphere or the ocean or both. The calculated biospheric reservoir uptake would have to be reduced by about 80% in order to achieve the global bomb radiocarbon balance. Such a reduction is very unlikely, as it strongly contradicts observations of bomb ^{14}C in wood (e.g., in tree rings: Levin *et al.*, 1985) and soils (Harrison *et al.*, 1993). A more likely, albeit tentative, explanation is that previous estimates of the oceanic bomb ^{14}C inventory compiled from the observations of the GEOSECS program (1973–78) (Broecker *et al.*, 1985) are too high by approximately 25%. This is slightly larger than the generally accepted uncertainty of this quantity. Because this explanation is inconsistent with a new assessment of the oceanic observations (Broecker *et al.*, 1995) the issue is still not fully resolved.

Observations of the distribution and inventories of bomb-produced radiocarbon constitute the key tracer either to calibrate simple ocean carbon models or to validate the upper ocean transport calculated by three-dimensional ocean carbon models. Therefore a downward revision of the oceanic ^{14}C inventory during the GEOSECS time period would have implications for our quantitative view of the global carbon cycle, requiring a smaller role for the oceans, and hence larger terrestrial uptake (or lower releases from land-use change). The implied changes in the carbon budget, although large (~0.5 GtC/yr), are within existing uncertainties and thus do not require any fundamental changes in understanding, but rather indicate the importance of adequate global sampling of critical variables.

2.2 Other Trace Gases And Atmospheric Chemistry

2.2.1 Introduction

Changes in the concentration of a number of gases (methane, nitrous oxide, the halocarbons and ozone) have resulted in a combined radiative forcing which is similar in magnitude to that of CO_2 since pre-industrial times. Their concentrations in the atmosphere depend on chemical processes, which regulate their removal rates (methane, nitrous oxide, halocarbons) and sometimes their production as well (ozone). The chemistry of the atmosphere is complex and the many reactions are closely inter-related.

In this section the discussion of the sources, sinks, lifetimes and trends of methane, nitrous oxide, halocarbons and ozone presented in IPCC (1994) is summarised and updated as is the discussion of the possible stabilisation of the atmospheric concentrations of these gases. Some of the discussion in IPCC (1994) is not updated here and the main points are now briefly summarised.

An intercomparison of tropospheric chemistry/transport models using a short-lived tracer showed how critical the model description of the atmospheric motions is, finding a high degree of consistency between three dimensional models, but distinctly different results among two dimensional models. This finding, illustrating the low degree of confidence we should have in numerical simulations involving gases such as tropospheric ozone and its precursors, still holds.

Our ability to model tropospheric ozone is not restricted solely by the limitations of our chemistry/transport models. Equally important is our lack of quantitative knowledge of the global sources and distribution of tropospheric ozone and its short-lived precursors (active nitrogen oxides (NO_x), hydrocarbons and carbon monoxide). For example, even the relative importance of the anthropogenic NO_x sources (transport of surface pollution out of the boundary layer, direct injection by aircraft) and natural sources (lightning, soils, stratospheric input) is not well known. Again, no major advances have been made in this field since IPCC (1994).

IPCC (1994) also described a model study which investigated the impact of a 20% increase in methane concentrations. Two main results were found:

(a) the chemical feedback of methane on tropospheric chemistry changes the methane removal rate by between –0.17% and –0.35% for each 1% increase in the methane concentration;

(b) increases of about 1.5 ppbv in tropospheric ozone occur in the tropics and summertime mid-latitudes, though there is a factor of 3 or more difference between models.

The first result was used to infer that the methane adjustment time was about 1.45 times the turnover time and the latter to estimate the ratio, about 0.25, of radiative forcing from the induced tropospheric ozone increase to that from the methane increase. These findings still hold.

2.2.2 Atmospheric Measurements and their Implications
2.2.2.1 Methane
Atmospheric methane (CH_4) has been increasing since the beginning of the 19th century; current levels, 1721 ppbv, are the highest ever observed, including ice core records

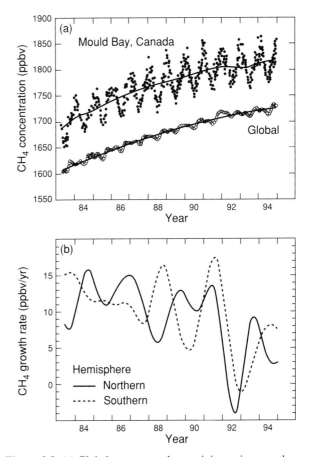

Figure 2.9: (a) Global average methane mixing ratios, together with the methane concentrations observed at Mould Bay, Canada. (b) The growth rates in average methane concentrations in the Northern and Southern Hemispheres (Dlugokencky *et al.*, 1994 a,c; E. Dlugokencky (NOAA) unpublished data).

that go back 160,000 years (Chappellaz *et al.*, 1990). That increase has not been regular. In the late 1970s methane grew at an annual rate of about 20 ppbv/yr (Blake & Rowland, 1988). More extensive atmospheric measurements have been made since 1984. These indicate lower growth rates through the 1980s, averaging about 13 ppbv/yr, declining through the decade to about 9 ppbv/yr in 1991 (Dlugokencky *et al.*, 1994a; Blake and Rowland, updated data).

During 1992/93, extremely low growth rates were observed (Dlugokencky *et al.*, 1994b), and during 1992 methane stopped growing at some locations. In 1994 (Figure 2.9) global methane growth rates recovered to about 8 ppbv/yr, close to the range of rates observed throughout the period 1984 to 1991 (13–9 ppbv/yr). Recent data for the high northern latitudes from Mould Bay, Canada show this pattern in growth rates (Figure 2.9). Atmospheric methane is still increasing albeit at a rate less than that observed over the previous decades.

There has been considerable speculation and modelling of the possible cause(s) of the 1992/93 anomaly. From methane isotope data, i.e., the relative abundance of $^{13}CH_4$ to $^{12}CH_4$, Lowe *et al.* (1994) suggest that decreased emissions of fossil fuels in the Northern Hemisphere (Dlugokencky *et al.*, 1994c) and decreased biomass burning in the tropics may have played significant roles. Schauffler and Daniel (1994) suggest that enhanced tropospheric–stratospheric exchange, caused by the additional stratospheric heating by the Mt. Pinatubo aerosol, could have been a contributory factor in 1992. Bekki *et al.* (1994) propose a mechanism related to Mt. Pinatubo, whereby significantly reduced stratospheric ozone levels following the eruption caused enhanced levels of UV to reach the troposphere (a known effect) and resulted in increased tropospheric hydroxyl radical (OH) levels and increased methane destruction (a model result). Hogan and Harriss (1994) have suggested that temperature effects on natural wetlands are a possible contributing cause. Many of these factors contributed to the observed methane anomaly in 1992/93, but at present it is not clear what their relative contributions are, whether they can fully explain the observed anomalies in concentration and isotopic composition, or even whether we have identified all of the important processes.

2.2.2.2 Nitrous oxide

Nitrous oxide is a major greenhouse gas because it has a long atmospheric lifetime (~120 years) and large radiative forcing about ~200 times that of carbon dioxide on a per molecule basis (IPCC, 1994). Nitrous oxide levels continue to grow in the global, background atmosphere. The 1993 growth rate (approximately 0.5 ppbv/yr) was lower than observed in the late 1980s-early 1990s (approximately 0.8 ppbv/yr). This lower growth rate has been tentatively associated with global cooling due to aerosols emitted by Mt. Pinatubo affecting N_2O soil emission rates (Bouwmann *et al.*, 1995), but the changes in N_2O growth rates are within the range of variability seen on decadal time-scales (Prinn *et al.*, 1990; Khalil and Rasmussen, 1992; Elkins *et al.*, 1994; Prather *et al.*, 1995; Prinn *et al.*, 1995a). Zander *et al.* (1994a) have updated trends in column abundance of N_2O from 1950 to 1990 that are broadly consistent with data obtained at remote surface stations.

2.2.2.3 Halocarbons

Chlorocarbons and bromocarbons strongly absorb infrared radiation (a direct warming) and also destroy ozone in the lower stratosphere (an indirect cooling since ozone is a strong greenhouse gas) (IPCC, 1994; WMO/UNEP, 1995).

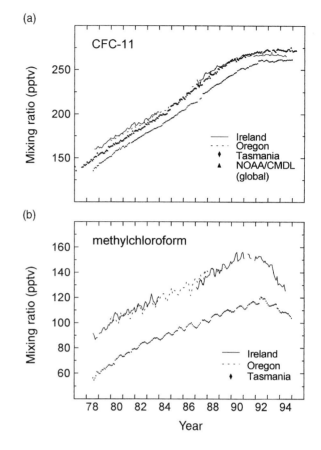

Figure 2.10: (a) Global average (NOAA) and selected station (AGAGE) mixing ratios of CFC-11 (Cunnold *et al.*, 1994; Prinn *et al.*, 1995a; AGAGE unpublished data). (b) Mixing ratios of methyl chloroform from selected AGAGE stations (Prinn *et al.*, 1995a,b; AGAGE unpublished data).

The slowdown in the growth rates of atmospheric chlorofluorocarbons, particularly CFC-11 and CFC-12 (Elkins *et al.*, 1993; Cunnold *et al.*, 1994) is consistent with the rapid phase-out of CFC consumption required by the Montreal Protocol and its Amendments (Copenhagen). Further data from the NOAA–CMDL and AGAGE networks (Elkins *et al.*, 1994; Prinn *et al.*, 1995a) show that CFC-11 global abundances probably stopped growing in early 1993, whereas CFC-12 growth rates declined from about 10 pptv/year (1991 to 1992) to 7 pptv/year (1992 to 1993). By mid-1994 global levels of CFC-12 had almost stopped growing and CFC-11 levels had started to decline (Figure 2.10).

These changes in atmospheric concentrations of CFC-11 and -12, clearly a response to large reductions in emissions, are probably in advance of those predicted to occur under the Montreal Protocol and its Amendments (see IS92a Scenario, Section 2.2.4). The inventory approach to the estimation of recent production and emissions of CFC-11 and -12 has become increasingly uncertain (Fisher and Midgley, 1994). The major uncertainty in the release of CFC-11 is now associated with release from closed cell foams while that for CFC-12 is the estimation of unreported production.

Methyl chloroform, a relatively short-lived species in the atmosphere (about 5 years) compared to the CFCs (about 50 years or more), is also a major source of stratospheric chlorine and so is regulated by the Montreal Protocol and its Amendments. Its concentration grew regularly at about 4–5%/yr until 1990, levelled off towards the end of 1991, and fell in 1994 by about 8% (Figure 2.10, Prinn *et al.*, 1995b) – the first reported global decrease in the atmospheric abundance of an anthropogenic chemical whose emissions are regulated by the Montreal Protocol and its Amendments. This dramatic reduction is consistent with the known emissions (Midgley and McCulloch, 1995), models of atmospheric transport and chemistry, including the control by tropospheric hydroxyl radicals (Prinn *et al.*, 1995b).

HCFCs (hydrochlorofluorocarbons) and HFCs (hydrofluorocarbons) are being used increasingly as interim (HCFC) and longer term (HFC) substitutes for CFCs, in response to the requirements of the Montreal Protocol and its Amendments. Their GWP values are typically less than the CFCs they replace, but are nevertheless high compared to carbon dioxide. Global data for HCFC-22 (Montzka *et al.*, 1993, 1994a; Elkins *et al.*, 1994; Figure 2.11) indicate a mean mixing ratio in 1994 of approximately 110 pptv with a constant growth rate (1992 to 1994) of about 5 pptv/yr. A new analysis of spectroscopic data confirms that from the mid-1980s to the early 1990s HCFC-22 levels grew at about 7%/yr in the Northern Hemisphere (Irion *et al.*, 1994; Zander *et al.*, 1994b). Other HCFCs such as HCFC-142b and -141b are being used increasingly as substitutes for CFC-12 and -11 in the production of closed cell foams (Elkins *et al.*, 1994; Montzka *et al.*, 1994b; Oram *et al.*, 1995; Schauffler *et al.*, 1995) (Figure 2.11). These gases show rapidly accelerating growth rates since 1993. In the remote Southern Hemisphere (Cape Grim, Tasmania), HCFC-142b increased from 0.2 pptv in 1978 to 3 pptv in 1993; and HCFC-141b, from 0.1 pptv in 1982 to 0.5 pptv in 1993 (Figure 2.11). These NOAA–CMDL data for 1994 give a mean concentration and growth rate of about 0.6 pptv/yr (21%/yr) for HCFC-142b, and 2 pptv/yr (67%/yr) for HCFC-141b. Analyses of both data sets indicate that industry estimates of emissions (AFEAS, 1995) may be low.

Simultaneous air-sea observations of methyl bromide in the east Pacific (40°N–55°S) and mid-Atlantic Oceans (53°N–45°S) during 1994 show that surface waters were

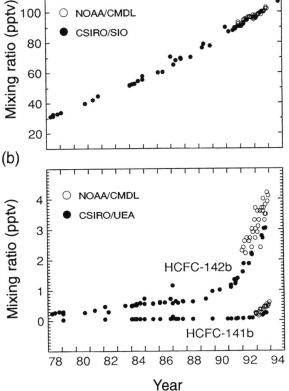

Figure 2.11: (a) HCFC-22 mixing ratios at Cape Grim, Tasmania (41°S), as observed in NOAA flasks (Montzka *et al.*, 1993, 1994a; Elkins *et al.*, 1994) and in the Cape Grim air archive (P. Fraser, unpublished data). (b) HCFC-141b and HCFC-142b mixing ratios at Cape Grim, as observed in NOAA flasks (Montzka *et al.*, 1994b; Elkins *et al.*, 1994) and in the Cape Grim air archive (Oram *et al.*, 1995).

undersaturated in all areas except for coastal and upwelling regions (Butler, 1994; Lobert *et al.*, 1995). The global mean atmospheric abundance is about 10 pptv, with an interhemispheric ratio of 1.3. Previously, the oceans were thought to be a net source of methyl bromide of about 35×10^9 g/yr (Khalil *et al.*, 1993), but extrapolation of these new measurements to the global oceans gives a net air-to-ocean sink of approximately 15×10^9 g/yr (Lobert *et al.*, 1995). The turnover time of atmospheric methyl bromide with respect to oceanic uptake and degradation is about 3–4 years, and the turnover time for all sinks is about 1.2 years, which includes the revised estimate for removal by OH (Section 2.2.3).

A re-analysis of six years of global halon-1211 and -1301 data shows these species are continuing to grow nearly linearly, in contrast with earlier reports (Butler *et al.*, 1992). These two halons represent about 30% of organic bromine in the remote atmosphere and hence of the ozone-depleting bromine that reaches the stratosphere. In 1993 the halon-1211 abundance was about 2 pptv growing at 0.1–0.2 pptv/yr, and that of halon-1301 was 3 pptv growing at 0.2 pptv/yr (Butler *et al.*, 1994; Elkins *et al.*, 1994; J. Butler, unpublished data). Production of halons has already been phased out under the Montreal Protocol and its amendments, and emissions depend on release of the currently existing "banks" of halons stored primarily in fire-fighting systems.

2.2.2.4 *Other perhalogenated species*

There is a group of trace gases, the exclusively anthropogenic perfluorinated species such as tetrafluoromethane (CF_4), hexafluorethane (CF_3CF_3) and sulphur hexafluoride (SF_6), that have very long atmospheric lifetimes (greater than 1000 years) and large GWP (Global Warming Potential) values (Ko *et al.*, 1993).

Tetrafluoromethane and hexafluorethane are by-products released to the atmosphere during the production of aluminium. Sulphur hexafluoride is principally used as a dielectric fluid in heavy electrical equipment. Comparatively small emissions of these species will accumulate and lead to radiatively significant atmospheric concentrations. Current concentrations of CF_4 and CF_3CF_3 probably exceed 70 and 2 pptv respectively (Prather *et al.*, 1995) and global average concentrations of SF_6 were 3–4 pptv, growing at 0.2 pptv/year in 1994 (Maiss and Levin, 1994; Maiss *et al.*, 1996; Figure 2.12).

2.2.2.5 *Tropospheric ozone*

There is observational evidence that tropospheric ozone (about 10% of the total-column ozone) has increased in the Northern Hemisphere (north of 20°N) over the past three decades (Logan, 1994; Harris *et al.*, 1995). The available measurements have been made primarily in industrialised countries so their geographic coverage is limited. Even so the trends are highly regional. They are smaller in the 1980s than in the 1970s and are negative at some locations (Tarasick *et al.*, 1995). European measurements at surface sites also indicate a doubling in the lower-tropospheric ozone concentrations since earlier this century. At the South Pole, a decrease has been observed since the mid-1980s. Elsewhere in the Southern Hemisphere, there are insufficient data to draw strong inferences. IPCC (1994) estimated that tropospheric ozone in the Northern Hemisphere has doubled, from 25 to 50 ppbv, since the pre-industrial. There is no new information to change or strengthen this tentative conclusion. The tropospheric ozone budget is discussed in Section 2.3.3.2.

2.2.2.6 *Stratospheric ozone*

Downward trends in total-column ozone continue to be observed over much of the globe. Decreases in ozone abundances of about 4–5% per decade at mid-latitudes in the Northern and Southern Hemispheres are observed by both ground-based and satellite-borne monitoring instruments. At mid-latitudes, the losses continue to be larger during winter/spring than during summer/fall in both hemispheres, and the depletion increases with latitude, particularly in the Southern Hemisphere. Little or no downward trends are observed in the tropics (20°N–20°S). While the current two-dimensional stratospheric models simulate the observed trends quite well during some seasons and latitudes, they underestimate the trends by factors of 1.5 to 3 in winter/spring at mid- and high latitudes. Several known atmospheric processes involving chlorine and bromine that affect ozone in the lower stratosphere are difficult to simulate numerically and have not been adequately incorporated into these models. A comprehensive review of issues pertaining to atmospheric ozone was recently completed (WMO/UNEP, 1995).

The average stratospheric ozone depletion over the last decade, as well as the large, but transient loss probably associated with Mt. Pinatubo's injection of sulphur dioxide (SO_2) into the stratosphere, are expected to have affected various chemical cycles in the troposphere. Several groups (Madronich and Granier, 1992; Bekki et al., 1994; Fuglesvedt et al., 1994) have modelled increases in tropospheric OH due to enhanced fluxes of solar ultraviolet

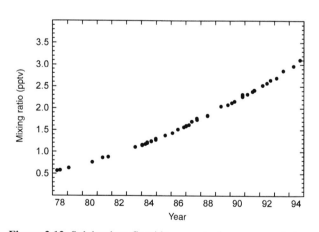

Figure 2.12: Sulphur hexafluoride concentrations measured at Cape Grim, Tasmania (41°S) from 1978 to 1994. The measurements were made at the University of Heidelberg on the CSIRO/Bureau of Meteorology air archive (Maiss *et al.*, 1996, M. Maiss, unpublished data).

radiation reaching the troposphere. Others have speculated on more involved mechanisms linking stratospheric changes to tropospheric effects: changes in cloud radiative properties (Toumi et al., 1994; Rodhe and Crutzen, 1995; Section 2.4.1.3) and changes in troposphere–stratosphere turnover (Kinnison et al., 1994; Schauffler and Daniel, 1994). While such mechanisms have been identified, there is as yet no consensus regarding the changes in tropospheric chemistry that occurred during this period.

2.2.2.7 Stratospheric water vapour

Stratospheric water vapour is a direct greenhouse gas (IPCC, 1990) as well as a critical chemical component of ozone chemistry (indirect effect). It is expected to increase as a result of increasing CH_4 abundances and also possibly from increases in tropopause temperatures that allow more water vapour to enter the stratosphere. Stratospheric water vapour measurements made over Boulder, Colorado (40°N) from 1981 to 1994 show increases that are statistically significant at some altitudes in the lower stratosphere (Oltmans and Hofmann, 1995). These increases are only partly explained by the concurrent increases in methane and may point to other long-term trends in the stratosphere. However, these results at a single site do not necessarily point to a global trend (see, for example, the different geographic trends in total ozone, WMO/UNEP (1992)).

2.2.2.8 Carbon monoxide

Carbon monoxide (CO), not itself an important greenhouse gas, is an important component of the oxidising capacity of the troposphere. It is short-lived, responds rapidly to changes in sources or its sink, the OH radical, and contributes to formation of tropospheric ozone. Intercomparison of CO standards at atmospheric levels from three laboratories (NOAA–CMDL, NASA-Langley and Fraunhofer-Institut) indicate general agreement to within 2% (Novelli et al., 1994a) and allow us to combine the different observational records. The recently reported global trend of −6 to −8%/yr since 1990 (IPCC 1994; Novelli et al., 1994b) has been observed in another global observational network but at a smaller rate of −2 to −3%/yr (Khalil and Rasmussen, 1994). At Mace Head, Ireland, however, small increases in CO (0.6%/yr) for 1990 to 1992 have been reported for pollution-free samples while decreases (−13%/yr) are found for air masses that originated over Europe (Derwent et al., 1994). These observations reflect the difficulty in determining global trends from short records for a gas such as CO. Several causes of these different "trends" have been postulated and

include reduced CO emissions from biomass burning and urban sources, increased OH-sink, possibly from enhanced stratospheric ozone depletion following the eruption of Mt. Pinatubo in 1991 (Bakwin *et al.*, 1994; Bekki *et al.*, 1994; Khalil and Rasmussen, 1994; Novelli *et al.*, 1994b). As in the case of the CH_4 anomaly in 1992/93, a single, dominant cause has not been quantitatively established. A recently published revision to one of the sources of CO (oxidation of isoprene and other hydrocarbons, Miyoshi *et al.*, 1994) is still consistent with the previous IPCC budget.

2.2.3 Chemical Lifetimes and Budgets

The atmospheric lifetimes of gases other than CO_2 are determined predominantly by chemical reactions. The term "lifetime" can refer to a specific process, a globally integrated loss (turnover time), or the duration of perturbations in the atmospheric concentration (adjustment time). These different definitions are discussed in the box. In the case of CH_4 and CO these gases' abundances influence OH concentrations and hence their own losses: adjustment times for additions of CH_4 or CO are longer than their respective turnover times (Prather, 1994).

There is little new information on the stratospheric losses of the greenhouse gases, and hence the recommended lifetimes for certain gases, e.g., N_2O and the CFCs, remain as in IPCC (1994). Fisher and Midgley (1994) have evaluated the empirical lifetime of CFC-11, 56 yr, from fitting observations to emissions, including uncertainty in the latter. Cunnold *et al.* (1994) analysed the same data with a different model and report a shorter lifetime of 44 yr. Allowing for uncertainties in the emissions and the absolute calibration, these are consistent with the IPCC lifetime of 50 yr. Morris *et al.* (1995) have evaluated the possible effects of ion reactions on the lifetimes of the fully fluorinated compounds and have shown that, even for the most extreme assumptions, the lifetimes of perfluoroalkanes and SF_6 are greater than 1000 years.

The lifetimes of all those gases that react primarily with tropospheric OH are revised downward relative to IPCC (1994) by about 10%. That revision includes CH_4 and other hydrocarbons, the HCFCs and other halogenated species that contain one or more hydrogen-atoms, and is based on a revised estimate of the mean global OH concentration. That in turn is based on a revision of the budget of methyl chloroform (Prinn *et al.*, 1995b) which serves as a reference to estimate mean global OH concentrations. A recent recalibration of the methyl chloroform absolute standard by AGAGE gave a value of 0.82 times that of their previous reported value (Prinn *et al.*, 1995b; 1992) lowering the estimate of the atmosphere CH_3CCl_3 burden by the same factor and lowering the turnover time of CH_3CCl_3 to 4.8 ± 0.2 years (Prinn *et al.*, 1995b) as compared to 5.7 ± 0.3 yr (Prinn *et al.*, 1992).

The new AGAGE calibration also removes much of the previous difference with the NOAA/CMDL calibration noted previously (Fraser *et al.*, 1994; IPCC, 1994). Using atmospheric models (Bloomfield *et al.*, 1994), an average of the two calibrations results in a recommended empirical

turnover time of 4.9 ± 0.4 yr (compared with 5.4 ± 0.6 in IPCC 1994) which is in close agreement with the estimate by Prinn *et al.* (1995b). After allowing for the loss to the stratosphere and the uptake by the oceans, the removal time for CH_3CCl_3 through reaction with troposphere OH is inferred to be 5.9 ± 0.7 yr. That value is used to recalculate the lifetimes of the other gases reacting with OH: for CH_4 see Table 2.2 and Section 2.2.3.1; for the other greenhouse and ozone-depleting gases see Table 2.2 and Section 2.5.

Table 2.2: *Lifetimes for radiatively active gases and halocarbons*

SPECIES		LIFETIME		CONCENTRATION (ppbv)		CURRENT GROWTH	RADIATIVE FORCING	
		Year	Uncert.	1992	pre-ind.	ppbv/yr	Wm⁻² /ppbv	Wm⁻²
NATURAL AND ANTHROPOGENICALLY INFLUENCED GASES								
carbon dioxide	CO_2	variable		356,000	278,000	1,600	1.8×10^{-5}	1.56
methane	CH_4@	12.2	25%	1714	700	8	3.7×10^{-4}	0.47
nitrous oxide	N_2O	120		311	275	0.8	3.7×10^{-3}	0.14
methyl chloride	CH_3Cl	1.5	25%	~0.6	~0.6	~0		0
methyl bromide	CH_3Br	1.2	32%	0.010	< 0.010	~0		0
chloroform	$CHCl_3$	0.51	300%	~0.012		~0	0.017	
methylene chloride	CH_2Cl_2	0.46	200%	~0.030		~0	0.03	
carbon monoxide	CO	0.25		50-150		~0		$
GASES PHASED OUT BEFORE 2000 UNDER THE MONTREAL PROTOCOL AND ITS AMENDMENTS								
CFC-11	CCl_3F	50	10%	0.268	0	+0.000**	0.22	0.06
CFC-12	CCl_2F_2	102		0.503	0	+0.007**	0.28	0.14
CFC-113	CCl_2FCClF_2	85		0.082	0	0.000**	0.28	0.02
CFC-114	$CClF_2CClF_2$	300		0.020	0		0.32	0.007
CFC-115	CF_3CClF_2	1700		< 0.01	0		0.26	<0.003
carbon tetrachloride	CCl_4	42		0.132	0	−0.0005**	0.10	0.01
methyl chloroform	CH_3CCl_3	4.9	8%	0.135#	0	−0.010**	0.05	0.007
halon-1211	$CBrClF_2$	20		0.007	0	.00015		
halon-1301	$CBrF_3$	65		0.003	0	.0002	0.28	
halon-2402	$CBrF_2CBrF_2$	20		0.0007	0			
CHLORINATED HYDROCARBONS CONTROLLED BY THE MONTREAL PROTOCOL AND ITS AMENDMENTS								
HCFC-22	$CHClF_2$	12.1	20%	0.100	0	+0.005**	0.19	0.02
HCFC-123	CF_3CHCl_2	1.4	25%		0		0.18	
HCFC-124	CF_3CHClF	6.1	25%		0		0.19	
HCFC-141b	CH_3CFCl_2	9.4	25%	0.002	0	0.001**	0.14	
HCFC-142b	CH_3CF_2Cl	18.4	25%	0.006	0	0.001**	0.18	
HCFC-225ca	$C_3HF_5Cl_2$	2.1	35%		0		0.24	
HCFC-225cb	$C_3HF_5Cl_2$	6.2	35%		0		0.28	

Table 2.2: *continued*

SPECIES		LIFETIME		CONCENTRATION (ppbv)		CURRENT GROWTH	RADIATIVE FORCING	
		Year	Uncert.	1992	pre-ind.	ppbv/yr	Wm^{-2} /ppbv	Wm^{-2}
PERFLUORINATED COMPOUNDS								
sulphur hexafluoride	SF_6	3200		0.032	0	+0.0002	0.64	0.002
perfluoromethane	CF_4	50000		0.070	0	+0.0012	0.10	0.007
perfluoroethane	C_2F_6	10000		0.004	0		0.23	
perfluoropropane	C_3F_8	2600			0		0.24	
perfluorobutane	C_4F_{10}	2600			0		0.31	
perfluoropentane	C_5F_{12}	4100			0		0.39	
perfluorohexane	C_6F_{14}	3200			0		0.46	
perfluorocyclobutane	$c\text{-}C_4F_8$	3200			0		0.32	
ANTHROPOGENIC GREENHOUSE GASES NOT REGULATED (PROPOSED OR IN USE)								
HFC-23	CHF_3	264	45%				0.18	
HFC-32	CH_2F_2	5.6	25%				0.11	
HFC-41	CH_3F	3.7					0.02	
HFC-43-10mee	$C_5H_2F_{10}$	17.1	35%				0.35	
HFC-125	C_2HF_5	32.6	35%				0.20	
HFC-134	CF_2HCF_2H	10.6	200%				0.18	
HFC-134a	CH_2FCF_3	14.6	20%				0.17	
HFC-143	CF_2HCH_2F	3.8	50%				0.11	
HFC-143a	CH_3CF_3	48.3	35%				0.14	
HFC-152a	CH_3CHF_2	1.5	25%				0.11	
HFC-227ea	C_3HF_7	36.5	20%				0.26	
HFC-236fa	$C_3H_2F_6$	209	50%				0.24	
HFC-245ca	$C_3H_3F_5$	6.6	35%				0.20	
HFOC-125e	CF_3OCHF_2	82	300%					
HFOC-134e	CHF_2OCHF_2	8	300%					
trifluoroiodomethane	CF_3I	< 0.005					0.38	

Notes: This table lists only the direct radiative forcing from emitted gases. The indirect effects due to subsequent changes in atmospheric chemistry, notably ozone (see below), are not included. The Wm^{-2} column refers to the radiative forcing since the pre-industrial, and the Wm^{-2}/ppbv column is accurate only for small changes about the current atmospheric composition (see Section 2.4 and IPCC (1994)). In particular, CO_2, CH_4 and N_2O concentration changes since pre-industrial times are too large to assume linearity; the formulae reported in IPCC(1990) are used to evaluate their total contribution.

A blank entry indicates that a value is not available. Uncertainties for many lifetimes have not been evaluated. The concentrations of some anthropogenic gases are small and difficult to measure. The pre-industrial concentrations of some gases with natural sources are difficult to determine. Radiative forcings are only given for those gases with values greater than 0.001 Wm^{-2}.

@Methane increases are calculated to cause increases in tropospheric ozone and stratospheric H_2O; these indirect effects, about 25% of the direct effect, are not included in the radiative forcings given here.

$ The direct radiative forcing due to changes in the CO concentration is unlikely to reach a few hundredths of a Wm^{-2} (see Section 2.4.1). The direct radiative forcing is hard to quantify (see Section 2.2.4).

** Gases with rapidly changing growth rates over the past decade, recent trends since 1992 are reported.

The change in CH_3Cl_3 concentration is due to the recalibration of the absolute standards used to measure this gas (see Section 2.2.3). Stratospheric ozone depletion due to halocarbons is about -2 % (globally) over the period 1979 to 1990 with half as much again occurring both immediately before and since; the total radiative forcing is thus now about -0.1 Wm^{-2}. Tropospheric ozone appears to have increased since the 19th Century over the northern mid-latitudes where few observational records are available; if over the entire Northern Hemisphere, tropospheric ozone increased from 25 ppb to 50 ppb at present, then the radiative forcing is about $+0.4$ Wm^{-2}.

2.2.3.1 The methane budget

The budget for CH_4 is given in Table 2.3. The estimate for the atmospheric turnover time of CH_4 has been revised to 8.6 ± 1.6 yr down from 10 ± 2 yr in the previous IPCC report. Two factors are responsible for that change: (a) a new estimate for the chemical removal rate (11% faster; see Section 2.2.3); and (b) the uptake by soils (included in the previous budget, but not in the calculation of the turnover time). The estimated error of the lifetime is based on two major uncertainties: that in the tropospheric turnover time of methyl chloroform, $\pm 12\%$, and that in the rate constants of the reactions of CH_4 and CH_3CCl_3 with OH. The new adjustment time for CH_4 is 12.2 ± 3 yr; its error also includes the uncertainty in the feedback factor, $\pm 17\%$ (see Table 2.4).

The revised removal rate implies that the global sink strength is 560 ± 100 $Tg(CH_4)/yr$, higher than the 1994 estimate. Given the uncertainties in the global strengths of the individual CH_4 sources, this implied atmospheric sink can still be matched by the estimated total sources (535 ± 125 $Tg(CH_4)/yr$) given in IPCC (1994).

CH_4 has many clearly identified chemical feedbacks. One already mentioned is that increases in atmospheric CH_4 concentrations, however small, will reduce the tropospheric concentrations of the hydroxyl radical, which in turn will increase the lifetime of CH_4 in the troposphere, and amplify the original perturbation in the CH_4 concentration (Prather, 1994). That feedback expresses itself in an adjustment time that is larger than the CH_4 turnover time. The ratio of adjustment time to turnover time depends on the exact formulation of the chemistry and varies between models. Table 2.4 presents such ratios along with the CH_4 global mean lifetimes (i.e., turnover) obtained from various models. The ratio, also called feedback factor in IPCC (1994), varies between 1.2 and 1.7 and is not correlated with the CH_4 lifetimes obtained by the models, which vary from 8.7 to 13.2 years. This suggests that the feedback factor does not depend on the CH_4 turnover time. Thus there is no reason to revise the ratio of adjustment time to turnover time used in IPCC (1994).

Another feedback loop has been suggested by King and Schnell (1994), who argue that increases in the

Table 2.3: *Estimated sources and sinks of methane for 1980 to 1990.*
(a) Observed atmospheric increase, calculated sinks and implied sources derived to balance the budget.

		Individual estimates	Total
Atmospheric increase			**37** (35–40)
Sinks of atmospheric CH_4:	tropospheric OH	**490** (405–575)	
	stratosphere	**40** (32–48)	
	soils	**30** (15–45)	
Total atmospheric sinks			**560** (460–660)
Implied sources (sinks + atmospheric increase)			**597** (495–700)

(b) Inventory of identified sources.

		Individual estimates	Total
Natural sources			**160** (110–210)
Anthropogenic sources:	Fossil fuel related	**100** (70–120)	
	Total biospheric	**275** (200–350)	
Total anthropogenic sources			**375** (300–450)
Total identified sources			**535** (410–660)

Notes: All figures are in $Tg(CH_4)/yr$. 1 Tg = 1 million million grams which is equivalent to 1 million tonnes. The total amount of CH_4 in the atmosphere is 4850 Tg.

The only change since IPCC (1994) is the estimate of loss through reaction with tropospheric OH. Estimates of individual sources (not shown) remain unchanged.

Table 2.4: *Methane adjustment time relative to turnover time from delta-CH4 simulations (IPCC, 1994)*

Model	Sensitivity@	Adjustment/turnover	Lifetime^ (yr)
U.Camb/2D	-0.17%	1.23 [#]	13.2
AER/2D	-0.18%	1.26 [#]	11.9
UEA/Harwell/2D	-0.20%	1.29 [§]	9.8
LLNL/3D	-0.22%	1.32 [#]	8.7
AER/2D	-0.26%	1.39 [§]	12.5
U.Oslo/3D	-0.34%	1.61 [§]	9.2
LLNL/2D	-0.35%	1.62 [§]	11.5

@ The %-change in CH4 loss frequency (= 1/lifetime) per 1% increase in CH4.

^ CH4 global mean turnover time.

Uses fixed CO concentrations.

§ Uses CO fluxes.

atmospheric CH_4 may inhibit the microbial uptake of CH_4 in soils via a process that couples soil methane and ammonia. That process would also amplify the original perturbation in the CH_4 concentrations. However, the effect of that mechanism is expected to be small, since the global sink strength for the microbial uptake of CH_4 by soils is small, about 30 $Tg(CH_4)$/yr.

If emissions were held constant at 1984–1994 levels (corresponding to current growth of 10 ppbv/yr), then methane levels would rise to about 1850 ppbv over the next 40 years. If emissions were cut by about 30 $Tg(CH_4)$/yr, about 8% of anthropogenic emissions (Table 2.3), then methane concentrations would remain at today's levels. These numbers are reduced from IPCC (1994), which used a current growth rate of 13 ppbv/yr and a longer adjustment time.

2.2.3.2 The tropospheric ozone and hydroxyl radical budget

Tropospheric ozone is one of the more reactive, radiatively active trace gases whose distribution is controlled by a complex interplay of chemical, radiative and dynamical processes. Ozone is transported down into the troposphere from the stratosphere at mid- and high latitudes, destroyed at the Earth's surface, produced by the photo-oxidation of CO, methane and other hydrocarbons in the presence of NO_x (NO_2 and NO), and destroyed by ultraviolet photolysis and reaction with HO_x (HO_2 and OH) radicals. The interplay between these processes results in local tropospheric turnover times for ozone varying from days to weeks. Hence, its concentration varies with latitude, longitude, altitude, season and time of day.

Most of the oxidation of long-lived radiatively active gases such as CH_4, CH_3Cl, CH_3Br, CH_3CCl_3 and HCFCs takes place in the tropics, where high UV and humidity promote the formation of OH from photolysis of ozone (Prather and Spivakovsky, 1990; Thompson, 1992). Ozone in the tropical troposphere thus plays a critical role in controlling the oxidising power of the atmosphere. An analysis of aircraft observations over the tropical South Atlantic (Jacob *et al.*, 1993b) indicates a close balance in the tropospheric column between photochemical production and loss of ozone, with net production in the upper troposphere balancing net loss in the lower troposphere. The production of ozone is driven by NO_x originating from continental sources (biomass burning, lightning, soils). Transport from the stratosphere is negligible as a source of tropospheric ozone or NO_x in this particular region. By extrapolation, Jacob *et al.* (1993b) proposed a general mechanism for the origin of ozone in the tropical troposphere: convection over the continents injects NO_x to high altitudes, leading to net ozone production in the upper troposphere; eventually the ozone subsides over the oceans and net loss takes place in the lower troposphere, closing the ozone cycle. The proposed mechanism implies a great sensitivity of the oxidising power of the atmosphere to the NO_x emissions from tropical continents. Unfortunately there has been no real increase either in our confidence in numerical simulations of global tropospheric ozone or in the measurements of ozone, particularly in the tropics and the Southern Hemisphere, needed to test the simulations.

The photolysis of ozone and its reactions with HO_x radicals, in addition to controlling the ozone lifetime and budget, are important components of the fast photochemical balance of the sunlit troposphere. This balance establishes a small steady state concentration of the highly reactive hydroxyl (OH) radical which control the removal of most trace gases including methane, carbon monoxide, NO_x and a wide range of organic compounds. The global tropospheric concentration of hydroxyl radicals thus determines the oxidising capacity of the troposphere, the build-up of radiatively active gases and the flux of some ozone-depleting halocarbons to the stratosphere.

There is increasing observational evidence that we understand and are able to measure and model local concentrations of OH (e.g., Poppe *et al.*, 1994; Wennberg *et al.*, 1994; Müller *et al.*, 1995). However, our knowledge of the global OH distribution is still limited by our ability to model accurately the large range of conditions that determine the OH concentration (e.g., O_3, NO_x, hydrocarbons, cloud cover). The global and hemispheric mean value of OH can be derived from analyses of observations of methyl chloroform and ^{14}CO. Analysis of atmospheric ^{14}CO data has been used to argue for an asymmetry in the mean tropospheric OH between hemispheres, with about 20% more ^{14}CO (and hence 20% less average OH) in the Northern Hemisphere (Mak *et al.*, 1994). Similar results come from the 12-box atmospheric model fitting of methyl chloroform (Prinn *et al.*, 1995b), but some 3-D models fit the observed CH_3CCl_3 without significant asymmetry in OH. More evidence is needed to be certain of any hemispheric asymmetry in tropospheric OH. The low uncertainty in OH levels ($\pm 12\%$) is limited, however, to globally integrated quantities. Confidence in the seasonality is limited.

2.2.3.3 Aircraft and ozone

The climatic impact of aircraft, due to emissions of particles, water vapour and NO_x, remains a serious concern in research and assessment (e.g., Graedel, 1994; Schumann and Wurzel, 1994; Stolarski and Wesoky, 1995). The potential impact of a supersonic fleet operating in the stratosphere has been studied and reviewed extensively. Stratospheric ozone depletion (a negative radiative forcing) due to emission of NO_x is currently predicted to be small, less than 1%, provided a new generation of engines with NO_x emission indices of 3 to 8g (NO_2) per kg of fuel can be built (Wahner *et al.*, 1995).

Climatic effects due to the addition of H_2O and sulphur to the stratosphere need to be assessed, and additional uncertainties remain about the ability of the 2-D assessment models to simulate the accumulation of exhaust products. Effects of the subsonic fleet, which adds NO_x primarily to the upper troposphere, centre on the ozone increases expected in the upper troposphere (positive radiative forcing) and on the modification of clouds through the emission of particles and contrails. The former effect depends strongly on the background concentrations of NO_x, O_3 and H_2O in the upper troposphere (Ehhalt and Rohrer, 1994), for which observations are extremely sparse. As a result the ability of chemical transport models to simulate these background concentrations in the upper troposphere remains untested, and the modelled impact of aircraft NO_x on ozone remains quite uncertain (Derwent, 1994; Ehhalt and Rohrer, 1994; Hauglustaine *et al.*, 1994).

2.2.4 IS92 Scenarios

The trace gas scenarios and consequent radiative forcing from the IS92 scenarios (IPCC 1992 and Appendices) are used in Chapter 6 to illustrate the implications of different emission pathways. There are still considerable uncertainties in the chemical models used to simulate possible future atmospheres, and we have chosen here to use the simple consensus on current time-scales and budgets given here and in IPCC (1994) to calculate future concentrations corresponding to the IS92 emission scenarios. The projections of trace gas concentrations from 1990 to 2100 using IS92 scenarios for CH_4, N_2O, tropospheric O_3, CFCs and chlorocarbons, HCFCs, stratospheric O_3 and HFCs are given in Tables 2.5 a-e.

We adopt turnover and adjustment times for the greenhouse gases as described in IPCC (1994) and updated in Table 2.2. For CH_4, the chemical feedback of its concentration on its own lifetime has been included as a sensitivity factor: the atmospheric loss rate (i.e., not including soil loss) decreases by 0.30% for every 1% increase in the methane concentration above 1700 ppbv. Thus as CH_4 increases from 1700 to 4100 ppbv, the global turnover time (including soil loss) increases from 8.6 to 11.0 yr. Since the IS92 emissions for CH_4 and N_2O are inconsistent with observations and these lifetimes, they were adjusted so that the concentrations in the 1990s matched the observations. This constant offset from the original IS92 scenarios is effectively a redefinition of the natural, baseline emissions. No attempt has been made to adjust the IS92 scenarios to match the 1992/93 anomaly. Changes in tropospheric O_3 are those induced by changes in CH_4 alone: those from NO_x changes, which may be equally important, are neglected since they cannot be quantified with confidence. The tropospheric ozone

Table 2.5a: IS92 scenarios for methane concentrations (ppbv)

Year	IS92a	IS92b	IS92c	IS92d	IS92e	IS92f
1990	1700	1700	1700	1700	1700	1700
1995	1749	1749	1743	1743	1753	1751
2000	1810	1810	1787	1787	1824	1817
2005	1882	1882	1832	1831	1910	1897
2010	1964	1964	1880	1878	2007	1995
2015	2052	2052	1931	1926	2112	2104
2020	2145	2145	1984	1975	2224	2221
2025	2242	2242	2038	2026	2341	2344
2030	2343	2343	2088	2074	2463	2474
2035	2450	2450	2129	2117	2593	2608
2040	2561	2561	2165	2157	2729	2748
2045	2676	2676	2195	2194	2870	2891
2050	2793	2793	2224	2230	3014	3038
2055	2905	2905	2243	2257	3156	3188
2060	3003	3003	2246	2267	3291	3342
2065	3092	3092	2239	2267	3421	3499
2070	3175	3175	2224	2259	3548	3659
2075	3253	3253	2204	2246	3673	3822
2080	3328	3328	2181	2230	3797	3987
2085	3402	3402	2155	2211	3921	4155
2090	3474	3474	2127	2190	4044	4324
2095	3545	3545	2098	2168	4167	4495
2100	3616	3616	2069	2146	4291	4669

The budget lifetime is 8.6 yr; there are 2.78 $Tg(CH_4)$/ppbv.

The IS92 fluxes are inconsistent with the current lifetime and growth rate, and thus 69 $Tg(CH_4)$/yr have been added to all IS92 fluxes.

Table 2.5b: *IS92 scenarios for N$_2$O concentration (ppbv)*

Year	IS92a	IS92b	IS92c	IS92d	IS92e	IS92f
1990	310	310	310	310	310	310
1995	314	314	314	314	314	314
2000	319	318	318	318	319	319
2005	323	323	323	323	323	323
2010	328	328	327	327	328	328
2015	333	333	332	332	334	334
2020	339	338	337	336	339	339
2025	344	343	342	341	345	345
2030	350	349	346	346	351	351
2035	355	354	351	351	357	357
2040	361	360	355	356	363	363
2045	366	365	360	360	370	369
2050	371	371	364	365	376	374
2055	376	376	368	369	382	380
2060	382	381	371	373	388	386
2065	386	386	374	377	394	392
2070	391	391	377	380	400	397
2075	396	395	380	383	405	403
2080	400	400	383	386	411	408
2085	404	404	385	389	417	414
2090	409	408	387	392	422	419
2095	413	412	389	394	428	425
2100	417	416	391	396	433	430

The budget lifetime is 120 yr; there are 4.81 Tg(N$_2$O)/ppbv.

The IS92 fluxes are inconsistent with current lifetime and growth rate and thus 3.4 Tg(N)/yr have been added to all IS92 fluxes.

Table 2.5c: *IS92 scenarios for tropospheric ozone (ppbv)#*

Year	IS92a	IS92b	IS92c	IS92d	IS92e	IS92f
1990	0.0	0.0	0.0	0.0	0.0	0.0
1995	0.2	0.2	0.2	0.2	0.2	0.2
2000	0.5	0.5	0.4	0.4	0.5	0.5
2005	0.8	0.8	0.6	0.6	0.9	0.9
2010	1.2	1.2	0.8	0.8	1.3	1.3
2015	1.5	1.5	1.0	1.0	1.8	1.8
2020	2.0	2.0	1.2	1.2	2.3	2.3
2025	2.4	2.4	1.5	1.4	2.8	2.8
2030	2.8	2.8	1.7	1.6	3.4	3.4
2035	3.3	3.3	1.9	1.8	3.9	4.0
2040	3.8	3.8	2.0	2.0	4.5	4.6
2045	4.3	4.3	2.2	2.2	5.1	5.2
2050	4.8	4.8	2.3	2.3	5.8	5.9
2055	5.3	5.3	2.4	2.4	6.4	6.5
2060	5.7	5.7	2.4	2.5	7.0	7.2
2065	6.1	6.1	2.4	2.5	7.6	7.9
2070	6.5	6.5	2.3	2.5	8.1	8.6
2075	6.8	6.8	2.2	2.4	8.7	9.3
2080	7.2	7.2	2.1	2.3	9.2	10.1
2085	7.5	7.5	2.0	2.2	9.8	10.8
2090	7.8	7.8	1.9	2.2	10.3	11.5
2095	8.1	8.1	1.8	2.1	10.9	12.3
2100	8.4	8.4	1.6	2.0	11.4	13.1

Global mean change, assuming increase due solely to CH_4 increase (IPCC, 1994). Possible effects from changes in NO_x, hydrocarbons, etc. are not included.

Table 2.5d: *Copenhagen-like scenario adopted for chlorocarbons used with all IS92 options.*

(i) Tropospheric mixing ratios (all units pptv, except for dO$_3$ which is the global mean ozone depletion in per cent).

Year	CFC-11	CFC-12	CFC-113	CFC-114	CFC-115	CCl$_4$	CH$_3$CCl$_3$	HCFC-22	HCFC-141b	HCFC-123	tropCl[@]	dO$_3$%
1970*	60	120	2	1	0	105	35	10	0	0	1563	0.0
1975*	115	205	6	2	1	111	60	25	0	0	2027	-0.3
1980*	173	295	15	4	2	118	85	50	0	0	2541	-1.2
1985*	222	382	30	8	4	126	110	70	0	0	3044	-2.0
1990#	263	477	77	19	5	133	133	91	2	0	3643	-3.1
1995	291	532	92	20	7	133	122	148	5	4	3922	-3.5
2000	289	545	97	20	8	120	61	241	13	9	3844	-3.4
2005	267	526	93	20	9	108	30	292	16	8	3644	-3.1
2010	242	501	88	20	9	96	12	299	16	6	3401	-2.7
2015	219	477	83	20	9	85	4	248	13	3	3140	-2.2
2020	198	454	78	19	9	76	2	178	8	1	2889	-1.8
2025	179	433	74	19	9	67	1	119	5	0	2670	-1.5
2030	162	412	70	19	8	60	0	79	3	0	2489	-1.2
2035	147	392	66	18	8	53	0	52	2	0	2334	-0.9
2040	133	373	62	18	8	47	0	35	1	0	2200	-0.7
2045	120	356	58	18	8	42	0	23	1	0	2081	-0.5
2050	109	339	55	17	8	37	0	15	0	0	1976	-0.3
2055	98	322	52	17	8	33	0	10	0	0	1880	-0.1
2060	89	307	49	17	8	29	0	7	0	0	1793	0.0
2065	81	292	46	17	8	26	0	4	0	0	1714	0.0
2070	73	278	44	16	8	23	0	3	0	0	1641	0.0
2075	66	265	41	16	8	20	0	2	0	0	1575	0.0
2080	60	252	39	16	8	18	0	1	0	0	1513	0.0
2085	54	240	366	15	8	16	0	1	0	0	1456	0.0
2090	49	229	34	15	8	14	0	1	0	0	1403	0.0
2095	44	218	32	15	8	14	0	0	0	0	1355	0.0
2100	40	207	31	15	8	11	0	0	0	0	1309	0.0
life-time (yr)	50	102	85	300	1700	42	4.9	12.1	9.4	1.4		
factor (Gg/pptv)		22.6	20.8	32.5	29.7	27.1	25.3	22.0	14.9	26.3	20.1	

(ii) *Annual flux^ (kton/yr)*

Year	CFC-11	CFC-12	CFC-113	CFC-114	CFC-115	CCl$_4$	CH$_3$CCl$_3$	HCFC-22	HCFC-141b	HCFC-123
1990-94	250	330	132	6	13	80	520	319	26	52
1990-99	125	165	66	6	6	10	120	520	65	130
2000-04	25	30	13	1	1	8	54	484	60	121
2005-09	2	2	1	0	0	0	6	383	42	85
2010-14	0	0	0	0	0	0	0	182	23	46
2015-19	0	0	0	0	0	0	0	53	7	13
2020-24	0	0	0	0	0	0	0	3	0	1
2025-29	0	0	0	0	0	0	0	3	0	1
2030-	0	0	0	0	0	0	0	0	0	0

@ TropCl represents the amount of chlorine that enters the stratosphere and is represented here by the sum of the chlorine contained in the chlorocarbons listed here plus 600 pptv of natural CH$_3$Cl.

* Taken from WMO (1992).

Initial conditions from IPCC (1994).

^ Emissions through 1993 are based on AFEAS (1995) for CFCs and Midgley and McCulloch (1995) for methyl chloroform; those for the HCFCs assume the IS92 branching for the maximum flux allowed during phaseout of CFCs (3.1% of ODP weighted production); the CFCs are assumed to follow their maximum allowable, including a 3-year bank released over the 10-year period 1991 to 1999 and an additional 10% flux with a 10-year lag in phaseout as in Article 5.

Table 2.5e: *IS92 scenarios for HFC concentrations (pptv)*

year	HFC-125			HFC-134a			HFC-152a		
	IS92acf	IS92b	IS92de	IS92acf	IS92b	IS92de	IS92acf	IS92b	IS92de
1990	0	0	0	0	0	0	0	0	0
1995	0	0	1	8	8	9	0	0	2
2000	0	0	6	31	31	36	0	0	6
2005	0	0	14	65	66	75	0	0	12
2010	0	0	28	107	109	122	0	0	20
2015	1	1	46	152	157	175	1	1	28
2020	2	2	69	198	206	231	2	2	36
2025	5	5	94	246	256	289	4	4	47
2030	10	10	122	297	309	350	12	12	58
2035	22	22	153	355	370	417	23	23	75
2040	39	39	187	418	434	488	36	36	93
2045	62	62	223	484	502	562	48	48	110
2050	88	88	261	554	573	638	60	60	127
2055	115	115	299	618	638	708	66	66	139
2060	140	140	333	669	689	763	70	70	147
2065	162	162	364	712	731	809	73	73	155
2070	182	182	393	749	766	848	77	77	163
2075	200	200	420	781	796	882	80	80	171
2080	217	217	444	808	820	910	82	82	174
2085	231	231	464	827	838	930	82	82	175
2090	243	243	482	841	850	944	82	82	175
2095	253	253	497	852	859	955	82	82	176
2100	262	262	510	860	865	961	82	82	176
life-time (yr)		32.6			14.6			1.5	
factor (Gg/pptv)		20.6			17.5			10.5	

increase, +1.5 ppbv for a 20% increase in CH_4, is based on the IPCC (1994) analysis of the delta-CH_4 simulations.

A major simplification of the IS92 scenarios is that we have chosen to ignore the changing emissions of the short-lived gases: CO, VOC (volatile organic compounds) and NO_x. Although we would like to include a complete atmospheric simulation with changing emissions of the major greenhouse gases and also these short-lived gases, there are two compelling reasons why this cannot be done at this time. First, there is not a strong enough consensus on how to treat the short-lived gases in the currently available coarse resolution global models, and on how the chemical feedbacks couple these highly reactive species with CH_4. Second, the impact of these short-lived species depends critically on where and when these gases are emitted. For example, the IS92 scenarios do not differentiate between aircraft, urban combustion and diffuse agricultural NO_x emissions. The yield of ozone (a greenhouse gas) per emitted NO_x molecule depends critically on these local conditions (e.g., Chameides *et al.*, 1994; Schumann and Wurzel, 1994; Derwent, 1996). Thus until we have a consensus on the tropospheric chemistry models and until the scenarios for short-lived gases include spatially resolved emissions (and speciated in the case of VOC), the IS92 calculations cannot include these short-lived gases. This decision is not a recommendation to use the extremely simplified representation of atmospheric chemistry in integrated models (e.g., Prather, 1988; Wigley, 1995).

To illustrate the potential importance of these short-lived gases, Derwent (1996) used the UKMO 2-D model to calculate the CH_4 increases resulting from a full IS92a Scenario. The scenario includes the projected growth in NO_x, CO and VOC emissions (with additional assumptions about the location and speciation of emissions), and the model includes a fairly full set of chemical feedbacks on tropospheric O_3 and OH. As a result the increases in tropospheric O_3 and NO_x also change the oxidising capacity of the atmosphere and partially offset the CH_4 chemical feedback described above. The increases in CH_4 for this 2-D model are approximately 20% less than those in the simplified CH_4 feedback-only model used here.

In another simplification of the IS92 scenarios, the chlorine and bromine containing halocarbons are assumed to be controlled by the Montreal Protocol and its Amendments (London, 1990; Copenhagen, 1992), and thus we give a single scenario for the greenhouse gases and for the ozone depletion they cause. The Copenhagen '92 amendments do not define phaseout steps for individual HCFCs and some latitude is allowed in implementing the

schedule which is fixed overall. We have chosen to illustrate the effects of the ozone-depleting substances with a scenario described in Table 2.5 which tends towards an upper limit to chlorine levels. Ozone depletion is scaled to tropospheric chlorine loading. Brominated halocarbons also contribute to ozone loss but they are not included in the table because they are not significant greenhouse gases in terms of direct forcing because of their low concentrations. Also, if the chlorine-plus-bromine-equivalent loading is used, no significant difference in the scaled future ozone depletion is apparent. Based on satellite and ground-based trends (Harris *et al.*, 1995), ozone depletion between 1979 (2.44 ppbv chlorine loading) and 1990 (3.64 ppbv) is taken as –2% in the global mean, with –1% occurring prior to 1979. With the linear fit, this model assumes depletion began at approximately 1.8 ppbv tropospheric chlorine. The lag in time between troposphere and stratosphere, the slightly different reactivities of the chlorocarbons and the inclusion of bromocarbons, are important to the details of ozone depletion (e.g., WMO/UNEP, 1995), but are not significant for these radiative forcing scenarios.

2.3 Aerosols

2.3.1 Summary of 1994 Report and Areas of Development

Aerosols influence climate in two ways, directly through scattering and absorbing radiation and indirectly through modifying the optical properties and lifetime of clouds. IPCC (1994) concluded that uncertainties in the estimation of direct aerosol forcing arose from the limited information on the spatial and temporal distribution of aerosol particles as well as of the optical properties of the particles themselves. While a body of observations of sulphate aerosol was available, there were only limited data on a global scale for other aerosol components. Furthermore, the sensitivity of the optical properties to the size distribution of the particles, as well as to their chemical composition, made it difficult to relate the aerosol forcing to emissions in a simple manner. Model distributions based on the present sources and sinks along with empirically determined optical properties have been used to estimate direct forcing although it is recognised that changes in the sources, even if total emissions are conserved, are likely to give a different magnitude as well as geographical distribution of the forcing. Similarly, although some local studies had been made of the impact of aerosol particles on cloud optical properties, there had been few observational studies either of possible global

Table 2.6: *Source strength, atmospheric burden, extinction efficiency and optical depth due to the various types of aerosol particles (after IPCC (1994), Andreae (1995) and Cooke and Wilson (1996)).*

Source	Flux (Tg/yr)	Global mean column burden (mg m^{-2})	Mass extinction coefficient (hydrated) (m^2 g^{-1})	Global mean optical depth
Natural				
Primary				
Soil dust (mineral aerosol)	1500	32.2	0.7	0.023
Sea salt	1300	7.0	0.4	0.003
Volcanic dust	33	0.7	2.0	0.001
Biological debris	50	1.1	2.0	0.002
Secondary				
Sulphates from natural precursors, as $(NH_4)_2SO_4$)	102	2.8	5.1	0.014
Organic matter from biogenic VOC	55	2.1	5.1	0.011
Nitrates from NO_x	22	0.5	2.0	0.001
Anthropogenic				
Primary				
Industrial dust, etc.	100	2.1	2.0	0.004
Soot (elemental carbon) from fossil fuels	8	0.2	10.0	0.002
Soot from biomass combustion	5	0.1	10.0	0.001
Secondary				
Sulphates from SO_2 as $(NH_4)_2SO_4$	140	3.8	5.1	0.019
Biomass burning	80	3.4	5.1	0.017
Nitrates from NO_x	36	0.8	2.0	0.002

changes in cloud condensation nuclei (CCN) or ice nuclei, which might be attributed to anthropogenic aerosol particles, or of the global influence of aerosol particles on cloud optical properties. The problem of the influence of aerosol particles on cloud lifetime and extent, and hence on the spatially and temporally averaged cloud optical thickness, was recognised but could not be quantified.

Several areas of research have been developed since the preparation of IPCC (1994) and these are discussed in this update. In particular, model studies of aerosol distribution and the consequent direct forcing have been extended and now include non-sulphate aerosol. Forcing by soot aerosol, and the optical depth due to soil dust, including natural and anthropogenic components, have

been estimated. (We define soot as the light-absorbing aerosol produced by incomplete combustion of carbon-based fuels, as distinct from non-absorbing organic aerosols from, for example, biomass combustion. Some authors use the term "black carbon" for this component.) In the context of indirect forcing, recent work has suggested a possible role for organic particles as CCN. Furthermore, recent field measurements have provided further data on the relationships between aerosol particles and cloud droplet concentration. As discussed by Jonas *et al.* (1995), the climatic influence of stratospheric aerosol due to volcanic eruptions is comparatively well understood so that the modelled and observed cooling following the eruption of Mt. Pinatubo are in good agreement.

2.3.2 *Modelling of Tropospheric Aerosol Distributions*

Whereas estimates of the radiative forcing of greenhouse gases can be based on measured concentrations of these gases in the atmosphere, the forcing due to aerosol particles at the present time has to be calculated from model spatial distributions. This is because the aerosol particles have a short lifetime in the atmosphere and are therefore much less uniformly distributed than the more long-lived greenhouse gases. The physical and radiative properties of aerosols are more variable and less well characterised than the equivalent properties of greenhouse gases. While there are many interactions between different chemical species in aerosols it has proved useful to consider the different species as independent.

The simplest way to estimate the average concentration of aerosol particles is to apply a box model in which the global rate of emission is multiplied by an average turnover time, estimated from the efficiency of the removal processes, to yield a global burden. Such first-order estimates have been made for anthropogenic sulphate (Charlson *et al.*, 1990; Charlson *et al.*, 1992) and soot (Penner, 1995). Radiative forcings estimated from the aerosol loadings predicted using such models tend to be overestimates, compared with those based on more detailed models, partly because box models neglect the spatial correlation that exists between aerosol concentration and cloudiness since large industrial emissions are in relatively cloudy regions. Forcing estimates which include explicit consideration of the spatial variability of the aerosol distribution, cloudiness and solar radiation have to be based on three-dimensional models of the global aerosol distribution. Such models have been developed for aerosol sulphate (Langner and Rodhe, 1991; Taylor and Penner, 1994; Feichter *et al.*, 1996; Pham *et al.*, 1995), for soot (Penner *et al.*, 1993; Cooke and Wilson, 1996), and for mineral dust (Tegen and Fung, 1994).

Most estimates of the radiative forcing due to anthropogenic sulphate aerosols have been based on the sulphate distributions calculated by Langner and Rodhe (1991) using the MPI–Mainz MOGUNTIA model. In order to illustrate the sensitivity of the result to assumptions about the rate of oxidation of sulphur dioxide (SO_2) to sulphate in clouds, one of the key processes in the sulphur cycle, Langner and Rodhe (1991) made two separate simulations. One simulation was based on immediate oxidation of SO_2 as soon as it encounters cloud, the other assumed a delayed rate of oxidation corresponding to some degree of oxidant limitation. The former case, referred to by the authors as the "standard" case (it might also have been named the "fast oxidation" case) gave a global

sulphate burden of 0.77 TgS compared with 0.55 TgS for the "slow oxidation" case. The forcing estimates by Charlson *et al.* (1991), Kiehl and Briegleb (1993) and Haywood and Shine (1995) are all based on the "slow oxidation" simulation of Langner and Rodhe (1991). Had the "fast oxidation" case been used, the forcing would have been proportionally increased (Kiehl and Rodhe, 1995). The simulations by Pham *et al.* (1995) and Feichter *et al.* (1996) using different models gave global sulphate burdens of 0.8 and 0.61 TgS respectively. The high value of Pham *et al.* (1995) is partly due to their assumption of a considerably higher anthropogenic SO_2 emission (92 TgS/yr) than the other studies (about 70 TgS/yr).

The average turnover time of aerosol sulphate, defined as the ratio of the global burden to the total rate of removal, falls in the range 4–5 days in all of the model simulations referred to above. The parametrization of precipitation scavenging of aerosol sulphate, which is the dominant removal pathway, and of the oxidation of SO_2 to sulphate, are probably the two largest sources of uncertainty in the sulphate simulations. From the additional model simulations, largely independent of the Langner and Rodhe (1991) study, that have been published during the past few years, there is some basis for estimating the uncertainty range for the global sulphate burden of roughly 0.5 to 0.8 TgS. However, more global observations are needed to verify these model calculations.

Other anthropogenic gaseous precursors of aerosol particles include nitrogen oxides and volatile organic compounds (VOCs) from fossil fuel combustion. Although their contribution to radiative forcing may well be more substantial than indicated in Table 2.6 – some studies (e.g., Diederen *et al.*, 1985) suggest that in highly polluted regions aerosol nitrate may be as important as aerosol sulphate for scattering visible light – no attempt has been made to estimate their global optical depth based on detailed modelling.

The study by Cooke and Wilson (1996) of soot is based on the MOGUNTIA model, as used by Langner and Rodhe (1991) in their sulphate study. The global source strength of soot from industrial and biomass combustion was estimated to be about 13 TgC/yr and the global burden 0.26 Tg. This model study provides the first consistent estimate of the global distribution of soot from which its direct radiative forcing can be estimated, for example using the approach of Haywood and Shine (1995).

The Tegen and Fung (1994) study of mineral dust is based on a source strength of 3000 Tg/yr, distributed between four size categories. The estimated turnover time of these categories ranged from 13 days for the clay fraction with an effective radius of 0.7 μm to 1 hour for sand with an effective radius of 38 μm. No attempt was made to separate out an anthropogenic component of the mineral dust and therefore it is not possible, from these results, to estimate the climate forcing by this aerosol component. However, recent studies by Tegen and Fung (1995) based on analysis of satellite observations suggest that a substantial amount (possibly 30–50%) of the soil dust burden may be influenced by human activities. No attempt has been made to estimate the global radiative forcing due to this anthropogenic component.

Table 2.6, adapted from IPCC (1994), contains revised and updated quantities for carbonaceous aerosols (organics and soot) including separate entries for fossil fuel and biomass combustion sources of carbonaceous aerosols. As in IPCC (1994), the main conclusions to be drawn from this table are:

(a)   Natural sources represent around 90% of the total mass emission to the atmosphere of aerosols and their precursors; but

(b)   Anthropogenic emissions are estimated to result in almost half of the global mean aerosol optical depth because of their size and optical properties, the major contribution arising from SO_2.

It is important to recognise that some fraction of the soot aerosols from biomass combustion is natural; however it is included here as anthropogenic, owing to a lack of detailed knowledge of the actual frequency of occurrence of natural fires initiated, for example, by lightning. Andreae (1995) estimates that biomass burning has increased by a factor of two or three since the mid-19th century, although large uncertainties remain concerning the history of the source strength of soot from biomass burning.

Caution should be exercised when considering global aerosol loadings and especially their trends with time. Figure 2.13 shows the Northern Hemisphere and regional source strengths of SO_2 (expressed as sulphur), and illustrates the lack of uniform trends and the current rapid growth of the Asian source.

2.3.3 *Optical Characteristics of Aerosols*
Sulphate aerosol is a large anthropogenic contributor to optical depth and is the best understood and most easily quantified of the key aerosol types. The second most important anthropogenic aerosol particle type is carbonaceous; taken together the two typically comprise over 80% of the sub-micrometre aerosol mass in industrial

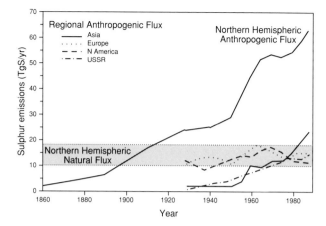

Figure 2.13: Natural and fossil fuel combustion sources of SO_2 in the Northern Hemisphere (after Dignon and Hameed, 1989; 1992).

regions (Jonas *et al.*, 1995). In contrast to the predominantly scattering effect of sulphate aerosol, carbonaceous aerosol absorbs solar radiation. While the uncertainty regarding the direct radiative forcing of sulphates has been narrowed somewhat (Section 2.4), increased recognition has been given to the role of carbonaceous aerosols, both organic compounds and soot. Neither the sources nor the ambient-air properties of these carbon-containing aerosols have been adequately characterised. As a result, the relative uncertainties in the calculated radiative forcing by carbonaceous aerosols are even larger than for the sulphates. In addition, sampling, measurement and chemical analysis methods for carbonaceous aerosols have not been standardised, and molecular characterisation is difficult and seldom attempted. Thus much of the present data set on carbonaceous aerosols must be viewed as not being truly appropriate for calculation of direct radiative forcing. Further, the lack of molecular information and the simultaneous lack of detailed information on water solubility and surface chemical properties make it difficult to address properly the role of carbonaceous particles as CCN. Nonetheless, some progress has been made towards bounding the direct forcing by both soot and organic aerosols, while empirical evidence has emerged that implicates organic aerosol particles in the process of cloud droplet formation (Novakov and Penner, 1993).

Referring back to Table 2.6, the source strengths of carbonaceous aerosol particles (organic as well as soot) are based on emission factors relating the mass of aerosol produced to the mass of carbon burned in the fuel. Some of the data on emission factors were obtained in the laboratory and some were derived from atmospheric sampling. While in a small number of cases the particle size distribution of the aerosol was measured, in most experiments only the total aerosol mass was measured. For the case of organic aerosol particles, the suggested size distribution is similar to that of sulphates (with the mass mainly in the 0.1 – 1.0 μm diameter range) so that there should be reasonable similarity between the scattering efficiencies of organic and sulphate aerosol (Penner *et al.*, 1992). However, since the data on soot emissions have not been acquired with any standardisation of sampling or analysis techniques, the emission factors must be regarded as preliminary estimates. Many of the soot emission data have been obtained from studies relating to inhalation and human health effects and these studies involve sampling of particles up to about 10 μm in diameter. As can be seen from Figure 2.14, the absorption efficiency of spherical soot particles is calculated to be a strong function of particle size, with little absorption occurring for particles larger than about 1 μm equivalent diameter. The dependence of absorption by non-spherical particles on their equivalent aerodynamic diameter is not well known. This, together with the fact that the atmospheric residence time of particles larger than 1 μm is much shorter than that for small particles, suggests that the use of existing emission data is uncertain and likely to overestimate both particle mass concentration and light absorption. Specific size-resolved sampling is needed to delineate the amount of soot emitted as particles smaller than 1 μm; in the absence of optical characterisation of soot from different sources, the existing data should be regarded more as upper limits on soot emissions than as current best estimates.

In spite of the large uncertainties concerning the composition and magnitude of the source strength of carbonaceous aerosols, the single scattering albedo $\omega = \sigma_{sp}/(\sigma_{sp} + \sigma_{ap})$ can be, and has been, measured (σ_{sp} and σ_{ap} are the scattering and absorption cross-sections of the aerosol particles). The results show finite but highly variable light absorption by soot (Waggoner *et al.*, 1981; White, 1990; Haywood and Shine, 1995). Table 3.3 of IPCC (1994) gives a range $0.8 \leq \omega \leq 0.95$ for low humidity polluted continental air, and higher values, $0.9 \leq \omega \leq 1.0$ for clean continental or remote marine conditions. While few data exist for biomass combustion smoke, values in the range from 0.5 to greater than 0.9 have been reported with a mean value for low relative humidity aerosol of 0.83 ±0.11 (Radke *et al.*, 1991). Because these quantities were measured using heated or highly desiccated samples, the ambient values of single scattering albedo would be expected to be considerably higher due to hygroscopic growth, giving a relative increase of σ_{sp} compared with

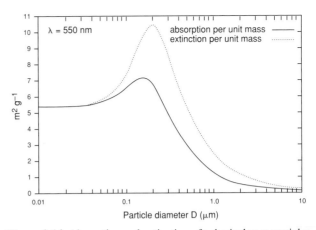

Figure 2.14: Absorption and extinction of spherical soot particles per unit mass as a function of particle size for light with a wavelength of 550 nm. The refractive index was assumed to be 1.773–0.626i and the calculations by S.F. Marshall (pers. comm.) assume Mie scattering.

σ_{ap}. Due to the large uncertainty ranges of both the source strength of soot and the single scattering albedo of the aerosol particles, it is necessary at present to adopt the precedent of Haywood and Shine (1995) who concluded that the role of light absorption by soot is uncertain but that the range of the climate forcing which results is bounded (see Section 2.4).

As pointed out in IPCC (1994) there are difficulties associated with the determination of the scattering efficiency of sulphate aerosol. The difficulties have not been fully resolved (Hegg *et al.*, 1993, 1994; Anderson *et al.*, 1994).

Some measurements of aerosol properties in China have been made (e.g., Shi *et al.*, 1994; Wu and Chen, 1994), but it remains difficult to evaluate adequately the effects of the aerosol particles in the Asian region, where the source strengths are increasing rapidly.

2.3.4 *Influence of Aerosols on Clouds*

The suggestion that aerosol particles might modify the cloud droplet size distribution and hence the radiative properties of clouds, on a global scale, was supported in IPCC (1994) by the satellite observations of Han *et al.* (1994) which showed evidence of systematic differences in the effective radius of the cloud droplets between the Northern and Southern Hemispheres, with smaller droplets in the Northern Hemisphere. More recently this hypothesis has been supported by observations of seasonal coherence of CCN and cloud optical depths at Cape Grim, Tasmania (Boers *et al.*, 1994).

The problem of quantifying the relationship between aerosol source strengths and the droplet size distribution in low-level water cloud is complex because of the number of processes involved, including chemical processes in the emissions and the interaction between cloud dynamics and microphysics. Jonas *et al.* (1995) showed examples of empirical relationships only between aerosol particle number concentrations and droplet concentrations. Studies by Jones *et al.* (1994), Boucher and Lohmann (1995) and Hegg *et al.* (1993) have provided further insight into the importance of the exact form of this relationship when estimating the possible range of indirect radiative forcing due to sulphate aerosol, but it is clear that no universal relationship exists which could be applied in all regions. Calculations by Chuang and Penner (1995) have also demonstrated the impact of aerosol particles produced by oxidation of SO_2 in the modification of the cloud droplet population. It is clear, however, that the present results are not sufficient to provide global relationships which might be used with confidence in calculations of global forcing.

Observations of cloud properties by Alkezweeny *et al.* (1993) demonstrate the impact of anthropogenic aerosol particles on cloud optical properties. Recent field experiments off the coast of California (MAST, the Monterey Area Ship Track experiment) and in the Arctic have provided further evidence for the impact of aerosol particles. Preliminary analyses of these experiments, and other studies (e.g., Hindman and Bodowski, 1994; Hindman *et al.*, 1995), have largely confirmed the impact of non-absorbing aerosol particles in increasing the albedo of clouds through their effect on the droplet concentration. The results emphasise that the sensitivity of clouds to aerosol particles is highly variable and is reduced in regions of high natural aerosol loading. There remain, however, some systematic differences between calculated and observed values of infrared albedo which may be due to the inhomogeneous nature of the cloud layers (Hayasaka *et al.*, 1994). Whether such inhomogeneities affect the sensitivity of cloud optical properties to droplet concentration has not been determined: there is some evidence that the optical properties of inhomogeneously mixed clouds are less sensitive to droplet concentration than more uniform clouds (Novakov *et al.*, 1994).

Experiments by Perry and Hobbs (1994) have also confirmed earlier results suggesting that regions close to clouds may act as favoured regions for the production of particles. These results again point to the complexity of the relationship between emissions and CCN concentrations. Kulmala *et al.* (1995) have suggested that the presence of NO_x may also modify the nucleation process but measurements to illustrate its importance have not yet been made. Calculations by Jensen and Toon (1994) have suggested

that conclusions concerning the sensitivity of cloud properties to aerosol particle concentration which relate to low level water cloud may not easily be transferred to the impact on upper tropospheric ice clouds. Their calculations assume that homogeneous nucleation of ice is the dominant mechanism by which ice crystals are produced in these clouds and that nucleation is relatively insensitive to the number of aerosol particles, although Sassen *et al.* (1995) provided observations that suggest the modification of cirrus by incursion of stratospheric aerosol particles. The extent to which aerosol particles influence the optical properties of ice clouds, especially ice clouds formed at lower levels, has still to be determined.

The problem of quantifying the impact of aerosol particles on cloud lifetime, through their impact on precipitation processes, was mentioned in IPCC (1994) although there were few quantitative results that could be used to assess the global impact of such changes which directly affect the mean optical thickness of the clouds. The problem was highlighted by Pincus and Baker (1994) who suggested that the impact of changes in the mean optical thickness might be comparable with the radiative effect of changing the droplet size distribution at cloud top. The problem is more acute for thin clouds in which precipitation is light but critical to the evolution of the cloud.

2.4 Radiative Forcing

The changes in greenhouse gas and aerosol concentrations reported in previous sections lead to a perturbation of the planetary radiation budget which is referred to here as "radiative forcing". It is the purpose of this section to report on calculations of the radiative forcing since pre-industrial times due to a number of mechanisms and to indicate the degree of confidence in these estimates.

The detailed rationale for using radiative forcing was given in IPCC (1994). It gives a first-order estimate of the potential climatic importance of various forcing mechanisms. The radiative forcing drives the climate to respond but, because of uncertainties in a number of feedback mechanisms (see Chapter 4), the radiative forcing can be calculated with more confidence than the consequent climate response. There are, however, limits to the utility of radiative forcing as neither the global mean radiative forcing, nor its geographical pattern, indicate properly the likely three-dimensional pattern of climate response; the general circulation models discussed in Chapters 5 and 6 are the necessary tools for the evaluation of climate response.

This section limits itself to mechanisms which, via their

interaction with solar or thermal infrared radiation, act to drive climate change. Changes in atmospheric water vapour, cloudiness or surface albedo which are a response to a radiative forcing and which, in turn, act to modify the climate response via their radiative effects, are considered in Chapter 4; similarly, recent studies of our fundamental understanding of the interaction of clouds with radiation are discussed in Chapter 4.

The definition of radiative forcing adopted in previous IPCC reports (1990, 1992, 1994) has been the perturbation to the net irradiance (in Wm^{-2}) at the tropopause after allowing for stratospheric temperatures to re-adjust (on a time-scale of a few months) to radiative equilibrium, but with the surface and tropospheric temperature and atmospheric moisture held fixed. This stratospheric adjustment is of crucial importance in evaluating properly the radiative forcing due to changes in stratospheric ozone and is important at the 5–10% level for changes in some other greenhouse gases; it is of less importance for changes in aerosol concentrations (see IPCC 1994). In this section, the reported radiative forcings due to greenhouse gas changes (including ozone) include the stratospheric temperature adjustment; all other forcings do not and are thus "instantaneous" values.

2.4.1 Greenhouse gases

Estimates of the adjusted radiative forcing due to changes in the concentrations of the so-called well-mixed greenhouse gases (CO_2, CH_4, N_2O and the halocarbons) since pre-industrial times remain unchanged from IPCC (1994); the forcing given there is 2.45 Wm^{-2} with an estimated uncertainty of 15%. CO_2 is by far the most important of the gases, contributing about 64% of the total forcing. We know of no new estimates of the forcing due to changes in tropospheric ozone since pre-industrial times and retain the estimate of between 0.2 and 0.6 Wm^{-2}.

Evans and Puckrin (1995) report measurements of the contribution of carbon monoxide to the greenhouse effect. The global mean radiative forcing due to the changes in carbon monoxide concentration is unlikely to exceed a few hundredths of a Wm^{-2}; its role in modifying tropospheric ozone is likely to be of greater importance for climate.

The basic physical understanding of the ways in which greenhouse gases absorb and emit thermal infrared radiation (e.g., Goody and Yung, 1989) is supported by abundant observations of the spectrally resolved infrared emission by the clear-sky atmosphere (e.g., Kunde *et al.*, 1974; Lubin, 1994). Barrett (1995)'s suggestion that greenhouse gases are unable to emit significant amounts of infrared radiation is contradicted by these observations.

2.4.1.1 Halocarbon radiative forcing

In IPCC (1994) it was noted that the radiative forcing per mass (or molecule) of a number of halocarbons was based on unpublished material for which details were not available. Since then Pinnock *et al.* (1995) have reported the integrated absorption cross-sections and radiative forcing of eighteen hydrohalocarbons, a number of which have not previously been reported. In addition, they examined the dependence of the radiative forcing on a number of assumptions concerning the effects of clouds and overlapping species. For most gases, the values reported by Pinnock *et al.* (1995), relative to CFC-11, are within 15% of the values given in Table 4.3 of IPCC (1994); since this is within the likely error of the estimates, we do not amend our earlier recommendations. For two gases, HFC-32 and HFC-236fa, more substantial discrepancies are found. We favour results from models for which details are available, and hence, in Table 2.7, the values for HFC-32 and HFC-236fa are updated. In addition, values for HFC-41 and HFC-272ca, which were not reported in IPCC (1994), are given.

A significant issue raised by Pinnock *et al.* (1995) concerns the absolute forcing due to CFC-11. A value of 0.22 Wm^{-2}/ppbv has been used in IPCC reports since IPCC (1990). Pinnock *et al.* obtain a value some 20% higher. Part of the problem may relate to the range of measurements of the absorption cross-section reported in the literature (see e.g., Li and Varanasi, 1994); however, the precise source of the discrepancy has not been established. If the Pinnock *et al.* (1995) value is confirmed by other work, the radiative forcing contribution of CFC-11 would have to be increased. Since in IPCC reports the radiative forcing due to other halogenated compounds has been reported relative to CFC-11, their absolute contribution would also have to be increased.

Roehl *et al.* (1995) reported absorption cross-sections and radiative forcings for 5 perfluorocarbons. The Roehl *et al.* values are favoured here as they are now published and form a consistent set of values for the perfluorocarbon series. The values are given in Table 2.7. It should be noted that there is a substantial discrepancy (30%) between the new value for C_2F_6 and the IPCC (1994) value; the source of this discrepancy has not been established.

2.4.1.2 Radiative forcing due to stratospheric ozone changes

The radiative forcing due to changes in ozone is more difficult to calculate than those of the other greenhouse gases for a number of reasons. First, ozone changes cause a significant change in both solar and thermal infrared radiation. Second, the effect of stratospheric temperature change as a consequence of ozone loss in the lower stratosphere significantly modifies the radiative forcing. Finally, uncertainty in the spatial distribution of the ozone loss, in particular in the vertical, introduces significant uncertainties in the consequent radiative forcing. These issues were discussed in detail in IPCC (1994) who concluded that the adjusted radiative forcing as a result of decreases in stratospheric ozone was about -0.1 Wm^{-2} with a factor of 2 uncertainty.

Molnar *et al.* (1994) calculated the radiative forcing and the equilibrium surface temperature response due to stratospheric ozone loss in the 1980s with a two-dimensional radiative-convective seasonal climate model. Using ozone data for broad latitude belts (90°–30°N, 30°N–30°S and 30°–90°S) their calculations reaffirm the sensitivity of the forcing to uncertainties in the vertical profile of ozone depletion (see IPCC 1994); Molnar *et al.* (1994) find uncertainties in the forcing of up to 50%. Their climate response results suggest that the surface cooling due to stratospheric ozone loss in the 1980s offsets about 30% of the warming due to well-mixed greenhouse gas increases over the same period. The offset is about a factor of two greater than that expected from radiative forcing calculations alone and appears to be due to the representation of the meridional transport of heat in the model. Because of a lack of other published studies, especially using three dimensional models, this result must be viewed with caution.

The reduction in stratospheric chlorine and bromine loading which will result from the Montreal Protocol and its Amendments is expected to lead to a recovery of the stratospheric ozone layer over the next century (WMO/UNEP, 1995). During the 1980s the ozone loss partially offsets the forcing due to the well-mixed greenhouse gases, but in the early decades of the next century the ozone recovery constitutes a positive radiative forcing that acts to enhance the effect of the well-mixed greenhouse gases. Solomon and Daniel (1996) have estimated the consequences of the recovery on radiative forcing using the IPCC (1990) Business-as-usual Scenario. The transition from ozone loss to ozone recovery changes the decadal increment of the greenhouse gas plus ozone forcing from about 0.44 Wm^{-2}/decade for the 1980s to about 0.59 Wm^{-2}/decade for the period 2000 to 2009. If the ozone changes are ignored, the increment of the forcing is 0.52 Wm^{-2}/decade for the 1980s and 0.56 Wm^{-2}/decade for 2000 to 2009. Hence the changes in stratospheric ozone lead to a significant transient acceleration of the greenhouse gas radiative forcing. This acceleration may be even greater, if the ozone forcing is enhanced by the effects of the depletion on tropospheric constituents via the

Table 2.7: *Radiative forcings due to halocarbons for a per unit mass and a per molecule increase in atmospheric concentration relative to CFC-11. The table shows direct forcings only. The values here are for gases for which the IPCC (1994) values have been updated or for gases not hitherto reported. The absolute forcing due to CFC-11 is taken from IPCC (1990) and is 0.22 ΔX Wm^{-2} where ΔX is the perturbation to the volume mixing ratio of CFC-11 in ppbv.*

Gas		ΔF per unit mass relative to CFC-11	ΔF per molecule relative to CFC-11	Source
* HFC-32	CH_2F_2	1.32	0.50	Pinnock *et al.* (1995)
+ HFC-41	CH_3F	0.44	0.11	Pinnock *et al.* (1995)
* HFC-236fa	$CF_3CH_2CF_3$	0.90	1.00	Pinnock *et al.* (1995)
+ HFC-272ca	$CF_2HCH_2CH_3$	0.57	0.33	Pinnock *et al.* (1995)
* CF_4		0.73	0.47	Roehl *et al.* (1995)
* C_2F_6		1.04	1.05	Roehl *et al.* (1995)
* C_3F_8		0.80	1.11	Roehl *et al.* (1995)
+ C_4F_{10}		0.80	1.40	Roehl et al. (1995)
+ C_5F_{12}		0.85	1.79	Roehl *et al.* (1995)
* C_6F_{14}		0.83	2.08	Roehl *et al.* (1995)

* Denotes value is amended from IPCC (1994) value.

+ Denotes gas not previously reported in IPCC reports.

increased penetration of UV (see Section 2.4.1.3). Solomon and Daniel (1996) show that if the ozone forcing were to be doubled as a consequence of the tropospheric changes, the forcing in the 1980s would be 0.36 Wm^{-2}/decade, increasing to about 0.61 Wm^{-2}/decade for 2000 to 2009.

2.4.1.3 Impact of ozone depletion on other radiatively active species

Depletion of stratospheric ozone can affect the distributions of other radiatively active species (see Section 2.2). Thus, the net radiative forcing due to ozone-depleting gases such as CFCs and halons depends not only upon the induced ozone depletion but also upon the subsequent chemical changes that can follow from stratospheric ozone depletion. Most notable among these is the dominant role of stratospheric ozone in controlling the transmission of the ultraviolet radiation to the troposphere and, in turn, the tropospheric OH concentrations that regulate the abundances of gases such as methane, HFCs, HCFCs and tropospheric ozone (e.g., Fuglestvedt *et al.*, 1994).

Madronich and Granier (1992) emphasised that through this mechanism ozone depletion could play a role in determining methane trends. The modelling study of Bekki *et al.* (1994) provides support for this mechanism by examining the role of enhanced ozone depletion following the eruption of Mt. Pinatubo upon the reduced methane growth rates observed at that time. However, other processes may also have affected the observed methane response as noted in Section 2.2. Bekki *et al.* (1994) also emphasised that the observed global ozone depletion of about 3% would be expected to produce an indirect negative radiative forcing due to methane decreases and associated tropospheric ozone decreases that is about 30–50% of the forcing due to the ozone depletion alone. Thus this study suggests a total radiative forcing associated with ozone depletion that could be 30–50% more negative than estimates of this forcing that neglect the impact of ultraviolet transmission changes upon methane and tropospheric ozone. Because the bulk of methane destruction normally takes place in tropical latitudes while ozone depletion maximises in mid- and high latitudes, the details of such estimates are dependent on the uncertainties in the latitudinal distribution of ozone depletion, as well as tropospheric transport characteristics.

Toumi *et al.* (1994) discussed another indirect chemical mechanism that could make the total radiative forcing due to ozone depletion even more negative. They pointed out that increases in tropospheric OH should be expected to increase the rate of oxidation of SO$_2$ to form sulphuric acid, which in turn provides the source of cloud condensation nuclei and hence has the potential to enhance the negative radiative forcing associated with aerosols; however, the importance of this mechanism has been challenged by Rodhe and Crutzen (1995). Toumi *et al.* (1994) used the same simple approach as that taken by Charlson *et al.* (1992) to relate increases in SO$_2$ oxidation to cloud albedo. They found that the observed decreases in ozone could have led to a negative radiative forcing due to this mechanism as large as 40–800% of the radiative forcing due to the ozone depletion itself. The many uncertainties associated with estimates of radiative forcing due to sulphate-induced cloud changes are discussed in Section 2.3.4. Toumi *et al.* (1994) also emphasised the uncertainties associated with the lack of convective transport in their two-dimensional model.

In spite of large uncertainties, the two studies taken together suggest that the total radiative forcing due to ozone depletion including indirect chemical effects relating to tropospheric OH is likely to exceed that obtained from the ozone depletion alone, perhaps by a factor of two or more. Any radiative forcing associated with methane changes will already be implicitly accounted for in the estimates of the forcing due to changes in the well-mixed greenhouse gases. The forcing resulting from changes in cloud properties would be in addition to the radiative forcing of –0.1 Wm^{-2} (with a factor of two uncertainty) due to stratospheric ozone change proposed in IPCC (1994). However, given the uncertainties in characterising the ozone-induced effect on clouds, we believe it to be too premature to suggest a revised value for the forcing (or the error bars) associated with stratospheric ozone change.

2.4.1.4 Other climate–chemistry interactions

A recent study of the impact of enhanced emission of pollutants in the Asian region on key chemical species using a 3-D chemical transport model suggests increases of free tropospheric ozone up to 30% for a doubling of NO$_x$ emissions during the summer months (Berntsen *et al.*, 1996). These increases in tropospheric ozone cause a local radiative forcing of 0.5 Wm^{-2} which is about 50% of the negative radiative forcing in the same region due to the direct effect of sulphate aerosols.

Hauglustaine *et al.* (1994) used a 2-D dynamical-radiation-chemistry model to estimate the effect of current day aircraft emissions of NO$_x$ on tropospheric ozone and, hence, on radiative forcing (although it should be noted that there are limitations in using 2-D models in such studies, see Section 2.2 and IPCC (1994)). The resulting radiative forcing is small. The global and annual mean radiative forcing due to aircraft NO$_x$ emissions is only

+0.02 Wm^{-2}, compared to their estimate of 0.1 Wm^{-2} from surface sources; the aircraft effect is greatest in northern mid-latitudes, reaching +0.08 Wm^{-2} in summer. One effect of these ozone increases is to increase the model's tropospheric OH concentrations; the subsequent decrease in methane lifetimes (and hence concentration) is found to offset the positive forcings by about 30%.

2.4.2 *Tropospheric Aerosols*

Aerosols can directly influence the radiation budget by scattering/absorption (this is referred to as the "direct" aerosol effect); they can also influence it by altering the structure and radiative properties of clouds (the "indirect" aerosol effect).

There is a large uncertainty range particularly for the radiative forcing due to the effect of aerosols on cloud properties. As also emphasised in IPCC (1994), the aerosol radiative forcing is spatially very inhomogeneous so that a direct comparison of the global mean forcing with the forcing due to other mechanisms might not give a complete picture. Nevertheless, the global mean value remains a useful single-parameter method of comparing different aerosol forcing estimates.

It should also be recognised that aerosol forcing estimates tend to be "partial derivatives": i.e., the forcing that would result from that aerosol component alone. Because of interactions between different aerosol components, the radiative properties of multi-component aerosols are not necessarily the same as the sum of the individual components (as will be discussed for soot aerosols in Section 2.4.2.2); the same consideration applies to the effect of aerosols on cloud properties.

A number of authors have reported a long-term decline in solar radiation reaching the surface at some locations using routine observations of global solar radiation since the 1950s (e.g., Liepert *et al.*, 1994; Stanhill and Moreshet, 1994; Li *et al.*, 1995). There are several potential factors that could cause such a change. These include the increased scattering from aerosols and changes in water vapour and cloud amounts. Thus the decline may be a direct indication of aerosol forcing but it may also be an indication of water vapour and/or cloud feedbacks as a consequence of recent climate change. Such observations may, nevertheless, provide useful additional information with which to constrain model estimates in the future.

2.4.2.1 *Direct forcing due to sulphate aerosols resulting from fossil fuel emissions and smelting*

A number of new estimates of the direct radiative forcing due to sulphate aerosols arising from fossil fuel burning and smelting have become available. As discussed in Section 2.3, the estimates of the direct forcing due to sulphate aerosols can differ because of significant differences in the source strength or representation of chemistry in the models used to construct sulphate aerosol climatologies. The Langner and Rodhe (1991) climatology derived using the MOGUNTIA model is available for a "standard" case and a so-called "slow oxidation" case in which the oxidation of SO$_2$ in clouds is reduced. The slow-oxidation case yields sulphate burdens in better agreement with more detailed models (see Section 2.3.2); the sulphate burden in the standard case is 40% higher for the present day. The newer Pham *et al.* (1995) climatology has a burden 45% higher than the Langner and Rodhe slow-oxidation case, due to a combination of faster oxidation and higher source strengths.

Kiehl and Rodhe (1995) have computed the direct radiative forcing due to sulphate aerosol using the Pham *et al.* (1995) climatology. They obtained a global and annual mean forcing of −0.66 Wm^{-2}. During July the magnitude of the aerosol forcing was a maximum at −11 Wm^{-2} in central Europe and −7.2 Wm^{-2} over Eastern China. These values are sufficient to make the sulphate plus greenhouse gas forcing since pre-industrial times substantially negative in these regions.

Boucher and Anderson (1995) examined the dependence of the radiative forcing on the size and chemical form of the sulphate aerosol, including the humidity dependence. For reasonable changes in these parameters they find a variation in the global mean forcing of only ±20%. They emphasise that the effect of changing size and composition of aerosols must be included self-consistently in all parameters that affect the radiative forcing calculation, as there is a tendency for the different effects to compensate. Kiehl and Briegleb (1993) had earlier shown a similar result for variations in particle size only. Boucher and Anderson's study incorporates the Langner and Rodhe (1991) "standard" sulphate climatology into the Laboratoire de Météorologie Dynamique du CNRS (LMD) GCM. Sulphate optical properties incorporate laboratory measurements of humidity-dependent sulphate particle growth. They compute a "base case" global mean radiative forcing of −0.29 Wm^{-2}, which is similar to the estimate of Kiehl and Briegleb (1993) but implies a 21% lower forcing per unit mass of sulphate, since Kiehl and Briegleb use the "slow oxidation" sulphate climatology. This difference is primarily due to the incorporation by Boucher and Anderson of the dependence of the asymmetry factor on humidity.

Chuang *et al.* (1994) have coupled a tropospheric chemistry model to the Livermore/NCAR Community

Climate Model (CCM1) to estimate the sulphate forcing for both direct scattering by the aerosols and the effect of the sulphate on cloud droplet effective radii. They obtain a global mean forcing of -0.92 Wm^{-2} and attribute -0.45 Wm^{-2} of this to the direct effect. This direct forcing is significantly smaller than the earlier direct radiative forcing estimate of -0.9 Wm^{-2} by the same group (Taylor and Penner, 1994). Most of this difference is due to the use of a humidity-dependent scattering coefficient in Chuang et al. (1994); Taylor and Penner (1994) had previously shown that use of such a humidity dependence led to a global mean forcing of -0.6 Wm^{-2}. The remainder of the difference is attributed to the overprediction of sulphate concentrations in Europe in Taylor and Penner (1994).

Haywood and Shine (1995) modified the simple radiation model used by Charlson et al. (1991) by incorporating a geographically and seasonally varying surface albedo; they used the aerosol optical properties appropriate for 0.7 μm (instead of 0.55 μm), which Blanchet (1982) showed gave irradiances in reasonable agreement with more rigorous multi-spectral calculations. They compared the clear sky forcing using two sulphate data sets – the "slow-oxidation" case of Langner and Rodhe (1991) and the Penner et al. (1994) sets. Both sets yielded a global mean forcing of about -0.33 Wm^{-2} although there was a marked difference in the seasonal variation of this forcing, with the Penner et al. set giving the more marked variation.

Direct forcing due to the scattering by aerosols in cloudy skies is generally smaller than in clear skies. Chuang et al. (1994) attribute about 25% of the direct forcing to cloudy regions. Boucher and Anderson (1995) get 22% and report that Kiehl and Briegleb's (1993) results give an estimated 40%. All of these values appear to have been obtained using a simplified global mean analysis and indicate the need for a more detailed calculation of the clear and cloudy sky contributions. The contribution of the cloudy sky forcing is ignored in simpler models such as Charlson et al. (1991) and Haywood and Shine (1995).

If the Langner and Rodhe (1991) source strengths and slow oxidation case are used, there appears to be some convergence in estimates of the direct effect of sulphate aerosols on the basis of the above estimates; this indicates that the uncertainties in radiative transfer calculations are likely to be less than the uncertainties in the column burden of sulphate aerosols. A central value of -0.4 Wm^{-2}, with a factor of 2 uncertainty is suggested (i.e., a range from -0.2 to -0.8 Wm^{-2}); in IPCC (1994) the range was given as -0.25 to -0.9 Wm^{-2}. Although this is a small change, we believe it is justified, as the upper limit (-0.9 Wm^{-2}) was

based on the Taylor and Penner (1994) estimate which, as discussed above, has been revised.

2.4.2.2 Soot aerosols

Much of the work on aerosol radiative forcing to date has concentrated on the role of sulphate aerosols. However, soot aerosols absorb radiation at solar wavelengths, so increases in the amount of soot may result in a positive radiative forcing. The quantitative understanding of the effects of soot on radiative forcing is still poor. Size-segregated climatologies of the distribution of soot are not currently available (see Section 2.3.2). Uncertainties in the size-dependent optical properties of soot particles pose a major complication in assessing the radiative forcing.

In the atmosphere, soot may be mixed with other components (Ogren, 1982), such as the non-absorbing sulphates, and the manner of the mixing has a significant bearing on the optical characteristics of the composite aerosol. Conceptual models of aerosol mixtures have been proposed, ranging from external mixtures (where the sulphate and soot aerosols are separate entities) to internal mixtures (e.g., concentric spheres with soot as the core or the shell of an aerosol (Ackerman and Toon, 1981)). In general, calculations demonstrate that when an absorber is internally mixed with a scattering particle the degree of absorption by the composite particle can be enhanced considerably (e.g., Heintzenberg, 1978; Ackerman and Toon, 1981). Thus, the quantitative aspects concerning the amount of radiation scattered and absorbed, besides being governed by the amount of the absorbing substance, also depend on the nature of the mixture and the optical mixing rules. Soot particles are often non-spherical, further adding to the uncertainty in their radiative properties (Chýlek et al., 1981). A recent theoretical development (Chýlek et al., 1995) has focused on developing general solutions for the radiative effect of soot present in sulphate aerosols; this study suggests that for high soot/sulphate mass ratios, a composite particle can cause a considerable reduction in the negative radiative forcing due to sulphate aerosol only.

Haywood and Shine (1995) have explored the possible effects of soot from fossil fuel sources on the clear-sky radiation budget using a simple radiation model (see Section 2.4.2.1). They generate soot distributions by making the simple assumption that it is a fixed mass fraction of the sulphate aerosol distributions given by Langner and Rodhe (1991) and Penner et al. (1994); thus, they neglect the effects of differences in the sources and life cycles of soot and sulphate which may lead to significantly different spatial distributions. The global mean forcing ranged from $+0.03$ Wm^{-2} for a soot/sulphate mass ratio of 0.05 using an external mixture to $+0.24$ Wm^{-2} for a soot/sulphate mass ratio of 0.1 using an internal mixture; these can be compared with their calculated sulphate only forcing of -0.33 Wm^{-2}. Since the effect of soot is highest over high albedo surfaces, its impact was greatest in the Northern Hemisphere; the ratio of Northern to Southern Hemisphere forcing was reduced from about 4 for sulphate only to 1.3 for the soot/sulphate mass ratio of 0.1 and an internal mixture.

Much work needs to be done to explore the validity of the assumptions made in this study and to extend it to cloudy regions. There are two distinct ways that clouds affect the soot forcing. First, soot placed above clouds with high albedos is able to cause a greater positive forcing than the same amount of soot over a low albedo surface. Second, if soot exists as an internal mixture with cloud liquid water, this could enhance the absorption considerably (see e.g., Danielson *et al.*, 1969; Chýlek *et al.*, 1984; Chýlek and Hallett, 1992). Sulphate aerosols lead to less solar radiation being absorbed by the Earth/atmosphere system as a whole and by the surface; soot aerosols increase the absorption by the system but decrease that reaching the surface. Therefore, the overall climate effect may differ from that with sulphate aerosols. It is clear that components of aerosol other than sulphate (including organics and soot) need to be considered if the total effects of aerosols are to be established.

Although it is very preliminary, we adopt an estimate for the global mean forcing due to soot aerosols of approximately $+0.1$ Wm^{-2} since pre-industrial times, with an uncertainty of at least a factor of 3.

2.4.2.3 Other aerosol types and sources
The forcing estimates described above are mainly for aerosols produced from fossil fuel burning and smelting. Biomass burning is a source of aerosols with potential for significant radiative forcing. We are unaware of any significant developments in this area and retain the IPCC (1994) estimate for the global mean radiative forcing since 1850 of -0.2 Wm^{-2} with a factor of 3 uncertainty. This uncertainty has been underlined by the work of Chýlek and Wong (1995) who have shown the direct radiative forcing to be dependent on the aerosol size distribution.

As discussed in Section 2.3.2, Tegen and Fung (1995) have estimated the visible optical depth of mineral dust from land surfaces as a result of human activity. To evaluate the resultant radiative forcing will require account of the modulating effects of cloud and the dependence of the forcing on the albedo of the underlying surface. Robock and Graf (1994) have also drawn attention to the fact that

in the pre-industrial period this, and other effects related to land surface modification, might have contributed a significant forcing.

2.4.2.4 Effect of aerosols on cloud properties
The effects of aerosols on changing the radiative properties of clouds were discussed in IPCC (1994). Two effects were identified: the effect on cloud albedo due to decreases in the droplet effective radius and the consequent effect of this decrease on cloud liquid water content and possibly cloud cover. The difficulties in quantifying the indirect effect of aerosols were stressed (see also the review by Schwartz and Slingo (1996)). As discussed in Section 2.3.4, there has been no significant increase in our understanding of the indirect effect since IPCC (1994). The purpose of this section is to report on recent estimates of the indirect radiative forcing due to aerosols; modelling studies continue to use simplified relationships between aerosol concentrations and droplet sizes. Thus, the similarity in the results from different groups need not imply added confidence in the estimates.

Boucher and Lohmann (1995) have used two GCMs (LMD and ECHAM) to estimate the indirect effect of anthropogenic sulphate aerosols on cloud albedo, by relating sulphate aerosol loading to cloud droplet number concentration. The sulphate aerosol distribution is derived using the MOGUNTIA model, as used by Langner and Rodhe (1991), and uses the standard oxidation rate. A range of different assumptions was made on the sulphate cloud droplet number concentration relationship: the global mean results fell in the range -0.5 to -1.5 Wm^{-2}, with a Northern Hemisphere – Southern Hemisphere ratio of between 2 and 4; the smaller ratios are from the ECHAM GCM. The methods used in this study are similar to those of Jones *et al.* (1994) reported in IPCC (1994); they obtained a global mean forcing of -1.3 Wm^{-2} with a North/South ratio of 1.6.

Chuang *et al.* (1994) (see Section 2.4.2.1) included a simplified representation of the effect of sulphate aerosol on cloud properties in their GCM study. They attributed -0.47 Wm^{-2} of their model derived forcing to the effect of sulphate on cloud albedo.

A different approach has been pursued by Boucher (1995) who used the low cloud droplet effective radii derived from satellite measurements by Han *et al.* (1994). Han *et al.* (1994) found that the mean droplet radius in the Northern Hemisphere was 11 mm, which was 0.7 mm lower than in the Southern Hemisphere. Boucher attributed this difference to anthropogenic aerosol production and computed a forcing of between -0.6 and -1.0 Wm^{-2}

averaged between 0 and 50°N. Boucher acknowledges the many simplifications inherent in this analysis; for example, other studies (such as Jones *et al.*, 1994) have derived a significant Southern Hemisphere forcing because the cleaner clouds there are more susceptible to changes in aerosol amounts. The method also assumes the interhemispheric differences are solely related to human activity. Boucher's values can be interpreted as the difference in anthropogenic forcing between the hemispheres. An estimate of the global mean forcing can then be made by applying the ratio of the forcing between the two hemispheres derived in the GCM studies and by assuming that Boucher's value for 0 to 50°N is representative of the entire Northern Hemisphere. Using an interhemispheric ratio of 3 together with Boucher's values yields a global mean forcing of between -0.6 and -1.0 Wm^{-2}; lower interhemispheric ratios would yield more negative global mean forcings.

Pincus and Baker (1994) have explored the possible effect of changes in droplet size on cloud thickness due to the suppression of precipitation. They use a simple model of the marine boundary layer. They find that the effect of precipitation suppression on cloud albedo can make the radiative forcing due to droplet effective radii changes alone more negative by between 50 and 200%. The applicability of such a model on larger scales is unclear but the conclusions reinforce those of the Boucher *et al.* (1995) GCM study.

These studies continue to indicate that the effect of aerosols on cloud droplet effective radius may be substantial although it remains very uncertain. We retain the range of 0 to -1.5 Wm^{-2} suggested in IPCC (1994). Our quantitative understanding is so limited at present that no mid-range estimate is given.

2.4.3 Stratospheric Aerosols

The volcanic eruption of Mt. Pinatubo in 1991 initiated a major global scale radiative forcing. The forcing was transient in nature reaching its most negative global mean value (-3 to -4 Wm^{-2}) in early 1992 (Hansen *et al.*, 1992; updated by Hansen *et al.*, 1995). Since that time, the stratospheric aerosol optical depths have declined and latest observations (e.g., Jäger *et al.*, 1995) indicate that the negative radiative forcing associated with Pinatubo aerosols has largely been removed. McCormick *et al.* (1995) and Hansen *et al.* (1995) have presented reviews of the atmospheric effects, including the radiative forcing, of the Pinatubo eruption. The climatic consequences of Pinatubo aerosols are discussed in Chapter 3.

Robock and Free (1995) use ice core acidity and sulphate records to deduce a new index of volcanic activity since 1850; they use 8 high latitude Northern Hemisphere cores, 5 high latitude Southern Hemisphere cores and 1 tropical core in their analysis. They compare their results with previous volcanic indices, including those of Khmelevtsov *et al.* (1996) and Sato *et al.* (1993). There is a general agreement between the records. In particular, even though there may have been no cumulative century scale trends in volcanic aerosol loading, there have been significant shorter time-scale variations in loading. Figure 2.15 shows estimates of the variation of the visible optical depth from Sato *et al.* (1993) and Robock and Free (1995). The associated radiative forcing is shown for the Sato *et al.* (1993) data (using the simple parametrization that the forcing in Wm^{-2} is -30 times the visible optical depth (see Lacis *et al.* (1992)). The forcing is not shown for the Robock and Free (1995) data as these are more representative of high latitudes and give added weight to the influence of high latitude eruptions; they are, therefore, less representative of global mean conditions. The period from about 1850 to 1920 was characterised by frequent eruptions of possible climatic significance; 1920 to 1960 was a period of reduced aerosol loading; and since 1960 the aerosol loading has on average again been higher. The decadal mean radiative forcings due to volcanic aerosols in the stratosphere may have varied by as much as 1.5 Wm^{-2} since 1850; hence such a variation can be large compared to the decadal-scale variation in any other known forcing.

These studies continue to support the conclusion that volcanic activity may be important in explaining some of the interdecadal variation in surface temperature during the instrumental record.

Changes in cirrus cloud properties as a result of the incursion of stratospheric aerosols into the upper troposphere are a possible source of additional radiative forcing, as mentioned in IPCC (1994) (see also Section 2.3.4). Sassen *et al.* (1995) and Wang *et al.* (1995) have provided further case-study evidence of cirrus modification by this mechanism. The extent of the effect, and even the sign of the resulting radiative forcing, remain unclear.

2.4.4 Solar Variability

IPCC (1994) reviewed direct observations of solar variability since 1978, presented correlative and theoretical studies of possible solar-climate connections, and described methods used to infer possible solar changes prior to the period of direct observations. Evidence suggests that the solar output was significantly lower during the Maunder Minimum (mainly in the 17th century), while recent observations and theoretical

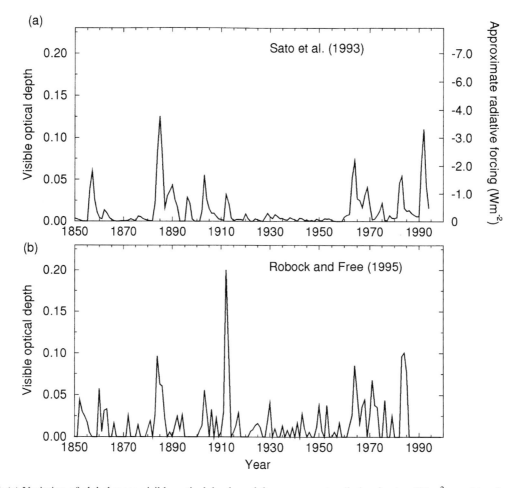

Figure 2.15: (a) Variation of global mean visible optical depth, and the consequent radiative forcing (Wm^{-2}) resulting from stratospheric aerosols of volcanic origin from 1850 to 1993, as estimated by Sato *et al.* (1993). The radiative forcing has been estimated using the simple relationship given in Lacis *et al.* (1992) where the radiative forcing is –30 times the visible optical depth. (b) Variation of visible optical depth from the ice core volcanic index of Robock and Free (1995) for 1850 to 1985. The Robock and Free (1995) index has been normalised so that their Northern Hemisphere mean agrees with the Sato *et al.* Northern Hemisphere optical depth for the mid-1880 Krakatau peak. Since Robock and Free's analysis is largely based on high latitude measurements, it is not likely to be representative of the global mean.

calculations imply that the radiative forcing due to changes in the Sun's output over the past century has been considerably smaller than anthropogenic forcing.

Estimates of the distant past and future role of solar variability for global radiative forcing are based upon observations and models of solar physics. For a concise review of solar processes and predictive models, see NRC (1994); solar-climate relationships have also been reviewed in Nesme-Ribes (1994). Empirical parametrizations have been developed to relate solar output fluctuations due to dark sunspots and bright faculae to indices such as the solar He I 1083 nm line (see, e.g., Lean *et al.*, 1992; Foukal, 1994a), while others have also considered possible changes in the solar diameter and rotation rate (e.g., Nesme-Ribes *et*

al., 1993) and comparisons of the behaviour of our Sun to other stars (Lockwood *et al.*, 1992). It is important to consider the age (Radick, 1994) and physical properties (e.g., photospheric magnetic structures) of such comparison stars and their relationship to the contemporary sun (Foukal, 1994b).

A recent paper by Zhang *et al.* (1994) helps to bracket the range of variability observed in sun-like stars and hence the likely past and future variability of our Sun. They noted that empirical models based upon sunspots and faculae do not account for all irradiance variations observed over an activity cycle (see also NRC (1994)) and base their correlation on an observed relationship between brightness and excess chromospheric emission, using the Ca II H and

K (396.8 nm and 393.3 nm) flux as the index. Using 33 sun-like stars, they estimate to 95% confidence that the solar brightness increase between the Maunder Minimum and the decade of the 1980s was likely to be 0.4 ± 0.2%. The lower limit of 0.2% (equivalent to a radiative forcing of 0.48 Wm^{-2}) agrees with the estimates of Lean *et al.* (1992) and Hoyt and Schatten (1993), while the upper range of 0.6% is in reasonable agreement with Nesme-Ribes *et al.* (1993) and corresponds to a radiative forcing of 1.4

Wm^{-2}. The radiative forcing since 1850 is likely to be no more than 50% of that since the Maunder Minimum (see IPCC 1994). This study thus provides support for the conclusions reached in IPCC (1994) that solar variations of the past century are highly likely to have been considerably smaller than the anthropogenic radiative forcings and expands that conclusion to show that the variations in solar output over the coming century are unlikely to exceed those observed since the Maunder Minimum.

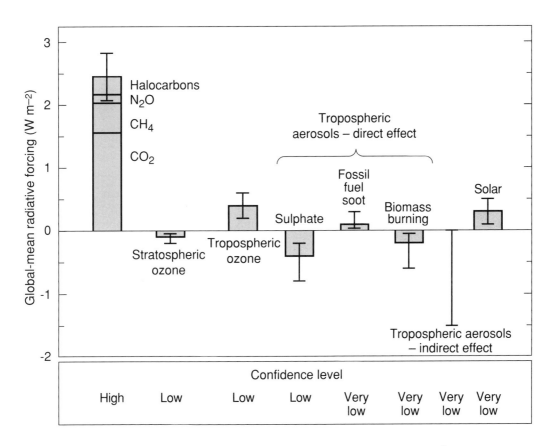

Figure 2.16: Estimates of the globally and annually averaged anthropogenic radiative forcing (in Wm^{-2}) due to changes in concentrations of greenhouse gases and aerosols from pre-industrial times to the present day and to natural changes in solar output from 1850 to the present day. The height of the rectangular bar indicates a mid-range estimate of the forcing whilst the error bars show an estimate of the uncertainty range, based largely on the spread of published values; our subjective confidence that the actual forcing lies within this error bar is indicated by the "confidence level". The contributions of individual gases to the direct greenhouse forcing is indicated on the first bar. The indirect greenhouse forcings associated with the depletion of stratospheric ozone and the increased concentration of tropospheric ozone are shown in the second and third bar respectively. The direct contributions of individual tropospheric aerosol components are grouped into the next set of three bars. The indirect aerosol effect, arising from the induced change in cloud properties, is shown next; our quantitative understanding of this process is very limited at present and hence no bar representing a mid-range estimate is shown. The final bar shows the estimate of the changes in radiative forcing due to variations in solar output. The forcing associated with stratospheric aerosols resulting from volcanic eruptions is not shown, as it is very variable over this time period; Figure 2.15 shows estimates of this variation. Note that there are substantial differences in the geographical distribution of the forcing due to the well-mixed greenhouse gases (CO$_2$, N$_2$O, CH$_4$ and the halocarbons) and that due to ozone and aerosols, which could lead to significant differences in their respective global and regional climate responses (see Chapter 6). For this reason, the negative radiative forcing due to aerosols should not necessarily be regarded as an offset against the greenhouse gas forcing.

2.4.5 Summary

Figure 2.16 shows our best estimates of the global mean radiative forcing due to a number of mechanisms since pre-industrial times. As was emphasised in IPCC (1994), the global mean forcing must be used with caution. In particular, the offset between positive and negative contributions to the global mean forcing does not take into account the fact that the regional distributions of the radiative forcing due, for example, to greenhouse gases and sulphate aerosols, are quite different. Thus even if, by coincidence, the global mean forcing were to be zero, due to a fortuitous cancellation between global mean forcings of opposite signs, this must not be taken to imply the absence of regional or possibly even global scale climate change.

It was also stressed in Section 4.7.1 of IPCC (1994) that different forcing mechanisms have different levels of confidence associated with them. The forcing due to the well-mixed greenhouse gases (CO_2, CH_4, N_2O and the CFCs) have the highest confidence, as they are based on well-characterised observations of the changes in gas concentration since pre-industrial times. On the other hand, estimates of the changes in concentrations of tropospheric ozone and tropospheric aerosols rely largely on model estimates which still have significant uncertainties. In Figure 2.16 the error bar indicates an estimate of the uncertainty range, based largely on the spread in the available literature; our subjective confidence that the actual forcing lies within this error bar is indicated by the "confidence level".

Our estimates of the contributions to radiative forcing since pre-industrial times due to changes in the well-mixed greenhouse gases are unchanged from those given in IPCC (1994) (i.e., a total of 2.45 Wm^{-2}); this remains the best determined radiative forcing. The estimates of the radiative forcing due to changes in ozone have also remained unchanged; these are 0.2 to 0.6 Wm^{-2} for tropospheric ozone and -0.1 Wm^{-2} (with a factor of 2 uncertainty) for stratospheric ozone. The most significant issue to be raised in work since IPCC (1994) concerns the consequences of stratospheric ozone depletion; the increased penetration of ultraviolet radiation into the troposphere may have a significant impact on tropospheric chemistry and a number of constituents of climatic relevance; as estimates of these effects are so preliminary, the IPCC (1994) forcing due to stratospheric ozone change, and the associated confidence level, are unaltered.

There has been a slight reduction in the uncertainty associated with our estimate of the range of the direct radiative forcing due to sulphate aerosols; the forcing is now estimated to be -0.4 Wm^{-2}, with a factor of two

uncertainty. In Figure 2.16, the contributions of different aerosol types to the direct aerosol forcing have been separated for clarity, although, due to interaction between them, the total direct aerosol forcing may not be the sum of the individual components. The contribution of soot aerosols is given as $+0.1$ Wm^{-2} with an uncertainty range of at least a factor of 3, although our confidence in this value is very low. The contribution of biomass burning remains at the IPCC (1994) value, with a central value of -0.2 Wm^{-2} with a factor of 3 uncertainty. Thus the total direct aerosol forcing is estimated to be -0.5 Wm^{-2} with a factor of 2 uncertainty.

The quantification of the effect of aerosols on cloud properties remains difficult. Although there have been further studies of the effect of aerosols on cloud properties (the indirect effect) the estimated range of 0 to -1.5 Wm^{-2} proposed in IPCC (1994) has not been changed. Since our quantitative understanding of this process is so limited at present, no mid-range estimate is given.

The estimate for the forcing due to changes in solar output since 1850 remains unchanged from the IPCC (1994) value of $+0.1$ to $+0.5$ Wm^{-2}. The available evidence indicates that natural variations in the radiative forcing, due to volcanic eruptions and changes in solar output, may have been important in determining some of the decadal scale variations in global climate over the past 150 years. Nevertheless, the cumulative radiative forcing due to human activity remains large compared to these and, on this evidence, this radiative forcing would be expected to have played a more significant role in determining the long-term trends in climate over the past 150 years.

It was noted in IPCC (1994) that changes in surface albedo could, potentially, contribute to radiative forcing; because of the problems discussed in IPCC (1994), it is still not possible to produce a reliable quantification of the contribution.

2.5 Trace Gas Radiative Forcing Indices

2.5.1 Introduction

Policymakers may need a means of estimating the relative radiative effects of various greenhouse gases. Chapter 5 of IPCC (1994) discussed in detail the current state of knowledge of trace gas radiative forcing indices used for this purpose. The definition and general limitations of a simple index were reviewed in Section 5.1 of that report. The choice of CO_2 as the reference molecule for calculations of the relative Global Warming Potential (GWP) and the resulting sensitivity of GWPs to incomplete knowledge of the response time of CO_2 were the subjects of Section 5.2.1. The sensitivities of GWPs to possible changes in the future composition of the Earth's atmosphere (Section 5.2.2.1) and future water vapour and clouds (Section 5.2.2.2) were shown to be of order 20%. GWPs representing the direct radiative forcing of 38 chemical species were given, including hydro-fluorocarbons, hydrochlorofluorocarbons, chlorofluoro-carbons, perfluorocarbons, N_2O and CH_4 (Section 5.2.4). The need to consider indirect effects on GWPs was also discussed (Section 5.2.5), and the indirect effects on the methane GWP were specifically estimated (Section 5.2.5.2). A typical uncertainty of these GWPs is ±35% relative to the CO_2 reference. The product of GWP and estimated current emissions for primary greenhouse gases (one approximate measure of the commitment to future radiative forcing from contemporary emissions) was presented for selected gases (Section 5.2.6). Choices of time horizon and the use of GWPs in policy formation were also considered.

In the short time since IPCC (1994) there have been a limited number of new scientific studies that expand upon knowledge of GWPs and their uncertainties; it is our goal to summarise briefly these new findings here. They pertain to three specific subjects:

(i) *net GWPs for ozone-depleting gases.* In IPCC (1994), only the *direct* GWPs for ozone-depleting gases (i.e., "warming" or positive radiative forcing) were presented. This current update describes an estimate of the *net* (i.e., direct "warming" and indirect "cooling" or negative radiative forcing) GWPs and uncertainties for ozone-depleting gases.

(ii) *updated or new GWPs for a number of key species due to improved or new estimates of atmospheric lifetimes (Section 2.2) or relative radiative forcing per unit mass (Section 2.4).* Three new gases have been added to the suite of GWP estimates presented below and the GWPs of several others have been improved.

Table 2.8: Net GWPs per unit mass emission for halocarbons including indirect effects for time horizons from 1990 (adapted from Daniel et al., 1995).

compound	Time Horizon = 2010 (i.e., after 20 years)					Time Horizon = 2090 (i.e., after 100 years)				
	Uncertainty in α		Uncertainty in cooling		Direct	Uncertainty in α		Uncertainty in cooling		Direct
	min	max	min	max		min	max	min	max	
CFC-11	2100	2900	1200	2900	4900	1300	1700	540	2100	3800
CFC-12	6400	6800	6000	6800	7800	6600	6800	6200	7100	8100
CFC-113	3300	3700	2800	3800	4900	3100	3300	2600	3600	4800
HCFC-22	3600	3700	3500	3700	4000	1300	1300	1300	1400	1500
HCFC-142b	3700	3800	3600	3800	4100	1600	1700	1600	1700	1800
H-1301	−97600	−22700	−31400	−14100	6100	−85400	−22400	−30700	−14100	5400
HCFC-141b	920	1200	660	1200	1800	270	350	170	370	600
CH_3Cl_3	−710	−420	−1000	−400	300	−220	−140	−320	−130	100
CCl_4	−1500	−550	−2600	−500	1900	−1500	−1100	−2400	−650	1400
HCFC-123	120	170	60	170	300	30	50	20	50	90
HCFC-124	1300	1400	1300	1400	1500	410	420	390	430	470

The CO_2 function from the Bern model was used for the decay function (slightly revised from IPCC (1994) as discussed in Section 2.1) and future concentrations of CO_2 were assumed to be constant. Lifetimes for HCFCs and CH_3Cl_3 are revised based upon new information on CH_3Cl_3 calibration discussed in detail in Section 2.2. The effects of the uncertainties in the relative efficiency of bromine versus chlorine (a) in ozone loss and in the magnitude of the ozone cooling are shown separately (see text). Direct GWPs are shown for comparison.

(iii) *expanded understanding of the difficulties in calculating GWPs for some chemicals.* A number of new studies have explored some of the scientific challenges related to the chemical production of tropospheric ozone, which is a key factor in estimates of GWPs that include indirect effects for several trace gases. These challenges are briefly reviewed here regarding their implications for the calculation of reliable GWPs.

2.5.2 Net GWPs for Ozone-Depleting Gases

Chlorofluorocarbons (CFCs) and halons effectively absorb infrared radiation, thereby contributing to the positive radiative forcing induced by greenhouse gases. However, their role in depleting stratospheric ozone also leads to a negative radiative forcing, particularly for ozone losses near the tropopause (Lacis *et al.*, 1990; Ramaswamy *et al.*, 1992). Ramaswamy *et al.* (1992) concluded that the globally averaged *decrease* in radiative forcing due to stratospheric ozone depletion including both infrared and solar effects represents an indirect effect that approximately balanced the globally averaged *increase* in direct radiative forcing due to halocarbons during the decade of the 1980s. Therefore, the *net* GWPs for ozone-depleting gases should consider both direct and indirect terms, together with their inherent uncertainties.

Daniel *et al.* (1995) estimated the indirect effects of ozone depletion upon the GWPs for halocarbons. They assumed that the indirect and direct radiative effects of halocarbons can be compared to one another in a globally averaged sense, an assumption that is being tested with two- and three-dimensional models (see Chapter 8 of WMO/UNEP, 1995; Chapter 4 of IPCC, 1994; Section 2.4 of this report; Molnar *et al.*, 1994). Daniel *et al.* (1995) combined estimates of the negative radiative forcing due to ozone depletion for the 1980s with an evaluation of past and future ozone loss contributions for each halocarbon. They assumed that the indirect negative forcing for each halocarbon depends linearly upon its contribution to stratospheric active chlorine or bromine release and examined the net radiative forcing that can be attributed to each gas. The resulting net GWPs are, of course, smaller than the earlier values consisting of the direct (positive) component alone (e.g., IPCC, 1994). However, a primary conclusion of their study was that some gases, such as the CFCs, are likely still to be a net positive forcing agent, while for compounds such as the halons and CH_3Br, the situation is reversed. This is due to the enhanced chemical effectiveness of brominated compounds compared to chlorinated species for ozone loss (see Chapter 13 of

WMO/UNEP, 1995). Carbon tetrachloride (CCl_4) and methyl chloroform (CH_3CCl_3), while not as effective as the bromocarbons for ozone destruction, contain several chlorine atoms per molecule and release them readily in the stratosphere, making them relatively effective ozone destroyers (and hence "cooling agents") as well.

Insofar as significant ozone loss likely occurs only for stratospheric active chlorine levels above a certain threshold, the total negative radiative forcing caused by any halocarbon depends upon the abundances of others and cannot be specified independent of past abundances and future scenarios (see Daniel *et al.*, 1995). This implies that GWPs for halocarbons based upon the indirect effects estimated for injection of an infinitesimally small amount of added gas cannot be used to calculate directly the total radiative impact of the true amount of that gas in the Earth's atmosphere; this limitation is similar to that for methane discussed in IPCC (1994). Further, the net GWPs for ozone depletors refer to an explicit period from a chosen start date (e.g., the present) to the end of the time horizon.

Daniel *et al.* (1995) considered the following uncertainties in deriving the net GWPs for halocarbons: (i) likely variations in the scenario for future concentrations of ozone-depleting gases (from WMO/UNEP, 1995), (ii) uncertainties in the globally averaged relative efficiency of bromine for ozone loss as compared to chlorine (α, assumed to lie between 40 and 200), and (iii) uncertainties in the magnitude of the cooling in the lower stratosphere due to uncertainties in the ozone loss profile (estimated by Schwarzkopf and Ramaswamy (1993) to be about $\pm30\%$). They found that the GWPs were not very sensitive to the adopted range of possible scenarios for future concentrations of halocarbons nor to the exact values of the threshold assumed. However, the GWPs for bromocarbons were found to be extremely sensitive to the chosen value of α, while those for CFCs were quite sensitive to the adopted uncertainty in the negative radiative forcing in the 1980s. Table 2.8 shows the uncertainty range for net halocarbon GWPs over 20- and 100-year time horizons from these sensitivity studies and compares them to GWPs for the direct effect only (adapted from Daniel *et al.* (1995) for the CO_2 denominator used here; see Section 2.1).

In addition to the indirect GWP component introduced by the ozone loss itself, there are other recognised indirect effects that may be associated with halocarbon GWPs. In the past year, several authors have noted that the impact of changing UV radiation due to ozone depletion upon OH and hence tropospheric chemistry could play a role in determining the total radiative cooling effect of ozone loss

Table 2.9: *Global Warming Potential (mass basis) referenced to the updated decay response for the adopted carbon cycle model (see Section 2.1) and future CO_2 atmospheric concentrations held constant at current levels. Typical uncertainties are about ±35%.*

Species	Chemical formula	Lifetime and reference	Global Warming Potential (Time Horizon)		
			20 years	100 years	500 years
CO_2	CO_2	Bern model, revised	1	1	1
HFC-23	CHF_3	264 (a)	9100	11,700	9800
HFC-32	CH_2F_2	5.6 (a)	2100	650	200
HFC-41	CH_3F	3.7 (a)	490	150	45
HFC-43-10mee	$C_5H_2F_{10}$	17.1 (a)	3000	1,300	400
HFC-125	C_2HF_5	32.6 (a)	4600	2,800	920
HFC-134	$C_2H_2F_4$	10.6 (a)	2900	1,000	310
HFC-134a	CH_2FCF_3	14.6 (a)	3400	1,300	420
HFC-152a	$C_2H_4F_2$	1.5 (a)	460	140	42
HFC-143	$C_2H_3F_3$	3.8 (a)	1000	300	94
HFC-143a	$C_2H_3F_3$	48.3 (a)	5000	3800	1400
HFC-227ea	C_3HF_7	36.5 (a)	4300	2900	950
HFC-236fa	$C_3H_2F_6$	209 (a)	5100	6300	4700
HFC-245ca	$C_3H_3F_5$	6.6 (a)	1800	560	170
Chloroform	$CHCl_3$	0.51 (a)	14	4	1
Methylene chloride	CH_2Cl_2	0.46 (a)	31	9	3
Sulphur hexafluoride	SF_6	3200 (b)	16300	23900	34900
Perfluoromethane	CF_4	50000 (b)	4400	6500	10000
Perfluoroethane	C_2F_6	10000 (b)	6200	9200	14000
Perfluoropropane	C_3F_8	2600 (b)	4800	7000	10100
Perfluorobutane	C_4F_{10}	2600 (b)	4800	7000	10100
Perfluoropentane	C_5F_{12}	4100 (b)	5100	7500	11000
Perfluorohexane	C_6F_{14}	3200 (b)	5000	7400	10700
Perfluorocyclobutane	$c\text{-}C_4F_8$	3200 (b)	6000	8700	12700
Methane**	CH_4	12.2±3 (a)	56	21	6.5
Nitrous oxide	N_2O	120 (c)	280	310	170
Trifluoroiodomethane	CF_3I	<0.005 (c)	<3	<1	<1

**The GWP for methane includes indirect effects of tropospheric ozone production and stratospheric water vapour production, as in IPCC (1994). The updated adjustment time for methane is taken from the discussion in Section 2.2 of this report.

(a) Based upon the revised CH_3Cl_3 lifetime from Prinn et al. (1995b) and updated chemical kinetic data where appropriate from JPL(1994), see Section 2.2 and Table 2.2. Also includes updated information relating to radiative forcing per molecule from Section 2.4 based upon Pinnock et al. (1995), where appropriate.

(b) See Section 2.4 and Roehl et al. (1995) for discussion of radiative forcing per molecule for these gases. Lifetimes as in IPCC (1994).

(c) As in IPCC (1994).

(see, e.g., Bekki *et al.*, 1994 and Toumi *et al.*, 1994) and hence would be a factor in the magnitude of halocarbon GWPs. The magnitudes of the negative radiative forcing associated with these newly identified mechanisms are currently considered quite uncertain (particularly that related to the OH role in indirect aerosol formation, see Section 2.4 of this report and Rodhe and Crutzen (1995)). Enhanced negative radiative forcing would also be associated with the large changes in ozone inferred from some satellite measurements near the tropical tropopause (see Chapter 1 of WMO/UNEP, 1995). These currently ill-quantified processes were not considered by Daniel *et al.* (1995) and are not included here. All of them would act to decrease the net halocarbon GWPs if they prove to be significant.

If improved understanding shows that the ozone depletion-related negative radiative forcing is larger than currently estimated, it would tend to reduce the GWPs for all ozone-depleting gases. However, this would not change the general conclusions regarding *relative* GWPs of ozone-depletors as indicated in Table 2.8. For example, if the magnitude of negative radiative forcing associated with ozone loss were to prove to be larger by a factor of two (see Toumi *et al.*, 1994), it would imply that the net GWPs for halons would become even more negative, that for CFC-11 would be close to zero, and that for CFC-12 would be reduced, but would still be positive. On the other hand, some studies suggest that the direct radiative forcings per molecule of halocarbons may be about 20% larger than current estimates (see Pinnock *et al.*, 1995). If the direct radiative forcings per molecule of halocarbons have indeed been underestimated by this amount, the net GWPs for halocarbons would increase. In no case would the increase in GWP exceed the increase in direct forcing, but would be considerably smaller for those gases (such as halons) with large negative contributions. The uncertainties in positive radiative forcing contributions are presently thought to be smaller than those relating to negative radiative forcing described above. The numbers presented in Table 2.8 should be considered illustrative of the relative values of GWPs among ozone-depleting gases, but their absolute magnitudes are presently subject to considerable quantitative uncertainties (at least ±50%). Current scientific evidence suggests that they are unlikely to be very much more positive than the values indicated, but could be more negative.

2.5.3 *Updated GWPs For Other Gases*
As was the practice in IPCC (1994), the same relative radiative forcings per unit mass are used for both the radiative forcing estimates of Section 2.4 and the GWPs presented here; they are summarised in Table 2.7 (Section 2.4.1.1). Note that overlapping absorptions are included (see Section 2.4 and IPCC, 1994). Sensitivities to possible future changes in CO_2, water vapour and clouds were shown to be of order ±20% in IPCC (1994). Since IPCC (1994), the molecular parameters for CF_4, C_2F_6, C_3F_8, C_4F_{10}, C_5F_{12}, and C_6F_{14} were reported in a consistent study by Roehl *et al.* (1995) and are considered the most appropriate to use now. In addition, Pinnock *et al.* (1995) presented a detailed and consistent laboratory and radiative transfer study of the radiative forcing per unit mass of a number of hydrofluorocarbons (HFCs). Improved radiative forcings per unit mass for HFC-32 and HFC-236fa are available from that work, and the forcing for HFC-41 was presented for the first time. Table 2.9 presents the resulting updates and additions.

The indirect contributions to the methane GWP due to tropospheric ozone and stratospheric water vapour production are included here, and the best estimate of the contributions to the total methane GWP from those sources remain the same as in IPCC (1994). The decay response for CO_2 has also been re-evaluated to refine the representation of the carbon cycle in Section 2.1. As in IPCC (1994), the decay response of the Bern carbon cycle model is used here for GWP calculations. Lifetimes are taken from Section 2.2 (Table 2.2) and include updated kinetic rate constants as well as lifetimes for methyl chloroform and related gases based on the study of Prinn *et al.* (1995b).

2.5.4 *Recent Studies Relating to the Challenges in Calculating GWPs for Some Chemical Species*
GWPs are simple, globally averaged indices with many limitations, as emphasised in Chapters 2 and 5 of IPCC (1994). As noted, key factors in GWP calculations include atmospheric lifetimes and distributions, as well as radiative properties. For a gas whose chemical lifetime is shorter than the time-scale for mixing in the troposphere (order of months within a hemisphere and a year globally), the distribution (and in turn, the chemical lifetime) can be strongly sensitive to local sources, sinks and transport. Current models are limited in their ability to reproduce faithfully the global transport and chemistry of short-lived gases in the troposphere, in part because of difficulties in representing physical and chemical processes occurring on relatively small spatial scales compared to the model resolution. These processes include the vertical transport due to deep convection (as shown by the [222]Rn model intercomparison described in Chapter 2 of IPCC (1994)) and boundary layer chemistry and transport. Accurate

representation of precipitation and liquid phase chemistry associated with clouds and rain also represent major challenges for two- and three-dimensional global models. In addition, aerosol optical effects on radiative forcing are spatially variable, insofar as they depend upon factors such as surface albedo. Due to these and other limitations (see Section 2.3 and Chapter 3 of IPCC (1994)), GWPs for the short-lived sulphur gases that are thought to contribute to negative radiative forcing through aerosol formation cannot be estimated with confidence at present and therefore were included neither in IPCC (1994) nor are they in this assessment.

Several anthropogenic gases are believed to contribute to the chemical production of tropospheric ozone which is central in the calculation of many of the indirect components of GWPs. Tropospheric ozone is a radiatively important gas (particularly in the upper troposphere), and there is evidence that its abundance has increased significantly in the past century in many locations because of human activities (see IPCC, 1994; Derwent, 1994). Emissions of nitrogen oxides (NO_x), non-methane hydrocarbons, carbon monoxide and methane are believed to play prominent roles in such trends. It would be highly desirable to determine the indirect GWPs associated with ozone production for these gases. However, with the important exception of methane, all of these compounds have very short atmospheric lifetimes of days or months, making confidence in model calculations of their spatial distributions and source/sink relationships considerably lower than those for long-lived gases (IPCC, 1994). The chemical production of tropospheric ozone and oxidative properties of the atmosphere are strongly sensitive to the NO_x abundance and emission rates; some authors have discussed these chemical non-linearities in detail (see e.g., Liu *et al.*, 1987; Sillman *et al.*, 1990; Thompson, 1992; Kleinman, 1994). The impact of emissions of short-lived gases upon the radiative forcing due to tropospheric ozone changes depends upon factors including the location of the emission and model-calculated OH and NO_x distributions, which in turn depend upon the uncertain convection, boundary layer, and chemical processes discussed above.

A number of recent studies have examined the sensitivity of tropospheric ozone production to emissions of short-lived gases using a range of models of varying resolution and illustrate the challenge of calculating GWPs for such gases. We briefly review their findings here. Thompson *et al.* (1994) examined the regional budget of carbon monoxide over the central USA and its role in ozone production using a detailed high-resolution model including deep convection based upon direct cloud observations. They concluded that deep convection over the central USA acts as a "chimney" for transport of carbon monoxide and other ozone precursors to the free troposphere and they estimated a gross production of ozone in the free troposphere of about 1 Gmol/day due to carbon monoxide transport and subsequent chemistry. Jacob *et al.* (1993a,b) used a continental-scale photochemical model including a subgrid nested scheme to arrive at a similar value of 1.2 Gmol/day (see the discussion of Jacob *et al.* in Thompson *et al.*, 1994). Jacob *et al.* (1993b) also found that direct export of ozone produced at low altitudes by pollution in the USA accounted for about 4.3 Gmol/day while NO_x export led to about 4.0 Gmol/day of free tropospheric ozone increase. All of these processes could contribute significantly to the hemispheric ozone budget as evidenced by comparison with the cross-tropopause transport of ozone over the entire Northern Hemisphere in summer of about 18–28 Gmol/day estimated by Jacob *et al.* (1993b). The studies of Jacob *et al.* (1993a,b) and Thompson *et al.* (1994) taken together underline the importance of convection and suggest that models with sufficient resolution and representation of physical processes may be able to predict key elements of tropospheric chemistry. Studies of other regions such as Europe and Asia (e.g., Berntsen *et al.*, 1996) show similar results. However, these model studies are constrained only by limited observations, and liquid phase chemistry in clouds is also important for some gases, (e.g., Lelieveld and Crutzen, 1994). Hence their uncertainties remain large.

Sensitivity studies show some of the non-linearities involved in ozone formation. Jacob *et al.* (1993b) found that reducing NO_x emissions by 50% from 1985 levels would reduce rural ozone concentrations over the eastern USA by about 15%, while reducing hydrocarbon emissions by 50% would have less than a 4% effect except in the largest urban plumes. Strand and Hov (1994) presented a global two-dimensional chemistry transport model that includes a parameterized treatment of convection (Strand and Hov, 1993). These authors found that a global 50% reduction in anthropogenic nitrogen oxide emissions led to about twice as large a reduction in ozone production as the same reduction in emissions of anthropogenic volatile organic compounds. While the findings of Strand and Hov (1994) are qualitatively similar to those reported by Jacob *et al.* (1993b), the differences between them directly illustrate the difficulty of quantitative estimation of global GWPs for such gases.

In contrast, the much longer lifetime of methane (in excess of several years) leads to a nearly well-mixed global tropospheric distribution and a lesser dependence of its

GWP upon these factors. Further, much of the chemical destruction of methane occurs in relatively clean regions in the tropics, improving confidence in the calculation of the methane adjustment time and ozone effects as compared to more short-lived gases (Chapter 2 of IPCC, 1994). Accordingly, the indirect GWP for methane was deemed estimatable in IPCC (1994), but not those for much shorter lived gases.

References

Ackerman, T.P. and O.B. Toon, 1981: Absorption of visible radiation in atmosphere containing mixtures of absorbing and nonabsorbing particles. *Appl. Opt.*, **20**, 3661–3668.

AFEAS (Alternative Fluorocarbons Environmental Acceptability Study), 1995: *Historic Production, Sales and Atmospheric Release of HCFC-142b*, SPA-AFEAS, Washington, DC, USA, pp. 1–14.

Alcamo, J., A. Bouwman, J. Edmonds, A. Grübler, T. Morita and A. Sugandhy, 1995: An Evaluation of the IPCC IS92 Emission Scenarios. In: *Climate Change 1994: Radiative Forcing of Climate Change and An Evaluation of the IPCC IS92 Emission Scenarios.*, J.T. Houghton, L.G. Meira Filho, J. Bruce, Hoesung Lee, B.A. Callander, E. Haites, N. Harris and K. Maskell (eds.), Cambridge University Press, Cambridge, UK.

Alkezweeny, A.J., D. Burrows and C. Grainger, 1993: Measurements of cloud-droplet size distributions in polluted and unpolluted stratiform clouds. *J. Appl. Met.*, **32**, 106–115.

Anderson, T.L., R.J. Charlson, W.H. White and P.H. McMurry, 1994: Comment on "Light scattering and cloud condensation nucleus activity of sulfate aerosol measured over the Northeast Atlantic Ocean" by D.A. Hegg *et al.*, *J. Geophys. Res.*, **99**, 25947–25949.

Andreae, M.O., 1995: Climate effects of changing atmospheric aerosol levels. In: *Future Climate of the World: A modelling perspective, World Survey and Climatology, Vol.XVI.*, A. Henderson-Sellers (ed.)., Elsevier, Amsterdam.

Bacastow, R.B., 1976: Modulation of atmospheric carbon dioxide by the Southern Oscillation. *Nature*, **261**, 116–118.

Bakwin, P., P. Tans and P. Novelli, 1994: Carbon monoxide budget in the Northern Hemisphere. *Geophys. Res. Lett.*, **21**, 433–436.

Barrett, J., 1995: The roles of carbon dioxide and water vapour in warming and cooling the Earth's troposphere. *Spectrochimica Acta*, **51A**, 415–417.

Bekki, S., K.S. Law and J.A. Pyle, 1994: Effects of ozone depletion on atmospheric CH_4 and CO concentrations. *Nature*, **371**, 595–597.

Berntsen, T., I.S.A. Isaksen, W.C. Wang and X.Z. Liang, 1996: Impacts from increased anthropogenic sources in Asia on tropospheric ozone and climate: A global 3-D model study. *Tellus*, (In press).

Blake, D.R. and F.S. Rowland, 1988: Continuing worldwide increase in tropospheric methane, 1978–1987. *Science*, **239**, 1129–1131.

Blanchet, J.P., 1982: Application of Chandrasekhar mean to aerosol optical properties. *Atmosphere-Ocean*, **20**, 189–206.

Bloomfield, P, 1994: Inferred lifetimes. In: *NASA Report on Concentrations, Lifetimes and Trends of CFCs, Halons and Related Species*, J. Kaye, S. Penkett and F. Ormond (eds.), NASA Reference Report No. **1339**, 3.1–3.23.

Boden, T.A., R.J. Sepanski and F.W. Stoss, 1991: Trends '91: A compendium of data on global change. Oak Ridge National Laboratory, *ORNL/CDIAC*-46.

Boers, R., G.P. Ayers and J.L. Gras, 1994: Coherence between the seasonal variation in satellite derived cloud optical depth and boundary layer CCN concentrations at a mid-latitude southern hemisphere station. *Tellus*, **46B**, 123–131.

Boucher, O., 1995: GCM estimate of the indirect aerosol forcing using satellite-retrieved cloud droplet effective radii. *J. Climate*, **8**, 1403–1409.

Boucher, O. and T.L. Anderson, 1995: GCM assessment of the sensitivity of direct climate forcing by anthropogenic sulfate aerosols to aerosol size and chemistry. *J. Geophys. Res.* **100**, 26061–26092.

Boucher, O. and U. Lohmann, 1995: The sulfate-CCN-cloud albedo effect. A sensitivity study with two general circulation models. *Tellus*, **47B**, 281–300.

Boucher, O., H. Le Treut and M.B. Baker, 1995: Precipitation and radiation modelling in a GCM: Introduction of cloud microphysical processes. *J. Geophys. Res.* **100**, 16395–16414.

Bouwman, A., K. van der Hoek and J. Olivier, 1995: Uncertainties in the global source distribution of nitrous oxide. *J. Geophys. Res.*, **100**, 2785–2800.

Broecker, W.S. and T.-H. Peng, 1994: Stratospheric contribution to the global bomb radiocarbon inventory: Model vs. observation. *Global Biogeochem. Cycles*, **8**, 377–384.

Broecker, W.S., T.-H. Peng and R. Engh, 1980: Modelling the carbon system. *Radiocarbon*, **22**, 565–598.

Broecker, W.S., T.-H. Peng, G. Ostlund and M. Stuiver, 1985: The distribution of bomb radiocarbon in the ocean. *J. Geophys. Res.*, **90**, 6953–6970.

Broecker, W.S., S. Sutherland, W. Smethie, T.-H. Peng and G. Oestlund, 1995: Oceanic radiocarbon: Separation of the natural and bomb components. *Global Biogeochem. Cycles*, **9**, 263–288.

Butler, J., 1994: The potential role of the ocean in regulating atmospheric methyl bromide. *Geophys. Res. Lett.*, **21**, 185–188.

Butler, J., J. Elkins, B. Hall, S. Cummings and S. Montzka, 1992: A decrease in the growth rates of atmospheric halon concentrations. *Nature*, **359**, 403–405.

Butler, J., J. Elkins, B. Hall, S. Montzka, S. Cummings, P. Fraser and L. Porter, 1994: Recent trends in the global atmospheric mixing ratios of halon-1301 and halon-1211. In: *Baseline 91*, A. Dick and J. Gras (eds.), 29–32, Bureau of Meteorology/CSIRO.

Chameides, W.L., P.S. Kasibhatala, J. Yienger and H. Levy II,

1994: Growth of continental-scale metro-agro-plexes, regional ozone pollution and world food production. *Science*, **264**, 74–77.

Chappellaz, J., J.M. Barnola, D. Raynaud, Y.S. Korotkevich and C. Lorius, 1990: Ice-core record of atmospheric methane over the past 160,000 years. *Nature*, **345**, 127–131.

Charlson, R.J., J. Langner and H. Rodhe, 1990: Sulfate aerosol and climate. *Nature*, **348**, 22.

Charlson, R.J., J. Langner, H. Rodhe, C.B. Leovy and S.G. Warren, 1991: Perturbation of the northern hemisphere radiative balance by backscattering from anthropogenic sulphate aerosols. *Tellus*, **43A-B**, 152–163.

Charlson, R. J., S.E. Schwartz, J.M. Hales, R.D. Cess, J.A. Coakley Jr., J.E. Hansen and D.J. Hoffman, 1992: Climate forcing by anthropogenic aerosols, *Science*, **255**, 423–430.

Chuang, C.C. and J.E. Penner, 1995: A study of the relationship between anthropogenic sulfate and cloud droplet nucleation. *Preprints, AMS Conf. on Cloud Physics, Dallas, January 1995*, pp. 493–497.

Chuang, C.C., J.E. Penner, K.E. Taylor and J.J. Walton, 1994: Climate effects of anthropogenic sulfate: simulation from a coupled chemistry, climate model. *Preprints of the Conference on Atmospheric Chemistry, Nashville, Tennessee, January 1994*, American Meteorological Society, Boston USA, pp. 170–174.

Chýlek, P. and J. Hallett, 1992: Enhanced absorption of solar radiation by cloud droplets containing soot particles in their surface. *Quart. J. R. Met. Soc.*, **118**, 167–172.

Chýlek, P. and J. Wong, 1995: Effect of absorbing aerosols on global radiation budget. *Geophys. Res. Lett.*, **22**, 929–931.

Chýlek, P., V. Ramaswamy, R. Pinnick and R. Cheng, 1981: Optical properties and mass concentration of carbonaceous aerosols. *Appl. Opt.*, **20**, 2980–2985.

Chýlek, P., V. Ramaswamy and R.J. Cheng, 1984: Effect of graphitic carbon on the albedo of clouds. *J. Atmos. Sci.*, **41**, 3076–3084.

Chýlek, P., G. Videen, D. Ngo, R.G. Pinnick and J.D. Klett, 1995: Effect of black carbon on the optical properties and climate forcing of sulfate aerosols. *J. Geophys. Res.*, **100**, 16325–16332.

Ciais, P., P.P. Tans, J.W.C. White, M. Trolier, R.J. Francey, J.A. Berry, D.R. Randall, P.J. Sellers, J.G. Collatz and D.S. Schimel, 1995a: Partitioning of ocean and land uptake of CO_2 as inferred by $\delta^{13}C$ measurements from the NOAA Climate Monitoring and Diagnostics Laboratory Global Air Sampling Network. *J. Geophys. Res.*, **100D**, 5051–5070.

Ciais, P., P.P. Tans, M. Trolier, J.W.C. White and R.J. Francey, 1995b: A large northern hemisphere terrestrial CO_2 sink indicated by $^{13}C/^{12}C$ of atmospheric CO_2. *Science*, **269**, 1098–1102.

Conway, T.J., P.P. Tans, L.S. Waterman, K.W. Thoning, D.R. Buanerkitzis, K.A. Maserie and N. Zhang, 1994: Evidence for interannual variability of the carbon cycle from the NOAA/CMDL global air sampling network. *J. Geophys. Res.*, **99D**, 22831–22855.

Cooke, W.F. and J.J.N. Wilson, 1996: A global black carbon aerosol model. *J. Geophys. Res.* (submitted).

Cooper, D.L., T.P. Cunningham, N.L.Allan and A. McCulloch, 1992: Tropospheric lifetimes of potential CFC replacements – rate co-efficients for reaction with the hydroxyl radical. *Atmos. Env.* **26A**, 1331–1334.

Cunnold, D., P. Fraser, R. Weiss, R. Prinn, P. Simmonds, B. Miller, F. Alyea and A. Crawford, 1994: Global trends and annual releases of CFC-11 and -12 estimated from ALE/GAGE and other measurements from July 1978 to June 1991. *J. Geophys. Res.*, **99**, 1107–1126.

Daniel, J., S. Solomon and D. Albritton, 1995: On the evaluation of halocarbon radiative forcing and global warming potentials. *J. Geophys. Res.*, **100**, 1271–1285.

Danielson, R.E., D.R. Moore and H.C. Van de Hulst, 1969: The transfer of visible radiation through clouds. *J. Atmos. Sci.*, **26**, 1078–1087.

Derwent, R. 1994: The estimation of global warming potentials for a range of radiatively active gases. In: *Non–CO_2 Greenhouse Gases*, J. van Ham, L.J.H.M. Janssen and R.J. Swart, (eds.), Kluwer Academic Publishers, Dordrecht, pp. 289–299.

Derwent, R.G., 1996: The influence of human activities on the distribution of hydroxyl radicals in the troposphere. *Phil. Trans. Roy. Soc. Lond.*, A, **354**, 1–30.

Derwent, R., P. Simmonds and W. Collins, 1994: Ozone and carbon monoxide measurements at a remote maritime location, Mace Head, Ireland, from 1990 to 1992. *Atmos. Env.*, **28**, 2623–2637.

Diederen, H.S.M.A., R. Guicherit and J.C.T. Hollander, 1985: Visibility reduction by air pollution in the Netherlands. *Atmos. Env.*, **19**, 377–383.

Dignon, J. and S. Hameed, 1989: Global emissions of nitrogen and sulfur oxides from 1860 to 1980. *JAPCA*, **39**, 180–186.

Dignon, J. and S. Hameed, 1992: Emission of nitrogen oxides and sulfur oxides from the former Soviet Union. *Ambio*, **21**, 481–482.

Dlugokencky, E., P. Lang and K. Masarie, 1994a: Flask measurements of methane. In: *CMDL No. 22 Summary Report 1993*, J. Peterson and R.Rossen, (eds.), U.S. Department of Commerce/NOAA/ERL, 22–24 .

Dlugokencky, E.J., K.A. Masarie, P.M. Lang, P.P. Tans, L.P. Steele and E.G. Nisbet, 1994b: A dramatic decrease in the growth rate of atmospheric methane in the Northern Hemisphere during 1992. *Geophys. Res. Lett.*, **21**, 45–48.

Dlugokencky, E.J., L.P. Steele, P.M. Lang and K.A. Masarie, 1994c: The growth rate and distribution of atmospheric methane. *J. Geophys. Res.*, **99**, 17021–17043.

Ehhalt, D.E. and F. Rohrer, 1994: The impact of commercial aircraft on tropospheric ozone, *Proc. 7th Priestley Conference*, 24–27 June 1994, Lewisburg, Pennsylvania.

Elkins, J., T. Thompson, T. Swanson, J. Butler, B. Hall, S. Cummings, D. Fisher and A. Raffo, 1993: Decrease in the growth rates of atmospheric chlorofluorocarbons-11 and -12. *Nature*, **364**, 780–783.

Elkins, J., J. Butler, S. Montzka, R. Myers, T. Baring, S. Cummings, G. Dutton, J. Giliga, A. Hayden, J. Lobert, T. Swanson, D. Hurst and C. Volk, 1994: Nitrous Oxide and Halocarbon Division Report. In: *CMDL No. 22 Summary Report 1993*, J. Peterson and R.Rossen (eds.) US Department of Commerce/NOAA/ERL, pp. 72–91.

Enting, I.G., 1995: Analysing the conflicting requirements of the framework convention on climate change. *Climatic Change*, **31**, 5–18.

Enting, I.G., T.M.L. Wigley and M. Heimann, 1994: Future emissions and concentrations of carbon dioxide: Key ocean/atmosphere/land analyses. *CSIRO Division of Atmospheric Research Technical Paper No. 31*, 120 pp.

Eswaran, H., E. Van den Berg and P. Reich, 1993: Organic carbon in soils of the world. *Soil Sci. Soc. America J.*, **57**, 192–194.

Evans, W.F.J. and E. Puckrin, 1995: An observation of the greenhouse radiation associated with carbon monoxide. *Geophys. Res. Lett.*, **22**, 925–928.

Feely, R.F., R. Wanninkhof, C.E. Cosca, P.P. Murphy, M.F. Lamb and M.D. Steckley, 1995: CO_2 distributions in the equatorial Pacific during the 1991–92 ENSO event. *Deep-Sea Research* II, **42**, 365–386.

Feichter, J., E. Kjellström, H. Rodhe, F. Dentener, J. Lelieveld and G-J. Roelofs, 1996: Simulation of the tropospheric sulfur cycle in a global climate model. *Atmos. Env.*, (In press).

Fisher, D. and P. Midgley, 1994: Uncertainties in the calculation of atmospheric releases of chlorofluorocarbons. *J. Geophys. Res.*, **99**, 16643–16650.

Foukal, P., 1994a: Stellar luminosity variations and global warming. *Science*, **247**, 556–558.

Foukal, P., 1994b: Response to Radick (1994). *Science*, **266**, 1073.

Francey, R.J., P.P. Tans, C.E. Allison, I.G. Enting, J.W.C. White and M. Trolier, 1995: Changes in oceanic and terrestrial carbon uptake since 1982. *Nature*, **373**, 326–330.

Fraser, P., S. Penkett, M. Gunson, R. Weiss and F.S. Rowland, 1994: Measurements. In: *Concentrations, Lifetimes and Trends of CFCs, Halons and Related species*, J. Kaye, S. Penkett and F. Ormond (eds.), NASA Report No. 1339, pp. 1.1–1.68.

Fuglestvedt, J.S., J.E. Jonson and I.S.A. Isaksen, 1994: Effects of reduction in stratospheric ozone on tropospheric chemistry through changes in photolysis rates. *Tellus*, **46B**, 172–192.

Goody, R.M. and Y.L. Yung, 1989: *Atmospheric Radiation*. Oxford University Press.

Graedel, T.E., D. Cariolle, M.A. Geller, J.L. Kerrebrock, D.H. Lister, K. Mauersberger, S.A. Penkett, U. Schmidt, S.E. Schwartz and S. Solomon, 1994: *Atmospheric Effects of Stratospheric Aircraft, an evaluation of NASA's interim assessment*. National Academy Press, Washington DC, USA.

Han, Q., W.B. Rossow and A.A. Lacis, 1994: Near-global survey of effective droplet radii in liquid water clouds using ISCCP data. *J. Climate*, **7**, 465–497.

Hansen, J., A. Lacis, R. Ruedy and M. Sato, 1992: Potential climate impact of Mt. Pinatubo eruption. *Geophys. Res. Lett.*, **19**, 215–218.

Hansen, J., M. Sato, R. Ruedy, A. Lacis, K. Asamoah, S. Borenstein, E. Brown, B. Cairns, G. Caliri, M. Campbell, B. Curran, S. de Castro, L. Druyan, M. Fox, C. Johnson, J. Lerner, M.P. McCormick, R. Miller, P. Minnis, A. Morrison, L. Pandolfo, I. Ramberran, F. Zaucker, M. Robinson, P. Russell, K. Shah, P. Stone, I. Tegen, L. Thomason, J. Wilder and H. Wilson, 1996: A Pinatubo climate modelling investigation. To appear in *The Effects of Mt. Pinatubo on the Atmosphere and Climate*, NATO ASI Series Volume, Subseries I, "*Global Environmental Change*, G. Fiocco, D. Fua and G. Visconti (eds.), Springer Verlag, Berlin.

Harris, N.R.P., G. Ancellet, L. Bishop, D.J. Hofmann, J.B. Kerr, R.D. McPeters, M. Préndez, W. Randel, J. Staehelin, B.H. Subbaraya, A. Volz-Thomas, J.M. Zawodny and C.S. Zerefos, 1995: Ozone Measurements. In: *Scientific Assessment of Ozone Depletion: 1994, Global Ozone Research and Monitoring Project Report No. 37*, World Meteorological Organisation, Geneva, pp. 1.1–1.54.

Harrison, K., W. Broecker and G. Bonani, 1993: A strategy for estimating the impact of CO_2 fertilisation on soil carbon storage. *Global Biogeochem. Cycles*, **7**, 69–80.

Hauglustaine, D.A., C. Granier, G.P. Brasseur and G. Megie, 1994: Impact of present aircraft emissions of nitrogen oxides on tropospheric ozone and climate forcing. *Geophys. Res. Lett.*, **21**, 2031–2034.

Hayasaka, T., M. Kuji and M. Tanaka, 1994: Air truth validation of cloud albedo estimated from NOAA advanced very high resolution radiometer data. *J. Geophys. Res.*, **99**, 18685–18694.

Haywood, J.M. and K.P. Shine, 1995: The effect of anthropogenic sulfate and soot aerosol on the clear sky planetary radiation budget. *Geophys. Res. Lett.*, **22**, 603–606.

Hegg, D.A., R.J. Ferek and P.V. Hobbs, 1993: Light scattering and cloud condensation nucleus activity of sulfate aerosol measured over the Northeast Atlantic Ocean. *J. Geophys. Res.*, **98**, 14887–14894.

Hegg, D.A., R.J. Ferek and P.V. Hobbs, 1994: Reply to comment by Anderson *et al. J. Geophys. Res.*, **99**, 25951–25954.

Heintzenberg, J., 1978: Light scattering parameters of internal and external mixtures of soot and non-absorbing material in atmospheric aerosols. *Proc. Conf. on Carbonaceous Particles in the Atmosphere*, Berkeley, California, 1978, pp. 278–281.

Hesshaimer, V., M. Heimann and I. Levin, 1994: Radiocarbon evidence suggesting a smaller oceanic CO_2 sink than hitherto assumed. *Nature*, **370**, 201–203.

Hindman, E.E. and R. Bodowski, 1994: A marine stratus layer modified by ship-produced CCN and updrafts. *Preprints, 6th WMO Scientific Conference on Weather Modification, Sienna*.

Hindman, E.E., W.M. Porch, J.G. Hudson and P.A. Durkee, 1995: Ship-produced cloud lines of 13 July 1991. *Atmos. Env.*, **28**, 3393–3403.

Hogan, K. and R. Harriss, 1994: Comment on "A dramatic decrease in the growth rate of atmospheric methane in the

Northern Hemisphere during 1992" by E. J. Dlugokencky *et al. Geophys. Res. Lett.*, **21**, 2445–2446.

Hoyt, D.V. and K.H. Schatten, 1993: A discussion of plausible solar irradiance variations, 1700–1992. *J. Geophys. Res.*, **98** (A11), 18895–18906.

INPE, 1992: *Deforestation in Brazilian Amazonia.* Instituto Nacional de Pesquisas Especiais, Sao Paulo, Brazil.

IPCC, 1990: *Climate Change: The IPCC Scientific Assessment,* J.T. Houghton, G.J. Jenkins and J.J. Ephraums (eds.). Cambridge University Press, Cambridge, UK.

IPCC, 1992: *Climate Change 1992: The Supplementary Report to the IPCC Scientific Assessment,* J.T. Houghton, B.A. Callander and S.K. Varney (eds.). Cambridge University Press, Cambridge, UK.

IPCC, 1994: *Climate Change 1994: Radiative Forcing of Climate Change and an Evaluation of the IPCC IS92 Emission Scenarios,* J.T. Houghton, L.G. Meira Filho, J. Bruce, Hoesung Lee, B.A. Callander, E.F. Haites, N. Harris and K. Maskell (eds.). Cambridge University Press, Cambridge, UK.

Irion, F., M. Brown, G. Toon and M. Gunson, 1994: Increase in atmospheric HCFC-22 over southern California from 1985–1990. *Geophys. Res. Lett.*, **21**, 1723–1726.

Jacob, D.J., J.A. Logan, R.M. Yevich, G.M. Gardner, C.M. Spivakovsky, S.C. Wofsy, J.W. Munger, S. Sillman, M.J. Prather, M.O. Rodgers, H. Westberg and P.R. Zimmerman, 1993a: Simulation of summertime ozone over North America. *J. Geophys. Res.*, **98**, 14797–14816.

Jacob, D.J., J.A. Logan, G.M. Gardner, R.M. Yevich, C.M. Spivakovsky, S.C. Wofsy, S. Sillman and M.J. Prather, 1993b: Factors regulating ozone over the United States and its export to the global atmosphere. *J. Geophys. Res.,* **98**, 14817–14826.

Jäger, H., O. Uchino, T. Nagai, T. Fujimoto, V. Freudenthaler and F. Homburg, 1995: Ground-based remote sensing of the decay of the Pinatubo eruption cloud at 3 northern hemispheric sites. *Geophys. Res. Lett.*, **22**, 607–610.

Jain, A.K., H.S. Kheshgi, M.I. Hoffert and D.J. Wuebbles, 1995: Distribution of radiocarbon as a test of global carbon cycle models. *Global Biogeochem. Cycles*, **9**, 153–166.

Jensen, E.J. and O.B. Toon, 1994: Ice nucleation in the upper troposphere: sensitivity to aerosol number density, temperature, and cooling rate. *Geophys. Res. Lett.*, **21**, 2019–2022.

Jonas, P.R., R.J. Charlson and H. Rodhe, 1995: Aerosols. In: *Climate Change 1994: Radiative Forcing of Climate Change and an Evaluation of the IPCC IS92 Emission Scenarios,* J.T. Houghton, L.G. Meira Filho, J. Bruce, Hoesung Lee, B.A. Callander, E. Haites, N. Harris and K. Maskell (eds). Cambridge University Press, Cambridge, UK.

Jones, A., D.L. Roberts and A. Slingo, 1994: A climate model study of the indirect radiative forcing by anthropogenic sulphate aerosols. *Nature,* **370**, 450–453.

Joos, F., M. Bruno, R. Fink, U. Siegenthaler, T. Stocker, C. Le Quéré and J.L. Sarmiento, 1996: An efficient and accurate representation of complex oceanic and biospheric models of anthropogenic carbon uptake. *Tellus* (In press).

JPL, 1994: Chemical Kinetics and Photochemical Data for Use in Stratospheric Modelling, *Jet Propulsion Laboratory report 94–26.*

Kaduk, J. and M. Heimann, 1994: The climate sensitivity of the Osnabrück Biosphere Model on the ENSO timescale. *Ecological Modelling*, **75/76**, 239–256.

Keeling, C.D., S.C. Piper and M. Heimann, 1989a: A three-dimensional model of atmospheric CO_2 transport based on observed winds: 4. Mean annual gradients and interannual variations. In: *Aspects of Climate Variability in the Pacific and Western Americas, Geophysical Monograph 55,* D.H. Peterson (ed.), American Geophysical Union, Washington, DC, pp. 305–363.

Keeling, C.D., R.B. Bacastow, A.F. Carter, S.C. Piper, T.P. Whorf, M. Heimann, W.G. Mook and H. Roeloffzen, 1989b: A three-dimensional model of atmospheric CO_2 transport based on observed winds: 1. Analysis and observational data. In: *Aspects of Climate Variability in the Pacific and Western Americas. Geophysical Monograph 55,* D.H. Peterson (ed.), American Geophysical Union, Washington, DC, pp. 165–236.

Keeling, C.D., T.P. Whorf, M. Wahlen and J. van der Plicht, 1995: Interannual extremes in the rate of rise of atmospheric carbon dioxide since 1980. *Nature*, **375**, 666–670.

Khalil, M.A.K. and R.A. Rasmussen, 1992: The global sources of nitrous oxide. *J. Geophys. Res.*, **97**, 14651–14660.

Khalil, M.A.K. and R.A. Rasmussen, 1994: Global decrease of atmospheric carbon monoxide concentration. *Nature*, **370**, 639–641.

Khalil, M.A.K., R.A. Rasmussen and R. Gunawardena, 1993: Atmospheric methyl bromide: trends and global mass balance. *J. Geophys. Res.*, **98**, 2887–2896.

Khmelevtsov, S.S., D.I. Busygina and Y.G. Kaufman, 1996: Modelling the spatial and temporal distribution and optical characteristics of stratospheric aerosols of volcanic origin. *J. Geophys. Res.* (Submitted).

Kiehl, J.T. and B.P. Briegleb, 1993: The relative role of sulfate aerosols and greenhouse gases in climate forcing. *Science*, **260**, 311–314.

Kiehl, J.T. and H. Rodhe, 1995: Modeling geographical and seasonal forcing due to aerosols. In *Aerosol Forcing of Climate*, R.J. Charlson and J. Heintzenberg (eds.), John Wiley, Chichester, pp. 281–296.

King, G. and S. Schnell, 1994: Effect of increasing atmospheric methane concentration on ammonium inhibition of soil methane consumption. *Nature*, **370**, 282–284.

Kinnison, D.E., K.E. Grant, P.S. Connell, D.A. Rotman and D.J. Wuebbles, 1994: The chemical and radiative effects of the Mount Pinatubo eruption. *J. Geophys. Res.*, **99**, 25705–25731.

Kleinman, L., 1994: Low and high NO_x tropospheric photochemistry. *J. Geophys. Res.*, **99**, 16831–16838.

Ko, M., N. Sze, W-C. Wang, G. Shia, A. Goldman, F. Murcray, D. Murcray and C.Rinsland, 1993: Atmospheric sulfur hexafluoride: sources, sinks and greenhouse warming. *J. Geophys. Res.*, **98**, 10499–10507.

Komhyr, W.D., R.H. Gammon, T.B. Harris, L.S. Waterman, T.J. Conway, W.R. Taylor and K.W. Thoning, 1985: Global atmospheric CO_2 distribution and variations from 1968–1982 NOAA/GMCC CO_2 flask sample data. *J. Geophys. Res.*, **90**, 5567–5596.

Kulmala, M., R. Korhonen, A. Laaksonen and T. Vesala, 1995: Changes in cloud properties due to NO_x emissions. *Geophys. Res. Lett.*, **22**, 239–242.

Kunde, V.G., B.J. Conrath, R.A. Hanel, W.C. Maguire, C. Prabhakara and V.V. Salomonson, 1974: The Nimbus-4 Infrared Spectroscopy Experiment, 2. Comparison of observed and theoretical radiances from 425 to 1450 cm^{-1}. *J. Geophys. Res*, **79**, 777–784.

Lacis, A.A., D.J. Wuebbles and J.A. Logan, 1990: Radiative forcing by changes in the vertical distribution of ozone. *J. Geophys. Res.*, **95**, 9971–9981.

Lacis, A., J. Hansen and M. Sato, 1992: Climate forcing by stratospheric aerosols. *Geophys. Res. Lett.*, **19**, 1607–1610.

Langner, J. and H. Rodhe, 1991: A global three-dimensional model of the global sulfur cycle. *J. Atmos. Chem.*, **13**, 255–263.

Lean, J., A. Skumanich and O. White, 1992: Estimating the Sun's radiative output during the Maunder minimum. *Geophys. Res. Lett.*, **19**, 1591–1594.

Lelieveld, J. and P.J. Crutzen, 1994: Role of deep cloud convection in the ozone budget of the troposphere. *Science*, **264**, 1759–1761.

Levin, I., B. Kromer, H. Schoch-Fischer, M. Bruns, M. Münnich, D. Berdau, J.C. Vogel and K.O. Münnich, 1985: 25 years of tropospheric ^{14}C observations in Central Europe. *Radiocarbon*, **27**, 1–19.

Li, X., X. Zhou, W. Li and L. Chen, 1995: The cooling of Sichuan province in recent 40 years and its probable mechanisms. *Acta Meteorologica Sinica*, **9**, 57–68.

Li, Z. and P. Varanasi, 1994: Measurements of the absorption cross-sections of CFC-11 at conditions representing various model atmospheres. *J. Quant. Spectrosc. Radiat. Transfer*, **52**, 137–144.

Liepert, B., P. Fabian and H. Grassl, 1994: Solar radiation in Germany – observed trends and an assessment of their causes. Part I: Regional approach. *Beitr. Phys. Atmosph.*, **67**, 15–29.

Liu, S.C., M. Trainer, F.C. Fehsenfeld, D.D. Parrish, E.J. Williams, D.W. Fahey, G. Huebler and P.C. Murphy, 1987: Ozone production in the rural troposphere and the implications for regional and global ozone distributions. *J. Geophys. Res.*, **92**, 4191–4207.

Lobert, J., J. Butler, S. Montzka, L. Geller, R. Myers and J. Elkins, 1995: A net sink for atmospheric methyl bromide in the east Pacific Ocean. *Science*, **267**, 1002–1005.

Lockwood, G.W., B.A. Skiff, S.L. Baliunas and R.R. Radick, 1992: Long-term solar brightness changes estimated from a survey of Sun-like stars. *Nature*, **360**, 653–655.

Logan, J.A., 1994: Trend in the vertical distribution of ozone: an analysis of ozonesonde data. *J. Geophys. Res.*, **99**, 25553–25585.

Lowe, D., C. Brenninkmeijer, G. Brailsford, K. Lassey and A.

Gomez, 1994: Concentration and ^{13}C records of atmospheric methane in New Zealand and Antarctica: evidence for changes in methane sources. *J. Geophys. Res.*, **99**, 16913–16925.

Lubin, D., 1994: Infrared radiative properties of the maritime Antarctic atmosphere. *J. Climate*, **7**, 121–140.

Madronich, S. and C. Granier, 1992: Impact of recent total ozone changes on tropospheric ozone photodissociation, hydroxyl radicals and methane trends. *Geophys. Res. Lett.*, **19**, 465–467.

Maiss, M. and I. Levin, 1994: Global increase in sulfur hexafluoride observed in the atmosphere. *Geophys. Res. Lett.*, **21**, 569–572.

Maiss, M., P. Steele, R. Francey, P. Fraser, R. Langenfelds, N. Trivett and I. Levin, 1996: Sulfur hexafluoride: a powerful, new atmospheric tracer, *Atmos. Env.* (in press)

Mak, J., C. Brenninkmeijer and J. Tamaresis, 1994: Atmospheric ^{14}CO observations and their use for estimating carbon monoxide removal rates. *J. Geophys. Res.*, **99**, 22915–22922.

McCormick, M.P., L.W. Thomason and C.R. Trepte, 1995: Atmospheric effects of the Mt. Pinatubo eruption. *Nature*, **373**, 399–404.

Midgley, P. and A. McCulloch, 1995: The production and global distribution of emissions to the atmosphere of 1,1,1-trichloroethane (methyl chloroform). *Atmos. Env.*, **29**, 1601–1608.

Miyoshi, A., S. Hatakeyama and N. Washida, 1994: OH radical-initiated photooxidation of isoprene: an estimation of global CO production. *J. Geophys. Res.*, **99**, 18779–18787.

Molnar, G.I., M.K W. Ko, S. Zhou and N.D. Sze, 1994: Climatic consequences of observed ozone loss in the 1980s: Relevance to the greenhouse problem. *J. Geophys. Res.*, **99**, 25755–25760.

Montzka, S., R.Myers, J. Butler, J. Elkins and S. Cummings, 1993: Global tropospheric distribution and calibration scale of HCFC-22. *Geophys. Res. Lett.*, **20**, 703–706.

Montzka, S., M. Nowick, R. Myers, J. Elkins, J. Butler, S. Cummings, P. Fraser and L. Porter 1994a: NOAA-CMDL chlorodifluoromethane observations at Cape Grim. In: *Baseline 91*, A. Dick and J. Gras (eds.), Bureau of Meteorology/CSIRO, pp. 25–28.

Montzka, S., R. Myers, J. Butler and J. Elkins, 1994b: Early trends in the global tropospheric abundance of hydrochlorofluorocarbon-141b and -142b. *Geophys. Res. Lett.*, **21**, 2483–2486.

Morris, R., T. Miller, A. Viggiano, J. Paulson, S. Solomon and G. Reid, 1995: Effects of electron and ion reactions on atmospheric lifetimes of fully fluorinated compounds. *J. Geophys. Res.*, **100**, 1287–1294.

Müller, M., A. Kraus and A. Hofzumahaus, 1995: $O_3O(^1D)$ photolysis frequencies determined from spectroradiometric measurements of solar actinic UV-radiation: comparison with chemical actinometer measurements. *Geophys. Res. Lett.*, **22**, 679–682.

Neftel, A., H. Oeschger, T. Staffelbach and B. Stauffer, 1988: CO_2 record in the Byrd ice core 50,000–5,000 years BP. *Nature*, **331**, 609–611.

Nesme-Ribes, E., 1994: *The Solar Engine and Its Influence on Terrestrial Atmosphere and Climate.* NATO ASI Series, Springer, New York.

Nesme-Ribes, E., E.N. Ferreira, R. Sadourny, H. Le Treut and Z.X. Li, 1993: Solar dynamics and its impact on solar irradiance and the terrestrial climate. *J. Geophys. Res., 98*, 18923–18935.

Novakov, T. and J.E. Penner, 1993: Large contribution of organic aerosols to cloud condensation nuclei concentrations. *Nature, 365*, 823–826.

Novakov, T., C. Rivera-Carpio, J.E. Penner and C.F. Rogers, 1994: The effect of anthropogenic sulfate aerosols on marine droplet concentrations, *Tellus, 46B*, 132–141.

Novelli, P., J. Collins, R. Myers, G. Sachse and H. Scheel, 1994a: Re-evaluation of the NOAA/CMDL carbon monoxide reference scale and comparisons with CO reference gases at NASA-Langley and at the Fraunhofer-Institut. *J. Geophys. Res., 99*, 18779–18787.

Novelli, P., K. Masarie, P. Tans and P. Lang, 1994b: Recent changes in atmospheric carbon monoxide. *Science, 263*, 1587–1590.

NRC, 1994: *Solar Influences on Global Change. A National Research Council report.* National Academy Press, Washington DC, USA.

Oeschger, H., U. Siegenthaler and A. Guglemann, 1975: A box-diffusion model to study the carbon dioxide exchange in nature. *Tellus, 27*, 168–192.

Ogren, J. A., 1982: Deposition of particulate elemental carbon from the atmosphere. In: *Particulate carbon: Atmospheric Life Cycle*, G.T. Wolff and R.L. Klimisch (eds.), Plenum Press, New York, 379–391.

Oltmans, S.J. and D.J. Hofmann, 1995: Increase in lower stratospheric water vapour at a mid-latitude northern hemisphere site from 1981 to 1994. *Nature, 374*, 146–149.

Oram, D., P. Fraser, C. Reeves and S. Penkett, 1995: Measurements of HCFC-142b and HCFC-141b in the Cape Grim air archive: 1978–1993. *Geophys. Res. Lett., 22*, 2741–2744.

Penner, J.E., 1995: Carbonaceous aerosols influencing atmospheric radiation: Black and organic carbon. In: *Aerosol Forcing of Cimate*, R.J. Charlson and J. Heintzenberg (eds.). John Wiley and Sons Ltd., Chichester, pp.91–108.

Penner, J.E., R.E. Dickinson and C.A. O'Neill, 1992: Effects of aerosol from biomass burning on the global radiation budget. *Science, 256*, 1432–1434.

Penner, J.E., H. Eddleman and T. Novakov, 1993: Towards the development of a global inventory for black carbon emissions. *Atmos. Env., 27A*, 1277–1295.

Penner, J.E., C.S. Atherton and T.E. Graedel, 1994: Global emissions and models of photochemically active compounds. To appear in *37th OHOLO Conference Series*, Plenum Press, New York.

Perry, K.D. and P.V. Hobbs, 1994: Further evidence for particle nucleation in clear air adjacent to marine cumulus clouds. *J. Geophys. Res., 99*, 22803–22818.

Pham, M., G. Mégie, J.F. Müller, G. Brasseur and C. Granier, 1995: A three-dimensional study of the tropospheric sulfur cycle. *J. Geophys. Res., 100*, 26061–26092.

Pincus, R. and M.B. Baker, 1994: Effect of precipitation on the albedo susceptibility of clouds in the marine boundary layer. *Nature, 372*, 250–252.

Pinnock, S., M.D. Hurley, K.P. Shine, T.J. Wallington and T.J. Smyth, 1995: Radiative forcing of climate by hydrochlorofluorocarbons and hydrofluorocarbons. *J. Geophys. Res., 100*, 23227–23238.

Poppe, D. J. Zimmermann, R. Bauer, T. Brauers, D. Brüning, J.Callies, H.-P. Dorn, A. Hofzumahaus, F.-J. Johnen, A. Khedim, H.Koch, R. Koppmann, H. London, K.-P. Müller, R. Neuroth, C. Plass-Dülmer, U. Platt, F. Rohrer, E.-P. Röth, J. Rudolph, U.Schmidt, M. Wallasch and D. Ehhalt, 1994: Comparison of measured OH concentrations with model calculations. *J. Geophys. Res., 99*, 16633–16642.

Post, W.M., J. Pastor, P.J. Zinke and A.G. Stangenberger, 1985: Global patterns of soil nitrogen storage. *Nature, 317*, 613–616.

Potter, C.S., J.T. Randerson, C.B. Field, P.A. Matson, P.M. Vitousek, H.A. Mooney and S.A. Klooster, 1993: Terrestrial ecosystem production: A process model based on global satellite and surface data. *Global Biogeochem. Cycles, 7*, 811–841.

Prather, M.J., 1988: *An Assessment Model for Atmospheric Composition*, NASA Conf. Publ. CP-3203, 64 pp.

Prather, M.J., 1994: Lifetimes and eigenstates in atmospheric chemistry. *Geophys. Res. Lett., 21*, 801–804.

Prather, M.J. and C.M. Spivakovsky, 1990: Tropospheric OH and the lifetimes of hydrochlorofluorocarbons (HCFCs). *J. Geophys. Res., 95*, 18723–18729.

Prather, M., R. Derwent, D. Ehhalt, P. Fraser, E. Sanhueza and X. Zhou, 1995: Other trace gases and atmospheric chemistry. In: *Climate Change 1994: Radiative Forcing of Climate Change and an Evaluation of the IPCC IS92 Emission Scenarios,* J.T. Houghton, L.G. Meira Filho, J. Bruce, Hoesung Lee, B.A. Callander, E. Haites, N. Harris and K. Maskell (eds.), Cambridge University Press, Cambridge, UK.

Prinn, R., D. Cunnold, R. Rasmussen, P. Simmonds, F.Alyea, A.Crawford, P. Fraser and R. Rosen, 1990: Atmospheric emissions and trends of nitrous oxide deduced from 10 years of ALE-GAGE data. *J. Geophys. Res., 95*, 18369–18385.

Prinn, R., D. Cunnold, P. Simmonds, F. Alyea, R. Boldi, A. Crawford, P. Fraser, D. Gutzler, D. Hartley, R. Rosen and R. Rasmussen, 1992: Global average concentration and trend for hydroxyl radicals deduced from ALE/GAGE trichloroethane (methylchloroform) data for 1978–1990. *J. Geophys. Res., 97*, 2445–2461.

Prinn, R., R. Weiss, P. Fraser, P. Simmonds, F. Alyea and D. Cunnold, 1995a: DOE-CDIAC World Data Center (Internet: cdp@ornl.gov) dataset no. DB-1001.

Prinn, R., R. Weiss, B. Miller, J. Huang, F. Alyea, D. Cunnold, P. Fraser, D. Hartley and P. Simmonds, 1995b: Atmospheric trends and lifetime of trichloroethane and global average hydroxyl radical concentrations based on 1978–1994 ALE/GAGE measurements. *Science*, **269**, 187–192.

Radick, R.R., 1994: Stellar variability and global warming. *Science*, **266**, 1072.

Radke, L.F., D.A. Hegg, P.V. Hobbs, J.D. Nance, J.H. Lyons, K.K. Laursen, R.E. Weiss, P.J. Riggan and D.E. Ward, 1991: Particulate and trace gas emission from large biomass fires in North America. In:*Global Biomass Burning*, J.S. Levine (ed.), MIT Press, pp. 209–224.

Ramaswamy, V., M.D. Schwarzkopf and K.P. Shine, 1992: Radiative forcing of climate from halocarbon-induced global stratospheric ozone loss. *Nature*, **355**, 810–812.

Revelle, R. and H.E. Suess, 1957: Carbon dioxide exchange between atmosphere and ocean and the question of an increase of atmospheric CO_2 during the past decades. *Tellus*, **9**, 18–27.

Robock, A. and M.P. Free, 1995: Ice cores as an index of global volcanism from 1850 to the present. *J. Geophys. Res.*, **100**, 11549–11568.

Robock, A. and H.F. Graf, 1994: Effects of preindustrial human activities on climate. *Chemosphere*, **29**, 1087–1099.

Rodhe, H. and P. Crutzen, 1995: Climate and CCN. *Nature*, **375**, 111.

Roehl, C.M., D. Boglu, C. Brühl and G.K. Moortgat, 1995: Infrared band intensities and global warming potentials of CF_4, C_2F_6, C_3F_8, C_4F_{10}, C_5F_{12} and C_6F_{14}. *Geophys. Res. Lett.*, **22**, 815–818.

Sarmiento, J.L., C. Le Quéré and S.W. Pacala, 1995: Limiting future atmospheric carbon dioxide. *Global Biogeochem. Cycles*, **9**, 121–137.

Sassen, K., D.O'C. Starr, G.G. Mace, M.R. Poellot, S.H. Melfi, W.L. Eberhard, J.D. Spinhirne, E.W. Eloranta, D.E. Hagen and J. Hallett, 1995: The 5–6 December 1991 FIRE IFO II jet stream cirrus case study: possible influences of volcanic aerosols. *J. Atmos. Sci.*, **52**, 97–123.

Sato, M., J.E. Hansen, M.P. McCormick and J.B. Pollack 1993: Stratospheric aerosol optical depths, 1850–1990. *J. Geophys. Res*, **98**, 22987–22994.

Schauffler, S.M. and J.S. Daniel, 1994: On the effects of stratospheric circulation changes on trace gas trends. *J. Geophys. Res.*, **99**, 25747–25754.

Schauffler, S.M., W.H. Pollock, E.L. Atlas, L.E. Heidt and J.S. Daniel, 1995: Atmospheric distributions of HCFC-141b. *Geophys. Res. Lett.*, 22, 819–822.

Schimel, D.S. and E. Sulzman, 1995: Variability in the earth climate system: Decadal and longer timescales. *Reviews of Geophysics*, Supplement July 1995, 873–882.

Schimel, D.S., B.H. Braswell Jr., E.A. Holland, R. McKeown, D.S. Ojima, T.H. Painter, W.J. Parton and A.R. Townsend, 1994: Climatic, edaphic and biotic controls over carbon and turnover of carbon in soils. *Global Biogeochem. Cycles*, **8**, 279–293.

Schimel, D., I. Enting, M. Heimann, T.M.L. Wigley, D. Raynaud, D. Alves and U. Siegenthaler, 1995: CO_2 and the carbon cycle. In: *Climate Change 1994: Radiative Forcing of Climate Change and an Evaluation of the IPCC IS92 Emission*

Scenarios, J.T. Houghton, L.G. Meira Filho, J. Bruce, Hoesung Lee, B.A. Callander, E. Haites, N. Harris and K. Maskell (eds.). Cambridge University Press, Cambridge, UK.

Schumann, E. and D. Wurzel, eds., 1994: Impact of Emissions from Aircraft and Spacecraft Upon the Atmosphere. *Proc. International Scientific Colloquium*, 18–20 Apr 1994, Koln, DLR Report 94–06, 496 pp.

Schwartz, S.E. and A. Slingo, 1996: Enhanced shortwave cloud radiative forcing due to anthropogenic aerosols. *Clouds, Chemistry and Climate, Proceedings of NATO Advanced Research Workshop*, P. Crutzen and V. Ramanathan (eds.), Springer, Heidelberg, pp.191–236.

Schwarzkopf, M.D. and V. Ramaswamy, 1993: Radiative forcing due to ozone in the 1980s: dependence on altitude of ozone change. *Geophys. Res. Lett.*, **20**, 205–208.

Shi, G.Y., X.B. Fan, J.D. Guo, L. Xu, R.M. Hu, J.P. Chen and B. Wang, 1994: Measurements of the atmospheric aerosols' optical properties in HEIFE area. *Proc. Int. Conf. on HEIFE*, Nov 8-12, Kyoto, Japan, pp. 642–647.

Siegenthaler, U. 1990: El Niño and atmospheric CO_2. *Nature*, **345**, 295–296.

Siegenthaler, U. and F. Joos, 1992: Use of a simple model for studying oceanic tracer distributions and the global carbon cycle. *Tellus*, **44B**, 186–207.

Siegenthaler, U. and H. Oeschger, 1987: Biospheric CO_2 emissions during the past 200 years reconstructed by deconvolution of ice core data. *Tellus*, **39B**, 140–154.

Siegenthaler, U. and J.L. Sarmiento, 1993: Atmospheric carbon dioxide and the ocean. *Nature*, **365**, 119–125.

Sillman, S., J.A. Logan and S.C. Wofsy, 1990: The sensitivity of ozone to nitrogen oxides and hydrocarbons in regional ozone episodes. *J. Geophys. Res.*, **95**, 1837–1851.

Solomon, S. and J.S. Daniel, 1996: Impact of the Montreal Protocol and its amendments on the rate of change of global radiative forcing. *Clim. Change, 32*, 7–17.

Stanhill, G. and S. Moreshet 1994: Global radiation climate change at seven sites remote from surface sources of pollution. *Clim. Change*, **26**, 89–103.

Starr, C., 1993: Atmospheric CO_2 residence time and the carbon cycle. *Energy*, **18**, 1297–1310.

Stolarski, R.S. and H.L. Wesoky, 1995: *The Atmospheric Effects of Stratospheric Aircraft: A Fourth Program Report*, NASA Ref. Publ., 1359, 246 pp.

Strand, A. and O. Hov, 1993: A two-dimensional zonally averaged transport model including convective motions and strategy for numerical solution. *J. Geophys. Res.*, **98**, 9023–9037.

Strand, A. and O. Hov, 1994: A two-dimensional global study of tropospheric ozone production. *J. Geophys. Res.*, **99**, 22877–22895.

Suess, H.E., 1955: Radiocarbon concentration in modern wood. *Science*, **122**, 415–417.

Tarasick, D.W., D.I Wardle, J.B. Kerr, J.J. Bellefleur and J. Davies, 1995: Tropospheric ozone trends over Canada: 1980–1993. *Geophys. Res. Lett.*, **22**, 409–412.

Taylor, K.E. and J.E. Penner, 1994: Response of the climate system to atmospheric aerosols and greenhouse gases. *Nature*, **369**, 734–737.

Tegen, I. and I. Fung, 1994: Modelling of mineral dust in the atmosphere: Sources, transport and optical thickness. *J. Geophys. Res.*, **99**, 22897–22914.

Tegen, I. and I. Fung, 1995: Contribution to the atmospheric mineral aerosol load from land surface modification. *J. Geophys. Res.*, **100**, 18707–18726.

Thompson, A.M., 1992: The oxidising capacity of the Earth's atmosphere: probable past and future changes. *Science*, **256**, 1157–1165.

Thompson, A.M., K.E. Pickering, R.R. Dickerson, W.G. Ellis Jr., D.J. Jacob, J.R. Scala, W-K. Tao, D.P. McNamara, J. Simpson, 1994: Convective transport over the central United States and its role in regional CO and ozone budgets. *J. Geophys. Res.*, **99**, 18703–18711.

Toumi, R., S. Bekki and K.S. Law, 1994: Indirect influences of ozone depletion on climate forcing by clouds. *Nature*, **372**, 348–351.

Vloedbeld, M. and R. Leemans, 1993: Quantifying feedback processes in the response of the terrestrial carbon cycle to global change: The modelling approach of IMAGE-2. *Water, Air and Soil Pollution*, **70**, 615–628.

Volk, T., 1989: Effect of the equatorial Pacific upwelling on atmospheric CO_2 during the 1982–83 El Niño. *Global Biogeochem. Cycles*, **3**, 267–279.

Waggoner, A.P., R.E. Weiss, N.C. Ahlquist, D.S. Covert, S. Will and R.J. Charlson, 1981: Optical characteristics of atmospheric aerosols. *Atmos. Env.*, **15**, 1891–1909.

Wahner, A., M.A. Geller, F. Arnold, W.H. Brune, D.A. Cariolle, A.R. Douglass, C. Johnson, D.H. Lister, J.A. Pyle, R. Ramaroson, D. Rind, F. Rohrer, U. Schumann and A.M. Thompson, 1995: Subsonic and supersonic aircraft emissions. In: *Scientific Assessment of Ozone Depletion: 1994, Global Ozone Research and Monitoring Project Report No. 37*, World Meteorological Organisation, Geneva, pp. 11.1–11.32.

Wang, P-H., P. Minnis and G.K. Yue, 1995: Extinction coefficient (1 µm) properties of high-altitude clouds from solar occultation measurements (1985–1990): Evidence of volcanic aerosol effect. *J. Geophys. Res.*, **100**, 3181–3199.

Wennberg, P.O., R.C. Cohen, R.M. Stimpfle, J.P. Koplow, J.G. Anderson, R.J. Salawitch, D.W. Fahey, E.L. Woodbridge, E.R. Keim, R.S. Gao, C.R. Webster, R.D. May, D.W. Toohey, L.M. Avallone, M.H. Profitt, M. Loewenstein, J.R. Podolske, K.R. Chan, and S.C. Wofsy, 1994: Removal of stratospheric O_3 by radicals: in situ measurements of OH, HO_2, NO, NO_2, ClO and BrO. *Science*, **266**, 398–404.

White, W., 1990: The contribution of fine particle scattering to total extinction. §4.1–4.4 in *Visibility: Existing and historical conditions – causes and effects*. Acid deposition science and technology report **24**, US Government Printing Office.

Wigley, T.M.L., 1993: Balancing the global carbon budget. Implications for projections of future carbon dioxide concentration changes. *Tellus*, **45B**, 409–425.

Wigley, T.M.L., 1995: Global-mean temperature and sea level consequences of greenhouse gas concentration stabilization. *Geophys. Res. Lett.*, **22**, 45–48.

Wigley, T.M.L., R. Richels and J.A. Edmonds, 1995: Economic and environmental choices in the stabilization of CO_2 concentrations: choosing the "right" emissions pathway. *Nature*, **379**, 240–243.

Winguth, A.M.E., M. Heimann, K.D. Kurz, E. Maier-Reimer, U. Mikolajewicz and J. Segeschneider, 1994: El Niño Southern Oscillation related fluctuations of the marine carbon cycle. *Global Biogeochem. Cycles*, **8**, 39–63.

WMO/UNEP, 1992: *Scientific Assessment of Ozone Depletion: 1991*. Global Ozone Research and Monitoring Project Report No. **25**, World Meteorological Organization, Geneva.

WMO/UNEP, 1995: *Scientific Assessment of Ozone Depletion: 1994*. Global Ozone Research and Monitoring Project Report No. **37**. World Meteorological Organization, Geneva.

Wong, C.S., Y.-H. Chan, J.S. Page, G.E. Smith and R.D. Bellegay, 1993: Changes in equatorial CO_2 flux and new production estimated from CO_2 and nutrient levels in Pacific surface waters during the 1986/87 El Niño. *Tellus*, **45B**, 64–79.

Wu, D. and W. Chen, 1994: Intra-annual variation features of mass distribution and water soluble composition distribution of atmospheric aerosols over Guangzhou, *Acta Meteorologica Sinica*, **52**, 499–505.

Zander, R., D.H. Ehhalt, C.P. Rinsland, U. Schmidt, E. Mahieu, M.R. Gunson, C.B. Farmer, M.C. Abrams and M.K.W. Ko, 1994a: Secular trend and seasonal variability of the column abundance of N_2O above the Jungfraujoch station determined from IR solar spectra. *J. Geophys. Res.*, **99**, 16745–16756.

Zander, R., E. Mahieu, Ph. Demoulin, C. Rinsland, D. Weisenstein, M. Ko, N. Sze and M. Gunson, 1994b: Secular evolution of the vertical column abundances of HCFC-22 in the earth's atmosphere inferred from ground-based IR solar observations at the Jungfraujoch and at Kitt Peak, and comparison with model observations. *J. Atmos. Chem.*, **18**, 129–148.

Zhang, Q., W.H. Soon, S.L. Baliunas, G.W. Lockwood, B.A. Skiff and R.R. Radick, 1994: A method of determining possible brightness variations of the Sun in past centuries from observations of solar-type stars. *Astrophys. J.*, **427**, L111–L114.

3

Observed Climate Variability and Change

N. NICHOLLS, G.V. GRUZA, J. JOUZEL, T.R. KARL, L.A. OGALLO, D.E. PARKER

Key Contributors:
J.R. Christy, J. Eischeid, P.Ya. Groisman, M. Hulme, P.D. Jones, R.W. Knight

Contributors:
J.K. Angell, S. Anjian, P.A. Arkin, R.C. Balling, M.Yu. Bardin, R.G. Barry, W. BoMin, R.S. Bradley, K.R. Briffa, A.M. Carleton, D.R. Cayan, F.H.S. Chiew, J.A. Church, E.R. Cook, T.J. Crowley, R.E. Davis, N.M. Datsenko, B. Dey, H.F. Diaz, Y. Ding, W. Drosdowsky, M.L. Duarte, J.C. Duplessy, D.R. Easterling, W.P. Elliott, B. Findlay, H. Flohn, C.K. Folland, R. Franke, P. Frich, D.J. Gaffen, V.Ya. Georgievsky, B.M. Ginsburg, V.S. Golubev, J. Gould, N.E. Graham, D. Gullet, S. Hastenrath, A. Henderson-Sellers, M. Hoelzle, W.D. Hogg, G.J. Holland, L.C. Hopkins, N.N. Ivachtchenko, D. Karoly, R.W. Katz, W. Kininmonth, N.K. Kononova, L.V. Korovkina, G. Kukla, C.W. Landsea, S. Levitus, T.J. Lewis, H.F. Lins, J.M. Lough, T.A. McMahon, L. Malone, J.A. Marengo, E. Mekis, A. Meshcherskya, P.J. Michaels, E. Mosley-Thompson, S.E. Nicholson, J. Oerlemans, G. Ohring, G.B. Pant, T.C. Peterson, N. Plummer, F.H. Quinn, E.Ya. Ran'kova, V.N. Razuvaev, E.V. Rocheva, C.F. Ropelewski, K. Rupa Kumar, M.J. Salinger, B. Santer, H. Schmidt, E. Semenyuk, I.A. Shiklomanov, M. Shinoda, I.I. Soldatova, D.M. Sonechkin, R.W. Spencer, N. Speranskaya, A. Sun, K.E. Trenberth, C. Tsay, J.E. Walsh, B. Wang, K. Wang, M.N. Ward, S.G. Warren, Q. Xu, T. Yasunari

CONTENTS

SUMMARY

Has the climate warmed?

- The estimate of warming since the late 19th century has not significantly changed since the estimates in IPCC (1990) and IPCC (1992), although the data have been reanalysed, and more data are now available. Global surface temperatures have increased by about 0.3 to 0.6°C since the late-19th century, and by about 0.2 to 0.3°C over the last 40 years (the period with most credible data). The warming has not been globally uniform. Some areas have cooled. The recent warming has been greatest over the continents between 40° and 70°N.

- The general, but not global, tendency to reduced diurnal temperature range over land, at least since the middle of the 20th century, noted in IPCC (1992), has been confirmed with more data (representing more than 40% of the global land mass). The range has decreased in many areas because nights have warmed more than days. Cloud cover has increased in many of the areas with reduced diurnal temperature range. Minimum temperature increases have been about twice those in maximum temperatures.

- Radiosonde and Microwave Sounding Unit observations of tropospheric temperature show slight overall cooling since 1979, whereas global surface temperature has warmed slightly over this period. There are statistical and physical reasons (e.g., short record lengths; the different transient effects of volcanic activity and El Niño-Southern Oscillation) for expecting different recent trends in surface and tropospheric temperatures. After adjustment for these transient effects, which can strongly influence trends calculated from short periods of record, both tropospheric and surface data show slight warming since 1979. Longer term trends in the radiosonde data, since the 1950s, have been similar to those in the surface record.

- Cooling of the lower stratosphere since 1979 is shown by both Microwave Sounding Unit and radiosonde data (as noted in IPCC, 1992), but is larger (and probably exaggerated because of changes in instrumentation) in the radiosonde data. The current (1994) global stratospheric temperatures are the coolest since the start of the instrumental record (in both the satellite and radiosonde data).

- As predicted in IPCC (1992), relatively cool surface and tropospheric temperatures, and a relatively warmer lower stratosphere, were observed in 1992 and 1993, following the 1991 eruption of Mt. Pinatubo. Warmer surface and tropospheric temperatures reappeared in 1994. Surface temperatures for 1994, averaged globally, were in the warmest 5% of all years since 1860.

- Further work on indirect indicators of warming such as borehole temperatures, snow cover, and glacier recession data, confirm the IPCC (1990) and (1992) findings that they are in substantial agreement with the direct indicators of recent warmth. Variations in sub-surface ocean temperatures have been consistent with the geographical pattern of surface temperature variations and trends.

- As noted in IPCC (1992) no consistent changes can be identified in global or hemispheric sea ice cover since 1973 when satellite measurements began. Northern Hemisphere sea ice extent has, however, been generally below average in the early 1990s.

Has the climate become wetter?

- There has been a small positive (1%) global trend in precipitation over land during the 20th century, although precipitation has been relatively low since about 1980. Precipitation has increased over land in high latitudes of the Northern Hemisphere, especially during the cold season, concomitant with temperature increases. A step-like decrease of precipitation occurred after the 1960s over the subtropics and tropics from Africa to Indonesia, as temperatures in this region increased. The various regional changes are consistent with changes in streamflow, lake levels, and soil moisture (where data are available and have been analysed).

- There is evidence to suggest increased precipitation over the central equatorial Pacific Ocean, in recent decades, with decreases to the north and south. Little can be said about precipitation changes elsewhere over the ocean.

- Northern Hemisphere snow cover extent has been consistently below the 21 year (1974-1994) average since 1988. Snow-radiation feedback has amplified spring-time warming over mid- to high latitude Northern Hemisphere land areas.

- Evaporation appears to have decreased since 1951 over much of the former Soviet Union, and possibly also in the USA. Evaporation appears to have increased over the tropical oceans (although not everywhere).

- The evidence still suggests an increase of atmospheric water vapour in the tropics, at least since 1973, as noted in IPCC (1992).

- In general, cloud amount has increased over the ocean in recent decades, with increases in convective and middle and high-level clouds. Over many land areas, cloud increased at least up to the 1970s. IPCC (1990) and (1992) also reported cloud increases.

Has the atmospheric/oceanic circulation changed?

- The behaviour of the El Niño-Southern Oscillation (ENSO), which causes droughts or floods in many parts of the world, has been unusual since the mid-1970s and especially since 1989. Since the mid-1970s, warm (El Niño) episodes have been relatively more frequent or persistent than the opposite phase (La Niña) of the phenomenon. Recent variations in precipitation over the tropical Pacific and the surrounding land areas (e.g., the relatively low

rainfall over the subtropical land areas in the last two decades) are related to this behaviour in the El Niño-Southern Oscillation, which has also affected the pattern and magnitude of surface temperatures.

Has the climate become more variable or extreme?

- The data on climate extremes and variability are inadequate to say anything about global changes, but in some regions, where data are available, there have been decreases or increases in extreme weather events and variability.

- Other than the few areas with longer term trends to lower rainfall (e.g., the Sahel), little evidence is available of changes in drought frequency or intensity.

- There have been few studies of variations in extreme rainfall events and flood frequency. In some areas with available data there is evidence of increases in the intensity of extreme rainfall events, but no clear, large-scale pattern has emerged.

- There is some evidence of recent (since 1988) increases in extreme extra-tropical cyclones over the North Atlantic. Intense tropical cyclone activity in the Atlantic has decreased over the past few decades although the 1995 season was more active than recent years. Elsewhere, changes in observing systems and analysis methods confound the detection of trends in the intensity or frequency of extreme synoptic systems.

- There has been a clear trend to fewer extremely low minimum temperatures in several widely separated areas in recent decades. Widespread significant changes in extreme high temperature events have not been observed.

- There have been decreases in daily temperature variability in recent decades, in the Northern Hemisphere mid-latitudes.

Is the 20th century warming unusual?

- Northern Hemisphere summer temperatures in recent decades appear to be the warmest since at least about 1400 AD, based on a variety of proxy records. The warming over the past century began during one of the colder periods of the last 600 years. Data prior to 1400 are too sparse to allow the reliable estimation of global mean temperature. However, ice core data from several sites around the world suggest that 20th century temperatures are at least as warm as any century since at least about 1400, and at some sites the 20th century appears to have been warmer than any century for some thousands of years.

- Large and rapid climatic changes affecting the atmospheric and oceanic circulation and temperature, and the hydrologic cycle, occurred during the last ice age and during the transition towards the present Holocene period. Changes of about 5°C occurred on time-scales of a few decades, at least in Greenland and the North Atlantic.

- Temperatures have been far less variable during the last 10,000 years (the Holocene), relative to the previous 100,000 years. Based on the incomplete observational and palaeoclimatic evidence available, it seems unlikely that global mean temperatures have increased by 1°C or more in a century at any time during the last 10,000 years.

3.1 Introduction

Observed climate change and variability are considered in this chapter by addressing six commonly asked questions related to the detection of climate change and sensitivity of the climate to anthropogenic activity. The questions are:

Has the climate warmed?
Has the climate become wetter?
Has the atmospheric/oceanic circulation changed?
Has the climate become more variable or extreme?
Is the 20th century warming unusual?
Are the observed trends internally consistent?

The conclusions from observations depend critically on the availability of accurate, complete, consistent series of observations. That conclusions regarding trends cannot always be drawn does not necessarily imply that trends are absent. It could reflect the inadequacy of the data, or the incomplete analysis of data. Karl *et al.* (1995a) demonstrate that, for many of the climate variables important in documenting, detecting, and attributing climate change, the data are not at present good enough for rigorous conclusions to be reached. This especially applies to global trends of variables with large regional variations, such as precipitation. The final section of this chapter attempts to indicate our confidence in the various trends observed in the less than complete data available.

3.2 Has the Climate Warmed?

3.2.1 Background

IPCC (1990) concluded that, on a global average, surface air and sea temperature had risen by between 0.3°C and 0.6°C since the mid-19th century. IPCC (1992) confirmed this. The recent warming is re-examined here, using updated and improved data, including the diurnal asymmetry of the warming, and the geographical and vertical structure of the warming. Conventional temperature observations are supplemented by indirect evidence and by satellite-based data.

Temperatures in this chapter are expressed relative to the 1961-90 averages, rather than the 1951-80 period used in IPCC (1990) and (1992). The new reference period has been used because it contains more sea surface temperature data, more data from currently operating land stations, and because it is the official reference period used by the World Meteorological Organisation and member states. Where series from IPCC (1992) are reproduced here they are expressed relative to the 1961 to 1990 mean, not the 1951 to 1980 mean used in IPCC (1992).

3.2.2 Surface Temperature

3.2.2.1 Land-surface air temperature

The land-surface air temperature data base developed by Jones *et al.* (1986a,b) and Jones (1988), and used by IPCC (1990) and IPCC (1992), has been substantially expanded and reanalysed (Jones, 1994a). Coverage in recent decades has benefited most from this expansion. The resulting global anomaly time-series are very similar, on decadal time-scales, to those in the previous (IPCC, 1990, 1992) analyses (Figure 3.1a). The small differences in the time-series are mainly due to the addition of a few extra Australian stations which resulted in warmer late 19th century temperatures, relative to the earlier time-series. The 19th century Australian temperatures may be biased warm relative to modern recordings, because of different methods of exposure (Nicholls *et al.*, 1996). Figure 3.1b shows time-series of global land temperatures updated from Hansen and Lebedeff (1988) and Vinnikov *et al.* (1990), for comparison with the Jones (1994a) time-series.

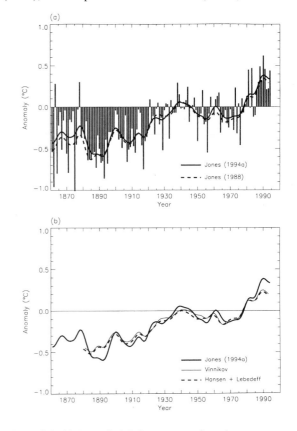

Figure 3.1: (a) Annual global average surface air temperature anomalies (°C) for land areas, 1861 to 1994, relative to 1961 to 1990. Bars and solid curve from Jones (1994a); dashed curve from Jones (1988). The smoothed curves were created using a 21-point binomial filter. (b) As (a) but updated from Hansen and Lebedeff (1988) – dashed line; Vinnikov *et al.* (1990) – thin line. Thick solid line is from Jones (1994a), as in (a).

IPCC (1992) cited results of Jones *et al.* (1990) which indicated that urbanisation influences have yielded, on average, a warming of less than 0.05°C during the 20th century in their data over the global land. In specific regions, however, urbanisation influences may be significant. Portman (1993) found that the average warming of 0.22°C between 1954 and 1983 over northern China in the gridded data of Jones *et al.* (1986a) was nearly as large as the 0.25°C warming in this period averaged over seven uncorrected large urban stations; Portman's adjustments using rural stations, however, suggested that there was in fact a regional cooling of about 0.05°C. The urbanisation warming trends were greatest in spring. Differences from Jones *et al.* (1990) and Wang *et al.* (1990), who found a smaller or no urbanisation warming trend in eastern China, may have resulted from the different, though overlapping, region covered. Hulme *et al.* (1994) note that urbanisation cannot fully account for the significant warming trends over eastern Asia. Christy and Goodridge (1994) found that the five longest term Californian stations used by Hansen and Lebedeff (1988) all had more positive temperature trends in 1910–89 than the median of 112 Californian stations. Moberg and Alexandersson (1995) found evidence of urban warming affecting apparent trends in Sweden. On the other hand, Barros and Camilloni (1994) note that the urbanisation effect in Buenos Aires relative to the surrounding rural environment has decreased since the 1950s. Remote sensing techniques hold promise for the eventual worldwide estimation of urban temperature bias (Johnson *et al.*, 1994).

Concerns about a possible link between climate change and desertification indicate the need for monitoring of arid region temperatures. An analysis of the Jones (1994a) gridded temperature anomalies for dryland areas of the world (UNEP, 1992) yielded a significantly greater trend for 1901–1993 (0.62°C) than for the land areas as a whole (0.44°C). Trends for dry lands were greatest in Central and North America (nearly 0.8°C) and least in South America (a little over 0.3°C). They were also significantly greater in the least arid subclass (nearly 0.8°C) than in the most arid subclass (around 0.3°C). The total arid area is only about 14% of the land surface. Even if all the greater warming in these areas was due to desertification, this would only have contributed a few hundredths of a degree to the observed global warming.

3.2.2.2 Sea surface temperature

A combined physical-empirical method (Folland and Parker, 1995) has been used to estimate improved adjustments to ships' sea surface temperature (SST) data obtained up to 1941, to compensate for heat losses from uninsulated (mainly canvas) or partly-insulated (mainly wooden) buckets. The time-series of global SST anomalies is very similar to that in IPCC (1992) (Figure 3.2). The differences compared with IPCC (1992) mainly result from an improved formulation of the heat transfers affecting wooden buckets and an improved climatology in the Southern Ocean. This climatology incorporates, for 1982 onwards, satellite-based data, using a Laplacian blending technique (Reynolds, 1988) to remove overall biases relative to the in situ data (Parker *et al.*, 1995). In addition, following Folland and Parker (1995), slightly fewer wooden buckets are assumed than was the case in IPCC (1992), and the SST data in the analysis of Jones and Briffa (1992) are no longer incorporated. The data used to construct the night-time marine air temperature (NMAT) series in Figure 3.2 largely avoid daytime heating of ships' decks and were corrected independently of SST from the mid-1890s onwards (Bottomley *et al.*, 1990; Folland and Parker, 1995). The NMAT data confirm the trends in SST.

Many historical in situ data remain to be digitised and incorporated into the data base, to improve coverage and reduce the uncertainties in our estimates of marine climatic variations. Annual hemispheric and global SST anomalies since 1950 appear to be insensitive to the spatial averaging technique used, to within about 0.02°C (Parker, 1994a).

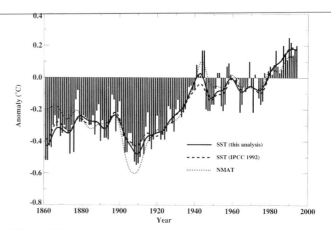

Figure 3.2: Annual global sea surface temperature anomalies (SST, bars and solid curve) and night marine air temperature (NMAT, dotted curve), 1861 to 1994, relative to 1961 to 1990 (°C). The smoothed curves were created using a 21-point binomial filter. The dashed curve is sea surface temperature from IPCC (1992), adjusted to be relative to 1961 to 1990.

But hemispheric and global trends of SSTs calculated over a decade or so are subject to large biases and sampling errors owing to incomplete data coverage, especially in the 19th century and during the period of the World Wars (Karl *et al.*, 1994). The spatial coherence and temporal persistence of seasonal 5° area SST anomalies have been used to estimate the random and sampling errors of decadal SST anomalies (Parker *et al.*, 1994). At the beginning of the century, the estimated combined random and sampling standard errors were generally less than 0.1°C *in those 5° areas with sufficient data to perform an analysis* (about 60% of the world's oceans).

3.2.2.3 *Land and sea combined*

Comparisons of island air temperatures with SSTs (e.g., Folland and Salinger, 1996) have demonstrated that the two independent fields exhibit similar decadal- and century-scale variations. Figures 3.3a to c show annual time-series of anomalies of combined land-surface air temperature and SST for the hemispheres and globe since 1861. These series, shown as bars and solid curves, differ slightly from those presented by IPCC (1992) which are summarised by the dashed curves, because of the above-mentioned improvements in both land and sea temperatures. Thus, the new curves are typically up to 0.05°C warmer before 1900, but differences are smaller thereafter.

In accord with IPCC (1990), global average land-surface air temperature anomalies exceeded SSTs by around 0.1°C in the 1920s and up to 0.2°C in recent years, but were up to 0.3°C lower in the 1880s (Figure 3.3d). Real atmospheric circulation changes, as well as some instrumental uncertainties, are likely to be the cause of this variation (Parker *et al.*, 1994).

A variety of tests (Parker *et al.*, 1994) suggests that the sampling uncertainty on the smoothed curves in Figure 3.3 is possibly within 0.05°C since the 1880s. However, these findings take no account of unsampled regions so a trend uncertainty of 0.1°C/century may be more realistic (Karl *et al.*, 1994). The instrumental bias uncertainty is assessed by Parker *et al.* (1994) as less than 0.1°C for land, around 0.1°C for SST, and less than 0.15°C for the combination, since the 1880s, consistent with preliminary estimates in IPCC (1992). In addition, an urbanisation uncertainty of less than 0.05°C is estimated (Section 3.2.2.1). The overall uncertainty remains about 0.15°C and leads to an estimate of 0.3 to 0.6°C for global near-surface warming since the late 19th century, unchanged from the estimates of IPCC (1990, 1992).

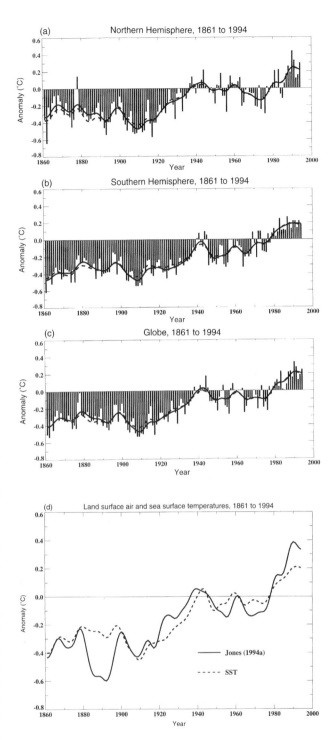

Figure 3.3: Combined annual land-surface air and sea surface temperature anomalies (°C) 1861 to 1994, relative to 1961 to 1990 (bars and solid smoothed curves): (a) Northern Hemisphere; (b) Southern Hemisphere; (c) Globe. The dashed smoothed curves are corresponding results from IPCC (1992), adjusted to be relative to 1961 to 1990. (d) Global land-surface air temperature (solid line) and SST (broken line).

Figure 3.4 shows the change of surface temperature, for the four seasons as well as for the annual averages, from the period 1955–1974 to the period 1975–1994. This second period has seen unusual ENSO activity (Section 3.4.2). Note, in accord with the results in IPCC (1990) and IPCC (1992), the recent warming over the mid-latitude Northern Hemisphere continents in winter and spring, and year-round cooling in the north-west North Atlantic and mid-latitudes over the North Pacific.

Zonal mean anomalies of combined land-surface air temperature (from Jones, 1994a) and SST (from Parker *et al.*, 1995), shown in Figure 3.5, confirm that the recent warmth is greatest in mid-latitudes of the Northern Hemisphere. This can be compared with the warm period of the mid-20th century, where the greatest warmth was in the higher latitudes of the Northern Hemisphere. The recent warm period also exhibits higher temperatures in the Southern Hemisphere. It should be noted that the anomalies in Figure 3.5 are less reliable in data sparse areas and periods, e.g., before 1900, during the World Wars, and in the Southern Ocean.

Figure 3.4 shows that maximum recent warming has been in winter over the high mid-latitudes of the Northern Hemisphere continents. Tropospheric data show no statistically significant Arctic-wide temperature trends for the 1958–86 period or during the 1950-90 period over the Arctic Ocean (Kahl *et al.*, 1993a,b). Significant warming since the 1950s along the west coast of the Antarctic Peninsula is not representative of continental Antarctica (Raper *et al.*, 1984; King, 1994), where weaker warming has been observed. However, expeditionary records (Jones, 1990) and ice temperature profiles (Nicholls and Paren, 1993) suggest 20th century warming for various parts of Antarctica, and Jones (1995a) reported an Antarctic warming trend from 1957 to 1994, although temperatures were low in 1993 and 1994. All of this warming occurred before the early 1970s.

3.2.2.4 Changes in the diurnal temperature range

An analysis of worldwide quality-selected station monthly average maximum (day) and minimum (night) temperatures (Horton, 1995) now covers about 41% of land areas. It confirms the report of IPCC (1992) and the findings of Karl *et al.* (1993a) that worldwide increases in minimum land-surface air temperature since 1950 have been about twice those in the maximum (Figure 3.6).

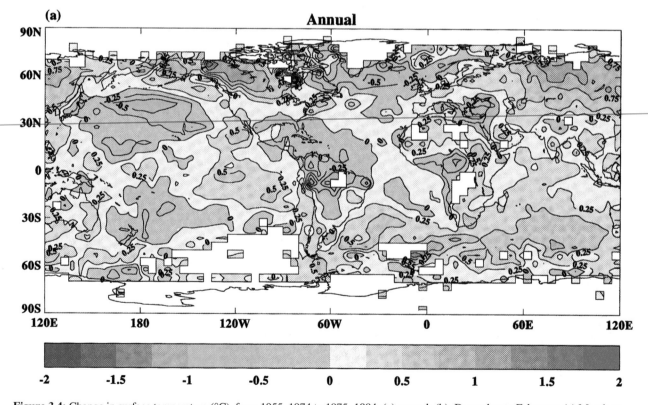

Figure 3.4: Change in surface temperature (°C), from 1955–1974 to 1975–1994: (a) annual; (b) December to February; (c) March to May; (d) June to August; (e) September to November. Sea surface temperatures are from Parker *et al.* (1995) and land-surface air temperatures are from Jones (1994a).

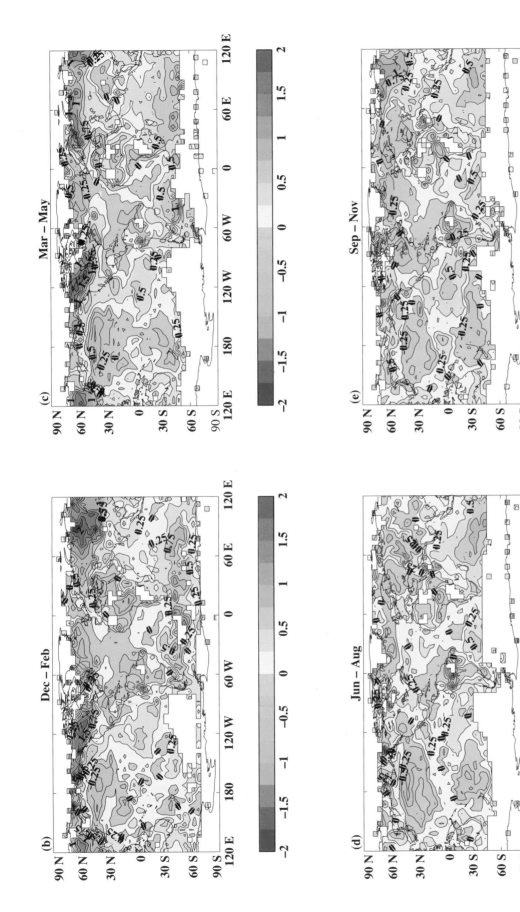

Figure 3.4: *cont.*

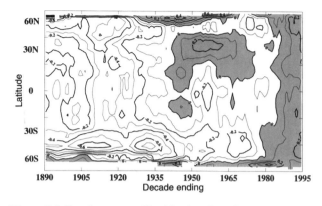

Figure 3.5: Zonal mean combined land-surface air and sea surface temperature anomalies (°C), 1881 to 1994, relative to 1961 to 1990. Values are 10-year running means. Data are updates of Parker *et al.* (1995) for SST and Jones (1994a) for land.

Additional areas with decreasing diurnal temperature range (DTR) have been identified (e.g., Jones, 1995b). The decrease of the DTR in several areas has been most pronounced in the Northern Hemisphere autumn (Karl *et al.*, 1993a; Kukla and Karl, 1993; Horton, 1995). The observed decreases in the DTR have been found to relate to increases of cloud cover (Plantico *et al.*, 1990; Henderson-Sellers, 1992; Dessens and Bücher, 1995; Jones, 1995b; Kaas and Frich, 1995; Plummer *et al.*, 1995; Salinger *et al.*,

1995b). The decrease of the DTR has mostly but not always been due to the faster rise of the night temperatures. The general increases in minimum (overnight) temperature occurred in rural areas as well as cities, and could not therefore simply reflect an increasing urban heat island effect. In the south-western part of the former Soviet Union, DTR has decreased because of depressed daily maxima (Razuvaev *et al.*, 1995).

Some coastal and island areas (Horton, 1995; Salinger *et al.*, 1993, 1995b), India (Rupa Kumar *et al.*, 1994), and alpine stations (Weber *et al.*, 1994) have shown no long-term decrease in annual average DTR. No systematic changes of minima or maxima and no general warming has been observed in the Arctic over the last 50 years or so (Kahl *et al.*, 1993a,b; Michaels *et al.*, 1995; Ye *et al.*, 1995), though regions with reduced DTR in recent decades do include Alaska and central Siberia (Figure 3.6).

3.2.3 Tropospheric and Lower Stratospheric Temperatures

Climate monitoring has focused on measurements at the Earth's surface; only in the past few decades has the atmosphere above the surface been regularly measured. Regular upper-air measurements by balloon ascents in scattered locations began in the 1940s and observations from satellites generally only began in the 1970s.

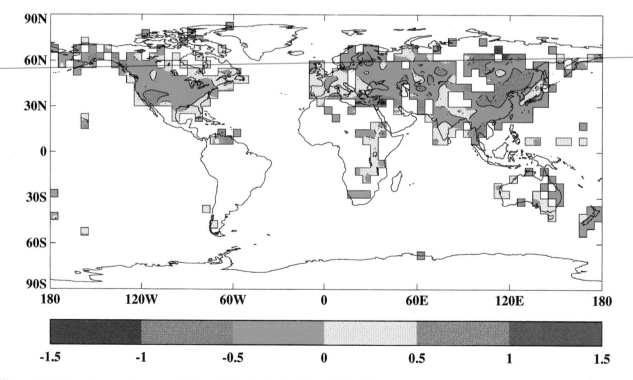

Figure 3.6: Diurnal temperature range 1981 to 1990 relative to 1951 to 1980 (°C).

3.2.3.1 Data reliability

IPCC (1992) presented radiosonde data since 1964, whereas here data from 1958 are presented. There are doubts, however, about the earlier data. Although the radiosonde coverage was adequate from 1958 in the Northern Hemisphere, it has only been adequate in the Southern Hemisphere since 1964. "Global" values from radiosondes from before 1964 are, therefore, somewhat suspect, because of the reduced geographical coverage.

Compilations for monitoring long-term changes of temperature aloft using radiosondes have been made by Angell (1988 and updates) and by Oort and Liu (1993). However, the sensors which measure temperature and humidity in radiosondes have undergone many changes in the past 40 years, and these changes have had significant effects on estimates of long-term trends (Gaffen, 1994; Parker and Cox, 1995). Humidity is important both on its own account (Section 3.3.7) and because it affects air density, which is often used as a surrogate for temperature. Elliott *et al.* (1994) calculate that spurious "drying" of the mid-troposphere (850-300 hPa) due to improved radiosonde sensors has led to a spurious cooling since 1958 of 0.05 to 0.1°C/decade. This spurious cooling is primarily related to sensor changes in the 1970s. An additional aspect of uncertainty is related to the non-uniform geographic distribution of stations since vast areas over the oceans, particularly in the Southern Hemisphere, are not monitored at all (Trenberth and Olson 1991), and coverage has declined over the oceans and parts of the tropics (Parker and Cox, 1995).

While information on changes in radiosonde sensors is rarely complete, Gaffen (1994) reported on the effects of documented changes in the temperature sensors and found that at least 43% of the stations used by Angell have clear heterogeneities. Oort and Liu's data will also have been affected. The net effect of these heterogeneities was to impose a spurious cooling trend (in addition to that from the humidity sensors noted above) on the time-series since 1958. The magnitude of the tropospheric impact appears to be small. However, the lower stratospheric time-series was found to have been significantly affected, particularly in the tropics and Southern Hemisphere, so as to bias the trend calculation to be cooler than actual. In addition Parker and Cox (1995) suggest that early radiosonde balloons were more likely to burst in cold conditions, biasing the incomplete data towards warmth, and giving a spurious cooling trend in the lower stratosphere as balloons were strengthened. Angell's global lower-stratospheric radiosonde time-series cooled since 1979 at about 0.6°C/decade relative to global satellite (Microwave Sounding Units (MSUs) on polar orbiting satellites) data. The relative cooling is less if only MSU data collocated with the radiosondes (rather than global MSU data) are used in the comparison, and trends in the Oort and Liu data (using more stations than Angell, but available only up to 1989) are more similar to those in the MSU, so the greater cooling in Angell's data appears to be due to undersampling of the global atmosphere (Christy, 1995), as well as to radiosonde inhomogeneities.

The MSU data, available since 1979, have exhibited high precision and global coverage for the temperature of deep layers in the troposphere and stratosphere (Spencer and Christy, 1992b, 1993). Biases because of water vapour and cloud changes have been shown to be minimal (Spencer *et al.*, 1990), although some uncertainty remains (Prabhakara et al., 1995). In addition, Spencer and Christy (1992a,b; 1993), Christy and Drouilhet (1994), and Christy and Goodridge (1995) used comparisons between satellites and with radiosondes to show that instrumental drift was insignificant until late 1991.

There has, however, been a spurious warming of 0.04°C/yr in NOAA 11[1] tropospheric data since 1990, owing to a drift in orbit times. An estimate of this bias, along with biases affecting the dynamic range of NOAA 12 MSU temperatures, has been removed from results shown here (Christy *et al.*, 1995).

3.2.3.2 Tropospheric trends

Figure 3.7a shows time-series of tropospheric temperature anomalies calculated from MSU data and radiosondes, and the global surface temperature for comparison. The global MSU tropospheric trend from 1979 to May 1995 was –0.06°C/decade, and that for the seasonal radiosonde data for the same period was –0.07°C/decade. However, if the transient effects of volcanoes and the El Niño-Southern Oscillation (which can bias trends calculated from short periods of data) are removed from the various time-series, positive trends become evident (e.g., 0.09°C/decade for MSU), in closer accord with surface data (Christy and McNider, 1994). Jones (1994b) calculated residual global trends (after removal of volcanic and ENSO transients) for the 1979–93 period of 0.09°C/decade from MSU data, 0.10°C from 850-300 hPa radiosonde temperature updated from Angell (1988), and 0.17°C from combined land-surface air temperature and SST data. The differences between these trends were about half the differences between trends in the raw data (i.e., without the removal of the transient El Niño-Southern Oscillation and volcanic

---

[1] NOAA 11 and NOAA 12 are USA weather satellites, launched in September 1988 and May 1991 respectively.

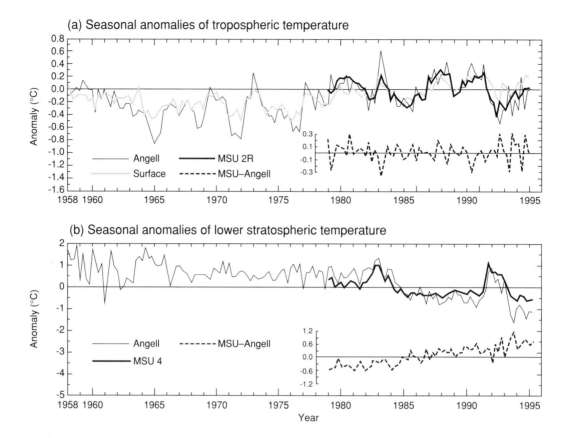

Figure 3.7: (a) Seasonal global temperature anomalies (°C), relative to 1979–1994 average, for the 850-300 hPa layer from radiosondes (Angell, 1988 and updates, light solid line) and for the troposphere from MSU 2R (Spencer and Christy, 1992b and updates, heavy solid line). Dashed line (inset) is MSU 2R minus radiosondes. Surface temperatures (shaded line) have been added for comparison with the tropospheric temperatures. (b) As in (a), but for the 100–50 hPa layer from radiosondes (Angell, 1988 and updates, light solid line) and for the lower stratosphere from MSU 4 (Spencer and Christy, 1993 and updates, heavy solid line). Dashed line (inset) is MSU 4 minus radiosonde. (Note: MSU 2R and MSU 4 are channels of the MSU instrument designed to sample the lower troposphere and stratosphere respectively.)

effects). So, apparent differences between surface and tropospheric trends for 1979–1993 appear to be partly a result of the greater influence of volcanic eruptions and ENSO on tropospheric temperatures. Hansen *et al.* (1995) also demonstrate that natural variability can account for some of the apparent differences between surface and lower tropospheric data.

For the longer period 1958 to 1993, Jones (1994b) found that the unadjusted and adjusted global trends from radiosonde and surface data were all between 0.08°C and 0.11°C/decade, reflecting the fact that longer-term trends are less likely to be biased by transient volcanic and ENSO influences. The unadjusted radiosonde trend to May 1995 was 0.09°C/decade. The similarity of the trends since the late 1950s in the tropospheric and surface temperatures is evident in Figure 3.7a.

3.2.3.3 Lower stratospheric trends

Lower stratospheric temperatures (17–22 km or 100–50 hPa) have demonstrated marked changes over the past few decades (Figure 3.7b). Sudden warmings were caused by infrared-absorbing aerosols from volcanoes (Nyamuragira 1981 with El Chichon 1982, and Mt. Pinatubo 1991). From 1979 through May 1995, the MSU global trend was –0.34°C/decade. The greater cooling in the radiosonde data is at least partly due to changes in the radiosondes and undersampling of the global atmosphere, for the reasons outlined in Section 3.2.3.1. A similar rate of cooling (–0.36°C/decade) has probably occurred since 1964, based on radiosonde data (Christy, 1995). Labitzke and van Loon (1994) examined daily historical analyses of the Northern Hemisphere lower stratosphere and concluded that the cooling was greatest at a height of 18–20 km at all latitudes in summer, and in the polar regions also in early winter.

Lower stratospheric trends are discussed in more detail in IPCC (1994) and the 1994 Ozone Assessment (WMO, 1994). The current (1994) global stratospheric temperatures are the coolest since the start of the instrumental record in both the satellite and radiosonde data.

3.2.4 Subsurface Ocean Temperatures

The deep ocean has a much lower noise level of temperature variability than at the surface, and hence long-term changes are more readily identified. Recent studies (e.g., Bindoff and Church, 1992; Bindoff and McDougall, 1994) have demonstrated that changes in water mass properties (salinity/temperature relationships) are even easier to detect.

Studies are beginning to indicate large-scale and coherent changes in subsurface conditions that appear to be related to changes, at the surface, in the formation regions of the relevant water masses. These lead to large, coherent regions of overlying layers of warming and cooling in the water column. The geographical variations in recent surface temperature changes (Figure 3.4) are linked to the subsurface changes. The surface changes can mix and advect downward with a strong component moving along surfaces of constant density.

In the Southern Hemisphere, Salinger *et al.* (1995b) report a basin-wide, depth-averaged warming of 0.3°C between 100 m and 800 m over the entire width of the Indian Ocean at 30°S over the past 20 years. Below this there was a depth-averaged cooling of up to 0.1°C. Bindoff and Church (1992) found warming in the upper 800 m at 43°S in the south-west Pacific between 1967 and 1989. Bindoff and McDougall (1994) inferred that the change between 300 m and 700 m resulted from warming at the surface where the water masses were formed during contact with the atmosphere. It appears that the upper ocean warming of these two basins is a reflection of the surface warming at the surface at latitudes of 40–50°S. The cooling at greater depth in the Indian Ocean appears to be a reflection of surface cooling at higher latitudes where there is poor surface coverage.

Recent transocean sections in the North Pacific are also beginning to reveal changes in subsurface temperatures reflecting surface temperature changes. Off California (a region of surface warming), subsurface temperatures have increased uniformly by 0.8°C in the upper 100 m in the past 42 years, and have risen significantly down to 300 m. Cooling of the surface of the central and western North Pacific is seen to depths of several hundred metres (Antonov, 1993).

In the North Atlantic, warming has occurred at 24°N in the subtropical gyre, with warming of up to 0.15°C between 800 m and 2500 m between 1957 and 1981 (Roemmich and Wunsch, 1984). A similar warming ensued between 1981 and 1992 (Parrilla *et al.*, 1994). The overall warming between 1957 and 1992 was very uniform across the Atlantic, averaging just over 0.3°C at 1100 m. Warming of 0.2°C has been observed since the late 1950s in the subtropical gyre at 1750 m near Bermuda (Levitus *et al.*, 1995).

Strong cooling at the surface of the subpolar North Atlantic connects with the subsurface ocean in the Labrador Sea, where a 0.9°C cooling of Labrador Sea Water has occurred between 1970 and 1995. This water mass is currently colder, fresher, and larger than ever before recorded, in observations extending back to the 1930s (Read and Gould, 1992; Lazier, 1995; Rhines and Lazier, 1995).

These subsurface variations are consistent with circulation and surface temperature variations.

3.2.5 Indirect Measures

A variety of indirect measures can provide supporting evidence of changes and variations in temperature. In general, however, these measures are affected by other factors as well as temperature, so care needs to be taken in determining the extent to which they provide supporting evidence of changes deduced from direct measurements of temperature.

3.2.5.1 Retreat of glaciers

Mountain glaciers provide integrated climatic evidence which is complementary to instrumental meteorological measurements (Haeberli, 1995). Measurements of glacier length extend back to the 19th century in many regions, and even to 1600 AD in Europe. Length reduction of mountain glaciers provides easily detectable evidence from cold regions that fast and worldwide climatic change has taken place over the past century (Haeberli *et al.*, 1989). The factors affecting glaciers, and a description of observed changes over the past century, are discussed in Chapter 7. The 20th century glacier retreat is consistent with a warming in alpine regions of 0.6–1.0°C (Schwitter and Raymond, 1993; Oerlemans, 1994).

3.2.5.2 Borehole temperatures

Underground temperatures, from boreholes in otherwise undisturbed locations, have been used to derive ground surface temperature (GST) histories which are found to be in accord with instrumental or proxy surface air temperature data. Low-frequency surface temperature variations occurring in the past 300 years can be detected in a borehole over 600 metres deep.

Warming has been observed in most boreholes in New England (Pollack and Chapman, 1993) and Canada (Beltrami and Mareschal, 1992; Wang *et al.*, 1992; Wang *et al.*, 1994), and the increase in GST suggested for the last century is very similar to the observed change in regional air temperatures during the same period (e.g., Beltrami and Mareschal, 1991). In temperate western North America (up to 62°N in the Yukon), however, the GST has remained relatively constant over the last century (e.g., Lewis and Wang, 1993) even though most glaciers have retreated. Further north, GSTs have increased over the last century on the Alaskan north slope (Judge *et al.*, 1983; Lachenbruch and Marshall, 1986) and in Canada's north-eastern Arctic Islands (Taylor, 1991), and there has been a retreat of the permafrost (Kwong and Gan, 1994). There are indications of increasing GSTs in the northern USA prairies (Gosnold *et al.*, 1992) and the Canadian prairies (Majorowicz, 1993). Cermak *et al.* (1992) have reported climatic warming in Cuba of 2-3°C over the last 200–300 years. Some borehole sites have indicated no recent change or cooling, e.g., in Utah (Chapman *et al.*, 1992). Deming (1995) assessed all the North American studies and concluded that all averages inferred from groups of boreholes revealed warming (ranging between 0.3 and 4.0°C) since the 19th century. The warming appears to have started in the middle of the 19th century in the eastern half of North America, whereas the warming in the west appears to have started near the beginning of the 20th century or even later.

In France, Mareschal and Vasseur (1992) made independent analyses of two boreholes and derived similar GST histories with peak warmth around 1000 AD, cooling to a minimum at 1700 AD, and warming starting at 1800 AD (see Section 3.6.2). Borehole temperatures also indicate increasing GST over the last century or two in Australia (Hyndman *et al.*, 1969) and the Ukrainian Shield (Diment, 1965), but an influence of deforestation was suspected. Where no ice was present in western Siberia, analyses indicate the most recent warming started 400 years ago in the south, and much earlier in the north (Duchkov and Devyatkin, 1992). Borehole measurements in New Zealand (Whiteford, 1993; Whiteford *et al.*, 1994) indicate cooling to a minimum around 1800 AD, followed by warming averaging 0.9°C over the past century.

Borehole temperature studies have also been conducted on ice sheets, and, especially those in Arctic regions, clearly show a warming since a cool period in the 1800s (e.g., Cuffey *et al.*, 1994).

3.2.5.3 Sea ice extent and mass

Neither hemisphere has exhibited significant trends in seaice extent since 1973 when satellite measurements began

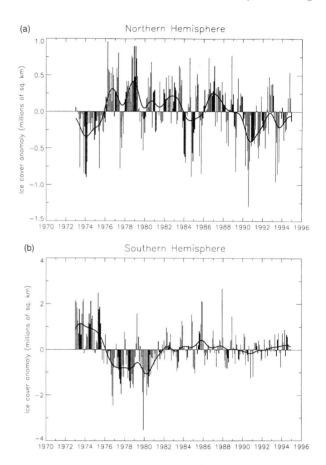

Figure 3.8: Sea ice extent anomalies relative to 1973–1994 for (a) the Northern Hemisphere and (b) the Southern Hemisphere. Data from NOAA (USA). Smooth lines generated from a 128-point binomial filter applied to the monthly anomalies. Heavy bars represent December–February in the Northern Hemisphere or June–August in the Southern Hemisphere.

(Figure 3.8). There has been below average extent in the Northern Hemisphere in the early 1990s, except for the second half of 1992 when the atmosphere was cooler. Coverage in the Southern Hemisphere has remained close to average. Jones (1995a) noted that the lack of sea ice variations around Antarctica seems unconnected to regional temperatures. For instance, sea ice did not increase during 1993 and 1994, despite low temperatures.

Sea ice total mass may be a more sensitive indicator of climate change than extent alone. Limited data on Arctic ice thickness, from upward sonar profiling from submarines and from moored subsurface sonar instruments, show large interannual variability but no trends from 1979 to 1990 (McLaren *et al.*, 1992).

3.2.5.4 Coral bleaching

Bleaching of coral reefs may result from high temperatures

or from other environmental stresses, e.g., pollution episodes. IPCC (1992) found the increased incidence of bleaching to be consistent with oceanic warming. Goreau and Hayes (1994) have shown that coral bleaching events since 1979 have been closely associated with warm-season sea surface temperature anomalies of 1°C or more. Bleaching events were, as a consequence, more prevalent in 1983, 1987 and 1991 in association with El Niño warming events, and scarce in 1992 following the eruption of Mt. Pinatubo and the subsequent cooling. Published records of coral reef bleaching from 1870 to the present suggest that the scale of bleaching since 1979 has not been observed previously (Glynn, 1993).

3.2.6 Mt. Pinatubo in the Temperature Record

Following the eruption of Mt. Pinatubo in June 1991, clear-sky solar radiation incident at the Earth's surface was reduced by about 3% for several months because of aerosol scattering (Dutton and Christy, 1992). Global tropospheric temperature anomalies cooled until August 1992 and then warmed in an irregular fashion (Figure 3.7a; see also Section 5.3.1.2). In the 16-year MSU lower troposphere time-series the coolest seasonal anomalies of the entire record occurred for the Northern Hemisphere in summer 1992 and for the Southern Hemisphere in summer 1992–3. Temperatures have increased since then, to levels similar to those prior to the eruption.

Seasonal series of land-surface air temperatures indicate that the cooling in 1992, which probably resulted from the Mt. Pinatubo eruption, was particularly marked in summer and autumn in the Northern Hemisphere. Robock and Mao (1995) found that this maximum cooling pattern, in the Northern Hemisphere summer of the year after the eruption, agrees with that after the five other largest volcanic eruptions since 1883. The winter warming pattern over the Northern Hemisphere continents in 1991–92 appears to be an indirect, dynamical response to the tropical stratospheric warming after the eruption (Graf *et al.*, 1993).

The rapid fall of globally averaged lower stratospheric temperature, as the aerosols dispersed, from the Mt. Pinatubo-induced peak in September 1991 to the minimum in late November 1993, represents a temperature change of about 2°C in just 28 months, with a 1.5°C cooling of seasonal averages (Figure 3.7b).

3.2.7 Possible Shift of Phase of the Annual Temperature Cycle

An analysis of monthly mean Central England Temperatures (CET) over the period 1659–1990 by Thomson (1995) indicated that there has been an observable phase shift in the mean annual cycle, implying, for example, a shift of peak

summer temperature from about 20 July to about 25 July. In addition, the rate of this phase shift appeared to have increased after 1940. Thomson also found such a shift in some, but not all, Northern Hemisphere stations with long records. Emslie and Christy (1995), however, could not confirm the apparent change in the phase trend in the CET. In addition, Karl *et al.* (1995b) concluded that the vagaries of weather could produce the phase shifts found by Thomson (1995) over century-scale periods.

3.2.8 Summary of Section 3.2

Global surface temperatures have warmed by 0.3–0.6°C since the late 19th century, with the greatest warming over the continents between 40°N and 70°N. A general, but not global, tendency to reduced diurnal range has been confirmed. Minimum temperatures have, in general, increased faster than maximum temperatures. As predicted in IPCC (1992) relatively cool surface and tropospheric temperatures, and a relatively warmer lower stratosphere, were observed in 1992 and 1993, following the 1991 eruption of Mt. Pinatubo. Warmer temperatures reappeared in 1994, with the global surface temperature for the year being in the warmest 5% of years since 1860, about 0.27°C higher than the 1961–90 mean.

Long-term variations in tropospheric temperatures, measured by radiosondes, have been similar to those in the surface record since the 1950s. More work has been done in comparing recent radiosonde and Microwave Sounding Unit observations of tropospheric temperature with surface temperatures, since IPCC (1992). Radiosonde and Microwave Sounding Unit observations of tropospheric temperature show slight overall cooling since 1979, whereas global surface temperature has warmed slightly over this period. After adjustment for the transient effects of volcanic activity and ENSO (which can bias trends calculated from short periods of record), both tropospheric and surface data show slight warming since 1979.

Indirect measures of temperature, such as borehole temperatures and glacier data, are in substantial agreement with the direct indicators of recent warmth. Sea ice extent does not show noticeable trends over the 22 years of available satellite data, although Northern Hemisphere sea ice has been below average in the early 1990s. Subsurface ocean measurements appear to be consistent with the geographical pattern of surface temperature changes.

3.3 Has the Climate Become Wetter?

3.3.1 Background

An enhanced greenhouse effect may lead to changes in the hydrologic cycle such as increased evaporation, drought, and precipitation. Unfortunately, our ability to determine the current state of the global hydrological cycle, let alone changes in it, is hampered by inadequate spatial coverage, inhomogeneities in climate records, poor data quality, and short record lengths. Nonetheless, some new aspects of changes and variations of the hydrological cycle have begun to emerge.

3.3.2 Precipitation

3.3.2.1 Land

IPCC (1990) concluded that precipitation over land was generally increasing in the extra-tropical areas, with a tendency for rainfall declines in the subtropics. It was noted in IPCC (1992) that precipitation over land is generally underestimated, typically by 10 to 15%, and that progressive improvements in instrumentation have introduced artificial, systematic increases in estimates of precipitation, particularly where snow is common. Nonetheless, the most reliable and useful measurements of multi-decadal precipitation variations are still these station data.

Figure 3.9a shows the change in precipitation from 1955–1974 to 1975–1994, expressed as a percentage of the 1955 to 1974 precipitation. The recent low rainfall in the Sahel is evident. Longer term trends in precipitation are

(a)

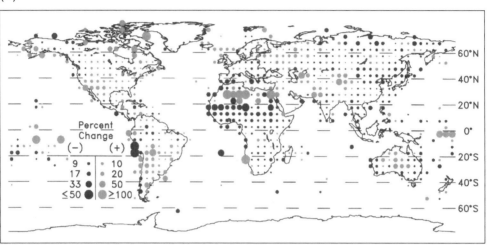

(b)

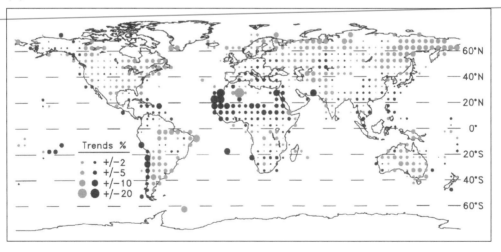

Figure 3.9: Precipitation changes based on the mean change from two data sets: "Hulme" (Hulme 1991; Hulme *et al.*, 1994) and the "Global Historical Climate Network-GHCN" (Vose *et al.*, 1992; Eischeid *et al.*, 1995). (a) Change in precipitation over land from 1955–1974 to 1975–1994, for 5° latitude by 5° longitude grid cells. Change expressed as percentage of 1955 to 1974 mean. (b) Trend in precipitation over land from 1900 to 1994. Average of trends from Hulme *et al.* (1994) and Eischeid *et al.* (1991) data sets. Trend (%/decade) expressed as percentage of the 1961 to 1990 mean from Hulme and 1951 to 1980 mean from GHCN. Magnitudes of the trends are depicted by the areas of the circles; shaded circles show increases; solid circles show decreases.

shown in Figure 3.9b. The 20th century rise in precipitation at high latitudes is clear, along with the decreases over the northern tropics of Africa. These trends are supported by regional and country studies, some of which have taken into account time-varying biases of precipitation measurements.

Groisman *et al.* (1991) evaluated information on the history of gauges and observational procedures over the former Soviet Union. Using this information they adjusted the measured precipitation for changes in instrumentation. Their adjusted data still show substantial increases of annual precipitation (~10%/100yr). The increase has been larger during winter than in summer. Much of the increase has been due to increases in the first half of the 20th century with a tendency for reduced precipitation in some areas since the middle of the century.

In North America annual precipitation has increased (Karl *et al.*, 1993b; Groisman and Easterling, 1994). The increase in the contiguous USA is most apparent after 1950 and is in large part due to increases during the autumn (September to November). Positive trends are apparent throughout much of northern Canada and Alaska during the past 40 years (Groisman and Easterling, 1994). Twentieth century station data over southern Canada and the northern USA (45–55°N) indicate increases of precipitation of 10–15%, with increases of about 5% averaged across the contiguous USA. Increases are more prevalent in the eastern two-thirds of North America (Findlay *et al.*, 1994; Lettenmaier *et al.*, 1994).

Precipitation changes in Europe (not including the former Soviet Union European countries) during the past century are latitudinally dependent. Time-varying measurement biases have been addressed in several countries (Hanssen-Bauer and Førland, 1994), where they are likely to be most serious (Karl *et al.*, 1993b). Overall, the data suggest an increase of precipitation in northern Europe and a decrease in southern Europe during the 20th century. In northern Europe (north of 55°N) Hanssen-Bauer *et al.* (1990) and Jónsson (1994) report an increase of precipitation since the 1960s for the Norwegian Arctic islands and Iceland while increasing precipitation has also been observed over the Faroe Islands (Frich, 1994). Denmark has had similar changes (Brázdil, 1992). West of the Scandinavian mountains increases have been reported (Frich, 1994; Hanssen-Bauer and Førland, 1994), but not in Finland (Heino, 1994). There is strong seasonality associated with the increases as they are in large part due to the changes during the autumn. In central Europe no significant positive trends have been observed over the past century (Brázdil, 1992). In southern Europe decreases tend to dominate (e.g., Palmieri *et al.*, 1991; Dahlström, 1994), but with some evidence for increased autumn precipitation over the maritime portions of south-western Europe (Mendes, 1994).

A long (1757–1992) record of areal average precipitation over Scotland has been analysed (Smith, 1995). This data set is believed to be consistent back to at least 1869. Annual precipitation has increased significantly, especially since the late 1970s, although summer rainfall has decreased. The recent increased precipitation is the largest sustained anomaly in the record.

Time-varying precipitation biases in the tropics and subtropics are not as severe as those in the mid- and high latitudes where higher wind speeds, smaller droplet size, and frozen precipitation all contribute to higher biases. The 20th century increase in precipitation in high latitudes is not matched by increases through the continental tropical and subtropical climates. Precipitation decreases, especially since mid-century, dominate large regions of the tropics and subtropics from North Africa east to Southeast Asia and Indonesia (Figures 3.9 and 3.10). Many of these areas with recent decreases are areas where droughts usually accompany El Niño episodes. So the decreases probably reflect the influence of the recent relatively frequent El Niño episodes (Section 3.4.2).

Consistent continental-scale trends of precipitation are not apparent through Central and South America. In part, this is due to the ENSO phenomenon, the influence of which varies across the continent. The relatively frequent El Niño episodes during the past few decades has led to decadal-scale precipitation variations in some areas. For example, precipitation has decreased over the western slopes of Central America, where the influence of ENSO is clearly evident, but March-May rainfall in the Parana-Paraguay River Basin of south-central South America has increased.

Australian and New Zealand precipitation is also heavily influenced by ENSO, and is characterised by large inter-annual fluctuations. There is evidence of an increase of summer rainfall (Nicholls and Lavery, 1992) in much of eastern Australia after the 1940s (with clusters of wet years in the 1950s and 1970s), although the north-east has suffered from frequent droughts in the last decade, associated with the relatively frequent El Niño events. The last four decades have seen a decrease in eastern Australian annual rainfall. There has been a long-term sustained decrease of winter rainfall in the far south-west, which has been linked to regional circulation changes (Allan and Haylock, 1993; see Section 3.4.4) and perhaps exacerbated by local changes in vegetation (Lyons *et al.*, 1993). Many

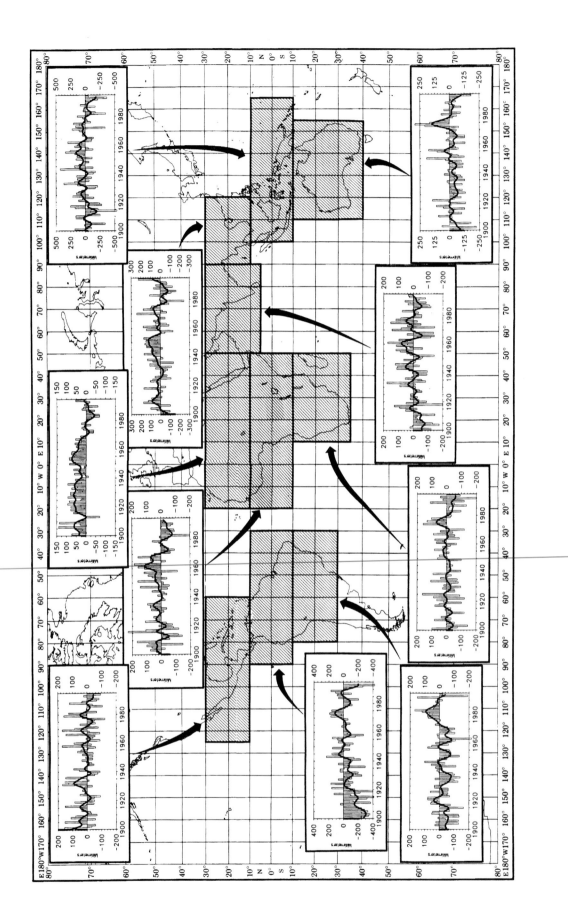

Figure 3.10: Variations of tropical and subtropical land-surface precipitation anomalies based on the average of the anomalies relative to 1961–90 from Hulme and GHCN. Smooth curves generated from nine-point binomial filters of the annual anomalies.

areas in the north and east of New Zealand have been drier since the mid-1970s, with areas in the south and west wetter (Salinger *et al.*, 1992).

One of the most complex and important precipitation systems in the Northern Hemisphere is the Asian summer monsoon. There has been a notable absence of very heavy monsoon onset (May–June) rains along the Yangtze River Valley of China ([†]Chen *et al.*, 1992) since about 1970, although total warm season rainfall has changed little. Indian summer monsoon rainfall reveals high interannual and decadal variability (Sontakke *et al.*, 1992, 1993; Parthasarathy *et al.*, 1992, 1994). The data indicate two wet periods of about 30 years (~1871–1900 and 1931–60) and two drier 30-year periods (1901–30 and 1961–90). Figure 3.10 indicates low rainfall through much of south and south-east Asia in the last few decades. At least some of this decrease reflects the influence of the recent frequent/intense El Niño episodes.

Throughout Africa marked variations in rainfall are evident (Figures 3.9 and 3.10). Rainfall in Sahelian West Africa from the late 1960s to 1993 was well below the amounts received earlier in the century, and over the last few decades has been about half that of the wet 1950s. Similar dry periods occurred during the historical and recent geologic past (Nicholson, 1994a,b; see Section 3.6.2). The recent period exhibits some differences in character from previous dry periods, namely a tendency towards continental-scale dryness (Nicholson, 1995). In recent years precipitation in eastern Northern Africa and the Arabian Peninsula has returned to levels more typical of the first-half of the century. Sahel rainfall in 1994 was greater than in any year since the 1960s.

Concerns about a possible link between climate change and desertification indicate the need for monitoring of arid region precipitation. The only large arid region with a strong trend to lower rainfall during the 20th century is Northern Africa, where rainfall has been low over the last few decades (Figure 3.10).

There have been several attempts to construct long-term instrumental worldwide precipitation time-series for hemispheric and global land areas (Bradley *et al.*, 1987; Diaz, et al., 1989; Eischeid *et al.*, 1991; Hulme, 1991; Hulme *et al.*, 1994). Figure 3.11 shows zonal average annual precipitation over land, in large latitudinal bands. There has been a substantial decrease of precipitation over the Northern Hemisphere subtropics primarily over the last three decades. Increases have occurred in the mid- and high latitudes. Since bias corrections have not yet been applied to the North American data in either the Eischeid *et al.* (1995) or the Hulme (1991) data sets these trends are likely

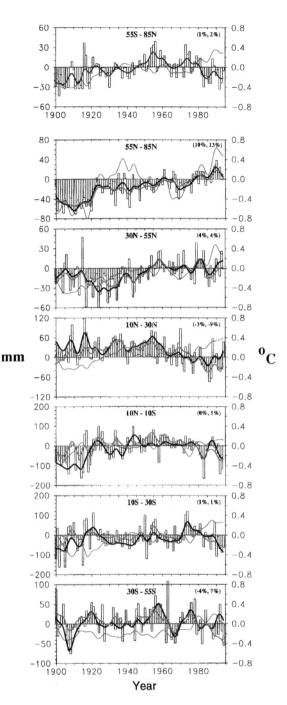

Figure 3.11: Annual and smoothed precipitation anomalies and smoothed temperature anomalies from 1961 to 1990 mean. Smooth curves were created using nine-point binomial filters of annual precipitation (blue – Hulme; black – GHCN; the green line in the top panel is the average of Hulme and GHCN), and temperature (red – Jones, 1994a). Annual bars are the average of Hulme and GHCN. Top panel is for the band 55°S to 85°N. Values in parentheses in upper right corner of each panel are the trend (%/century) of precipitation for GHCN (based on 1951 to 1980 means) and Hulme (1961 to 1990 means), respectively.

to be exaggerated. Based on the regional and country-wide studies where adjustments have been made however, the sign of the mid- to high latitude trend is not in doubt. Precipitation trends over land in the tropics and Southern Hemisphere are not statistically significant, although they indicate decreased precipitation in recent years.

In summary, the best evidence available suggests there has been a small positive (1%) global trend in precipitation over land during the 20th century, although precipitation has been relatively low since the late 1970s (Figure 3.11).

3.3.2.2 Ocean

Measurement of changes of precipitation over the oceans presents extreme difficulties (IPCC, 1990, 1992). Multi-decadal surface-based observations from gauges are limited to very small atolls. Space-based observations, although providing more or less continuous coverage, have several limitations, not least of which is that measurements to directly estimate precipitation are not available until 1979, with the operation of the Micro-Wave Sounding Unit (MSU) aboard NOAA polar orbiting satellites. As a result, researchers have had to resort to the use of data such as real-time cloud imaging, atmospheric temperature profiling, etc. to infer changes of precipitation.

Three methods (Garcia, 1981; Arkin and Ardanuy, 1989; Spencer, 1993) have been used to infer decadal variability of precipitation over the oceans from space-based instruments. These are: identifying Highly Reflective Clouds (HRC); measuring cloud top temperatures from Outgoing Long-wave Radiation (OLR); and the use of the Microwave Sounding Unit (MSU). The HRC data is limited by poor resolution of the diurnal cycle, the subjective nature of the technique, variations from satellite to satellite, and spatial sampling primarily confined to the tropics and subtropics. The OLR data are limited to those areas where precipitation is dominated by cold cloud top temperatures, again principally the tropics and subtropics. Additionally, they are based on observations from a number of instruments on successive satellites, and temporal inhomogeneities associated with satellite drift and instrument calibrations have affected the data set. Therefore, any estimate of precipitation change based on the HRC and OLR data sets is embedded with internal inhomogeneities. Being a more direct measure of rainfall, the MSU data avoid some of the limitations of the HRC and OLR. Nonetheless, these measurements are spatially limited, restricted to unfrozen ocean areas, liquid precipitation, and suffer from the low spatial resolution of the instrument, as well as diurnal sampling inadequacies (Negri *et al.*, 1994).

In IPCC (1990) evidence was presented to suggest that the OLR data analysed by Nitta and Yamada (1989) indicated an increase of tropical precipitation. A re-examination of the biases inherent in these data led Chelliah and Arkin (1992) and IPCC (1992) to conclude that a substantial portion of the increase of precipitation since 1974 was not climate-induced because of OLR biases due to changes in equatorial crossing time and uncertainties in sensor calibration (Gadgil *et al.*, 1992). However, new analyses by Graham (1995) using satellite data, a coarse Tropical Pacific Ocean island data set, and model-derived moisture flux convergence based on observed SSTs suggest an increase of tropical Pacific precipitation between two six-year periods before and after 1976, the time of the commencement of a decadal-scale fluctuation in the atmospheric circulation over the Pacific Ocean (Trenberth and Hurrell, 1994; see Section 3.4.3). Salinger *et al.* (1995a) analysed trends of precipitation from tropical Pacific atolls and islands. Central and eastern equatorial Pacific rainfall increased in the mid-1970s, concomitant with the change to increased frequency of El Niño episodes and the observed decreased precipitation in the south-west Pacific and in some land areas affected by ENSO (Figure 3.10).

3.3.3 Concomitant Changes of Precipitation and Temperature

Strong low-frequency relationships exist between temperature and land-based precipitation, on regional space-scales. Figure 3.11 indicates that the relationship between multi-decadal fluctuations of precipitation and temperature is latitudinally dependent. For the mid- to high latitudes of the Northern Hemisphere simultaneous increases of temperature and precipitation have been observed, but for the tropics, subtropics, and mid-latitudes of the Southern Hemisphere warmer temperatures have been associated with decreases of precipitation. This latter characteristic is a dominant feature of the recent increase of global temperature beginning in the mid-1970s.

3.3.4 Snow Cover, Snowfall, and Snow Depth

The lack of long-term homogeneous data is the major obstacle to evaluation of decadal-scale changes in land-surface cryospheric variables. Recent analyses however, have made some progress toward improving our understanding of changes in these variables.

Northern Hemisphere land-surface snow cover extent has decreased in recent years (1988–1994) of the 21-year period-of-record of satellite data (Robinson *et al.*, 1993; Groisman *et al.*, 1994a). The deficit of snow has been particularly apparent in spring (Figure 3.12). The snow cover of summer and autumn of these recent years has also been low, while winter snow cover exhibits less apparent decadal changes. The annual mean extent of snow cover has decreased by about 10% during the past 21 years over the Northern Hemisphere. Percentage decreases have been similar over North America and Asia (Groisman *et al.*, 1994b). The decrease in snow cover extent is closely coupled to an increase in temperature (Figure 3.12). The warming has been strong in spring over the northern land areas (Chapman and Walsh, 1993; Parker *et al.*, 1994). This is also reflected in earlier lake ice melting as reported by Skinner (1993), earlier snowmelt related floods on rivers in west-central Canada (Burn, 1994) and California (Dettinger and Cayan, 1995), and reduced duration of river ice over the former Soviet Union (Ginsburg, *et al.* 1992; Soldatova, 1993). Groisman *et al.* (1994a,b) have shown that snow-radiation feedback can account for up to 50% of the springtime April-May warming over the Northern Hemisphere land areas since the early 1970s. These results help explain why the increase of surface air temperatures over the Northern Hemisphere land areas has been more significant in spring than in other seasons.

While satellite remote sensing is being explored as a source of data on snow depth water equivalent, station data provide the only information on snowfall and snow depth over periods longer than about two decades. Station measurements indicate that annual snowfall has increased over the period 1950–90 by ~20% over northern Canada (north of 55°N) and by ~11% over Alaska (Groisman and Easterling, 1994). Total precipitation has increased in all these regions (Section 3.3.2.1); in southern Canada and the northern USA however, the increase of precipitation has been accompanied by higher ratios of liquid to solid precipitation as the temperatures have also increased. A small decrease of snowfall has been observed in the 45–55°N latitudinal belt. Decadal summaries of snowfall measurements in China have been compiled by Li (1987), who found a decrease of snowfall over China during the 1950s followed by an increase during the 1960s and 1970s.

There have been several analyses of changes in snow depth over Europe and Asia. Snow depth responds to both changes of precipitation and temperature. A synthesis of 20th century station snow depth measurements for the former Soviet Union (Meshcherskaya *et al.*, 1995) indicates decreases (~14%) of snow depth during February over the European portion of the former Soviet Union with a smaller decrease over the Asian sectors of the former Soviet Union where snow depth has actually increased since the 1960s.

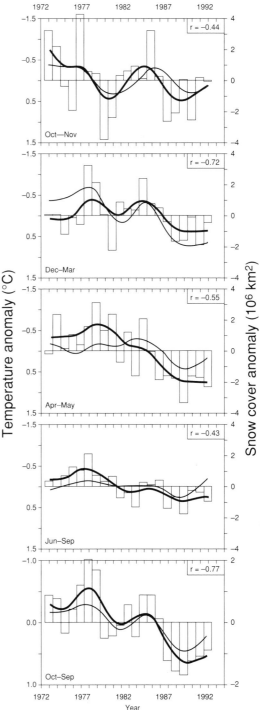

Figure 3.12: Seasonal and annual variations of Northern Hemisphere land-surface snow cover extent (Greenland excluded) and the surface temperature over regions of transient snow cover. Yearly anomalies (shown as bars) are given for snow cover extent. Smooth curves were created using nine-point binomial filters for yearly snow cover (thick) or temperature anomalies (thin) with scale reversed. "r" indicates correlation between annual values. Note bottom panel is for "snow year" (October–September).

3.3.5 Land-surface and Subsurface Water

Changes of precipitation and/or evaporation may lead to changes in runoff or soil moisture storage. Thus changes in streamflow, lake levels, levels of inland seas, and soil moisture may provide information about changes in the hydrological cycle. Data on many of these aspects of the hydrological cycle are suspect because of human influences, such as building of dams to regulate streamflow, so it is important to develop an internally consistent scenario of change.

3.3.5.1 Streamflow

Historical records of 142 rivers throughout the world with more than 50 years of historical data and drainage areas larger than 1000 km^2 were analysed for trends by Chiew and McMahon (1993, 1995). Although statistically significant trends and changes in means were detected at several locations, they were not always consistent within regions. No clear evidence of wide-spread change in the annual streamflow and peak discharges of rivers in the world was found.

Analyses from a number of streamflow gauging stations on major drainage basins across South America do not reveal general increases or decreases of streamflow. There are however, a number of drainage basins that depict important decadal-scale climate variations (Marengo, 1995). Reduced streamflow beginning in the 1970s in Colombia is consistent with similar behaviour in Pacific Central America. The Chicama and Chira River Basins in north-west Peru also reflect large interdecadal variations. Streamflow of the Rio Negro at Manaus is closely related to rainfall in north-west Amazonia (Marengo, 1992, 1995). Consistent trends are absent from the record, although there appears to be a recent (since 1973) minor short-term

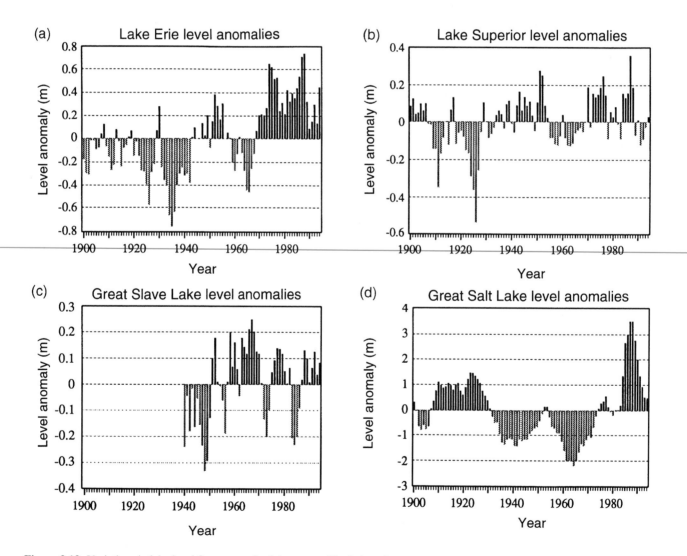

Figure 3.13: Variations in lake level for some major lakes across North America.

increase of streamflow, probably related to an increase in rainfall associated with increased convection over Amazonia as reported by Chu *et al.*, (1994) who also found little change in convective activity over areas of deforestation in south-western Amazonia. Generally, for South America, no clear unidirectional trend of streamflow is evident for the records analysed to date. A strong signal of the El Niño-Southern Oscillation is evident in several regions. These variations often dominate decadal-scale variability of streamflow. The lack of an overall trend is consistent with the precipitation variations.

In North America, the autumn increase of precipitation in the contiguous USA is also reflected in an increase in streamflow. Lins and Michaels (1994) found statistically significant increases in natural streamflow during the autumn and winter in nearly all regions of the contiguous USA (1948-88). The most significant (statistically) increases were generally found from the Rocky Mountains eastward to the Atlantic.

3.3.5.2 *Lake levels*

Large lakes and inland seas can serve as useful indicators of climate variability and change. With large surface areas and limited outflow capacity, these natural reservoirs filter out short-term variability and respond to longer term variations in the hydrologic cycle. However, local effects often confound the use of lake levels to monitor climate variations.

The historically small variations of levels of the North American Great Lakes are mainly due to natural changes in lake levels (Magnuson *et al.*, 1995). The lakes have been in a regime of high water levels since the late 1960s culminating in record high levels in 1986 (Changnon, 1987). A major drop occurred in response to an intense drought in 1988, but lake levels continue to remain above the long-term mean for Lakes Michigan, Huron, and Erie. Lake Superior has been slightly below its long-term average since 1988 (Figure 3.13). Levels of the Great Salt Lake in Utah are somewhat similar to those in the Great Lakes, but the Lake has recently been regulated with the introduction of pumps to relieve the historic peak water levels of 1986–87. The Great Slave Lake, located in the Northwest Territories of Canada has also continued to be at above average levels for the past several years compared to the relatively low levels observed during the 1940s. Although there has been regulation on one of the rivers flowing into the Great Slave Lake since 1968, comparison of the water levels of the Great Slave and Great Bear Lakes suggests it has had limited influence on annual or decadal fluctuations.

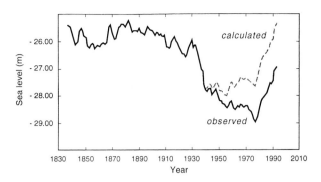

Figure 3.14: Annual mean sea level of the Caspian Sea as observed (solid line) and as calculated without local anthropogenic changes in land-use (dashed line) (updated after Shiklomanov, 1976).

The Caspian Sea is the largest closed water body in the world. Its sea level has been measured since 1837 (Figure 3.14) and has fluctuated by nearly 4 m. Sea level was quite stable until the 1930s when it fell 1.6 m in six years. The level abruptly began to rise in 1977, from record low values, perhaps the lowest values since the 14th century. The main contribution to the rapid decrease of sea level in the 1930s was reduced runoff into the Sea, but the most recent increase is attributed to a combination of reduced evaporation (Section 3.3.6.1; Figure 3.15), increases in runoff, and an engineered blockage of sea water into the Kara Bogaz Gol (which has since been reopened; Golitsyn *et al.*, 1990; Golitsyn, 1995).

Since 1965 lake levels in Northern Hemisphere Africa have declined sharply (Grove, 1995). The early 1960s were relatively wet, but the low Sahel/Sudan precipitation since then (Figure 3.10) led to Lake Chad shrinking from its highest level and extent in the 1960s to about one tenth of the area by the 1980s. This decline was due to decreased rainfall, and not the result of excessive extraction of water for irrigation.

3.3.5.3 *Soil moisture*

There are few long-term data sets of soil moisture measurements. One example is the so-called "water-balance" network within the former Soviet Union, at some places for more than 30 years (Robock *et al.*, 1995; Vinnikov *et al.*, 1995). In this network, total soil moisture measurements have been made over natural vegetation soil using a thermostat-weighing technique (Vinnikov and Yeserkepova, 1991). The data from the European part of the former Soviet Union exhibit a general increase in soil moisture from the 1970s to the 1980s, consistent with the increase of precipitation observed over the European portion of the former Soviet Union, and a reduction of evaporation (Section 3.3.6.1).

3.3.6 Evaporation

Systematic long-term measurements of surface evaporation have been carried out over many continental areas, but few analyses of their trends have been completed. Changes in evaporation over the ocean have proven to be particularly difficult to monitor, and are prone to time-varying biases.

3.3.6.1 Land

Estimates of evaporation over land have been obtained from pan evaporimeters. Although these measurements can neither be considered as evaporation from lakes or reservoirs, nor as evaporation from ground, they can be used as a composite index that characterises the annual and seasonal water and heat balance between the water surface of the evaporimeter and the atmosphere. Trends in such an

evaporation index from the former Soviet Union have recently been analysed (Golubev and Zmeikova, 1991) for a network of 190 stations. Although some measurements date back to the late 19th century, changes in instrumentation, observational procedures, and data management practices make the data inhomogeneous prior to 1951. The analysis indicates a reduction of evaporation since 1951 during the warm season (the freeze-free period) over the European and Siberian (north of 55°N) portions of the former Soviet Union (Figure 3.15), with little change in central Asia and Kazakhstan. The rapid decrease in evaporation in the European sector in 1976 corresponds with the timing of the decrease in the diurnal temperature range (Section 3.2.2.4 and Karl *et al.*, 1993a; Peterson *et al.*, 1995) over the former Soviet Union. In the USA, where

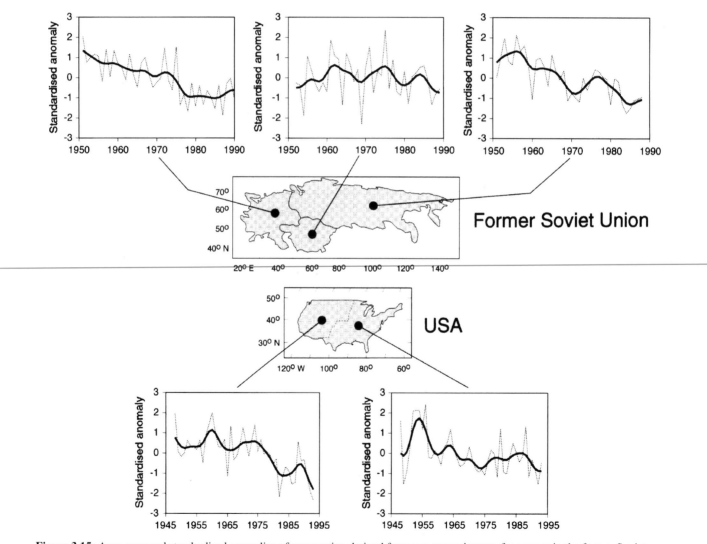

Figure 3.15: Area-averaged standardised anomalies of evaporation derived from pan evaporimeters for sectors in the former Soviet Union in the warm period of the year (updated from Golubev and Zmeikova, 1991), and for the USA (Peterson *et al.*, 1995). The dashed lines represent interannual variations and the smoothed curves suppress variations on time-scales of a decade or less.

data from a network of over 300 stations are available but have yet to be thoroughly inspected for inhomogeneities, an overall decrease in evaporation is also observed (Figure 3.15; Peterson *et al.*, 1995) that closely resembles the decrease in the diurnal temperature range.

3.3.6.2 Ocean

The oceans provide a source of moisture for the atmosphere and subsequent precipitation over land. Estimating long-term changes of oceanic evaporation is very difficult, and all estimates must be viewed with considerable caution due to time varying biases in the data. Bulk aerodynamic parametrizations are the only practical means of comprehensively calculating evaporation over regions as large as ocean basins and over multi-decadal time-scales. Routine marine weather observations of wind, temperature, and humidity, among other elements, are available over broad sectors of the oceans for the past four decades. These observations allow estimation of evaporation based on the sea surface saturation specific humidity, the specific humidity of the overlying air, and the near-surface wind speed.

Marine weather observation procedures were not designed to provide long-term homogeneous climate data, and several problems exist that can obscure climate-induced changes. Marine data show an increase of wind speed (Cardone *et al.*, 1990) over the four decades since World War II, believed to be largely artificial, involving changes from Beaufort scale sea state estimates of wind speed to anemometer observations (Ramage, 1987). Posmentier *et al.* (1989) argued an increase in wind is inconsistent with observed trends in tropical sea level and sea surface temperature, which in the eastern tropical Pacific have increased rather than decreased as would be expected under an increasing trade wind regime. The notion of no global increase in wind speed is consistent with the calculations of Ward (1992, 1994) who used sea level pressure gradients to estimate changes in wind speed. Even after removing the apparent wind bias however, evaporation was estimated to have increased by more than 0.5 mm/day over the tropical oceans (10°S to 14°N) during the last four decades (Flohn *et al.*, 1992). Graham (1995) used model calculations to suggest that increases of evaporation should also have occurred in the tropical Pacific, even without any increase of wind speed.

Weare (1984) and Cayan (1992) indicate that anomalous evaporation is influenced about equally by fluctuations in wind speed and fluctuations in vertical moisture gradients. Furthermore, the calculations of evaporation trends over much of the Atlantic indicate that evaporation has increased both because of vertical moisture gradients and increases in wind speed. That is, evaporation trends would still be positive even without any change in wind speed. Increases in evaporation are found in both winter and summer.

Other areas of the oceans are known to exhibit multi-decadal changes in evaporation. For example, there was a transition from warm to cool conditions in the western and central North Pacific associated with a cluster of winters with deep Aleutian Lows beginning in the winter of 1976–77 (Douglas *et al.*, 1982; Venrick *et al.*, 1987; Trenberth, 1990; Miller *et al.*, 1994; Trenberth and Hurrell, 1994). Since such transitions involve a redistribution of the paths of storms across the Pacific basin there are large-scale changes in the wind, temperature, and humidity with concomitant changes of evaporation. The magnitude of these changes over the North Pacific basin is estimated to be between 0.2 and 0.4 mm/day, with evaporation increasing over the western North Pacific, but decreasing over the Gulf of Alaska.

3.3.7 Water Vapour

Water vapour is the most abundant greenhouse gas and makes the largest contribution to the natural greenhouse effect. Half of all the moisture in the atmosphere is between sea level and 850 hPa and less than 10% is above 500 hPa. The amount in the stratosphere is probably less than 1% of the total. Despite the small amount of water vapour in the stratosphere, changes there may lead to significant changes in the radiative forcing of climate (Oltmans and Hofmann, 1995).

Monitoring atmospheric moisture presents many difficulties. Because the residence time of water vapour in the atmosphere is short, about 10 days, its distribution is horizontally, as well as vertically, heterogeneous. Therefore observations at many places and levels are necessary for adequate representation of climatically important changes. Measurement problems also make detecting trends of water vapour difficult. Most of the information about tropospheric water vapour comes from radiosonde measurements taken for daily weather forecasts. Unfortunately, this instrumentation is not capable of accurate measurements at the low temperatures and very low moisture conditions of the upper troposphere, stratosphere, and polar regions of the planet. Another serious problem with radiosonde data stems from the changes of instruments and reporting practices that have occurred through the period of record. About a dozen different manufacturers supply radiosondes to the world's weather services, some with quite different sensor designs.

These factors make it difficult to separate climate changes from changes in measurement programs (Elliott and Gaffen, 1991). Our inability to account for these inhomogeneities limits assessments of the change in atmospheric moisture to the last 20 years, thereby making it difficult to distinguish between transient phenomena, such as ENSO, and long-term change. Nevertheless, with careful attention to these problems, some estimates of tropospheric water vapour can be obtained.

Elliott *et al.* (1995) found increases in precipitable water from 1973 to 1993 over all of North America except northern and eastern Canada, where it decreased slightly. The positive trends tended to increase with decreasing latitude, with maximum values of about 3 mm/decade. The increases as a percentage of the annual mean also tended to be greater at lower latitudes, approaching a maximum of 8%/decade. The regions of moisture increase were generally also regions with rising temperatures over the same period and the regions of moisture decrease were generally regions with decreasing temperatures. There was a strong (but not universal) tendency for the moisture increases to be accompanied by increases in relative humidity.

There have been several analyses of changes in water vapour over the tropical oceans using radiosonde data. Hense *et al.* (1988) report an upward trend of moisture in the 700–500 hPa layer in the western Pacific from 1965 to 1986, but the data prior to 1973 are not necessarily homogeneous because of a change in instrumentation in 1973 (Elliott and Gaffen, 1991). Elliott *et al.*, (1991) document a moisture increase in the equatorial Pacific from 1973-86 and Gaffen *et al.* (1991) using more stations also found an increase of specific humidity on the order of 10% from 1973 to 1986, but the scatter of data is such that this value is only a rough estimate of the order-of-magnitude of the change. Much of the change occurred during a relatively brief interval, beginning about 1977 (see Section 3.4.3). An increase in tropical western Pacific moisture was also found by Gutzler (1992) at four tropical island stations. There were increases in humidity throughout the troposphere with the largest increases near the surface. The study indicates an increase in water vapour of about 6%/decade for the period 1973–1988.

Gaffen *et al.* (1992) computed trends of water vapour for some 35 radiosonde stations. Most of these stations, especially in the western tropical Pacific, showed an increase in precipitable water during the period 1973–90. The largest and most statistically significant trends were at tropical stations, where increases as large as 13%/decade were found.

Oltmans and Hofmann (1995) analysed stratospheric and upper tropospheric water vapour measurements (10–26

km) over Boulder, Colorado during the period 1981–94. Over this short period they found positive trends of water vapour in the stratosphere, but statistically significant increases, of between 0.5 and 1%/yr, were confined to the layers between 16 and 24 km. Data from other locations must be analysed and over longer time intervals however, before these results can be generalised.

3.3.8 Clouds

In addition to providing evidence of an enhanced hydrological cycle, changes of cloud distribution can have very important climatic feedbacks. Changes in clouds modify both incoming and outgoing radiation depending on their level, amount, vertical extent, and composition.

3.3.8.1 Land

There have not been any major new analyses of large-scale changes in land-based observations of cloud cover published since the IPCC (1990) report. Henderson-Sellers (1992) has, however, summarised her analyses of changes in cloud cover over Australia, Europe, India and North America. Added confidence has been attached to these analyses because the diurnal temperature range, which is quite sensitive to changes of cloud cover (Karl *et al.*, 1993a), has decreased in many of the regions with a reported increase in cloud cover. Moreover, Karl *et al.* (1995e) report a step-like increase in cloud amount over the former Soviet Union around 1976 consistent with the step-like decreases of evaporation (in the European sector of the former Soviet Union, Figure 3.15) and of the diurnal temperature range. In addition, analyses by Environment Canada (1995) generally support an increase in cloud amount throughout Canada since 1953, in particular over the lower Great Lakes region during autumn. There have been increases of cloud cover over Australia and the USA since at least the mid-20th century. In at least some areas (e.g., Australia), however, cloud cover over land may have decreased since about 1980 (Henderson-Sellers, 1992).

3.3.8.2 Ocean

Marine weather observations of clouds have been analysed. Satellite observations have also been used, but owing to instrumental changes and the shortness of their record, cannot be used with any degree of confidence to assess multi-year trends or changes.

Fifty million routine weather observations from ships were analysed by Warren *et al.* (1988), who computed average total cloud cover and the amounts of six cloud types for each of the four seasons for the 30 year period 1952 to 1981. During the 1952 to 1981 period London *et al.* (1991) reported a global ocean average increase of cloud cover of 0.7%. This increase was concentrated at low latitudes (20°N–20°S), and the cloud types contributing most to the increase were cumulonimbus and cirrus. The increase in cumulonimbus was nearly 3% and in cirrus about 1%, at the expense of cumulus clouds which decreased by about 1%. This data set has recently been updated through 1991, and indicates that cloud cover over the oceans has continued to increase.

Interannual variations of marine stratus and stratocumulus in the subtropics and mid-latitudes are inversely correlated with SST (Norris and Leovy, 1994). The 30-year trends of marine stratus and stratocumulus exhibit the same geographical patterns as do SSTs (cf. Parker *et al.*, 1994): increases in cloud amount over areas with decreasing temperatures in the northern mid-latitude oceans, and decreases with warming off the west coast of North America. Analysis of changes in cloud types (Parungo *et al.*, 1994) suggests an increase in altostratus and altocumulus cloud amount centred in the Northern Hemisphere mid-latitudes (30°N–50°N). Over the 30-year period 1952 to 1981 these mid-level clouds increased in frequency from 20 to 26%. This large apparent increase requires investigation of possible time-varying biases.

3.3.9 Summary of Section 3.3.

Precipitation has increased over land in high latitudes of the Northern Hemisphere, especially during the autumn. A decrease of precipitation occurred after the 1960s over the subtropics and tropics from Africa to Indonesia, as temperatures increased. Many of the changes in precipitation over land are consistent with changes in streamflow, lake levels, and soil moisture. There is evidence to suggest increased precipitation over the equatorial Pacific Ocean (near the dateline) in recent decades, with decreases to the north and south. Based on the available data, there has been a small positive (1%) global trend in precipitation over land during the 20th century, although precipitation has been low since about 1980.

Evaporation potential has decreased (since 1951) over much of the former Soviet Union (and possibly also in the USA). Evaporation appears to have increased over the tropical oceans (although not everywhere). There is evidence suggesting an increase of atmospheric water vapour in the tropics, at least since 1973. In general, cloud amount has increased both over land (at least up to the end of the 1970s) and the ocean. Over the ocean increases in both convective and middle and high level clouds have been reported.

3.4 Has the Atmospheric/Oceanic Circulation Changed?

3.4.1 Background

The atmospheric circulation is the main control behind regional changes in wind, temperature, precipitation, soil moisture and other climatic variables. Variations in many of these variables are strongly related through large-scale features of the atmospheric circulation, as well as through interactions involving the land and ocean surfaces (IPCC, 1990). Two well-known examples of such large-scale features are the El Niño-Southern Oscillation and the North Atlantic Oscillation, both of which are closely related to climatic fluctuations in many areas. Evidence of associated changes or variations in the atmospheric circulation may enhance confidence in the reality of observed changes and variations in the climate variables in these areas.

Comprehensive, long-term monitoring of changes and variations in the El Niño-Southern Oscillation and North Atlantic Oscillation, and the atmospheric circulation in general, requires the analysis of meteorological and oceanic fields on a routine basis. Analyses of such fields are performed every day by national meteorological services. However, the analysis schemes, their observational basis, and even the theoretical meteorology underlying the analysis have all changed dramatically over the decades (Trenberth, 1996). Thus, these analyses are of only limited use for analysis of variations and changes in circulation over extended periods. Where such analyses are used, their results need to be confirmed by analysis based solely upon high-quality station data. The lack of homogeneity in the atmospheric analyses restricted the variety of atmospheric circulation features IPCC (1990) decided could be examined with confidence. The same problem occurred in IPCC (1992), and again here only a small number of the more reliable circulation features are examined, namely the El Niño-Southern Oscillation and the North Atlantic Oscillation. The discussion of a possible recent increase in the intensity of the winter atmospheric circulation over the extra-tropical Pacific and Atlantic (IPCC, 1990) is also updated.

Ocean circulation variations and change can also be important determinants of climate variations, so their routine monitoring is also necessary. However, little information is available, as yet, on variations and change in ocean circulation on the time-scales of relevance to this chapter.

3.4.2 El Niño-Southern Oscillation

The El Niño-Southern Oscillation (ENSO) phenomenon is the primary mode of climate variability in the 2–5 year time band. Release of latent heat associated with ENSO episodes affects global temperature (Pan and Oort, 1990; Graham, 1995), and associated changes in oceanic upwelling influence atmospheric CO_2 levels (Keeling *et al.*, 1989). It is important to assess the stability of the ENSO, i.e., to determine how long it has been operating, and whether it exhibits longer-term variations in its influence.

As instrumental records for the El Niño-Southern Oscillation extend back only to the late 19th century, it is necessary to use various proxy approaches to derive the history of El Niño-Southern Oscillation events. In addition to analyses of historical records (Quinn, 1992), corals, tree rings, tropical ice cores, and varve sediments have also been employed to derive temporal histories of ENSO variability (Diaz and Markgraf, 1992; Cole *et al.*, 1993; Diaz and Pulwarty, 1994). Whetton *et al.* (1995) demonstrate that proxy records of ENSO from different regions do not always correspond with each other. One coral record (Dunbar *et al.*, 1994) is from the Galapagos Islands, where a time-series of oxygen isotope measurements indicates significant variations in the El Niño-Southern Oscillation. Tree ring and other records also indicate temporal variations in some of the ENSO periods (Diaz and Markgraf, 1992).

Instrumental records have also been examined to search for possible changes in the El Niño-Southern Oscillation. The Southern Oscillation Index (SOI), a simple index of the El Niño-Southern Oscillation based on surface atmospheric pressures at Darwin and Tahiti, is plotted in Figure 3.16. El Niño (or warm) events are associated with large negative excursions of the SOI and warmer than normal temperatures in the central and east equatorial Pacific. La Niña episodes (or cold events, when the east equatorial Pacific is cool) are associated with positive values of the SOI. The dominance of time-scales of 2–5 years in the El Niño-Southern Oscillation is apparent in this figure. The periodicity of El Niño-Southern Oscillation has varied since 1950. Between 1950 and 1965 the phenomenon exhibited a period of about five years. Since 1965 the period has shortened somewhat, closer to four years. A similar pattern of change in frequency was evident in a central equatorial Pacific coral record of the El Niño-Southern Oscillation (Cole *et al.*, 1993). El Niño events occurred more often around the start of the 20th century than during the 1960s and 1970s (Anderson, 1992).

Allan (1993) and Karoly *et al.* (1996) have documented decadal-scale variations in the El Niño-Southern Oscillation and its teleconnections over the last century.

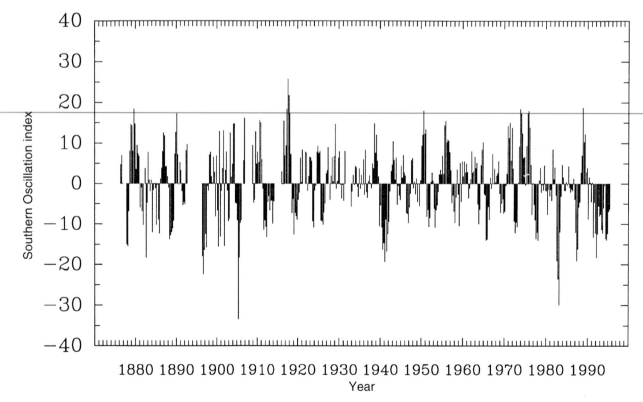

Figure 3.16: Seasonal values of the Troup Southern Oscillation Index (SOI), March–May 1876 through March–May 1995. SOI only plotted when both Darwin and Tahiti data available.

The most robust and coherent signals were evident in the late 19th century, and since the 1940s. ENSO activity was weaker and more fragmented in the intervening period. Wang (1995) found that the characteristics of the onset of the Pacific-wide warming associated with an El Niño changed after the mid-1970s.

Indeed, there appears to have been a rather distinct change in ENSO in 1976/77 (Figure 3.16). Since then, there have been relatively more frequent El Niño episodes, with only rare excursions into the other extreme (La Niña episodes) of the phenomenon (Trenberth and Hurrell, 1994), and the SOI has tended to be negative for extended periods. During this period sea surface temperatures in the central and equatorial Pacific have tended to remain anomalously high, relative to previous decades. As well, precipitation over land in many areas where dry conditions usually accompany El Niño episodes (e.g., Indonesia, north-east Australia) has been low.

The recent ENSO behaviour, and especially the consistent negative SOI since 1989, appears to be unusual in the context of the last 120 years (the instrumental record). However, some data for Tahiti is missing from before 1935, so it is difficult to compare the length of the post-1989 period of negative SOI with earlier periods. It may be, for instance, that the period of negative SOI values starting around 1911 was comparable with the recent period. Trenberth and Hoar (1996) examined the recent El Niño-Southern Oscillation behaviour using Darwin data only. The Darwin data are continuous and therefore allow a strict comparison of the post-1989 behaviour with that of earlier periods, although by themselves they may not provide a complete representation of the El Niño-Southern Oscillation. The Trenberth and Hoar analysis indicates that the post-1989 behaviour is very unusual. From the Darwin data, there has been no period in the last 120 years with such an extended period of negative SOI. Trenberth and Hoar fitted statistical models to the 1882 to 1981 Darwin data, to determine how unusual the recent behaviour has been. They concluded that the 1990 to 1995 behaviour had a probability of natural occurrence of about once in 2,000 years. Allan and D'Arrigo (1996), however, present evidence that episodes with similarly persistent ENSO characteristics have occurred prior to the period of instrumental data.

3.4.3 Northern Hemisphere Circulation

In the past few years considerable work has been done to document decadal and longer time-scale variability regionally in several parts of the globe, especially the North Pacific (IPCC, 1990; Trenberth and Hurrell, 1994;

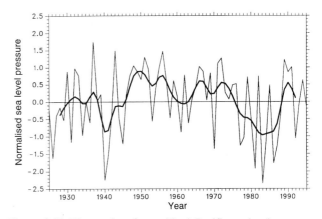

Figure 3.17: Time-series of mean North Pacific sea level pressures averaged over 30–65°N, 160°E to 140°W for the months November to March beginning in 1925 and smoothed with a low pass filter (thick line). Updated from Trenberth and Hurrell (1994).

Hurrell, 1995). Substantial changes in the North Pacific atmosphere and ocean began about 1976. This is illustrated in Figure 3.17 which shows a time-series of mean sea level pressures for the five winter months November to March, averaged over most of the extra-tropical North Pacific (updated from Trenberth and Hurrell, 1994). This index is closely related to changes in the intensity of the Aleutian low pressure centre, and to a pattern of atmospheric circulation variability known as the Pacific-North American (PNA) pattern (Wallace and Gutzler, 1981). The abrupt shift in pressure after 1976 is evident. Observed significant changes in the atmospheric circulation involving the PNA teleconnection pattern were described in IPCC (1990). Associated changes occurred in surface winds. The storm tracks and associated synoptic eddy activity shifted southward. This regime appears to have ended about 1989.

The Pacific decadal time-scale variations have been linked to recent changes in the frequency and intensity of El Niño versus La Niña events and it has been hypothesised that the decadal variation has its origin in the tropics (Trenberth and Hurrell, 1994). While some analyses of observations (Deser and Blackmon, 1995; Zhang *et al.*, 1995) show that this link is not linear, observational studies by Kawamura (1994), and Lau and Nath (1994) have shown that the decadal variation in the extratropics of the Pacific is closely tied to tropical sea surface temperatures (SSTs) in the Pacific and Indian Oceans. Several aspects of the decadal-scale fluctuation beginning around 1976 have been simulated with atmospheric models using specified SSTs (Kitoh, 1991; [+]Chen *et al.*, 1992; Miller *et al.*, 1994; Kawamura *et al.*, 1995). These studies also suggest that the changes over the North Pacific are substantially controlled by the anomalous SST forcing from the tropics. Yamagata and Masumoto (1992) have suggested that the decadal time-scale variations may also involve the Asian monsoon system.

The North Atlantic Oscillation (NAO) is of interest because of its relationship to regional precipitation and temperature variations. The NAO is a large-scale alternation of atmospheric pressure between the North Atlantic regions of subtropical high pressure (centred near the Azores) and subpolar low surface pressure (extending south and east of Greenland). The state of the NAO determines the strength and orientation of the poleward pressure gradient over the North Atlantic, and hence the speed and direction of the mid-latitude westerlies across that ocean. These, in turn, affect the tracks of European-sector low-pressure storm systems (Lamb and Peppler, 1991). One extreme of the NAO coincides with strong westerlies across the North Atlantic, cold winters in western Greenland and warm ones in northern Europe, while the other NAO extreme is associated with the opposite pattern. In addition, precipitation over Europe is related to the NAO (Hurrell, 1995). The recent decade-long winter dry conditions over southern Europe and the Mediterranean, and the wet anomalies from Iceland eastward to Scandinavia, are related to the recent persistent positive phase of the NAO (Hurrell, 1995).

IPCC (1990) used the smoothed standardised difference of December-February atmospheric pressure between Ponta Delgada, Azores, and Stykkisholmur, Iceland, to describe the variations in the westerly flow in the North Atlantic, and thus the NAO. Figure 3.18 updates IPCC (1990) and reveals strong westerlies since 1989. It also highlights an apparent near century time-scale variation in the westerly flow with peaks around 1910 and at the present. The amplitude of this variation is as large as one standard deviation of the interannual variations. Flohn *et al.* (1992), using meteorological analyses, reported a deepening of quasi-stationary cyclones in the Northern Atlantic and in the Northern Pacific leading to an intensification of the extra-tropical circulation, most pronounced during winter, supporting the suggestions of intensification in Figures 3.17 and 3.18.

In the North Atlantic, decadal period fluctuations are superposed on a longer time-scale variation (Deser and Blackmon, 1993; Kushnir, 1994; Schlesinger and Ramankutty, 1994), which is reflected in the global mean temperature variations. The short-period mode, i.e., irregular fluctuations averaging about 9 years in length before about 1945 and about 12 years thereafter, is

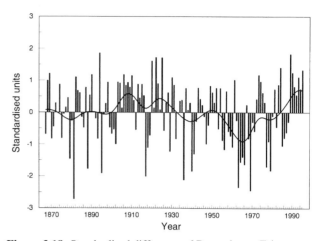

Figure 3.18: Standardised difference of December to February atmospheric pressure between Ponta Delgada, Azores, and Stykkisholmur, Iceland, 1867 to 1995. Smooth curve was created using a 21-point binomial filter. Updated from IPCC (1990).

characterised by a dipole pattern in SSTs and surface air temperatures, with anomalies of opposite polarity east of Newfoundland and off the south-east coast of the USA. Distinctive sea level pressure and wind patterns have been identified with this short-period mode and it seems that the SSTs result largely from the winds, with cooler-than-normal SSTs accompanied by stronger-than-normal winds.

The lower frequency mode is identified as a temperature fluctuation with largest amplitude in the North Atlantic and surrounding continents where the period appears to vary regionally from 50 to 88 years. Relative to the record from 1850 on, it corresponds to a much colder period from the late 1800s to about 1920, much warmer than normal from about 1930 to 1960, and another cooler period in the 1970s and 1980s (Kushnir, 1994; Schlesinger and Ramankutty, 1994). The sea level pressures and surface winds associated with this pattern have been explored by Deser and Blackmon (1993) and Kushnir (1994) who find an anomalous cyclonic circulation during the warmer years and anticyclonic during the cooler years, centred in the mid-Atlantic near 40°N.

It seems likely that relationships among the atmosphere and ocean variables include the patterns of dynamical coupling between the atmosphere and ocean, which may extend outside the Atlantic domain. It is likely that there are changes in the ocean thermohaline circulation such as those seen in models with time-scales of roughly 50 years (Delworth *et al.*, 1993). The thermohaline circulation is greatly influenced by density fluctuations at high latitudes which apparently arise mainly in the fresh water availability which alters salinity, and originates from changes in precipitation, runoff, and evaporation. These kinds of fluctuations may also be related to the abrupt climate changes seen in recent Greenland ice core analyses (Section 3.6.3).

Some reported variations in temperature and precipitation have been related to observed regional variations in atmospheric circulation. Analysis of trends in 700 hPa heights (O'Lenic, 1995) provides some evidence linking the regional circulation anomalies to the trends in surface temperature over the past 40 years, e.g., negative trend in the North Pacific and positive over western Canada. Ward (1992, 1994) demonstrated that the recent low rainfalls in the Sahel are also related to circulation anomalies. IPCC (1990) pointed out that changes in SSTs on a global scale played a major role in the reduction of Sahel precipitation. Lare and Nicholson (1994) have again reiterated the potential role of land-surface changes in the prolonged dry period in the Sahel, but Shinoda and Kawamura (1994a) suggest that changes in the SSTs of the Indian Ocean and the tropical North and South Atlantic modulate different aspects of the Sahel rainfall. Specifically, the contrasting anomalies of North and South Atlantic SSTs play a major role in the year-to-year latitudinal displacement of the rainbelt and the strong warming trend of the SSTs in the Indian Ocean (Figure 3.4) is associated with large-scale subsidence over the rainbelt. A new analysis of the interannual and decadal variations of the summer rains over the Sahel (Shinoda and Kawamura, 1994b) however, suggests that the decreased rainfall from the 1950s to the 1980s primarily resulted from weakened convection over the entire monsoonal rainbelt, rather than a systematic displacement of its central position.

One aspect of atmospheric circulation variations considered recently is the possible impact of volcanic eruptions. The eruption of Mt. Pinatubo (Section 3.2.6) led to studies which suggested, based on geographical variations in temperature after major eruptions, that heating of the tropical stratosphere resulting from volcanic aerosols drives an enhanced zonal wind (Groisman, 1992; Robock and Mao, 1992; Graf *et al.*, 1993). This stronger wind advects warmer maritime air over the continents, leading to warming, in the Northern Hemisphere winter, in higher latitudes. Such variations are limited to a year or two after a major eruption. So, only a protracted change in the level of volcanic activity could be expected to lead to a long-term atmospheric circulation change.

3.4.4 *Southern Hemisphere Circulation*

Over the Southern Hemisphere, analysis of decadal variability is difficult because of the paucity of data. Nevertheless careful analysis of station data together with the available analyses has revealed real decadal-scale climate changes (van Loon *et al.*, 1993; Hurrell and van Loon, 1994). A change in the atmospheric circulation is evident in the late 1970s with sea level pressures in the circumpolar trough generally lower in the 1980s than in the 1970s and with the changes most pronounced in the second half of the year, so that the tropospheric polar vortex remained strong into November. This was associated with a delayed breakdown in the stratospheric polar vortex and the ozone deficit in the Antarctic spring.

A near-global atmospheric pressure data set (Allan, 1993) and other data have been used to examine longer term variations in circulation in the Indian Ocean, since late last century (Allan *et al.*, 1995; Salinger *et al.*, 1995b). Circulation patterns in the periods 1870 to 1900 and 1950 to 1990 were more similar to each other than to those in the 1900 to 1950 period. These variations were most apparent in the mid- to high latitudes of the Southern Hemisphere. A more meridional long-wave pattern occurred in the early and late periods, and a more zonal flow regime occurred in the 1900 to 1950 period. This research suggests that the vigour of the atmospheric circulation has increased since the 1950s, particularly around the anticyclones in the southern Indian Ocean during summer, and Australia in winter. These high pressure features were found to wax and wane, but show no distinct displacements in the latitude of central pressure.

Coherent changes in the amplitude of the long-wave troughs to the south-west of Australia and in the Tasman Sea/New Zealand region have occurred during spring and summer (Salinger *et al.*, 1995b). On a decadal time-scale the troughs in the two regions have varied out of phase during winter. Most of the longer term trends are associated with the Tasman Sea/New Zealand trough during summer and autumn. Pressure in the south-west Australian winter trough has increased since the 1960s, and this has led to a decrease in the winter rainfall in the far south-west (Nicholls and Lavery, 1992; Allan and Haylock, 1993).

3.4.5 *Summary of Section 3.4*

Evidence that the El Niño-Southern Oscillation has varied in period, recurrence interval, and area and strength of impact is found in historical instrumental data and in palaeoclimatic data. The cause of these variations is not known. The rather abrupt change in the El Niño-Southern Oscillation, and atmospheric circulation, around 1976/77, noted in IPCC (1990), has continued. Since then, there have been relatively more frequent El Niño episodes, with only rare excursions into the other extreme (La Niña episodes) of the phenomenon. This ENSO behaviour, and especially its behaviour since 1989, is unusual in the context of the last 120 years (the instrumental record). At least some of the recent fluctuations in rainfall in the tropics and subtropics (Figure 3.10) appear to be related to this unusual ENSO behaviour.

Anomalous atmospheric circulation regimes persisted over the North Pacific from 1976 to about 1988, and in the North Atlantic since 1989. Many long-lived temperature and precipitation anomalies are now known to be associated with regional anomalous atmospheric circulation features.

There remain problems with the monitoring of changes and variations in atmospheric circulation, because of doubts about the long-term consistency of analysis techniques, observations, and even the meteorological theories underlying analysis techniques. Such doubts restrict the number of fields which can be examined confidently for evidence of real change and variation.

3.5 Has the Climate Become More Variable or Extreme?

3.5.1 Background

Concerns are often expressed that the climate may be more variable (i.e., more droughts and extended wet periods) or extreme (i.e., more frequent severe weather events) now than in the past. A possible source of confusion in any discussion of variability is that variability of the climate can be defined and calculated in several ways. Apparently contradictory conclusions may be reached from the different definitions. For instance, using the standard deviation or a similar measure to determine variability will, in certain circumstances, result in different conclusions from the use of interperiod differences (e.g., first differences of annual means) as a measure of variability. Care needs to be taken in the comparison of variabilities calculated and defined in different ways. Analysis of how the frequency and intensity of extreme weather events (e.g., tornadoes) have varied over time is also difficult, because of inhomogeneities in the data. Comparisons over longer periods, such as comparing late 20th century extremes with those of the late 19th century, are especially problematic because of changes in instrumentation and analysis methods, as well as site and exposure changes.

3.5.2 Climate Variability

3.5.2.1 Temperature
Parker *et al.* (1992) show plots of the low-pass filtered variances of the daily central England temperature series

since the mid-18th century. There is no clear evidence of trends in the variance over this very long record. Although there are interdecadal differences in variability, no tendency to higher variance is evident in recent decades.

Karl *et al.* (1995c) searched for evidence of changes in temperature variability in the USA, the former Soviet Union, China and Australia. They defined temperature variability as the absolute difference in temperature between two adjacent periods of time (1-day to annual). In the Northern Hemisphere their results indicate a tendency towards decreased temperature variability on intraseasonal (1-day to 5-day) time-scales. The decrease is strongest in the USA and China. The decreased variability at short time-scales could reflect a decrease in baroclinicity or the frequency of air mass change, suggesting that the Northern Hemisphere circulation may have become less variable through the 20th century. The interannual variability changes showed no consistent pattern across the three Northern Hemisphere areas. There was little consistency in patterns of change in temperature variability for Australia since 1961, apart from statistically significant decreases in interannual variability in September-November, for minimum, maximum, and average temperature, especially in temperate regions.

Parker *et al.* (1994) calculated the interannual variability of seasonal surface temperature anomalies for the 20-year period 1974 to 1993 for each calendar season separately, and compared it with the period 1954 to 1973. There was evidence of only a small global increase of variability in the later period. The spatially averaged ratios of the variances between the two 20-year periods, for the four seasons, were between 1.04 and 1.11. This suggests surface temperature variability increased slightly over the period of analysis. The analysis did reveal regions with strongly enhanced variability in the later period. Variance was enhanced over central North America between December and August. The El Niño events of 1982/3, 1986/7, and 1991/2 produced enhanced variance in the eastern tropical Pacific in 1978 to 1992 relative to 1951 to 1980.

In summary, temperature shows no consistent, global pattern of change in variability. Regional changes have occurred, but even these differ with the time-scale considered.

3.5.2.2 Precipitation and related moisture indices
Means and standard deviations of Indian summer monsoon rainfall, calculated using a 31-year moving window, over the period 1871 to 1993, were examined by Sontakke *et al.* (1992). No trend was apparent in variability over the entire period. However, higher variability is generally associated

with drier periods, such as that observed since about 1970. Droughts have increased in frequency in Sub-Saharan Africa over the past few decades. This partly reflects the downward trend in mean rainfall (Section 3.3.2.1), but there were also changes in variability. Hulme (1992) examined changes in relative variability over Africa between 1931 to 1960 and 1961 to 1990 and found that areas of increased variability outweighed areas of reduced variability. The Sahel showed increased rainfall variability, especially in the east where increases were over 5% and in some places over 15%.

Changes in variability of the diurnal temperature range may provide some suggestive information regarding changes of variability of precipitation or cloudiness, since increased cloud cover can lead to warmer overnight temperatures, and lower daytime temperatures, in some regions. Karl *et al.* (1995c) examined changes in the variability of the diurnal range of temperature in the USA, the former Soviet Union, and China. They found decreases in the variability of the diurnal range, especially in the former Soviet Union. Since the diurnal range of temperature is related to cloud cover and humidity, decadal decreases in the variability of the diurnal range may reflect a decrease in variability of cloud cover, soil moisture, humidity, or wind.

Droughts have been relatively frequent since the late 1970s in some of the areas where drought usually accompanies El Niño events (e.g., north-east Australia). This presumably reflects the relatively frequent El Niño events during this period (see Section 3.4.2). Few analyses of the frequency of floods have been conducted. A study of floods in Sweden (Lindström, 1993) found no convincing evidence of trends through the 20th century, although the 1980s had larger floods than usual and the 1970s had few high floods. Wigley and Jones (1987) reported a higher than usual frequency of extreme dry summers and wet springs over England and Wales in 1976 to 1985. Gregory *et al.* (1991) updated the Wigley and Jones data to 1989 and found no evidence of a continuation of this tendency.

3.5.2.3 Atmospheric circulation

Born (1995) and Born and Flohn (1995) investigated atmospheric circulation variability using interdiurnal (i.e., day-to day) changes of surface pressure and geopotential in the Atlantic sector. Since 1970 there appears to have been a trend to increased variability on this time-scale. They also investigated maritime winds in the Atlantic between 60°N and 30°S, by examining trends in constancy (the resultant wind speed as a percentage of scalar wind speed). The trend at most latitudes from 1949 to 1989 was negative,

indicating increasing variability. A trend towards lower variability (increasing constancy) was found only in winter between 45°N and 60°N. Doubts remain about the long-term consistency of the analyses from which these results were derived. Trenberth (1996) notes some of the changes in the systems used to analyse atmospheric circulation.

The results of Karl *et al.* (1995c) suggest that these findings cannot be extrapolated to infer increases in atmospheric circulation variability over the USA or much of Asia. Karl *et al.* found evidence of decreased variability of temperature at the interdiurnal time-scale in the USA, former Soviet Union and China. Since the variability of the atmospheric circulation provides the dominant forcing of interdiurnal temperatures in these regions, it seems unlikely that the interdiurnal variability of atmospheric circulation has increased there.

Changes in the mean atmospheric circulation over the North Pacific (Section 3.4.3) in the late 1970s were accompanied by a southward shift in storm tracks and the associated synoptic eddy activity (Trenberth and Hurrell, 1994). The incidence of cold outbreaks increased across the plains of North America at this time. Such shifts could be interpreted as changes in atmospheric circulation variability, at least on a regional scale.

The variability of the El Niño-Southern Oscillation has been examined by Wang and Ropelewski (1995) who noted a general rise in the level of variability from the earliest records to the 1910s, followed by a period of 20 to 40 years of no change, with a suggestion of increased variability over recent decades. ENSO variability was greater when the sea surface temperature was generally higher. However, the results were sensitive to the sea surface temperature analysis technique. Nonetheless, their conclusion was consistent with that of Rasmusson *et al.* (1994), who found a relative minimum in surface pressure variance and precipitation variability during the periods of relatively low mean sea surface temperature.

3.5.3 *Extreme Weather Events*

3.5.3.1 *Tropical cyclones*

Atlantic hurricane (tropical cyclone) activity over the period 1970 to 1987 was less than half that in the period 1947 to 1969 (Gray, 1990). A similar quiet period occurred at the same time in the western North Pacific, suggestive of a decrease in the number of very intense tropical cyclones. Bouchard (1990) and Black (1992), however, demonstrated that this apparent change in intensity in the western North Pacific was an artefact, due to a change around 1970 in the method used to derive wind estimates from pressure estimates. When a consistent method for determining wind estimates was used throughout the period of record, the pre-1970 data were statistically indistinguishable from the post-1970 data. Landsea (1993) suggested that the Atlantic hurricane intensity record was probably also biased. Winds were 5 kt higher before 1970, compared with hurricanes with the same minimum pressure after 1970. After adjusting for this bias a substantial downward trend in intense hurricane activity is still apparent. There remains a possibility that not all the bias in the Atlantic records has been removed, because of the remarkable drop in apparent intensity from 1970, the same year the artificial drop in frequency of intense typhoons appeared in the north-west Pacific. However, this step decrease was also observed in the frequency of storms hitting the USA. These storms were categorised by using minimum sea level pressure recorded at landfall. Such observations should not be as suspect as observations over the ocean. Also, the decrease in hurricanes appears to reflect a relationship between hurricane activity and Sahel rainfall (Landsea and Gray, 1992; Goldenberg and Shapiro, 1996). Finally, hurricane activity is also weaker during El Niño episodes. The relatively more frequent El Niño episodes since the mid-1970s (see Section 3.4.2) would, therefore, have led to a tendency for weak hurricane activity. There are strong grounds, therefore, for concluding that the decrease in Atlantic intense hurricane activity is real (Landsea *et al.*, 1996). However, Karl *et al.* (1995d) found no trend in the numbers of hurricanes crossing the USA coast since 1900, although numbers were higher during the middle decades of the century. The 1995 Atlantic hurricane season was more active than recent years.

In the north-east and south-west Pacific the number of cyclones appears to have increased (Thompson *et al.*, 1992; Landsea and Gray, 1995), although doubts must be expressed about the consistency of the observation systems over this period. In the south-east Pacific, where tropical cyclone activity is usually associated with El Niño events, the frequency of occurrence appears to have increased in recent decades. Again, the quality of the long-term cyclone data base in this region is suspect. In the Australian region much of the apparent long-term variations in cyclone activity appear to be the result of changes in observing systems or analysis techniques (Nicholls, 1992). Raper (1993) noted that little confidence could be placed in apparent long-term trends (except perhaps in the North Atlantic) because of doubts about the consistency of the data.

Doubts about the quality and consistency of the data on maximum wind speeds in most cyclone basins preclude convincing analysis of how peak cyclone intensity might have changed in recent decades. Only in the Atlantic do the

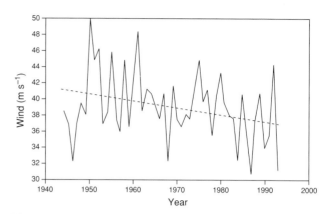

Figure 3.19: Time-series of mean annual maximum sustained wind speed attained in Atlantic hurricanes (Landsea *et al.*, 1996). Linear trend shown as dashed line.

data seem of sufficient quality to allow such an analysis. Figure 3.19 shows the mean maximum sustained wind speed attained each year in Atlantic hurricanes (Landsea *et al.*, 1996). Mean maximum wind speed appears to have decreased. However, the peak intensity reached by the strongest hurricane each year has shown no trend in the five-decade period.

3.5.3.2 *Extra-tropical storminess*

There is some evidence of recent increased storminess around the North Atlantic, although there are doubts about the consistency of the meteorological analyses from which this increase has been derived. Hand-drawn analyses of Atlantic surface pressure patterns, produced since 1956 by Seewetteramt, Hamburg, have been used to analyse the annual occurrence of extra-tropical cyclones (also known as low pressure systems or depressions) with minimum pressures below 950 hPa (Dronia, 1991). Little trend was found, except for an abrupt increase in frequency since 1988. Stein and Hense (1994), using daily grid-point pressure data, deduced that the North Atlantic winters since 1988/89 had been affected by a higher frequency of extreme low pressure systems than at any time since 1880. Schinke (1993) used once-daily analyses (USA analyses 1939 to 1964; thereafter German analyses) to count the number of storms with pressures below certain thresholds. This study revealed a substantial increase in the number of severe storms in the North Atlantic area, in the early 1970s. The mean central pressure showed a decreasing trend. Stein and Hense (1994) also deduced an increase in the numbers of extreme lows in the early 1970s, but concluded that this increase does not appear to be significant, when considered in the context of the observed interannual variability.

Schmidt and von Storch (1993) suggest that local studies with homogenous data bases may provide more definite answers, at least for specific regions, than the use of analyses with time-varying biases (Section 3.4.1). They used daily air pressure observations at three stations in the south-east North Sea to calculate the annual distributions of daily geostrophic wind speeds and concluded that the frequency of extreme storms in this area has not changed in the past 100 years. Von Storch *et al.* (1993) analysed a long time-series of "severe storm days" on Iceland, based on local wind observations, and found little trend in the number of severe storm days, (apart from an artificial change in storm frequency around 1949, because of changes in observing practices). Numbers of storms were quite low during the 1980s. Von Storch *et al.* (1993) also examined high water levels at Hoek van Holland, after removing the effects of tides and sea level changes. The resultant time-series should reveal storm-related surge heights. No trend was found in the frequency of extreme surge heights.

Bardin (1994) reported an abrupt increase in the frequency of extra-tropical depressions in the central North Pacific, in the late 1970s. For the Northern Hemisphere as a whole, Bardin found a decrease in cyclone frequency beginning in the second half of the 1980s, after a period of about a decade with higher numbers of cyclones. The size and intensity of cyclones were approximately constant until 1980, and had increased since then. Serreze *et al.* (1993) found increasing numbers of cyclones and anticyclones over the Arctic between 1952 and 1989.

Agee (1991) combined data from three previous studies of cyclone and anticyclone frequency around North America to examine trends. He found evidence of increases in extra-tropical cyclone numbers between 1905 and 1940 and decreasing numbers of both cyclones and anticyclones between about 1950 and 1980. Agee found some evidence of an increase in cyclone numbers between 1980 and 1985, but his data did not extend into more recent years. Temporal variations in winter storm disasters (Changnon and Changnon, 1992) closely matched the trends in cyclone numbers for North America. Davis and Dolan (1993) found a similar variation in the occurrences of mid-latitude cyclones over eastern North America. Cyclones were less frequent between the mid-1960s and the mid-1970s, relative to the period from 1942. Since the mid-1970s the yearly number has increased but has not consistently reached pre-1965 levels. The prevalence of the most destructive storms has been erratic, but has increased overall. Seven of the eight most intense storms that have developed in the past 50 years occurred in the last 25 years.

One Southern Hemisphere region where atmospheric depressions appear to have increased in number in recent decades is along the east coast of Australia. Hopkins and Holland (1995) determined the numbers of east coast cyclones with an objective method of specification based on a consistent set of observing stations spread along the coast. They found an upward trend in the numbers of cyclones between 1958 and 1992.

In summary, the evidence on changes in extra-tropical synoptic systems is inconclusive. There is no clear evidence of any uniform increase. Evidence from some areas suggests no change during the 20th century; in some other areas there is some evidence of change.

3.5.3.3 *Intense rainfalls*

Long-term rainfall observing sites can be used to examine changes in the frequency of intense 24-hour rainfall totals. Iwashima and Yamamoto (1993) did this, for 55 stations in Japan and for 14 stations in the contiguous USA, by determining the decades in which the three highest daily rainfall totals were recorded. A trend towards higher frequencies of extreme rainfalls in recent decades was evident. Such a trend was also found by Karl *et al.* (1995c) who analysed the trends in the percentage of total seasonal and annual precipitation occurring in heavy daily rainfall events (days with rainfalls exceeding 50.8 mm) over the USA (1911 to 1992), the former Soviet Union (1935 to 1989), and China (1952 to 1989). A significant trend to increased percentages of rainfall falling in heavy events is evident in the USA, largely due to a strong increase in extreme rainfall events during the warm season. A similar trend is also apparent in heavy rainfall events calculated from 3-day total rainfalls. Trends to more extreme rainfall events were not apparent in the other regions. A trend to increased annual rainfall around the periphery of the North Atlantic, since early this century, also appears to reflect an increase in the number of heavy rainfall events (Frich, 1994).

Rakhecha and Soman (1994) examined annual extreme rainfalls in the time-scale of 1–3 days at 316 stations well-distributed across India, for the period 1901 to 1980. Most annual extreme rainfall records were free of trend or autocorrelation. The extreme rainfall series over the west coast north of 12°N and at some stations to the east of the Western Ghats over the central parts of the Peninsula showed a statistically significant increasing trend. Stations over the southern Peninsula and over the lower Ganga valley exhibited a decreasing trend.

In Australia, the relationship between trends in mean rainfall and trends in intensity of rain events is complex (e.g., Nicholls and Kariko, 1993; Yu and Neil, 1991, 1993), except in the tropical regions (Lough, 1993; Suppiah and Hennessey, 1996). Lough (1993) found no evidence of a trend to more intense rainfall, or for greater numbers of heavy rain days between 1921 and 1987, in general over north-east Australia. Suppiah and Hennessey (1995), however, reported an increase in the frequency of extreme rainfall events at a majority of stations in northern Australian summers since 1910. Few of the increases were statistically significant.

Few studies have reported changes in occurrence of hail. Dessens (1994) reported a substantial increase in hail fall severity in France during summer. This increase was related to an increase in the summer mean minimum temperature which is significantly correlated with hail occurrence in France.

3.5.3.4 Extreme temperatures

One extreme that might be anticipated to change in frequency if mean temperatures increased is the frequency of frosts or freezes. Decreases over the past few decades in the frequency of extreme low minimum temperatures or the length of the frost season have been reported for several widely separated locations (Salinger *et al.*, 1990; Karl *et al.*, 1991; Watkins, 1991; Bootsma, 1994; Palecki, 1994; Cooter and LeDuc, 1995; Stone *et al.*, 1996). Cooter and LeDuc, for instance found that the frost-free season in the north-eastern USA now begins 11 days earlier than 30 years ago. However, not all the stations and regions examined in these studies exhibited a decrease in frost occurrence.

Changes in frequency of extreme maximum (daytime) temperatures were less consistent than was the case for minimum temperatures. Across the USA a spatially complicated pattern of variations in extreme maximum temperatures is evident, with no evidence of a country-wide increase in extremes (Karl *et al.*, 1991; Balling and Idso, 1990; DeGaetano *et al.*, 1994; Henderson *et al.*, 1994). Karl *et al.* found evidence of increases in extreme seasonal minimum temperatures in the contiguous USA and the former Soviet Union, but little or no change in extreme seasonal maximum temperatures. Plummer (1996) found that, in Australia, extreme minimum temperatures have been increasing at a similar rate to average temperatures but that trends in the extreme maximum temperatures were smaller.

In some areas there is evidence of decadal variations in the frequency of occurrence of regional cold outbreaks. For instance, changes in Pacific Ocean storm tracks since about 1976 (Trenberth and Hurrell, 1994) have been implicated in higher incidences of regional cold outbreaks, from the late 1970s to the mid-1980s, across the plains of North America, ultimately leading to an increased frequency of major freezes affecting Florida (Rogers and Rohli, 1991; Downton and Miller, 1993).

3.5.3.5 Tornadoes, thunderstorms, dust storms, and fire weather

The final type of extreme weather event considered here consists of events normally subject to visual reports. Identification of trends in such data is likely to be problematic, because of doubts about consistency of observer behaviour. There is little or no evidence of consistent increases in such events. For example, Ostby (1993) found no evidence of increased occurrence of strong or violent tornadoes in the USA, although the numbers of reports of less severe tornadoes appears to have increased, perhaps due to increased population, eagerness in reporting, or improved reporting procedures. Grazulis (1993) reported a drop in damaging tornadoes in the 1980s over the USA.

There is some evidence of an increase in thunderstorms in the global tropics, from observations of trends in cloudiness (London *et al.*, 1991). The amount of cumulonimbus clouds, often associated with thunderstorms, increased over the tropical oceans between 1952 and 1981 partly at the expense of cumulus clouds (Section 3.3.8.2). This suggests an increase in thunderstorm activity in the tropics. Increases in summer extreme rainfall events in the USA reported by Karl *et al.* (1995c), and in tropical Australia during summer (Suppiah and Hennessey, 1995) also may reflect an increase in thunderstorm activity. Overall, however, land based tropical rainfall has declined in the last few decades (see Figure 3.11). A large percentage of rainfall is associated with thunderstorm activity, so the decrease in rainfall might reflect a drop in thunderstorm activity, at least over tropical land areas. Karl *et al.* (1995c) did not find increases in summer extreme rainfall events in China or the former Soviet Union.

Goudie and Middleton (1992) examined time-series of dust storms for many parts of the world. They found no global pattern of dust storm frequency trend, and concluded that, in the absence of regional-scale human activities (e.g., changes in agricultural practices), the major factor affecting numbers of dust storms is rainfall. So a change in rainfall might be expected to lead to changes in the frequency of dust storms.

Balling *et al.* (1992) examined variations in wildfire data in the Yellowstone National Park between 1895 and 1990. The area burnt is positively related to summer temperature and negatively related to summer and antecedent precipitation. Summer temperatures have been increasing in this area through the 20th century, while antecedent precipitation has been decreasing. Balling *et al.* concluded that there has been a significant trend to a set of climatic conditions favouring the outbreak of wildfires.

3.5.4 Summary of Section 3.5

There has been no consistent trend in interannual temperature variability in recent decades. In some areas variability on shorter time-scales has decreased. Few regions have been examined for evidence of changes in interannual variability of rainfall. The areas examined have not exhibited a consistent pattern. Trends in intense rainfalls are not globally consistent, although in some areas (Japan, the USA, tropical Australia) there is some evidence of increases in the intensity or frequency of extreme events. There has been a clear trend to fewer low temperatures and frosts in several widely-separated areas in recent decades. Widespread significant changes in extreme high temperature events have not been observed, even in areas where the mean temperatures have increased.

There are grounds for believing that intense tropical cyclone activity has decreased in the North Atlantic, the one tropical cyclone region with apparently consistent data over a long period. Elsewhere, apparent trends in tropical cyclone activity are most likely due to inconsistent analysis and observing systems. Doubts in the consistency of meteorological analyses also confound the estimation of trends in extra-tropical cyclones. In some regions (e.g., over the USA, the east coast of Australia, the North Atlantic) there is some evidence suggestive of recent increases. However some other highly reliable records in particular regions (e.g., the German Bight) do not exhibit any trends to increased storminess.

Overall, there is no evidence that extreme weather events, or climate variability, has increased, in a global sense, through the 20th century, although data and analyses are poor and not comprehensive. On regional scales there is clear evidence of changes in some extremes and climate variability indicators. Some of these changes have been toward greater variability; some have been toward lower variability.

3.6 Is the 20th Century Warming Unusual?

3.6.1 Background

To understand recent and future climatic change, it is necessary to document how climates have varied in the past, i.e., the space- and time-scales of natural climate variability. Such information is necessary, for instance, to determine whether the changes and variations documented earlier in this chapter are likely to reflect natural (rather than human-induced) climate variability. Some measure of this natural variability can be deduced from instrumental observations, but these are restricted, for the most part, to less than 150 years. Longer records of climate variations are required to provide a more complete picture of natural climate variability against which anthropogenic influences in the observed climate record can be assessed. Changes in the thermohaline circulation are one example of internal (natural) variability in the climate system affecting temperature trends. Evidence of such changes has been found in some ice-age records, although the evidence is more muted for fluctuations of the last 10,000 years.

IPCC (1990) provided a broad overview of the climates of the past 5,000,000 years, and a more detailed presentation of three periods suggested as possible analogues of a greenhouse-enhanced world. IPCC (1992) concentrated on the climate of the past 1000 years, a topic which this chapter also examines, because of the recent increase in data for this period, and because of its relevance to the current climate. The data from the last 1000 years are the most useful for determining the scales of natural climate variability. Rapid natural climate changes also provide information on the sensitivity/stability of the climate system of relevance in projections of climate change. Recent evidence of such changes over the last 150,000 years is also assessed here.

Climates from before the recent instrumental era must be deduced from palaeoclimatic records. These include tree rings, pollen series, faunal and floral abundances in deep-sea cores, isotope analysis from coral and ice cores, and diaries and other documentary evidence. The difficulty of determining past climates can be illustrated by the case of tropical sea surface temperatures (SSTs) around the time of the last glacial maximum (around 22,000-18,000 years ago). The lack of consensus on this topic provides an indication of the problems of using proxy data. The topic is important partly because of the possible use of glacial data in validating climate models, and relating past climates to radiative forcings (Crowley, 1994). The relationship between modern and past terrestrial and sea surface temperatures is still one of the major uncertainties involved in understanding the climate during the last ice age. Terrestrial surface temperature estimates from snow line and pollen data (e.g., Webster and Streten, 1978; Seltzer, 1992), SSTs from coral (e.g., Beck *et al.*, 1992; Aharon *et al.*, 1994; Guilderson *et al.*, 1994), and from ice cores in the Andes (Thompson *et al.*, 1995) do not concur with estimates from other sources such as oxygen isotope, planktonic foraminiferal and pollen records and the temperature dependent saturation of long-chain alkenones from algae (e.g., Ohkouchi *et al.*, 1994; Sikes and Keigwin, 1994; Thunnell *et al.*, 1994). The land-based data (and some estimates from coral) suggest that ice age SSTs in the tropics were about 5°C lower than present, while the other sources suggest about 1–2°C. These inconsistencies cause problems in the use of palaeoclimatic data to validate climate models, and need to be resolved.

3.6.2 *Climate of the Past 1000 Years*

Various methods based on historical, ice core, tree-ring, lake level and coral data have been used to reconstruct the climate of the last millennium (Cook, 1995). For example, a significant number of annually resolved, precisely dated temperature histories from tree rings are available. However, these records are still too sparse to provide a complete global analysis, and must, in general be interpreted in a regional context. In addition, they usually only reflect changes in warm-season (growing season) temperatures. Tree-ring records frequently represent interannual and decadal time-scale climate variability with good fidelity, as indicated by comparison with recent instrumental records. However, the extent to which multi-decadal, century, and longer time-scale variability is expressed can vary, depending on the length of individual ring-width or ring-density series that make up the chronologies, and the way in which these series have been processed to remove non-climatic trends. In addition, the possible confounding effects of carbon dioxide fertilisation needs to be taken into account when calibrating tree-ring data against climate variations. Coral records are available from regions not represented by tree rings and usually have annual resolution. However, none extends back more than a few hundred years. The interpretation of ice core records from polar ice sheets and tropical glaciers may be in some cases limited by the noise inherent to snow depositional processes, especially during this period when climate changes were rather small (by comparison, for instance, with the large rapid changes discussed in the next section). On the other hand, they can give an unambiguous record of accumulation change, on an annual basis. All these forms of data can and have been used to provide information regarding climate variations of the past 1000 years.

There are, for this last millennium, two periods which have received special attention, the Medieval Warm Period and the Little Ice Age. These have been interpreted, at times, as periods of global warmth and coolness, respectively. Recent studies have re-evaluated the interval commonly known as the Medieval Warm Period to assess the magnitude and geographical extent of any prolonged warm interval between the 9th and 14th centuries (Hughes

and Diaz, 1994). The available evidence is limited (geographically) and is equivocal. A number of records do indeed show evidence for warmer conditions at some time during this interval, especially in the 11th and 12th centuries in parts of Europe, as pointed out by Lamb (1965, 1988). There are also indications of changes in precipitation patterns and associated droughts both in California and Patagonia during medieval time (Stine, 1994). However other records show no such evidence, or indicate that warmer conditions prevailed, but at different times. This rather incoherent picture may be due to an inadequate number of records or a bias in the geographical and seasonal representation in the available data (Briffa and Jones, 1993; Jones and Briffa, 1996), and a clearer picture may emerge as more and better calibrated proxy records are produced. However, at this point, it is not yet possible to say whether, on a hemispheric scale, temperatures declined from the 11–12th to the 16–17th century. Nor, therefore, is it possible to conclude that global temperatures in the Medieval Warm Period were comparable to the warm decades of the late 20th century.

The term Little Ice Age is often used to describe a 400-500 year long, globally synchronous cold interval, but studies now show that the climate of the last few centuries was more spatially and temporally complex than this simple concept implies (Jones and Bradley, 1992). It was a period of both warm and cold climatic anomalies that varied in importance geographically. For the Northern Hemisphere as a whole, the coldest intervals of summer temperature were from 1570 to 1730 (especially 1600 to 1609) and during most of the 19th century, though individual records show variations in this basic pattern. Warmer conditions were more common in the early 16th century and in most of the 18th century, though for the entire hemisphere, conditions comparable to the decades from 1920 onward (the time when instrumental records become more reliable and widely available) have not been experienced for at least several hundred years. Temperatures from boreholes (Section 3.2.5.2) also suggest that the present temperatures in parts of North America and perhaps elsewhere may be warmer than the last few hundred years. Regional temperatures do not all, of course, conform to this pattern. For instance, spring temperatures during the period 1720 to 1770 appeared to be warmer than the 20th century in parts of China (Hameed and Gong, 1994). However, despite the spatial and temporal complexity, it does appear that much of the world was cooler in the few centuries prior to the present century.

Bradley and Jones (1993, 1995) developed a Northern Hemisphere summer temperature reconstruction since 1400 from 16 palaeoclimatic records (Figure 3.20). Recent decades (the recent instrumental record is also plotted on this figure) appear to be warmer than any extended period since 1400, and the warming since the late 19th century is unprecedented in this record. Regionally, however, the recent warming is not always exceptional. For instance, in the Swedish Torneträsk tree-ring series, the 20th century warming appears as a relatively minor event (Briffa *et al.*, 1990; 1992). In this region, in both the 1400s and, especially, the 900–1100 interval, summer conditions appear to have been warmer than today. Tree-ring width series from the northern Urals (Graybill and Shiyatov, 1992) indicate that in this area 20th century summers have been somewhat warmer than average. A recent analysis, using tree-ring density data, has attempted to reproduce more of the century time-scale temperature variability in this region (Briffa *et al.*, 1995). This shows that the 20th century was clearly the warmest in the last 1000 years in this region, though shorter warmer periods occurred, for example, in the 13th and 14th centuries. The Californian Cirque Peak ring-width cool-season temperature history (Graybill and Funkhouser, 1994) indicates unusual warmth during most of the 20th century, but with recent cooling (contrary to instrumental observations in this area). The nearby Campito Mountain ring-width warm-season temperature history (LaMarche, 1974) agrees well with the Cirque Peak record, e.g., the 20th century appears to have been unusually warm, while the 17th century was generally cool. North-west Alaska has been unusually warm during the 20th century (Jacoby *et al.*, 1996).

In the Southern Hemisphere, the records from Australasia show evidence of recent unusual warmth, especially since 1960, but some from South America do not. Thus the Northern Patagonia and Rio Alerce ring-width records from Argentina (Villalba, 1990; Boninsegna, 1992) and the Chilean Lenca ring-width temperature history (Lara and Villalba, 1993) show no clear indication of 20th century warming, in accord with local instrumental records (Villalba, 1990). Recently analysed ice cores from the north-central Andes (Thompson *et al.*, 1995) indicate that temperatures were cool in the 200–500 years before the present. Strong warming has dominated the last two centuries in this region. Temperatures in the last two centuries appear higher than for some thousands of years in this area. New Zealand tree rings reproduce the warming observed instrumentally since about 1950, although there is also a suggestion of warmer periods early in the 18th and 19th centuries (Salinger *et al.*, 1994), near a time of maximum mountain glacier ice volumes in the Southern Alps of the country. The Tasmanian Lake Johnston history

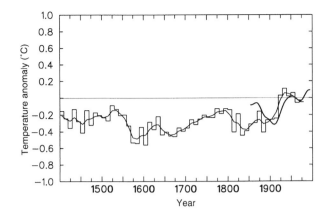

Figure 3.20: Decadal summer temperature index for the Northern Hemisphere, from Bradley and Jones (1993), up to 1970-1979. The record is based on the average of 16 proxy summer temperature records from North America, Europe and east Asia. The smooth line was created using an approximately 50-year Gaussian filter. Recent instrumental data for Northern Hemisphere summer temperature anomalies (over land and ocean) are also plotted (thick line). The instrumental record is probably biased high in the mid-19th century, because of exposures differing from current techniques (e.g., Parker, 1994b).

(Cook *et al.*, 1991) shows recent anomalous warming, especially since 1960 (Cook *et al.*, 1992).

Temperature records derived from coral supplement the records from tree rings. A Galapagos Islands coral temperature record for 1607 to 1982 has been derived by Dunbar *et al.* (1994) who found a very high correlation with directly measured local annual SST for 1961 to 1982. Coral-estimated SSTs have fallen significantly throughout the 20th century in this location and may now be at their lowest since 1650. This is in broad agreement with the lack of recent warming in the South American tree-ring studies. However, there has not been a general cooling in the eastern tropical Pacific this century (Bottomley *et al.*, 1990), so the Galapagos cooling may be a local phenomenon. Australian Great Barrier Reef temperature histories since 1583 have been derived from corals distributed along the reef from 10–30°S (Lough *et al.*, 1996). These coral-based SSTs have been relatively warm for most of the 20th century. This warm period was preceded by below-average SSTs in the 1850 to 1900 period. An earlier warm period is indicated around 1830 to 1840, but with less certainty. The pre-1800 period was probably a time of conspicuously below-average SSTs. Similar results were obtained from the Abraham Reef record (Druffel and Griffin, 1993), on the south-western tip of the Great Barrier Reef.

Just as important as temperature are variations in the hydrological cycle during the last millennium. Global-scale records do not exist, but useful information has been derived for several continents. Especially relevant, because of the recent low rainfall in the Sahel (Section 3.3.2.1), is the evidence from Africa. From the 10th to 13th centuries the evidence suggests a much wetter climate over much of North Africa (Nicholson, 1978). After a drier regime during the 14th and 15th centuries, resembling the 20th century climate, North Africa again reverted to relatively wet conditions during the 16th to the late 18th centuries. Interspersed in this moderately wet era were extreme droughts in the Sahel in the 1680s and the 1740s to 1750s. Continental-scale dryness began in the 1790s and climaxed in a severe drought in the 1820s and 1830s. These two decades were likely drier than current conditions throughout the continent, except in the Sahel where current conditions are similar to those in the 1820 to 1830 period. The late 19th century again saw a rise in African rainfall that subsided around 1900. Rainfall increased during the 1950s and 1960s in North and Equatorial Africa, followed by a sharp decrease in the Sahel. The conditions during the last 100 years have been, for the most part, substantially drier than the rest of the last millennium. Of particular note are the dry conditions in the Sahel between 1968 and 1993. Nicholson (1989) has documented several previous dry regimes also lasting on the scale of a decade or two, particularly the 1680s, 1740s and 1750s, and 1820s and 1830s, that mirrored the recent period in the Sahel. So, the recent past has exhibited periods of comparable dryness to the last few decades.

The climate of the past 1000 years provides opportunities to determine the spatial and temporal scales of natural climate variability required for detection of climate change, and the causes of the natural climate variations. A relatively small number of records from key regions could help discriminate the relative importance of different phenomena (Crowley and Kim, 1993), in the creation of decadal-century time-scale variability. For instance, volcanism would influence land more than ocean, North Atlantic thermohaline processes may have unique climate signatures, records from the equatorial Pacific could be used to monitor decadal-scale variations in that region (e.g., variations in the El Niño-Southern Oscillation – see Section 3.4.2). Some of the palaeoclimatic records already exist. Figure 3.21 (after Thompson *et al.*, 1993) shows temperature estimates from ice cores in several widespread locations. These show warming this century, but the warming appears to be more significant in some locations than others. As noted above, geographical

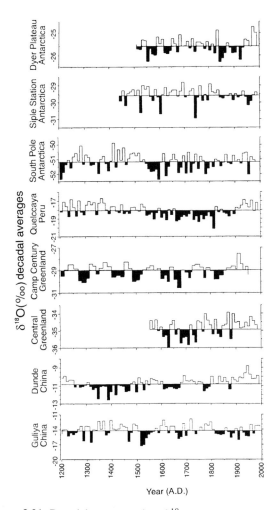

Figure 3.21: Decadal averages of the $\delta^{18}O$ records from ice cores in several widespread locations (after Thompson *et al.*, 1993). The shaded areas represent isotopically more negative (cooler) periods relative to the individual record means.

differences also exist in the warming estimated from tree-ring chronologies. The development and combination of more high-quality records such as those in Figure 3.21 are needed, if a more complete understanding of the scales and causes of natural climate variability is to be gained. Overall, however, it appears that the 20th century has been at least as warm as any century since at least 1400 AD. In at least some areas, the recent period appears to be warmer than has been the case for a thousand or more years (e.g., Briffa *et al.*, 1995; Thompson *et al.*, 1995). Alpine glacier advance and retreat chronologies (Wigley and Kelly, 1990) suggest that in at least alpine areas, global 20th century temperatures may be warmer than any century since 1000 AD, and perhaps as warm as during any extended period (of several centuries) in the past 10,000 years. Crowley and Kim (1995) estimate the variability of global mean

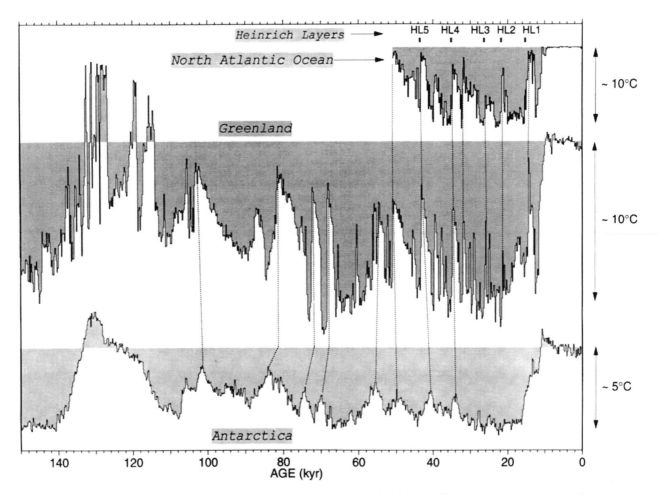

Figure 3.22: Reconstructed climate records showing rapid changes in the North Atlantic and in Greenland; the corresponding events (indicated by thin dashed vertical lines) are damped in the Antarctic record. Temperature changes are estimated from the isotopic content of ice (Greenland and Antarctica) and from faunal counts (North Atlantic). HL1 to HL5 indicate sedimentary "Heinrich" layers. Figure adapted from Jouzel et al. (1994).

temperature on century time-scales, over the past millennium, as less than ±0.5°C.

3.6.3 Rapid Cimate Changes in the Last 150,000 Years

The warming of the late 20th century appears to be rapid, when viewed in the context of the last millennium (see above, and Figures 3.20, 3.21). But have similar, rapid changes occurred in the past? That is, are such changes a part of the natural climate variability? Large and rapid climatic changes did occur during the last ice age and during the transition towards the present Holocene period which started about 10,000 years ago (Figure 3.22). Those changes may have occurred on the time-scale of a human life or less, at least in the North Atlantic where they are best documented. Many climate variables were affected: atmospheric temperature and circulation, precipitation

patterns and hydrological cycle, temperature and circulation of the ocean.

Much information about rapid climatic changes has recently been obtained either from a refined interpretation of existing records or from new ice, ocean and continental records from various parts of the world. Of particular significance are those concerning the North Atlantic and adjacent continents such as the GRIP (Dansgaard *et al.*, 1993) and GISP 2 (Grootes *et al.*, 1993) central Greenland ice cores, numerous deep-sea core records from the North Atlantic, and continental records (lake sediments, pollen series, etc.) from Western Europe and North America. These records provide descriptions of the last glacial period and the following deglaciation. The observed rapid changes are often large in magnitude, and thus there is considerable confidence in their reality.

There is evidence from these records of rapid warming ~11,500 calendar years ago. Central Greenland temperatures increased by ~7°C in a few decades (Dansgaard *et al.*, 1989; Johnsen *et al.*, 1992; Grootes *et al.*, 1993). There are indications of an even more rapid change in the precipitation pattern (Alley *et al.*, 1993) and of rapid reorganisations in the atmospheric circulation (Taylor *et al.*, 1993a; Mayewski *et al.*, 1993). Changes in SST, associated with sudden changes in oceanic circulation, also occurred in a few decades, at least in the Norwegian Sea (~5°C in fewer than 40 years; Lehman and Keigwin, 1992). During the last 20,000 years surface water salinity and temperature have exhibited parallel changes that resulted in reduced oceanic convection in the North Atlantic and in reduced strength of the global conveyor belt ocean circulation (Duplessy *et al.*, 1992). There was also a hiatus in the warming in the Southern Hemisphere during the last deglaciation. The subsequent warming was much less abrupt than in the Northern Hemisphere (Jouzel *et al.*, 1995). Similar behaviour occurred in New Zealand (Suggate, 1990; Denton and Hendy, 1994; Salinger, 1994).

There is also firm evidence of rapid warm-cold oscillations during the last glacial (Dansgaard-Oeschger events) in the central Greenland records (Johnsen *et al.*, 1992). Rapid warmings of ~5–7°C in a few decades were followed by periods of slower cooling and then a generally rapid return to glacial conditions. About 20 such intervals (interstadials) lasting between 500 and 2000 years occurred during the last glacial period (Dansgaard *et al.*, 1993). Their general progression was similar to the rapid changes in North Atlantic deep-sea core records. The most prominent of these interstadials may be associated with the sedimentary "Heinrich" layers interpreted as reflecting massive iceberg discharge from Northern Hemisphere ice sheets (Bond *et al.*, 1993; Mayewski *et al.*, 1994; Bond and Lotti; 1995), see Figure 3.22. These discharges occurred at the end of the cooling cycles and were followed by abrupt shifts to warmer SSTs.

During the last glacial period, rapid changes were also felt over at least parts of the continents as suggested by recently obtained records for Western Europe, North America, and China (Grimm *et al.*, 1993; Guiot *et al.*, 1993; Porter and An, 1995). Moreover, a significant increase in methane concentration is associated with the warm interstadials (Chappellaz *et al.*, 1993) that may be due to variations in the hydrological cycle at low latitudes. This suggests that the interstadials were at least hemispheric in their extent. Moreover, Greenland interstadials lasting more than 2000 years have weaker and smoother counterparts in Antarctica (Bender *et al.*, 1994;

Jouzel *et al.*, 1994). Keigwin *et al.* (1994) and McManus *et al.* (1994) found that Atlantic deep-water circulation and surface temperatures exhibited high variability through much of the last glacial, and that these variations were closely related to fluctuations in the Greenland ice cores. Keigwin *et al.* (1994) also report evidence from terrestrial palaeoclimate data suggesting that the short duration events since about 100,000 years ago were at least hemispheric, possibly global in extent.

The GRIP central Greenland ice core suggests that climate instability may also have been present during the last interglacial (Eemian). This period (~115,000–130,000 years ago) was locally warmer in Greenland than the present climate by up to 4°C but may have been interrupted by a series of changes that began extremely rapidly and lasted from decades to centuries. Questions about the validity of interpretation of the GRIP Eemian changes arise because there are significant differences between the GRIP and the GISP 2 records for the Eemian whereas the two cores, only 28 km apart, are in excellent agreement throughout the Holocene and the last glacial period. The bottom 10% of each core contains deformational features that are likely to have disturbed their stratigraphy (Taylor *et al.*, 1993b; Alley *et al.*, 1995). Comparison of the composition of entrapped air with the undisturbed Antarctic air composition record shows that mixing of ice of different origin has affected the bottom part of the GISP 2 core (Bender *et al.*, 1994). Until such an approach has been exploited for GRIP the possibility that disturbances simply change the apparent timing of events within the Eemian cannot be excluded for this core.

Evidence from other sources does little, at this stage, to determine whether or not these apparent Eemian rapid climate changes are real. Keigwin *et al.* (1994) and McManus *et al.* (1994) find little evidence of variability during the Eemian. Thouveny *et al.* (1994) and Field *et al.* (1994), however, report that pollen and rock magnetism data from western Europe show somewhat similar changes to the GRIP ice core record although the abrupt change found by Field *et al.*, may be a statistical artefact (Aaby and Tauber, 1995). Cortijo *et al.* (1992) also found rapid cooling in the ocean record during the Eemian. This may suggest that climate changes during the Eemian were fundamentally different in scale and area of influence, compared to more recent events.

These rapid events are relevant to understanding current climate because they affect on the human time-scales important climatic variables on a large geographical scale. However, at least some of these abrupt changes have been attributed to the instability of an ice sheet which does not exist in today's world. The relevance of past abrupt events to present and future climate would be more convincing if the suggested high climate variability in the Eemian was confirmed.

Abrupt regional events also occurred in the past 10,000 years (e.g., Berger and Labeyrie, 1987). These changes, however, have been smaller and smoother than those during the previous glacial period. Such events are perhaps more relevant to the estimation of the possible speed of natural climate variations in the current climate, because of the closer similarity of boundary conditions (e.g., continental positions and ice sheets) to the current situation. It seems unlikely, given the smaller regional changes, that global mean temperatures have varied by 1°C or more in a century at any time during the last 10,000 years (e.g., Wigley and Kelly, 1990).

3.6.4 Summary of 3.6

Palaeoclimatic data are needed to better estimate natural climate variability to help resolve the climate change detection issue (Chapter 8), and are useful for the evaluation of climate models (Chapter 5).

Large and rapid climatic changes occurred during the last ice age and during the transition towards the present Holocene period. Some of those changes may have occurred on time-scales of a few decades, at least in the North Atlantic where they are best documented. They affected atmospheric and oceanic circulation and temperature, and the hydrologic cycle. There are suggestions that similar rapid changes may have also occurred during the last interglacial (the Eemian) period, but this requires confirmation. The recent (20th century) warming needs to be considered in the light of evidence that rapid climatic changes can occur naturally in the climate. However, temperatures have been far less variable during the last 10,000 years (i.e., during the Holocene).

Recent studies have demonstrated that the two periods commonly known as the Medieval Warm Period and the Little Ice Age were geographically more complex than previously believed. It is not yet possible to say whether, on a hemispheric scale, temperatures declined from the 11–12th to the 16–17th century. However, it is clear that the period of instrumental record began during one of the cooler periods of the past millennium.

Two views of the temperature record of the last century are possible if this record is viewed with the longer perspective provided by the palaeoclimatic data (Figure 3.20). On the one hand, the long-term change of temperature could be interpreted as showing a gradual increase from the late 16th century, interrupted by cooler conditions in the 19th century. Alternatively, one could argue that temperatures fluctuated around a mean somewhat lower than the 1860 to 1959 average (punctuated by cooler intervals in the late 16th, 17th and 19th centuries) and then underwent pronounced, and unprecedented (since 1400) warming in the early 20th century. Whichever view is considered, mid-late 20th century surface temperatures appear to have been warmer than any similar period of at least the last 600 years (Figures 3.20, 3.21). In at least some regions 20th century temperatures have been warmer than any other century for some thousands of years.

3.7 Are the Observed Trends Internally Consistent?

Estimates of changes in some important climatic variables examined earlier in this chapter, for the instrumental period are provided in Figure 3.23. The periods over which the changes have been estimated differ for the different variables, reflecting the availability of credible data. The information in the figure is meant to provide only a gross picture of climate variations during the instrumental record. In particular, it should not be interpreted as implying that any changes in the climate during the instrumental period have been linear or even necessarily monotonic. Examination of the earlier parts of this chapter will reveal more information, including the general absence of simple, linear trends over extended periods, for all the variables included in Figure 3.23.

Not all the climatic variables exhibiting substantial variations are represented in Figure 3.23, for simplicity. For example, changes in extreme events are not indicated on the schematic. There has been a clear trend to fewer frosts and low temperatures in several widely separated areas in recent decades. One recent circulation change is well-documented: the rather abrupt change in ENSO and atmospheric circulation, around 1976/77, noted in IPCC (1990). El Niño episodes have been relatively frequent since that time, with only rare excursions into La Niña episodes (e.g., 1988/89). There have been substantial prolonged precipitation and circulation anomalies associated with this extended period.

The pattern of change represented in Figure 3.23 is internally consistent. All the forms of data used to examine climate change and variability suffer from problems of quality and consistency, so conclusions reached on the basis of just one form of data must always be somewhat suspect. However, the internal consistency shown in Figure 3.23, of a warm late-20th century world, accompanied by some changes in the hydrologic cycle, confirms the validity of the messages from the individual forms of data. So, the

(a)   **Temperature indicators**

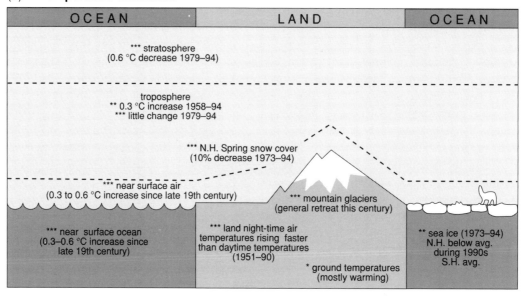

(b)   **Hydrological indicators**

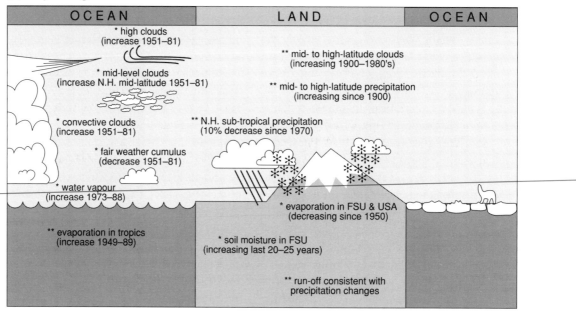

Asterisk indicates confidence level (i.e., assessment): * high, ** medium, * low**

Figure 3.23: (a) Schematic of observed variations of temperature. (b) Schematic of observed variations of the hydrologic cycle.

observed warming cannot be attributed to urbanisation since it is also found in ocean temperatures and reflected in indirect temperature measurements. The increased 40–70°N precipitation is reflected in increased streamflow.

Despite this consistency, it should be clear from the earlier parts of this chapter that current data and systems are inadequate for the complete description of climate change. Virtually every monitoring system and data set requires better data quality and continuity. New monitoring systems, as well as improvements on current systems and studies to reduce quality problems from historical data, are required. Such improvements are essential, if we are to answer conclusively the questions posed in this chapter. Enormous amounts of meteorological data have been collected and archived over the past century. Even greater amounts will be collected, using new observing systems, in the future. The old and new observations will need to be combined carefully, and comprehensive efforts made to reduce the influence of time-varying biases in all the data, and to protect the integrity of long-established, high-quality observing systems and sites, if we are to obtain more accurate and complete estimates of observed climate change and variability.

Conventional meteorological data, both now and in the past, were collected for weather prediction and for the description of the current climate. They require considerable work to ensure that they are useful for monitoring climate variability and change. Studies are required to ensure that adequate corrections can be made for changes in instrumentation, exposure, etc. The reduction of such problems in the future, as well as the past, will require the protection of high-quality climate observing sites.

Probably no single climate element has been studied more than near-surface temperature over land. Unfortunately, the present rate of decline of global data acquisition and exchange across the Global Telecommunications System threatens estimates of near-surface global temperature change (Karl *et al.*, 1995a). Historical temperature data are plagued by inhomogeneities from changes in instrumentation, exposure, site-changes, and time-of-observation bias. Many of these problems can be overcome by thorough comparisons between stations and with the help of metadata (documentation regarding site and instrumentation changes etc.). Few countries, however, provide adequate support for such activities. Changes in instrumentation in recent times continue to pose problems. Karl *et al.* (1995a) note, for instance, that the introduction of a new maximum-minimum thermistor in the USA co-operative observing network has introduced a systematic bias. Oceanographic data suffer from a lack of continuity. Conclusive detection and attribution of global climate change will require an ongoing homogenous, globally representative climate record. This needs to be given high priority in the design and maintenance of meteorological and oceanographic monitoring systems.

References

Aaby, B. and H. Tauber, 1995: Eemian climate and pollen. *Nature*, **376**, 27–28.

Agee, E.M., 1991: Trends in cyclone and anticyclone frequency and comparison with periods of warming and cooling over the Northern Hemisphere. *J. Climate*, **4**, 263–267.

Aharon, P., C.W. Wheeler, J.M.A. Chappell and A.L. Bloom, 1994: Oxygen isotopes and Sr/Ca coral records from core Huon-1 in New Guinea document a strong Younger Dryas

resonance in the western tropical Pacific. *EOS, Abstracts 1994 Spring Meeting*, pp 205.

Allan, R.J., 1993: Historical fluctuations in ENSO and teleconnection structure since 1879: Near-global patterns. *Quaternary Australasia*, **11**, 17–27.

Allan, R.J. and R.D. D'Arrigo, 1996: 'Persistent' ENSO sequences: How unusual was the recent El Niño? *Holocene*, (submitted).

Allan, R.J. and M.R. Haylock, 1993: Circulation features associated with the winter rainfall decrease in southwestern Australia. *J. Climate*, **6**, 1356–1367.

Allan, R.J., J.A. Lindesay and C.J.C. Reason, 1995: Multidecadal variability in the climate system over the Indian Ocean region during the austral summer. *J. Climate*, **8**, 1853–1873.

Alley, R.B., D.A. Meese, C.A. Shuman, A.J. Gow, K.C. Taylor, P.M. Grootes, J.W. C. White, M. Ram, E.D. Waddington, P.A. Mayewski and G.A. Zielinski, 1993: Abrupt increase in Greenland snow accumulation at the end of the Younger Dryas event. *Nature*, **362**, 527–529.

Alley, R.B., A.J. Gow, S.J. Johnson, K. Kiefstuhl, D.A. Meese and Th. Thorsteinsson, 1995: Comparison of deep ice cores. *Nature*, **373**, 393–394.

Anderson, R.Y., 1992: Long-term changes in the frequency of occurrence of El Niño events. In: *El Niño: Historical and Paleoclimatic Aspects of the Southern Oscillation*, H.F. Diaz and V. Markgraf (eds.), Cambridge University Press, Cambridge, UK, pp 193–200.

Angell, J.K., 1988: Variations and trends in tropospheric and stratospheric global temperatures, 1958-87. *J. Climate*, **1**, 1296–1313.

Antonov, J.I., 1993: Linear trends of temperature at intermediate and deep layers of the North Atlantic and the North Pacific Oceans: 1957–1981. *J. Climate*, **6**, 1928–1942.

Arkin, P.A. and P.E. Ardanuy, 1989: Estimating climatic-scale precipitation from space: A review. *J. Climate*, **2**, 1229–1238.

Balling, R.C. and S.E.Idso, 1990: Effects of greenhouse warming on maximum summer temperatures. *Agricultural and Forest Meteorology*, **53**, 143–147.

Balling, R.C., G.A. Meyer and S.G. Wells, 1992: Climate change in Yellowstone National Park: Is the drought-related risk of wildfires increasing? *Clim. Change*, **22**, 35–45.

Bardin, M.Yu., 1994: Parameters of cyclonicity at 500 mb in the Northern Hemisphere extratropics. In: *Proc. XVIII Climate Diagnostics Workshop, Boulder, Co.*, NTIS, US Dept. of Commerce, Sills Building, 5285 Port Royal Road, Springfield, VA 22161, U.S.A., pp 397.

Barros, V. and I. Camilloni, 1994: Urban-biased trends in Buenos Aires' mean temperature. *Clim. Res.*, **4**, 33–45.

Beck, J.W., R.L. Edwards, E. Ito, F.W. Taylor, J. Recy, F. Rougerie, P. Joannot and C. Henin, 1992: Sea surface temperature from coral skeletal strontium/calcium ratios. *Science*, **257**, 644–647.

Beltrami, H. and J.-C. Mareschal, 1991: Recent warmings in eastern Canada inferred from geothermal measurements. *Geophys. Res. Lett.*, **18**, 605–608.

Beltrami, H. and J.-C. Mareschal, 1992: Ground temperature histories for central and eastern Canada from geothermal measurements: little ice age signature. *Geophys. Res. Lett.*, **19**, 689–692.

Bender, M., T. Sowers, M.L. Dickson, J. Orchards, P. Grootes, P.A. Mayewski and M.A. Meese, 1994: Climate connection between Greenland and Antarctica during the last 100,000 years. *Nature*, **372**, 663–666.

Berger, W.H.,and L.D. Labeyrie (eds.), 1987: *Abrupt Climatic Change: Evidence and Implications.* Reidel, Dordrecht, pp. 425.

Bindoff, N.L. and J.A. Church, 1992: Warming of the water column in the southwest Pacific Ocean. *Nature*, **357**, 59–62.

Bindoff, N.L. and T.J. McDougall, 1994: Diagnosing climate change and ocean ventilation using hydrographic data. *J. Phys. Oceanog.*, **24**, 1137–1152.

Black, P.G., 1992: Evolution of maximum wind estimates in typhoons. *ICSU/WMO International Symposium on Tropical Cyclone Disasters, October 12–16, 1992, Beijing.*

Bond, G.C. and R. Lotti, 1995: Iceberg discharges into the North Atlantic on millennial time scales during the last glaciation. *Science*, **267**, 1005–1010.

Bond, G., W.S. Broecker, S.J. Johnsen, J. Mc Manus, L.D. Labeyrie, J.Jouzel and G. Bonani, 1993: Correlations between climate records from North Atlantic sediments and Greenland ice. *Nature*, **365**, 143–147.

Boninsegna, J.A., 1992: South American dendroclimatological records. In: *Climate Since AD 1500,* R.S. Bradley and P.D. Jones (eds.), Routledge, London, pp 446–462.

Bootsma, A., 1994: Long term (100 yr) climatic trends for agriculture at selected locations in Canada. *Clim. Change,* **26**, 65–88.

Born, K., 1995: Tropospheric warming and changes in weather variability over the Northern Hemisphere during the period 1967–1991. *Meteorology and Atmospheric Physics,* (submitted).

Born, K. and H. Flohn, 1995: The detection of changes in baroclinicity and synoptic activity on the Northern Hemisphere for the period 1967-91using two data sets. *Meteorologische Zeitschrift,* (submitted).

Bottomley, M., C.K. Folland, J. Hsiung, R.E. Newell and D.E. Parker, 1990: *Global Ocean Surface Temperature Atlas (GOSTA).* Joint Meteorological Office/Massachusetts Institute of Technology Project. Project supported by US Dept of Energy, US National Science Foundation and US Office of Naval Research. Publication funded by UK Depts of Energy and Environment, HMSO, London. 20 + iv pp and 313 Plates.

Bouchard, R.H., 1990: A climatology of very intense typhoons: or where have all the super typhoons gone? In: *1990 Annual Tropical Cyclone Report*, Joint Typhoon Warning Center, Guam, pp 266–269.

Bradley, R.S. and P.D. Jones, 1993: Little Ice Age summer temperature variations: their nature and relevance to recent global warming trends. *The Holocene*, **3**, 367–376.

Bradley, R.S. and P.D. Jones., 1995: Recent developments in studies of climate since AD 1500. In: *Climate Since AD 1500* (second edition), R.S. Bradley and P.D. Jones (eds.), Routledge, London, pp. 666–679.

Bradley, R.S., H.F. Diaz, J.K. Eischeid, P.D. Jones, P.M. Kelly and C.M. Goodess, 1987: Precipitation fluctuations over northern hemisphere land areas since the mid-19th Century. *Science*, **237**, 171–275.

Brázdil, R., 1992: Fluctuation of atmospheric precipitation in Europe. *GeoJournal 27*, 275–291,

Briffa, K.R. and P.D. Jones, 1993: Global surface air temperature variations during the twentieth century: Part 2, implications for large-scale high frequency paleoclimatic studies. *Holocene*, **3**, 77–88.

Briffa, K.R., T.S. Bartholin, D. Eckstein, P.D. Jones, W. Karlen, F.H. Schweingruber and P. Zetterberg, 1990: A 1,400-year tree-ring record of summer temperatures in Fennoscandia. *Nature*, **346**, 434–439.

Briffa, K.R., P.D. Jones, T.S. Bartholin, D. Eckstein, F.H. Schweingruber ,W. Karlen, P. Zetterberg, and M. Fronen, 1992: Fennoscandian summers from AD 500: Temperature changes on short and long timescales. *Clim. Dyn.,* **7**, 111–119.

Briffa, K.R., P.D. Jones, F.H. Schweigruber, S.G. Shiyatov and E.R. Cook, 1995: Unusual twentieth-century summer warmth in a 1,000-year temperature record from Siberia. *Nature*, **376**, 156–159.

Burn, D.H, 1994: Hydrologic effects of climate change in west-central Canada. *J. Hydrol.*, **160**, 53–70.

Cardone, V.J., J.G. Greenwood and M.A. Cane, 1990: On trends in historical marine data. *J. Clim.ate*, **3**, 113–127.

Cayan, D.R., 1992: Latent and sensible heat flux anomalies over the northern oceans: The connection to monthly atmospheric circulation. *J. Climate*, **5**, 354–369.

Cermak, V., L. Bodri and J. Safanda, 1992: Underground temperature fields and changing climate: evidence from Cuba. *Global and Planetary Change*, **97**, 325–337.

Changnon, S. 1987: Climatic fluctuation and record-high levels of Lake Michigan. *Bull. Am. Met. Soc.*, **68**, 1394–1402.

Changnon, S.A. and J.M. Changnon, 1992: Temporal fluctuations in weather disasters: 1950–1989. *Clim. Change,* **22**, 191–208.

Chapman, D.S., T.J. Chisholm, and R.N. Harris, 1992: Combining borehole temperature and meteorological data to constrain past climate change. *Global and Planetary Change,* **6**, 269–281.

Chapman, W.L. and J.E. Walsh, 1993: Recent variations of sea ice and air temperature in high latitudes. *Bull. Am. Met .Soc.*, **74**, 33–47.

Chappellaz, J., T. Blunier, D. Raynaud, J.M. Barnola, J. Schwander and B. Stauffer, 1993: Synchronous changes in atmospheric CH_4 and Greenland climate between 40 and 8 kyr BP. *Nature*, **366**, 443–445.

Chelliah, M. and P. Arkin, 1992: Large-scale interannual variability of monthly outgoing long wave radiation anomalies over the global tropics. *J. Climate,* **5**, 371–389.

†Chen, L-X., M. Dong and Y.-N. Shao, 1992: The characteristics of interannual variations of the East Asian monsoon. *J. Met. Soc. Japan*, **70**, 397–421.

+Chen, T.C., H. van Loon, K.D. Wu and M.C. Yen, 1992: Changes in the atmospheric circulation over the North Pacific-North America area since 1950. *J. Met. Soc. Japan*, **70**, 1137–1146.

Chiew, F.H.S. and T.A. McMahon, 1993: Detection of trend or change in annual flow in Australian rivers. *Int. J. Climatology*, **13**, 643–653.

Chiew, F.H.S. and T.A. McMahon, 1995: Trends and changes in historical annual streamflow volumes and peak discharges of rivers in the world. *Proc. Int. Congress on Modelling and Simulation, November 1995, Newcastle, Australia* (in press).

Christy, J.R., 1995: Temperature above the surface layer. *Clim. Change,* (in press).

Christy, J.R. and S.J. Drouilhet, 1994: Variability in daily, zonal mean lower-stratospheric temperatures. *J. Climate*, **7**, 106–120.

Christy, J.R. and J.D. Goodridge, 1995: Precision global temperatures from satellites and urban warming in non-satellite data. *Atmos. Env.*, **29**, 1957–1961.

Christy, J.R. and R.T. McNider, 1994: Satellite greenhouse warming. *Nature*, **367**, 325.

Christy, J.R., R.W. Spencer and R.T. McNider, 1995: Reducing noise in the MSU daily lower tropospheric global temperature data set. *J. Climate*, **8**, 888–896.

Chu, P-S., Z.-P. Yu and S. Hastenrath, 1994: Detecting climate change concurrent with deforestation in the Amazon basin: Which way has it gone? *Bull. Am .Met .Soc.*, **75**, 579–583.

Cole, J.E., R.G. Fairbanks and G.T. Shen, 1993: Recent variability in the Southern Oscillation: Isotopic results from a Tarawa Atoll coral. *Science*, **260**, 1790–1793.

Cook, E.R., 1995: Temperature histories from tree rings and corals. *Clim. Dyn.,* **11**, 211–222.

Cook, E.R., T. Bird, M. Peterson, M. Barbetti, B. Buckley, R. D'Arrigo, R. Francey and P.Tans, 1991: Climatic change in Tasmania inferred from a 1089-year tree-ring chronology of Huon pine. *Science*, **253**, 1266–1268.

Cook, E.R., T. Bird, M. Peterson, M. Barbetti, B. Buckley, R. D'Arrigo and R. Francey, 1992: Climatic change over the last millennium in Tasmania reconstructed from tree-rings. *Holocene*, **2**, 205–217.

Cooter, E.J. and S.K. LeDuc, 1995: Recent frost date trends in the northeastern United States. *Int. J. Climatology*, **15**, 65–75.

Cortijo, E., J.C. Duplessy, L. Lubeyrie, H. Leclaire, J. Duprat, and T.C.E. Van Weering, 1992: Eemian cooling in the Norwegian Sea and North Atlantic Ocean preceding ice-sheet growth. *Nature*, **372**, 446–449.

Crowley, T.J., 1994: Pleistocene temperature changes. *Nature*, **371**, 664.

Crowley, T.J. and K-Y.Kim, 1993: Towards development of strategy for determining the origin of decadal-centennial scale climate variability. *Quat. Sci. Rev.*, **12**, 375–385.

Crowley, T.J. and K-Y.Kim, 1995: Comparison of longterm greenhouse projections with the geologic record. *Geophys. Res. Lett.*, **22**, 933–936.

Cuffey, K.M., R.B. Alley, P.M. Grootes, J.F. Bolzan and S. Anandakrishnan, 1994: Calibration of the $\delta^{18}O$ isotopic paleothermometer for central Greenland, using borehole temperatures. *J. Glaciology*, **40**, 341–349.

Dahlström, B. 1994: Short term fluctuations of temperature and precipitation in Western Europe. In: *Climate Variations in Europe,* R.Heino (ed.), Academy of Finland, Publication 3/94, pp 30–38.

Dansgaard, W., J.W.C. White and S.J. Johnsen, 1989: The abrupt termination of the Younger Dryas. *Nature*, **339**, 532–534.

Dansgaard, W., S. J. Johnsen, H.B. Clausen, D. Dahl-Jensen, N.S. Gunderstrup, C.U. Hammer, C.S. Hvidberg, J.P. Steffensen, A. Sveinbjörnsdottir, J. Jouzel and G. Bond, 1993: Evidence for general instability of past climate from a 250-kyr ice-core record. *Nature*, **364**, 218–220.

Davis, R.E. and R. Dolan, 1993: Nor'easters. *American Scientist*, **81**, 428–439.

DeGaetano, A.T., K.L. Eggleston and W.W. Knapp, 1994: Trends in extreme temperature events in the northeastern United States. In: *Preprints, Sixth Conference on Climate Variations, January 23-28, 1994*, Nashville, Tennessee, American Meteorological Society, pp. 136–139.

Delworth, T., S. Manabe and R.J. Stouffer, 1993: Interdecadal variations of the thermohaline circulation in a coupled ocean-atmosphere model. *J. Climate*, **6**, 1993–2011.

Deming, D., 1995: Climatic warming in North America: Analysis of borehole temperatures. *Science*, **268**, 1576–1577.

Denton, G.H. and C.H. Hendy, 1994: Younger Dryas age advance of Franz Josef glacier in the southern alps of New Zealand. *Science*, **264**, 1434–1437.

Deser, C. and M.L. Blackmon, 1993: Surface climate variations over the North Atlantic ocean during winter: 1900-1989. *J. Climate*, **6**, 1743–1753.

Deser, C. and M.L. Blackmon, 1995: On the relationship between tropical and North Pacific sea surface temperature variations. *J. Climate*, **8**, 1677–1680.

Dessens, J., 1994: Climatic response to a differential change in minimum and maximum temperatures. In: *Abstracts, AGU Western Pacific Geophysics Meeting, EOS*, June 21, 1994.

Dessens, J. and A. Bücher, 1995: Changes in minimum and maximum temperatures at the Pic du Midi in relation with humidity and cloudiness, 1882–1984. *Atmos. Res.*, **37**, 147–162.

Dettinger, M.D. and D.R. Cayan, 1995: Large-scale atmospheric forcing of recent trends toward early snowmelt runoff in California. *J. Climate*, **8**, 606–623.

Diaz, H.F. and V. Markgraf, 1992: *El Niño: Historical and Paleoclimatic Aspects of the Southern Oscillation*, Cambridge University Press, Cambridge, UK, pp 476.

Diaz, H.F. and R.S. Pulwarty, 1994: An analysis of the time scales of variability in centuries-long ENSO-sensitive records in the last 1000 years. *Clim. Change,* **26**, 317–342.

Diaz, H.F., R.S. Bradley and J.K. Eischeid, 1989: Precipitation fluctuation over global land areas since the late 1800s. *J. Geophys. Res.*, **94**, 1195–1240.

Diment, W.H., 1965: Comments on paper by E.A. Lubimova, 'Heat flow in the Ukrainian Shelf in relation to recent tectonic movement'. *J. Geophys. Res.*, **70**, 2466–2467.

Douglas, A.V., D.R. Cayan and J.Namias, 1982: Large-scale changes in North Pacific and North American weather patterns in recent decades. *Month. Wea. Rev.*, **110**, 1851–1862.

Downton, M.W. and K.A. Miller, 1993: The freeze risk to Florida citrus. Pt II: Temperature variability and circulation patterns. *J. Climate*, **6**, 364–372.

Dronia, H., 1991: On the accumulation of excessive low pressure systems over the North Atlantic during the winter seasons (November to March) 1988/89 to 1990/91. *Die Witterung im übersee*, **39**, 27.

Druffel, E.R.M and S. Griffin, 1993: Large variations of surface ocean radiocarbon: evidence of circulation changes in the southwestern Pacific. *J. Geophys. Res.*, **98**, 20249–20259.

Duchkov, A.D. and V.N. Devyatkin, 1992: Reduced geothermal gradients in the shallow West-Siberian Platform. *Global and Planetary Change*, **6**, 245–250.

Dunbar, R.B., G.M. Wellington, M.Colgan and P.W.Glynn, 1994: Eastern Pacific sea surface temperature variability since 1600 AD: The $\delta^{18}O$ record of climatic variability in Galapagos corals. *Paleoceanography*, **9**, 291–315.

Duplessy, J.C., L. Labeyrie, M. Arnold, M. Paterne, J. Duprat and T.C.E. Van Weering, 1992: North Atlantic sea surface salinity and abrupt climatic changes. *Nature*, **358**, 485–488.

Dutton, E.G. and J.R. Christy, 1992: Solar radiative forcing at selected locations and evidence for global lower tropospheric cooling following the eruptions of El Chichon and Pinatubo. *Geophys. Res. Lett.*, **19**, 2313–2316.

Eischeid, J.K., H.F. Diaz, R.S. Bradley and P.D. Jones, 1991: *A comprehensive precipitation data set for global land areas.* U.S. Department of Energy Report No. DOE/ER-69017T-H1, Washington, DC, 81 pp.

Eischeid, J.K., C.B. Baker, T.R. Karl and H.F. Diaz, 1995: The quality control of long-term climatological data using objective data analysis. *J. Appl. Meteor.*, **34**, 2787–2795.

Elliott, W.P. and D.J. Gaffen, 1991: On the utility of radiosonde humidity archives for climate studies. *Bull. Am . Met. Soc.*, **72**, 1507–1520.

Elliott, W.P., M.E. Smith and J.K. Angell, 1991: On monitoring tropospheric water vapor changes using radiosonde data. In: *Greenhouse-Gas-Induced Climatic Change: A Critical Appraisal of Simulations and Observations*, M.E. Schlesinger (ed.), Elsevier, Amsterdam, pp 311–328.

Elliott, W.P., D.J. Gaffen, J.D.W. Kahl and J.K. Angell, 1994: The effect of moisture on layer thicknesses used to monitor global temperatures. *J. Climate*, **7**, 304–308.

Elliott, W.P., R.J. Ross and D.J. Gaffen, 1995: Water vapor trends over North America. *Sixth Symposium on Global Change Studies, Amer. Meteor. Soc., Preprints*, pp 185–186.

Emslie, A.G. and J.R. Christy, 1995: The role of astronomical influences in driving long-term trends in terrestrial temperature, *Geophys. Res. Lett.*, (submitted).

Environment Canada, 1995: *The state of Canada's climate: monitoring variability and change.* State of Environment Rep. No. 95-1, Minister of Supply and Services, Canada, 52 pp.

Field, M.H., B. Huntley and H. Mſller, 1994: Eemian climate fluctuations observed in a European pollen record. *Nature*, **371**, 779–783.

Findlay, B.F., D.W. Gullett, L. Malone, J. Reycraft, W.R. Skinner, L. Vincent and R. Whitewood, 1994: Canadian national and regional standardized annual precipitation departures. In: *Trends '93: A Compendium of Data on Global Change*, T.A. Boden, D.P. Kaiser, R.J. Sepanski and F.W. Stoss (eds.), ORNL/CDIAC-65, Carbon Dioxide Information Analysis Center, Oak Ridge National Laboratory, Oak Ridge, U.S.A., pp. 800–828.

Flohn, H., A. Kapala, H.R. Knoche, and H. Mächel, 1992: Water vapour as an amplifier of the greenhouse effect: new aspects. *Meteorol. Zeitschrift N.F.*, **1**, 122–138.

Folland, C.K. and D.E. Parker, 1995: Correction of instrumental biases in historical sea surface temperature data. *Quart. J. R. Met. Soc.*, **121**, 319–367.

Folland, C.K. and M.J. Salinger, 1996: Surface temperature trends and variations in New Zealand and the surrounding ocean, 1871–1993. *Int. J. Climatology*, **15**, 1195–1218.

Frich, P., 1994: Precipitation trends in the North Atlantic European region. In: *Climate Variations in Europe*, R.Heino (ed.), Academy of Finland, Publication 3/94, pp 196–200.

Gadgil, S., A. Guruprasad and J. Srinivasan, 1992: Systematic bias in the NOAA outgoing longwave radiation dataset? *J. Climate*, **5**, 867–875.

Gaffen, D.J., 1994: Temporal inhomogeneities in radiosonde temperature records. *J. Geophys. Res.*, **99**, 3667–3676.

Gaffen, D.J., T.P. Barnett and W.P. Elliott, 1991: Space and time scales of global tropospheric moisture, *J. Climate*, **4**, 989–1008.

Gaffen, D.J., W.P. Elliott and A. Robock, 1992: Relationships between tropospheric water vapor and surface temperature as observed by radiosondes. *Geophys. Res. Let.*, **19**, 1839–1842.

Garcia, O., 1981: A comparison of two satellite rainfall estimates for GATE. *J. Appl. Meteor.*, **20**, 430–438.

Ginsburg, B.M., C.N. Polyakova and I.I. Soldatova, 1992: Centennial variations of the ice appearance dates on river and their relation to climate changes. *Meteorologia i gidrologia*, **12**, 71–79.

Glynn, P.W., 1993: Coral reef bleaching: Ecological perspectives. *Coral Reefs*, **12**, 1–17.

Goldenberg, S.B. and L.J. Shapiro, 1996: Physical mechanisms for the relationships between El Niño, West African rainfall, and North Atlantic major hurricanes. *J. Climate*, (in press).

Golitsyn, G.S., 1995: The Caspian sea level as a problem of diagnosis and prognosis of the regional climate change. *Isvestia – Atmospheric and Oceanic Physics*, **31**, 385–391.

Golitsyn, G.S., A.V. Dzuba, A.G. Osipov and G.N. Panin, 1990: Regional climate changes and their impacts on the Caspian Sea level rise. *Doklady USSR Acad. Sci.*, **313**, 5, 1224–1227.

Golubev, V.S. and I.C. Zmeikova, 1991: Long-term changes of evaporation conditions in the Aral area. In: *Environmental Monitoring in the Aral Sea Basin.*.Yu.A. Izrael and Yu.A. Anokhin (eds.), Gidrometeoizdat, Saint-Petersburg, 80–86.

Goreau, T.J. and R.L. Hayes, 1994: Coral bleaching and ocean "hot spots". *Ambio*, **73**, 176–180.

Gosnold, W.D., A.M. Farrow and H.J. Pollack, 1992: Microclimate effects on the climate record in the geothermal gradient, *EOS, 1992 Fall Meeting Supplement*, 70.

Goudie, A.S. and N.J. Middleton, 1992: The changing frequency of dust storms through time. *Clim. Change,* **20**, 197–225.

Graham, N.E., 1995: Simulation of recent global temperature trends. *Science*, **267**, 666–671.

Graf, H.-F., I. Kirchner, A. Robock and I. Schult, 1993: Pinatubo eruption winter climate effects: model versus observations. *Clim. Dyn.* **9**, 81–93.

Graybill, D.A. and S.G. Shiyatov, 1992: Dendroclimatic evidence from the northern Soviet Union. In: *Climate Since AD 1500*, R.S. Bradley and P.D. Jones (eds.), Routledge, London, pp 393–414.

Graybill, D.A. and G. Funkhouser, 1994: Dendroclimatic reconstructions during the past millennium in the southern Sierra Nevada and Owens Valley, California. In: *Southern California Climate: the last 2000 Years*, M. Rose and P. Wigand (eds.), Natural History Museum of Los Angeles County, Los Angeles (in press).

Gray, W.M., 1990: Strong association between West African rainfall and U.S. landfall of intense hurricanes. *Science*, **249**, 1251–1256.

Grazulis, T.P., 1993: A 110-year perspective of significant tornadoes. In: *The Tornado: Its Sructure, Dynamics, Prediction, and Hazards*, C. Church, D. Burgess, C. Doswell, and R. Davies-Jones (eds.), American Geophysical Union, pp 467–474.

Gregory, J.M., P.D. Jones, and T.M. Wigley, 1991: Precipitation in Britain: An analysis of area-averaged data updated to 1989. *Int. J. Climatology*, **11**, 331–345.

Grimm, E.C., G.L. Jacobson, W.A. Watts, B.C.S. Hansen and K.A. Maasch, 1993: A 50000-year record of climate oscillations from Florida and its temporal correlation with the Heinrich Events. *Science*, **261**, 198–200.

Groisman, P.Ya, 1992: Possible regional climate consequences of the Pinatubo eruption. *Geophys. Res. Lett.*, **19**, 1603–1606.

Groisman, P.Ya and D.R. Easterling, 1994: Variability and trends of precipitation and snowfall over the United States and Canada. *J. Climate*, **7**, 184–205.

Groisman, P.Ya, V.V. Koknaeva, T.A. Belokrylova and T.R. Karl, 1991: Overcoming biases of precipitation measurement: A history of the USSR experience. *Bull. Am. Met. Soc.*, **72**, 1725–1733.

Groisman, P.Ya, T.R. Karl, R.W. Knight and G.L. Stenchikov,

1994a: Changes of snow cover, temperature, and the radiative heat balance over the Northern Hemisphere. *J. Climate*, **7**, 1633–1656.

Groisman, P.Ya, T.R. Karl, and R.W. Knight, 1994b: Observed impact of snow cover on the heat balance and the rise of continental spring temperatures. *Science*, **263**, 198–200.

Grootes, P.M., M. Stuiver, J.W.C. White, S. Johnsen and J. Jouzel, 1993: Comparison of oxygen isotope records from the GISP2 and GRIP Greenland ice cores. *Nature*, **366**, 552–554.

Grove, A.T., 1995: African river discharge and lake levels in the twentieth century. In: *The Limnology, Climatology and Paleoclimatology of the East African Lakes*, T.C. Johnson and E.Odada (eds.), Gordon and Breach, London (in press).

Guilderson, T.P., R.G. Fairbanks, and J.L. Rubenstone, 1994: Tropical temperature variations since 20,000 years ago: Modulating interhemispheric climate change. *Science*, **263**, 663–665.

Guiot, J., J.L. de Beaulieu, R. Cheddad, F. David, P. Ponel and M. Reille, 1993: The climate in western Europe during the last glacial interglacial cycle derived from pollen and insect remains. *Palaeogeography, Palaeoclimatology,* **103**, 73–93.

Gutzler, D., 1992: Climatic variability of temperature and humidity over the tropical western Pacific. *Geophys. Res. Lett*, **19**, 1595–1598.

Haeberli, W., 1995: Glacier fluctuations and climate change detection – operational elements of a worldwide monitoring strategy. *Bulletin World Meteorological Organization*, **44**, 23–31.

Haeberli, W., P. Müller, P. Alean and H. Bösch, 1989: Glacier changes following the Little Ice Age: a survey of the international data basis and its perspectives. In: *Glacier Fluctuations and Climatic Change*, J. Oerlemans (ed.), Kluwer Academic Publishers, Dordrecht, pp. 77–101.

Hameed, S. and G. Gong, 1994: Variation of spring climate in lower-middle Yangtse River valley and its relation with solar-cycle length. *Geophys. Res. Lett.*, **21**, 2693–2696.

Hansen, J. and S. Lebedeff, 1988: Global surface temperatures: update through 1987. *Geophys. Res. Lett.*, **15**, 323–326.

Hansen, J., H. Wilson, M. Sato, R. Ruedy, K. Shah and E. Hansen, 1995: Satellite and surface temperature data at odds? *Clim. Change*, **30**, 103–117.

Hanssen-Bauer, I. and E. Førland, 1994: Homogenizing of long Norwegian precipitation series. *J. Climate*, **7**, 1001–1013.

Hanssen-Bauer, I., M.K. Solës and E.L. Steffensen, 1990: *The Climate of Spitsbergen*. DNMI-Rapport 39/90 KLIMA, 40pp.

Heino, R., 1994: *Climate in Finland During the Period of Meteorological Observations*. Finnish Meteorological Institute Contributions. No. 12. 209pp.

Henderson, K.G., G.E. Faiers, J.M. Grymes and R.A. Muller, 1994: Temporal variation in temperature and dew point in the Southern Region, In: *Preprints, Sixth Conference on Climate Variations, January 23-28, 1994*, Nashville, Tennessee, American Meteorological Society, pp. 146–147.

Henderson-Sellers, A., 1992: Continental cloudiness changes this century. *GeoJournal*, **27.3**, 255–262.

Hense, A., P. Krahe and H. Flohn, 1988: Recent fluctuations of tropospheric temperature and water vapor content in the tropics. *Meteorol. Atmos. Phys.*, **38**, 215–227.

Hopkins, L.C. and G.J. Holland, 1995: Australian east-coast cyclones and heavy rain days: 1958–1992. *J. Climate*, (submitted).

Horton, E.B., 1995: Geographical distribution of changes in maximum and minimum temperatures. *Atmos. Res.*, **37**, 102–117.

Hughes, M. and H.F. Diaz, 1994: Was there a Medieval Warm Period, and if so, where and when? *Clim. Change*, **26**, 109–142.

Hulme, M., 1991: An intercomparison of model and observed global precipitation climatologies. *Geophys. Res. Lett.*, **18**, 1715–1718.

Hulme, M., 1992: Rainfall changes in Africa: 1931–1960 to 1961–1990. *Int. J. Climatology*, **12**, 685–690.

Hulme, M., Z-C. Zhao and T. Jiang, 1994: Recent and future climate change in East Asia. *Int. J. Climatology*, **14**, 637–658.

Hurrell, J .W., 1995: Decadal trends in the North Atlantic Oscillation and relationships to regional temperature and precipitation. *Science*, **269**, 676–679.

Hurrell, J.W. and H. van Loon, 1994: A modulation of the atmospheric annual cycle in the Southern Hemisphere. *Tellus*, **46A**, 325–338.

Hyndman, R., J. Jaeger, and J. Sass, 1969: Heat flow measurements on the southeast coast of Australia. *Earth Planet. Sci. Lett.*, **7**, 12–16.

IPCC, 1990: *Climate Change, The IPCC Scientific Assessment*, J.T. Houghton, G.J. Jenkins and J.J. Ephraums (eds.), Cambridge University Press, Cambridge, UK, 365 pp.

IPCC, 1992: *Climate Change, 1992: The Supplementary Report to the IPCC Scientific Assessment*, J.T. Houghton, B.A. Callander and S.K. Varney (eds.), Cambridge University Press, Cambridge, UK, 198 pp.

IPCC, 1994: *Climate Change 1994. Radiative Forcing of Cimate Change and an Evaluation of the IPCC IS92 Emissions Scenarios*. J.T. Houghton, L.G. Meira Filho, J. Bruce, Hoesung Lee, B.A. Callander, E. Haites, N. Harris and K. Maskell (eds.), Cambridge University Press, Cambridge, UK, 339 pp.

Iwashima, T. and R. Yamamoto, 1993: A statistical analysis of the extreme events: Long-term trend of heavy daily precipitation. *J. Met. Soc. Japan*, **71**, 637–640.

Jacoby, G.C., R. D'Arrigo and B. Luckman, 1996: In: *Climatic Variations and Forcing Mechanisms of the Last 2000 Years*, P.D. Jones, R.S. Bradley and J. Jouzel (eds.), Springer Verlag, Heidelburg (in press).

Johnsen, S.J., H. Clausen, W. Dansgard, K. Fuhrer, N.S. Gunderstrup, C.U. Hammer, P. Iverssen, J. Jouzel, B. Stauffer and J.P. Steffensen, 1992: Irregular glacial interstadials recorded in a new Greenland ice core. *Nature*, **359**, 311–313.

Johnson, G.L., J.M. Davis, T.R. Karl, A.L. McNab, K.P. Gallo, J.D. Tarpley and P. Bloomfield, 1994: Estimating urban temperature bias using polar-orbiting satellite data. *J. Appl. Met.*, **33**, 358–369.

Jones, P.D., 1988: Hemispheric surface air temperature variations: recent trends and an update to 1987. *J. Climate*, **1**, 654–660.

Jones, P.D., 1990: Antarctic temperatures over the present century – a study of the early expedition record. *J. Climate*, **3**, 1193–1203.

Jones, P.D., 1994a: Hemispheric surface air temperature variations: a reanalysis and an update to 1993. *J. Climate*, **7**, 1794–1802.

Jones, P.D., 1994b: Recent warming in global temperature series. *Geophys. Res. Lett.*, **21**, 1149–1152.

Jones, P.D., 1995a: Recent variations in mean temperature and the diurnal temperature range in the Antarctic. *Geophys. Res. Lett.* **22**, 1345–1348.

Jones, P.D., 1995b: Maximum and minimum temperature trends in Ireland, Italy, Thailand, Turkey and Bangladesh. *Atmos. Res.*, **37**, 67–78.

Jones, P.D. and R.S. Bradley, 1992: Climatic variations over the last 500 years. In: *Climate Since AD 1500*, R.S. Bradley and P.D. Jones (eds.), Routledge, London, pp 649–665.

Jones, P.D. and K.R. Briffa, 1992: Global surface air temperature variations during the twentieth century. *Holocene*, **2**, 165–179.

Jones, P.D. and K.R. Briffa, 1996: What can the instrumental record tell us about longer timescale paleoclimatic reconstructions? In: *Climatic variations and forcing mechanisms of the last 2000 years*, P.D. Jones, R.S. Bradley, and J. Jouzel (eds.), Springer Verlag, Heidelburg (in press).

Jones, P.D., S.C.B. Raper, R.S. Bradley, H.F. Diaz, P.M. Kelly and T.M.L. Wigley, 1986a: Northern Hemisphere surface air temperture variations, 1851–1984. *J. Climate Appl. Met.*, **25**, 161–179.

Jones, P.D., S.C.B. Raper, R.S. Bradley, H.F. Diaz, P.M. Kelly and T.M.L. Wigley, 1986b: Southern Hemisphere surface air temperature variations, 1851–1984. *J. Climate Appl. Met.*, **25**, 1213–1230.

Jones, P.D., P.Ya. Groisman, M. Coughlan, N. Plummer, W.-C. Wang and T.R. Karl, 1990: Assessment of urbanization effects in time series of surface air temperature over land. *Nature*, **347**, 169–172.

Jónsson, T., 1994: Precipitation in Iceland 1857–1992. In: *Climate Variations in Europe* , R. Heino (ed.), Academy of Finland, Publication 3/94, pp. 183–188.

Jouzel, J., C. Lorius, S.J. Johnsen and P. Grootes, 1994: Climate instabilities: Greenland and Antarctic records. *C.R. Acad. Sci., Paris*, **319**, Serie II, 65–77.

Jouzel, J., R. Vaikmae, J.R. Petit, M. Martin, Y. Duclos, M. Stievenard, C. Lorius, M. Toots, M.A. Melieres, L.H. Burckle, N.I. Barkov and V.M. Kotyakov, 1995: The two-step shape and timing of the last deglaciation in Antarctica. *Clim. Dyn.* **11**, 151–161.

Judge, A.S., G. Cheng, T.E. Osterkamp, M. Smith and J. Gray, 1983: Climate change and geothermal regime. In: *Permafrost: Fourth International Conference-Final Proceedings*, 137–159. National Academy Press, Washington, DC.

Kaas, E. and P. Frich, 1995: Diurnal temperature range and cloud cover in the Nordic countries: observed trends and estimates for the future. *Atmos. Res.*, **37**, 211–228.

Kahl, J.D.W., M.C. Serreze, R.S. Stone, S. Shiotani, M. Kisley and R.C. Schell, 1993a: Tropospheric temperature trends in the Arctic: 1958–1986. *J. Geophys. Res.*, **98**, 12825–12838.

Kahl, J.D., D.J. Charlevoix, N.A. Zaitseva, R.C. Schnell and M.C. Serreze, 1993b: Absence of evidence for greenhouse warming over the Arctic Ocean in the past 40 years. *Nature*, **361**, 335–337.

Karl, T.R., G. Kukla, V.N. Razuvayev, M.J. Changery, R.G. Quayle, R.R. Heim, D.R. Easterling and Cong Bin Fu, 1991: Global warming: evidence for asymmetric diurnal temperature change. *Geophys. Res. Lett.*, **18**, 2253–2256.

Karl, T.R., P.D. Jones, R.W. Knight, G. Kukla, N. Plummer, V. Razuvayev, K.P. Gallo, J. Lindseay, R.J. Charlson and T.C. Peterson, 1993a: A new perspective on recent global warming: Asymmetric trends of daily maximum and minimum temperature. *Bull. Am. Met. Soc.*, **74**, 1007–1023.

Karl, T.R., P.Y. Groisman, R.W. Knight and R.R. Heim, Jr., 1993b: Recent variations of snow cover and snowfall in North America and their relation to precipitation and temperature variations. *J. Climate*, **6**, 1327–1344.

Karl, T.R., R.W. Knight and J.R. Christy, 1994: Global and hemispheric temperature trends: Uncertainties related to inadequate spatial sampling, *J. Climate*, **7**, 1144–1163.

Karl, T.R., V.E. Derr, D.R. Easterling, C. Folland, D.J. Hofmann, S. Levitus, N. Nicholls, D. Parker and G.W. Withee, 1995a: Critical issues for long-term climate monitoring. *Clim. Change*, **31**, 185–221.

Karl, T.R., P.D. Jones, and R.W. Knight, 1995b: Testing for bias in the climate record. *Science*, (in press).

Karl, T.R., R.W. Knight and N. Plummer, 1995c: Trends in high-frequency climate variability in the twentieth century. *Nature*, **377**, 217–220.

Karl, T.R., R.W. Knight, D.R. Easterling and R.G. Quayle, 1995d. Trends in U.S. climate during the twentieth century. *Consequences*, **1**, 3–12.

Karl, T.R., R.W. Knight, G. Kukla and J. Gavin, 1995e: Evidence for radiative effects of anthropogenic aerosols in the observed climate record. In: *Aerosol Forcings of Cimate*, R.J. Charlson and J. Heintzenberg (eds.), John Wiley and Sons Ltd., Chichester.

Karoly, D.J., P. Hope and P.D. Jones, 1996: Decadal variations of the Southern Hemisphere circulation. *Int. J. Climatology*, (in press).

Kawamura, R., 1994: A rotated EOF analysis of global sea surface temperature variability with interannual and interdecadal scales. *J. Phys. Oceanogr.*, **24**, 707–715.

Kawamura, R., M. Sugi and N. Sato, 1995: Interdecadal and interannual variability in the northern extratropical circulation simulated with the JMA global model. *J. Climate*, (in press).

Keeling, C.D., R.B. Bascastow, A.F. Carter, S.C. Piper, T.P. Wholf, M. Heimann, W.G.Mook and H. Roeloffzen, 1989: A three-dimensional model of atmospheric CO_2 transport based on observed winds: Analysis of observational data. In: *Aspects of Cimate Variability in the Pacific and the Western Americas*, D.H.Peterson (ed.), *Geophys. Mono.*, **55**, 165–236, AGU, Washington DC.

Keigwin, L.D., W.B. Curry, S.J. Lehman and S. Johnsen, 1994: The role of the deep ocean in North Atlantic climate change between 70 and 130 kyr ago. *Nature*, **371**, 323–326.

King, J.C., 1994: Recent climate variability in the vicinity of the Antarctic Peninsula. *Int. J. Climatolology*, **14**, 357–369.

Kitoh, A., 1991: Interannual variations in an atmospheric GCM forced by the 1970-1989 SST. PtII: Low frequency variability of the wintertime Northern Hemisphere extratropics. *J. Met. Soc. Japan*, **69**, 271–291.

Kukla, G and T.R. Karl, 1993: Nighttime warming and the greenhouse effect. *Environmental Science and Technology*, **27**, 1468–1474.

Kushnir, Y., 1994: Interdecadal variations in North Atlantic sea surface temperature and associated atmospheric conditions. *J. Climate*, **7**, 141–157.

Kwong, Y.T.J. and T.Y. Gan, 1994: Northward migration of permafrost along the Mackenzie Highway and climatic warming. *Clim. Change*, **26**, 399–419.

Labitzke, K. and H.van Loon, 1994: Trends of temperature and geopotential height between 100 and 10 hPa on the Northern Hemisphere. *J. Met. Soc. Japan*, **72**, 643–652.

Lachenbruch, A.H. and B.V. Marshall, 1986: Changing climate; geothermal evidence from permafrost in the Alaskan Arctic. *Science*, **234**, 689–696.

LaMarche, V.C., Jr., 1974: Paleoclimatic inferences from long tree-ring records. *Science*, **183**, 1043–1048.

Lamb, H.H., 1965: The medieval warm epoch and its sequel. *Palaeogeography, Palaeoclimatology, Palaeoecology*, **1**, 13–37.

Lamb, H.H., 1988. Climate and life during the Middle Ages, studied especially in the mountains of Europe. In: *Weather, Climate and Human Affairs*, H.H. Lamb, Routledge, London, pp.40–74.

Lamb, P.J. and R.A. Peppler, 1991: West Africa. In: *Teleconnections Linking Worldwide Climate Anomalies*, M.H. Glantz, R.W. Katz, and N. Nicholls (eds.), Cambridge Univ. Press, Cambridge, UK, 535 pp.

Landsea, C.W., 1993: A climatology of intense (or major) Atlantic hurricanes. *Mon. Weath. Rev.*, **121**, 1703–1713.

Landsea, C.W. and W.M. Gray, 1992: The strong association between Western Sahelian monsoon rainfall and intense Atlantic hurricanes. *J. Climate*, **5**, 435–453.

Landsea, C.W. and W.M. Gray, 1996: Northern hemispheric surface temperature and tropical cyclone activity associations. *Int. J. Climatology* (in press).

Landsea, C.W., N. Nicholls, W.M. Gray, and L.A. Avila, 1996: Quiet early 1990s continues trend of fewer intense Atlantic hurricanes. *Geophys. Res. Lett.*, (submitted).

Lara, A. and R.Villalba, 1993: A 3620-year temperature record from Fitzroya cupressoides tree rings in southern South America. *Science*, **260**, 1104–1106.

Lare, A.R. and S.E. Nicholson, 1994: Contrasting conditions of surface water balance in wet and dry years as a possible land surface-atmosphere feedback mechanism in the West African Sahel. *J. Climate*, **7**, 653–668.

Lau, N.-C. and M.J. Nath, 1994: A modeling study of the relative roles of tropical and extratropical SST anomalies in the variability of the global atmosphere-ocean system. *J. Climate*, **7**, 1184–1207.

Lazier, J.R.N., 1995: The salinity decrease in the Labrador Sea over the past thirty years. In *Natural Climate Variability on Decade to Century Time Scales*, D.G. Martinson *et al.* (eds.), National Research Council, National Academy Press, Washington DC, (in press).

Lehman, S.J. and L.D. Keigwin, 1992: Sudden changes in North Atlantic circulation during the last deglaciation. *Nature*, **356**, 757–762.

Lettenmaier, D., E.F. Wood and J.R. Wallis, 1994: Hydro-climatological trends in the continental United States, 1948-88. *J. Climate*, **7**, 586–607.

Levitus, S., J. Antonov, X. Zhou, H. Dooley, K. Selemenov and V. Tereschenkov, 1995: Decadal-scale variability of the North Atlantic Ocean. In: *Natural Climate Variability on Decade to Century Time Scales*, D.G. Martinson *et al.* (eds.), National Research Council, National Academy Press, Washington DC, (in press).

Lewis, T.J. and K. Wang, 1993: Different patterns of climatic change in eastern and western Canada: Evidence in borehole temperature measurements. *EOS, 1993 Fall Meeting Supplement,* 103.

Li, P., 1987: Seasonal snow resources and their fluctuations in China. In: *Large-Scale Effects of Seasonal Snow Cover*, B.E. Goodison, R.G. Barry and J. Dozier (eds.), IAHS Press, Wallingford, UK, 93–104.

LindstrÖm, G., 1993: *Floods in Sweden – Trends and Occurrence*. SMHI RH No. 6, May 1993, 83 pp.

Lins, H.F. and P.J. Michaels, 1994: Increasing streamflow in the United States. *EOS*, **75**, 281–286.

London, J., S.G. Warren and C.J. Hahn, 1991: Thirty-year trend of observed greenhouse clouds over the tropical oceans. *Adv. Space Res.*, **11(3)**, 45–49.

Lough, J.M., 1993: Variations of some seasonal rainfall characteristics in Queensland, Australia: 1921-1987. *Int. J. Climatology*, **13**, 391–409.

Lough, J.M., D.J. Barnes and R.B. Taylor, 1996: The potential of massive corals for the study of high-resolution climate variations in the past millenium. In: *Climatic Variations and Forcing Mechanisms of the last 2000 Years*, P.D.Jones, R.S.Bradley, and J.Jouzel (eds.), Springer Verlag, Heidelburg (in press).

Lyons, T.J., P. Schwerdtfeger, J.M. Hacker, I.J. Foster, R.C.G. Smith and H. Xinmei, 1993: Land-atmosphere interaction in a semiarid region: the bunny fence experiment. *Bull. Amer. Met. Soc.*, **74**, 1327–1334.

Magnuson, J.J, R.A. Assel, C.J. Bowser, B.T. DeStasio, P.J.

Dillon, J.R. Eaton, E.J. Fee, L.M. Mortsch, F.H. Quinn, N.T. Roulet, and D.W. Schindler, 1995: Regional assessment of freshwater ecosystems and climate change in North America, Laurentian Great Lakes and precambrian Shield. *J. Hydrologic Processes*, (in press).

Majorowicz, J.A., 1993: Climate change inferred from analysis of borehole temperatures: first results from Alberta Basin, Canada. *Pageoph.*, **140**, 655–666.

Marengo, J .A., 1992: Interannual variability of surface climate in the Amazon basin. *Int. J. Climatology,* **12**, 853–863.

Marengo, J., 1995: Variations and change in South American streamflow. *Clim. Change,* **31**, 99–117.

Mareschal, J.-C. and G. Vasseur, 1992: Ground temperature history from two deep boreholes in central France. *Global and Planetary Change*, **6**, 185–192.

Mayewski, P.A., L.D. Meeker, S. Whitlow, M.S. Twicker, M.C. Morrison, R.B. Alley, P. Bloomfield and K. Taylor, 1993: The atmosphere during the Younger Dryas. *Science*, **262**, 195–197.

Mayewski, P.A., L.D. Meeker, S. Whitlow, M.S. Twicker, M.C. Morrison, P.M. Grootes, G.C. Bond, R.B. Alley, D.A. Meese, A.J. Gow, K.C. Taylor, M. Ram and M. Wunkes, 1994: Changes in atmospheric circulation and ocean ice cover over the North Atlantic during the last 41,000 years. *Science*, **263**, 1747–1751.

McLaren, A.S., J.E. Walsh, R.H. Bourke, R.L. Weaver and W. Wittman, 1992: Variability in sea ice thickness over the North Pole from 1979 to 1990. *Nature*, **358**, 224–226.

McManus, J.F., G.C. Bond, W.S. Broecker, S.J. Johnsen, L. Labeyrie and S. Higgins, 1994: High-resolution climate records from the North Atlantic during the last interglacial. *Nature*, **371**, 326–329.

Mendes, J.C., 1994: Climatic variability of precipitation in Portugal. In: *Climate Variations in Europe*. R.Heino (ed.), Academy of Finland, Publication 3/94, pp 189–195.

Meshcherskaya, A.V., I.G. Belyankina and M.P. Golod, 1995: Monitoring of the snow depth over the main grain-growing zone of the former Soviet Union during the instrumental period. *Izvestiya of the Russ. Acad. of Sci.*, Geog. Series, **4**, 101–111.

Michaels, P.J., P.C. Knappenberger and D.A. Gay, 1995: Predicted and observed long night and day temperature trends. *Atmos. Res.*, **37**, 257–266.

Miller, A.J., D.R. Cayan, T.P. Barnett, N.E. Graham and J.M. Oberhuber, 1994: Interdecadal variability of the Pacific Ocean: Model response to observed heat flux and wind stress anomalies. *Clim. Dyn.*, **9**, 287–301.

Moberg, A. and H. Alexandersson, 1995: Urban bias trend in the Swedish part of gridded temperature data. *Int. J. Climatology*, (submitted).

Negri, A.J., R.F. Adler, E.J. Nelkin and G.J. Huffman, 1994: Regional rainfall climatologies derived from special sensor microwave Imager (SSM/I) data. *Bull. Am. Met. Soc.*, **75**, 1165–1182.

Nicholls, K.W. and J.G. Paren, 1993: Extending the Antarctic meteorological record using ice-sheet temperature profiles. *J. Climate*, **6**, 141–150.

Nicholls, N., 1992: Recent performance of a method for forecasting Australian seasonal tropical cyclone activity. *Aust. Met. Mag.*, **40**, 105–110.

Nicholls, N. and B. Lavery, 1992: Australian rainfall trends during the twentieth century. *Int. J. Climatology*, **12**, 153–163.

Nicholls, N. and A. Kariko, 1993: East Australian rainfall events: Interannual variations, trends, and relationships with the Southern Oscillation. *J. Climate*, **6**, 1141–1152.

Nicholls, N., R. Tapp, K. Burrows and D. Richards, 1996: Historical thermometer exposures in Australia, *Int. J. Climatology*, (in press).

Nicholson, S.E., 1978: Climatic variations in the Sahel and other African regions during the past five centuries. *J. Arid. Environ.*, **1**, 3–24.

Nicholson, S.E., 1989: Long-term changes in African rainfall. *Weather*, **44**, 46–56.

Nicholson, S.E., 1994a: A review of climate dynamics and climate variability in eastern Africa. *The Limnology, Climatology and Paleoclimatology of the East African Lakes*, T.C. Johnson and E. Odada (eds.) Gordon and Breach, London, in press.

Nicholson, S.E., 1994b: Recent rainfall fluctuations in Africa and their relationship to past conditions. *Holocene*, **4**, 121–131.

Nicholson, S.E., 1995: Variability of African rainfall on interannual and decadal time scales. In *Natural Climate Variability on Decade-to-Century Time Scales*, D. Martinson, K. Bryan., M. Ghil, T. Karl, E. Sarachik, S. Sorooshian, and L. Talley (eds.), USA National Academy Press, in press.

Nitta, T. and S. Yamamada, 1989: Recent warming of the tropical sea surface and its relationship to the Northern Hemisphere circulation. *J. Met. Soc. Japan*, **67**, 375–383.

Norris, J.R. and C.B. Leovy, 1994: Interannual variability in stratiform cloudiness and sea surface temperature. *J. Climate*, **7**, 1915.

O'Lenic, E., 1995: The relationship between low-and high-frequency variability of Northern Hemisphere 700-hPa height. *J. Climate*. (in press).

Oerlemans, J., 1994: Quantifying global warming from the retreat of glaciers. *Science*, **264**, 243–245.

Ohkouchi, N., K. Kawamura, T. Nakamura and A. Taira, 1994: Small changes in the sea surface temperature during the last 20,000 years: Molecular evidence from the western tropical Pacific. *Geophys. Res. Lett*, **21**, 2207–2210.

Oltmans, S.J. and D.J. Hofmann, 1995: Increase in lower-stratospheric water vapour at a mid-latitude Northern hemisphere site from 1981 to 1994. *Nature*, **374**, 146–149.

Oort, A.H. and H. Liu, 1993: Upper-air temperature trends over the globe, 1958-1989. *J. Climate*, **6**, 292–307.

Ostby, F.P., 1993: The changing nature of tornado climatology. In: *Preprints, 17th Conference on Severe Local Storms, October 4-8, 1993, St Louis, Missouri*, pp. 1–5.

Palecki, M.A., 1994: The onset of spring in the eastern United States during the 20th century. *Preprints, Sixth Conference on Climate Variations, January 23-28, 1994*, Nashville, Tennessee, American Meteorological Society, pp. 108–109.

Palmieri, S., A.M. Siani and A.D. Agostina, 1991: Climate fluctuations and trends in Italy within the last 100 years. *Ann. Geophysicae*, **9**, 769–776.

Pan, Y.H. and A.H. Oort, 1990: Correlation analyses between sea surface temperature anomalies in the eastern equatorial Pacific and the world ocean. *Clim. Dyn.*, **4**, 191–205.

Parker, D.E., 1994a: Long term changes in sea surface temperatures. *Proc. Air and Waste Management Assoc. Conf. on Global Climate Change, Phoenix, Arizona, 5-8 April 1994*, C.V. Mathar and G. Steislaed (eds.), pp 102–113.

Parker, D.E., 1994b: Effects of changing exposures of thermometers at land stations. *Int. J. Climatology*, **14**, 1–31.

Parker, D.E. and D.I. Cox, 1995: Towards a consistent global climatological rawinsonde data-base. *Int. J. Climatolology*, **15**, 473–496.

Parker, D.E., T.P. Legg and C.K. Folland, 1992: A new daily central England temperature series. *Int. J. Climatology*, **12**, 317–342.

Parker, D.E., P.D. Jones, C.K. Folland and A. Bevan, 1994: Interdecadal changes of surface temperatures since the late 19th century. *J. Geophy. Res.*, **99**, 14373–14399.

Parker, D.E., C.K. Folland and M. Jackson, 1995: Marine surface temperature: observed variations and data requirements. *Clim. Change*, **31**, 559–600.

Parrilla, G., A. Lavin, H. Bryden, M. Garcia and R. Millard, 1994: Rising temperatures in the subtropical North Atlantic Ocean over the past 35 years. *Nature*, **369**, 48–51.

Parthasarathy, B., K. Rupa Kumar and D.R. Kothawale, 1992: Indian summer monsoon rainfall indices: 1871-1990. *Meteorol. Magaz.*, **121**, 174–186.

Parthasarathy, B., A.A. Munot and D.R. Kothawale, 1994: All-India monthly and seasonal rainfall series: 1871–1993. *Theor. Appl. Climatol.*, **49**, 217–224.

Parungo, F., J.F. Boatman, H. Sievering, S. Wilkison and B.B. Hicks, 1994: Trends in global marine cloudiness and anthropogenic sulfur. *J. Climate*, **7**, 434–440.

Peterson, T.C., V.S. Golubev and P.Ya. Groisman, 1995: Evaporation losing its strength. *Nature*, **377**, 687–688.

Plantico, M.S., T.R. Karl., G. Kukla and J. Gavin, 1990: Is recent climate change across the United States related to rising levels of anthropogenic greenhouse gases? *J. Geophys. Res.*, **95**, 16617–16637.

Plummer, N., 1996: Temperature variability and extremes over Australia: Part 1 – Recent observed changes, *Aust. Meteorol. Mag.*, (submitted).

Plummer, N., Z. Lin and S. Torok, 1995: Recent changes in the diurnal temperature range over Australia. *Atmos. Res.*, **37**, 79–86.

Pollack, H.N. and D.S. Chapman, 1993: Underground records of changing climate. *Scientific American*, **268**, 44–50.

Porter, S.C. and An Zhisheng, 1995: Correlation between climate events in the North Atlantic and China during the last glaciation. *Nature*, **375**, 305–308.

Portman, D.A., 1993: Identifying and correcting urban bias in regional time series: surface temperature in China's northern plains. *J. Climate*, **6**, 2298–2308.

Posmentier, E.S., M.A. Cane and S.E. Zebiak, 1989: Tropical Pacific trends since 1960. *J. Climate*, **2**, 731–736.

Prabhakara, C., J.C. Nucciarone and J.M. Yoo, 1995: Examination of global atmospheric temperature monitoring with satellite microwave measurements: 1) Theoretical considerations. *Clim. Change*, **30**, 349–366.

Quinn, W.H., 1992: A study of Southern Oscillation – related climatic activity for AD 622-1900 incorporating Nile River flood data. In: *El Niño: Historical and Paleoclimatic aspects of the Southern Oscillation*, H.F. Diaz and V. Markgraf (eds.), Cambridge University Press, Cambridge, UK 476pp.

Rakhecha, P.R. and M.K. Soman, 1994: Trends in the annual extreme rainfall events of 1 to 3 days duration over India. *Theor. Appl. Climatol.*, **48**, 227–237.

Ramage, C.S. 1987: Secular change in reported surface wind speeds over the ocean. *J. Clim. Appl. Meteorol.*, **26**, 525–528.

Raper, S.C.B., 1993: Observational data on the relationships between climatic change and the frequency and magnitude of severe tropical storms. In: *Climate and Sea Level Change: Observations, Projections and Implications*, R.A. Warrick, E.M. Barrow, and T.M.L. Wigley (eds.), Cambridge University Press, Cambridge, UK, pp 192–212.

Raper, S.C.B., T.M.L. Wigley, P.R. Mayes, P.D. Jones and M.J. Salinger, 1984: Variations in surface air temperatures, part 3: The Antarctic, 1957-1982. *Month. Weath. Review*, **112**, 1341–1353.

Rasmusson, E.M., X.L. Wang and C.F. Ropelewski, 1994: Secular variability of the ENSO cycle. In: *Natural Climate Variability on Decade to Century Time Scales*, D.G. Martinson *et al.* (eds.), National Research Council, National Academy Press, Washington DC, in press.

Razuvaev, V.N., E.G. Apasova, O.N. Bulygina and R.A. Martuganov, 1995: Variations in the diurnal temperature range in the European region of the former USSR during the cold season. *Atmos. Res.*, **37**, 45–51.

Read, J .F. and W.J. Gould, 1992: Cooling and freshening of the subpolar North Atlantic Ocean since the 1960s. *Nature*, **360**, 55–57.

Reynolds, R.W., 1988: A real-time global sea surface temperature analysis. *J. Climate*, **1**, 75–86.

Rhines, P.B. and J.R.N. Lazier, 1995: Deep convection and climate change in the Labrador Sea. *Nature*, (submitted).

Robinson, D.A., K.F. Dewey and R.R. Heim, Jr., 1993: Global snow cover monitoring: An update. *Bull. Am. Met. Soc.*, **74**, 1689–1696.

Robock, A. and Jianping Mao, 1992: Winter warming from large volcanic eruptions. *Geophys. Res. Lett.*, **12**, 2405–2408.

Robock, A.and Jianping Mao, 1995: The volcanic signal in surface temperature observations. *J. Climate*, **8**, (in press).

Robock, A., K.Ya. Vinnikov, C.A. Schlosser, N.A. Speranskaya and Y. Xue, 1995: Use of midlatitude soil moisture and meteorological observations to validate soil moisture simulations with biosphere and bucket models. *J. Climate*, **8**, 15–35.

Roemmich, D. and C. Wunsch, 1984: Apparent changes in the climatic state of the deep North Atlantic Ocean. *Nature*, **307**, 447–450.

Rogers, J.C. and R.V. Rohli, 1991: Florida citrus freezes and polar anticyclones in the Great Plains. *J. Climate*, **4**, 1103–1113.

Rupa Kumar, K., K. Krishna Kumar and G.B. Pant, 1994: Diurnal asymmetry of surface temperature trends over India. *Geophys. Res. Lett.*, **21**, 677–680.

Salinger, M.J., 1994: New Zealand climate in the last 25000 years. In: *Paleoclimates and Cimate Modelling. Proceedings of the National Science Strategy Committee for Climate Change Workshop, Royal Society of New Zealand Miscellaneous Series* **29**, 15–16.

Salinger, M.J., A.B. Mullan, A.S. Porteous, S.J. Reid, C.S. Thompson, L.A. Coutts and E. Fouhy, 1990: *New Zealand Cimate Extremes: Recent Trends*. New Zealand Meteorological Service Report to new Zealand Ministry for the Environment, 33 pp.

Salinger, M.J., R. McGann, L. Coutts, B. Collen and E. Fouhy, 1992: *Rainfall Trends in New Zealand and Outlying Islands, 1920–1990. South Pacific Historical Climate Network*, National Institute of Water and Atmospheric Research Report, 33 pp.

Salinger, M.J., J. Hay, R. McGann and B. Fitzharris, 1993: Southwest Pacific temperatures: diurnal and seasonal trends. *Geophys. Res. Lett.*, **20**, 935–938.

Salinger, M.J., J.G. Palmer, P.D. Jones and K.R. Briffa, 1994: Reconstruction of New Zealand climate indices back to AD 1731 using dendroclimatic techniques: Some preliminary results. *Int. J. Climatology*, **14**, 1135–1149.

Salinger, M.J., R.E. Basher, B.B. Fitzharris, J.E. Hay, P.D. Jones, J.P. MacVeigh and I. Schmidely-Leleu, 1995a: Climate trends in the southwest Pacific. *Int. J. Climatology*, **15**, 285–302.

Salinger, M.J., R. Allan, N. Bindoff, J. Hannah, B. Lavery, L. Leleu, Z. Lin, J. Lindesay, J.P. MacVeigh, N. Nicholls, N. Plummer and S. Torok, 1995b: Observed variability and change in climate and sea level in Oceania. In *Proc. GREENHOUSE '94*, (in press).

Schinke, H., 1993: On the occurrence of deep cyclones over Europe and the North Atlantic in the period 1930-1991. *Beitr. Phys. Atmosph.*, **66**, 223–237.

Schlesinger, M.E. and N. Ramankutty, 1994: An oscillation in the global climate system of period 65-70 years. *Nature*, **367**, 723–726.

Schmidt, H., and H. von Storch, 1993: German Bight storms analysed. *Nature*, **365**, 791.

Schwitter, M.P. and C.F. Raymond, 1993: Changes in the longitudinal profiles of glaciers during advance and retreat. *J. Glaciology*, **39**, 582–590.

Seltzer, G.O., 1992: Late Quaternary glaciation of the Cordifera

Real, Bolivia. *J. Quat. Sci.*, **7**, 87–98.

Serreze, M.C., J.E. Box, R.G. Barry and J.E. Walsh, 1993: Characteristics of Arctic synoptic activity, 1952–1989. *Meteorol. Atmos. Phys.*, **51**, 147–164.

Shiklomanov, I.A., 1976: *Hydrological Problems Related to the Caspian Sea Problems*. Gidrometeoizdat, Leningrad, 78pp., in Russian.

Shinoda, M. and R. Kawamura, 1994a: Tropical rainbelt circulation, and sea surface temperatures associated with the Sahalian rainfall trend. *J. Met. Soc. Japan*, **72**, 341–359.

Shinoda, M. and R. Kawamura, 1994b: Tropical African rainbelt and global sea surface temperatures: Interhemispheric comparison. In: *Proc. Intl. Conf. on Monsoon Variability and Prediction*. Trieste, Italy, 9-13 May 1994, 288–295.

Sikes, E.L. and L.D. Keigwin, 1994: Equatorial Atlantic sea surface temperature for the last 30 kyr: a comparison of U37', d180, and foraminiferal assemblage temperature estimates. *Paleoceanography*, **9**, 31–45.

Skinner, W.R., 1993: Lake ice conditions as a cryospheric indicator for detecting climate variability in Canada. In: *Snow Watch '92, Detection Strategiess for Snow and Ice, Proc. Int'l Workshop on Snow and Ice Cover and the Climate System*, Barry, R.G., B.E. Goodison, and E.F. LeDrew (eds.), World Data Center A for Glaciology, Boulder, Co, 204–240.

Smith, K., 1995: Precipitation over Scotland 1757-1992: Some aspects of temporal variability. *Int. J. Climatology*, **15**, 543–556.

Soldatova, I .I, 1993: Secular variations in river "break-up" data and their relationship with climate variation. *Meteorologia i gidrologia*, **9**, 89–96.

Sontakke, N.A., G.B. Pant and N. Singh, 1992: *Construction and analysis of All-India summer monsoon rainfall series for the longest instrumental period: 1813-1991*. Research Report No. 053, IITM, Pune, India, 18 pp.

Sontakke, N.A., G.B. Pant and N. Singh, 1993: Construction of All-India summer monsoon rainfall series for the period 1844-1991. *J. Climate*, **6**, 1807–1811.

Spencer, R.W., 1993: Global oceanic precipitation from the MSU during 1979-91 and comparisons to other climatologies. *J. Climate*, **6**, 1301–1326.

Spencer, R.W. and J.R. Christy, 1992a: Precision and radiosonde validation of satellite gridpoint temperature anomalies, Part I: MSU channel 2. *J. Climate*, **5**, 847–857.

Spencer, R.W. and J.R. Christy, 1992b: Precision and radiosonde validation of satellite gridpoint temperature anomalies, Part II: A tropospheric retrieval and trends during 1979-90. *J. Climate*, **5**, 858–866.

Spencer, R.W. and J.R. Christy, 1993: Precision lower stratospheric temperature monitoring with the MSU: Validation and results 1979-91. *J. Climate*, **6**, 1194–1204.

Spencer, R.W. J.R. Christy and N.C. Grody, 1990: Global atmospheric temperature monitoring with satellite microwave measurements: method and results 1979-84. *J. Climate*, **3**, 1111–1128.

Stein, O. and A. Hense, 1994: A reconstructed time series of the number of extreme low pressure events since 1880. *Meteorol. Zeitschrift*, **N.F. 3**, 43–46.

Stine, S., 1994: Extreme and persistent drought in California and Patagonia during medieval times. *Nature*, **369**, 546–549.

Stone, R., N. Nicholls and G. Hammer, 1996: Frost in NE Australia: Trends and influence of phases of the Southern Oscillation. *J. Climate,* (in press)

Suggate, R.P., 1990: Late Pliocene and Quaternary glaciations of New Zealand. *Quaternary Science Reviews*, **9**, 175–197.

Suppiah, R. and K.J. Hennessy, 1996: Trends in the intensity and frequency of heavy rainfall in tropical Australia and links with the Southern Oscillation. *Aust. Meteorol. Mag.*, (in press).

Taylor, A.E., 1991: Holocene paleoenvironmental reconstruction from deep ground temperatures: a comparison with paleoclimate derived from the $\delta^{18}O$ record in an ice core from the Agassiz Ice Cap, Canadian Arctic Archipelago. *J. Glaciology*, **37**, 209–219.

Taylor, K.C., G.W. Lamorey, G.A. Doyle, R.B. Alley, P.M. Grootes, P.A. Mayewski, J.W.C. White and L.K. Barlow, 1993a: The "flickering switch" of late Pleistocene climate change. *Nature*, **361**, 432–436.

Taylor, K.C., C.U. Hammer, R.B. Alley, H.B. Clausen, D. Dahl-Jensen, A.J. Gow, N.S. Gundestrup, J. Kipfstuhl, J.C. Moore and E.D. Waddington, 1993b: Electrical conductivity measurements from the GISP2 and GRIP Greenland ice cores. *Nature*, **366**, 549–552.

Thompson, C., S. Ready and X. Zheng, 1992: *Tropical Cyclones in the South West Pacific: November 1979 to May 1989*. National Institute of Water and Atmospheric Research Report. 35 pp.

Thompson, L.G., E. Mosley-Thompson, M. Davis, P.N. Lin, T. Yao, M. Dyergerov and J. Dai, 1993: Recent warming: ice core evidence from tropical ice cores with emphasis on central Asia. *Global and Planetary Change*, **7**, 145–156.

Thompson, L.G., E. Mosley-Thompson, M.E. Davis, P.-N. Lin, K.A. Henderson, J. Cole-Dai, J.F. Bolzan and K.-B. Liu, 1995: Late glacial stage and holocene tropical ice core records from Huascarçn, Peru. *Science*, **269**, 46–50.

Thomson, D.J., 1995: The seasons, global temperature, and precession. *Science*, **268**, 59–68.

Thouveny, N., J-L. de Beaulieu, E. Bonifay, K.M. Creer, J. Guiot, M. Icole, S. Johnsen, J. Jouzel, M. Reille, T. Williams and D. Williamson, 1994: Climate variations in Europe over the past 140 kyr deduced from rock magnetism. *Nature*, **371**, 503–506.

Thunnell, R., D. Anderson, D. Gellar and Qingmin Miao, 1994: Sea-surface temperature estimates for the tropical western Pacific during the last glaciation and their implications for the Pacific warm pool. *Quat. Res.*, **41**, 255–264.

Trenberth, K., 1990: Recent observed interdecadal climate changes in the Northern Hemisphere. *Bull. Am. Met. Soc.*, **71**, 988–993.

Trenberth, K.E., 1996: Atmospheric circulation climate changes. *Clim. Change*, (in press).

Trenberth, K.E. and J.G. Olson, 1991: Representativeness of a 63-station network for depicting climate changes. In:

Greenhouse-Gas-Induced Climatic Change: A Critical Appraisal of Simulations and Observations, M.E. Schlesinger (ed.). Elsevier, Amsterdam, pp 249–259.

Trenberth, K.E. and J.W. Hurrell, 1994: Decadal atmosphere-ocean variations in the Pacific. *Clim. Dyn.* **9**, 303–319.

Trenberth, K.E.and T.J. Hoar, 1996: The 1990-1995 El Niño-Southern Oscillation Event: Longest on record. *Geophys. Res. Lett.* **23**, 57.

UNEP, 1992: *World Atlas of Desertification*. Edward Arnold, London.

Van Loon, H., J.W. Kidson and A.B. Mullan, 1993: Decadal variation of the annual cycle in the Australian dataset. *J. Climate*, **6**, 1227–1231.

Venrick, E.L., J.A. McGowan, R.D. Cayan and T. Hayward, 1987: Climate and chlorophyll a: Long term trends in the central North Pacific Ocean. *Science*, **238**, 70–72.

Villalba, R., 1990: Climatic fluctuations in northern Patagonia during the last 1000 years as inferred from tree-ring records. *Quaternary Research*, **34**, 346–360.

Vinnikov, K. Ya. and I.B. Yeserkepova, 1991: Soil moisture: empirical data and model results. *J. Climate*, **4**, 66–79.

Vinnikov, K.Ya., P.Ya. Groisman and K.M. Lugina, 1990: Empirical data on contemporary global climate changes (temperature and precipitation). *J. Climate*, **3**, 662–677.

Vinnikov, K.Ya., A. Robock, N.A. Speranskaya and C.A. Schlosser, 1995: Scales of temporal and spatial variability of midlatitude soil moisture. *J. Geophy. Res.* (in press).

Von Storch, H., J. Guddak, K.A. Iden, T. Jùnson, J. Perlwitz, M. Reistad, J. de Ronde, H. Schmidt and E. Zorita, 1993: *Changing Statistics of Storms in the North Atlantic?*, Report No. 116, Max-Planck-Institut für Meteorologie, Hamburg, 18 pp.

Vose, R.S., R.L. Schmoyer, P.M. Steurer, T.C. Peterson, R. Heim, T.R. Karl and J.Eischeid, 1992: *The Global Historical Climatology Network: Long-term monthly temperature, precipitation, sea level pressure, and station pressure data.* Report ORNL/CDIAC-53, NDP-041 (Available from Carbon Dioxide Information Analysis Center, Oak Ridge National Laboratory, Oak Ridge, Tennessee.)

Wallace, J.M. and D.S. Gutzler, 1981: Teleconnections in the geopotential height field during the Northern hemisphere winter. *Mon. Weath. Rev.*, **109**, 784–811.

Wang, B., 1995: Interdecadal changes in El Niño onset in the last four decades. *J. Climate*, **8**, 267–285.

Wang, K., T.J. Lewis and A.M. Jessop, 1992: Climatic changes in central and eastern Canada inferred from deep borehole temperature data. *Global and Planetary Change*, **6**, 129–141.

Wang, K., T.J. Lewis, D.S. Belton,and P-Y. Shen, 1994: Differences in recent ground surface warming in eastern and western Canada: Evidence from borehole temperatures. *Geophys. Res. Lett.*, **21**, 2689–2692.

Wang, W.C., Z. Zeng and T.R. Karl, 1990: Urban heat islands in China. *Geophys. Res. Lett.*, **17**, 2377–2380.

Wang, X.L. and C.F. Ropelewski, 1995: An assessment of ENSO-scale secular variability. *J. Climate*, **8**, 1584–1599.

Ward, M.N., 1992: Provisionally corrected surface wind data,

worldwide ocean-atmosphere surface fields and Sahelian rainfall variability. *J. Climate*, **5**, 454–475.

Ward, M.N., 1994: Tropical North African rainfall and worldwide monthly to multi-decadal climate variations. Ph.D. thesis, Reading University, UK.

Warren, S.G., C.J. Hahn, J. London, R.M. Chervin, and R.L. Jenne, 1988: *Global Distribution of Total Cloud Cover and Cloud Type Amounts over the Ocean*. NCAR Technical Note TN-317+STR, 42 pp + 170 maps.

Watkins, C., 1991: The annual period of freezing temperatures in central England: 1850-1989. *Int. J. Climatology*, **11**, 889–896.

Weare, B.C., 1984: Interannual moisture variations near the surface of the tropical Pacific Ocean. *Quart. J. R. Met. Soc.*, **110**, 489–504.

Weber, R.O., P. Talkner and G. Stefanicki, 1994: Asymmetric diurnal temperature change in the Alpine region. *Geophys. Res. Lett.*, **21**, 673–676.

Webster, P.J.and N.A. Streten, 1978: Late Quaternary ice age climates of tropical Australasia: Interpretations and reconstructions. *Quat. Res.*, **10**, 279–309.

Whetton, P., R. Allan and I. Rutherford, 1995: Historical ENSO teleconnections in the eastern hemisphere: Comparison with latest El Niño series of Quinn. *Clim.Change* (in press).

Whiteford, P., 1993: Evaluation of past climate using borehole temperature measurements. *Weather and Climate*, **13**, 3–8.

Whiteford, P.C., R.G. Allisand R.H. Funnel, 1994: Past climate from borehole temperature measurements. In: *Paleoclimates and Climate Modelling. Proceedings of the National Science Strategy Committee for Climate Change Workshop, Royal Society of New Zealand Miscellaneous Series*, **29**, 37–38.

Wigley, T.M.L. and P.D. Jones, 1987: England and Wales precipitation: A discussion of recent changes in variability and an update to 1985. *J. Climatology*, **7**, 231–246.

Wigley, T.M.L. and P.M. Kelly, 1990: Holocene climatic change, ^{14}C wiggles and variations in solar irradiance, *Phil. Trans. R. Soc. Lond.*, **A 330**, 547–560.

WMO/UNEP, 1995: *Scientific Assessment of Ozone Depletion: 1994*. Global Ozone Research and Monitoring Project Report No. **37**. World Meteorological Organization, Geneva.

Yamagata, T. and Y. Masumoto, 1992: Interdecadal natural climate variability in the western Pacific and its implication in global warming. *J. Met. Soc. Japan.*, **70**, 167–175.

Ye, H., L.S. Kalkstein and J.S. Greene, 1995: The detection of climate change in the Arctic: An updated report. *Atmos. Res.*, **37**, 163–173.

Yu, B. and D.T. Neil, 1991: Global warming and regional rainfall: the difference between average and high intensity rainfalls. *Int. J. Climatology*, **11**, 653–661.

Yu, B. and D.T. Neil, 1993: Long-term variations in regional rainfall in the south-west of Western Australia and the difference between average and high intensity rainfalls. *Int. J. Climatology*, **13**, 77–88.

Zhang, Y., J.M. Wallace and N. Iwasaka, 1995: Is climate variability over the North Pacific a linear response to ENSO? *J. Climate* (in press).

4

Climate Processes

R.E. DICKINSON, V. MELESHKO, D. RANDALL, E. SARACHIK, P. SILVA-DIAS, A. SLINGO

Contributors:
A. Barros, O. Boucher, R. Cess, T. Charlock, L. Dümenil, A. Del Genio, R. Fu, P. Gleckler, J. Hansen, R. Lindzen, K. McNaughton, J. McWilliams, E. Maier-Reimer, G. Meehl, M. Miller, D. Neelin, E. Olaguer, T. Palmer, C. Penland, R. Pinker, D. Rind, V. Ramaswamy, A. Robock, M. Salby, C. Senior, M. Schlesinger, H.P. Schmid, Q.Q. Shao, K. Shine, H. Sundqvist, A. Vogelmann, A.J. Weaver

CONTENTS

SUMMARY

This chapter assesses the processes in the climate system that are believed to contribute the most to the uncertainties in current projections of greenhouse warming. Many of these processes involve the coupling of the atmosphere, ocean, and land through the hydrological cycle. Continued progress in climate modelling will depend on the development of comprehensive data sets and their application to improving important parametrizations. The large-scale dynamical and thermodynamical processes in atmospheric and oceanic models are well treated and are one of the strengths of the modelling approach. As previously indicated in IPCC (1990, 1992), the radiative effects of clouds and their linkages to the hydrological cycle remain a major uncertainty for climate modelling. The present report, however, goes into much more detail than the past reports in summarising the many facets of this question, recent progress in understanding different feedbacks, and in the development of climate model parametrization treating these processes. This is now a very active research area with much that has been accomplished since IPCC (1990).

Current climate models are highly sensitive to cloud parametrizations, and there are not yet satisfactory means for evaluating the correctness of such treatments. Progress will require improved understanding, observational data sets, sub-grid scale parametrization, and improved modelling of the distribution of atmospheric water in its vapour, liquid, and solid forms, and will not be achieved quickly. Sub-grid scale parametrizations are especially difficult to improve. The determination of cloud-dependent surface radiative and precipitation fluxes is a significant source of uncertainty for both land-surface and ocean climate modelling, making attempts to assess regional climate change problematic.

Clear-sky feedbacks involving changes in water vapour distribution and lapse rates are also uncertain but their global sum varies little between models. The processes determining the distribution of upper tropospheric water vapour are still poorly understood. Water vapour feedback in the lower troposphere is undoubtedly positive, and the preponderance of evidence also points to it being positive in the upper troposphere.

A large-scale dynamical framework for treating the ocean component of climate models is now being used for climate change projections. Sub-grid scale parametrizations in these models are important for surface energy exchange and the thermohaline circulation. A new parametrization for interior mixing appears promising in providing an improved simulation of the global thermocline. Both high latitude and tropical elements of ocean climate models involve important and still inadequately represented processes. In high latitudes, coupling to sea ice models and deep convection are especially important. In the tropics, ocean models need to simulate the large-scale sea surface temperature variability of the El Niño-Southern Oscillation systems.

Climate model treatments of land processes have advanced rapidly since the last assessment. However, there are lags in the validation of these models, in the development of required data sets, in an adequate assessment of how sub-grid scale processes should be represented, and in their implementation in models for climate change projection. None of the land parametrizations include a physically based and globally validated treatment of runoff.

4.1 Introduction to Climate Processes

Climate processes are all the individual physical processes that separately contribute to the overall behaviour of the climate system. They are also the interactions and feedbacks among the individual processes that determine the response of the climate system to external forcing, including the response to global anthropogenic forcing. There are a myriad of such climate processes. This report is focused by the present community experience with numerical modelling. The success of numerical simulations of future climate change hinges on the adequate inclusion of all the climate processes that are responsible for determining the behaviour of the system. A triage approach is used here; processes are treated lightly that are already included adequately in numerical models or those for which there is little or no evidence supporting their importance. The present report emphasises those processes known to contribute substantially to the uncertainties of current numerical simulations of long-term climate response. These all are physical processes. Chemical and biological components are treated in Chapters 2, 9 and 10 respectively, and at present would be de-emphasised by the triage approach for lack of substantial effort to include them in climate models.

A globally averaged temperature increase is the response most easily related to global greenhouse forcing. Current climate models project global temperature increases that vary by over a factor of two to three for a given forcing scenario. These differences are directly attributable to the treatment of cloud processes, their links to the hydrological cycle and their interaction with radiation in the models. Indeed, individual models can give this range of answers, depending on changes only in the cloud parametrization (Senior and Mitchell, 1993). Thus, all the likely contributors to the ultimate representations of cloud processes in climate models need emphasis, as do other aspects of the hydrological cycle and model dynamics that interact with cloud evolution and properties.

Another current focus for climate process studies is the evolving interpretation of the past climate record. Most notable is that of the observed warming of about half a degree over the last century. This is about a factor of two smaller than conventional estimates according to models including only trace gas increases but is consistent with the inclusion of the expected effects of sulphate aerosols.

Observations over continental surfaces show that the warming has been more pronounced at night than during the day (*cf.* Chapter 3). Hansen *et al.* (1995) have hypothesised from a large number of numerical simulations with a simple sector General Circulation Model (GCM)

that the observed global trends and continental predominance of night-time warming can be explained if clouds and aerosol have increased over continental surfaces with a global average radiative cooling of about half that of the warming by greenhouse gases increase. For the same direct radiative effects, a cloud increase would have a larger impact on diurnal range than would dry aerosols, since clouds not only cool during daytime but warm the surface at night. GCM calculations by Mitchell *et al.* (1995) indicate that the direct radiative effect of aerosols alone can only partially explain the observed decrease in diurnal range. Further progress will require understanding the time history and distribution of anthropogenic aerosol as well as its effects on cloud properties and its consequent direct and indirect radiative effects.

As mentioned above, ocean surface temperatures, and hence much of climate variability and change are strongly influenced by net energy exchanges between the oceans and the atmosphere. In high latitudes, sea ice is of major significance for modifying these energy exchanges. The possible rapid variations of the thermohaline circulation, first suggested by Bryan (1986), indicate that rearrangements of heat in the ocean can occur relatively quickly so that surface temperatures can respond rapidly to changes in the thermohaline circulation. Further, the thermohaline circulation is particularly sensitive to changes in the high latitude temperature and hydrology; variability on decadal-to-millennial time-scales can arise. On somewhat shorter time-scales, coupled atmosphere-ocean processes in the tropics are a major source of interannual climate variability. Therefore, dynamically active oceans are included in coupled models for a comprehensive evaluation of the response to increases in the radiatively active gases.

Land-surface processes are also now highlighted for several reasons. The land surface is readily modified on the large-scale by human activities. Such modifications may have important regional consequences; historical land modifications may have had larger regional climate impacts up to now than has had greenhouse gas warming. Land-surface processes probably are of lesser importance for the future globally averaged temperature response to greenhouse warming than are cloud processes, but how climate changes over land is of the greatest practical importance to humans and depends substantially on land-surface processes. Furthermore, land-surface processes strongly affect the overlying atmospheric hydrological cycle including clouds, so they arguably will be of major importance as feedbacks for determining changes in regional climate patterns.

This chapter identifies and assesses important processes for incorporation into climate models for projecting climate change and highlights major gaps in our understanding of these processes. These processes occur in the atmosphere, ocean and land surface and involve their coupling through the hydrological cycle.

4.2 Atmospheric Processes

The processes of large-scale dynamics, thermodynamics and mass balance in the atmospheric and oceanic components of the climate models are now included in models with considerable confidence. This is not to imply that the modelling of large-scale circulation and temperature structure can be viewed as completely successful, since it depends not only on the relatively robust components but also on weak elements. Diagnostic comparisons with observations of potential vorticity transport and mixing could help improve confidence in the treatments of large-scale dynamics. Stable dynamical modes in which the coupled atmosphere-ocean system may respond as part of climate change are inadequately understood (e.g., Palmer, 1993). The hydrological cycle and radiation budget are so intimately coupled that they should not be treated separately. Therefore, wherever possible in the following sections, these two aspects are discussed together. Clear-sky water vapour feedback is first examined, then the various feedbacks attributed to clouds that affect the Earth-atmosphere radiative balance, then the additional issues of surface radiative fluxes, and then coupling to precipitation processes. The intention here is not to describe all the physical processes operating in the atmosphere, as that would be far too broad an approach, but rather to focus on those processes that are especially relevant to the various feedbacks, as identified in studies of global warming with climate models.

4.2.1 Water Vapour Amounts

A positive water vapour feedback was hypothesised in the earliest simulations of global warming with simple radiative-convective models (Manabe and Wetherald, 1967). It arises for water vapour near the surface from the strong dependence of the saturation vapour pressure on temperature, as given by the Clausius-Clapeyron equation. Increases in temperature are thus expected to lead to increases in the atmospheric water vapour mixing ratio. Since water vapour is the most important greenhouse gas, such increases in water vapour enhance the greenhouse effect; that is, they reduce the thermal infrared (long-wave) flux leaving the atmosphere-surface system, providing a positive feedback amplifying the initial warming. This feedback operates in all the climate models used in global

warming and other studies. However, intuitive arguments for it to apply to water vapour in the upper troposphere are weak; observational analyses and process studies are needed to establish its existence and strength there.

Changes in the vertical decrease of temperature with altitude change surface temperature for a given radiative balance and are known as "lapse rate feedback". Changes in lapse rate act as an additional feedback that can also be substantial and that generally oppose the water vapour feedback. The sum of the water vapour and lapse rate feedbacks comprise the clear-sky feedback. Cess *et al.* (1990) show that the magnitude of the clear-sky feedback is very similar in a wide range of models. However, the partitioning between lapse rate and water vapour feedback may vary substantially between models (Zhang *et al.*, 1994) depending on how convection is parametrized and may depend on the time-scale of the climate change or climate fluctuations considered (Bony *et al.*, 1995).

The consensus view that water vapour provides a strong positive feedback has been challenged by Lindzen (1990), who emphasised the sensitivity of the water vapour feedback to poorly understood processes, such as the profile of cumulus detrainment and the related distribution of water vapour in the upper troposphere. Water vapour is physically most closely controlled by temperature in the lower troposphere, and by transport processes in the upper troposphere. Both regions contribute comparably to the water vapour greenhouse effect.

How upper tropospheric water vapour is distributed and varies with other climate parameter variations is best studied with satellite data (e.g., Soden and Bretherton, 1994). Soden and Fu (1995) relate climatic variations of upper tropospheric water vapour to clear-sky long-wave radiation and to moist convection indicated by the International Satellite Cloud Climatology Project (ISCCP), over a five-year period. These are shown to be highly correlated in the tropical half of the world but uncorrelated outside the tropics. Thus, these data support the conventional view that in the tropics, water vapour is supplied to the upper troposphere primarily by moist convection. They find that this result is maintained, even averaging over the whole tropical belt of 30°S-30°N, and so conclude that the net effect of convection is moistening even allowing for compensating regions of subsidence. Chou (1995) in a case study with two months of data (April 1985 and 1987) over 100°W–100°E infers a somewhat contradictory conclusion that increased convection in the tropics leads to a net reduction in the atmospheric clear-sky greenhouse effect and hence a net drying.

Soden and Fu show that occurrence of tropical convection in the Geophysical Fluid Dynamics Laboratory (GFDL) GCM has a very similar correlation with upper tropospheric water vapour as observed, suggesting that the model captures the essential upward transport processes of water in the tropics. Sun and Held (1995), on the other hand, show that for specific humidities averaged over the tropics, the correlation with surface values declines with height to much smaller values for observed data than in the same model. As they discuss, lack of radiosonde coverage over the eastern and central Pacific may throw into question their observational analyses.

Detailed process studies of cumulus detrainment and of the water budget in mesoscale cumulus convection are also being made through field experiments, to provide a more solid physical basis for testing the GCM treatments of water vapour. Lau *et al.* (1993) used the Goddard Cumulus Ensemble model (GCEM) to investigate the water budget of tropical cumulus convection. Their results on changes in temperature and water vapour induced by surface warming are in agreement with those from GCMs which use only crude cumulus parametrizations.

The details of water vapour feedback in the extra-tropical upper troposphere are also poorly characterised observationally. In the extra-tropics, the relative contributions to upper tropospheric water vapour of lateral transport from the tropics, versus upward transport by large-scale motions or by moist convection are poorly known. Lacis and Sato (1993) showed, in the Goddard Institute for Space Science (GISS) GCM, that the water vapour feedback was almost as strong at middle and high latitudes as it was at low latitudes. Pierrehumbert and Yang (1993) have emphasised the potential small-scale complexity of latitudinal exchanges of water vapour by large-scale eddies at high latitudes, and Kelly *et al.* (1991) have shown some observational evidence for temperature-dependent large-scale high latitude exchanges of water vapour creating a hemispheric asymmetry in upper tropospheric water vapour at these latitudes. Del Genio *et al.* (1994) find, for the GISS model, that large-scale eddies dominate the seasonal variation of upper troposphere water vapour outside the tropical rainbelt.

Feedback from the redistribution of water vapour remains a substantial uncertainty in climate models. That from the lower troposphere seems least controversial. Much of the current debate has been addressing feedback from the tropical upper troposphere, where the feedback appears likely to be positive. However, this is not yet convincingly established; much further evaluation of climate models with regard to observed processes is needed.

Somewhat independent of feedbacks affecting top of the atmosphere long-wave fluxes, water vapour in the lower troposphere affects the long-wave contribution to surface radiation. This feedback has lately been emphasised as contributing somewhat to a reduction in diurnal temperature range with increasing water vapour concentrations from global warming (Mitchell *et al.*, 1995).

4.2.2 Cloud Amounts

The first cloud feedback to be studied in detail in global models involves changes in cloud amounts. These can be extremely complicated because of the many different types of clouds, whose properties and coverage are controlled by many different physical processes, and which affect the radiation budget in many different ways. In the global and annual mean, clouds have a cooling effect on the present climate (the surface and atmosphere), as evaluated from the Earth Radiation Budget Experiment (ERBE) and other satellite measurements. That is, the 31 Wm^{-2} enhancement of the thermal greenhouse effect is exceeded by a 48 Wm^{-2} increase in the reflection of short-wave radiation to space (Ramanathan *et al.*, 1989). But there are large variations in the net cloud forcing with geography and cloud type; indeed some clouds contribute a net warming. For low clouds, the reflected short-wave dominates so that an increase in amount would cool the climate and be a negative feedback on global warming. But thin tropical cirrus clouds are much colder than the underlying surface and act more to enhance the greenhouse effect, so an increase in the amount of this cloud type would be a positive feedback.

Climate model simulations of global warming have found the tropical troposphere to become higher and, in most models, the amounts of high cloud to increase. This increase (which depends on the somewhat uncertain changes of water vapour and relative humidity) enhances the greenhouse effect and produces a positive feedback (e.g., Wetherald and Manabe, 1988; Mitchell and Ingram, 1992). However, its importance relative to other changes in cloud amounts and the detailed changes in the three dimensional cloud distribution, varies significantly between models.

Until recently, climate models have used cloud prediction schemes based largely on presumed relationships between cloud amount and relative humidity and giving cloud amount as the only parameter. Such schemes introduce or remove condensed water instantaneously in amounts that are prescribed or depend only on temperature. Many modelling groups are now moving to prognostic cloud water variables that explicitly determine the amount of liquid water in each grid cell (e.g., Sundqvist, 1978; Le Treut and Li, 1988; Sundqvist *et al.*, 1989; Roeckner *et al.*, 1990; Smith, 1990; Ose, 1993; Tiedtke, 1993; Del Genio *et al.*, 1995; Fowler *et al.*, 1996). The predicted cloud water may vary more smoothly, persisting for hours after the agencies that formed it have ceased (as is especially true for cirrus); it can be used to determine interactively the optical properties of the clouds, as well as the precipitation rate. A prognostic cloud water variable thus improves a model's physical basis, but not without considerable difficulties, including proper representation of mixed-phase clouds (e.g., Senior and Mitchell 1993) and various numerical issues.

Some features of this new approach for predicting clouds are:

- Prediction of the condensed water and its partitioning between liquid and ice (Sundqvist, 1978; Li and Le Treut, 1992). Separate treatment of the liquid and ice cloud particles is important, because they undergo significantly different microphysical and thermodynamical processes and have different optical properties. The transition of water to ice is a critical process and empirically dealt with in some of the latest models.

- Action of cumulus clouds as liquid and/or ice sources for the stratiform clouds in some models (e.g., Ose, 1993; Tiedtke, 1993; Del Genio *et al.*, 1995; Fowler *et al.*, 1996). This important physical link between two types of cloud systems and its explicit incorporation into GCMs may mark a significant step forward in cloud parametrization.

- Physically based parametrization of the various cloud microphysical processes, such as the evaporation of cloud water and ice to water vapour in subsaturated air, conversion of cloud water and ice to rain and snow in supersaturated air, and parametrization of cloud droplets and ice crystals depending on cloud condensation nuclei (CCN) and ice nuclei. Figure 4.1 schematically illustrates some of these processes. For example, cumulus detrainment can produce small ice crystals that combine to make larger falling snowflakes. The snow falling through a warm lower atmosphere will melt to become rain, although ice and liquid may coexist in a range of temperatures whose width is somewhat uncertain. The efficiency of the so-called Bergeron-Findeisen mechanism in this range of coexistence has a decisive impact on how effective the release of precipitation is in the cloud. As a consequence, this mechanism strongly influences the resulting amount of cloud water, and indirectly the optical properties of the cloud.

- Physically based parametrization of the cloud optical properties and fractional cloud amount. It is far from obvious how to determine the cloudy fraction of a GCM grid-box; this cloud fraction depends on the sub-grid scale distribution of the water and ice contents (e.g., Kristjánsson, 1991; Kvamstø, 1991). The dependence of cloud optical properties on sub-grid scale spatial heterogeneity and mixed phases remains an important problem.

The new generation of cloud parametrization has led to some improvements in simulations of the Earth's radiation budget (e.g., Senior and Mitchell, 1993; Del Genio *et al.*, 1995; Fowler and Randall, 1996). Comparisons of the global distribution of simulated cloud water and ice concentrations against observations are problematic. Although satellite observations of the macroscopic distribution of total-column liquid water content are available over the oceans (Njoku and Swanson, 1983; Prabhakara *et al.*, 1983; Greenwald *et al.*, 1993), these show substantial differences, possibly a result of different algorithms (Lin and Rossow, 1994). Also, observations of macroscopic ice water content are lacking. Field data for local cloud ice measurements have been obtained in regional experiments such as the First ISCCP Regional Experiment (FIRE) (Heymsfield and Donner 1990) and the International Cirrus Experiment (ICE) (Raschke *et al.*, 1990), but these are very difficult to convert into macroscopic averages.

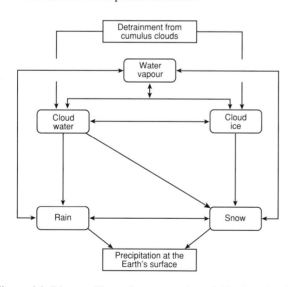

Figure 4.1: Diagram illustrating prognostic variables in a cloud microphysics scheme, and the processes that affect them.

High clouds

High clouds are very effective in trapping outgoing long-wave radiation, and so tend to warm the Earth. The net radiative effect on climate of anvils and cirrus clouds associated with deep convection in the tropics is near zero (e.g., Ramanathan *et al.*, 1989) because short-wave cooling and long-wave warming nearly cancel each other. The solar cooling acts mainly at the Earth's surface; however, how much is absorbed rather than reflected by clouds is currently controversial, as discussed in Section 4.2.6. Long-wave warming by anvils and cirrus clouds associated with deep convection in the tropics acts mainly on the atmosphere and can influence the general circulation of the atmosphere (Slingo and Slingo, 1988; Harshvardhan *et al.*, 1989).

Recent work has led to simple ice crystal scattering parametrizations for climate models (Ebert and Curry, 1992; Fu and Liou, 1993). Improvements may be needed to account for ice crystal size and shape effects, especially as climate models develop the capability to compute crystal sizes and realistic distributions of the ice water path. Comparison of cirrus properties from ISCCP with those generated by GCM parametrization using observed large-scale dynamic and thermodynamic fields from operational analyses shows encouraging agreement in the spatial patterns of the cirrus optical depths (Soden and Donner, 1994). However, determining cirrus modification of net radiative fluxes to the accuracies desirable for climate models may require accuracies in measurement of cirrus cloud temperature, ice water content and/or scattering properties that, in some cases, are beyond current observational and computational abilities (Vogelmann and Ackerman, 1996), suggesting that improved accuracies in these will be needed for climate studies.

Middle clouds

Frontal cloud systems are major sources of precipitation and cloud cover in the mid-latitudes. Climate models are incapable of explicitly resolving frontal circulations, although they do resolve "large-scale" cloud cover and rain associated with these systems. However there is a dearth of observations over the ocean and generally in the Southern Hemisphere to validate the microphysical parametrizations used to calculate the radiative and precipitation processes with these mid-level clouds (Ryan, 1996).

Low clouds

The marine stratocumulus clouds that commonly occur on the eastern sides of the subtropical oceans (e.g., Hanson, 1991) are important for their solar reflection (e.g., Slingo,

1990), and are, at present, under-predicted by many atmospheric GCMs. Similar clouds occur in the Arctic in summer (e.g., Herman and Goody, 1976), as well as over the mid-latitude oceans (Klein and Hartmann, 1993). They must be simulated successfully in order to obtain realistic sea surface temperature (SST) distributions in coupled atmosphere-ocean models (e.g., Robertson *et al.*, 1995).

Low clouds, such as marine stratocumulus, are favoured by strong capping temperature inversions (e.g., Lilly, 1968; Randall, 1980; Klein and Hartmann, 1993) as, for example, when a subsidence inversion associated with a subtropical high pressure cell confines moisture evaporated from the ocean within a thin, cool marine layer. At the same time, the radiative cooling associated with the clouds helps to maintain such inversions by lowering the temperature of the cloudy air. Turbulent entrainment, driven in part by radiative destabilisation, also maintains the inversion. Under suitable conditions, an external perturbation that reduces the SST favours a change in cloudiness that further reduces the SST (e.g., Hanson, 1991).

The positive feedback maintaining low clouds may be suppressed by various dynamical processes. Mesoscale circulations are forced in the marine boundary layer by the strong cloud-top radiative cooling that, in turn, breaks stratiform clouds into mesoscale cloud patches (Shao and Randall, 1996). Such broken clouds allow more solar radiation to warm the sea surface. Furthermore, temperature, moisture, and the subsidence rate of the subsiding air above the inversion may be changed to weaken the inversion (Siems *et al.*, 1990), thus increasing the likelihood of the stratiform clouds breaking up (Deardorff, 1980; Randall, 1980). Still controversial are details of the criteria and relative importance of other mechanisms, such as drizzle, absorbed sunlight, and entrainment decoupling that might detach planetary boundary layer (PBL) clouds from the surface (Kuo and Schubert, 1988; MacVean and Mason, 1990; Siems *et al.*, 1990).

Since subsiding air in the subtropics is connected to the sinking branch of a Hadley cell, subtropical boundary layer cloud properties may be partly determined by remote processes. A model with inadequate dynamical coupling to other regions may exaggerate the strength of the low cloud feedback and have a tendency to produce excessive low-level cloudiness. Coupled atmosphere-ocean models may be particularly susceptible because they have the ability to produce negative SST anomalies in response to increases in low-level cloudiness. On the other hand, over-emphasis of coupling to other regions may lead the feedback to another positive loop. Miller and Del Genio (1994) found in a version of the GISS GCM that the reduced low cloud amount led to stronger surface solar radiation, therefore a warming of the subtropical sea surface, and a further reduction in low cloud amount. The initial reduction of low cloud amount was due to the temperature changes at remote grid points, apparently weakening the inversion. They found this mechanism to give oscillations with periods on the order of a few decades. They are appropriately cautious about concluding that the feedbacks in the GISS model also operate in the real climate system, but suggest that positive cloud feedback could enhance variability.

Arctic clouds

Clouds are the dominant modulators of the Arctic radiation climate, affecting sea ice characteristics such as temperature, albedo and ice volume, (Curry and Ebert 1990, 1992), and so indirectly possibly altering the rate of sea ice transport to the North Atlantic. Sea ice and related high latitude physical processes including Arctic cloud formation are incorporated into current models with many unverified assumptions, so that the reliability of the simulated Arctic climate change scenarios is not high and the model-to-model differences are not surprising.

Ingram *et al.* (1989) used the UK Meteorological Office climate model to investigate the sea ice feedback on greenhouse warming. They performed sensitivity tests with prescribed, fixed sea ice distributions and compared their results with those of a simulation in which the sea ice distribution was permitted to change in response to climate change. They found that cloud-ice feedbacks were very important; the clouds obscure surface albedo exchanges, thus minimising their effects.

4.2.3 Cloud Water Content

As already noted, the climate models used for the first studies of global warming ignored the possibility of changes in cloud-water content, but are now beginning to include cloud formulations which explicitly predict the liquid- and ice-water content. The distinction between liquid-water and ice is important, not only for thermodynamic reasons but also because of differing radiative properties. The need to include formulations of cloud water and ice comes in part from the strong dependence of the short-wave, and for high clouds, long-wave radiative properties on water content and the possibility that the latter will change during global warming. Rennó *et al.* (1994) suggest that climate is sensitive to cloud microphysical processes and in particular, precipitation efficiencies.

Liquid water feedback might be substantially negative, as was emphasised by the simple radiative-convective model study carried out by Somerville and Remer (1984). Subsequently, a different sign for this feedback was inferred in a GCM by Roeckner *et al.* (1987) through a large increase in the long-wave greenhouse effect from thin cirrus clouds. The complexity of such cloud feedbacks is further illustrated by the studies of Li and Le Treut (1992) and Taylor and Ghan (1992).

In climate change studies with the UK Meteorological Office (UKMO) GCM, Senior and Mitchell (1993) found that in a warmer atmosphere, water clouds (lasting longer because of slower fall rates) replace ice clouds in the mixed phase region and hence cloud amount, especially at low and mid-levels, is increased. This "change of phase" feedback led to larger cloud amounts and to an increased short-wave cloud cooling with a warming climate. Consequently climate sensitivity was reduced with a 2.8°C warming in response to a doubling of CO_2, in comparison with an older relative humidity dependent cloud scheme where the global mean warming was 5.4°C. The inclusion of interactive cloud radiative properties further reduced the global mean warming to 1.9°C. In regions where cloud amount increased, the optical depth also increased. The net cloud feedback in this experiment was negative.

Changes in the long-wave cloud feedback depend on the representation of clouds in the model (Senior and Mitchell, 1995). A sensitivity experiment, in which the assumed statistical distribution of cloud water in a grid box is changed, produced a slightly improved validation of radiative fluxes against ERBE data (Barkstrom, 1984), but increased the climate sensitivity of the model from 3.4°C to 5.5°C (Senior and Mitchell, 1995). The effect of the change was to reduce high cloud amount for a given cloud water content. The reduced high cloud amount led to a much smaller "change of phase" feedback and so a higher climate sensitivity.

Several of the GCMs incorporate new cloud microphysics parametrizations for stratiform clouds and consequently produce weak negative cloud feedbacks in the climate sensitivity experiments reported by Cess *et al.* (1996). For example, the new version of the GISS GCM (Del Genio *et al.*, 1995) contains a prognostic cloud water parametrization and incorporates interactive cloud optical properties. When subjected to globally uniform increases and decreases of SST, the new model's climate sensitivity is only about half that of the earlier GISS model, and its cloud feedback is slightly negative as opposed to the substantial positive cloud feedback in the earlier model. This change in the cloud feedback is largely a result of a dramatic increase in both the cloud cover and cloud water content of tropical cirrus anvil clouds in the warmer climate. This result is sensitive to the type of climate experiment conducted. Tests with an SST perturbation that reduces the tropical Pacific SST gradient and weakens the Walker circulation give a higher climate sensitivity (Del Genio *et al.*, 1995).

The crude nature of current parametrization of detrainment of cumulus ice and the sensitivity to this term as discussed above, suggest that this may be an important area for future work in cloud parametrization. Low-level clouds may not produce the large negative feedback that is characteristic of models with only temperature-dependent cloud water because increasing cloud water content in a warmer climate could be offset by a decreasing geometric thickness of these clouds. This is consistent with the behaviour of satellite-derived optical thickness in the ISCCP data set (Tselioudis *et al.*, 1992). Thus the sign of the cloud liquid-water feedback in the real climate system is still unknown. Further study will be needed to reach the goal to determine the overall sign and magnitude of the real world's cloud feedback, as cloud feedback varies considerably with cloud type and geography and presumably with time. Nevertheless, the results from the new models provide valuable new insight into the physical issues that must be confronted before this goal can be achieved.

4.2.4 Cloud Particle Size

Studies of the effects of possible changes in cloud particle size on global warming integrations are at an early stage, but mechanisms by which changes might occur have been suggested. Cloud drops are formed by condensation on submicron diameter hygroscopic aerosol particles, which are present throughout the troposphere. The concentrations of CCN are highly variable and are influenced by air pollution as well as by natural processes as addressed in Chapters 3 and 4 of IPCC (1994) and Sections 2.3 and 2.4 of this report. The number of nuclei available has a strong effect on the number of cloud particles formed, and this in turn affects both the cloud optical properties and the likelihood of precipitation. Changes in the CCN concentration might alter the number of cloud drops that can grow, and hence alter the mean radius for a given liquid-water path. Some parametrization are attempting to represent the effects of nuclei availability on cloud particle number density and cloud particle size.

Twomey *et al.* (1984) noted that pollution produced by burning fossil fuels consists not only of carbon dioxide (CO_2) but also sulphur dioxide (SO_2), the gaseous precursor of sulphate aerosols, and that cloud drop sizes may be sensitive to changes in sulphate aerosol concentrations. Increased pollution could therefore increase the number of these CCN, increase the number of cloud drops, and hence reduce the mean particle size (provided that the water content does not change). This would increase the cloud optical thickness and hence the cloud albedos, leading to a cooling influence on climate. Efforts to include effects of sulphate aerosols from pollution in climate model cloud parametrization have been reported by Ghan *et al.* (1993, 1995), Jones *et al.* (1994), Boucher *et al.* (1995) and Boucher and Lohmann (1995). Kiehl (1994) argues that the difference between continental and oceanic cloud droplet sizes needs to be accounted for in determining climate model cloud albedos.

Charlson *et al.* (1987) suggested that the primary source of sulphate aerosol and hence CCNs over the ocean involves biological organisms, so that much of the sulphate aerosol over the remote oceans comes not from pollution but from dimethylsulphide (DMS), excreted by marine phytoplankton (discussed further in Section 10.3.4). Charlson *et al.* discussed the possibility of a regulation of the climate system by such marine organisms. Attempts to establish the details of processes by which this regulation might occur have not been successful, and there is no observational evidence that DMS sources would change with climate change.

Relatively shallow clouds may develop drizzle. This drizzle production depends on cloud depth, cloud liquid-water content and CCN distribution, as well as on cloud dynamics and lifetime. Drizzle depletes the cloud liquid-water, and reduces the cloud reflectivity. Below the cloud the drizzle may evaporate and cool the air, thus possibly leading to a decoupling of the cloud from the air below. If future anthropogenic emissions of SO_2 increase, thus leading to more CCN, then cloud droplets may become more numerous and smaller. This will likely impede drizzle production and possibly lead to longer cloud lifetimes.

4.2.5 Model Feedback Intercomparisons

The intercomparison reported by Cess *et al.* (1990) of climate models' response to changing the SST by $\pm 2°C$ provided a snapshot of the feedbacks operating in GCMs at that time. The models agreed in the magnitudes of the clear-sky feedbacks, as noted earlier, but cloud feedbacks varied considerably and were responsible for a threefold variation in the overall climate sensitivity between the participating models. This exercise has recently been repeated by Cess *et al.* (1996), who find that the disparity between the models has been reduced significantly (Figure 4.2). The most notable change is the removal of the largest (positive) values of cloud feedback. A detailed analysis of the reasons for this convergence has not yet been made. A small, overall cloud feedback can result from a cancellation of much larger feedbacks of opposite signs in the long-wave and short-wave regions of the spectrum. These components of the overall feedback vary considerably between the models. In addition, several modelling centres have produced a wide range of cloud feedbacks from slightly different cloud parametrizations in the same model. Hence, the convergence found by Cess *et al.* (1996) may not represent a reduction in the uncertainty of the magnitude of the cloud feedback. The idealised SST perturbation applied in these comparisons is very different from the more complex patterns obtained in coupled model simulations of global warming, and consequently the cloud feedback produced by a model forced by the $\pm 2°C$ SST perturbation might be completely different from that found in global warming simulations (Senior and Mitchell, 1993). These intercomparisons may provide only limited guidance as to the cloud feedback to be expected during greenhouse warming.

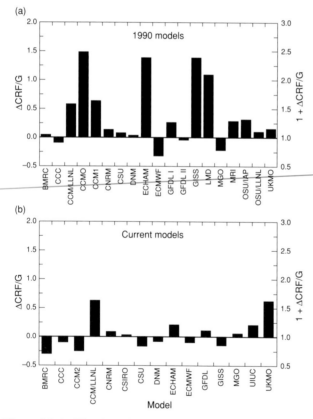

Figure 4.2: (a) The cloud feedback parameter, $\Delta CRF/G$, as produced by the 19 atmospheric GCMs used in the Cess *et al.* (1990) study, where ΔCRF is the Wm^{-2} due to cloud changes and G is the overall Wm^{-2} change, both as a result of the prescribed SST change. (b) The same as (a) but for the Cess *et al.* (1996) study.

A further intercomparison that gives insights into the impact of differences between cloud simulations in models is provided by Gleckler *et al.* (1995). They analysed the ocean energy transports implied by the ensemble of GCMs participating in the Atmospheric Model Intercomparison Project (AMIP) (Gates, 1992). The models were run for ten simulated years, using prescribed seasonally and interannually varying SST and sea ice distributions, as observed for the years 1979 to 1988. The models calculated the net radiation at the top of the atmosphere and the net energy flux across the Earth's surface.

The pattern of ten-year-averaged net radiation at the top of the atmosphere implies a pattern of total energy transport inside the system, since there is a net energy input in some parts of the world, and a net energy output in other parts, giving a global total very close to zero. This energy transport is accomplished by the circulation of the atmosphere and of the oceans. A pattern of ocean meridional energy transports is implied by the ten-year averages of the net ocean surface energy flux for each atmospheric GCM.

The implied ocean energy transports T_O are represented by the thin lines in the upper panel of Figure 4.3. The grey stippling shows the range of observationally derived upper and lower bounds of T_O (*cf.* Trenberth and Solomon, 1994, for the most recent observational study). Most of the simulations imply ocean energy transports that differ markedly from those inferred from observations, particularly in the Southern Hemisphere, where the implied T_O for many of the models is towards the equator.

Gleckler *et al.* (1995) determined the total meridional energy transport by the atmosphere and ocean combined, denoted by T_{A+O}, from the simulated ten-year averages of the top-of-the-atmosphere net radiation for each model. They then determined the simulated atmospheric energy transport, T_A by subtracting the ocean transport from the sum of the ocean and atmosphere transport. Finally, they computed a "hybrid" value of T_O, denoted by $T_{O\ hybrid}$, by subtracting each model's simulated T_A from the ERBE-observed T_{A+O}. This "hybrid" combines the simulated T_A with the observed T_{A+O}. The results for $T_{O\ hybrid}$ are shown in the lower panel of Figure 4.3. On the whole, the various curves for $T_{O\ hybrid}$ bear a much closer resemblance to the observations than do the model curves, indicating that the simulated atmospheric meridional energy transports are relatively realistic in most cases. Evidently, improved cloudiness parametrization and improved simulations of the effects of clouds on the radiation budget are needed to improve oceanic forcing in coupled atmosphere-ocean models.

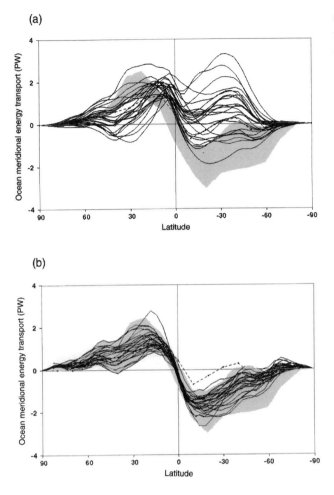

Figure 4.3: Ocean meridional energy transport from models and observations (a) model results from a range of atmosphere GCMs, derived from the ten-year averages of their implied net surface energy flux (thin lines); bounds on observed ocean transport (shaded area); model results from Semtner and Chervin (1992) in a numerical simulation of the general circulation of the oceans, forced with the observed atmospheric climate (dashed line). (b) As in (a), except that the thin lines show the "hybrid"ocean transport (see text).

4.2.6 Coupling of Clouds with the Surface

The effects of clouds on net solar radiation at the top of the atmosphere are largely mirrored in the effects of clouds on surface solar radiative fluxes. The uncertainties in the input of solar radiation to ocean and land models are a major source of uncertainties in determining the regional and global response to increasing greenhouse gases. Long-wave radiation would appear to be a smaller source of error, except in high latitudes, because of the atmosphere's large opacity to long-wave, even without clouds. In any

case, the surface solar fluxes are more easily estimated from remote sensing since long-wave fluxes depend on cloud bases which are not easily seen from space.

Clouds are sensitive to the changes of both atmospheric circulation and the surface boundary conditions. Small changes in the surface boundary layer can cause substantial differences in convective clouds over the tropical oceans (e.g., Fu *et al.*, 1994). However, a change of large-scale circulation can modify atmospheric conditional instability and thus clouds, even without a change at the surface (Lau *et al.*, 1994; Fu *et al.*, 1996). Because the prediction of future cloud changes depends on the correctness of these responses in GCM cloud schemes, more stringent tests, using satellite and *in situ* observations, are needed to ensure that the observed sensitivities of clouds to the changes of atmospheric circulation and surface conditions are adequately simulated in GCMs.

Consideration of ocean surface-atmosphere interactions have led to a controversial hypothesis. Figure 4.4 schematically illustrates what is known as the Thermostat Hypothesis (Ramanathan and Collins, 1991). If a positive SST perturbation leads to an increase in surface evaporation and moisture convergence (Lindzen and Nigam, 1987) then the increased moisture supply induces more convection, which leads to the formation of more high, bright clouds, which reflect more solar energy back to space. The resulting reduction in the solar radiation absorbed at the sea surface thus acts to dampen the postulated positive SST perturbation. The initial perturbation might be the climatological differences between east and west Pacific, or warming of the east Pacific associated with El Niño, or overall SST increases associated with greenhouse warming. Similar mechanisms would not necessarily apply to all these situations.

To support their idea, Ramanathan and Collins presented observational evidence that SST fluctuations associated with El Niño are accompanied by changes in the solar cloud radiative forcing (CRF) that would tend to dampen the SST fluctuations regionally. Where the ocean warms, the solar radiation reaching the sea surface diminishes, and where the ocean cools, the increased solar radiation tends to warm it. Ramanathan and Collins argued that convection and high bright clouds increase when the SST increases to about 30°C. They suggested that the increased solar cloud forcing associated with deep convection might act to prevent much higher SSTs.

Although Ramanathan and Collins explicitly discussed only regional climatological effects, their paper has been widely interpreted as suggesting that the global surface temperature of the Earth may also be limited in this way.

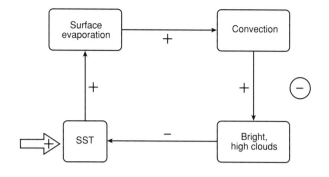

Figure 4.4: The "Thermostat Hypothesis" of Ramanathan and Collins (1991). An external perturbation leads to an increase in the SST, either locally or globally. The signs in the diagram are "+" for amplification of the next term in the loop and "−" for reduction. Increased SST promotes stronger evaporation and moisture convergence, which then lead to more vigorous convection. The convection generates high, bright clouds, which reduce the insolation of the ocean, thus counteracting the external perturbation. This is, therefore, a negative feedback and indicated by the circled "−".

This Thermostat Hypothesis has been very controversial and remains an active research topic. It has been criticised by Wallace (1992), Hartmann and Michaelson (1993) and Lau *et al.* (1994) for failure to recognise the importance of regional effects associated with large-scale dynamics, and also for under-emphasising the tendency of surface evaporation to cool the oceans. Several critical papers have emphasised a more conventional view of the tropical energy balance (e.g., Pierrehumbert, 1995). Fu *et al.* (1992) have argued on the basis of satellite data that the strong regional cloud radiative forcing anomalies associated with El Niño average to near zero over the tropics as a whole. The reality of locally negative short-wave cloud radiative forcing anomalies in response to local positive SST anomalies in the central tropical Pacific is apparent, but the importance of such short-wave cloud radiative forcing anomalies relative to other processes, and also their importance for the globally averaged surface temperature, are still in dispute.

Miller and Del Genio (1994) found, with their version of the GISS GCM, that a negative tropical evaporation anomaly resulted in a warming of the SST, leading to enhanced convection and rainfall. This convection decreased the solar radiation incident on the sea surface, and, not unlike the Thermostat Effect envisioned by Ramanathan and Collins (1991), dampened the initial warming of the sea surface. Spectral analysis of the model

results showed that this mechanism led to oscillations with periods on the order of years up to a decade.

Three recent papers argue that clouds absorb much more solar radiation than current physical understanding and radiation codes would allow. Suggestions that some clouds absorb more short-wave radiation than can be accounted for on the basis of the known radiative properties of water and ice have appeared at various times in the literature over several decades (see the review by Stephens and Tsay, 1990). This possibility has been invoked by Ramanathan *et al.* (1995), in the context of the West Pacific warm pool. According to their results, the effect of clouds on short-wave radiation at the surface needs to be 50% more than that at the top of the atmosphere, whereas current radiation schemes predict an enhancement of less than 20%. The implication is that the clouds are absorbing several times more short-wave radiation than previously believed. Ramanathan *et al.* do accept that, by making "extreme (but plausible)" changes to their numbers, it is possible to close the heat budget without recourse to anomalous absorption. However, the companion paper by Cess *et al.* (1995) claims that the anomalous absorption is a global phenomenon, showing evidence from a wide variety of locations to support the 50% enhancement required by Ramanathan *et al.* This enhancement was also obtained by Pilewski and Valero (1995) in an analysis of observations from research aircraft in the tropics.

A global enhancement of the magnitude proposed by Ramanathan is in conflict with many other documented studies with research aircraft where the measured absorption was not substantially different from the theoretically predicted value (Stephens and Tsay, 1990). Additionally, a critical examination of the anomalous absorption papers has revealed major flaws in the analysis methods which appear to invalidate the conclusions (Stephens, 1995). Hayasaka *et al.* (1995) have concluded "that the anomalous absorption pointed out by aircraft observations in previous studies does not exist". An extensive survey of surface and satellite data (Li *et al.*, 1995a; Whitlock *et al.*, 1995) suggests that short-wave budgets are unlikely to be in error by more than 15 Wm^{-2}. Li *et al.* (1995b) carried out extensive analyses of data "following the same methodologies" as Cess *et al.* and Ramanathan *et al.* They do not find anomalous absorption except possibly in the tropics where the data are most uncertain. The disagreement with Cess *et al.* is attributed to the use of different data sets, not due to different methodologies. Thus, at present, the evidence is weak for the claim that clouds absorb substantially more short-wave radiation than is predicted by models.

4.2.7 Precipitation and Cumulus Convection

Cumulus convection provides very rapid mass, energy, and momentum exchanges between the lower and upper troposphere. Much of the precipitation that falls to Earth is produced during this convective overturning, and a substantial fraction of the cloudiness in the tropics is produced by cumulus convection, either directly in the cumulus clouds themselves, or indirectly in the cirrus and other debris that cumuli generate. The proper treatment of these processes in climate models is still far from established. Errors in treating momentum exchange may seriously affect surface winds in the tropics and hence coupling to ocean models.

Cumulus convection is a manifestation of a buoyancy-driven instability that occurs when the vertical decrease of temperature is sufficiently rapid (i.e., when the "lapse rate" of temperature is sufficiently strong) and, at the same time, sufficient moisture is available. Because of the latter condition, cumulus instability is often called "conditional instability". The degree to which buoyancy forces can drive cumulus convection thus depends on both the lapse rate and the humidity. The time-scale for convective release is on the order of an hour – very short, compared to the multi-day time-scale of large-scale weather systems. This disparity of time-scales implies that ensembles of cumulus clouds must stay nearly in balance with large-scale weather systems. If a large-scale motion system or surface heating tries to promote cumulus instability, convection releases the instability restoring the system to a near-neutral state almost as rapidly as the instability is generated.

"Large-scale precipitation" refers to a somewhat old-fashioned but still widely used parametrization forming stratiform clouds such as cirrus or stratus and the accompanying precipitation that occurs when the mean state relative humidity reaches or tries to exceed a threshold value, such as 100%. Although generally larger than cumulus clouds, these systems are still typically sub-grid scale in climate models. Some such schemes accordingly include a sub-grid scale distribution of humidities, so that precipitation occurs with mean state relative humidities <100%. Relative humidities exceeding the threshold can be produced, for example, by large-scale rising motion which leads to adiabatic cooling and a decrease of the saturation mixing ratio. The excess humidity is typically assumed to condense and fall out as precipitation. In prognostic schemes, it may also be stored as liquid water. In many models, the falling precipitation is permitted to evaporate or partially evaporate on the way down.

Precipitation and convection are coupled to atmospheric radiative cooling in several ways. Slingo and Slingo (1988) discuss a positive feedback between the horizontal gradients of atmospheric radiative warming/cooling associated with localised high clouds produced by deep convection and the large-scale rising motion associated with the convection. In convectively active regions, long-wave radiation is trapped by anvils and cirrus produced by convective detrainment, and so the long-wave radiative cooling of the atmospheric column is reduced, and may even be transformed into a heating. The convectively active column is consequently radiatively warmed relative to the surrounding, convectively inactive regions, reinforcing the latent heating. The combination of these two heatings, together with the radiative cooling in the surrounding radiatively inactive regions, amplifies, on the average, the rising motion in the convectively active column.

Evidently, the strengths of the cloud feedbacks on precipitation must be further quantified. They do not occur in isolation, but coexist not only with each other, but also with many other powerful processes that can affect weather and climate. Idealised numerical experiments with GCMs can be designed to focus on such feedbacks in relative isolation, and so are particularly well suited to investigating their relative strengths.

The distribution of modelled precipitation and its changes with climate change needs more extensive validation as it is a major coupling link to both the land hydrological cycle and oceanic buoyancy forcing of the thermohaline circulation. High latitude precipitation is especially important for the latter.

4.2.8 Assessment of the Status of Moist Processes in Climate Models

In the previous IPCC report the radiative feedbacks of clouds were identified as a major source of uncertainty for modelling future climate change. Considerable research efforts have addressed this issue since the last assessment and have further reinforced this conclusion. They have also added considerably to our understanding of the complexity of this issue. Some conclusions from these studies are:

- Different cloud parametrizations in current GCMs give a wide range of radiative feedbacks, affecting global and regional energy balances and the occurrence and intensities of atmospheric precipitation. These are derived from plausible physical assumptions and parametrizations, but the issue is extremely complex and many assumptions or approximations are made. Models including only cloud amount feedbacks have indicated that these could amplify global warming. Feedbacks involving cloud liquid-water and phase could also have major impacts on the global energy balance. At present, it is not possible to judge even the sign of the sum of all cloud process feedbacks as they affect greenhouse warming, but it is assessed that they are unlikely either to be very negative or to lead to much more than a doubling of the response that would occur in their absence. Improved treatments are vigorously being pursued.

- The cloud feedback processes are intimately linked to the atmospheric hydrological cycle, and can only be simulated satisfactorily if there is a comprehensive treatment of water in all its phases – vapour, liquid and ice. Many of these cloud and linked hydrological processes occur on scales not resolved by current GCMs. The sub-grid scale parametrizations treating these processes should be physically based and carefully evaluated with observational data.

- Inadequate simulation of cloud amounts and optical properties in GCMs contributes major errors to the simulation of surface net radiation, and thereby introduces errors in simulation of regional ocean and land temperatures. Uncertainties in the simulation of changes in these properties with climate change have a major impact on confidence in projections of future regional climate change.

- There is an important, but poorly understood, linkage of cloud optical properties to the CCN distribution. Inclusion of this linkage for models of climate change will require an improved description of the time and spatially varying distribution of global sulphate and other aerosols, as well as detailed microphysical treatments of cloud droplet size distributions.

- There is a consensus among different GCMs as to the sign and magnitude of clear-sky feedbacks but not for water vapour feedbacks alone. With these clear-sky feedbacks but with fixed cloud properties these GCMs would all report climate sensitivities in the range 2-3°C. There is no compelling evidence that the water vapour feedback is anything but the positive feedback indicated by the models. However,

the partitioning between water vapour and lapse rate feedback is not well established, and the processes maintaining water vapour in the upper troposphere are poorly understood.

4.3 Oceanic Processes

The ocean covers about 70% of the surface area of the Earth, has most of the thermal inertia of the atmosphere-ocean-land-ice system, is a major contributor to total planetary heat transport, and is the major source of atmospheric water vapour. Its interaction with the atmosphere through its surface quantities occurs through: SST, sea ice extent and thickness, surface albedo over ice-covered and ice-free regions, sea surface salinity and the partial pressure of CO_2 at the surface (pCO_2). It is a major component of the climate system in determining the mean (annually averaged) climate, the annual variations of climate, and climate variations on time-scales as long as millennia. While it is only through ocean surface variations that the atmosphere and land can be affected, these surface variations, in turn, depend on the thermal and saline coupling between the deeper ocean and the surface. Thus, the thermal and saline structure and variations in the deeper ocean must be simulated in order to determine surface variations on long time-scales. In general (except in those few regions of deep convection and other water mass transformation regions), the longer the time-scale of interest, the greater the depth of ocean that communicates with the surface. In turn, the ocean is driven by fluxes from the atmosphere of heat, momentum, and fresh water at the surface of the ocean, so that the only consistent way of simulating the evolution of the climate is through coupled atmosphere-ocean models.

The first atmosphere-ocean coupled models studying greenhouse warming concentrated on the sensitivity and response of climate to sudden and transient changes of radiatively active gases by using well-mixed (slab) oceans of fixed depth with no (or specified) transport of thermal energy. While such simplifications are useful for understanding atmospheric responses and for qualitative estimates of certain ocean responses, work over the last few years has indicated that the ocean circulation itself is sensitive to changes in forcing at the surface. Thus as the radiatively active gases increase, the ocean circulation may change and these changes may affect the mean climate and its variability. These changes may be significant, so that only coupled models that include the relevant parts of the ocean circulation are capable of simulating the entire range of possible climatic responses.

The ocean, like the atmosphere, has complex internal processes that must be parametrized. It has its own unique properties of boundaries and a density (buoyancy) structure that is affected by salt as well as temperature. Although venting of heat from fissures in the deep ocean may contribute to its circulation (Riser, 1995), to a good first approximation the ocean is driven entirely at the surface by the input of heat, fresh water and momentum fluxes from the atmosphere.

4.3.1 Surface Fluxes

Because the inertia (mechanical, thermal, and chemical) of the ocean is large compared to that of the atmosphere, changes in ocean surface properties for the most part occur relatively slowly. These slow changes depend on the surface fluxes of heat, momentum, fresh water, and CO_2 from the atmosphere. Models of the mean circulation of the ocean are sensitive to changes in heat flux (Maier-Reimer *et al.*, 1993) and fresh water flux (Mikolajewicz and Maier-Reimer, 1990; Weaver *et al.*, 1993).

The momentum input at the surface of the ocean, as wind stress, depends mostly on the winds near the surface and on the wave response. The heat input to the ocean surface consists of latent and sensible heat exchange between the ocean and the atmosphere, (depending on near surface winds, air temperature, and humidity), state of the sea, and radiative inputs (which depend on the overlying atmospheric column). The fresh water input to the surface of the ocean is composed of the difference between precipitation and evaporation, of runoff from land, and of the difference between the melting and freezing of ice (Schmitt, 1994). The carbon dioxide flux into the ocean depends on pCO_2, the concentration of CO_2 in the atmosphere, and on the near-surface winds. The pCO_2 is in turn controlled by oceanic transport processes, geochemical processes, and upper ocean biotic processes. The possibility of large errors in incident solar flux and latent heat flux are especially of concern for climate modelling; such errors may account for much of the flux adjustment needed for many models.

4.3.2 Processes of the Surface Mixed Layer

The near surface ocean is usually well-mixed by sub-grid scale processes involving stirring by the wind and by its convection. Therefore, quantities at the surface are determined by their mixed-layer values. The mixed-layer temperature is determined by heat fluxes at the surface, by mixing and advection of temperature horizontally, by the depth of the mixed layer, and by the entrainment heat flux at the bottom of the mixed layer (the interface between the near-surface turbulence and the relatively non-turbulent ocean interior). Skin effects at the atmospheric interface may give departures from mixed-layer values, especially under low winds. The interior ocean affects the surface

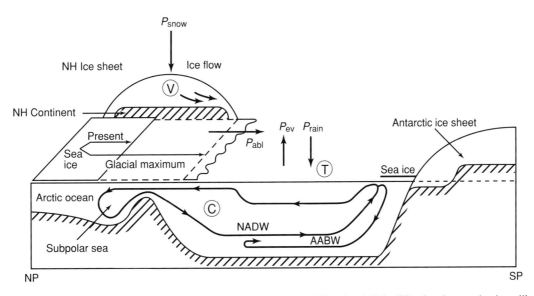

Figure 4.5: Diagram of an Atlantic meridional cross-section from North Pole (NP) to South Pole (SP), showing mechanisms likely to affect the thermohaline circulation on various time-scales. The change in hydrologic cycle, expressed in terms of water fluxes, $P_{rain} - P_{evaporation}$, for the ocean and water fluxes, $P_{snow} - P_{ablation}$, for the snow and ice, is due to changes in ocean temperature. Deep-water formation in the North Atlantic Subpolar Sea (North Atlantic Deep Water: NADW) is affected by changes in ice volume and extent (V), and regulates the intensity of the thermohaline circulation (C); changes in Antarctic Bottom Water (AABW) formation are neglected in this approximation. The thermohaline circulation affects the system's temperature (T) and is also affected by it (Ghil and McWilliams, 1994).

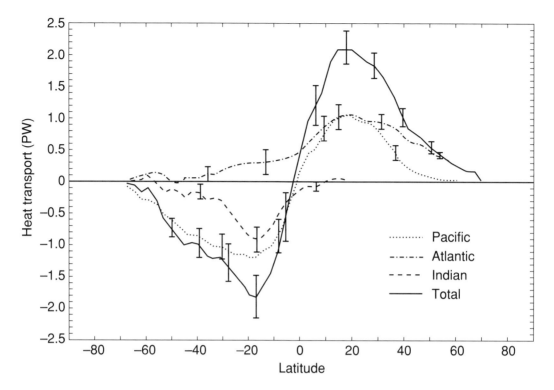

Figure 4.6: The poleward ocean heat transports in each ocean basin and summed over all oceans as calculated indirectly from energy balance requirements using ERBE for top of the atmosphere radiative fluxes and ECMWF data for atmospheric energy fluxes (from Trenberth and Solomon, 1994).

ocean only through this entrainment at the bottom of the mixed layer. Entrainment depends on the strength of the mixed-layer turbulence, on the motion of the bottom of the mixed layer itself, and on upwelling in the interior through the bottom of the mixed layer. Sterl and Kattenberg (1994) examine the effects on an ocean model of a mixed-layer parametrization and suggest that wind-stirring has important consequences not captured in the present ocean GCMs.

4.3.3 Wind Driven and Thermohaline Ocean Circulation

The wind driven circulation is that directly driven by the wind stress: because it is slow and large-scale in the interior of the ocean, it can conveniently be expressed by the conservation of vorticity (the so-called Sverdrup balance) which says that the vertically integrated meridional flow, where not affected by lateral boundaries and bottom topography, is given directly by the curl of the wind stress. This is an absolute constraint that would exist whatever the internal stratification of the ocean.

The thermohaline circulation is driven by changes in sea water density arising from changes in temperature versus salinity. Its functioning in the Atlantic Ocean is illustrated schematically in Figure 4.5. Formation of sea ice increases

the salinity of adjacent unfrozen water. At low temperatures and relatively high salinity, cold dense waters sink convectively and spread throughout the oceanic depths, thereby maintaining the stable vertical stratification. Warmer surface waters flow toward these sinking regions and are cooled along their journey by heat fluxes from the ocean to the atmosphere. The sinking regions are highly restricted in area: "deep waters" are formed only in the North Atlantic Greenland, Norwegian, Iceland, and Labrador seas. The world's "bottom waters" are formed only in restricted regions of the Southern Oceans near the coast of Antarctica in the Weddell and Ross Seas. Waters from these sinking regions spread at depth to fill the world's entire basin and thereby help maintain the vertical stratification even in oceans where no deep sinking exists. These processes and the thermohaline circulation may change on various time-scales.

In the North Atlantic the combination of warm surface water flowing northward and cold water flowing southward at depth gives a net northward heat transport which has been estimated by direct measurements to be about 1 PW at 24°N (see Bryden, 1993 and references therein) and verified by indirect methods to be about the same (Figure 4.6, from Trenberth and Solomon, 1994). Deep cold waters flow southward across the equator from the North Atlantic and thereby imply a northward heat flux even in the South Atlantic. The Pacific is less saline than the Atlantic, is bounded further south by Alaska, and has no deep water formation. Its transport is more nearly symmetrical about the equator and similarly has a magnitude at 24°N of about 1 PW (Bryden *et al.*, 1991 and Figure 4.6). The properties of the Atlantic and Pacific rapidly interact through the Antarctic Circumpolar Current.

A coupled simulation (Manabe and Stouffer, 1988) has shown that during periods of no thermohaline circulation, the high northern Atlantic would be ice-covered to south of Iceland and be much colder than now. Palaeo-records indicate that oscillations in the thermohaline circulation leading to warming and cooling at high latitudes prevailed during the last glacial interval (Bond *et al.*, 1993). If the thermohaline circulation were to weaken with the expected larger inputs of fresh water to high latitudes during global warming, the net effect would be to either weaken the warming in high latitudes and amplify it in lower latitudes, and/or to make the thermohaline circulation and, hence latitudinal temperature gradients, more variable as discussed in Section 4.3.9.

Ocean only models are useful for isolating the ocean processes that may be present in more comprehensive coupled models of the climate although such may be considerably altered when coupled to the atmosphere. The idea that the stability and variability of the thermohaline circulation depends on the relative strength of high latitude thermal to fresh water forcing was introduced in coarse resolution ocean GCMs by Weaver *et al.* (1991, 1993). A steady thermohaline circulation can only exist if the fresh water input by high latitude precipitation, runoff, and ice melt is balanced by the fresh water export by that same thermohaline circulation. As the high latitude fresh water flux is increased, the ability of the thermohaline circulation to remove the fresh water is limited and the thermohaline circulation may have multiple equilibrium solutions (Stommel, 1961; Bryan 1986; Marotzke, 1988) and large variability. Such variability on decadal (Weaver and Sarachik, 1991) to millennial (Winton and Sarachik, 1993, Winton, 1993) time-scales has been demonstrated in ocean models with relatively large fresh water stochastic forcing. While the imposed boundary conditions on temperature have come into question, the mechanism is physically plausible and may survive the transition to a responsive atmosphere. If so, the implications are considerable: in a warmer world with warmer high latitudes and a stronger hydrologic cycle the thermohaline circulation could become less stable and more variable.

Recent simulations of the coupled transient response to increases of radiatively active gases that reach twice the pre-industrial concentration of CO_2 (Manabe and Stouffer, 1994) have indicated that the thermohaline circulation first weakens and then returns. The experiments that reach four times the pre-industrial concentration of CO_2 (in 140 years) have the thermohaline circulation weaken and stay weak for up to 500 years. While this model has a severe flux correction which stabilises the thermohaline circulation, it raises the important question of the response of the thermohaline circulation to changes in the greenhouse forcing and the subsequent effects of this response on climate.

Very high resolution models are being used to explicitly examine small-scale orographic, topographic and eddy processes in the ocean (e.g., Semtner and Chervin 1988, 1992). These models are useful for studying relatively short-lived phenomena and have given impressive simulations of the wind driven annual variation of heat transports in the Atlantic (Bönning and Herrmann, 1994) but, because of the huge computational overhead, can only be run for relatively short periods of model time. Thus, they cannot yet be coupled to model atmospheres for use in climate simulations nor to examine the thermohaline circulation. Such simulations must still use coarser-resolution models with parametrized eddies.

4.3.4 Ocean Convection

Given the surface fluxes and a formulation of mixed-layer processes near the surface, ocean general circulation models have to solve the advective equations for temperature and salt in the presence of convective overturning and stable ocean mixing. Convection arises when the density stratification becomes unstable and when relatively salty water is cooled to low temperature and becomes so dense that a water column becomes unstable. Convection homogenises the column and allows overturning to occur (see Killworth (1983) for a complete review).

Convection occurs over scales of a few km and is therefore hard to observe (but see MEDOC, 1970). Detailed simulations of individual convective elements in neutral stratification show the process to be complex and rife with small-scale features (Jones and Marshall, 1993; Legg and Marshall, 1993). Nevertheless, it is important to represent this process in ocean models with resolution coarser than the scales on which convection is known to occur. Simple convective adjustment parametrization (e.g., Cox, 1984) has long been used with improvement developed over time (e.g., Yin and Sarachik, 1994).

Parametrization of deep convection based on the detailed simulations of convective elements cited above are under development but it is not yet known how this will affect the simulation of the thermohaline circulation.

4.3.5 Interior Ocean Mixing

Parametrization of the mixing of stably stratified water in the interior of the ocean is crucial for the simulation of ocean circulation. Traditionally, eddy mixing coefficients have been used with values of order $1 cm^2/s$ for vertical diffusion and viscosity and order $10^7 cm^2/s$ for horizontal diffusion. These diffusion values are sometimes derived from tracer experiments, but more often they are selected to ensure numerical stability of the simulation. Parcels tend to stay and move on surfaces of constant density, which are predominantly horizontal in the interior ocean, and only small values of cross-density diffusion are expected. A tracer experiment (Ledwell *et al.*, 1993) has recently indicated that the correct vertical diffusion coefficient for the ocean interior is closer to $0.1 cm^2/s$, an order of magnitude smaller than often used. Vertical mixing in the regions of lateral boundary currents or perhaps sea mounts is likely to be larger. Large *et al.* (1995) review a new scheme for mixed-layer dynamics and for ocean vertical mixing.

Further insights into turbulent diffusion in the ocean depend on a detailed knowledge of the precise mechanisms of that mixing. Mesoscale eddies, for example, are plentiful in the ocean and their finer scales would imply enhanced mixing through the production of frontal type gradients in temperature and tracers. A recent parametrization (Gent and McWilliams, 1990; Danabasoglu *et al.*, 1994) of these eddies follows an approach pioneered in stratospheric tracer modelling; that is, the mixing is primarily accomplished by advection by the "residual circulation" induced by the eddies rather than by the direct effects of the eddies themselves. Experiments with this parametrization show striking improvements in simulation of the depth and sharpness of the global thermocline and of the meridional heat transport.

4.3.6 Sea Ice

In high latitudes, sea ice is a major modulator of energy exchange between ocean and atmosphere. It is an insulating and highly reflecting surface. In the Arctic, much of the ice lasts permanently through the summer so that it has a substantially larger impact on surface albedos per unit area than do continental snow surfaces.

In winter, sea ice controls the transfer of heat from the relatively warm ocean to the cold atmosphere. The sea ice-cover is normally not complete. Even in the Central Arctic, 1-2% of open water permanently exists in the winter and a larger fraction in summer. Observation and model simulations indicate that surface sensible heat fluxes from the open water are one to two orders of magnitude larger than from the surface of the pack ice in the Arctic Basin (e.g., Meleshko *et al.*, 1991). Albedo of the sea ice-covered region also depends on sea ice concentration. Given the significant low-frequency variability of sea ice-cover and the persistence of its anomalies, its variable control of heat fluxes may significantly affect the atmospheric circulation and climate of the mid-latitude regions.

In the Northern Hemisphere, sea ice reaches its minimum extent in September when it covers $8.5 \times 10^6 km^2$. It attains a maximum extent of $15 \times 10^6 km^2$ in March. Its interannual variability varies from $1.1 \times 10^6 km^2$ in winter to $1.8 \times 10^6 km^2$ in summer, mainly in the marginal zones, and depending on atmospheric circulation and oceanic currents (Parkinson and Cavalieri, 1989).

Many GCMs and other climate models still treat the sea ice-cover as a single slab and do not take into account the always present but randomly distributed open waters ("leads"). Climate simulations with GCMs that incorporate sea ice inhomogeneities in a single grid box (sea ice concentration), show additional and substantial heating of the atmosphere in winter (Kattsov *et al.*, 1993; Groetzner *et al.*, 1994). This heating amounts to $10 Wm^{-2}$ over the polar cap of 60°N–90°N, increasing surface air temperature by 3.2°C over the same region. The warming is confined to the lower troposphere by the high stability of the polar atmosphere. The leads have a comparable heating effect to that produced by the observed sea ice anomalies (Kattsov *et al.*, 1993).

The largest surface air temperature increase with greenhouse warming is expected in the high latitudes of the Northern Hemisphere, because of the large atmospheric stability and the positive feedbacks between sea ice albedo and the surface temperature of the mixed ocean layer. These feedbacks are inadequately characterised, in part because of uncertain cloud feedbacks at the ice margins. Climate models that do not account for open waters over the ice-covered ocean probably overestimate the effect of the sea ice albedo in summer and underestimate the ocean cooling in winter. Changing snow cover may also feed back on ice thickness (Ledley, 1993).

The distribution of leads and thickness of sea ice is complex, of small scale and depends substantially on the dynamics of the ice as forced by wind and water drag. The required drag from the atmospheric model is, furthermore, not always of the correct strength and direction. Flato and Hibler (1992) have proposed a practical method for sea ice dynamics that is now being used in several GCMs.

4.3.7 The El Niño-Southern Oscillation as a Climate Process

The El Niño-Southern Oscillation (ENSO) is the coupled atmosphere-ocean phenomenon (Figure 4.7) wherein the normally cool, dry, eastern and central tropical Pacific becomes warmer, wetter, and with a lower sea level pressure every few years. The entire global tropics warm by about 1°C, and, by virtue of the large fraction of global area covered by the tropics, affect the globally averaged surface temperature on seasonal-to-interannual time-scales (e.g., Yulaeva and Wallace, 1994). The ENSO cycle affects the distribution and concentration of global CO$_2$ (e.g., Feely *et al.*, 1987 and Chapter 10). Major progress in

understanding, simulating, and predicting ENSO has been made in the last 10 years (e.g., Battisti and Sarachik, 1995).

A number of recent papers have shown that:

- The recent land temperature record over the last few decades can be modelled by an atmospheric GCM forced by the observed record of global SST (Kumar *et al.*, 1994; Graham, 1995), and,

- These results are dominated by tropical rather than mid-latitude SST (Graham *et al.*, 1994; Lau and Nath, 1994; Smith, 1994).

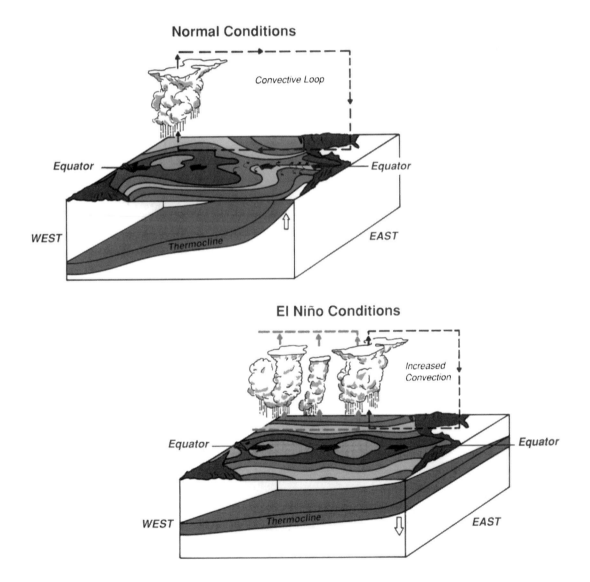

Figure 4.7: Schematic illustration of the differences in tropical climate between normal and El Niño conditions. For the latter, the thermocline becomes less tilted, SSTs increase in the eastern Pacific and regions in the central and eastern Pacific see increased convection.

These studies indicate that the unusual warmth of the tropical SST since the mid 1970s, during the warm phase of ENSO, has imprinted an unusual warmth on the entire global circulation. ENSO is evidently a major contributor to the natural variability of SST in the tropics, and arguably, also globally.

Warm phases of ENSO coincide with a warming of the tropical troposphere of close to 1°C. The net effect of ENSO is to warm the surface waters of the eastern Pacific. Ocean models that try to simulate the tropical Pacific SST in response to a repeating annual cycle of surface fluxes tend to simulate too cold an eastern Pacific. The simulation where the ocean is forced by long records of observed fluxes is more difficult and has not yet been done, though it is likely that such more realistic forcing would ameliorate the cold eastern Pacific problem.

Interestingly, all coupled model simulations (Mechoso *et al.*, 1995) of eastern Pacific SST show a result which contrasts with the ocean-only simulations: an eastern Pacific that is too warm, a feature attributed to the inadequacy of stratus cloud simulations in the atmospheric component of the coupled model (Koberle and Philander, 1994). To the extent that tropical Pacific temperatures matter for global climate and to the extent that ENSO variability dominates SST and land temperatures in the tropics (Wallace, 1995), it is clear that ENSO must be considered a vital part of the global climate system and should be accurately simulated. In order to correctly simulate ENSO, the meridional resolution at the equator must be a fraction of a degree in order to simulate wave processes and the meridional extent of the upwelling, both crucial. To date, no coupled model used for projecting the response to greenhouse warming has such resolution.

There is one final and intriguing possibility that the above cited papers imply: the possibility that global warming not only affects ENSO by affecting the background state (Graham *et al*, 1995; and a contrary view by Knutson and Manabe, 1994, in a coarse resolution GCM) but that indeed much of the effects of greenhouse warming might be modulated through changes in the magnitude and regularity of the warm and cold phases of ENSO.

4.3.8 *Assessment of the Status of Ocean Processes in Climate Models*

- Comprehensive ocean GCMs, coupled to the atmosphere through fluxes of energy, momentum, and fresh water, are required for assessment of the rate, magnitude and regional distribution of climate change.

- The large-scale dynamics of current ocean models seem reasonably realistic, but are not completely validated, in part because of a dearth of appropriate observations. An outstanding question to be resolved is the response of the thermohaline circulation in coupled models to increased high latitude inputs of fresh water. This question needs to be answered to assess how latitudinal gradients of SST may respond to global warming. Current suggestions point to either a large increase in or a highly variable latitudinal temperature gradient as possible responses.

- Fluxes at the ocean-atmosphere interface in coupled models have not yet been fully examined. In some cases there may be serious errors – for example, surface radiative fluxes depending on inadequately parametrized cloud processes, and high latitude inputs of fresh water depending on poorly characterised changes in the atmosphere-hydrological cycle.

- Oceanic models still use relatively crude parametrizations of sub-grid scale processes for near surface and interior mixing and for deep convection. New parametrizations for interior mixing associated with mesoscale eddies are likely to improve the simulated depth and sharpness of the global thermocline.

- Details of sea ice treatments in GCMs are still questionable, although some improvements are being examined. The role of sea ice in climate change is especially uncertain because of poorly known interface feedbacks; that is, overlying clouds modifying radiation, surface wind stress, ocean currents and changes in oceanic heat transport underneath the sea ice.

- ENSO processes have major effects on the tropical climate system, with a strong impact on hydrological processes and surface temperatures on interannual time-scales. Some coupled atmosphere-ocean models appear to give reasonable simulations of this system and show promise for providing useful predictions. However, the current generation of models used for projection of greenhouse gas response do not satisfactorily simulate ENSO processes, in part because the spatial resolution required to do so is not computationally feasible for century-long climate simulations.

4.4 Land-surface processes

Fluxes of heat and moisture between land and the atmosphere are central to the role of land processes in the climate system. These fluxes determine the overlying distributions of atmospheric temperature, water vapour, precipitation, and cloud properties. Atmospheric inputs of precipitation and net radiative heating are crucial for determining land-surface climate (climate over land is of greatest practical importance) and in turn are modified by land process feedbacks. Solar fluxes at the surface are currently highlighted as being significantly in error compared to observations, in some and perhaps most climate models due to the inadequate treatments of clouds (e.g., Garratt, 1994; Ward, 1995). These comparisons are being made possible by the recent availability of satellite-derived surface solar fluxes (e.g., Pinker *et al.*, 1995; Whitlock *et al.*, 1995).

Compared to the ocean, the land's relatively low heat capacity and limited capacity for water storage lead to strong diurnal variations in surface conditions and direct local responses to radiative and precipitation inputs. These limited storage capacities combined with the heterogeneous nature of the underlying soils, vegetation, and slope (e.g., Figure 4.8) imply potentially large heterogeneities in sensible and latent fluxes which may drive mesoscale atmospheric effects.

4.4.1 Soil-Vegetation-Atmosphere Transfer Schemes

Sensitivity studies with GCMs have shown that the treatment of land in climate models has major effects on the model climate and especially near the land surface (e.g., Koster and Suarez, 1994). The schemes to represent land in climate models are called soil-vegetation-atmosphere transfer schemes (SVATs). Important elements

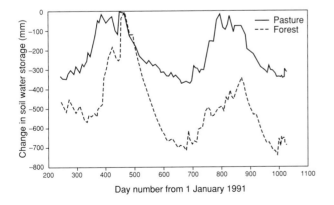

Figure 4.8: The seasonal variation in soil moisture storage for a tropical forest in Rondonia, Brazil (dashed line), compared to an adjacent pasture (solid line) (Institute of Hydrology, 1994).

of these schemes are their storage reservoirs and their mechanisms for the exchange with the atmosphere of water and thermal energy. Water storage occurs in soil reservoirs and in some models also as fast time-scale canopy or surface terms. The water intercepted by the canopy may be stored or evaporated. This store generally holds about 1 mm of water. The soil store, usually in the range of 50 to 500 mm of water, depends on soil porosity, wilting point for vegetation, soil drainage rates and especially the depth from which water can be extracted from the soil. This depth is associated with the rooting depth of vegetation and the root distribution.

Canopy transpiration is a physiological process depending on water transfer from the soil through roots, stems and leaves. The canopy resistance measures the effectiveness of this moisture transfer. It is primarily the integrated stomatal resistances. The transpiration, as mediated by the stomates, is limited by the supply of water from the roots and atmospheric conditions of demand. Neglect of this canopy resistance, now included in the SVAT models, has perhaps been the largest source of error in the older "bucket" models. The importance of canopy resistance is illustrated by two independent studies with GCM simulations of the effect of doubling stomatal resistance within the model SVATs (Henderson-Sellers *et al.*, 1995; Pollard and Thompson, 1995). The computed effects are largest for forests, and hence the largest areas affected are the boreal and tropical forests. In the boreal forests, summertime evapotranspiration (ET) is reduced by at least 20%, and surface air temperature increased by up to several degrees.

Surface roughness is the basis for determining the aerodynamic drag coefficient, C_D, for a surface. In the early GCMs, C_D for land was specified as 0.003, a typical value for short vegetation and for conditions of neutral stability. To achieve adequate accuracy, it is necessary to represent drag coefficients in terms of surface similarity theory, where transfer coefficients for momentum, heat and moisture are determined from a roughness length, Z_0 and the thermal stability of the near surface air. It may be necessary to distinguish between coefficients for momentum, heat and moisture. In particular, all sub-grid scale roughness elements and topography may contribute to momentum transfer but it is likely that only those on the scale of individual vegetation elements contribute to heat and moisture transfer. Schmid and Bünzli (1995) suggest a new approach for scaling roughness elements to model resolution and emphasise the importance of surface texture. The largest departures in newer models from the earlier ones are over forests, where C_D can readily exceed 0.01.

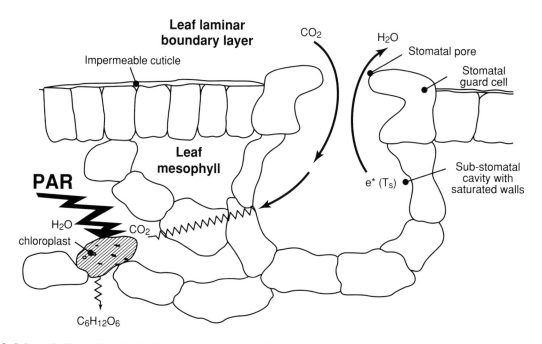

Figure 4.9: Schematic illustration of a leaf cross-section showing links between stomatal gas exchange (CO_2, H_2O) and photosynthesis. The stomatal conductance is related to the area-averaged value of the stomatal pore width which is of the order of 10 μm. The stomatal pores are under active physiological control and appear to act so as to maximise the influx of CO_2 for photosynthesis for a minimum loss of leaf water. Thus photosynthesis and transpiration are dependent on PAR flux, atmospheric CO_2 concentration, humidity, temperature, and soil moisture (Sellers et al., 1992). $e^*(T_s)$ is the saturation vapour pressure at the leaf surface.

Parametrization of vegetation properties related to canopy architecture determines significant features of the treatment of vegetation for evapotranspiration. The total surface of photosynthesising leaves and stem surfaces influences canopy resistance and transfer of heat and moisture from the canopy to the atmosphere. The flux of photosynthetically active radiation (PAR) normal to leaf surfaces, as required for stomatal parametrization, depends on canopy and leaf architecture. Furthermore, the net radiative loading over the surface of a given canopy element depends on these properties. Because of the large effect that canopy resistance has on SVAT models, they are sensitive to such details.

Recently, some canopy and soil schemes have included the uptake, storage and release of carbon through carbon dioxide exchanges with the atmosphere as illustrated in Figure 4.9. These sub-schemes will be increasingly important as physical models are coupled to biogeochemical models (*cf.* Chapters 2 and 9).

Vegetation cover and properties are now, for the most part, included in climate models as prescribed from inadequate observations. However, interactions between vegetation and climate may have significant effects. On seasonal time-scales, such interaction includes the effects of drought on vegetation cover, and on longer time-scales,

possible changes in structure, e.g., transition between forest and grassland.

Canopy albedo determines the fraction of incident solar radiation that is absorbed. Current model parametrizations of surface albedo are largely inferred from some limited surface measurements for various kinds of vegetation canopies. Satellites are, in principle, the only means of globally establishing surface albedos. Because albedo can change substantially with vegetation cover it has been a major parameter in studies of the response of regional climate to land-use change (Section 4.4.3).

Runoff depends on soil moisture, properties of the incident precipitation, and characteristics of soils and topography. A theoretical foundation exists for the local vertical infiltration of water in soil, given soil hydraulic properties. In reality, runoff rarely results from precipitation exceeding maximum infiltration. Rather, much more often lateral down-slope flows carry soil water to low regions where the water table reaches the surface and into streams. However, there are intrinsic difficulties in parametrizing slope effects to determine runoff in a climate model. Furthermore, soil properties are highly heterogeneous both horizontally and vertically so that specifying them as constants over a model grid square or in a soil column is questionable. Changes of soil hydraulic properties with depth may strongly affect runoff. The recent intercomparisons through the Project for the Intercomparison of Land-surface Parametrization Schemes (PILPS) (Shao *et al.*, 1995) indicate a wide range of runoff rates between different land models, leading to substantial differences in annual average evapotranspiration. The current lack of a physically based and adequately validated treatment for runoff may be the biggest single obstacle to achieving an SVAT adequate for climate modelling.

In cold regions and seasons, processes involving snow cover and soil freezing become important for surface energy and water balances. Depending on its depth, snow masks some part of the underlying surface (e.g., Robinson and Kukla, 1985 and Baker *et al.*, 1991). Atmospheric models provide snow to the surface in liquid water equivalent, depending on criteria for transition between rain and snow. The surface model must determine the snow's density, temperature, albedo, and spatial heterogeneities. Long-term data records of snow cover and other surface conditions allow validation of the parametrization of snow processes in climate models (e.g., Foster *et al.*, 1996; Yang *et al.*, 1996). New treatments of snow processes in climate models have been proposed by Loth *et al.* (1993) and Lynch-Stieglitz (1995).

Although our understanding of how to model land processes has advanced considerably, there has been a substantial lag in implementing this understanding in models of greenhouse gas response. Some models with detailed land process treatments are now being used for such studies, but it is not yet possible to assess how future climate projections are influenced by these treatments. Current intercomparisons of off-line models by PILPS indicate a considerable divergence of results between different land-surface models for the same prescribed forcing. Improved criteria for accuracy and validation will be needed before the current conceptual improvements of land-surface models can be translated into increased confidence in climate change projection.

4.4.2 *Questions of Spatial Heterogeneity*

One of the common criticisms of present treatments of land processes in climate models is their failure to include many of the essential aspects of sub-grid scale heterogeneity. Heterogeneity is manifested in the precipitation and radiative inputs and in modelling the land processes themselves. The issue of precipitation heterogeneity (Milly and Eagleson, 1988) has been addressed through a simple statistical model in several GCMs, as reviewed by Thomas and Henderson-Sellers (1991). In this approach, precipitation is assumed to occur over some fraction of a model grid square, within which the precipitation is assumed to have an exponential distribution of intensities. Simple runoff models having a non-linear dependence on precipitation are integrated across the distribution of precipitation intensities to provide a grid-square runoff. This approach has been generalised to interception (Shuttleworth, 1988). Pitman *et al.* (1990) have demonstrated, for prescribed atmospheric forcing, that the partitioning between ET and runoff in a land model can be very sensitive to the fractional area of precipitation. On the other hand, Dolman and Gregory (1992), who allow for atmospheric feedbacks with a 1-D model, find very little sensitivity of average ET to assumptions about the rainfall distribution, but that the partitioning between interception versus evaporation and hence short time-scale rates of ET, can vary widely with assumptions about the precipitation distribution.

All of the above approaches assume variability of precipitation but retain a homogeneous water storage. However spatial variability of storage can also have a substantial effect (Wood *et al.*, 1992). Entekhabi and Eagleson (1989) have developed separate statistical models for precipitation and soil moisture that have been tested in the GISS GCM (Johnson *et al.*, 1993). They find a large variation in ET and runoff depending on the assumed model. Eltahir and Bras (1993) have generalised Shuttleworth's approach to a statistical description of interception. In particular, they assume a statistical distribution of leaf water stores and find, for prescribed atmospheric forcing, that interception changes substantially but that little change from homogeneous conditions would be realised if the leaf water stores were assumed uniform.

Another issue is the inclusion of heterogeneities in land-surface cover (Avissar and Pielke, 1989; Seth *et al.*, 1994) and hence inferred parameters such as roughness. At least three scales need to be considered. On a very fine scale, canopy air interacts between different surfaces, and surface roughness lengths cannot be associated with individual surfaces. Koster and Suarez (1992) refer to this scale of heterogeneity as "mixture" and for a given roughness give a simple model for deriving a total canopy temperature and water vapour from individual elements. This type of heterogeneity needs to be provided as part of the overall land cover description.

On a somewhat coarser scale, surfaces independently interact with an overlying homogeneous atmosphere. Koster and Suarez (1992) refer to this as "mosaic", and Shuttleworth (1988) as "disordered" heterogeneity. This is the scale, for example, of typical agricultural fields or small stands of forest. Finally, at scales of at least a few tens of kilometres and certainly at the scales of GCM resolution, each surface has a different PBL overlaying it. Shuttleworth (1988) refers to these as "ordered" heterogeneity. Mesoscale circulations on this scale may substantially add to boundary layer fluxes (e.g., Pielke *et al.*, 1991). They may also modify processes of clouds and convection in ways not accounted for by grid box-mean information.

Another issue of heterogeneity is in the distribution of incident radiation. Sub-grid scale clouds, perhaps associated with precipitation, may be important. For example, if precipitation occurs over some fraction of a grid square and spatial variation of surface wetness is included, but radiation is assumed homogeneous, the estimated evaporation from the wet surfaces will be excessive. Besides clouds, surface slope can be a major cause of heterogeneity in the amounts of absorbed surface solar radiation (Avissar and Pielke, 1989), as well as determining further heterogeneities in clouds and precipitation. Barros and Lettenmaier (1994) review the role of orography in triggering clouds and precipitation.

Some aspects of heterogeneity can be treated with relatively straightforward approaches including how to determine an average over a wide range of surface types with different characteristics. Raupach and Finnigan (1995) have considered energetic constraints on areally averaged energy balances in heterogeneous regions. In some cases, particularly if surface characteristics do not differ strongly, surface parameters can be aggregated (Claussen, 1990; Blyth *et al.*, 1993). For instance, albedo can simply be linearly averaged, (when the underlying surface is otherwise fairly homogeneous) whereas for roughness length a more complex aggregation is necessary (Taylor, 1987; Mason, 1988; Claussen, 1990).

The calculation of regional surface fluxes is made difficult by the non-linear dependence between fluxes and driving mean gradients. For example, where parts of the area are snow-covered with surface temperatures held at the freezing point, the mean vertical temperature structure over the area may imply a downward heat flux but, because the transfer coefficients are larger in the snowfree, statically unstable part of the area, the actual mean heat flux can be upward. Moreover, if surface types strongly differ, "parameter aggregation" becomes unfeasible. For example, definition of an aggregated soil temperature diffusivity does not make sense if parts of the area consist of bare soil, where heat conduction is diffusive, and of open water, where heat can be advected horizontally as well as vertically. In these more complex landscapes, "flux aggregation" is preferred (Avissar and Pielke, 1989; Claussen, 1991). Flux aggregation implies the computation

of surface fluxes for each type in a grid box separately. Consequently, a regional surface flux is obtained by a linear average. Flux and parameter aggregation may be combined or use can be made of the intermediate approach of averaging exchange coefficients (Mahrt, 1987). The various aggregation methods require specification of a "blending height" where the aggregated information is matched to the overlying GCM grid squares (Claussen, 1990, 1991; Wood and Mason, 1991; Dolman, 1992; Blyth *et al.*, 1993). For ordered heterogeneity, this concept is less useful since the blending height would be above the surface layer. Practical application of these averaging procedures to global models will require high resolution global data sets on vegetation properties from future satellite sensors.

The aggregation or averaging methods may effectively treat the prescribed distribution of surface heterogeneities, but are not readily generalised to the dynamic interactions with sub-grid scale atmospheric inputs of water and radiation. Alternatively, a model grid box can be subdivided into sub-elements with both distinct surface characteristics and distinct atmospheric inputs. The "tile" or "mosaic" scheme by Koster and Suarez (1992) emphasises representation of the land heterogeneity. Different approaches may be needed to efficiently treat the heterogeneous distribution of atmospheric inputs; these atmospheric inputs may provide the largest overall departure in results from that inferred for homogeneous conditions. The possible importance of mesoscale circulation effects as a function of the spatial scale of individual homogeneous element on surface and boundary layer fluxes still needs to be assessed, although some preliminary work has been done (e.g., Zeng and Pielke, 1995) to parametrize mesoscale and turbulent fluxes in the boundary layer over inhomogeneous surfaces.

4.4.3 Sensitivity to Land-use Changes

Processes at the land surface influence the atmosphere through fluxes of heat and moisture into the PBL. These in turn affect atmospheric stability and the occurrence of precipitation and cloud radiative effects. Betts *et al.* (1993, 1994) have shown a close coupling between errors in a model's surface parametrization and its PBL, clouds, and moist convection. Sensitivity studies with GCMs have looked at the question of the possible effect of land cover modifications. These studies have indicated major surface influences on the atmospheric hydrological processes. This should perhaps not be too surprising since variations in ocean surface latent heat fluxes driven by small changes in oceanic surface temperatures have long been known to have large regional climate effects. However, the feedbacks of land-surface processes to the atmosphere are more complex and still far from understood.

Recent studies of the sensitivity of the Amazon Basin climate to a change from forest to grassland have been published, e.g., by Nobre *et al.* (1991), Henderson-Sellers *et al.* (1993) and Lean and Rowntree (1993). The response of climate to deforestation in the Amazon has been found to be sensitive to the specification of surface properties such as albedo (Dirmeyer and Shukla, 1994) and surface roughness (Sud *et al.*, 1995). A contrasting question is the effect of changing a semi-arid grassland such as the Sahel in Africa to a desert (e.g., Xue and Shukla, 1993). Reductions in absorbed solar energy due to higher surface albedos reduce ET but it is not clear how the change in precipitation should be related to change in ET. Many of the simulations have shown precipitation to reduce substantially more than ET, which implies a reduction in the convergence of moisture from the ocean.

The sum of various regional climate changes from land-use may contribute an overall effect on global climate. Estimates of global radiative forcing depending on albedo change are usually small compared to energy flux change from greenhouse gases. However, effects of changing land surface on the ratio of sensible to latent fluxes and on precipitation are not accounted for in such estimates and may be significant not only regionally but globally.

4.4.4 Assessment of the Current Status of Land Processes in Climate Change Simulations

- Land processes, driven by incoming solar radiation and precipitation, play an important role in the determination of near-surface climate, surface temperature, soil moisture, etc., and hence regional climates. Biases and uncertainties in the surface energy balance, and radiation and water budgets, are a significant source of error in simulations of regional climate.

- Detailed treatments of land processes are now available, replacing the previously used bucket models, and are being incorporated into numerous climate models. These treatments are substantial conceptual improvements, but it is premature to judge how they will modify or improve our confidence in climate change simulations. A number of issues must be resolved before their inclusion in climate models may be viewed as satisfactory. These include: the reasons for the current divergences in

answers between conceptually similar models in the PILPS model intercomparisons; the relative importance of various sub-grid scale processes, their interaction, and how to represent those that are important in the model; what level of observational detail is needed to prescribe the land properties needed by the models; and how these observations will be made available.

- Modelling of runoff has large uncertainty in global models, there are no convincing treatments of the scaling of the responsible processes over the many orders of magnitude involved, and in high latitudes of the effects of frozen soils. Global data on soils, topography, and water holding capacities at the relevant scales will be urgently needed to make progress on this issue.

- The question of the averaging of heterogeneous land surfaces has been clarified and reasonable approaches proposed.

- Future models should begin to address the issue of how land-surface characteristics might change with climate and CO_2 change and how important that is as a feedback. Sensitivity studies of the impact of doubling stomatal resistance indicates some climatic effects over forest regions comparable to those anticipated from global warming over the next century. How the integrated stomatal resistance for global vegetation might change with changing climate and CO_2 concentrations is largely unknown.

References

Avissar, R. and R.A. Pielke, 1989: A parametrization of heterogeneous land surfaces for atmospheric numerical models and its impact on regional meteorology. *Mon. Wea. Rev.,* **117**, 2113–2136.

Baker, D.G., R.H. Skaggs and D.L. Ruschy, 1991: Snow depth required to mask the underlying surface. *J. Appl. Met.,* **30**, 387–392.

Barkstrom, B.R., 1984: The Earth Radiation Budget Experiment (ERBE). *Bull. Am. Met. Soc.,* **65**, 1170–1185.

Barros, A.P. and D.P. Lettenmaier, 1994: Dynamic modeling of orographically induced precipitation. *Rev. Geophys.,* **32**, 265–284.

Battisti, D.S. and E.S. Sarachik, 1995: Understanding and Predicting ENSO. U.S. National Report to International Union of Geodesy and Geophysics 1991-1994, *Reviews of Geophysics,* Supplement, 1367–1376.

Betts, A.K., J.H. Ball and A.C.M. Belijaars, 1993: Comparison between the land surface response of the ECMWF model and the FIFE-1987 data. *Quart. J. R. Met. Soc.,* **199**, 975–1001.

Betts, A.K., J.H. Ball, A.C.M. Beljaars, M.J. Miller and P. Viterbo, 1994: Coupling between land-surface, boundary-layer parametrizations and rainfall on local and regional scales: lessons from the wet summer of 1993. Fifth Conference on Global Change Studies, Nashville, TN, Jan. 23-28, 1994, American Meteorological Society, Boston, MA.

Blyth, E.M., A.J. Dolman and N. Wood, 1993: Effective resistance to sensible and latent heat flux in heterogeneous terrain. *Quart. J. R. Met. Soc.,* **119**, 423–442.

Bond, G., W. Broecker, S. Johnsen, J. McManus, L. Labeyrie, J. Jouzel and G. Bonani, 1993: Correlations between climate records from North Atlantic sediments and Greenland ice. *Nature,* **365**, 143–147.

Bönning, C.W.and P. Herrmann, 1994: Annual cycle of poleward heat transport in the ocean: Results from high-resolution modeling of the north and equatorial Atlantic. *J. Phys. Oceanogr.,* **24**, 91–107.

Bony, S., J.P. Duvel and H. Le Treut, 1995: Observed dependence of the water vapor and clear sky greenhouse effect on sea surface temperature. Comparison with Climate Warming Experiments. *Clim. Dyn.* **11** (5), 307–320.

Boucher, O. and U. Lohmann, 1995: The sulfate-CCN-cloud albedo effect: A sensitivity study with two general circulation models. *Tellus.,* **47B**, 281–300.

Boucher, O., H. Le Treut, and M.B. Baker, 1995: Precipitation and radiation modelling in a GCM: Introduction of cloud microphysical processes. *J. Geophys. Res.,* **100**, 16395–16414.

Bryan, F., 1986: High latitude salinity effects and interhemispheric thermohaline circulations. *Nature,* **305**, 301–304.

Bryden, H.L., 1993: Ocean heat transport across 24N latitude. In: *Interactions between global climate subsystems: the legacy of Hann.* Geophysical Monograph #75, American Geophysical Union, 65–75.

Bryden, H.L., D.H. Roemmich and J.A. Church, 1991: Ocean heat transport across 24N in the Pacific. *Deep Sea Res.,* **38**, 297–324.

Cess, R.D., G.L. Potter, J.P. Blanchet, G.J. Boer, A.D. Del Genio, M. Deque, V. Dymnikov, V. Galin, W.L. Gates, S.J. Ghan, J.T. Kiehl, A. Lacis, H. Le Treut, Z.-X. Li, X.-Z. Liang, B.J. McAvaney, V.P. Meleshko, J.F.B. Mitchell, J.-J. Morcrette, D.A. Randall, L. Rikus, E. Roeckner, J.F. Royer, U. Schlese, D.A. Sheinin, A. Slingo, A.P. Sokolov, K.E. Taylor, W.M. Washington, R.T. Wetherald, I. Yagai and M.-H. Zhang, 1990: Intercomparison and interpretation of climate feedback processes in 19 atmospheric general circulation models. *J. Geophys. Res.,* **95**, 16601–16615.

Cess, R.D., M.H. Zhang, P. Minnis, L. Corsetti, E.G. Dutton, B.W. Forgan, D.P. Garber, W.L. Gates, J.J. Hack, E.F. Harrison, X. Jing, J.T. Kiehl, C.N. Long, J.-J. Morcrette, G.L. Potter, V. Ramanathan, B. Subasilar, C.H. Whitlock, D.F. Young and Y. Zhou, 1995: Absorption of solar radiation by clouds: Observations versus models. *Science,* **267**, 496–498.

Cess, R.D., M. H. Zhang, G. L. Potter, V. Alekseev, H. W. Barker, E. Cohen-Solal, R. A. Colman, D. A. Dazlich, A. D. Del Genio, M. R. Dix, V. Dymnikov, M. Esch, L. D. Fowler, J. R. Fraser, V. Galin, W. L. Gates, J. J. Hack, W. J. Ingram, J. T. Kiehl, H. Le Treut, K. K.-W. Lo, B. J. McAvaney, V. P. Meleshko, J.-J. Morcrette, D. A. Randall, E. Roeckner, J.-F. Royer, M. E. Schlesinger, P. V. Sporyshev, B. Timbal, E. M. Volodin, K. E. Taylor, W. Wang and R. T. Wetherald, 1996: Cloud feedback in atmospheric general circulation models: An update. Submitted to *J. Geophys. Res.*

Charlson, R.J., J.E. Lovelock, M.O. Andreae and S.G. Warren, 1987: Oceanic phytoplankton, atmospheric sulphur, cloud albedo and climate. *Nature,* **326**, 655–661.

Chou, M.-D., 1995: Coolness in the tropical Pacific during an El Niño episode. *J. Clim.,* **7**, 1684–1692.

Claussen, M., 1990: Area averaging of surface fluxes in a neutrally stratified, horizontally inhomogeneous atmospheric boundary layer. *Atmos. Env.,* **24a**, 1349–1360.

Claussen, M., 1991: Estimation of areally-averaged surface fluxes. *Boundary-layer Meteor.,* **54**, 387–410.

Cox, M., 1984: A primitive equation three-dimensional model of the ocean. GFDL Ocean Group Tech. Rept. No. 1., GFDL.

Curry, J.A. and E. Ebert, 1990: Sensitivity of the thickness of Arctic sea ice to the optical properties of clouds. *Ann. Glaciol.,* **14**, 43–46.

Curry, J.A. and E. Ebert, 1992: Annual cycle of radiative fluxes over the Arctic Ocean: Sensitivity to cloud optical properties. *J. Climate,* **5**, 1267–1280.

Danabasoglu, G., J.C. McWilliams and P.R. Gent, 1994: The role of mesoscale tracer transports in the global ocean circulation. *Science,* **264**, 1123–1126.

Deardorff, J.W., 1980: Cloud-top entrainment instability. *J. Atmos. Sci.,* **37**, 131–147.

Del Genio, A.D., W. Kovari Jr. and M.-S. Yao, 1994: Climatic implications of the seasonal variation of upper troposhere water vapor. *Geophys. Res. Lett.* **21**, 2701–2704.

Del Genio, A.D., M.-S. Yao, W. Kovari and K.K.-W. Lo, 1995: A prognostic cloud water parameterization for global climate models. *J. Climate,* in press.

Dirmeyer, P.A. and J. Shukla, 1994: Albedo as a modulator of climate response to tropical deforestation. *J. Geophys. Res.,* **99**, 20863–20877.

Dolman, A.J., 1992: A note on areally-averaged evaporation and the value of the effective surface conductance. *J. Hydrology,* **138**, 583–589.

Dolman, A.J. and D. Gregory, 1992: The parametrization of rainfall interception in GCMs. *Quart. J. R. Met. Soc.,* **188**, 455–467.

Ebert, E.E.and J.A. Curry, 1992: A parameterization of ice cloud optical properties for climate models. *J. Geophys. Res.,* **97**, 3831–3836.

Eltahir, E.A.B. and R.L. Bras, 1993: A description of rainfall interception over large areas. *J. Climate*, **6**, 1002–1008.

Entekhabi, D. and P.S. Eagleson, 1989: Land-surface hydrology parametrization for atmospheric general circulation models including subgrid-scale spatial variability. *J. Climate*, **2**, 816–831.

Feely, R.A., R.H. Gammon, B.A. Taft, P.E. Pullen, L.S. Waterman, T. J. Conway, J.F. Gendron and D.P. Wisegarver, 1987: Distribution of chemical tracers in the eastern equatorial Pacific during and after the 1982-1983 El Niño/Southern Oscillation event. *J. Geophys. Res.*, **92**, 6545–6558.

Flato, G.M. and W.D. Hibler III, 1992: Modeling pack ice as a cavitating fluid. *J. Phys. Oceanogr.*, **22**, 626–651.

Foster, J., G. Liston, R. Koster, R. Essery, H. Behr, L. Dümenil, D. Verseghy, S. Thompson, D. Pollard and J. Cohen, 1996: Snow cover and snow mass intercomparisons from general circulation models and remotely-sensed data sets. *J. Climate*, in press.

Fowler, L.D. and D.A. Randall, 1996: Liquid and ice cloud microphysics in the CSU general circulation model. Part 2: Simulation of the Earth's radiation budget. *J. Climate*, in press.

Fowler, L.D., D.A. Randall and S.A. Rutledge, 1996: Liquid and ice cloud microphysics in the CSU general circulation model. Part 1: Model description and simulated microphysical processes. *J. Climate*, in press.

Fu, Q. and K.N. Liou, 1993: Parameterization of the radiative properties of cirrus clouds. *J. Atmos. Sci.*, **50**, 2008–2025.

Fu, R., A.D. Del Genio, W.B. Rossow and W.T. Liu, 1992: Cirrus-cloud thermostat for tropical sea surface temperatures tested using satellite data. *Nature*, **358**, 394–397.

Fu, R., A. D. Del Genio and W. B. Rossow, 1994: Influence of ocean surface conditions on atmospheric vertical thermodynamic structure and deep convection. *J. Climate*, **7**, 1092–1108.

Fu, R., W.T. Liu and R.E. Dickinson, 1996: Response of tropical clouds to the interannual variation of sea surface temperature. *J. Climate*, in press.

Garratt, J.R., 1994: Incoming shortwave fluxes at the surface – a comparison of GCM results with observations. *J. Climate*, **7**, 72–80.

Gates, W.L., 1992: AMIP: The Atmospheric Model Intercomparison Project. *Bull. Am. Met. Soc.*, **73**, 1962–1970.

Gent, P.R. and J.C. McWilliams, 1990: Isopycnal mixing in ocean circulation models. *J. Phys. Oceanogr.*, **20**, 150–155.

Ghan, S.J., C.C. Chuang and J.E. Penner, 1993: A parametrization of cloud droplet nucleation. Part I: Single aerosol type. *Atmos. Res.*, **30**, 198–221.

Ghan, S.J., C.C. Chuang, R.C. Easter and J.E. Penner, 1995: A parametrization of cloud droplet nucleation. Part II: Multiple aerosol types. *Atmos. Res.*, **36**, 39–54.

Ghil, M. and J. McWilliams, 1994: Workshop tackles oceanic thermohaline circulation. *EOS, Transactions, American Geophysical Union*, **75**, 42, 493–498.

Gleckler, P.J., D.A. Randall, G. Boer, R. Colman, M. Dix, V. Galin, M. Helfand, J. Kiehl, A. Kitoh, W. Lau, X.-Z. Liang, V. Lykossov, B. McAvaney, K. Miyakoda, S. Planton and W. Stern, 1995: Interpretation of ocean energy transports implied by atmospheric general circulation models. *Geophy. Res. Lett.*, **22**, 791–794.

Graham, N.E., 1995: Simulation of recent global temperature trends. *Science*, **267**, 666–671.

Graham, N.E., T.P. Barnett, R. Wilde, M. Ponater and S. Shubert, 1994: On the roles of tropical and midlatitude SST in forcing interannual to interdecadal variability in the winter Northern Hemisphere circulation. *J. Climate*, **7**, 1416–1441.

Graham, N.E., T.P. Barnett, M.A. Cane and S.E. Zebiak, 1995: Greenhouse warming and its possible effect on El Niño. *Geophys.Res. Lett.*, submitted.

Greenwald, T.J., G.L. Stephens, T.H. Vonder Haar and D.L. Jackson, 1993: A physical retrieval of cloud liquid water over the global oceans using Special Sensor Microwave/Imager (SSM/I) observations. *J. Geophys. Res.*, **98**, 18471–18488.

Groetzner, A., R. Sausen and M. Claussen, 1994: The impact of sub-grid inhomogeneities on the performance of the atmospheric general circulation model ECHAM. *Clim.Dyn.* in press.

Hansen, J., M. Sato and R. Ruedy, 1995: Long-term changes of the diurnal temperature cycle: Implications about mechanisms of global climate change. *Atmos. Res.*, **37**, 175–209.

Hanson, H.P., 1991: Marine stratocumulus climatologies. *Int. J. Climatol.*, **11**, 147–164.

Harshvardhan, D.A. Randall, T.G. Corsetti and D.A. Dazlich, 1989: Earth radiation budget and cloudiness simulations with a general circulation model. *J. Atmos. Sci.*, **46**, 1922–1942.

Hartmann, D.L. and M.L. Michaelson, 1993: Large-scale effects on the regulation of tropical sea surface temperature. *J. Climate*, **6**, 2049–2062.

Hayasaka, T., N. Kikuchi and M. Tanaka, 1995: Absorption of solar radiation by stratocumulus clouds: Aircraft measurements and theoretical calculations. *J. Appl. Met.*, **34**, 1047–1055.

Henderson-Sellers, A., R.E. Dickinson, T.B. Durbidge, P.J. Kennedy, K. McGuffie and A.J. Pitman, 1993: Tropical deforestation: Modeling local- to regional-scale climate change. *J. Geophys. Res.*, **98**, 7289–7315.

Henderson-Sellers, A., K. McGuffie and C. Gross, 1995: Sensitivity of global climate model simulations to increased stomatal resistance and CO_2 increases. *J.Climate*, **8**, 1738–1756.

Herman, G. and R. Goody, 1976: Formation and persistence of summertime arctic stratus clouds. *J. Atmos. Sci.*, **33**, 1537–1553.

Heymsfield, A.J. and L.J. Donner, 1990: A scheme for parameterizing ice-cloud water content in general circulation models. *J. Atmos. Sci.*, **47**, 1865–1877.

Ingram, W.J., C.A. Wilson and J.F.B. Mitchell, 1989: Modeling climate change: An assessment of sea ice and surface albedo feedbacks. *J. Geophys. Res.*, **94**, 8609–8622.

Institute of Hydrology, 1994: *Results of Project ABRACOS*, an 18-page booklet.

IPCC, (Intergovernmental Panel on Climate Change) 1990: *Climate Change: The IPCC Scientific Assessment*, J.T. Houghton, G.J. Jenkins and J.J. Ephraums (eds.). Cambridge University Press, Cambridge, UK, 365 pp.

IPCC, 1992: *Climate Change 1992: The Supplementary Report to the IPCC Scientific Assessment*, J.T. Houghton, B.A. Callander and S.K. Varney (eds.). Cambridge University Press, Cambridge, UK.

IPCC, 1994: *Climate Change 1994: Radiative Forcing of Climate Change a*nd an Evaluation of the IPCC 1993 Emission Scenarios, J.T. Houghton, L.G. Meira Filho, J. Bruce, Hoesung Lee, B. A. Callander, E. Haites, N. Harris and K. Maskell (eds). Cambridge University Press, Cambridge, UK.

Johnson, K.D., D. Entekhabi and P.S. Eagleson, 1993: The implementation and validation of improved land-surface hydrology in an atmospheric General Circulation Model. *J.Climate*, **6**, 1009–1026.

Jones, A., D.L. Roberts and A. Slingo, 1994: A climate model study of indirect radiative forcing by anthropogenic sulphate aerosols. *Nature*, **370**, 450–453.

Jones, H. and J. Marshall, 1993: Convection with rotation in a neutral ocean: a study of open-ocean deep convection. *J. Phys. Oceanogr.*, **23**, 1009–1039.

Kattsov, V.M., V.P. Meleshko, A.P. Sokolov and V.A. Lubanskaya, 1993: Role of sea ice in maintenance of thermal regime and atmospheric circulation of the Northern Hemisphere. *Meteor. Gidrologia*, **12**, 5–24.

Kelly, K.K., A.F. Tuck and T. Davies, 1991: Wintertime asymmetry of upper tropospheric water between the Northern and Southern Hemispheres. *Nature*, **353**, 244–247.

Kiehl, J.T., 1994: Sensitivity of a GCM climate simulation to differences in continental versus maritime cloud drop size. *J. Geophys. Res.*, **99**, 23107–23116.

Killworth, P.D., 1983: Deep convection in the world ocean. *Revs. Geophys.* **21**, 1–26.

Klein, S.A. and D.L. Hartmann, 1993: The seasonal cycle of low stratiform clouds. *J. Climate*, **6**, 1587–1606.

Knutson, T.R. and S. Manabe, 1994: impact of increased CO_2 on simulated ENSO-like phenomena. *Geophys. Res. Lett*, **21**, 2295–2298.

Koberle, C. and S.G.H. Philander, 1994: On the processes that control seasonal variations of sea surface temperature in the tropical Pacific Ocean. *Tellus*, **46A**, 481–496.

Koster, R.D. and M.J. Suarez, 1992: Modelling the land surface boundary in climate models as a composite of independent vegetation stands. *J. Geophys. Res.*, **97**, 2697–2715.

Koster, R.D. and M.J. Suarez, 1994: The components of a 'SVAT' scheme and their effects on a GCM's hydrological cycle. *Advan. Water Res.*, **17**, 61–78.

Kristjánsson, J.E., 1991: Cloud parameterization at different horizontal resolutions. *Quart. J. R. Met. Soc.*, **117**, 1255–1280.

Kumar, A., A. Leetmaa and M. Ji, 1994: Simulation of atmospheric variability induced by sea surface temperatures and implications for global warming. *Science,* **266**, 632–634.

Kuo, H. and W.H. Schubert, 1988: Stability of cloud-topped boundary layers. *Quart. J. R. Met. Soc.*, **114**, 887–916.

Kvamstø, N.G., 1991: An investigation of relations between stratiform fractional cloud cover and other meteorological parameters in numerical weather prediction models. *J. Appl.Met.*, **30**, 200–216.

Lacis, A.A. and M. Sato, 1993: GCM feedback assessment with a 2-D radiative-convective-dynamic equilibrium model. Preprints, *4th Symposium on Global Change Studies*, American Meteorological Society, Anaheim, CA 198–202.

Large, W.G., J.C. McWilliams and S.C. Doney, 1995: Oceanic vertical mixing: A review and a model with nonlocal boundary layer parameterizations. *Revs. Geophys.*, **32**, 363–403.

Lau, K.M., C.H. Sui and W.K. Tau, 1993: A preliminary study of the tropical water cycle and its sensitivity to surface warming. *Bull. Am. Met. Soc.*, **74**, 1313–1321.

Lau, K.M., C.-H. Sui, M.-D. Chou and W.K. Tao, 1994: An inquiry into the cirrus-cloud thermostat effect for tropical sea surface temperature. *Geophys. Res. Lett.*, **21**, 12, 1157–1160.

Lau, N.C. and M.J. Nath, 1994: A modeling study of the relative roles of tropical and extratropical SST in the variability of the global atmosphere-ocean system. *J. Climate*, **7**, 1184–1207.

Le Treut, H. and Z.-X. Li, 1988: Using Meteosat data to validate a prognostic cloud generation scheme. *Atmos. Res.*, **21**, 273–292.

Lean, J. and P.R. Rowntree, 1993: A GCM simulation of the impact of Amazonian deforestation on climate using an improved canopy representation. *Quart. J. R. Met. Soc.*, **199**, 509–530.

Ledley, T.S., 1993: Variations in snow on sea ice: A mechanism for producing climate variations. *J. Geophys. Res.*, **98**, 10401–10410.

Ledwell, J.R., A.J. Watson and C.S. Law, 1993: Evidence for slow mixing across the pycnocline from an open-ocean tracer-release experiment. *Nature*, **364**, 701–703.

Legg, S. and J. Marshall, 1993: A heton model of the spreading phase of open-ocean deep convection. *J. Phys. Oceanogr.*, **23**, 1040–1056.

Li, Z.-X. and H. Le Treut, 1992: Cloud-radiation feedbacks in a general circulation model and their dependence on cloud modelling assumptions. *Clim. Dyn.*, **7**, 133–139.

Li, Z., C.H. Whitlock and T.P. Charlock, 1995a: Assessment of the global monthly mean surface insolation estimated from satellite measurements using Global Energy Balance Archive data. *J. Climate*, **8**, 315–328.

Li, Z., H.W. Barker and L. Moreau, 1995b: The variable effect of clouds on atmospheric absorption of solar radiation. *Nature*, **376**, 486–490.

Lilly, D.K., 1968: Models of cloud-topped mixed layers under a strong inversion. *Quart. J. R. Met. Soc.*, **94**, 292–309.

Lin, B. and W.B. Rossow, 1994: Observations of cloud liquid water path over oceans: Optical and microwave remote sensing methods. *J. Geophys. Res.*, **99**, 20909–20927.

Lindzen, R.S., 1990: Some coolness concerning global warming. *Bull. Am. Met. Soc.*, **71**, 288–299.

Lindzen, R.S. and S. Nigam, 1987: On the role of sea surface temperature gradients in forcing low-level winds and convergence in the tropics. *J. Atmos. Sci.*, **44**, 2418–2436.

Loth, B., H.-F. Graf and J.M. Oberhuber, 1993: Snow cover model for global climate simulations. *J. Geophys. Res.*, **98**, 10451–10464.

Lynch-Stieglitz, M., 1995: The development and validation of a simple snow model for the GISS GCM. *J. Climate*, **7**, 1842–1855.

MacVean, M.K. and P.J. Mason, 1990: Cloud-top entrainment instability through small-scale mixing and its parametrization in numerical models. *J. Atmos. Sci.*, **47**, 1012–1030.

Mahrt, L., 1987: Grid-averaged surface fluxes. *Mon. Wea. Rev.*, **115**, 1550–1560.

Maier-Reimer, E., U. Mikolajewicz and K. Hasselmann, 1993: Mean circulation of the Hamburg LSG OGCM and its sensitivity to the thermohaline surface forcing. *J. Phys. Oceanogr.*, **23**, 731–757.

Manabe, S. and R.T. Wetherald, 1967: Thermal equilibrium of the atmosphere with a given distribution of relative humidity. *J. Atmos. Sci.*, **24**, 241–259.

Manabe, S. and R.J. Stouffer, 1988: Two stable equilibria of a coupled ocean-atmosphere model. *J. Climate*, **1**, 841–866.

Manabe, S. and R.J. Stouffer, 1994: Multiple century response of a coupled ocean-atmosphere model to an increase of atmospheric carbon dioxide. *J. Climate*, **7**, 5–23.

Marotzke, J., 1988: Instabilities and multiple steady states of the thermohaline circulation. In: *Oceanic Circulation Models: Combining Data and Dynamics,* D.L.T. Anderson and J. Willebrand (eds.), NATO ASI Series, Kluwer, 501–511.

Mason, P.J., 1988: The formation of areally-averaged roughness lengths. *Quart. J. R. Met. Soc.*, **114**, 399–420.

Mechoso, C.R., & 19 authors, 1995: The seasonal cycle over the tropical Pacific in coupled ocean-atmosphere general circulation models. *Mon. Wea. Rev.* **123**, 2825–2838.

MEDOC Group, 1970: Observations of formation of deep water in the Mediterranean Sea. *Nature*, **227**, 1037–1040.

Meleshko, V.P., B.E.Sneerov, A.R.Sokolov and V.M.Kattsov, 1991: Sea ice anomaly impact on surface heat fluxes and atmospheric circulation as evaluated by the MGO GCM. *Report of the Workshop on Polar Radiation Fluxes and Sea-Ice Modeling,* Bremerhaven, Germany, November 1990, WMO/TD No. 442.

Mikolajewicz, U. and E. Maier-Reimer, 1990: Internal secular variability in an ocean general circulation model. *Clim. Dyn.*, **4**, 145–156.

Miller, R.L. and A.D. Del Genio, 1994: Tropical cloud feedbacks and natural variability of climate. *J. Climate*, **7**, 1388–1402.

Milly, P.C.D. and P.S. Eagleson, 1988: Effect of storm scale on surface runoff volume. *Water Resour. Res.*, **24**, 620–624.

Mitchell, J.F.B. and W.J. Ingram, 1992: Carbon dioxide and climate: Mechanisms of changes in cloud. *J. Climate*, **5**, 5–21.

Mitchell, J.F.B., R.A. Davis, W.J. Ingram and C.A. Senior, 1995: On surface temperature, greenhouse gases and aerosols: models and observations. *J. Climate*, **8**, 2364–2386.

Njoku, E.G. and L. Swanson, 1983: Global measurements of sea surface temperature, wind speed, and atmospheric water content from satellite microwave radiometry. *Mon. Wea. Rev.*, **111**, 1977–1987.

Nobre, C.A., P.J. Sellers and J. Shukla, 1991: Amazon deforestation and regional climate change. *J. Climate*, **4**, 957–988.

Ose, T., 1993: An examination of the effects of explicit cloud water in the UCLA GCM. *J. Meteor. Soc. Japan*, **71**, 93–109.

Palmer, T., 1993: A nonlinear dynamical perspective on climate change. *Weather*, **48**, 314–325.

Parkinson, C. and Cavalieri, 1989: Arctic sea ice 1973-1997: Seasonal, regional and interannual variability. *J. Geophys. Res.*, **94**, C10, 14499–14523.

Pielke, R.A., G.A. Dalu, J.S. Snook, T.J. Lee and T.G.F. Kittel, 1991: Nonlinear influence of mesoscale land use on weather and climate. *J. Climate*, **4**, 1053–1069.

Pierrehumbert, R.T., 1995. Thermostats, radiator fins and the local runaway greenhouse. *J. Atmos. Sci.*, **52**, 1784–1806.

Pierrehumbert, R.T. and H. Yang, 1993: Global chaotic mixing on isentropic surfaces. *J.Atmos. Sci.*, **50**, 2462–2480.

Pilewski, P. and F.P.J. Valero, 1995: Direct observations of excess solar absorption by clouds. *Science*, **267**, 1626–1629.

Pinker, R.T., R. Frouin and A. Li, 1995: A review of satellite methods to derive surface shortwave irradiance. *Remote Sens. Env.*, **51**, 108–124.

Pitman, A.J., A. Henderson-Sellers and Z.-L.Yang, 1990: Sensitivity of regional climates to localized precipitation in global models. *Nature*, **346**, 734–737.

Pollard, D. and S.L. Thompson, 1995: The effect of doubling stomatal resistance in a global climate model. *Global Planet. Change,* **10**, 1–4.

Prabhakara, C., I. Wang, A.T.C. Chang and P. Gloersen, 1983: A statistical examination of Nimbus-7 SMMR data and remote sensing of sea surface temperature, liquid water content in the atmosphere and surface wind speed. *J. Clim. Appl. Meteor.*, **22**, 2023–2037.

Ramanathan, V. and W. Collins, 1991: Thermodynamic regulation of ocean warming by cirrus clouds deduced from observations of the 1987 El Niño. *Nature*, **351**, 27–32.

Ramanathan, V., R.D. Cess, E.F. Harrison, P. Minnis, B.R. Barkstrom, E. Ahmad and D. Hartmann, 1989: Cloud-radiative forcing and climate: Results from the Earth Radiation Budget Experiment. *Science*, **243**, 57–63.

Ramanathan, V., B. Subasilar, G.J. Zhang, W. Conant, R.D. Cess, J.T. Kiehl, H. Grassl and L. Shi, 1995: Warm pool heat budget and shortwave cloud forcing: A missing physics? *Science*, **267**, 499–502.

Randall, D.A., 1980: Conditional instability of the first kind, upside-down. *J. Atmos. Sci.*, **37**, 125–130.

Raschke, E., J. Schmetz, J. Heitzenberg, R. Kandel and R. Saunders, 1990: The International Cirrus Experiment (ICE). A joint European effort. *ESA Journal*, **14**, 193–199.

Raupach, M. and Finnigan, 1995: Scale issues in boundary-layer meteorology: Surface energy balances in heterogeneous terrain. *Hydrol. Processes*, in press.

Riser, S.C., 1995: Geothermal heating and the large scale ocean circulation. Submitted to *Rev. Geophys.*

Rennó, N.O., K.A. Emmanuel and P.H. Stone, 1994: Radiative-convective model with an explicit hydrologic cycle 1. Formulation and sensitivity to model parameters. *J. Geophys. Res.*, **99**, 14429–14441.

Robertson, A.W., C.-C.Ma, C.R. Mechoso and M. Ghil, 1995: Simulation of the tropical-Pacific climate with a coupled ocean-atmosphere general circulation model. Part I: The seasonal cycle. *J. Climate*, **8**, 1178–1198.

Robinson, D.A. and G. Kukla, 1985: Maximum surface albedo of seasonally snow-covered lands in the Northern Hemisphere. *J. Climate*, **24**, 402–411.

Roeckner, E., U. Schlese, J. Biercamp and P. Loewe, 1987: Cloud optical depth feedbacks and climate modelling. *Nature*, **329**, 138–140.

Roeckner, E., M. Rieland and E. Keup, 1990: Modelling of clouds and radiation in the ECHAM model. *Proceedings of ECMWF/WCRP Workshop on Clouds, Radiation, and the Hydrologic Cycle*, ECMWF, Reading, UK, pp. 199–222.

Ryan, B.F., 1996: On global variations of precipitating layer clouds. *Bull. Amer. Met. Soc.*, in press.

Schmid, H.P. and B. Bünzli, 1995: The influence of surface texture on the effective roughness length. *Quart. J. R. Met. Soc.* **121**, 1–22.

Schmitt, R.W., 1994: The ocean freshwater cycle. *JSC Ocean Observing System Development Panel Background Report Number 4*, Texas A&M University, College Station, 40pp.

Sellers, P.J., F.G. Hall, G.Asrar, D.E. Strebel and R.E. Murphy, 1992: An overview of the First International Satellite Land Surface Climatology Project (ISLSCP) Field Experiment (FIFE). *J. Geophys. Res.*, **97**, 18345–18371.

Semtner, A.J. Jr. and R.M. Chervin, 1988: A simulation of the global ocean circulation with resolved eddies. *J. Phys. Oceanogr.*, **6**, 379–389.

Semtner, A.J. Jr. and R.M. Chervin, 1992: Ocean general circulation from a global eddy-resolving simulation. *J. Geophys. Res.*, **97**, 5493–5550.

Senior, C.A. and J.F.B. Mitchell, 1993: Carbon dioxide and climate: The impact of cloud parametrization. *J. Climate*, **6**, 393 – 418.

Senior, C.A. and J.F.B. Mitchell, 1995: Cloud feedbacks in the UKMO unified model. In: *Climate Sensitivity to Radiative Perturbations*, Le Treut (ed.) NATO I 34, Springer-Verlag.

Seth, A., F. Giorgi and R.E. Dickinson, 1994: Simulating fluxes from heterogeneous land surfaces: Explicit subgrid method employing the biosphere-atmosphere transfer scheme (BATS). *J. Geophys. Res.*, **99**, D9, 18651–18667.

Shao, Q. and D.A. Randall, 1996: Closed mesoscale cellular convection driven by cloud-top radiative cooling. *J. Atmos. Sci.* (In press)

Shao, Y., R.D. Anne, A. Henderson-Sellers, P. Irannejad, P.

Thornton, X. Liang, T.-H. Chen, C. Ciret, C. Desborough, O. Balachova, A. Haxeltine and A. Ducharne, 1995: *Soil Moisture Simulation*. A report of the RICE and PILPS workshop, Climatic Impacts Centre, Macquarie University, Sydney, Australia, November 1994.

Shuttleworth, W.J., 1988: Evaporation from Amazonian rain forest. *Proc. Roy. Soc. London*, **B233**, 321–346.

Siems, S.T., C.S. Bretherton, M.B. Baker, S. Shy and R.T. Breidenthal, 1990: Buoyancy reversal and cloudtop entrainment instability. *Quart. J. R. Met. Soc.*, **116**, 705–739.

Slingo, A., 1990: Sensitivity of the Earth's radiation budget to changes in low clouds. *Nature*, **343**, 49–51.

Slingo, A. and J.M. Slingo, 1988: The response of a general circulation model to cloud longwave radiative forcing. I: Introduction and initial experiments. *Quart. J. R. Met. Soc.*, **114**, 1027 – 1062.

Smith, I.N., 1994: A GCM simulation of global climate trends: 1950-1988. *J. Climate*, **7**, 732–744.

Smith, R.N.B., 1990: A scheme for predicting layer clouds and their water content in a general circulation model. *Quart. J. R. Met. Soc.*, **116**, 435–460.

Soden, B.J. and F.P. Bretherton, 1994: Evaluation of water vapour distribution in general circulation models using satellite observations. *J. Geophys. Res.*, **99**, 1187–1210.

Soden, B.J. and L.J. Donner, 1994: Evaluation of a GCM cirrus parameterization using satellite observations. *J. Geophys. Res.*, **99**, 14401–14413.

Soden, B.J. and R. Fu, 1995: A satellite analysis of deep convection, upper-tropospheric humidity, and the greenhouse effect. *J. Climate*, **8**, 2333–2351.

Somerville, R.C.J. and L.A. Remer, 1984: Cloud optical thickness feedbacks in the CO_2 climate problem. *J. Geophys. Res.*, **89**, 9668–9672.

Stephens, G.L., 1995: How much solar radiation do clouds absorb? *Science*, in press.

Stephens, G.L. and S.-C. Tsay, 1990: On the cloud absorption anomaly. *Quart. J. R. Met. Soc.* **116**, 671–704.

Sterl, A. and A. Kattenberg, 1994: Embedding a mixed layer model into an ocean general circulation model of the Atlantic: The importance of surface mixing for heat flux and temperature. *J. Geophys. Res.*, **99**, 14139–14157.

Stommel, H., 1961: Thermohaline convection with two stable regimes of flow. *Tellus*, **13**, 224–230.

Sud, Y.C., G.K. Walker, J.-H. Kim, G.E. Liston, P.J. Sellers and W.K.-M. Lau, 1995: *A simulation study fo the importance of surface roughness in tropical deforestation. Special Volume on TROPMET-94*, Indian Institute of Tropical Meteorology, Pune, India, in press.

Sun, D.-Z. and I. Held, 1995: A comparison of modeled and observed relationships between interannual variations of water vapor and temperature. *J. Climate*, in press.

Sundqvist, H., 1978: A parametrization scheme for non-convective condensation including prediction of cloud water content. *Quart. J. R. Met. Soc.*, **104**, 677–690.

Sundqvist, H., E. Berge and J.E. Kristjánsson, 1989: Condensation and cloud parameterization studies with a mesoscale numerical weather prediction model. *Mon. Wea. Rev.,* **117**, 1641–1657.

Taylor, K.E. and S.J. Ghan, 1992: An analysis of cloud liquid water feedback and global climate sensitivity in a general circulation model. *J. Climate,* **5**, 907–919.

Taylor, P.A., 1987: Comments and further analysis on effective roughness lengths for use in numerical three-dimensional models. *Boundary-Layer Meteor.,* **39**, 403–418.

Thomas, G. and A. Henderson-Sellers, 1991: An evaluation of proposed representations of subgrid hydrologic processes in climate models. *J. Climate,* **4**, 898–910.

Tiedtke, M., 1993: Representation of clouds in large-scale models. *Mon. Wea. Rev.,* **121**, 3040–3061.

Trenberth, K.E. and A. Solomon, 1994: The global heat balance: heat transports in the atmosphere and ocean. *Clim. Dyn.,* **10**, 107–134.

Tselioudis, G., W.B. Rossow and D. Rind, 1992: Global patterns of cloud optical thickness variation with temperature. *J. Climate,* **5**, 1484–1495.

Twomey, S.A., M. Piepgrass and T.L. Wolfe, 1984: An assessment of the impact of pollution on global cloud albedo. *Tellus,* **36B**, 356–366.

Vogelmann, A.M. and T.P. Ackermann, 1996: Relating cirrus cloud properties to observed fluxes – a critical assessment. *J. Atmos. Sci.,* in press.

Wallace, J.M., 1992: Effect of deep convection on the regulation of tropical sea surface temperature. *Nature,* **357**, 230–231.

Wallace, J.M., 1995: Natural and forced variability in the climate record. In *The Natural Variability of the Climate System on Decade-to-Century Time Scales,* D.G. Martinson, K. Bryan, M. Ghil, M.M. Hall, T.R. Karl, E.S. Sarachik, S. Sorooshian and L.D. Talley (eds.). National Academy Press, Washington DC, in press.

Ward, D.M., 1995: Comparison of the surface solar radiation budget derived from satellite data with that simulated by the NCAR CCM2. *J. Climate,* **8**, 2824–2842.

Weaver, A.J. and E.S. Sarachik, 1991: Evidence for decadal variability in an ocean general circulation model: An advective mechanism. *Atmosphere-Ocean,* **29**, 197–231.

Weaver, A.J., E.S. Sarachik and J. Marotzke, 1991: Freshwater flux forcing of decadal/interdecadal oceanic variability. *Nature,* **353**, 836–838.

Weaver, A.J., J. Marotzke, P.F. Cummins and E.S. Sarachik, 1993: Stability and variability of the thermohaline circulation. *J. Phys. Oceanogr.,* **23**, 39–60.

Wetherald, R.T. and S. Manabe, 1988. Cloud feedback processes in a general circulation model. *J. Atmos. Sci.,* **45**, 1397–1415.

Whitlock, C.H., T.P. Charlock, W.F. Staylor, R.T. Pinker, I. Laszlo, A. Ohmura, H. Gilgen, T. Konzelmann, R.C. DiPasquale, C.D. Moats, S.R. LeCroy and N.A. Ritchey, 1995: First global WCRP shortwave surface radiation budget data set. *Bull. Am. Met. Soc.,* **76**, 905–922.

Winton, M., 1993: Deep decoupling oscillations of the oceanic thermohaline circulation. In: *Ice in the Climate System,* W.R. Peltier (ed.), NATO ASI Series, I 12, Springer-Verlag.

Winton, M. and E.S. Sarachik, 1993: Thermohaline oscillations of an oceanic general circulation model induced by strong steady salinity forcing. *J. Phys. Oceanogr.,* **23**, 1389–1410.

Wood, E.F., D.P. Lettenmaier and V.G. Zartarian, 1992: A land-surface hydrology parametrization with subgrid variability for general circulation models. *J. Geophys. Res.,* **97**, 2717–2728.

Wood, N. and P.J. Mason, 1991: The influence of static stability on the effective roughness lengths for momentum and heat transfer. *Quart. J. R. Met. Soc.,* **117**, 1025–1056.

Xue, Y. and J. Shukla, 1993: The influence of land-surface properties on Sahel climate. Part I: Desertification. *J. Climate,* **6**, 2232–2245.

Yang, Z.-L., R.E. Dickinson, A. Robock and K.Y. Vinnikov, 1996: On validation of the snow sub-model of the Biosphere-Atmosphere Transfer Scheme with Russian snow cover and meteorological observational data. *J.Climate,* In press.

Yin, F.L. and E.S. Sarachik, 1994: A new convective scheme for ocean general circulation models. *J. Phys. Oceanogr.,* **24**, 1425–1430.

Yulaeva, E. and J.M. Wallace, 1994: The signature of ENSO in global temperature and precipitation fields derived from the Microwave Sounding Unit. *J. Climate,* 7, 1719–1736.

Zeng, X. and R.A. Pielke, 1995: Landscape-induced atmospheric flow and its parameterization in large-scale numerical models. *J. Climate,* **8**, 1156–1177.

Zhang, M.H., J.J. Hack, J.T. Kiehl and R.D. Cess, 1994: Diagnostic study of climate feedback processes in atmospheric general circulation models. *J. Geophys. Res.,* **99,** 5525–5537.

5

Climate Models – Evaluation

W.L. GATES, A. HENDERSON-SELLERS, G.J. BOER,
C.K. FOLLAND, A. KITOH, B.J. McAVANEY, F. SEMAZZI,
N. SMITH, A.J. WEAVER and Q.-C. ZENG

Contributors:
*J.S. Boyle, R.D. Cess, T.H. Chen, J. Christy, C.C. Covey, T.J. Crowley, U. Cubasch,
J. Davies, M. Fiorino, G. Flato, C. Fredericksen, F. Giorgi, P.J. Gleckler, J. Hack,
J. Hansen, G. Hegerl, R.X. Huang, P. Irannejad, T.C. Johns, J. Kiehl, H. Koide,
R.D. Koster, J. Kutzbach, S.J. Lambert, M. Latif, N.-C. Lau, P. Lemke, R.E. Livezey,
P.K. Love, N. McFarlane, K. McGuffie, G.A. Meehl, I. Mokhov, B.L. Otto-Bliesner,
T.N. Palmer, T.J. Phillips, A.J. Pitman, J. Polcher, G.L. Potter, S.B. Power, D. Randall,
P. Rasch, A. Robock, B.D. Santer, E. Sarachik, N. Sato, A.J. Semtner, J. Slingo,
I. Smith, R. Stouffer, K.R. Sperber, J. Syktus, M. Sugi, K.E. Taylor, S.F.B. Tett,
S. Tibaldi, W.-C. Wang, W.M. Washington, B.C. Weare, D.L. Williamson,
T. Yamagata, Z.-L. Yang, M.H. Zhang, R.H. Zhang, F.W. Zwiers*

CONTENTS

SUMMARY

Coupled climate models

- The most powerful tools available with which to assess future climate are coupled climate models, which include three-dimensional representations of the atmosphere, ocean, cryosphere and land surface. Coupled climate modelling has developed rapidly since 1990, and current models are now able to simulate many aspects of the observed climate with a useful level of skill. Coupled model simulations are most accurate at large space scales (e.g., hemispheric or continental); at regional scales skill is lower. The extent to which differences among coupled models may result in differences in their simulations of climate change is, however, not yet fully understood. More detailed and accurate simulations are expected as models are further developed and improved.

- The evaluation of coupled climate models is a challenging task for which a suitable formalism is only now being developed. Considerable progress has been made in evaluating the performance of the component atmospheric, oceanic, land-surface and sea ice models and their interactions.

- Flux adjustment and spin-up methodologies commonly used in coupled models affect the simulation of surface temperature, sea ice cover and the strength of the thermohaline circulation. The need for flux adjustment is expected to diminish as the component atmospheric, oceanic, land-surface and sea ice models are improved.

Component models of the atmosphere, land surface, ocean and sea ice

- The ability of current atmospheric models to simulate the observed climate varies with scale and variable, with the radiative effects of clouds remaining an area of difficulty. Given the correct sea surface temperature, most models simulate the observed large-scale climate with skill, and give a useful indication of some of the observed regional interannual climate variations or trends.

- Land-surface processes can be modelled more realistically than in 1990, although there continues to be wide disparity among current schemes. The evaluation of surface parametrizations such as soil moisture continues to be hindered by the lack of observations.

- Ocean and sea ice models portray the observed large-scale distribution of temperature, salinity and sea ice more accurately than in 1990. High resolution ocean models simulate mesoscale ocean eddies with striking realism, but their computational cost is presently prohibitive for coupled climate simulations.

The performance of climate models under other conditions

- Weather forecasting and palaeoclimate simulation continue to provide important tests of the realism of components of climate models under different initial and external conditions. Atmospheric forecast models show increased skill since 1990 due to improvements in parametrization, resolution and data assimilation schemes.

How can confidence in climate models be increased?

- The major areas of uncertainty in climate models concern clouds and their radiative effects, the hydrological balance over land surfaces and the heat flux at the ocean surface.

- The comprehensive diagnosis and evaluation of both component and coupled models are essential parts of model development, although the lack of observations and data sets is a limiting factor.

- In addition to data rescue and re-analysis efforts, a comprehensive global climate observing system is urgently needed.

5.1 What is Model Evaluation and Why is it Important?

In the evaluation of coupled models there is no single "figure of merit" that adequately represents the validity of a model's simulations or predictions, and a variety of performance measures are required. The evaluations reported in this chapter are of three types: evaluation of selected aspects of coupled models (Section 5.2), evaluation of the atmospheric, oceanic, land-surface and sea ice model components (Sections 5.3 and 5.4), and evaluation of the sensitivity of the links between these components (Section 5.5). There has been significant progress in all three types of model evaluation since IPCC (1990).

The concept of evaluation used here corresponds to the ideas of Oreskes *et al.* (1994) in that we are trying to demonstrate the degree of correspondence between models and the real world they seek to represent. We will therefore avoid use of the word "validation", although this term is commonly used in the sense of evaluation in the climate modelling community. The intent of this chapter is to summarise the performance of models as used for the simulation of the current observed climate, and to point out deficiencies and uncertainties with a view to further model improvement. The models' behaviour near the surface is stressed since surface and near-surface variables are those of greatest human interest and socio-economic importance. The use of coupled climate models for the simulation of climate change is considered in Chapter 6.

In recent years the systematic evaluation and intercomparison of climate models and their components has been a useful and popular mechanism for identifying common model problems. For example, Boer *et al.* (1992) identified a cold polar upper tropospheric bias of the then current atmospheric general circulation models, and Stockdale *et al.* (1993) showed that the then current ocean general circulation models produced sea surface temperatures (SSTs) that are too warm in the western tropical Pacific and too cold along the equator in the east-central Pacific. Neelin *et al.* (1992) found a wide variation in the ability of the then current coupled models to simulate interannual variability in the upper layers of the tropical Pacific, while Gates *et al.* (1993) reported significant differences in the extended control runs of four early coupled models. The evaluation of land-surface schemes in progress under the auspices of the Project for Intercomparison of Land surface Parametrization Schemes (PILPS) (Henderson-Sellers *et al.*, 1995) has identified important differences in energy and water partitioning at the surface. Comprehensive analysis and intercomparison of virtually all current global atmospheric models is being undertaken by the international Atmospheric Model Intercomparison Project (AMIP) (Gates, 1992). A comprehensive and systematic evaluation of coupled models based on global observational data is a goal yet to be achieved.

5.2 How Well Do Coupled Models Reproduce Current Climate?

5.2.1 Introduction

As discussed in Chapter 1, the coupled climate system comprises the atmosphere, ocean, cryosphere and land surface, and the models we consider here are "full" models in that they include the three-dimensional representation and interaction of these components on a global time-dependent basis. Such coupled models, in which other aspects of the climate system such as the chemical composition of the atmosphere and the surface vegetation are specified, provide the current scientific basis for understanding and simulating the climate system and its future changes. We note that none of the control simulations considered in this section include the direct effects of aerosols (see Chapter 6).

Simulations of the coupled system require integrations over 100–1000 simulated years (Cubasch *et al.*, 1992; Stouffer *et al.*, 1994), and computational cost restricts the model complexity and resolution that is currently possible. Table 5.1 lists the coupled model integrations that have been completed or are currently underway, together with some of their characteristics. The "model number" links our discussion of the simulation of current climate with that of climate change in Chapter 6. The first published intercomparison of the simulation of the current climate with coupled models was that of Gates *et al.* (1993), in which a number of basic parameters were examined in four models. Only one of these (MPI-LSG) appears in Table 5.1 and in Chapter 6. The remaining entries in the table represent new coupled models or new versions of previous models. Information from eleven of these coupled model integrations (BMRC, CCC, COLA, CSIRO, GFDL, GISS1, MPI-LSG, MPI-OPYC, MRI, NCAR and UKMO)[1] is given its first comparative analysis in this chapter.

5.2.2 Spin-up and Flux Adjustment

Time-scales for the atmosphere are on the order of weeks, those for the land surface and upper ocean extend to seasons, while those for the deep ocean are hundreds to

---

[1] See the Supplementary Table at the end of this chapter for definitions of these acronyms.

Table 5.1: *Coupled model control simulations.*[*]

Group	Model No.[†]	Country	AGCM Resolution	OGCM Resolution	Sea ice[@]	Flux correction[+]	Land-surface[$] scheme	Initial state[^]	Notes[§]
BMRC	1	Australia	R21 L9	3.2° × 5.6° L12	T	none	B	E	d
CCC	2	Canada	T32 L10	1.8° × 1.8° L29	T	H, W, T	BB	E	d
CERFACS	3	France	T42 L31	1° × 2° L20		none	B	E	
COLA	4	USA	R15 L9	3° × 3° L16	T	none	C r_s (SSiB)		d
CSIRO	5	Australia	R21 L9	3.2° × 5.6° L12	T/R	H, W, τ, T	C r_s (CSIRO)	E	d
GFDL	6	USA	R30 L14	2° × 2° L18	T/Dr	H, W	B	E	d
GISS	7	USA	4° × 5° L9	4° × 5° L13	T	none	C	I	d (1)
GISS	8	USA	4° × 5° L9	4° × 5° L16	T	none	C	E	(2)
IAP	9	China	4° × 5° L2	4° × 5° L20	T	H, W	B	E	NPOGA
LMD/OPA	10	France	3.6° × 2.4° L15	1° × 2° L20	T	none	B(SECHIBA)	E	
MPI	11	Germany	T21 L19	5.6° × 5.6° L11	T	H, W, τ, T	B	E	d E1/LSG
MPI E2/OPYC	12	Germany	T21 L19	2.8° × 2.8° L9	T/R	H, W, τ, T	C r_s (ECHAM)	E	d
MRI	13	Japan	4° × 5° L15	(0.5–2°) × 2.5° L21	T/Dr	H, W	B	E	d
NCAR	14	USA	R15 L9	1° × 1° L20	T/R	none	B	I	d
UCLA	15	USA	4° × 5° L9	1° × 1° L15	T/Dr	none	fixed wetness		NPOGA
UKMO	16	UK	2.5° × 3.8° L19	2.5° × 3.8° L20	T/Dr	H, W	C r_s (UKMO)	U	d

[*] The entries in the table apply to coupled atmosphere/ocean models for which a coupled control integration for current climate is under way. (See Supplementary Table for model identification).

[†] The model no. corresponds to the simulations listed in Table 6.3 and discussed in Chapter 6.

[@] T refers to "thermodynamic" and R to "dynamic" sea ice with rheology. Dr stands for "free drift" sea ice.

[+] H, W, τ, T stand for flux adjustment of heat, fresh water, surface stress and ocean surface temperature, respectively.

[$] B refers to a "simple bucket"; BB refers to "modified bucket"; C includes canopy processes; r_s denotes inclusion of stomatal resistance. The name of the land-surface scheme is given in parentheses.

[^] The method of initialising the coupled model is indicated by E for an equilibrium of the coupled system, U for equilibrium of the upper ocean and I for initial conditions specified from available observations.

[§] NPOGA: stands for "no polar ocean with global atmosphere" and reflects models which either exclude the polar oceans and ice by specifying climatological values or which control polar deep ocean quantities with a relaxation to observed values. E1 and E2 refer to the ECHAM1 and ECHAM2 AGCMs, (1) and (2) refer to two versions of the GISS model, and LSG and OPYC refer to ocean models. d indicates that data from this simulation are included in Section 5.2.

thousands of years. A coupled model integration could start from the observed three-dimensional distribution of atmosphere-ocean-land-surface variables, but this approach is not currently possible due to data limitations. An alternative is to integrate the coupled model from arbitrary initial conditions to a climatic (statistical) equilibrium under seasonally-varying forcing. Computational costs and long oceanic time-scales usually preclude this, and the typical experimental strategy is to "spin-up" separately the component atmosphere and ocean models before coupling. Such "initialisation" of the coupled system, however, can have a long-term effect on the simulated climate.

An atmospheric model requires the sea surface temperature (SST) as a lower boundary condition, while an ocean model requires the surface fluxes of energy, fresh water and momentum (or wind stress) as upper boundary conditions. After the ocean and atmosphere models are separately "spun up" with boundary conditions representing present-day conditions and are joined (possibly using flux adjustment), the coupled model may be integrated for a further spin-up or adjustment period. The state of the coupled system after such adjustment is then used to initiate both the control runs examined here and the climate change experiments examined in Chapter 6.

Ocean models present special spin-up problems since, when forced with specified fluxes, they typically arrive at an unrealistic state because atmosphere-ocean feedbacks are absent in this case. Feedback is usually introduced by adding a "relaxation" to observationally based reference upper ocean temperature and salinity fields. Three approaches are used among the coupled models in Table 5.1. They are: (1) integrate the ocean model for the several thousand years required for

---

Flux Adjustment: – A Modelling Dilemma

The atmosphere and ocean interact via fluxes of heat, momentum and fresh water. Local values of the heat flux are of order hundreds of Wm^{-2} and to first order the ocean is warmed in summer and cooled in winter, with the fluxes largely cancelling on the annual average. The relatively small residual heat flux, together with the fresh water flux, control the ocean's thermohaline circulation and the associated poleward ocean transports of heat and salt.

The coupling of ocean and atmosphere models can highlight discrepancies in the surface fluxes that may lead to a drift away from the observed climate. A systematic error of ± 5 Wm^{-2} in heat flux, which is smaller than the local observational error, could lead to a relatively large error in the annual average poleward oceanic heat transport. Modellers were made especially aware of this from the study by Gleckler *et al.* (1994) which showed that the surface fluxes in many uncoupled atmospheric models would imply a northward oceanic heat transport in the Southern Hemisphere in contrast with observational estimates there. This is a result of errors in the cloud radiative forcing in the models which allows excessive solar radiation to reach the surface.

Climate drift may be ameliorated by flux adjustment whereby the heat and fresh water fluxes (and possibly the surface stresses) are modified before being imposed on the ocean by the addition of a "correction" or "adjustment." The term "flux adjustment" is not meant to imply a knowledge of the "right" answer for the fluxes, since they are only imprecisely known. The flux adjustment terms are calculated from the difference between the modelled surface fluxes and those required to keep the model close to current climate. After running the model for a period suitable for the calculation of average flux adjustments, these terms are applied throughout the control and anomaly experiments. The alternative approach of not making any flux adjustments and accepting the resultant climate drift has been chosen by some modellers.

Kerr (1994) has highlighted this dilemma between the pragmatic need to conduct long runs with imperfect coupled models and the continuing desire to develop the best possible models on purely physical grounds. Some modellers have drawn an analogy with the early development of numerical weather prediction when ad hoc corrections were used to improve forecasts. Later improvements in model formulation and initialisation techniques made such corrections unnecessary.

This conflict between the use of a fully physically based model and the use of a non-physical flux adjustment is reduced if the response of the model to a perturbation is not seriously altered. The main purpose of the flux adjustment is to ensure that any perturbation, such as that due to increased CO_2, is applied about a realistic reference climate so that distortion of the major climate feedback processes is minimised (see Section 6.2.6). The need for flux adjustment in coupled climate models will decrease as models are further improved.

equilibrium under surface forcing, possibly using "acceleration" techniques (e.g., MPI, CCC, GFDL, BMRC, CSIRO, GISS2, MRI); (2) integrate the coupled model for a shorter period until the upper ocean is in quasi-equilibrium (e.g., UKMO); or (3) initialise the ocean model with the aid of three-dimensional ocean data (e.g., GISS1). In the absence of a climate equilibrium, an adjustment or drift of the simulated climate will almost always occur. The magnitude of this drift is generally least for the first spin-up approach and greatest for the last.

Inconsistencies between the surface fluxes developed in atmosphere and ocean models during spin-up, as a result of imperfections in either (or both) components, typically result in a drift away from a realistic current climate when the models are coupled. This is often ameliorated by flux adjustment (see Box on flux adjustment) (Manabe and Stouffer, 1988; Sausen *et al.*, 1988). Here, for example, the surface heat flux is modified for the ocean model by the addition of a "flux adjustment" that depends on the mean fluxes typically calculated by the separate ocean and atmospheric models in the spin-up phase. A similar adjustment may be made for the fresh water flux and surface wind stress. It is also possible to adjust the surface temperature of the ocean as seen by the atmospheric GCM, although this is less common. The types of flux adjustments used in current coupled model simulations are indicated in Table 5.1.

Flux adjustment is strictly justified only when the corrections are relatively small, and in fact the flux adjustments in some coupled models are comparatively large (Figure 5.5). The alternative is to avoid flux adjustment and to accept the resulting climate drift, a choice that is made in several of the models listed in Table 5.1. However, confidence in a coupled model's simulation of transient climate change is not improved if the climate drift is large and/or if the feedbacks are seriously distorted by flux adjustment.

5.2.3 Simulation of Mean Seasonal Climate of the Coupled System

Here we intercompare and evaluate the control runs for eleven of the coupled models of Table 5.1, some of which are used for the transient climate change simulations discussed in Chapter 6. We concentrate mainly on surface parameters that are particularly important for climate change and which reflect the "coupled" nature of the simulation. We consider December to February (DJF) and June to August (JJA) results in comparison with climatological observations in order to demonstrate the models' ability to simulate the change of the seasons. For each season we show both the average and the standard deviation of the models' simulations, the latter providing a measure of the models' scatter or disagreement.

5.2.3.1 Surface air temperature

Figures 5.1a and d give the DJF and JJA distributions of the average of the coupled models' simulated surface air temperature. Overall, the models successfully simulate the large-scale seasonal distribution of this important variable, so much so that the observed fields are not presented since they are very similar to the averaged model results.

Figures 5.1b and e show the differences between the model average and the "observed" surface air temperature. The differences are largest over the land, especially over mountains, reflecting uncertainties in both the observed and modelled values. (It should be recalled that these simulations do not include the direct effects of aerosols which could change the differences in some areas.) By contrast, differences over the ocean are rather small, with largest differences found in the Southern Ocean. This is particularly evident in the zonally averaged differences shown in Figures 5.1g and h, which indicates that some models have large climate drifts in this region.

The bottom panels, Figures 5.1c and f, show the disagreement or scatter among the models in terms of the standard deviation of the simulated values. Once again the largest values are over land and over the high latitude oceans. Table 5.2 shows that, on the global average the

Table 5.2: *Coupled model simulations of global average temperature and precipitation.*

	Surface air temperature (°C)		Precipitation (mm/day)	
	DJF	**JJA**	**DJF**	**JJA**
BMRC*	12.7	16.7	2.79	2.92
CCC	12.0	15.7	2.72	2.86
COLA*	12.6	15.5	2.64	2.67
CSIRO	12.1	15.3	2.73	2.82
GFDL	9.6	14.0	2.39	2.50
GISS (1)*	13.0	15.6	3.14	3.13
MPI (LSG)	11.0	15.2		
MPI (OPYC)	11.2	14.8	2.64	2.73
MRI	13.4	17.4	2.89	3.03
NCAR*	15.5	19.6	3.78	3.74
UKMO	12.0	15.0	3.02	3.09
Observed	12.4	15.9	2.74	2.90

* Models without flux adjustment.

Here the observed surface air temperature is from Jenne (1975) and the observed precipitation from Jaeger (1976).

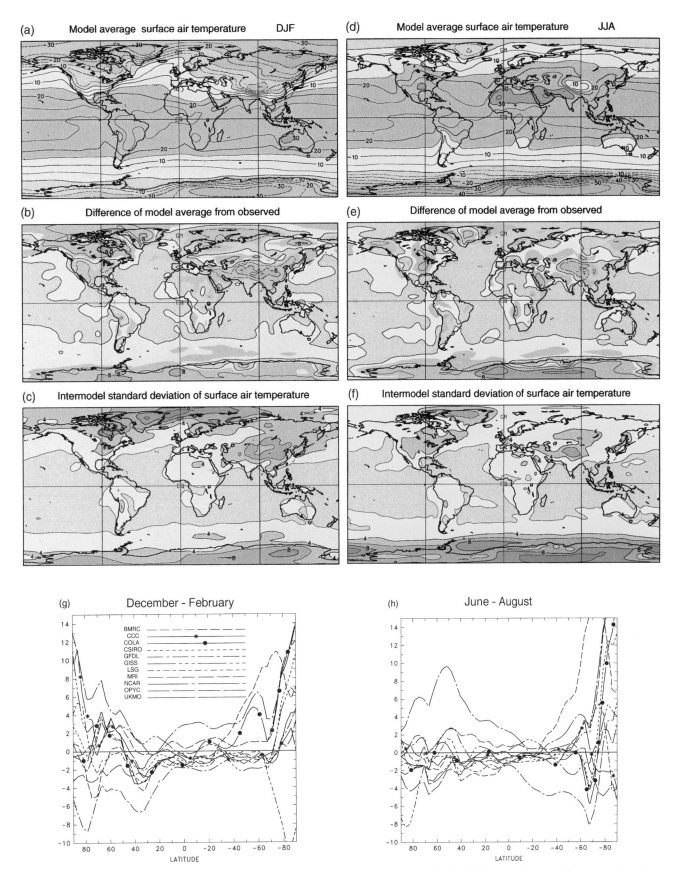

Figure 5.1: (a) The average surface air temperature simulated by nine coupled models, (b) the difference of the model average from the observed data of Jenne (1975) and (c) the intermodel standard deviation for DJF. (d-f) As in (a-c) but for JJA. (g) The zonally averaged difference of eleven coupled models' surface air temperature from observations for DJF. (h) As in (g) but for JJA. Units °C.

mean model surface air temperature agrees well with observations, but that individual models span a range of about 5°C. Three of the four warmest models are from simulations without flux adjustment, with the flux-adjusted models generally having a better simulation of surface air temperature over the oceans.

5.2.3.2 Precipitation rate

The precipitation rate is a measure of the intensity of the hydrological cycle and also influences the ocean thermohaline circulation through the fresh water flux. Figures 5.2a–c and d–f show the global distribution of the observed precipitation rate, the coupled models' average, and the intermodel standard deviation during DJF and JJA, respectively. The coupled models are generally successful in simulating the broad-scale structure of the observed precipitation, although some features of the actual precipitation are not well observed. The scatter among the models is comparatively large in the tropics where the precipitation rates are themselves largest, as seen in Figure 5.2g and h in terms of zonal averages. The globally averaged precipitation rate is given in Table 5.2, where we note that warmer mean temperatures are associated with larger precipitation rates as might be expected.

5.2.3.3 Mean sea level pressure

The seasonal average mean sea level pressure, displayed in Figure 5.3, is a measure of both dynamical and thermodynamical surface features and is linked to the surface wind stress that drives the ocean. Current coupled models succeed in representing the observed large-scale geographical distribution of this field rather well as shown in Figures 5.3a–f. The scatter among the models seen here and in Figures 5.3g and h may be partly due to differences in reduction to sea level, although the scatter at polar latitudes is a characteristic difficulty also seen in uncoupled atmospheric models (Boer *et al.*, 1992).

5.2.3.4 Snow and sea ice cover

The extent of snow and sea ice cover summarised in Tables 5.3 and 5.4 is a measure of coupled models' ability to simulate some aspects of the seasonal variation of the cryosphere (although the extreme values of these fields occur near the ends of the DJF and JJA periods). These data show that some models have an apparent excess of winter snow cover which tends to persist into summer. The ratio of summer to winter snow cover and the simulated sea ice cover vary greatly among the models, with some of the models without flux adjustment losing their sea ice in the Arctic summer and almost all their sea ice in the Southern

Hemisphere in both summer and winter. The apparent difficulties the models have in simulating the seasonal snow and ice cover could have consequences for the "albedo feedback" and the soil moisture and runoff in the climate change simulations of Chapter 6.

Table 5.3: *Coupled model simulations of Northern Hemisphere snow cover (10^6 km^2).*

	DJF (winter)	JJA (summer)
BMRC*	62.0	28.4
CCC	42.9	11.5
COLA*	53.4	12.1
CSIRO	37.5	11.6
GFDL	64.4	10.0
GISS (1)*	41.2	2.5
MPI (LSG)	51.4	5.8
MPI (OPYC)	54.6	16.7
MRI	34.3	2.9
NCAR*	41.4	2.1
UKMO	35.3	5.1
Observed (Matson *et al.*, 1986)	44.7	7.8

* Models without flux adjustment

Table 5.4: *Coupled model simulations of sea ice cover (10^6 km^2).*

	Northern Hemisphere		Southern Hemisphere	
	DJF (winter)	JJA (summer)	JJA (winter)	DJF (summer)
BMRC*	18.9	16.7	<1.0	<1.0
CCC	9.7	7.1	12.2	7.5
COLA*	9.3	1.6	4.0	3.7
CSIRO	16.6	14.5	21.1	18.9
GFDL	16.0	12.7	24.7	16.0
GISS (1)*	11.0	8.3	12.4	6.4
MRI	19.1	11.5	11.6	3.1
NCAR*	13.6	<1.0	5.3	<1.0
UKMO	10.2	5.3	18.0	5.9
Observed (Gloersen *et al.*, 1992)	14.5	11.5	16.0	7.0
(Ropelewski, 1989)	15.0	9.8	16.7	5.3

* Models without flux adjustment

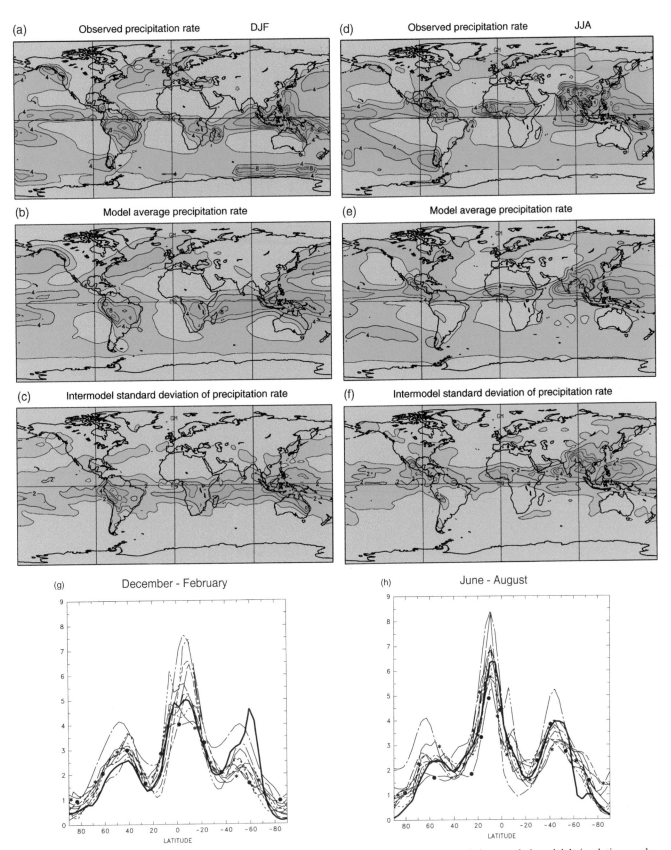

Figure 5.2: (a) The observed precipitation rate as estimated by Jaeger (1976), (b) the average of nine coupled models' simulations and (c) the intermodel standard deviation for DJF. (d-f) As in (a-c) but for JJA. (g) The zonally averaged precipitation rate from eleven coupled models and that from observations according to Jaeger (1976) for DJF (solid line). (h) As in (g) but for JJA. Units mm/day. (See Figure 5.1(g) for model identification.)

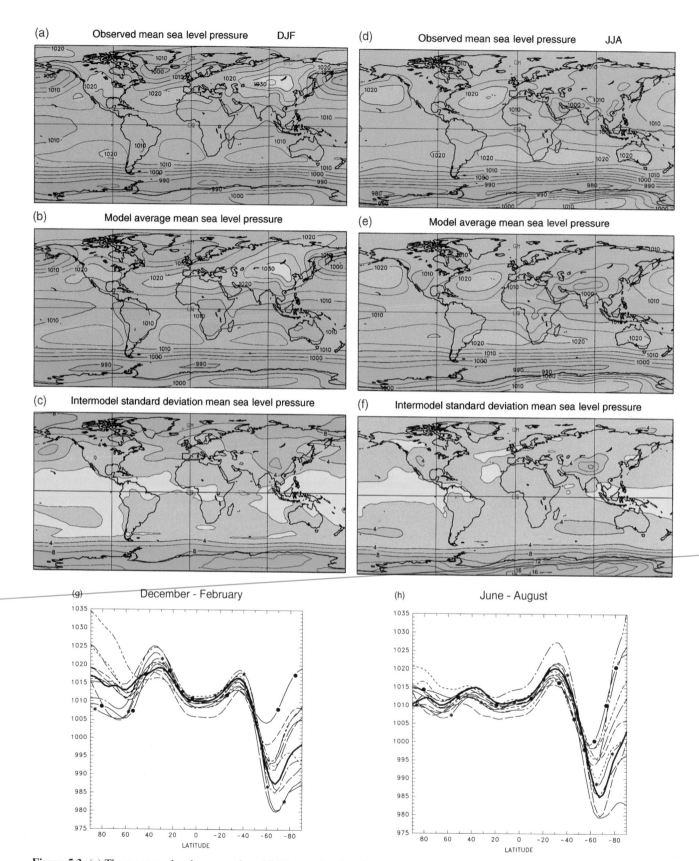

Figure 5.3: (a) The mean sea level pressure from NMC operational analyses during 1978 to1988, (b) the average of nine coupled models' simulations, and (c) the intermodel standard deviation for DJF.(d-f) Units hPa. (d-f) As in (a-c) but for JJA. (g) The zonally averaged mean sea level pressure from eleven coupled models and that from observations (solid line) for DJF. (h) As in (g) but for JJA. Units hPa. (See Figure 5.1(g) for model identification.)

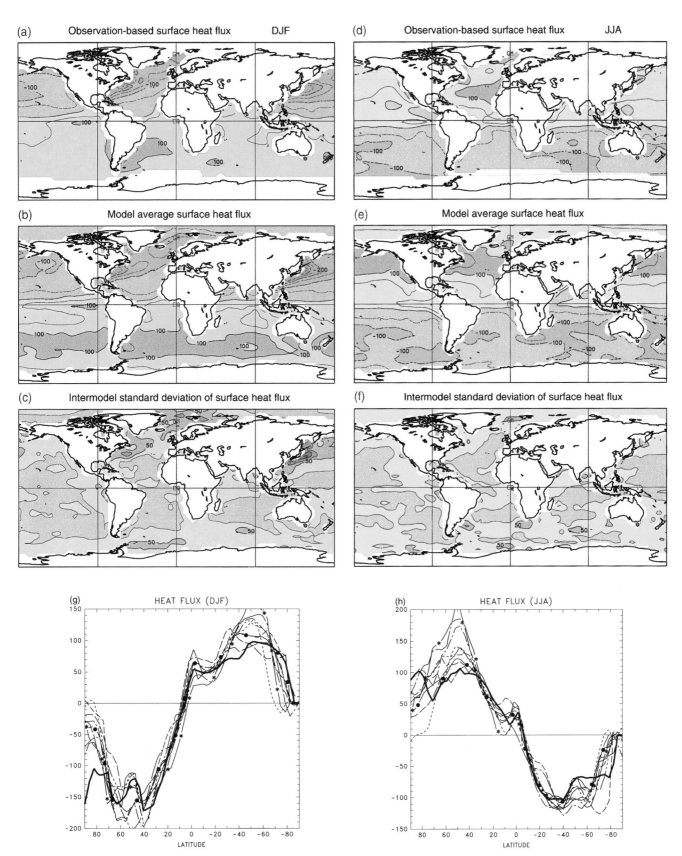

Figure 5.4: (a) The surface heat flux from observations (Esbensen and Kushnir, 1981), (b) the average of nine coupled models' simulations, and (c) the intermodel standard deviation for DJF. (d-f) As in (a-c) but for JJA. (g) The zonally averaged surface heat flux from nine coupled models and that from observations (solid line). (h) As in (g) but for JJA. Units Wm^{-2}. (See Figure 5.1(g) for model identification.)

5.2.3.5 Ocean surface heat flux and thermohaline circulation
The surface heat flux measures the energetic interaction of the atmosphere and ocean, and together with the surface fresh water flux, governs the strength of the ocean's thermohaline circulation. Figure 5.4 shows the coupled models' net surface heat flux, while Figure 5.5 shows the average heat flux adjustment in the models that use one. All the coupled models portray the ocean's absorption of heat in summer and its release in winter (Figures 5.4a–f), although there is a tendency for the modelled fluxes to exceed the observationally based estimates on the zonal average (Figure 5.4g and h). The heat flux adjustment, however, is seen to vary widely among the models that use it (Figure 5.5b and d; see also Table 5.1).

The thermohaline circulation in the oceans is important for the poleward heat transport in the climate system. Table 5.5 shows that only 4 of the 10 models considered have Atlantic thermohaline circulations with maxima in the "accepted" range of 13–18 Sv (10^6 m^3 s^{-1}) near 50°N (Schmitz and McCartney, 1993). These results suggest that flux adjustment and an equilibrium ocean model spin-up may be necessary (but not sufficient) conditions for thermohaline circulations in this range. No model without flux adjustment attains the accepted range, and of the four thermohaline circulations that are weak, only one is for a flux-adjusted model.

Table 5.5: *Coupled model simulations of the strength of the North Atlantic thermohaline circulation*[§].

	Strength (Sv) (10^6 m^3 s^{-1})	State	Flux correction	E
BMRC[*]	6	weak	no	yes
CCC	14	moderate	yes	yes
CSIRO	18	moderate	yes	yes
GFDL	17[†]	moderate	yes	yes
GISS (1)[*]	2	weak	no	no
GISS (2)[*]	26	strong	no	yes
MPI (LSG)	26[†]	strong	yes	yes
MRI	2	weak	yes	yes
NCAR[*]	2[†]	weak	no	no
UKMO	16	moderate	yes	yes

§   This table is intended to represent the strength of the meridional mass flux maximum near 50°N.

†   Values taken from Gates *et al.* (1993).

E   Indicates whether the model spin-up is to a long-term equilibrium state.

*   Models without flux adjustment.

5.2.4 Simulation of the Variability of the Coupled System
While much attention has been focused on the mean climate simulated by coupled models, the simulation of variability is at least as important in many considerations, including the detection of climate change (see Chapter 8).

5.2.4.1 Monthly and seasonal variability
The evaluation of coupled model variability on monthly to seasonal time-scales is illustrated by the work of Meehl *et al.* (1994), who calculated a measure of intermonthly standard deviation of lower tropospheric temperature from their coupled model and compared it with observations from the microwave sounding unit (MSU) at "periods greater than one month but less than twelve months". The simulated values were generally larger than those observed in the tropics and smaller in southern high latitudes. The model successfully simulated some of the features of the observed intermonthly variation, including a representation of the Madden-Julian Oscillation in the tropics and blocking events in the extratropics (see also Section 5.3.1.2).

5.2.4.2 Interannual and ENSO time-scale variability
Figure 5.6 displays the standard deviation of the annual mean surface air temperature from observations and from three coupled model simulations. There is qualitative agreement in that the models reproduce the larger variability observed over the continents and at high latitudes. The El Niño-Southern Oscillation-related variability in the eastern tropical Pacific, however, is generally underestimated in the models. The largest differences among these models is found over the tropical oceans where values differ by as much as a factor of 2.

The El Niño-Southern Oscillation (ENSO) represents one of the coupled system's dominant interannual variations. ENSO is thought to involve modes of variability that produce coupled anomalies in the eastern Pacific, and modes that involve internal ocean dynamics and wave reflection from boundaries requiring high resolution for their simulation (Schopf and Suarez, 1988; Neelin, 1991; Jin and Neelin, 1993a, b; Neelin and Jin, 1993).

Multi-decadal integrations with coarse-resolution global coupled models display interannual variability in the tropical Pacific that resembles some aspects of the observed variability associated with ENSO, although typically of lesser amplitude (Sperber *et al.*, 1987; Meehl, 1990; Lau *et al.*, 1992; Sperber and Hameed, 1991; Nagai *et al.*, 1992; Meehl *et al.*, 1993). Changes in such features as the mean thermocline depth, however, can produce different manifestations of interannual variability (Nagai *et al.*, 1992; Moore, 1995; Tett, 1995). Higher resolution but

December-February

June-August

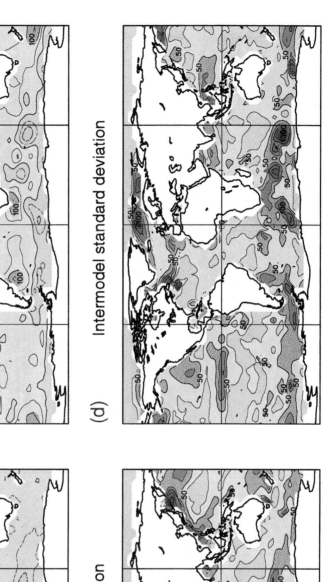

(a)   Model average heat flux correction

(c)   Model average heat flux correction

(b)   Intermodel standard deviation

(d)   Intermodel standard deviation

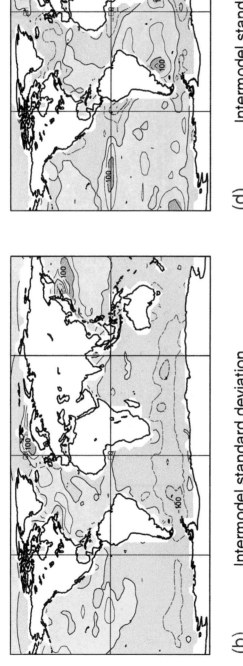

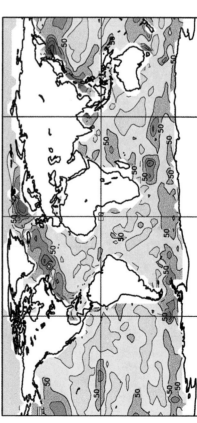

Figure 5.5: (a) The average surface heat flux adjustment for models which use such an adjustment (see Table 5.1) for DJF and (b) the corresponding intermodel standard deviations. Units Wm^{-2}. (c) and (d) As in (a) and (b) but for JJA.

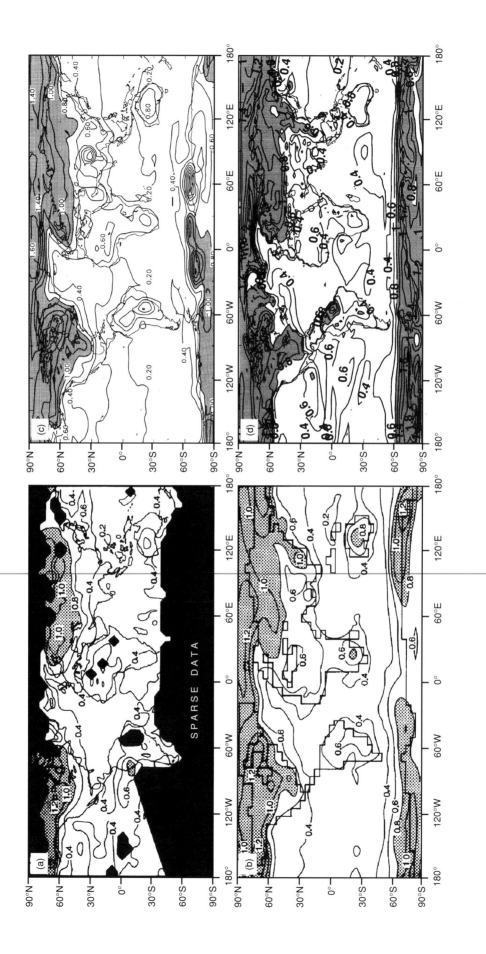

Figure 5.6: The standard deviation of the annual mean surface air temperature: (a) over 110 years of observations (Jones *et al.*, 1991), (b) over 1000 yr from the GFDL coupled model, (c) over 350 yr from the MPI/LSG coupled model, and (d) over 130 yr from the UKMO coupled model. Units °C.

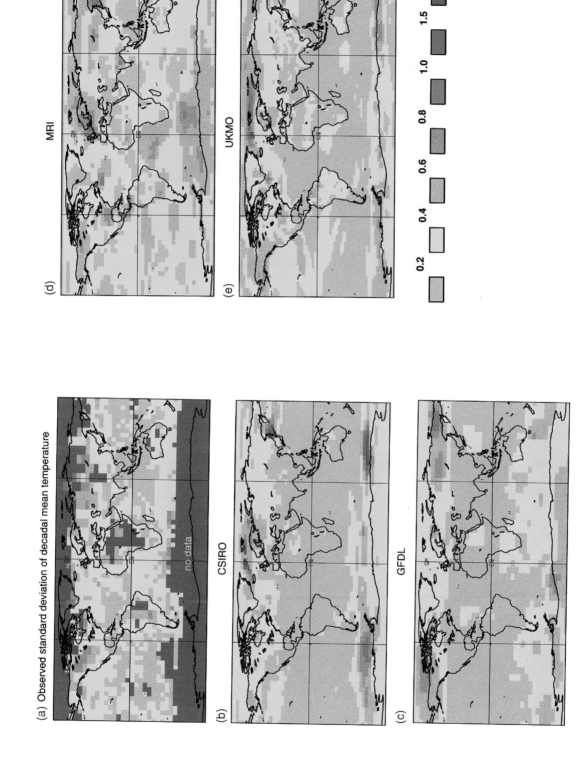

Figure 5.7: The standard deviation of the decadal mean surface temperature from observations (a) (Parker *et al.*, 1994) compared with simulations by the following coupled models: CSIRO (b), GFDL (c), MRI (d) and UKMO (e). Units °C.

limited domain coupled models have also been used successfully in ENSO forecast studies (Latif *et al.*, 1994), and global ENSO forecast models are under development (Ji *et al.*, 1994). A class of model that combines a global atmospheric model and a limited domain high-resolution ocean (usually over the tropical Pacific) has been used to study ENSO-type interannual variability over longer time periods (Philander *et al.*, 1992; Gent and Tribbia, 1993; Latif *et al.*, 1993), as well as to demonstrate linkages between the Indian monsoon and ENSO (Barnett *et al.*, 1989). These studies give some credibility to the simulated changes of interannual variability on ENSO time-scales associated with greenhouse gas increases (see Chapter 6). Improved simulations of ENSO variability are expected from coupled global models that better resolve equatorial ocean dynamics.

5.2.4.3 *Decadal and longer-term variability*
The extended coupled model integrations now becoming available allow investigation of longer time-scale variability. Although the surface air temperature has perhaps the "best" observed long time-series record (100-150 years), there are problems in comparing this record, which includes the effects of both internally generated and externally forced variability, with model simulations which include only the internally generated component (Santer *et al.*, 1995b; Hegerl *et al.*, 1996; Chapter 8). Such a comparison is also hindered by the effects of sparse observations during the early part of the record and at high latitudes (Karl *et al.*, 1993; Jones, 1994), and difficulties in correcting the systematic errors due to urbanisation and changes in measurement methods.

The standard deviation of decadal mean surface temperature is shown in Figure 5.7. This measure of long time-scale variability indicates that coupled models are capable of simulating the observed increase of interdecadal variability with latitude. The variability of decadal means associated with the ice edge can also be seen in some models. There are substantial unexplained differences among the models, however, and this must be kept in mind when interpreting the results of long-term transient experiments (see Chapter 6).

Long time-series of the global mean surface air temperature anomalies from the MPI/LSG and GFDL coupled models are shown in Figure 5.8, and similar records are available from other shorter simulations. The MPI model exhibits a large fluctuation in the first 250 years of the integration (which is related to sea ice fluctuations), while other models (e.g., BMRC and GISS1) show a slow climate drift over the first century. Coupled

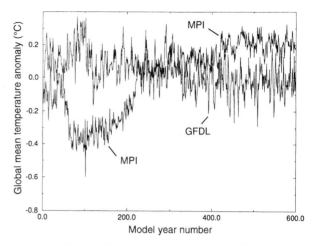

Figure 5.8: The globally averaged annual mean surface air temperature time-series from the GFDL and MPI coupled model simulations. Units °C, expressed as anomalies from the time mean.

models are able to simulate a range of natural or internal variability, whose correct simulation is important for the detection of climate change (see Chapter 8).

5.2.5 *Simulation of Regional Climate*
The key factors that affect the regional performance of global coupled models are their horizontal resolution and their physical parametrizations. Coarse resolution atmospheric models are unable to realistically portray the extent and height of mountains, with consequent distortions in their simulation of orographic precipitation on regional scales, while coarse resolution ocean models suffer similar distortions in their simulation of boundary currents. A GCM's resolution may also introduce systematic errors in the depiction of coastlines, with consequent effects on the simulation of regional circulation and temperature. With their present horizontal resolution of several hundred km (Table 5.1), the largest errors of coupled models' simulation of the observed seasonal temperature generally occur on the smaller resolved scales over the continents (Figure 5.1). Differences between models also tend to be greater at the regional scale (Figure 5.1). See Chapter 6 (Section 6.6.1 and Figure 6.32) where the regional simulations of precipitation and surface temperature by five coupled ocean-atmosphere models are compared with observations.

As the resolution of models increases, the parametrization of subgrid-scale physical processes such as convection, mixing, and surface fluxes may need to be reformulated to more adequately represent their interaction with the newly resolved scales of motion. The accurate portrayal of regional climate is also limited by the general increase of variability found on smaller scales (see Section 5.2.5).

Regional-scale climate can also be simulated by using a higher-resolution model over a limited area of the world (e.g., Giorgi *et al.*, 1994, as discussed further in Chapter 6). With adequate boundary conditions (or lateral forcing) provided either by observations or by a global model, and with careful placement of the boundaries of the area, such models can portray the climate at regional and local scales with more accuracy than that given by global climate models of conventional resolution. This indicates that when it is possible to use higher resolution in coupled global models, a general increase in accuracy may be expected, especially near mountains and coasts and in regions of high variability.

5.2.6 Summary

Coupled modelling is rapidly accelerating, and the present state may be summarised as follows:

(1)   The large-scale features of the current climate are well simulated on average by current coupled models;

(2)   Different coupled models simulate the current climate with various degrees of success, and this affects the confidence that can be placed in their simulations of climate change;

(3)   Flux adjustments are relatively large in the models that use them, but their absence affects the realism of the control climate and the associated feedback processes;

(4)   The analysis of variability is providing new opportunities for model evaluation.

5.3   How Well Do the Component Atmosphere, Land-surface, Ocean and Sea Ice Models Reproduce Current Climate?

5.3.1 Atmospheric General Circulation Models

5.3.1.1 Simulation of mean seasonal climate

5.3.1.1.1 Introduction

Atmospheric general circulation models (GCMs) have undergone many refinements in formulation and parametrization during more than four decades of continuous development and use for weather prediction and climate simulation. Evaluation of individual atmospheric GCMs used for climate simulation are widely available in the literature, although attempts to intercompare them under standard conditions have been undertaken only relatively recently. In general, there has been a progressive increase in the accuracy of atmospheric climate models' simulation of the observed average large-scale distribution of pressure, temperature and circulation, and of elements of the heat and hydrologic balances such as radiative fluxes and precipitation. There have also been progressive improvements in atmospheric

models' portrayal of regional mean climate (Chapter 6, Section 6.6) and of atmospheric variability on a variety of time-scales (Section 5.3.1.2).

Here we present a new assessment of the ability of current atmospheric GCMs to simulate the observed mean seasonal climate. For this purpose we use the emerging results of the international Atmospheric Model Intercomparison Project (AMIP) (Gates, 1992). Virtually all global atmospheric modelling groups[1] are participating in this project, which calls for a ten-year simulation under agreed conditions. The AMIP boundary conditions used by all participating atmospheric GCMs consist of the 1979 to 1988 sequence of observed monthly averaged distributions of sea surface temperature and sea ice, an atmosphere CO_2 concentration of 345 ppmv and a value of 1365 Wm^{-2} for the solar constant. We focus here on the AMIP models' simulation of selected variables that were not considered in the section on coupled models (Section 5.2). The actual climate is given by observational estimates that are believed to be the most representative of the AMIP decade, although in some cases other data are also used to provide global coverage. For this purpose both the models' results and the observed data were regridded to a common 4° latitude and 5° longitude grid, which is representative of the AMIP models' horizontal resolution.

5.3.1.1.2 Surface air temperature over land, precipitation and mean sea level pressure

In general, the AMIP models' simulation of these variables is similar to that shown for coupled models in Section 5.2. Although the prescription of sea surface temperature effectively determines the surface air temperature over the oceans, the AMIP models' performance provides reassuring evaluation of their ability to portray the summer and winter temperatures (not shown) over the continents. There are, however, relatively large regional discrepancies over continental interiors and over Antarctica. Current models' overall simulation of the surface air temperature is superior to that of previous generations of atmospheric GCMs.

The overall geographical distribution of the AMIP models' precipitation (not shown) resembles that observed, but there are notable deficiencies. The simulation of tropical precipitation shows a slight improvement over that given by Boer *et al.* (1992) for models of comparable resolution. Much of this increase in accuracy is thought to be due to improvements in the parametrization of penetrating cumulus convection, on which the low-latitude precipitation is

---

[1] See the Supplementary Table at the end of this chapter for a detailed list of institutions involved.

critically dependent. The seasonal sea level pressure distribution (not shown) reveals a slight improvement over that seen in Boer *et al.* (1992), and is a marked improvement over that of the earlier generation of models (Gates, 1987). At least part of this increase in accuracy may be due to increases in model resolution.

5.3.1.1.3 Tropospheric temperature and circulation

The three-dimensional temperature distribution (not shown) simulated by the AMIP models bears a close resemblance to that observed in both summer and winter. The most prominent discrepancies are a tendency for the models to be systematically too cold at lower levels in the tropics and in the upper troposphere and lower stratosphere in higher latitudes, and too warm in the tropical lower stratosphere. Similar but more pronounced errors were reported by Boer *et al.* (1992). The circulation accompanying the temperature distribution is shown in Figure 5.9 in terms of the zonally averaged zonal wind at 200 hPa. These data show that while the AMIP models generally simulate the zonal winds with considerable

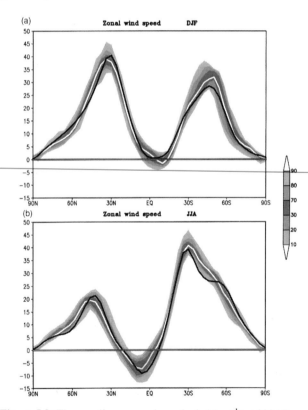

Figure 5.9: The zonally-averaged zonal wind (ms^{-1}) at 200 hPa as observed (black line) and as simulated by the AMIP models for (a) DJF and (b) JJA. The mean of the models' results is given by the full white line, and the 10, 20, 30, 70, 80 and 90 percentiles are given by the shading surrounding the model mean. The observed data are from Schubert *et al.* (1992).

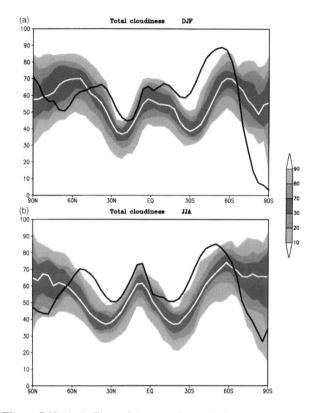

Figure 5.10: As in Figure 5.9 except for total cloudiness (%). The observational estimates are from ISCCP data for 1983 to 1990 (Rossow and Schiffer, 1991).

accuracy, there is a tendency to overestimate the strength of the Southern Hemisphere zonal winds near the tropopause, a feature that may be related to the models' limited vertical resolution. In general, the AMIP models' winds in the troposphere closely resemble those observed.

5.3.1.1.4 Cloudiness

Clouds exert a significant influence on the atmospheric heat budget through their reflection and absorption of radiation (Chapter 4, Section 4.2), and the treatment of cloud type, amount, height, optical properties and water content differ widely in current atmospheric GCMs. Figure 5.10 shows the distribution of the AMIP models' simulation for the zonally averaged total cloudiness during DJF and JJA, together with an estimate of the zonal average of the observed cloudiness. These data show that current models portray the large-scale latitudinal structure and seasonal change of the observed total cloud cover with only fair accuracy, and there is an apparent systematic underestimate of the cloudiness in low and middle latitudes in both winter and summer. In the higher latitudes the models appear to overestimate the observed cloudiness, especially over Antarctica, although here the difficulties in the satellite observing system's discrimination between cloud cover and a snow- or ice-covered surface should be kept in mind. The simulated total cloudiness may also reflect adjustments or tuning of the models' cloud properties made in most of the AMIP models.

5.3.1.1.5 Outgoing long-wave radiation and cloud radiative forcing

The long-wave radiation leaving the top of the atmosphere is the heat sink for the Earth-atmosphere system (Chapter 1, Section 1.2.1), and the comparison of its simulation with direct satellite measurements is an important evaluation of atmospheric models. Figure 5.11 shows the distribution of the AMIP models' simulation for the zonally averaged outgoing long-wave radiation (OLR) during DJF and JJA, along with the zonal average of the observed OLR. Assuming that the satellite measurements are representative of the AMIP decade, these data show that the mean of the model simulations is in good agreement with the observed estimate at all latitudes in both summer and winter. Since the minimum OLR in low latitudes is due to the presence of high and therefore cold convective

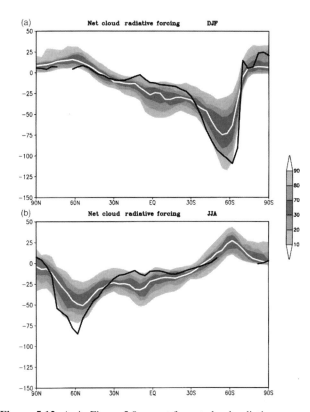

Figure 5.12: As in Figure 5.9 except for net cloud radiative forcing (Wm^{-2}). The observational estimates are from ERBE data for 1985–88 (Harrison *et al.*, 1990).

clouds, these results constitute an evaluation of the ability of current atmospheric GCMs to simulate tropical convection while maintaining a realistic overall thermal radiation budget.

Another important measure of the effects of clouds on the atmospheric radiation budget is the cloud radiative forcing, which is defined as the difference in the radiative flux at the top of the atmosphere when clouds are present and when they are absent. Since the effect of clouds (compared to clear sky) is to trap long-wave (or infrared) radiation in the Earth-atmosphere system, long-wave cloud radiative forcing is generally positive while short-wave cloud forcing is nearly always negative since clouds tend to reflect short-wave (or solar) radiation. The net radiative effect of clouds may therefore be either positive or negative. Figure 5.12 shows the distribution of the AMIP models' zonally averaged net cloud radiative forcing during DJF and JJA, together with the zonal average of the observational estimates from ERBE for the years 1985 to 1988. These data show that while the models have successfully reproduced the general large-scale structure of the observed cloud radiative forcing, there are a number of discrepancies. Perhaps the most obvious is the models'

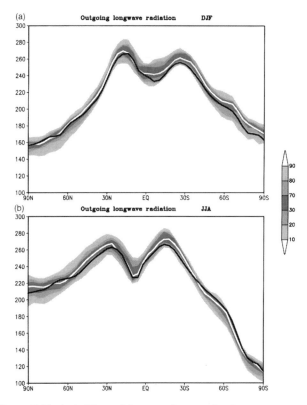

Figure 5.11: As in Figure 5.9 except for outgoing long-wave radiation (Wm^{-2}). The observational estimates are from Hurrell and Campbell's (1992) summary of NOAA polar orbiter data for the period 1979 to 1988.

systematic underestimate of the negative cloud radiative forcing between about 30°N and 30°S in both summer and winter, and their corresponding overestimate of the maximum negative forcing near 60° latitude in the summer hemisphere. These apparent errors are likely related to the models' treatment of cloud radiative processes, although the differences in how the modelled and observed cloud radiative forcing are calculated should be kept in mind.

5.3.1.1.6 Multivariate statistics

More insight into the nature of model differences is provided by multivariate statistics that measure the errors in the time mean and temporal variability relative to observations, as illustrated in Figure 5.13 for the mean sea level pressure in the Northern Hemisphere. These results show that in comparison to the models' natural variability (given by an ensemble of runs made with perturbed initial conditions, whose results (M vs M) all lie near the origin), the AMIP models have relatively large errors in their estimates of the temporal mean and variability (D vs M). The models' errors in simulating the time mean and temporal variability of sea level pressure are also somewhat larger than the observational uncertainties as

given by the D vs D results. Such statistics permit an assessment of the differences among models' errors without addressing formal questions of statistical significance (Katz, 1992), and emphasise the need for use of a variety of measures to adequately characterise model performance.

5.3.1.1.7 Summary

The AMIP results confirm that current atmospheric models generally provide a realistic portrayal of the phase and amplitude of the seasonal march of the large-scale distribution of pressure, temperature and circulation. In general, the models' largest discrepancies in the seasonal cycle of sea level pressure and surface air temperature occur in the higher latitudes, while the largest discrepancies in the seasonal cycle of precipitation are found in the tropics. Overall, atmospheric models have shown some improvement since 1990, and are markedly superior to the models in use in the 1970s and early 1980s.

The area-mean root-mean-square (RMS) difference between a simulated variable and an estimate of its observed value provides a useful measure of overall model performance. Table 5.6 shows the average RMS error of

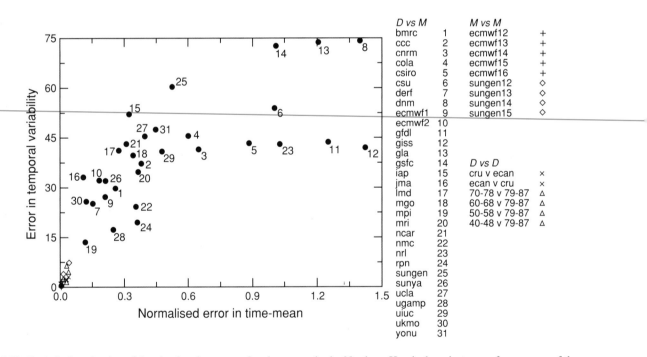

Figure 5.13: Statistical evaluation of the simulated mean sea level pressure in the Northern Hemisphere in terms of a measure of the error in temporal variability (ordinate) and a measure of the error in time means (abscissa). D vs M (data vs. model) refers to the results of the AMIP models (see Supplementary Table for model identification), M vs M (model vs. model) refers to the results of model ensembles with different initial conditions, and D vs D (data vs. data) refers to the results of intercomparisons of the CRU data (Jones, 1987) and ECMWF analyses. For details of the co-ordinates' statistics and data, see Santer and Wigley (1990), Wigley and Santer (1990) and Santer *et al.* (1995a).

the AMIP simulations for a number of common climate variables for both summer and winter in each hemisphere. Of particular note are the relatively large RMS deviations of the cloud radiative forcing (CRF) and of the net ocean surface heat flux in the summer hemisphere, which are illustrative of the need for flux adjustment in coupled climate models. We also note the relatively large RMS differences in the surface air temperature over land in the winter hemisphere, which emphasise the need for improved parametrization of the surface heat fluxes and land-surface processes. The simulation of clouds and their seasonal variation remains a major source of uncertainty in atmospheric models.

5.3.1.2 Simulation of variability and trends

Here we consider atmospheric GCMs' simulation of variability on a variety of time-scales and their portrayal of recent trends, as a supplement to the discussion of the mean seasonal climate given in Section 5.3.1.1.

Table 5.6: *Hemispheric mean seasonal root-mean-square differences between observations and the mean of the AMIP models.*

Variable	DJF NH	DJF SH	JJA NH	JJA SH
Mean sea level pressure (hPa)	1.4	1.4	1.3	2.4
Surface air temperature (°C) (over land)	2.4	1.6	1.3	2.0
Zonal wind (ms^{-1}) (at 200 hPa)	2.4	1.8	1.8	2.4
Precipitation (mm /day)	0.80	0.71	0.62	0.77
Cloudiness (%)	10	21	14	16
Outgoing long-wave radiation (OLR) (Wm^{-2})	2.8	3.2	2.9	5.5
Cloud radiative forcing (Wm^{-2})	9.1	20.5	16.2	6.5
Surface heat flux (Wm^{-2}) (over ocean)	22.5	27.3	30.5	17.2

See Supplementary Table for identification of the AMIP models. Observed data used are ECMWF analyses for mean sea level pressure (Trenberth, 1992) for 1979 to 1988, diagnoses of Schemm *et al.* (1992) and Schubert *et al.* (1993) for surface air temperature and precipitation over land for 1979 to 1988, MSU data of Spencer (1993) for precipitation over the oceans for 1979-1988, COADS data of Da Silva *et al.* (1994) for net surface heat flux over the ocean, and ECMWF TOGA analyses as summarised by Schubert *et al.* (1992) for the 200 hPa zonal wind for 1985 to 1991. The observed data sources for cloudiness, outgoing long-wave radiation and cloud radiative forcing are as in Figures 5.10, 5.11 and 5.12.

5.3.1.2.1 Diurnal and seasonal ranges

Recent observations have shown a reduction in the diurnal range of surface air temperature (nominally at 1.5 m) over the continents (Karl *et al.*, 1993; Horton, 1994; see also Chapter 3.3.2.4) and have suggested a reduction of the diurnal range of surface air temperature as carbon dioxide increases (Cao *et al.*, 1992; Mitchell *et al.*, 1995b). In general, the patterns of observed and modelled diurnal range are similar, though modelled values exceed those observed in high northern latitudes in January and over northern continents and deserts in July (e.g., Hansen *et al.*, 1994). The observed and modelled seasonal range of surface air temperature is shown in Figure 5.14 from the AMIP models, in which there is an overestimate of the seasonal amplitude of surface air temperature in the drier regions of the continents. There are, however, considerable differences among the models in higher latitudes.

5.3.1.2.2 Synoptic variability

Figure 5.15 shows the climatology of observed and modelled synoptic variability in DJF and JJA, as measured by the standard deviation of daily mean sea level pressure (MSLP) about each constituent monthly mean. Simulated values are from the AMIP models, while observations are derived from the Climate Analysis Center Diagnostic Data Base for 1979 to 1988. The models on average capture the broad pattern of the observed synoptic variability quite well, although the models' estimates are less than those observed in regions of maximum storminess. There is, however, considerable disagreement among the observed data themselves, and there are marked differences among the models away from the tropics (Hay *et al.*, 1992; Hulme *et al.*, 1993).

The Madden-Julian Oscillation (MJO) (Slingo and Madden, 1991) is a major feature of tropical variability on time-scales of about 30-60 days. Since the physical causes of the MJO are not fully understood (Rui and Wang, 1990), it is not surprising that analyses of near-equatorial winds at 200 hPa in the AMIP runs (Slingo *et al.*, 1995) show only a modest skill in reproducing the observed MJO magnitude and frequency.

5.3.1.2.3 Interannual variability

The ability of models to reproduce the observed mean interannual variability of climate is an important aspect of their performance, but in examining the ability of models to reproduce specific time sequences of interannual variability it must be borne in mind that not all interannual variations are forced (e.g., by SST) and that a fraction of this variability, often large in the extratropics, is internal to

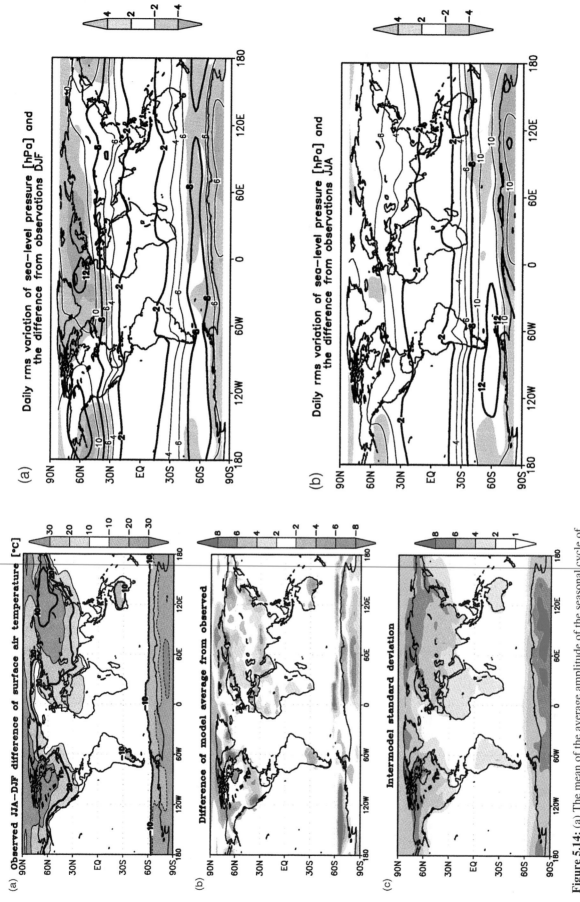

Figure 5.15: Mean of the daily RMS variability of mean sea level pressure simulated by the AMIP models during (a) DJF and (b) JJA for 1979 to 1988, and the difference (shaded in colour) of the modelled mean from the observed RMS variability using the ECMWF TOGA analyses for 1985 to 1991 (Schubert *et al.*, 1992). Units are hPa.

Figure 5.14: (a) The mean of the average amplitude of the seasonal cycle of surface air temperature as given by the difference between the JJA and DJF averages from the AMIP models. (b) The difference between the simulated mean seasonal amplitude and that from a merged observational estimate from Jones (1994), Schemm *et al.* (1992), Da Silva *et al.* (1994) and Willmott and Legates (1993). (c) The standard deviation of the average amplitude of the seasonal cycle of surface air temperature. Units are °C.

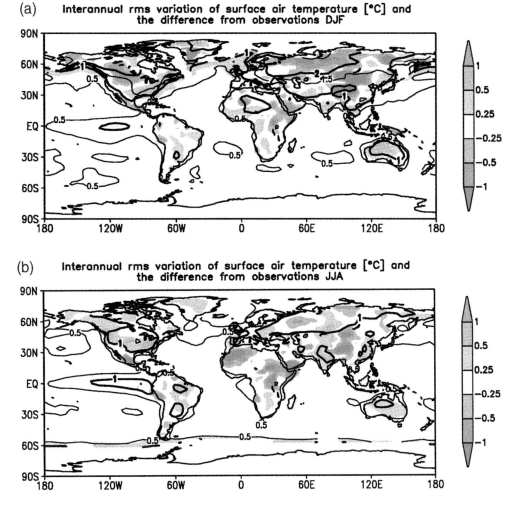

Figure 5.16: Mean of the interannual RMS variability of surface air temperature simulated by the AMIP models during (a) DJF and (b) JJA of 1979 to 1988, and the difference (shaded in colour) of the modelled mean from the observed RMS variability from the data of Schemm *et al.* (1992) and Schubert *et al.* (1993).

the atmosphere. Figure 5.16a shows the interannual standard deviation of mean surface air temperature for the AMIP models in DJF, forced with the observed monthly mean SST and sea ice extent, along with the differences from an observed data set for 1979 to 1988. Observations over land are from Schemm *et al.* (1992) and oceanic observations from a version of the COADS marine data set (Willmott and Legates, 1993). The broad pattern of observed interannual variability is portrayed well, with a maximum over the winter continental interiors. The oceanic maximum over the tropical eastern Pacific is forced by the imposed SST variations, and the differences from observations in Figure 5.16 over the oceans are generally small for this reason. Over high latitude continental interiors the modelled variability is typically too low. In JJA the modelled variability is also generally

too low over the North African desert, and in both seasons considerable variation exists among the models (Figure 5.16b). The blocking simulated in atmospheric models also generally shows too little interannual variability (Sausen *et al.*, 1993), as shown by the typical example given in Figure 5.17.

Figure 5.18 shows the ability of atmospheric models to reproduce the interannual variability of rainfall in north-east Brazil where the wet season rainfall is strongly influenced by tropical SST (Hastenrath, 1990; Ward and Folland, 1991; Sperber and Hameed, 1993). Figure 5.18b shows simulations of the north-east Brazil February-May rainfall anomalies by 22 AMIP models that correctly represent the observed teleconnection patterns with worldwide SST, while Figure 5.18a shows results from four models forced with observed SST for 1949 to 1993, some being ensemble averages. These results, together with an ensemble of integrations using the ECMWF model for 1979 to 1988 (not shown), indicate that current atmospheric models have the ability to simulate observed interannual rainfall variations in north-east Brazil reasonably well.

Another measure of the ability of atmospheric models to respond realistically to large-scale SST patterns is the Southern Oscillation Index (SOI), which measures variations in the coupled ocean-atmosphere system that include El Niño warming and La Niña cooling events in the tropical Pacific. Figure 5.19b shows the modelled and

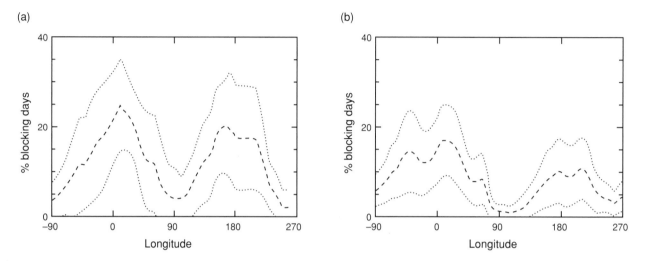

Figure 5.17: The observed mean (thick dashed line) and observed interannual standard deviation (thin dashed lines) of the Tibaldi and Molteni (1990) 500 hPa blocking index for DJF of 1960 to 1990 in the Northern Hemisphere (a), and that calculated from 6 runs of the UKMO atmospheric model forced by the GISST data set (Parker *et al.*, 1994) (b).

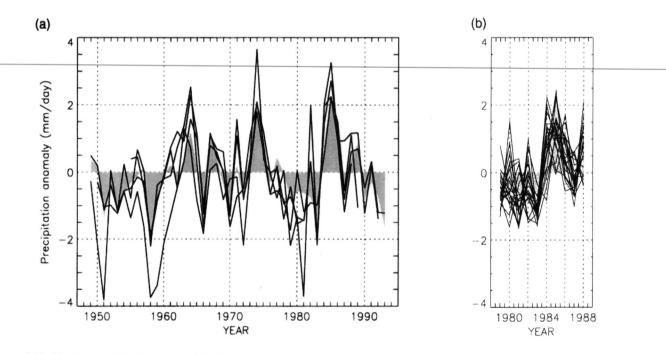

Figure 5.18: The observed March to May rainfall anomaly in north-east Brazil (approximately 4°S–10°S, 42.5°W–35°W) compared with that from (a) simulations during February to May of 1949 to 1993 by 4 models and (b) 22 selected AMIP models. Anomalies are relative to the 1979 to 1988 average. The observed data are from Hulme (1995).

(a)

(b)

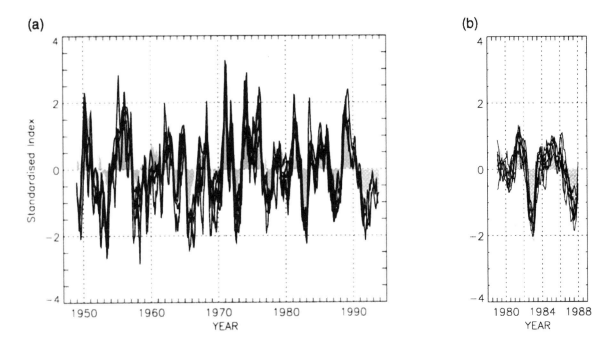

Figure 5.19: The observed Tahiti minus Darwin Southern Oscillation Index (SOI) compared with that from (a) the averages from 7 atmospheric model simulations for 1949 to 1993 and (b) 7 AMIP models. Anomalies are relative to the 1979 to 1988 average.

observed SOI indices for 1979 to 1988 from seven AMIP models, all of which capture the main features of the SOI variations. Simulated SOI indices forced by observed SST over several recent decades from versions of the atmospheric components of seven of the coupled models considered in Chapter 6 are shown in Figure 5.19a (Folland and Rowell, 1995). These results indicate that current atmospheric models respond to large-scale, large-amplitude tropical SST patterns in the Pacific with considerable skill (Smith, 1995).

A major fluctuation in global temperature occurred between 1991 and 1995 following the eruption of Mt. Pinatubo in the Philippine Islands in June 1991. Recent estimates indicate that net global radiative forcing at the top of the atmosphere peaked at about -3.5 Wm^{-2} a few months after the eruption, and then declined to small values by late 1994 (Parker *et al.*, 1996). Figure 5.20 shows the observed MSU temperature changes in the lower troposphere after the eruption as running seasonal averages (Parker *et al.*, 1996). Two predictions are also shown starting from different initial conditions made by Hansen *et al.* (1992) using a version of the GISS model with a simple diffusive ocean. The model captured the phase and magnitude of the overall changes in temperature rather well. In another study Graf *et al.* (1993) used the ECHAM2 model to successfully simulate the patterns of the

atmospheric circulation and surface air temperature anomalies that were observed in the Northern Hemisphere during the winter of 1991 to 1992 immediately following the Pinatubo eruption. These results indicate that models can respond realistically to a sudden but short-lived radiative forcing that is comparable to that following an instantaneous doubling of CO_2, and raise our confidence in atmospheric models' ability to portray longer-term climate changes. It should be noted, however, that the response of the deeper ocean has not been considered here.

5.3.1.2.4 Recent trends
Figure 5.21 shows the observed and modelled surface air temperature over the global land surface for the past several decades. The results from the UKMO model, a version of the CSIRO model (Smith, 1994), and three other models have been filtered to highlight trends on decadal time-scales, and were forced with observed SST and sea ice extent during 1949 to 1993, mostly by the GISST data set (Parker *et al.*, 1994). Although the shape of the observed warming trend is moderately well simulated, when a fixed CO_2 concentration is used only about half the observed amplitude of warming is reproduced (Folland and Rowell, 1995). When changing CO_2 is included in the UKMO model the warming increases, and when a representation of tropospheric aerosols is also included as

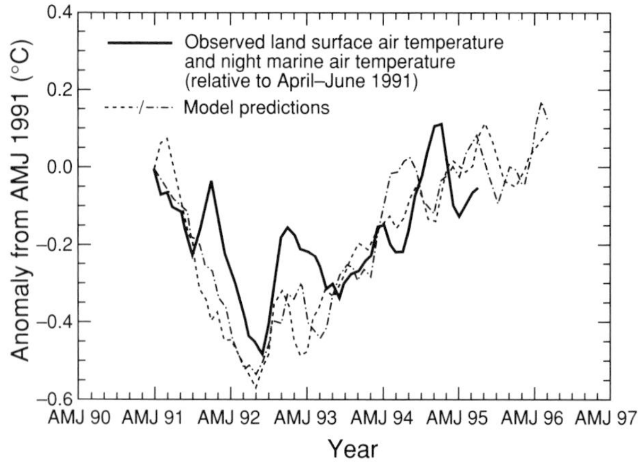

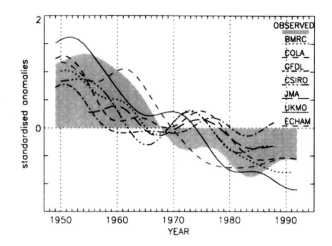

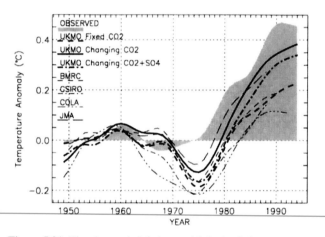

Figure 5.20: The predicted and observed changes in global land and ocean surface air temperature after the eruption of Mount Pinatubo, in terms of three-month running averages from April to June 1991 to March to May 1995 (Hansen *et al.*, 1992).

Figure 5.22: The observed and modelled standardised July to September rainfall for 1949 to 1993 in the Sahel from 7 models. Standardisation is done separately for each model for the common period 1955 to 1988.

Wolter, 1989; Folland *et al.*, 1991; Shinoda and Kawamura, 1994; Smith, 1994; Rowell *et al.*, 1995), though land-surface processes such as desertification may also be important (Xue and Shukla, 1993). Figure 5.22 shows the observed and modelled Sahel rainfall for the wet season July-September. All seven models (six of which are atmospheric components of coupled models) show decreasing trends in Sahel rainfall, though modelled trends are generally smaller than those observed. South of the Sahel the results are less consistent, suggesting that the models respond in different ways to decadally varying SST patterns in this region (the Soudan). This evidence and that discussed above indicates that current atmospheric general circulation models simulate some regional interannual climate variations and trends with useful skill, given the observed SST forcing.

Figure 5.21: The observed global annual 1.5m land air temperature for 1949 to 1993 from the Jones (1994) data set (shaded) and the corresponding modelled 1.5 m air temperature deviations from the 1950 to 1959 average for: (i) the average of 4 simulations with the UKMO model without progressive changes in radiative forcing, (ii) the average of 4 simulations with changing CO_2, (iii) the average of 4 models with a representation of tropospheric aerosols, and (iv) the averages from 4 other models with no changes in radiative forcing.

5.3.2 Simulation of Land-surface Processes

The surface climate over land is important for many aspects of human well-being. The land-surface components of coupled climate models range from rather simple schemes to complex representations of soil and vegetation. Increased realism in land-surface parametrizations has been shown to improve the representation of surface climates, but has also led to different sensitivities to atmospheric forcing, particularly in terms of precipitation and net surface radiation. The dearth of appropriate observations hinders land-surface parametrization improvement and model evaluation.

5.3.2.1 Why consider the land surface?

Models indicate that changes in the state of the continental surface affect other components of the climate system,

in Mitchell *et al.* (1995a) the warming trend is reduced but the spatial pattern of the warming is more consistent with observations (Chapter 8).

Widespread attention has been given to summer rainfall trends in sub-Saharan Africa, where wet conditions in the 1950s gave way to droughts in the 1970s and 1980s. These changes are now thought to have been influenced by changes in SST patterns (Lamb, 1978; Folland *et al.*, 1986;

including precipitation, and indirectly, sea ice distributions and ocean surface temperatures (Mintz, 1984; Bonan *et al.*, 1992). Soil moisture anomalies may also have far-reaching effects (Manabe and Delworth, 1990; Beljaars *et al.*, 1993; Koster and Saurez, 1996). In general, the simulated changes in the continental surface climate depend on the land-surface scheme employed as well as on the meteorological forcing from the atmosphere (Garratt *et al.*, 1993; Milly and Dunne, 1994; Shao *et al.*, 1995).

Land-surface processes partition the precipitation among evaporation, runoff, and temporary stores such as snow and soil moisture, and partition the net incident radiant energy between latent (evaporation) and sensible heat loss. The land surface also alters radiative fluxes through the albedo and surface temperature. Momentum exchange between the atmosphere and the land surface is determined by the representation of surface drag. The parametrization of these processes affects short-period meteorological behaviour (Walker and Rowntree, 1977; Rowntree and Bolton, 1983; Beljaars *et al.*, 1993) and the longer-term climate (Dickinson and Henderson-Sellers, 1988; Manabe and Delworth, 1990; Xue and Shukla, 1993; Milly and Dunne, 1994; McGuffie *et al.*, 1995).

In the 1970s and early 1980s, the representation of land-surface processes used prescribed albedos and roughness lengths and a simple "bucket" hydrology (precipitation filled the "bucket", evaporation emptied it, and runoff occurred when the "bucket" overflowed; Manabe, 1969). More recent land-surface parametrizations include representations of leaves, roots, multiple soil types and layers, an interception store, the dependence of albedo, roughness and stomatal conductance on the type of vegetation and its canopy geometry, and carbon uptake and respiration (Sellers, 1992; Dickinson *et al.*, 1993; Koster and Saurez, 1994). These land-surface schemes generally reduce the overestimation of land-surface evaporation given by "bucket" schemes in the tropics (Sellers, 1992; Scott *et al.*, 1996), and improve the simulation of the diurnal cycles of evaporative and other surface fluxes (Dickinson and Henderson-Sellers, 1988). The divergence among current land-surface parametrization schemes (Pitman *et al.*, 1993; Henderson-Sellers *et al.*, 1995), when combined with the different forcing supplied by different atmospheric models (Shuttleworth and Dickinson, 1989; Garratt, 1993), results in a wider range of land-surface hydroclimates than was reported in IPCC (1990). Although determining soil moisture or river discharge off-line in an impact model can produce different results from those determined in a GCM, it is difficult to discern the better schemes because adequate evaluation data are generally lacking.

5.3.2.2 *Land-surface fluxes*

Fluxes between the atmosphere and land may be poorly simulated because of inadequacies in the atmospheric and land-surface models themselves. For example, the bucket model is known to exaggerate daytime evaporation when energy and soil water are available (Dickinson and Henderson-Sellers, 1988), with a consequent distortion of the diurnal temperature range. Shuttleworth and Dickinson

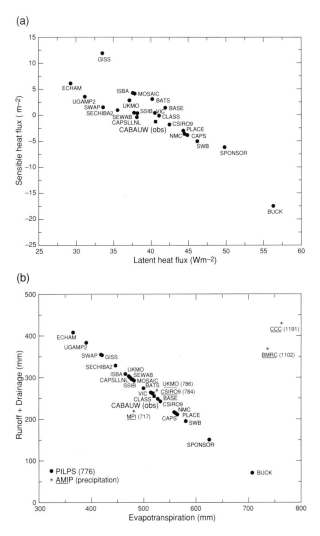

Figure 5.23: (a) The annually averaged latent and sensible heat fluxes predicted by the PILPS land-surface schemes. A single year's observations from Cabauw, the Netherlands, were used for as many annual cycles as was required for each land-surface scheme to conserve energy (≤ 3 Wm^{-2}) and water (≤ 3 mm/yr). (b) As in (a) but for the annual totals of evaporation and runoff plus drainage. The AMIP/PILPS models' 10-year mean values are the weighted average for the closest GCM grid point and as many of the surrounding eight grid points that as designated as land. (The AMIP precipitation totals are given in parentheses and differ from those prescribed for the off-line forcing.)

(1989) also showed that use of the Biosphere-Atmosphere Transfer Scheme (BATS) can lead to erroneous results when the incident surface solar radiation provided by the atmospheric GCM is poor (Garratt, 1993).

The Project for Intercomparison of Land-surface Parametrization Schemes (PILPS) (Henderson-Sellers *et al.*, 1993, 1995) is evaluating land-surface simulations. Figure 5.23a shows the uncoupled annually averaged partition of net incident radiant energy into sensible and latent heat with prescribed atmospheric fluxes at Cabauw, the Netherlands. The scatter in the energy partition is due to different calculations of the surface temperature and, to a lesser extent, to the differences in winter surface albedos; the range (from upper left to lower right) is the result of the use of different land-surface parametrizations. The land surface also interacts with the ocean through fresh water runoff, which has recently been improved in some models (Johnson *et al.*, 1993; Beljaars *et al.*, 1993; Viterbo and Beljaars, 1995; Polcher *et al.*, 1996) for the major humid river basins, but in other basins the models' poor precipitation (and perhaps evaporation) can overwhelm the improvement in the land-surface scheme (Miller *et al.*,

1994). Uncoupled simulations, in which the variation due to differences in atmospheric forcing is removed, also exhibit a wide variation in the partition of precipitation into runoff (Figure 5.23b); for example, ECHAM puts less than 50% of the precipitation into runoff, while a budget scheme such as BUCK (Robock *et al.*, 1995) puts more than 90% into runoff.

5.3.2.3 *Soil moisture simulation*

IPCC (1990) noted that "the representation (and validation) of soil moisture in current climate models is still relatively crude". Robock *et al.* (1995) find that both a "bucket" scheme and the simplified SiB (Simple Biosphere) scheme have serious deficiencies in their simulations of soil moisture in mid-latitudes. Figure 5.24 shows the annual cycle of soil moisture simulated in PILPS for the HAPEX-MOBILHY site in France (Shao and Henderson-Sellers, 1996). The range among the simulations is about 100 mm at the end of an equilibrated year, and is greatest in the summer when most schemes underestimate soil moisture. The differences in the simulation of soil moisture result in part from different runoff and drainage parametrizations

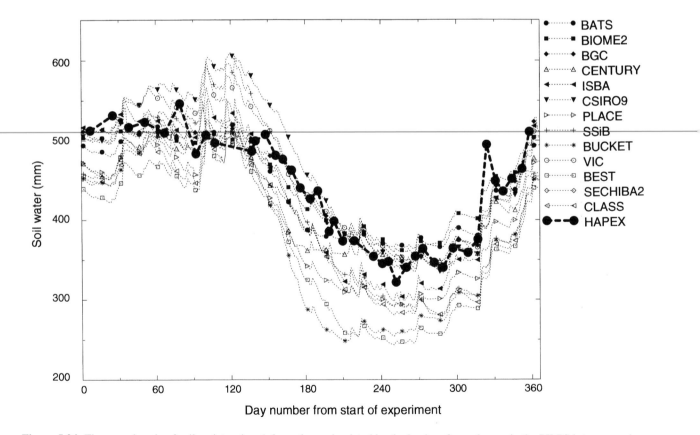

Figure 5.24: The annual cycle of soil moisture in a 1.6 m column simulated by the land-surface schemes in the PILPS intercomparison for the HAPEX-MOBILHY site in southern France (Shao *et al.*, 1995; Shao and Henderson-Sellers, 1996).

(see Figure 5.23b), and the bucket scheme with no stomatal resistance evaporates more than other schemes. Mitchell *et al.* (1990) also noted a proportionality between the seasonal variation of soil moisture in the control climate simulation and the soil moisture decrease in increased CO_2 experiments. These results emphasise the need for improved treatments of soil moisture and runoff, especially in view of the summer drying found over the mid-latitude continents in some greenhouse warming experiments with coupled models (see Chapter 6).

5.3.2.4 Summary

In summary, the general agreement found among the results of relatively simple land-surface schemes in 1990 has been reduced by the introduction of more complex parametrizations. While models can be tuned for particular locations, regional distributions depend upon the adequacy of the forcing meteorology and the antecedent conditions, both of which are difficult to establish. Use of site-specific parameters improves the simulation in more realistic models, but the lack of suitable observations is inhibiting the development and evaluation of universal land-surface schemes.

5.3.3 Oceanic General Circulation Models

5.3.3.1 Introduction

The oceanic components of present coupled models range in spatial resolution from around 1 to 4 horizontally and from 6 to 30 vertical levels. Current "high-resolution" ocean models have many more vertical levels (up to 60) and have resolutions as fine as $1/6^\circ$. The counterpart to meteorological weather occurs in the ocean in the form of eddies at scales of less than $1/2^\circ$. The high resolution models mentioned above explicitly resolve much of the mixing due to eddies but at great computational expense, while the lower resolution models typically used in climate simulations must use parametric representations of the mixing. Here the former class of model is referred to as eddy-resolving, and the latter as coarse resolution or non-eddy-resolving.

Many of the deficiencies in coarse resolution models noted by Gates *et al.* (1992) are only starting to be effectively addressed. These problems include: (i) the representation of geometry and bathymetry; (ii) the parametrization of sub-grid scale processes such as convection, mixing and mesoscale eddies; (iii) errors in surface forcing for ocean-alone simulations; (iv) a thermocline that is often too deep and too diffuse; (v) weak poleward heat transport; (vi) distortion of upper ocean and deep boundary currents; and (vii) temperature and salinity

errors in the deep waters. WOCE (World Ocean Circulation Experiment) (1993, 1994) and the recent studies of Covey (1995) and Weaver (pers. comm.) suggest that higher horizontal resolution, but not necessarily eddy-resolving, may redress some of these deficiencies.

5.3.3.2 Evaluation against present ocean climate

5.3.3.2.1 Coarse resolution models

The ocean stores heat and various chemical constituents at one place and transports them over large distances for release at some later time, perhaps many centuries later. England (1993) and England *et al.* (1993) have found that the GFDL ocean model can reproduce the observed major

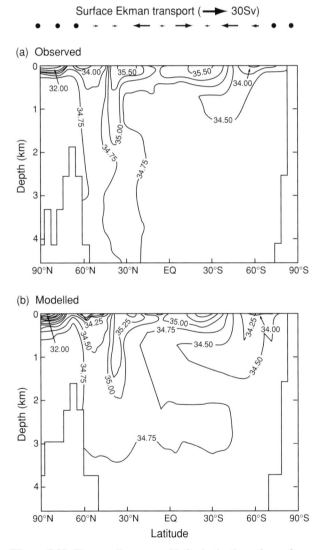

Figure 5.25: The zonally averaged latitude-depth sections of salinity (psu) in the World Ocean as observed (a) (Levitus *et al.*, 1994b) and as simulated (b). Contour interval is 0.25. Vectors at the top indicate the zonally integrated strength of the surface Ekman transport in Sverdrup (England *et al.*, 1993).

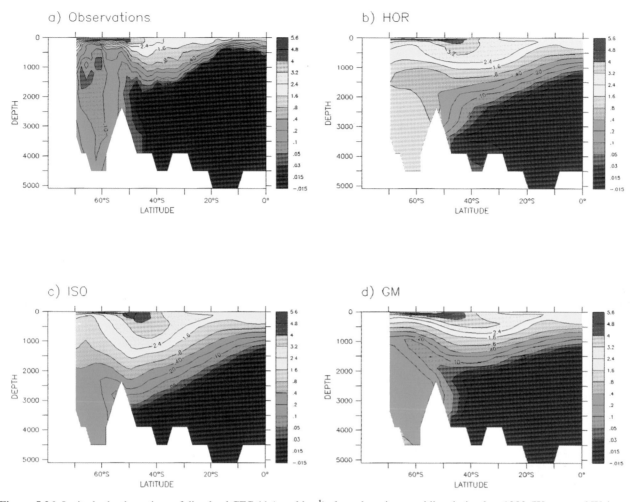

Figure 5.26: Latitude-depth sections of dissolved CFC-11 (pmol kg^{-1}) along the prime meridian during late 1983 (Warner and Weiss, 1992; Robitaille and Weaver, 1995; England, 1995) as given by (a) observations; (b) a control experiment with mixing in a Cartesian reference frame (HOR); (c) an experiment with mixing predominantly parallel to surfaces of constant density (ISO), and (d) an experiment with mixing configured according to the parametrization of Gent and McWilliams (1990) (GM).

features of the water masses of the world ocean (Figure 5.25). Such results increase our confidence that the time-scales and volumes of water involved in the conversion processes for heat and water (salinity) in climate change experiments are realistic.

Coarse resolution ocean models have also been widely used in the study of CO_2 uptake (Orr, 1993; Siegenthaler and Sarmiento, 1993), and Wallace (1995) has noted that a variety of box-diffusion ocean models converge on similar values for the net uptake of CO_2. For ocean GCMs, where the ocean carbon data are used primarily for evaluation, the consistency with simpler models increases confidence that the vertical and horizontal transport mechanisms of non-eddy-resolving models are reasonable in lower and middle latitudes. Results such as those shown in Figure 5.26b and c indicate that ocean models still have difficulty in

simulating observed tracer distributions, especially in regions where deep convection occurs. England (1995) and Robitaille and Weaver (1995) suggested that the model estimates of the CFC-11 uptake in the Southern Ocean were too large if the usual mixing schemes (Figure 5.26b and c) were employed, and that the role of the Southern Ocean in moderating climate change may be over-estimated. Improvements in the parametrization of eddy mixing have alleviated some of this problem as shown in Figure 5.26d, but errors still exist in the thermocline region.

Data assimilation experiments have also been conducted in recent years with coarse resolution ocean models similar to those used in climate simulation. Such methods provide an ideal environment for model evaluation since they yield not only an estimate of the distribution of oceanic variables based on data and model physics, but also quantify the expected errors in these estimates (Wunsch 1989; Bennett 1992; Tziperman and Bryan, 1993). For example, Marotzke and Wunsch (1993) found that it was not possible to make a steady meridional overturning in the North Atlantic consistent with both the model and observations, indicating a weakness in the model formulation and/or errors in the data.

5.3.3.2.2 Eddy-resolving models

It is presently not possible to incorporate an eddy-resolving ocean component in a coupled climate model because of the lack of computational resources. While such models provide an important means for testing processes and parametrizations, they are subject to the effects of spin-up in response to initial conditions and forcing to at least as great an extent as are coarse-resolution ocean models. The near-eddy-resolving global model of Semtner and Chervin (1988, 1992) has been integrated at $\frac{1}{4}^{\circ}$ resolution at NCAR and at $\frac{1}{6}^{\circ}$ at LANL (Los Alamos National Laboratory), while the WOCE Community Modelling Effort includes a high-resolution GFDL ocean GCM for the North Atlantic (Bryan and Holland 1989; Boning *et al.*, 1991) with resolution of 10-15 km and a high-resolution model on constant density surfaces (Campos and Bleck, 1992). The model developed for studies of the Antarctic

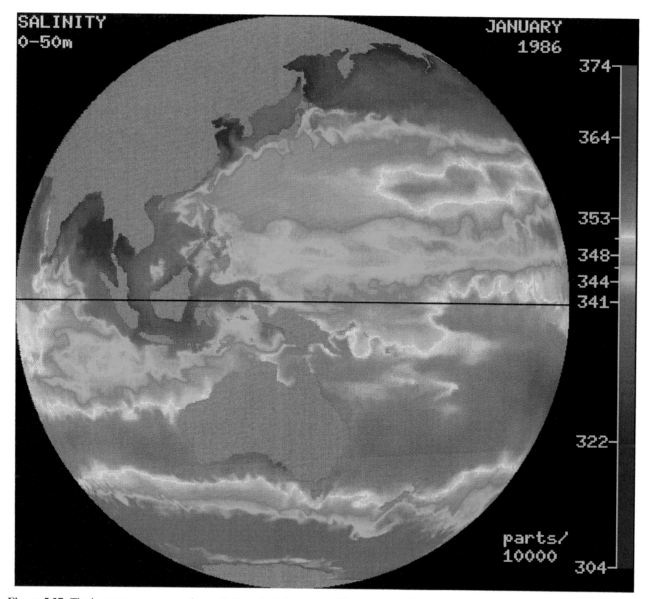

Figure 5.27: The instantaneous sea surface salinity field in January 1986 as simulated by the NCAR/Los Alamos eddy-resolving model (Semtner, 1994).

Circumpolar Current (FRAM Group, 1991) is being tested at comparable resolution in the UK Ocean Circulation and Climate Advanced Modelling (OCCAM) project.

Semtner (1994) finds that many of the known features of the global ocean circulation are reproduced by the sixth-degree global simulation at Los Alamos. In particular, this model provides support for the hypothesised routes of low-salinity warm water from the western Pacific to the North Atlantic via the Indonesian throughflow (Figure 5.27), which is a critical element of North Atlantic overturning. This result suggests that non-eddy-resolving ocean models without such a route may not realistically account for this overturning. Figure 5.28 shows analyses of the horizontal wavenumber spectra (representing the potential energy as a function of scale) at 400 m and 800 m at 24°N in the western Pacific Ocean, based on observations and on the NCAR and Los Alamos model integrations. The NCAR model matches the observed variability at 800 m but shows too little variability at 400 m, even at scales which are well resolved by the model. The finer-resolution Los Alamos model matches the observed temperature variability at both depths except at the shortest wavelengths. From these and other comparisons between observed and modelled spatial variability, it has been suggested that $\frac{1}{12}^{\circ}$ resolution will be needed in ocean models before the modelled variability converges to the observed spectra.

The TOPEX/POSEIDON satellite mission is producing data on the global sea surface elevation with an accuracy of ±4 cm. Such measurements provide a unique source of data for evaluating ocean circulation on small scales. Figure 5.29 shows the standard deviation of sea surface height from the $\frac{1}{6}^{\circ}$ simulation at LANL and from TOPEX altimetry data. The agreement between the observed and modelled distributions increases confidence in the realism of the model's simulations.

Convection and meridional circulation in the vicinity of the Antarctic have been identified as key processes for moderating the warming effect of enhanced greenhouse gas concentrations. In this context, the results of Döös (1994) and Döös and Webb (1994) are particularly important. Using the Fine Resolution Antarctic Model (FRAM Group, 1991) they showed that the wind-driven cell found around the latitude of the Drake Passage in coarse resolution models (the so-called Deacon Cell) is associated with changes in the depth of the density surfaces between South America and the interior of the ocean, and involves little transport through density surfaces. However, these results and those of Klinck (1994) from the CME model (Bryan and Holland, 1989) suggest that eddy-resolving models need to be integrated over the millennial periods characteristic of the ocean's

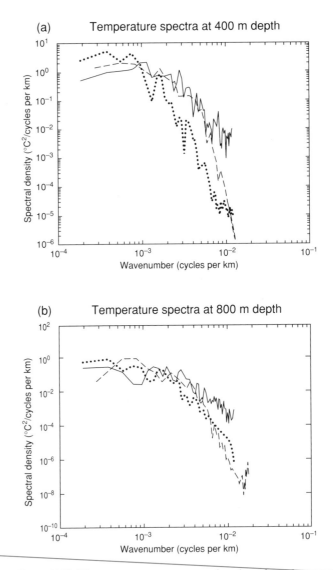

Figure 5.28: The temperature spectra as a function of zonal wavenumber at 24°N in the Pacific for (a) 400 m depth and (b) 800 m depth. The observed structure is shown as a solid line, and the results from the NCAR model are shown as a dotted line for $\frac{1}{4}^{\circ}$ by $\frac{2}{5}^{\circ}$ resolution and as a dashed line for $\frac{1}{6}^{\circ}$ resolution (Semtner, 1994).

thermal adjustment time before actual water mass conversion processes are accurately represented.

5.3.3.3 *Evaluation of temporal variability*

One of the characteristics of ocean circulation is that it appears to have significant energy on many time-scales; some of the variability is clearly externally forced (e.g., seasonal variations), some appears to be the result of atmosphere-ocean coupling (e.g., El Niño-Southern Oscillation), and some is due to internal ocean variability

(a)

Modelled surface height variability

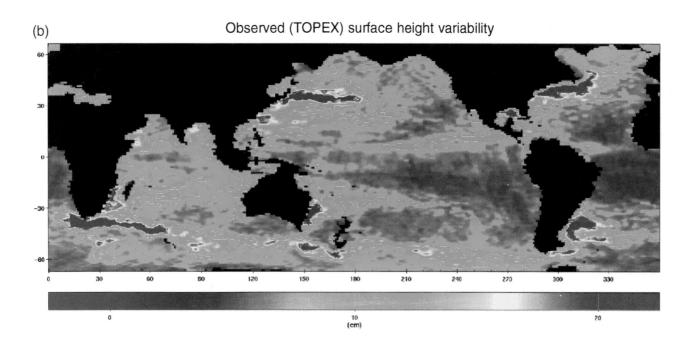

(b)

Observed (TOPEX) surface height variability

Figure 5.29: The standard deviation of sea surface height from (a) the $\frac{1}{6}^{\circ}$ by $\frac{1}{4}^{\circ}$ eddy-resolving simulation conducted at LANL (averaged over 1985 to 1994), and (b) TOPEX altimetry data averaged from Oct. 1992 to Oct. 1994 (Wunsch, 1994).

(e.g., eddies and internal fluctuations of the thermohaline circulation). Evaluation of the longer time-scales is difficult due to the lack of data, and even at interannual time-scales the available data are barely adequate.

5.3.3.3.1 Seasonal to interannual variability

The seasonal cycle in surface temperature and surface heat and moisture fluxes is accompanied by seasonal cycling in the upper ocean stratification (i.e., deep mixing in winter and a shallow thermocline in summer). Very few ocean models used in coupled simulations have adequate representation of these processes, and evaluation is difficult because of the paucity of suitable global data sets (see Levitus and Boyer, 1994).

Global ocean-only models have simulated realistic modes of interannual variability (Jacobs *et al.*, 1994; Zhang and Endoh, 1994), although it is clear that the model must possess adequate resolution in the tropical region in order to avoid distortion of the important equatorial Kelvin and Rossby waves. Philander *et al.* (1992) have noted that low-resolution ocean models may not be capable of representing the important mechanisms of ENSO, though in coupled systems they do possess ENSO-like oscillations (see Section 5.2.4.2).

5.3.3.3.2 Decadal and longer variability

While there has been considerable progress in the study of ocean variability at decadal and longer time-scales with stand-alone ocean models (Weaver and Hughes, 1992), the lack of suitable verifying data in the deep ocean has inhibited model evaluation. Long SST data sets have enabled extended atmospheric model integrations under realistic boundary conditions (Bottomley *et al.*, 1990; Parker *et al.*, 1994), but similar integrations for the ocean will require, at a minimum, an extended record of global wind stress and estimates of fresh water forcing, particularly at high latitudes. Nevertheless, decadal variability has been found in ocean models in basins where deep water formation occurs (Weaver and Sarachik, 1991a; Greatbatch and Zhang, 1994; Weaver *et al.*, 1994; Weisse *et al.*, 1994), and further analysis of such results might help to interpret some of the causes of decadal variability found in coupled models such as those of Delworth *et al.* (1993) (see Section 5.2.4.3). The latter authors have identified irregular oscillations of the North Atlantic thermohaline circulation in a long coupled model simulation, in which the circulation anomalies are driven by out-of-phase density anomalies in the sinking and rising regions. The spatial patterns of the associated sea surface temperature anomalies resemble an interdecadal pattern identified in

observations by Kushnir (1994). Another class of decadal variability, however, exists in mid-latitudes that arises from unstable ocean-atmosphere innteractions, as shown by Latif and Barnett (1994).

A fundamental period for ocean model variability also occurs on the century time-scale (Mikolajewicz and Maier-Reimer, 1990; Weaver *et al.*, 1993; Winton and Sarachik, 1993). That such modes exist in some climate models is not in dispute, although the lack of observational data leaves open the questions whether these unstable modes have counterparts in reality (Bond *et al.*, 1993) or whether counterparts exist in coupled climate models. A similar caveat applies to modelled variations of the thermohaline circulation on millennial time-scales (Marotzke, 1989; Weaver and Sarachik, 1991b; Wright and Stocker, 1991).

5.3.3.4 Issues in ocean model evaluation

5.3.3.4.1 Evaluating model processes

Observationalists have for some time recognised the importance of bathymetry in shaping the global exchange and conversion of water masses. Döscher *et al.* (1994) incorporated a detailed representation of the sills which control deep water flow in the North Atlantic, and showed that this improved the simulation of the circulation and heat transport. Recent studies of the mixing associated with mesoscale eddies (Gent and McWilliams, 1990; Gent et al., 1995) and that associated with convection (Send and Marshall, 1996) are motivated in part by recognised discrepancies between ocean climate models and observations. Danabasoglu et al. (1994) have shown promising results from experiments using a parametrization for mesoscale eddy-induced mixing. When such parametrization was used in an oceanic general circulation model, the thermocline became sharper, the deep ocean became colder, and the meridional transport of heat and salt increased (Figure 5.30). Robitaille and Weaver (1995) and England (1995) have also shown that this parametrization improves the ability of a global ocean model to reproduce observed CFC distributions (see Figure 5.26). All these features are improvements on the results obtained using traditional mixing schemes, and address some of the recognised weaknesses of ocean models used for climate studies.

5.3.3.4.2 Boundary forcing

The evaluation of ocean surface boundary conditions is critical, since the surface forcing ultimately determines the large-scale ocean circulation. There have been considerable improvements in the determination of the surface wind stress and surface heat and moisture fluxes

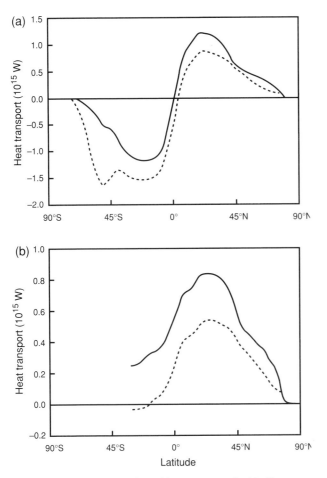

Figure 5.30: The mean northward heat transport in (a) all oceans and (b) in the Atlantic for experiments conducted with a global model using the Gent and McWilliams (1990) parametrization for sub-grid scale mixing (solid line) and the traditional lateral/vertical mixing scheme (dashed line) (Danabasoglu *et al.*, 1994).

(OOSDP, 1995), and atmospheric reanalysis projects should lead to further improvements.

5.3.3.4.3 Observations

In general, the ocean is so poorly observed and the instrumental record so incomplete that there are regions of the ocean for which no observations exist. While over one hundred years of instrumental data are available for the surface of the ocean in COADS (Slutz *et al.*, 1985), the situation for internal ocean data is much more restricted. A recent effort at "data archaeology" (Levitus and Boyer, 1994; Levitus *et al.*, 1994a,b) has expanded the total data record and should allow a better benchmark of the ability of ocean models to simulate the mean state of the ocean (WOCE, 1994).

A future ocean observing system designed specifically for climate purposes is being considered by the Global Climate Observing System (GCOS) and the Global Ocean Observing System (GOOS). Since there are now no continuous long-term measurement sites in the deep ocean, the establishment of such a system is important for the evaluation of climate models and for the measurement of the natural variability that is necessary for the unambiguous detection of a climate response to anthropogenic forcing (see Chapter 8).

5.3.3.5 Summary

The observational data sets on which the evaluation of ocean models rests have significant spatial and temporal shortcomings, with many regions of the global ocean not sampled and many others having only short sampling records. This means it is difficult to form a reliable estimate of the ocean climate and its variability. Evaluation of ocean models in the coastal region is critical for regional interpretation of projected changes in sea level, and the interactions between the open ocean and coastal regions are affected by the fresh water input to ocean models from continental runoff (Section 5.3.2). The need for flux adjustment is an explicit recognition of the inadequacies of the components of coupled climate models (Section 5.2.2), and evaluation of (and better parametrizations for) air-sea exchanges are necessary for the sustained improvement of ocean models.

In summary, evaluation of the climate simulation capability of ocean GCMs shows that they realistically portray the large-scale structure of the oceanic gyres and the gross features of the thermohaline circulation. The models' major deficiencies are the representation of mixing processes, the structure and strength of the western boundary currents, the simulation of the meridional heat transport, and the portrayal of convection and subduction. Some of these shortcomings may be related to the relatively coarse resolution of the ocean components of current coupled models.

5.3.4 Sea Ice Models

Although sea ice occupies only a small portion of the ocean surface, it has an important role in climate variation through its effects on the surface energy and moisture exchanges. The simplest sea ice model consists of a motionless uniform layer of ice whose thickness is governed by thermodynamics. More realistic models incorporate snow on the ice, fractional ice coverage, multi-layer ice, the effects of salinity, and, most importantly, sea ice dynamics. Because of the sea ice motion, melting may occur in places far away from the formation area. Averaged

(a)

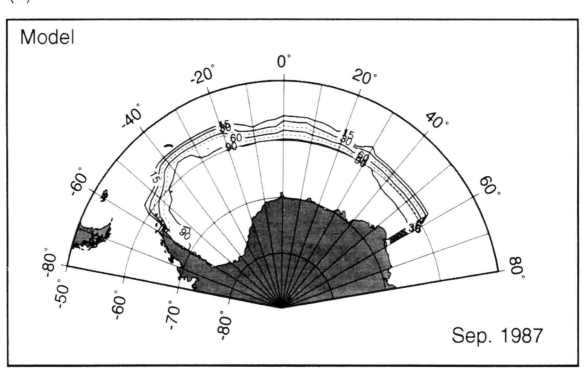

(b)

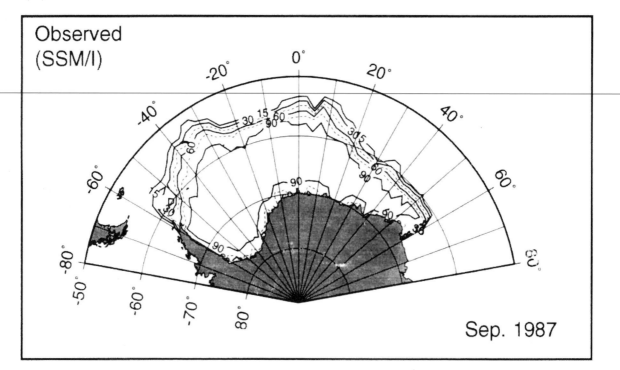

Figure 5.32: (a) Simulated and (b) observed sea ice concentrations (in %) in the Weddell Sea for September 1987 (Fischer and Lemke, 1994).

over a seasonal cycle, the net freezing rate in a particular area is rarely zero due to the divergence or convergence of the sea ice flow. The net freezing rates represent surface buoyancy fluxes, which in turn affect the density structure and baroclinic flow in the ocean. The characteristics of the sea ice motion depend strongly on the rheological characteristics of the sea ice as a plastic material, and also on the geometry of the coastlines and the atmospheric forcing fields, especially the wind.

Recent experiments with sea ice models with different internal ice stress parametrizations (Ip *et al.*, 1991; Flato and Hibler, 1992) have shown that dynamic/ thermodynamic models can describe the observed seasonal cycle of ice extent (Figure 5.31) and the location of the ice edge (Figure 5.32, opposite) quite realistically (Fischer and Lemke, 1994; Harder and Lemke, 1995). Comparison of observed and modelled buoy trajectories also indicates that the large-scale structure as well as the small-scale excursions of the sea ice flow are represented well. A proper description of the sea ice motion is especially important to obtain the correct buoyancy flux forcing for the ocean.

Evaluation of sea ice in the current generation of models is usually limited to comparison of the modelled and observed ice edge position (as available from satellites, e.g., Gloersen *et al.*, 1992). A common problem in coarse-resolution coupled climate models is an ice edge that is too far equatorward in the North Atlantic but not far enough equatorward in the Southern Ocean (Washington and Meehl, 1989; Henderson-Sellers and Hansen, 1995). Under these circumstances it is difficult to achieve an adequate simulation of the surface fresh water fluxes that are crucial to sea ice formation. Flux adjustment may be undertaken for ice-covered areas, although it is not as straightforward as over the open ocean because of strong feedback effects. (For instance, if the flux adjustment is negative the result may be an unstable growth of ice when the ice advances over a cooling ocean.) Observations of sea ice thickness are needed for evaluating the model treatment of thermodynamic growth and melting, ice dynamics, and the effects of ocean and atmospheric circulation on ice deformation; only a tentative climatology is available for the Arctic based on sparse submarine sonar observations (Bourke and Garrett, 1987) and drill-hole observations in the thinner ice of the Antarctic. Satellite-based ice concentration and some ice velocity fields from Arctic buoys are also available.

In summary, evidence indicates that incorporating sea ice dynamics reduces the modelled climate sensitivity (Hibler, 1984; Pollard and Thompson, 1994; Ramsden and Fleming, 1995), which underlines the need for

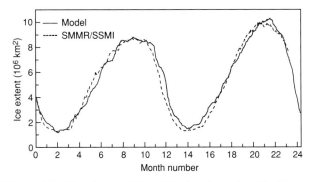

Figure 5.31: The observed (dashed line) and simulated (solid line) sea ice extent in the Weddell Sea for 1986 and 1987 (Fischer and Lemke, 1994).

increased attention to the sea ice in coupled climate models. In addition to verification data, accurate atmospheric forcing fields are necessary to obtain realistic ice simulations. Although dynamic/thermodynamic sea ice models are less sensitive to variations of atmospheric and oceanic boundary conditions than purely thermodynamic models, their accuracy requirements are nevertheless quite high.

5.4 How Well Do Models Perform Under Other Conditions?

Numerical weather analysis and prediction and palaeoclimate simulations provide opportunities for model evaluation under conditions that are different from those in the simulation of the current climate, and constitute an important additional source of confidence in models.

5.4.1 Weather Analysis and Prediction
There has been a progressive increase in the overall accuracy of operational (atmospheric) numerical weather prediction (NWP) models as measured in their daily forecasts, as illustrated in Figure 5.33. The sea level pressure forecast scores show a similar increase in accuracy in the 1980s, although the improvement has slowed in the 1990s. Some of this improvement is due to increased resolution and some to improvements in physical parametrizations and data assimilation schemes. The evaluation of NWP models can lead to the identification of deficiencies in the generally similar atmospheric models used in coupled climate simulations (Hollingsworth, 1993).

The reanalysis projects in progress at several operational centres with "frozen" state-of-the-art atmospheric models and advanced data assimilation systems will provide a multi-year data set of reasonable homogeniety that is suitable for climate model evaluation. Using data produced by operational NWP models, the intercomparison of modelled surface stress and precipitation has shown good agreement over the ocean, although there are significant intermodel differences in the estimated solar radiation and sensible/latent heat fluxes (CAS/JSC, 1994). It is anticipated that atmospheric water budgets, surface fluxes and oceanic heat transport estimates will be substantially improved from the results of reanalyses.

Evidence of the skill in predicting the onset and decay of atmospheric blocking has also been accumulating in the context of short- and medium-range weather prediction (Tibaldi and Molteni, 1990; Tibaldi *et al.*, 1994) and dynamical extended-range forecasting (Tracton *et al.*, 1989; Miyakoda and Sirutis, 1990; Tracton, 1990; Anderson, 1993) (See Figure 5.17). The prediction of blocking onset is on average very poor beyond a few days into a forecast. This appears to be related to the fact that blocking is associated with subtleties in the dynamical balances that maintain it, which make it critically dependent upon the accuracy of the initial conditions, model resolution and model parametrizations (Palmer *et al.*, 1990; Anderson, 1993; Ferranti *et al.*, 1994; Tibaldi *et al.*, 1994).

The simulation of tropical cyclones may be considered an indicator of a model's ability to portray extreme events. In an intercomparison of tropical cyclone track forecasts in the north-western Pacific by operational NWP models

(CAS/JSC, 1994; Muroi and Sato, 1994) it was found that tropical cyclones tend to dissipate earlier in the models than in the real atmosphere. Attempts to infer tropical cyclones in climate models have had some success (Broccoli and Manabe, 1990; Haarsma *et al.*, 1993), although the limited resolution of earlier models demands caution in any "prediction" of tropical cyclone changes in a warmer climate (Lighthill *et al.*, 1994). More recent studies with models of higher resolution appear to simulate tropical cyclone climatology more realistically (Bengtsson, 1994).

5.4.2 *Palaeoclimate Simulation*

Knowledge of the climate's response to the long-term changes in external forcing and in the configuration of the Earth's surface over geological time-scales provides a unique opportunity to increase our confidence in climate models. The relative scarcity of palaeoclimatic data, however, restricts their use in model evaluation to a portrayal of general climatic regimes rather than the detailed evaluation of models and processes afforded by modern observations.

The changes of the Earth's climate during the Holocene and since the last glacial maximum are large and comparatively well-documented, and data are available for most continents; the time control (based upon radiocarbon dating) is reasonable, and proxy estimates of

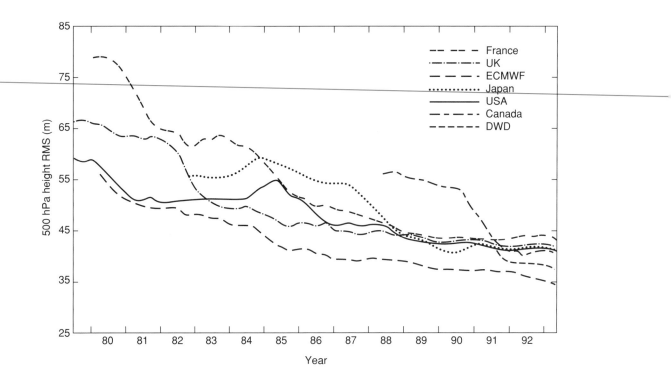

Figure 5.33: Evolution of the root-mean-square error of 72-hr forecasts of the 500 hPa height over the period 1979 to 1992 at operational NWP centres in France, UK, ECMWF, Japan, USA, Canada and Germany (DWD) (CAS/JSC, 1994).

palaeoclimatic conditions can be obtained from a variety of environmental records (see Chapter 3, Section 3.6). Using such data, atmospheric models have been used to simulate the climate of the last glacial maximum (IPCC, 1990; Joussaume, 1993). The glacial boundary conditions consist of the ice sheets' topography and sea ice, CO_2 levels, ocean surface temperature, and land albedo. Changes in the ice sheets themselves or in vegetation, however, are not simulated (see Chapters 7 and 9). The earlier ice age SST data sets (IPCC, 1990; Wright *et al.*, 1993) are now being expanded and revised for use in the newly organized Palaeoclimate Modelling Intercomparison Project (PMIP), which is focusing on simulations for the last glacial maximum and for 6000 years BP using atmospheric models with both fixed SST and mixed-layer oceans.

The widespread warmth and enhanced northern summer monsoon circulations of the mid-Holocene have also recently been simulated by applying appropriate changes in the orbital parameters (Mitchell, 1993; Phillips and Held, 1994). Modelling studies of pre-Pleistocene climates (i.e., those prior to about 1.8 million years ago) have concentrated on how to achieve warm, ice-free conditions year-round at high latitudes, especially over continental areas, and on the conditions necessary for the initiation, maintenance and demise of land ice (Rind and Chandler, 1991; Sloan and Barron, 1992; Barron *et al.*, 1993; Kutzbach and Ziegler, 1993). Some data, however, indicate that the tropical SST has remained relatively stable during times when CO_2 levels were probably higher (Crowley, 1993), while climate models predict increasing tropical SST with higher CO_2 (e.g., Cubasch *et al.*, 1992) (see Chapter 3).

5.5 How Well Do We Understand Model Sensitivity?

Issues relating to the equilibrium and transient response of models to changes in forcing due to increasing CO_2 and aerosols over the period of the observational record are discussed in Chapter 6. Although such sensitivity experiments are an important source of model evaluation, they are more appropriately considered in the context of model projections of climate change than in connection with the evaluation of the models' control or reference simulations. Therefore, only model sensitivity to changes in their formulation, boundary conditions and/or parametrizations are discussed here.

In general, any model sensitivity indicates an area of uncertainty where further research is needed, while a model's response to changes in external forcing indicates its potential for climate change. Some sensitivity experiments indicate that a particular formulation or

parameter value gives a better evaluation of the model, but the required observations are often not available. In other sensitivity experiments with idealised changes in forcing, such as those of Cess *et al.* (1990, 1991, 1993) and Randall *et al.* (1994), no systematic evaluation against observed data is generally possible, although such experiments highlight the importance of particular processes and illustrate the wide differences that can exist among models.

While a given parametrization scheme may perform well in "off-line" tests, interactions with other parametrizations and with dynamics may not result in an improved simulation in a coupled model. A particular scheme may also perform well in one model but perform poorly when used in a different model. The development of parametrizations that can be easily moved from one model to another would help to understand this behaviour, and would facilitate more systematic model evaluation (Roeckner *et al.*, 1995).

5.5.1 Sensitivity to Representation of Water Vapour

The distribution of moisture is characterised by large values near the surface and by low values in high latitudes and in the upper troposphere. This can lead to a number of computational difficulties, of which the appearance of non-physical "overshoots" and "undershoots" of moisture are the most severe (Rasch and Williamson, 1990). This has led some modelling groups to use semi-Lagrangian methods for moisture transport (Williamson and Rasch, 1993), in which the moisture is tracked by individual air parcels rather than being carried on a fixed numerical grid. Hogan and Rosmond (1991), Thuburn (1993), Navarra *et al.* (1994) and Boer (1995) have also reduced errors in the modelled moisture distribution by using special numerical techniques.

5.5.2 Sensitivity to Resolution

Debate continues over the most appropriate horizontal and vertical resolution to use in climate models. Kiehl and Williamson (1991), Boyle (1993) and Williamson *et al.* (1995) find noticeable differences in the quality of atmospheric simulations made with low horizontal resolution spectral models (such as R15 or T21) and those made with medium resolution (such as T42). While some climate statistics converge as the resolution is increased still further, others do not, especially when considered on a regional basis. The strong (and often implicit) dependence of a model's parametrizations on resolution makes it difficult to separate the purely dynamical and physical effects of resolution changes, and the climate of lower resolution models is particularly sensitive to the

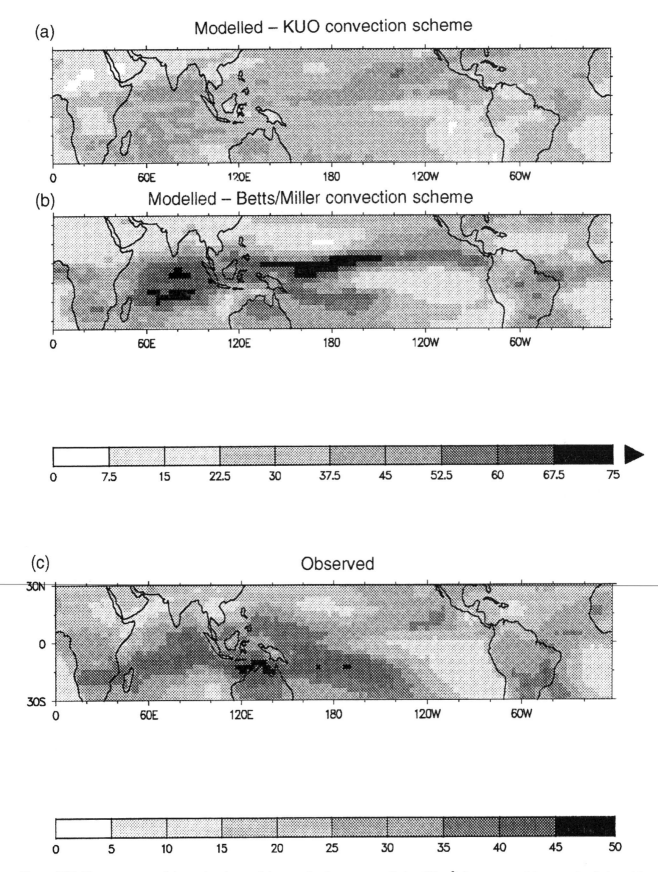

Figure 5.34: The square root of the total variance of the outgoing long-wave radiation (Wm^{-2}) for a perpetual January simulation with the UGAMP atmospheric GCM using the convection schemes of (a) Kuo (1974) and (b) Betts-Miller. (c) As in (a) and (b) but for the ensemble mean for December-February from NOAA AVHRR data during 1979 to 1991. (Betts and Miller, 1993).

formulation and strength of the sub-grid scale damping (Held and Phillipps, 1993). Studies on the impact of vertical resolution have also been conducted, and indicate that higher resolution near the surface and near the tropopause generally improves the simulations (Bushell and Gregory, 1994).

The sensitivity of ocean models to horizontal resolution is also an important issue in climate simulation with coupled models. Covey (1995) has shown that there is a resolution sensitivity of the annual mean meridional heat transport simulated with a global eddy-resolving model (Semtner and Chervin, 1992), while Washington *et al.* (1994) have demonstrated that a moderate-resolution (1 and 20 vertical levels) ocean model is capable of simulating many features of the corresponding eddy-resolving version (see Section 5.3.3.2.2). The effect of resolution on heat transport has also been recently considered by Fanning and Weaver (1995).

Atmospheric and ocean models are sometimes formulated with spatially variable horizontal resolution in order to provide more detail in areas of particular interest or importance, such as regions where there are generally large gradients of precipitation or circulation. This technique allows improved simulation in selected regions while permitting full interaction with the larger-scale climate, and is an alternative to the use of imbedded fine-mesh mesoscale models discussed in Chapter 6 (see also Section 5.2.5).

5.5.3 Sensitivity to Convection and Clouds

The sensitivity of atmospheric models to the convective parametrization has been investigated by several groups (Slingo *et al.*, 1994; Zhang and McFarlane, 1995). The use of alternative convective parametrizations is found to have a marked influence on a model's transient behaviour in the tropics as shown in Figure 5.34. In the UGAMP model, the Kuo scheme significantly underestimates the observed variability of the outgoing long-wave radiation, while there is a tendency for the Betts-Miller scheme to overestimate it. Recent studies by Wang and Schlesinger (1995) suggest, however, that with some adjustment of parameters (e.g., the threshold relative humidity) currently used convection schemes can be made to produce realistic intraseasonal oscillations.

The sensitivity of the simulated tropical precipitation to the choice of convective parametrization has also been studied by Numaguti (1993), Colman and McAvaney (1996) and McAvaney *et al.* (1995), among others. In general, the Kuo moisture convergence scheme is found to produce maxima of precipitation on both sides of the

equator in contrast to other convective parametrizations. Cess *et al.* (1995) have diagnosed a systematic error in the seasonal variation of the cloud radiative forcing in models that use a moisture convergence closure in their convective parametrization (see also Figure 5.12).

Climate models are also known to be sensitive to details of the parametrization of cloud microphysical processes. In particular, Fowler and Randall (1996) have found significant differences in the proportions of cloud liquid water and cloud ice in the CSU model depending upon the temperature range in which they are allowed to coexist, although the inclusion of cloud microphysics in the CSU model generally improves the simulation of cloud radiative forcing. Other studies show a marked sensitivity to the choice of cloud parametrizations (Mitchell *et al.*, 1989; Le Treut and Li, 1991; Boer, 1993; Senior and Mitchell, 1993; Washington and Meehl, 1993; Meehl and Washington, 1995; Kiehl, 1996), and a consensus on which parametrizations are the most appropriate has not yet been reached.

5.5.4 Sensitivity to Land-surface Processes

The sensitivity of the surface climate to the parametrization of the land surface has been considered in Section 5.3.2. In anticipation of their importance to climate change studies (see Chapter 6), here we note that recent studies by Thompson and Pollard (1996) show that the Land Surface Transfer (LSX) scheme produces smaller variations of soil moisture in a greenhouse simulation than does a bucket scheme. Henderson-Sellers *et al.* (1995) have also found that a doubling of stomatal resistance in the Biosphere-Atmosphere Transfer Scheme (BATS) significantly reduces the evaporative flux over the northern mid-latitude continents in summer. These results demonstrate the urgent need for further studies of the climate's sensitivity to land-surface processes in order to determine which schemes are the more realistic.

5.5.5 Sensitivity to Initial Conditions and Surface Boundary Conditions

It is well-known that numerical weather forecasts cannot be made with useful accuracy beyond a few weeks ahead, after which the solutions bear no resemblance to the actual sequence of events. This limit on forecasting skill is due to the inevitable growth of errors too small to detect in the initial and/or boundary conditions. Such sensitivity is termed "chaos", and is also present in the GCMs used for climate simulation. In the case of climate, however, since we are primarily interested only in the statistics of the models' simulation and not the sequence of individual solutions, a model's chaotic behaviour may be effectively removed by considering an ensemble of simulations made with different initial conditions. This technique is especially useful when considering interannual variability (see Section 5.3.1.2.3) or the transient response of a model to external forcing as in Chapter 6.

The sensitivity of the thermohaline circulation to the surface boundary conditions in coarse resolution ocean models has been explored by Birchfield (1989), Power and Kleeman (1993, 1994), Zhang *et al.* (1993), Mikolajewicz and Maier-Reimer (1994), Power *et al.* (1994, 1995) and Rahmstorf and Willebrand (1996), although the corresponding studies with eddy-resolving models have not yet been made. These studies indicate that more realistic formulations of the surface fluxes in ocean models generally increase the stability of the thermohaline circulation and generally improve the realism of the solutions.

5.5.6 *Summary*

Clouds, the hydrological cycle and the treatment of the land surface remain the largest areas of uncertainty in climate models, and are generally the cause of the largest intermodel differences in both control and sensitivity experiments. Although a large number of sensitivity studies have been conducted, it is still difficult to identify the best overall formulations and parametrizations to be used in climate models. Recent experience suggests that a higher horizontal resolution than that achieved in most earlier studies is needed in atmospheric models, although the minimum resolution requirements for ocean models remain to be determined.

Sensitivity studies have also suggested that a model's response to changes in external forcing may usefully be viewed as systematic changes in the amplitude and frequency of the model's characteristic or internal circulation regimes (or EOFs). This aspect is further discussed in Chapter 6 in connection with future climate projections.

5.6 How Can Our Confidence in Models be Increased?

Confidence in climate models depends partly upon their ability to simulate the current climate and recent climate changes, and partly upon the realistic representation of the physical processes that are important to the climate system. Current models of the atmosphere, ocean, land surface and sea ice may be rated as good to very good on these grounds, while coupled models may be rated as fair to good. This seeming discrepancy is due to the fact that most coupled climate models do not yet include the newer parametrizations and/or higher resolution used in the component models.

The development of more complete, more efficient, and more accurate coupled models has long been the aim of the climate modelling community, since it is generally believed that it is only through such models that we can gain a scientific understanding (and hence a reliable predictive capability) of climate and climate change. Although some modelling innovations at times appear to give puzzling results, and model evaluation has not always been carried out as systematically as it should be, the progressive development of more realistic parametrizations, the improvement of numerical techniques and resolution, and the inclusion of previously missing processes has in general led to better climate models. This faith in the fundamental soundness of the modelling approach does not deny the presence of significant errors in current models nor the utility of models known to be incomplete, but does provide confidence that these errors can and will be reduced through continuing modelling research.

As discussed more fully in Chapter 6, the successful simulation of some of the climate changes that have taken place during the past century has provided increased confidence in using coupled models for projections of future climate change. In particular, the recent coupled climate simulations of Mitchell *et al.* (1995a) and Hasselmann *et al.* (1995) have included the direct effect of sulphate aerosols along with changing levels of carbon dioxide concentration. While there are many uncertainties associated with both the implied radiative forcing due to aerosols and the climate sensitivity of the models used (and other important forcings such as those due to volcanic and other aerosols, solar variation, ozone and other greenhouse gases are yet to be included), a clear improvement is evident in the simulation of the observed change in the global surface air temperature and the zonal average of the vertical distribution of temperature over the last few decades (see Chapter 8).

Model evaluation needs to be carried out with an emphasis on those aspects and processes that are critical to model performance, such as cloud radiative effects and ocean surface fluxes. Co-ordinated model intercomparisons such as AMIP (Gates, 1992) or PILPS (Henderson-Sellers and Dickinson, 1992) are useful in this regard, since they serve to identify common errors and to document model performance and improvement. The evaluation and intercomparison of coupled models in particular needs to be intensified and broadened, and increased recognition needs to be given to comprehensive evaluation as an essential part of modelling.

The acquisition of new observational data for the evaluation of both model parametrizations and overall model performance must go hand-in-hand with model development. The long-term observational data sets that are being assembled by GEWEX from in-situ and satellite sources for the global distribution of radiation, cloudiness, water vapour, precipitation and land-surface properties provide unique opportunities for the evaluation of atmospheric models. Similarly, the data being assembled by TOGA, WOCE and other focused observational programs, including the emerging CLIVAR programme (CLIVAR, 1995) and the Global Climate Observing System (GCOS), should provide research-quality data that will be critical to the evaluation and further development of coupled climate models.

Supplementary Table: *Institutions Involved in the Atmospheric Model Intercomparison Project .*

The atmospheric general circulation models with which the AMIP simulations were made are from the institutions listed below. A documentation of the dynamical, physical and numerical properties of these models is given by Phillips (1994).
Bureau of Meteorology Research Centre (BMRC, Melbourne, Australia)

Canadian Centre for Climate (CCC, Victoria, Canada)

Centre National de Recherches Météorologiques (CNRM, Toulouse, France)

Center for Ocean, Land and Atmosphere (COLA, Calverton, USA)

Commonwealth Scientific and Industrial Research Organization (CSIRO, Melbourne, Australia)

Colorado State University (CSU, Ft. Collins, USA)

Dynamic Extended Range Forecast Group at the Geophysical Fluid Dynamics Laboratory (DERF, Princeton, USA)

Department of Numerical Mathematics of the Russian Academy of Sciences (DNM, Moscow, Russia)

European Centre for Medium Range Weather Forecasts (ECMWF, Reading, UK)

Geophysical Fluid Dynamics Laboratory (GFDL, Princeton, USA)

Goddard Institute for Space Studies (GISS, New York, USA)

Goddard Laboratory for Atmospheres (GLA, Greenbelt, USA)

Goddard Space Flight Center (GSFC, Greenbelt, USA)

Institute for Atmospheric Physics of the Chinese Academy of Sciences (IAP, Beijing, China)

Japan Meteorological Agency (JMA, Tokyo, Japan)

Laboratoire de Meteorologie Dynamique (LMD, Paris, France)

Main Geophysical Observatory (MGO, St. Petersburg, Russia)

Max Planck Institute for Meteorology (MPI, Hamburg, Germany)

Meteorological Research Institute (MRI, Tsukuba, Japan)

National Center for Atmospheric Research (NCAR, Boulder, USA)

National Meteorological Center (NMC, Washington DC, USA)

Naval Research Laboratory (NRL, Monterey, USA)

"Recherche en Prévision Numerique" group of Environment Canada (RPN, Quebec, Canada)

State University of New York at Albany (SUNYA, Albany, USA) "Genesis" model that is jointly administered by SUNYA and NCAR, the University of California at Los Angeles (UCLA, Los Angeles, USA)

University Global Atmospheric Modelling Project (UGAMP, Reading, UK)

University of Illinois at Urbana-Champaign (UIUC, Urbana-Champaign, USA)

United Kingdom Meteorological Office (UKMO, Bracknell, UK)

Yonsei University (YONU, Seoul, Korea)

References

Anderson, J.L., 1993: The climatology of blocking in a numerical forecast model. *J. Climate*, **6**, 1041–1056.

Barnett, T.P., L. Dumenil, U. Schlese, E. Roeckner and M. Latif, 1989: The effect of Eurasian snow cover on regional and global climate variations. *J. Atmos. Sci.*, **46**, 661–685.

Barron, E.J., P.J. Fawcett, W.W. Peterson, D. Pollard and S. Thompson, 1993: Model simulations of Cretaceous climates: The role of geography and carbon dioxide. *Phil. Trans. R. Soc. Lond., Series B*, **341**, 307–316.

Beljaars, A.C.M., P. Viterbo, M.J. Miller, A.K. Betts and J.H. Ball, 1993: A new surface boundary layer formulation at ECMWF and experimental continental precipitation forecasts. *GEWEX News*, **3**(3), 1, 5–8.

Bengtsson, L., 1994: Hurricane-type vortices in a general circulation model. Part I. *Max-Planck-Institute für Meteorologie, Hamburg, Report* No. **123**, 42 pp.

Bennett, A.F., 1992: *Inverse Methods in Physical Oceanography*. Cambridge monographs on Mechanics and Applied Mathematics, Cambridge University Press, New York, 346 pp.

Betts, A.K., and M.J. Miller, 1993: The Betts-Miller scheme. In: *The Representation of Cumulus Convection in Numerical Models of the Atmosphere*, K.A. Emanuel and D.J. Raymond, (eds.). American Meteorological Society, Boston.

Birchfield, G.E., 1989: A coupled ocean-atmosphere climate model: Temperature versus salinity effects on the thermohaline circulation. *Clim. Dyn.* **4**, 57–71.

Boer, G.J., 1993: Climate change and the regulation of the surface moisture and energy budgets. *Clim. Dyn.*, **8**, 225–239.

Boer, G.J., 1995: *A hybrid moisture variable suitable for spectral GCMs. Research Activities in Atmospheric and Oceanic Modelling*. Report No. 21, WMO/TD – No. 665, World Meteorological Organization, Geneva.

Boer, G.J., K. Arpe, M. Blackburn, M. Déqué, W.L. Gates, T.L. Hart, H. Le Treut, E. Roeckner, D.A. Sheinin, I. Simmonds, R.N.B. Smith, T. Tokioka, R.T. Wetherald and D. Williamson, 1992: Some results from an intercomparison of the climates simulated by 14 atmospheric general circulation models. *J. Geophys. Res.*, **97**, 12771–12786.

Bonan, G.B., D. Pollard and S.L. Thompson, 1992: Effects of boreal forest vegetation on global climate. *Nature*, **359**, 716–718.

Bond, G., W. Broecker, S. Johnsen, J. McManus, L. Labeyrie, J. Jouzel and G. Bonani, 1993: Correlations between climate records from North Atlantic sediments and Greenland ice. *Nature*, **365**, 143–147.

Boning, C.W., R. Döscher and R.G. Budich, 1991: Seasonal transport variations in the western subtropical North Atlantic: Experiments with an eddy-resolving model. *J. Phys. Oceanogr.*, **21**, 1271–1289.

Bottomley, M., C.K. Folland, J. Hsiung, R.E. Newell and D.E. Parker, 1990: *Global Ocean Surface Temperature Atlas "GOSTA"*. HMSO, London, 20 pp.+ iv, 313 plates.

Bourke, R.H. and R.P. Garrett, 1987: Sea-ice thickness distribution in the Arctic Ocean. *Cold Reg. Sci. and Tech.*, **13**, 259–280.

Boyle, J.S., 1993: Sensitivity of dynamical quantities to horizontal resolution for a climate simulation using the ECMWF (Cycle 33) Model. *J. Climate*, **6**, 796–815.

Broccoli, A.J. and S. Manabe, 1990: Can existing climate change models be used to study anthropogenic changes in tropical cyclone climate? *Geophys. Res. Lett.*, **17**, 1917–1920.

Bryan, F.O. and W.R. Holland, 1989: A high-resolution simulation of the wind- and thermohaline-driven circulation in the North Atlantic Ocean. In: *Proceedings of the 'Aha Huliko'a Hawaiian Winter Workshop on Parameterization of Small Scale Processes*, P Muller and D. Henderson (eds.), Hawaii Institute of Geophysics, Honolulu, pp 99-116. (Available from Hawaii Institute of Geophysics.)

Bushell, A.C. and D. Gregory, 1994: Sensitivity to vertical resolution in a climate model simulation of South West African marine stratocumulus. In: Research Activities in Atmospheric and Oceanic Modelling, Report No. 19, G.J. Boer (ed.), WMO/TD – No. 592, Geneva, 4.13–4.14.

Campos, J.D. and R. Bleck, 1992: A numerical study of the tropical Atlantic Ocean circulation with an isopycnic-coordinate circulation model – preliminary results. *Trans. Amer. Geophys. Union*, **73** (suppl.), 292.

Cao, H-X., J.F.B Mitchell and J.R Lavery, 1992: Simulated diurnal range and variability of surface temperature in a global climate model for present and doubled CO_2 climates. *J Climate*, 5, 920–943.

CAS/JSC, 1994: *Report of ninth session of the CAS/JSC Working Group on Numerical Experimentation, WMO/TD-No. 607*, WMO, Geneva.

Cess, R.D., G.L. Potter, J.P. Blanchet, G.J. Boer, A.D. Del Genio, M. Déqué, V. Dymnikov, V. Galin, W.L. Gates, S.J. Ghan, J.T. Kiehl, A.A. Lacis, H. Le Treut, Z.-X. Li, X.-Z. Liang, B.J. McAvaney, V.P. Meleshko, J.F.B. Mitchell, J.-J. Morcrette, D.A. Randall, L.J. Rikus, E. Roeckner, J.F. Royer, U. Schlese, D.A. Sheinin, A. Slingo, A.P. Sokolov, K.E. Taylor, W.M. Washington, R.T. Wetherald, I. Yagai and M.-H. Zhang, 1990: Intercomparison and interpretation of cloud-climate feedback processes in nineteen atmospheric general circulation models. *J. Geophys. Res.*, **95**, 16601–16615.

Cess, R.D., G.L. Potter, J.P. Blanchet, G.J. Boer, R. Colman, S.J. Ghan, J. Hansen, J.T. Kiehl, H. Le Treut, Z.-X. Li, X.-Z. Liang, B.J. McAvaney, V.P. Meleshko, J.F.B. Mitchell, J.-J. Morcrette, D.A. Randall, L.J. Rikus, E. Roeckner, U. Schlese, D.A. Sheinin, A. Slingo, A.P. Sokolov, K.E. Taylor, W.M. Washington, R.T. Wetherald and I. Yagai, 1991: Intercomparison and interpretation of snow-climate feedback processes in seventeen atmospheric general circulation models. *Science*, **253**, 888–892.

Cess, R.D., M.-H. Zhang, G.L. Potter, H. Barker, R.A. Colman, D.A. Dazlich, A.D. Del Genio, M. Esch, J.R. Fraser, V. Galin, W.L. Gates, J.J. Hack, J.T. Kiehl, A.A. Lacis, X.-Z. Liang, J.-F. Mahfouf, B.J. McAvaney, V.P. Meleshko, J.F.B. Mitchell,

J.-J. Morcrette, D.A. Randall, E. Roeckner, J.-F. Royer, D.A. Sheinin, A.P. Sokolov, K.E. Taylor, W.-C. Wang and R.T. Wetherald, 1993: Uncertainties in carbon dioxide radiative forcing in atmospheric general circulation models. *Science*, **262**, 1252–1255.

Cess, R.D., M.H. Zhang, P. Minnis, L. Corsetti, E.G. Dutton, B.W. Forgan, D.P. Garber, W.L. Gates, J.J. Hack, E.F. Harrison, X. Jing, J.T. Kiehl, C.N. Long, J.-J. Morcrette, G.L. Potter, V. Ramanathan, B. Subasilar, C.H. Whitlock, D.F. Young and Y. Zhou, 1995: Absorption of solar radiation by clouds: Observations versus models. *Science*, **267**, 496–499.

CLIVAR, 1995: *A study of climate variability and predictability, Initial Science Plan*. World Climate Research Programme, Geneva, 187pp.

Colman, R.A. and B.J. McAvaney, 1996: The sensitivity of the climate response of an atmospheric general circulation model to changes in convective parameterization and horizontal resolution. *J. Geophys. Res.* (in press).

Covey, C., 1995: Global ocean circulation and equator-pole heat transport as a function of ocean GCM resolution. *Clim. Dyn.*, **11**, 425–437.

Crowley, T.J., 1993: Geological assessment of the greenhouse effect. *Bull. Amer. Meteor. Soc.*, **74**, 2363–2373.

Cubasch, U., K. Hasselmann, H. Höck, E. Maier-Reimer, U. Mikolajewicz, B.D. Santer and R. Sausen, 1992: Time-dependent greenhouse warming computations with a coupled ocean-atmosphere model. *Clim. Dyn.*, **9**, 55–69.

Danabasoglu, G., J.C. McWilliams and P.R. Gent, 1994: The role of mesoscale tracer transports in the global ocean circulation. *Science*, **264**, 1123–1126.

Da Silva, A. M., C.C. Young and S. Levitus, 1994: *Atlas of Surface Marine Data 1994. Vol. 1: Algorithms and Procedures*. NOAA Atlas NESDIS 6, US Dept. Commerce, Washington, DC, 83 pp.

Delworth, T., S. Manabe and R.J. Stouffer, 1993: Interdecadal variations of the thermohaline circulation in a coupled ocean-atmosphere model. *J. Climate*, **6**, 1993–2011.

Dickinson, R.E., and A. Henderson-Sellers, 1988: Modelling tropical deforestation: a study of GCM land-surface parameterizations. *Quart. J. R. Met. Soc.*, **114(B)**, 439–462.

Dickinson, R.E., A. Henderson-Sellers and P.J. Kennedy, 1993: Biosphere-Atmosphere Transfer Scheme (BATS) Version 1e as coupled to the NCAR Community Climate Model. NCAR/TN-387+STR, National Center for Atmospheric Research, Boulder, CO, 80 pp.

Döös, K., 1994: Semianalytical simulation of the meridional cells in the Southern Ocean. *J. Phys. Oceanogr.*, **24**, 1281–1293.

Döös, K. and D.J. Webb, 1994: The Deacon cell and the other meridional cells in the Southern Ocean. *J. Phys. Oceanogr.*, **24**, 429–442.

Döscher, R., C.W. Böning and P. Herrmann, 1994: Response of circulation and heat transport in the North Atlantic to changes in thermohaline forcing in northern latitudes: A model study. *J. Phys. Oceanogr.*, **24**, 2306–2320.

England, M.H., 1993: Representing the global-scale water masses in ocean general circulation models. *J. Phys. Oceanogr.*, **23**, 1523–1552.

England, M.H., 1995: Using chlorofluorcarbons to assess ocean climate models. *Geophys. Res. Lett.* **22**, 3051–3054.

England, M.H., J.S. Godfrey, A.C. Hirst and M. Tomczak, 1993: The mechanism for Antarctic intermediate water renewal in a world ocean model. *J. Phys. Oceanogr.*, **23**, 1553–1560.

Esbensen, S.K. and Y. Kushnir, 1981: *The heat budget of the global ocean: An atlas based on estimates from surface marine observations*. Rep. No. 29, Climatic Research Institute, Oregon State University, Corvallis, 244 pp.

Ferranti, L., F. Molteni, C. Brankovic and T.N. Palmer, 1994: Diagnosis of extratropical variability in seasonal integrations of the ECMWF model. *J. Climate, 7*, 849–868.

Fischer, H. and P. Lemke, 1994: On the required accuracy of atmospheric forcing fields for driving dynamic-thermodynamic sea-ice models. In: *The Polar Oceans and Their Role in Shaping the Global Environment*, O.M. Johannessen, R.D. Muench, and J.E. Overland (eds.), Geophysical Monograph 85, American Geophysical Union, Washington, pp. 373–381.

Flato, G.M. and W.D. Hibler, 1992: Modelling sea-ice as a cavitating fluid. *J. Phys. Oceanogr.*, **22**, 626–651.

Folland, C.K. and D.P. Rowell (eds.), 1995: Workshop on simulations of the climate of the twentieth century using GISST (28-30 November 1994). *Hadley Centre CRTN 56*, Bracknell, UK, 111 pp.

Folland, C.K., J. Owen, M.N. Ward and A. Colman, 1991: Prediction of seasonal rainfall in the Sahel region using empirical and dynamical methods. *J. Forecasting*, **10**, 21–56.

Folland, C.K., D.E. Parker and T.N. Palmer, 1986: Sahel rainfall and worldwide sea temperatures 1901-85. *Nature*, **320**, 602–607.

Fowler, L.D. and D.A. Randall, 1996: Liquid and ice cloud microphysics in the CSU general circulation model. Part 3: Model sensitivity tests. *J. Climate* (in press).

FRAM Group (Webb, D.J., and collaborators), 1991: An eddy-resolving model of the Southern Ocean. *EOS*, **72**(15) 169, 174–175.

Garratt, J.R., 1993: Sensitivity of climate simulations to land-surface and atmospheric boundary-layer treatments – a review. *J. Climate*, **6**, 419–449.

Garratt, J.R., P.B. Krummel and E.A. Kowalczyk, 1993: The surface energy balance at local and regional scales – comparison of general circulation model results with observations. *J. Climate*, **6**, 1090–1109.

Gates, W.L., 1987: Problems and prospects in climate modeling. In: *Toward Understanding Climate Change*, Westview Press, Boulder, CO, 5–33.

Gates, W.L., 1992: AMIP: The Atmospheric Model Intercomparison Project. *Bull. Amer. Meteor. Soc.*, **73**, 1962–1970.

Gates, W.L., J.F.B. Mitchell, G.J. Boer, U. Cubasch and V.P. Meleshko, 1992: Climate modelling, climate prediction and model validation. In: *Climate Change, The Supplementary Report to the IPCC Scientific Assessment* , J.T. Houghton, B.A. Callander and S.K. Varney (eds.), Cambridge University Press, Cambridge, UK, 97–134.

Gates, W.L., U. Cubasch, G.A. Meehl, J.F.B. Mitchell and R.J. Stouffer, 1993: An intercomparison of selected features of the control climates simulated by coupled ocean-atmosphere general circulation models. *World Climate Research Programme WCRP-82, WMO/TD No. 574*, World Meteorological Organization, Geneva, 46 pp.

Gent, P.R. and J.C. McWilliams, 1990: Isopycnal mixing in ocean circulation models. *J. Phys. Oceanogr.*, **20**, 150–155.

Gent, P.R. and J.J. Tribbia, 1993: Simulation and predictability in a coupled TOGA model. *J. Climate*, **6**, 1843–1858.

Gent, P.R., J. Willebrand, T.J. McDougall and J.C. McWilliams, 1995: Parameterizing eddy-induced tracer transports in ocean circulation models. *J. Phys. Oceanogr.*, **25**, 463–474.

Giorgi, F., C.S. Brodeur and G.T. Bates, 1994: Regional climate change scenarios over the United States produced with a nested regional climate model. *J. Climate*, **7**, 375–399.

Gleckler, P.J., D.A. Randall, G. Boer, R. Colman, M. Dix, V. Galin, M. Helfand, J. Kiehl, A. Kitoh, W. Lau, X.-Z. Liang, V. Lykossov, B. McAvaney, K. Miyakoda and S. Planton, 1994: Cloud-radiative effects on implied oceanic energy transports as simulated by atmospheric general circulation models. *Report No. 15, PCMDI*, Lawrence Livermore National Laboratory, Livermore, CA, 13 pp.

Gloersen, P., W.J. Campbell, D.J. Cavalieri, J.C. Comiso, C.L. Parkinson and H.J. Zwally, 1992: Arctic and Antarctic sea-ice, 1978-1987: Satellite passive-microwave observations and analysis. *NASA SP-511*, 290 pp.

Graf, H.-F., I. Kirchner, A. Robock and I. Schult, 1993: Pinatubo eruption winter climate effects: Model versus observations. *Clim. Dyn.*, **9**, 81–93.

Greatbatch, R.J. and S. Zhang, 1994: An interdecadal oscillation in an idealized ocean basin forced by constant heat flux. *J. Climate*, **8**, 81–91.

Haarsma, R.J., J.F.B. Mitchell and C.A. Senior, 1993: Tropical disturbances in a GCM. *Clim. Dyn.*, **8**, 247–257.

Hansen, J., A. Lacis, R. Ruedy and M. Sato, 1992: Potential climate impact of Mount Pinatubo eruption. *Geophys. Res. Lett.*, **19**, 215–218.

Hansen, J., M. Sato and R. Ruedy, 1994: Long-term changes of the diurnal temperature cycle: Implications about mechanisms of global climate change. *Atmos. Res.*, **32**, 1–25.

Harder, M. and P. Lemke, 1995: Modelling the extent of sea-ice ridging in the Weddell Sea. *Nansen Centennial Volume*, AGU, (in press).

Harrison, E.F., P. Minnis, B.R. Barkstrom, V. Ramanathan, R.D. Cess and G.G. Gibson, 1990: Seasonal variation of cloud radiative forcing derived from the Earth Radiation Budget Experiment. *J. Geophys. Res.* **95**, 18687–18703.

Hasselmann, K., L. Bengtsson, U. Cubasch, G.C.Hegerl, H. Rodhe, E. Roeckner, H.v. Storch, R. Voss and J. Waskewitz,

1995: Detection of anthropogenic climate change using a fingerprint method. *Max-Planck Institute for Meteorology, Report No. 168*, Hamburg, 20pp.

Hastenrath, S., 1990: Prediction of north-east Brazil rainfall anomalies. *J. Climate*, **3**, 893–904.

Hay, L.E., G.J. McCabe, D.M. Wolock and M.A. Ayres, 1992: Use of weather types to disaggregate general circulation model predictions. *J. Geophys. Res.*, **97**, 2781–2790.

Hegerl, G.C., H.V. Storch, K. Hasselmann, B.D. Santer, U. Cubasch and P.D. Jones, 1996. Detecting anthropogenic climate change with an optimal fingerprint method. *J. Climate* (in press).

Held, I.M. and P.J. Phillipps, 1993: Sensitivity of the eddy momentum flux to meridional resolution in atmospheric GCMs. *J. Climate*, **6**, 499–507.

Henderson-Sellers, A. and R.E. Dickinson, 1992: Intercomparison of land-surface parameterizations launched. *EOS*, **73**(17), 195–196.

Henderson-Sellers, A. and A.-M. Hansen, 1995: *Climate Change Atlas*, Kluwer, Dordrecht, 159pp.

Henderson-Sellers, A., Z.-L. Yang and R.E. Dickinson, 1993: The Project for Intercomparison of Land-surface Parameterization Schemes (PILPS). *Bull. Amer. Meteor. Soc.*, **74**(7), 1335–1349.

Henderson-Sellers, A., A.J. Pitman, P.K. Love, P. Irannejad and T.H. Chen, 1995: The Project for Intercomparison of Land Surface Parameterization Schemes (PILPS): Phases 2 and 3. *Bull. Amer. Meteor. Soc.*, **76**, 489–503.

Hibler, W.D., III., 1984: The role of sea-ice dynamics in modeling CO_2 increases. In: *Climate Processes and Climate Sensitivity*, Geophysical Monograph 29, American Geophysical Union, pp. 238–253.

Hogan, T.R. and T.E. Rosmond, 1991: The description of the Navy operational global atmospheric prediction system's spectral forecast model. *Mon. Wea. Rev.*, **119**, 1786–1815.

Hollingsworth, A., 1993: Validation and diagnostics of atmospheric models. In *Proceedings of Validation of Models over Europe*, Volume I, 7-11 September 1992, ECMWF, Reading, UK.

Horton, E.B., 1994: The geographical distribution of changes in maximum and minimum temperature. *Proc NOAA USDOE MINIMAX Workshop*, G. Kukla *et al.*, (eds.) College Park, USDOE Conf. Proc. 25, pp. 179–198.

Hulme, M., 1995: Validation of large-scale precipitation fields in general circulation models. In: *Global Precipitation and Climate Change*, M. Dubois and F Desalmand, D Reidel (eds.), Dordrecht (in press).

Hulme, M., K.R. Briffa, P.D. Jones and C.A. Senior, 1993: Validation of GCM control simulations using indices of daily airflow types over the British Isles. *Clim. Dyn.*, **9**, 95–105.

Hurrell, J.W. and G.G. Campbell 1992: Monthly mean global satellite data sets available in CCM history tape format, *NCAR Tech. Note NCAR/TN-371+STR*, 94 pp.

Ip, C.F., W.D. Hibler and G.M. Flato, 1991: On the effect of rheology on seasonal sea-ice simulations. *Ann. Glaciol.*, **15**, 17–25.

IPCC (Intergovernmental Panel on Climate Change), 1990: *Climate Change: The IPCC Scientific Assessment*, J.T. Houghton, G.J. Jenkins and J.J. Ephraums (eds.), Cambridge University Press, Cambridge, UK.

Jacobs, G.A., H.E. Hurlburt, J.C. Kindle, E.J. Metzger, J.L. Mitchell, W.J. Teague and A.J. Wallcraft, 1994: Decade-scale trans-Pacific propagation and warming effects of an El Niño anomaly. *Nature*, **370**, 360–363.

Jaeger, L., 1976: Monatskarten des Niederschlags für die ganze Erde. *Ber. Deutschen Wetterdienstes*, Nr. 139, 38 pp.

Jenne, R.L., 1975: Data sets for meteorological research. *NCAR Technical Note, NCAR-TN/IA-111*, NCAR, Boulder, 194 pp.

Ji, M., A. Kumar and A. Leetmaa, 1994: A multiseason climate forecast system at the National Meteorological Center. *Bull. Amer. Meteor. Soc.*, **75**, 569–577.

Jin, F.-F. and J.D. Neelin, 1993a: Model of interannual tropical ocean-atmosphere interaction – a unified view. Part I: Numerical results. *J. Atmos. Sci.*, **50**, 3477–3503.

Jin, F.-F.,and J.D. Neelin, 1993b: Modes of interannual tropical ocean-atmosphere interaction – a unified view. Part III: Analytical results in fully coupled cases. *J. Atmos. Sci.*, **50**, 3524–3540.

Johnson, K.D., D. Entekhabi and P.S. Eagleson, 1993: The implementation and validation of improved land surface hydrology in an atmospheric general circulation model. *J. Climate*, **6**, 1009–1026.

Jones, P.D., 1987: The early twentieth century Arctic high-fact or fiction? *Clim. Dyn.*, **1**, 63–75.

Jones, P.D., 1994: Hemispheric surface air temperature variations: a reanalysis and an update to 1993. *J. Climate*, **7**, 1794–1802.

Jones, P.D., T.M.L. Wigley and G. Farmer, 1991: Marine and land temperature data sets: A comparison and a look at recent trends. In: *Greenhouse-Gas-Induced Climatic Change: A Critical Appraisal of Simulations and Observations*, M.E. Schlesinger (ed.), Elsevier, Amsterdam, pp.153–172.

Joussaume, S., 1993: Paleoclimatic tracers: An investigation using an atmospheric general circulation model under ice age conditions 1. Desert dust. *J. Geophys. Res.*, **98**, 2767–2805.

Karl, T.R., P.D. Jones, R.W. Knight, G. Kukla, N. Plummer, V. Rasuvayev, K.P. Gallo, J. Lindseay, R.J. Charlson and T.C. Peterson, 1993: A new perspective on recent global warming: Asymmetric trends of daily maximum and minimum temperature. *Bull. Amer. Meteor. Soc.*, **74**, 1007–1023.

Katz, R.W., 1992: Role of statistics in the validation of general circulation models. *Clim. Res.*, **2**, 35–45.

Kerr, R.A., 1994: Climate modeling's fudge factor under fire. *Science*, **265**, 1528.

Kiehl, J.T., 1996: Sensitivity of a GCM Climate simulation to differences in continental versus maritime cloud drop size. *J. Geophys. Res.* (in press).

Kiehl, J.T. and D.L. Williamson, 1991: Dependence of cloud amount on horizontal resolution in the National Center for Atmospheric Research Community Climate Model. *J. Geophys. Res.*, **96**, 10955–10980.

Klinck, J.M., 1994: Thermohaline structure of the CME. *WOCE Notes*, **6**(2), 4–8.

Koster, R.D. and M.J. Suarez, 1994: The components of a SVAT scheme and their effects on a GCM's hydrological cycle. *Adv. Water Resources*, **17**, 61–78.

Koster, R.D. and M.J. Suarez, 1996: The relative contributions of land and ocean processes to precipitation variability. *J. Geophys. Res.* (in press).

Kuo, H.L., 1974: Further studies of the parameterization of the influence of cumulus convection on large-scale flow. *J. Atmos. Sci.*, **31**, 1232–1240.

Kushnir, Y., 1994: Interdecadal variations in North Atlantic sea surface temperature and associated atmospheric conditions. *J. Climate*, **7**, 141–157.

Kutzbach, J.E., and A.M. Ziegler, 1993: Simulation of Late Permian climate and biomes with an atmosphere-ocean model: Comparisons with observations. *Phil. Trans. R. Soc. Lond.*, Series B, **341**, 327–340.

Lamb, P.J., 1978: Large-scale tropical Atlantic surface circulation patterns associated with Subsaharan weather anomalies. *Tellus*, **30**, 240–251.

Latif, M. and T.P. Barnett, 1994: Causes of decadal climate variability over the North Pacific and North America. *Science*, **266**, 634–637.

Latif, M., A. Sterl, E. Maier-Reimer and M.M. Junge, 1993: Climate variability in a coupled GCM. Part I: The tropical Pacific. *J. Climate*, 6, 5–21.

Latif, M., T.P. Barnett, M.A. Cane, M. Fluegel, N.E. Graham, H. von Storch, J.-S. Xu and S.E. Zebiak, 1994: A review of ENSO prediction studies. *Clim. Dyn.*, **9**, 167–179.

Lau, N.-C., S.G.H. Philander and M.J. Nath, 1992: Simulation of ENSO-like phenomena with a low-resolution coupled GCM of the global ocean and atmosphere. *J. Climate*, **5**, 284–307.

Le Treut, H., and Z.-X. Li, 1991: Sensitivity of an atmospheric general circulation model to prescribed SST changes: Feedback effects associated with the simulation of cloud optical properties. *Clim. Dyn.*, **5**, 175–187.

Levitus, S., and T.P. Boyer, 1994: *World Ocean Atlas 1994, Volume 4: Temperature*. NOAA/NESDIS E/OC21, Washington, DC, 117 pp.

Levitus, S., J. Antanov and T.P. Boyer, 1994a: Interannual variability of temperature at 125 m depth in the North Atlantic Ocean. *Science*, **266**, 96–99.

Levitus, S., R. Burgett, and T.P. Boyer, 1994b: *World Ocean Atlas 1994, Volume 3: Salinity*. NOAA/NESDIS E/OC21, Washington DC, 99pp.

Lighthill, J., G. Holland, W. Gray, C. Landsea, G. Craig, J. Evans, Y. Kurihara and C. Guard, 1994: Global climate change and tropical cyclones. *Bull. Amer. Meteor. Soc.*, **75**, 2147–2157.

Manabe, S., 1969: Climate and the ocean circulation: I, the atmospheric circulation and the hydrology of the Earth's surface. *Mon. Wea. Rev.*, **7**, 739–774.

Manabe, S., and R. Stouffer, 1988: Two stable equilibria of a coupled ocean-atmosphere model. *J. Climate*, **1**, 841–866.

Manabe, S., and T. Delworth, 1990: The temporal variability of soil wetness and its impact on climate. *Climatic Change*, **16**, 185–192.

Marotzke, J., 1989: Instabilities and multiple steady states of the thermohaline circulation. In: *Oceanic Circulation Models: Combining Data and Dynamics*, D.L.T. Anderson and J. Willebrand, eds., NATO ASI series, Kluwer, Dordrecht, pp. 501–511.

Marotzke, J. and C. Wunsch, 1993: Finding the steady state of a general circulation model through data assimilation: Application to the North Atlantic ocean. *J. Geophys. Res.*, **98** (C11), 20149–20167.

Matson, M., C.F. Ropelewski and M.S. Varnadore, 1986: An atlas of satellite-derived northern hemispheric snow cover frequency. *NOAA Atlas*. U.S. Department of Commerce, National Oceanic and Atmospheric Administration, Washington, DC, 75 pp.

McAvaney, B.J., R.R. Dahni, R.A. Colman and J.R. Fraser, 1995: The dependence of the climate senditivity on convective parametrization: Statistical evaluation. *Global and Planetary Change*, **10**, 181–200.

McGuffie, K., A. Henderson-Sellers, H. Zhang, T.B. Durbidge and A.J. Pitman, 1995: Global climate sensitivity to tropical deforestation. *Global and Planetary Change*, **10**, 97–128.

Meehl, G.A., 1990: Seasonal cycle forcing of El Niño-Southern Oscillation in a global coupled ocean-atmosphere GCM. *J. Climate*, **3**, 72–98.

Meehl, G.A. and W.M. Washington, 1995: Cloud albedo feedback and the super greenhouse effect in a global coupled GCM. *Clim. Dyn.*, **11**, 399–411.

Meehl, G.A., G.W. Branstator and W.M. Washington, 1993: Tropical Pacific interannual variability and CO_2 climate change. *J. Climate*, **6**, 42–63.

Meehl, G.A., M. Wheeler and W. M. Washington, 1994: Low-frequency variability and CO_2 climate change. *J. Climate*, **6**, 42–63.

Mikolajewicz, U. and E. Maier-Reimer, 1990: Internal secular variability in an ocean general circulation model. *Clim. Dyn.*, **4**, 145–156.

Mikolajewicz, U. and E. Maier-Reimer, 1994: Mixed boundary conditions in ocean general circulation models and their influence on the stability of the model's conveyor belt. *J. Geophys. Res.*, **99**, 22633–22644.

Miller, J.R., G.L. Russell and G. Caliri, 1994: Continental-scale river flow in climate models. *J. Climate*, **7**, 914–928.

Milly, P.C.D. and K.A. Dunne, 1994: Sensitivity of the global water cycle to the water-holding capacity of land. *J. Climate*, **7**(4), 506–526.

Mintz, Y., 1984: The sensitivity of numerically simulated climates to land-surface boundary conditions. In: *The Global Climate*, J. Houghton (ed.), Cambridge University Press, Cambridge, UK, 79–105.

Mitchell, J.F.B., 1993: Modelling of palaeoclimates: Examples from the recent past. *Phil. Trans. Roy. Soc. Lond.*, **341**(1297), 267–275.

Mitchell, J.F.B., C.A. Senior and W.J. Ingram, 1989: CO_2 and climate: A missing feedback? *Nature*, **341**, 132–134.

Mitchell, J.F.B., S. Manabe, T. Tokioka and V. Meleshko, 1990: Equilibrium climate change – and its implications for the future. In: *Climate Change, The IPCC Scientific Assessment*, J.T. Houghton, G.J. Jenkins and J.J. Ephraums (eds.), Cambridge University Press, Cambridge, UK, 131–172.

Mitchell, J.F.B., T.C. Johns, J.M. Gregory and S.F.B. Tett, 1995a: Climate response to increasing levels of greenhouse gases and sulphate aerosols. *Nature*, **376**, 501–504.

Mitchell, J.F.B, R.A. Davis, W.J. Ingram and C.A. Senior, 1995b: On surface temperature, greenhouse gases and aerosols: Models and observations. *J. Climate*, **8**, 2364–2386.

Miyakoda, K., and J. Sirutis, 1990: Subgrid scale physics in 1-month forecast. Part II: Systematic error and blocking forecast. *Mon. Wea. Rev.*, **118**, 1065–1081.

Moore, A.M., 1995: Tropical interannual variability in a global coupled GCM: Sensitivity to mean climate state. *J. Climate*, **8**, 807–828.

Muroi, C. and N. Sato, 1994: Intercomparison of tropical cyclone track forecast by ECMWF, UKMO and JMA operational global models. *JMA/NPD Tech. Rep., No 31*, JMA, Tokyo.

Nagai, T., T. Tokioka, M. Endoh and Y. Kitamura, 1992: El Niño-Southern Oscillation simulated in an MRI atmosphere-ocean coupled general circulation model. *J. Climate*, **5**, 1202–1233.

Navarra, A., W.F. Stern and K. Miyakoda, 1994: Reduction of the Gibbs oscillation in spectral model simulations. *J. Climate*, **7**, 1169–1183.

Neelin, J.D., 1991: The slow sea surface temperature mode and the fast-wave limit: Analytic theory for tropical interannual oscillations and experiments in a hybrid coupled model. *J. Atmos. Sci.*, **48**, 584–606.

Neelin, J.D., and F.-F. Jin, 1993: Modes of interannual tropical ocean-atmosphere interaction – a unified view. Part II: Analytical results in the weak-coupling limit. *J. Atmos. Sci.*, **50**, 3504–3522.

Neelin, J.D., M. Latif, M.A.F. Allaart, M.A. Cane, U. Cubasch, W.L. Gates, P.R. Gent, M. Ghil, C. Gordon, N.C. Lau, C.R. Mechoso, G.A. Meehl, J.M. Oberhuber, S.G., H. Philander, P.S. Schopf, K.R. Sperber, A. Sterl, T. Tokioka, J. Tribbia and S.E. Zebiak, 1992: Tropical air-sea interaction in general circulation models. *Clim. Dyn.*, **7**, 73–104.

Numaguti, A., 1993: Dynamics and energy balance of the Hadley circulation and the tropical precipitation zones: Significance of the distribution of evaporation. *J. Atmos. Sci.*, **50**, 1874–1887.

OOSDP (Ocean Observing System Development Panel), 1995: Scientific design for the common module of the Global Ocean Observing System and the Global Climate Observing System. U.S. WOCE Office, Washington, DC, 228 pp.

Oreskes, N., K. Shrader-Frechette and K. Belitz, 1994: Verification, validation and confirmation of numerical models in the Earth sciences. *Science*, 263, 641–646.

Orr, J.C., 1993: Accord between ocean models predicting uptake of ocean CO_2. *Water, Air and Soil Pollution*, 70, 465–481.

Palmer, T.N., C. Brankovic, F. Molteni and S. Tibaldi, 1990: Extended-range predictions with ECMWF models: Interannual variability in operational model integrations. *Quart. J. R. .Met. Soc.*, 116, 799–834.

Parker, D.E., P.D. Jones, C.K. Folland and A.C. Bevan, 1994: Interdecadal changes of surface temperature since the late nineteenth century. *J. Geophys. Res.*, 99, 14373–14399.

Parker, D.E., H. Wilson, P.D. Jones, J. Christy and C.K. Folland, 1996: The impact of Mount Pinatubo on worldwide temperatures. *Int. J. Climatol.* (in press).

Philander, S.G.H., R.C. Pacanowski, N.-C. Lau and M.J. Nath, 1992: Simulation of ENSO with a global atmospheric GCM coupled to a high-resolution tropical Pacific Ocean GCM. *J. Climate*, 5, 308–329.

Phillipps, P.J. and I.M. Held, 1994: The response to orbital perturbations in an atmospheric model coupled to a slab ocean. *J. Climate*, 7, 767–782.

Phillips, T.J., 1994: A summary documentation of the AMIP models. *Report No. 18, PCMDI*, Lawrence Livermore National Laboratory, Livermore, CA, UCRL-ID-116384, 343 pp.

Pitman, A.J., A. Henderson-Sellers, F. Abramopoulos, R. Avissar, G. Bonan, A. Boone, J.G. Cogley, R.E. Dickinson, M. Ek, D. Entekhabi, J. Famiglietti, J.R. Garratt, M. Frech, A. Hahmann, R. Koster, E. Kowalczyk, K. Laval, J. Lean, T.J. Lee, D. Lettenmaier, X. Liang, J.-F. Mahfouf, L. Mahrt, P.C.D. Milly, K. Mitchell, N. de Noblet, J. Noilhan, H. Pan, R. Pielke, A. Robock, C. Rosenzweig, S.W. Running, C.A. Schlosser, R. Scott, M. Suarez, S. Thompson, D. Verseghy, P. Wetzel, E. Wood, Y. Xue, Z.-.L. Yang and L. Zhang, 1993: Project for Intercomparison of Land-Surface Parameterization Schemes (PILPS). GEWEX Report, *IGPO Publication Series No. 7*, Washington, DC, 47 pp.

Polcher, J., K. Laval, L. Dumenil, J. Lean and P.R. Rowntree, 1996: Comparing three land surface schemes used in GCMs. *J. Hydrol.* (submitted).

Pollard, D. and S.L. Thompson, 1994: Sea-ice dynamics and CO_2 sensitivity in a global climate model. *Atmosphere-Ocean*, 32, 449–467.

Power, S.B. and R. Kleeman, 1993: Multiple equilibria in a global ocean general circulation model. *J. Phys. Oceanogr.*, 23, 1670–1681.

Power, S.B. and R. Kleeman, 1994: Surface heat flux parameterization and the response of ocean general circulation models to high-latitude freshening. *Tellus*, 46A, 86–95.

Power, S.B., A. Moore, D.A. Post, N.R. Smith and R. Kleeman, 1994: Stability of North Atlantic deep water formation in a global ocean general circulation model. *J. Phys. Oceanogr.*, 24, 904–916.

Power, S.B., R. Kleeman, R.A. Colman and B.J. McAvaney,

1995: Modelling the heat flux response to long-lived SST anomalies in the North Atlantic. *J. Climate*, 8, 2161–2180.

Rahmstorf, S. and J. Willebrand, 1996: The role of temperature feedback in stabilising the thermohaline circulation. *J. Phys. Oceanogr.* (in press).

Ramsden, D., and G. Fleming, 1995: Use of a coupled ice-ocean model to investigate the sensitivity of the Arctic ice cover to doubling atmospheric CO_2. *J. Geophys. Res.* 100, 6817–6828.

Randall, D.A., R.D. Cess, J.P. Blanchet, S. Chalita, R. Colman, D.A. Dazlich, A.D. Del Genio, V. Dymnikov, V. Galin, D. Jerrett, E. Keup, A.A. Lacis, H. Le Treut, X.-Z. Liang, B.J. McAvaney, J.F. Mahfouf, V.P. Meleshko, J.F.B. Mitchell, J.-J. Morcrette, P.M. Norris, G.L. Potter, L.J. Rikus, E. Roeckner, J.F. Royer, U. Schlese, D.A. Sheinin, A.P. Sokolov, K.E. Taylor, R.T. Wetherald, I. Yagai and M.-H. Zhang, 1994: Analysis of snow feedbacks in fourteen general circulation models. *J. Geophys. Res.*, 99, 20757–20771.

Rasch, P.J. and D.L. Williamson, 1990: Computational aspects of moisture transport in global models of the atmosphere. *Quart. J. R. Met. Soc.*, 116, 1071–1090.

Rind, D.,and M. Chandler, 1991: Increased ocean heat transports and warmer climate. *J. Geophys. Res.*, 96, 7437–7461.

Robitaille, D.Y. and A.J. Weaver, 1995: Validation of sub-grid scale mixing schemes using CFCs in a global ocean model. *Geophys. Res. Let.* 22, 2917–2920.

Robock, A., K.Y. Vinnikov, C.A. Schlosser, N.A. Speranskaya and Y. Xue, 1995: Use of midlatitude soil moisture and meteorological observations to validate soil moisture simulations with biosphere and bucket models. *J. Climate*, 8, 15–35.

Roeckner, E., T. Siebert and J. Feichter, 1995: Climatic response to anthropogenic sulfate forcing simulated with a general circulation model. *Proceedings of Dahlem workshop "Aerosol Forcing of Climate"*, Academic Press (in press).

Ropelewski, C.F., 1989: Monitoring large-scale cryosphere/ atmosphere interactions. *Adv. Space Res.*, 9, 213–218.

Rossow, W.B. and R.A. Schiffer, 1991: ISCCP cloud data products. *Bull. Amer. Meteor. Soc.*, 72, 2–20.

Rowell, D.P., C.K. Folland, K. Maskell and M.N. Ward, 1995: Variability of summer rainfall over Tropical North Africa (1906-1992): Observations and modelling. *Quart. J. R. Met. Soc.*, 121, 669–704.

Rowntree, P.R.and J.A. Bolton, 1983: Simulation of the atmospheric response to soil moisture anomalies over Europe. *Quart. J. R. Met. Soc.*, 109, 501–526.

Rui, H. and B. Wang, 1990: Development characteristics and dynamic structure of tropical intraseasonal convection anomalies. *J. Atmos. Sci.*, 47, 357–379.

Santer, B.D. and T.M.L. Wigley, 1990: Regional validation of means, variances and spatial patterns in GCM control runs. *J. Geophs. Res.*, 95, 829–850.

Santer, B.D., K.E. Taylor and L.C. Corsetti, 1995a: Statistical evaluation of AMIP model performance. In *Proc. First AMIP Sci. Conf.*, WCRP-92, WMO TD-No. 732, Geneva.

Santer, B.D., K.E. Taylor, T.M.L. Wigley, J.E. Penner, P.D. Jones and U. Cubasch, 1995b: Towards the detection and attribution of an anthropogenic effect on climate. Clim. Dyn. 12, 77–100.

Sausen, R., K. Barthel and K. Hasselmann, 1988: Coupled ocean-atmosphere models with flux corrections. *Clim. Dyn.*, **2**, 154–163.

Sausen, R., W. Koenig and F. Sielmann, 1993: Analysis of blocking events from observations and ECHAM model simulations. *Max-Planck-Institut für Meteorologie Rep. 11*, Hamburg.

Schemm, J.-K., S. Schubert, J. Terry and S. Bloom, 1992: Estimates of monthly mean soil moisture for 1979-1989. *NASA Tech. Memo. 104571*, Goddard Space Flight Center, Greenbelt, MD, 260 pp.

Schmitz, W.J. and M.S. McCartney, 1993: On the North Atlantic circulation. *Rev. Geophys.*, **31**, 29–49.

Schopf, P.S.,and M.J. Suarez, 1988: Vacillations in a coupled ocean-atmosphere model. *J. Atmos. Sci.*, **45**, 549–566.

Schubert, S.D., C.-Y. Wu, J. Zero, J.-K. Schemm, C.-K. Park and M. Suarez, 1992: Monthly means of selected climate variables from 1985 to 1989. *NASA Tech. Memo 104565*, Goddard Space Flight Center, Greenbelt, MD, 376 pp.

Schubert, S.D., R.B. Rood and J. Pfaendtner, 1993: An assimilated dataset for earth science applications. *Bull. Amer. Meteor. Soc.*, **74**, 2331–2342.

Scott, R., R.D. Koster, D. Entekhabi and M.J. Suarez, 1996: Effect of a canopy interception reservoir on hydrological persistence in a general circulation model. *J. Climate* (in press).

Sellers, P.J., 1992: Biophysical models of land surface processes. In: Climate System Modeling K.E. Trenberth (ed.), Cambridge University Press, Cambridge, UK, 451–490.

Semtner, A.J., 1994: Sixth-degree global ocean model. *U.S. WOCE Report 1994*, U.S. WOCE Office, Texas A&M University, College Station, TX, 34–36.

Semtner, A.J. and R.M. Chervin, 1988: A simulation of the global ocean circulation with resolved eddies. *J. Geophys. Res.*, **93**, 15502–15522.

Semtner, A.J. and R.M. Chervin, 1992: Ocean general circulation from a global eddy-resolving model. *J. Geophys. Res.*, **97**, 5493–5550.

Send, U. and J. Marshall, 1996: Integral effects of deep convection. *J. Phys. Oceanogr.* (in press).

Senior, C.A. and J.F.B. Mitchell, 1993: CO_2 and climate: The impact of cloud parameterization. *J. Climate*, **6**, 393–418.

Shao, Y. and A. Henderson-Sellers, 1996: Soil moisture simulation in land-surface parameterisation schemes. *Global and Planetary Change* (Special Issue) **13**, 11–27.

Shao, Y., R.D. Anne, A. Henderson-Sellers, P. Irannejad, C. P. Thornton, X. Liang, O. Balachova, A. Haxeltine, A. Durcharne, T.H. Chen, C. Ciret and C. Desborough, 1995: Soil moisture simulation. *A report of the RICE and PILPS Workshop, GEWEX/GAIM Report. IGPO Publication Series No. 14*, 179 pp.

Shinoda, M. and R. Kawamura, 1994: Tropical rainbelt, circulation, and sea surface temperatures associated with Sahelian rainfall trend. *J. Met. Soc. Japan*, **72**, 341–357.

Shuttleworth, J.W. and R.E. Dickinson, 1989: Comments on "Modelling tropical deforestation: a study of GCM land-surface parameterizations" by R.E. Dickinson and A. Henderson-Sellers. *Quart. J. R. Met. Soc.*, **115**, 1177–1179.

Siegenthaler, U. and J.L. Sarmiento, 1993: Atmospheric carbon dioxide and the ocean. *Nature*, **365**, 119–125.

Slingo, J.M. and R.A. Madden 1991: Characteristics of the tropical intraseasonal oscillation in the NCAR community model. *Quart. J. R. Met. Soc.*, **117**, 1129–1169.

Slingo, J., M. Blackburn, A. Betts, R. Brugge, K. Hodges, B. Hoskins, M. Miller, L. Steenman-Clark, and J. Thuburn, 1994: Mean climate and transience in the tropics of the UGAMP GCM: Sensitivity to convective parametrization. *Quart. J. R. Met. Soc.*, **120**, 881–922.

Slingo, J.M., K.R. Sperber, J.S. Boyle, J.-P. Ceron, M. Dix, B. Dugas, W. Ebisuzaki, J. Fyfe, D. Gregory, J.-F. Guerney, J. Hack, A. Harzallah, P. Inness, A. Kitoh, W.K.-M. Lau, B. McAvaney, R. Madden, A. Matthews, T.N. Palmer, C.-K. Park, D. Randall and N. Rennó, 1995: Intraseasonal oscillations in 15 AGCMs (Results from an AMIP diagnostic subproject). *WCRP-88, WMO/TD-No. 661*, WMO, Geneva, 32 pp.

Sloan, L.C., and E.J. Barron, 1992: A comparison of Eocene climate model results to quantified paleoclimatic interpretations. *Paleogeography, Paleoclimatology, Paleoecology*, **93**, 183–202.

Slutz, R.J., S.J. Lubker, J.D. Hiscos, S.D. Woodruff, R.L. Jenne, D.H. Joseph, P.M. Steurer and J.D. Elms, 1985: *Comprehensive Ocean-Atmosphere Data Set*, Release 1, Univ. Colorado/NOAA CIRES, Boulder, 255pp.

Smith, I.N., 1994: A GCM simulation of global climate trends, 1950-1988. *J. Climate*, **7**, 732–744.

Smith, I.N., 1995: A GCM simulation of global climate interannual variability. *J. Climate*, **8**, 709–718.

Spencer, R.W., 1993: Global oceanic precipitation from the MSU during 1979-91 and comparisons to other climatologies. *J. Climate*, **6**, 1301–1326.

Sperber, K.R., and S. Hameed, 1991: Southern Oscillation simulation in the OSU coupled upper ocean-atmosphere GCM. *Clim. Dyn.*, **6**, 83–97.

Sperber, K.R. and S. Hameed, 1993: Phase locking of Nordeste Precipitation with sea surface temperatures. *Geophys. Res. Lett.*, **20**, 113–116.

Sperber, K.R., S. Hameed, W.L. Gates and G.L. Potter, 1987: Southern Oscillation simulated in a global climate model. *Nature*, **329**, 140–142.

Stockdale, T., D. Anderson, M. Davey, P. Delecluse, A. Kattenberg, Y. Kitamura, M. Latif and T. Yamagata, 1993: Intercomparison of tropical ocean GCMs. *World Climate Research Programme WCRP-79, WMO/TD No. 545*, World Meteorological Organization, Geneva, 43 pp.

Stouffer, R.J., S. Manabe and K. Ya. Vinnikov, 1994: Model assessment of the role of natural variability in recent global warming. *Nature*, **367**, 634–636.

Tett, S., 1995: Simulation of El Niño/Southern Oscillation-like variability in a global AOGCM and its response to CO_2 increase. *J. Climate*, **8**, 1473–1502.

Thompson, S.L. and D. Pollard, 1996: A global climate model (GENESIS) with a land-surface-transfer scheme (LSX). Part 2: CO_2 sensitivity. *J. Climate* (in press).

Thuburn, J., 1993: Use of a flux-limited scheme for vertical advection in a GCM. *Quart. J. R. Met. Soc.*, **119**, 469–487.

Tibaldi, S. and F. Molteni, 1990: On the operational predictability of blocking. *Tellus*, **42A**, 343–365.

Tibaldi, S., E. Tosi, A. Navarra and L. Pedulli, 1994: Northern and Southern hemisphere seasonal variability of blocking frequency and predictability. *Mon. Wea. Rev.*, **122**, 1971–2003.

Tracton, M.S., 1990: Predictability and its relationship to scale interaction processes in blocking. *Mon. Wea. Rev.*, **118**, 1666–1695.

Tracton, M.S., K. Mo, W. Chen, E. Kalnay, R. Kistler and G. White, 1989: Dynamical extended range forecasting (DERF) at the National Meteorological Center. *Mon. Wea. Rev.*, **117**, 1606–1637.

Trenberth, K.E., 1992: Global analyses from ECMWF and atlas of 1000 to 10 mb circulation statistics. *NCAR Tech. Note NCAR/TN-373+STR*, 191 pp.

Tziperman, E., and K. Bryan, 1993: Estimating global air-sea fluxes from surface properties and from climatological flux data using an oceanic general circulation model. *J. Geophys. Res.*, **98**, 22629–22644.

Viterbo, P. and A.C.M. Beljaars, 1995: An improved land surface parameterization scheme in the ECMWF model and its validation. *J. Climate* , **8**, 2716–2748.

Walker, J. and P.R. Rowntree, 1977: The effect of soil moisture on circulation and rainfall in a tropical model. *Quart. J. R. Met. Soc.*, **103**, 29–46.

Wallace, D.W.R., 1995: An ocean observing system to monitor global ocean carbon inventories. *OOSDP Background Report No. 4*, US WOCE Office, Washington, DC, 59 pp.

Wang, W. and M.E. Schlesinger, 1995: Simulation of tropical intraseasonal oscillations by the UIUC GCM. *Proc. First AMIP Sci. Conf.*, WCRP-92, WMO TD-No. 732, Geneva.

Ward, M.N. and C.K. Folland, 1991: Prediction of seasonal rainfall in the north Nordeste of Brazil using eigenvectors of sea-surface temperature. *Int. J. Climatol.*, **11**, 711–743.

Warner, M.J. and R.F. Weiss, 1992: Chlorofluorocarbons in South Atlantic Intermediate Water. *Deep-Sea Res.*, **39**, 2053–2075.

Washington, W.M. and G.A. Meehl, 1989: Climate sensitivity due to increased CO_2: Experiments with a coupled atmosphere and ocean general circulation model. *Clim. Dyn.*, **4**, 1–38.

Washington, W.M.,and G.A. Meehl, 1993: Greenhouse sensitivity experiments with penetrative cumulus convection and tropical cirrus albedo effects. *Clim. Dyn.*, **8**, 211–223.

Washington, W.M., G.A. Meehl, L. VerPlank and T. Bettge, 1994: A world ocean model for greenhouse sensitivity studies: Resolution intercomparison and the role of diagnostic forcing. *Clim. Dyn.*, **9**, 321–344.

Weaver, A.J. and E.S. Sarachik, 1991a: Evidence for decadal variability in an ocean general circulation model: An advective mechanism. *Atmos.-Ocean.*, **29**, 197–231.

Weaver, A.J. and E.S. Sarachik, 1991b: The role of mixed boundary conditions in numerical models of the ocean's climate. *J. Phys. Oceanogr.*, **21**, 1470–1493.

Weaver, A.J. and T.M.C. Hughes, 1992: Stability and variability of the thermohaline circulation and its link to climate. In: *Trends in Physical Oceanography*, Council of Scientific Research Integration, Trivandrum, India, **1**, 15–70.

Weaver, A.J., J. Marotzke, P.F. Cummins and E.S. Sarachik, 1993: Stability and variability of the thermohaline circulation. *J. Phys. Oceanogr.*, **23**, 39–60.

Weaver, A.J., S.M. Aura and P.G. Myers, 1994: Interdecadal variability in a coarse resolution North Atlantic model. *J. Geophys. Res.*, **99**, 12423–12441.

Weisse, R., U. Mikolajewicz and E. Maier-Reimer, 1994: Decadal variability of the North Atlantic in an ocean general circulation model. *J. Geophys. Res.*, **99**, 12411–12421.

Wigley, T.M.L. and B.D. Santer, 1990: Statistical comparison of spatial fields in model validation, perturbation and predictability experiments. *J. Geophys. Res.*, **95**, 851–865.

Williamson, D.J. and P.J. Rasch, 1993: Water vapor transport in the NCAR CCM2. *Tellus*, **46A**, 34–51.

Williamson, D.J., J.T. Kiehl and J.J. Hack, 1995: Climate sensitivity of the NCAR Community Climate Model (CCM2) to horizontal resolution. *Clim. Dyn.*, **11**, 377–397.

Willmott, C.J. and R.D. Legates, 1993: A comparison of GCM-simulated and observed mean January and July surface air temperature. *J. Climate*, **6**, 274–291.

Winton, M. and E.S. Sarachik, 1993: Thermohaline oscillations induced by strong steady salinity forcing of ocean general circulation models. *J. Phys. Oceanogr.*, **23**, 1389–1410.

WOCE (World Ocean Circulation Experiment), 1993: Workshop on WOCE Data Assimilation. *WOCE International Project Office, Report No. 102/93*, Wormley, UK, 37 pp.

WOCE (World Ocean Circulation Experiment), 1994: WOCE Strategy for Ocean Modelling. *WOCE International Project Office, Report No. 112/94*, Wormley, UK, 35 pp.

Wolter, K., 1989: Modes of tropical circulation, Southern Oscillation and Sahel rainfall anomalies. *J. Climate*, **2**, 149–172.

Wright, D.G. and T.F. Stocker, 1991: A zonally averaged ocean model for the thermohaline circulation. Part 1: Model development and flow dynamics. *J. Phys. Oceanogr.*, **21**, 1713–1724.

Wright, H.E., J.E. Kutzbach, T. Webb, W.F. Ruddiman, F.A. Street-Perrott and P.J. Bartlein, eds., 1993: *Global Climates Since the Last Glacial Maximum*. University of Minnesota Press, Minneapolis, MN, 544 pp.

Wunsch, C., 1989: Tracer inverse problems. In: *Ocean Circulation Models: Combining Dynamics and Data*, D.L.T. Anderson and J. Willebrand (eds.), Kluwer Academic, Hingham, MA, pp. 1–7.

Wunsch, C., 1994: The TOPEX/POSEIDON data. *AVISO Altimetry Newsletter No. 3*, pp. 14–16.

Xue, Y. and J. Shukla, 1993: The influence of landsurface properties on Sahel climate. Part I: Desertification. *J. Climate*, **6**, 2232–2245.

Zhang, G.J. and N.A. McFarlane, 1995: Sensitivity of climate simulations to the parametrization of cumulus convection in the Canadian Climate Centre general circulation model. *Atmosphere-Ocean*, **33**, 407–446.

Zhang, R.H. and M. Endoh, 1994: Simulation of the 1986-1987 El Niño and 1988 La Niña events with a free surface tropical Pacific Ocean general circulation model. *J. Geophys. Res.*, **99**, 7743–7759.

Zhang, S., R.J. Greatbatch and C.A. Lin, 1993: A re-examination of the polar halocline catastrophe and implications for coupled ocean-atmosphere modelling. *J. Phys. Oceanogr.*, **23**, 287–299.

6

Climate Models – Projections of Future Climate

A. KATTENBERG, F. GIORGI, H. GRASSL, G.A. MEEHL, J.F.B. MITCHELL,
R.J. STOUFFER, T. TOKIOKA, A.J. WEAVER, T.M.L. WIGLEY

Contributors:
*P.A. Barros, M. Beniston, G. Boer, T.A. Buishand, R. Colman, J. Copeland, P.M. Cox,
A. Cress, J.H. Christensen, U. Cubasch, M. Deque, G. Flato, C. Fu, I. Fung, J. Garratt,
S. Ghan, H. Gordon, J.M. Gregory, P. Guttorp, A. Henderson-Sellers, K.J. Hennessy,
H. Hirakuchi, G.J. Holland, B. Horton, T. Johns, A. Jones, M. Kanamitsu, T. Karl,
D. Karoly, A. Keen, T. Kittel, T. Knutson, T. Koide, G. Können, M. Lal, R. Laprise,
R. Leung, A. Lupo, M. Lynch, C.-C. Ma, B. Machenhauer, E. Maier-Reimer,
M.R. Marinucci, B. McAvaney, J. McGregor, L.O. Mearns, N.L. Miller, J. Murphy,
A. Noda, M. Noguer, J. Oberhuber, S. Parey, H. Pleym, J. Raisanen, D. Randall,
S.C.B. Raper, P. Rayner, J. Roads, E. Roeckner, G. Russell, H. Sasaki, F. Semazzi,
C.A. Senior, S.V. Singh, C. Skelly, K. Sperber, K. Taylor, S. Tett, H. von Storch,
K. Walsh, P. Whetton, D. Wilks, F.I. Woodward, F. Zwiers*

Modelling Contributors: see tables

CONTENTS

SUMMARY

General circulation models (GCMs), and in particular coupled atmosphere-ocean general circulation models (AOGCMs), are the state-of-the-art tool for understanding the Earth's present climate, and for estimating the effects on past and future climate of various natural and human factors. This chapter focuses on the estimation of the effects on future climate of changes in atmospheric composition due to human activities. An important development since IPCC(1990) is the improved quantification of some radiative effects of aerosols, and climate projections presented here include, in addition to the effects of increasing greenhouse gas concentrations, some potential effects of anthropogenic aerosols.

Climate simulations using GCMs require substantial computer resources and it is not generally feasible to carry out separate simulations for a large number of forcing scenarios. In order to interpolate and extrapolate global mean projections from GCMs to a wider range of greenhouse gas and aerosol scenarios, simple upwelling diffusion-energy balance models are employed. These models are calibrated to give the same globally averaged temperature response as the global coupled GCMs. Since the amount of anthropogenic aerosols has most probably grown alongside the growth in fossil fuel use since pre-industrial times, the estimated historical changes of radiative forcing up to 1990 used in this report for global mean temperature projections include a component due to aerosols.

Projections of global mean temperature

Using the IS92 emission scenarios, projected global mean temperature changes were calculated up to 2100 assuming low (1.5°C), "best estimate" (2.5°C) and high (4.5°C) values of the climate sensitivity (similar to IPCC (1990)). Taking account of increases of greenhouse gas concentrations alone (i.e., assuming aerosol concentrations remain constant at 1990 levels) the models project an increase in global mean temperature relative to the present of between 1 and 4.5°C by 2100 for the full range of IPCC scenarios. These projections are lower than the corresponding projections presented in IPCC (1990) partly because of the inclusion of aerosols in the pre-1990 radiative forcing history and partly for other reasons, including revised understanding of the carbon cycle (see Chapter 2). Incorporating possible effects of future changes of anthropogenic aerosol concentrations implied by the IS92 scenarios leads to lower projections of temperature change of between 1°C and 3.5°C by 2100. In all cases these projections would represent a substantial warming of climate. Uncertainty in the projections is introduced by uncertainty in the climate sensitivity and by uncertainty in the radiative forcing scenarios.

Projections of continental scale climate change

Spatial patterns of climate change in recent publications tend to confirm and extend the 1990 results. With increasing greenhouse gases, the warming of the land is generally more than that of the oceans, similar to equilibrium simulations. There is a minimum warming around Antarctica and in the northern North Atlantic which is associated with deep oceanic mixing in those areas. The maximum annual mean warming occurs in high northern latitudes associated with reduced sea ice cover. The warming here is largest in late autumn and winter, but becomes negligible for a short period in summer. There is little seasonal variation of the warming in low latitudes or over the southern circumpolar ocean. The diurnal range of land temperature is reduced in most seasons and most regions.

Including the effects of aerosols in simulations of future climate leads to a somewhat reduced warming in middle latitudes of the Northern Hemisphere and the maximum winter warming in high northern latitudes is less extensive.

All models produce an increase in global mean precipitation. If the direct effect of sulphate aerosol forcing is taken into account, the total increase in global precipitation is smaller, as would be expected with the smaller net warming. Precipitation increases in high latitudes in winter and in most cases the increases extend well into mid-latitudes. In the tropics, the patterns of change vary from model to model, with shifts or changes in intensity of the main rainfall maxima. In general, changes in the dry subtropics are small. Recent coupled models (with increased greenhouse gases only) agree with earlier model results that mean south Asian monsoon rainfall could increase. In contrast the limited number of simulations with aerosols included show a reduction in monsoon precipitation relative to the current climate.

There is less confidence in simulated changes in soil moisture than in those of temperature. All models produce predominantly increased soil moisture in high northern latitudes in winter. Most models produce a drier surface in summer in northern mid-latitudes. This occurs most consistently over southern Europe and North America. Summer soil moisture in northern mid-latitudes increases when aerosol effects are included (in contrast to the decrease found in greenhouse gas-only simulations).

In response to increasing greenhouse gases, models show a decrease in the strength of the meridional circulation in the northern North Atlantic oceanic circulation, further reducing the strength of the warming around the North Atlantic. The increase in precipitation in high latitudes decreases surface salinity, inhibiting the sinking of water at high latitude which drives this circulation.

Projections of regional scale climate change

Recent estimates of regional climate changes, that did not take into account the effects of aerosols, for the year 2030 for five areas match the spread of the earlier results reported in IPCC (1990): temperature changes due to a doubling of CO_2 concentration varied between 0.6°C and 7°C and precipitation changes varied from –35% to +50% of control run values. Tropospheric aerosols arising from human activities, because of their uneven spatial distribution, are likely to greatly influence future regional climate change.

Small changes in the mean climate or climate variability can produce relatively large changes in the frequency of extreme events (defined as events where a certain threshold is surpassed); a small change in the variability has a stronger effect than a similar change in the mean.

A general warming would tend to lead to an increase in extremely high temperature events and a decrease in winter days with extremely low temperatures (e.g., frost days in some areas). With increasing greenhouse gas concentration, many models suggest an increase in the probability of intense precipitation. A number of simulations also show in some areas an increase in the probability of dry days and the length of dry spells (consecutive days without precipitation). Where mean precipitation decreases, the likelihood of drought increases. New results reinforce the view that variability associated with the enhanced hydrological cycle translates into prospects for more severe droughts and/or floods in some places and less severe droughts and/or floods in other places.

In the few analyses available, there is little agreement between models on changes in storminess that might occur in a warmer world. Conclusions regarding extreme storm events are obviously even more uncertain. The formation of tropical cyclones depends not only on sea surface temperature (SST), but also on a number of atmospheric factors. Although some models now represent tropical storms with some realism for present day climate, the state of the science does not allow assessment of future changes.

Associated with the mean increase of tropical sea surface temperatures (SSTs) as a result of increased greenhouse gas concentrations, there could be enhanced precipitation variability associated with El Niño-Southern Oscillation (ENSO) events, especially over the tropical continents. Several AOGCMs generate ENSO-like SST variability in their simulations of present-day climate, and continue to do so with increasing greenhouse gas concentrations.

The consequences for global mean temperature of some scenarios leading to stabilisation of greenhouse gas concentrations have been analysed. In each case the climate shows considerable warming during the 21st century. Stabilisation of greenhouse gas concentration does not lead to an immediate stabilisation of global mean temperature, which continues to rise for hundreds of years afterwards due to the long time-scales of the oceanic response.

6.1 Introduction

This chapter on model projections of future climate change is central to this IPCC assessment. It presents estimates of the likely response of the climate system (described in Chapter 1) to scenarios of greenhouse gas and aerosol emissions (which are discussed in Chapter 2). The projections presented here rely heavily on results from general circulation models (GCMs); the underlying climatic processes and feedbacks that provide the greatest challenges to modellers are outlined in Chapter 4. Chapter 5 (Model evaluation) gives an assessment of current models, and Chapter 3 outlines recent observed trends.

The projected magnitudes, rates and patterns of climate change from the present chapter (temperature, wind, precipitation, soil moisture, etc.) form the basis of the sea level projections of Chapter 7 and link directly to the problems of the detection of climate change and its attribution to radiative forcing by greenhouse gases and aerosols (Chapter 8). Climate change projections form a cornerstone of the IPCC Working Group II report, where possible impacts of climate change are assessed.

The chapter's sections are ordered according to scale, both in time and space, and hence in order of increased uncertainty. The methodology moves from GCM results through energy balance to regional climate models and indirect methods like (statistical) downscaling. The final section, 6.7, looks to "Reducing Uncertainties, Future Model Capabilities and Improved Climate Change Estimates".

Section 6.2, "Mean Changes in Climate Simulated by Three-dimensional Climate Models", assesses recent climate model projections for the next decades or century on the globally averaged and continental scales on which the models are most likely to be trustworthy (Figure 6.1). Most results are obtained using (equivalent) CO_2-only scenarios starting from present day conditions. Two studies start with mid-19th century conditions and also take into account the negative radiative forcing of sulphate aerosols (direct effect), though the results have not yet been fully analysed. These studies are a first step towards incorporating more realistic forcings into the models.

Section 6.3, "Global Mean Temperature Changes for the IPCC (1992) Emission Scenarios", presents results of climate projections that explore the response of the climate system to different emission scenarios. Due to the computing costs, this type of calculation cannot be done directly with GCMs. The upwelling diffusion-energy balance (UD/EB) models employed in this section contain important simplifications, but have the advantage of computational efficiency and physical transparency. The

UD/EB model employed in Section 6.3 has been calibrated to give a similar globally averaged response as global coupled climate models.

Section 6.4, "Simulated Changes of Variability Induced by Increased Greenhouse Gas Concentrations", discusses results from coupled atmosphere-ocean GCMs in more detail. Possible changes in variability from monthly to decadal time-scales show little consistency between different simulations, and there is less confidence in this type of result than in the large-scale changes and patterns described in Sections 6.2 and 6.3.

Sections 6.5 on "Changes in Extreme Events" and 6.6 on "Simulation of Regional Climate Change" assess climate model results for small time and spatial scales. Such results are potentially of great practical significance, but confidence in them is at present very low.

6.2 Mean Changes in Climate Simulated by Three-Dimensional Climate Models

By the time of writing of the 1992 IPCC Supplementary Report (IPCC 1992), only four centres had completed extended transient coupled experiments. An additional seven centres have now completed such experiments (Table 6.3: details of many of the models are given in Table 5.1 of Chapter 5), and the original four centres have completed new simulations with higher resolution and/or different scenarios. Here we assess the results from transient simulations with a gradual increase in CO_2; these largely confirm the findings in IPCC (1992). As noted in Section 2.4, forcing from sulphate aerosols may have a marked effect on spatial patterns of climate change, even if their global mean forcing is small relative to that from greenhouse gases. We also consider new transient simulations which indicate the likely additional effect on the magnitude and patterns of climate change of increases in concentrations of sulphate aerosols. A wider range of emission scenarios using simpler energy balance models is considered in Section 6.3. Finally, additional technical matters which are nevertheless pertinent to the interpretation of model results are considered.

6.2.1 *Simulation of Changes in Climate Since the Late 19th Century*

In Chapter 5, the ability of climate models to represent the current climate and its variability was assessed. Before considering predictions of future climate, we examine the ability of models to reproduce changes in climate over the period of the instrumental temperature record (from the latter half of the 19th century to present).

(a) **Temperature indicators**

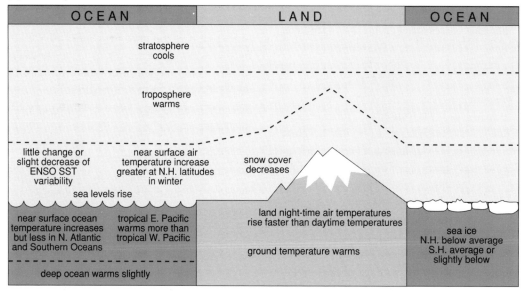

(b) **Hydrological indicators**

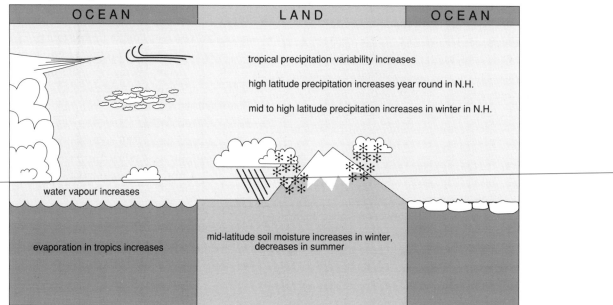

Figure 6.1: A qualitative description of modelled anthropogenic climate change for some temperature (a), and hydrological indicators (b).

6.2.1.1 Estimates of the equilibrium response due to the change in radiative forcing from the late 19th century

Because of the expense of running fully coupled atmosphere-ocean models, preliminary attempts to estimate the effect of past changes in radiative forcing have used atmospheric models coupled to a simple oceanic well mixed layer. This allows an estimate of the equilibrium response to be made with much less computing time required than for a full climate model. However, these calculations neglect the thermal inertia of the deep ocean, and so give an exaggerated response, particularly in regions of deep mixing (see Section 6.2.2). Note that the equilibrium global mean temperature response of a model to doubling atmospheric CO_2 is often referred to in this report as the model's "climate sensitivity".

Table 6.1: *Equilibrium global mean response to the increase in greenhouse gases and sulphate aerosol concentrations over the 20th century. The experimental designs in the four studies differ, so some of the entries are derived under specific assumptions defined below.*

	Study	Sensitivity to Doubling CO$_2$ (°C)	Direct Aerosol Forcing (Wm^{-2})	Temperature response of equilibrium due to aerosols (°C)	Temperature response of equilibrium to combined aerosol and CO$_2$ forcing since 1900 (°C)
S1	Roeckner *et al.* (1995)	2.8	−0.7	−0.9[*]	0.5
S2	Taylor and Penner (1994)	5.2[#]	−0.9	−0.9	0.6+
S3	Mitchell *et al.* (1995a)	5.2	−0.6	−0.8	1.6
S4	Le Treut *et al.* (1995)	3.9	−0.3 (direct) −0.8 (indirect)	−1.6[$]	0.0[$]

S1 and S3 use the aerosol distribution of Langner and Rodhe (1991) and represent aerosols as an increase in surface albedo. S2 derives the sulphate loading from a coupled atmosphere sulphur cycle model and includes an explicit radiative scattering treatment of aerosols.

[*]   Assuming 40% increase in CO$_2$ gives 50% of the warming due to doubling, and subtracting this value from the combined forcing experiment in the final column.

[#]   Assuming 25% increase in CO$_2$ gives 29% of the warming due to doubling.

[+]   Using a 25% increase in CO$_2$, whereas S1 and S3 use a 40% increase to allow for changes in all greenhouse gases.

[$]   The forcing used includes both the estimated direct and indirect forcing. In the last column, CO$_2$ was increased by 25%. Substantially higher sensitivities are found if a colder control simulation is used. Although the global mean temperature change in S4 is zero, the model gives a cooling in the Northern Hemisphere and a warming in the Southern Hemisphere.

Radiative forcing

IPCC (1990) estimated the changes in radiative forcing since pre-industrial times due to increases in well-mixed greenhouse gases. The change in radiative forcing since 1900 is approximately equivalent to that given by a 40% increase in CO$_2$. The equilibrium response to the explicit trace gases and an equivalent increase in CO$_2$ are broadly similar. Wang *et al.* (1992) did find some difference in the patterns of change when using changes in equivalent CO$_2$ instead of changes in the individual gases, but the difference is small compared with that between different models with identical forcing. The equilibrium distribution of warming due to increases in CO$_2$ was discussed extensively in the 1990 IPCC report.

In the last few years, quantitative estimates have also been made of the direct effect of changes in sulphate aerosol concentrations on the Earth's radiation balance (Chapter 2). The equilibrium response to the scattering of radiation by sulphate aerosols has been estimated using atmospheric models coupled to a simple model of the oceanic mixed layer (Table 6.1). Distributions of current sulphate aerosol loading due to human activity were calculated off-line (Langner and Rodhe, 1991) or in the climate model (Taylor and Penner, 1994). The estimated global mean radiative forcing ranges from −0.6 Wm^{-2} to −0.9 Wm^{-2} for scattering by sulphate aerosols (Table 6.1) compared with −0.8 to −0.2 Wm^{-2} (central value −0.4 Wm^{-2}) for the direct forcing estimated in Section 2.4.2.1. In studies S1 and S3, the aerosol forcing is approximated by an increase in surface albedo, following Charlson *et al.* (1991), though Mitchell *et al.* (1995a) found a high spatial correlation between the resulting forcing and that calculated using an explicit scattering scheme with the same aerosol distribution. Note also that the simulations in Table 6.1 ignore other radiative forcing factors including the indirect effect of sulphate aerosols (not S4), volcanic aerosols, biogenic aerosols, soot, ozone and solar variations. In Chapter 2, the net effect of the direct effect and these other factors is estimated to be probably a small cooling though there is large uncertainty both in the net global mean forcing and its geographical distribution.

Because of their short atmospheric lifetime, aerosol concentrations are largest near their source region. Hence, in contrast to the well-mixed greenhouse gases, they have the potential to produce strong regional changes in climate even if their global mean radiative forcing is relatively small. As a result, their relative importance is greater than might be deduced by considering only global mean quantities.

Mean surface air temperature

The direct sulphate aerosol forcing produces a cooling which is most pronounced over the northern mid-latitude continents and downwind oceanic regions. The cooling is also large in the vicinity of winter sea ice, even though the radiative forcing is zero at this time (the polar night). This is because the aerosol cooling is spread by the atmospheric circulation and amplified locally through temperature sea ice feedbacks as in CO_2 only simulations (see for example, Ingram *et al.*, 1989, Mitchell *et al.*, 1995a).

When combined with the increase in greenhouse gases since 1900, the aerosol effect is sufficient to produce areas of net cooling over regions of eastern North America, the Middle East and eastern China in two of the three experiments (S1, S2; see Table 6.1). In S2, it has been shown that the patterns correlate more closely with the observed changes than do simulations with increases in equivalent CO_2 only (Santer *et al.*, 1995; see also the section on transient simulations below and Chapter 8).

In energy balance models it is usually assumed that the sensitivity of global mean surface temperature to global mean aerosol forcing (defined here as temperature response/unit forcing) is the same as to global mean greenhouse gas forcing. There is some evidence that the sensitivity to aerosol forcing is greater than to CO_2 forcing, at least for small amounts of aerosol (Table 6.1). However, because of the large interdecadal variability of temperature and the small aerosol forcing generally used, longer experiments than those listed in Table 6.1 are required to demonstrate that the apparent difference is not an artefact of poor sampling. With a larger aerosol forcing, and hence a larger signal-to-noise ratio, Mitchell *et al.* (1995a) found that sensitivity to aerosol and greenhouse gas forcing was similar. This result suggests that using the same global mean climate sensitivity for greenhouse gases and direct sulphate aerosol forcing, as in Section 6.3, may not lead to serious error.

The range of estimates of forcing due to factors other than well-mixed greenhouse gases in Section 2.4.2 encompasses global mean forcings of up to –2 to –3 Wm^{-2}. The large range of uncertainty makes it difficult to validate models against the observational record, and to estimate the radiative forcing of these factors in the future. Many of the factors considered (sulphate aerosols, soot, tropospheric ozone) are concentrated over the northern mid-latitude continents. It is possible that the larger estimates of

negative radiative forcing would produce a temperature change in this region which is inconsistent with the observed record. For example, in S3, an additional sensitivity experiment with tripled aerosol forcing (giving a global mean forcing of –1.8 Wm^{-2}) and adding the current greenhouse gas forcing gave an unrealistically large area of cooling over the northern continents when compared with the observed changes this century. Thus, it may be possible to use models in this type of experiment to provide an upper limit to the net radiative cooling due to factors other than well-mixed greenhouse gases.

Diurnal range

In recent decades, a widespread decrease in diurnal temperature range has been reported over the main continents (Chapter 3). In IPCC (1990), a few studies reported a small decrease in diurnal range in equilibrium $2 \times CO_2$ experiments, though at that stage no detailed analysis had been performed. More recent studies show that decreases are associated with increased evaporative cooling and other features accompanying climate warming (Cao *et al.*, 1992, Hansen *et al.*, 1995). The changes vary with location and season, and from model to model, though decreases are predominant (for example Cao *et al.*, 1992; Cubasch *et al.*, 1995a; Hansen *et al.*, 1995; Mearns *et al.*, 1995b; Mitchell *et al.*, 1995a). In general, increases in diurnal range are associated with reductions in cloud or soil moisture.

The most obvious effect of aerosol forcing is a tendency to reduce diurnal range, since it reduces the amplitude of the diurnal cycle of insolation. However, simulations with aerosol forcing included indicate that other feedbacks can counteract this effect. Sulphate aerosols can cool climate leading to a reduction in evaporative cooling of the surface, and an increase in global mean diurnal range (Hansen *et al.*, 1995; Mitchell *et al.*, 1995a). In other words, the cooling reverses the processes that lead to a reduced diurnal range in CO_2 only experiments. In these two studies, it is only in regions of maximum aerosol loading that the reduction in the diurnal amplitude of solar heating due to the presence of aerosol is sufficient to produce a net reduction in diurnal range due to aerosols. One consequence of this finding, if correct, is that the observed reduction in diurnal range cannot necessarily be attributed to increases in sulphate aerosols.

The effect of CO_2 and of aerosols on diurnal range is illustrated in Table 6.2 and Figure 6.2. In equilibrium simulations with a 40% increase in CO_2 (which gives the increase in greenhouse gas forcing from 1990 to present) there is a widespread reduction in diurnal range (Figure 6.2b). For comparison, the observed change in diurnal

Table 6.2: *Changes in diurnal range of 1.5 m temperature averaged over seasons and the annual cycle. The simulated and observed values are averaged over the regions where observations are available (see Figure 6.2a). The observed data are from Horton (1995) and the simulations from S3 (see Table 6.1). The changes in greenhouse gas forcing (represented by an equivalent increase in CO_2) and direct sulphate aerosol forcing are those estimated to have occurred since 1900. Note that the observed changes are available only over the latter half of this period and are the difference between the mean for 1981 to 1990 and the mean for 1951 to 1980.*

Source	DJF*	MAM*	JJA*	SON*	YEAR
Simulated, increase in equivalent CO_2 since 1900	–0.46	–0.35	–0.08	–0.34	–0.29
Simulated, aerosol forcing and equivalent CO_2 increase since 1990	–0.43	–0.27	–0.16	–0.32	–0.27
Observations of recent change (1981 to 1990 mean less 1951 to 1980 mean)	–0.28	–0.17	–0.19	–0.36	–0.19

* DJF = December, January, February; MAM = March, April, May; JJA = June, July, August; SON = September, October, November.

range from the mean for 1950 to 1980 to the mean for 1980 to 1990 is shown in Figure 6.2a. (Observed data from the beginning of the century are unreliable, but the period corresponds to when most of the increase in radiative forcing occurred and so allows qualitative comparison). The addition of aerosol effects (S3, Figure 6.2c) leads to slightly larger reductions in diurnal range in parts of mid-latitudes. The seasonal variation of the changes in diurnal range is similar to that observed, with smaller reductions occurring in northern spring and summer (Table 6.2). The observed changes are discussed in Section 3.2.2.4. Hansen *et al.* (1995) find that increases in CO_2 and aerosols alone are insufficient to explain the observed reduction in diurnal range, which they can only explain by additionally increasing cloud.

In summary, models in general produce decreases in diurnal range over much of the mid-latitude continents when forced with increases in aerosols, but the simulated magnitude varies from model to model, and the relative importance of different physical mechanisms leading to the observed diurnal range is uncertain. Additional uncertainties in radiative forcing (Chapter 2) and the observed changes (Chapter 3) mean that changes in diurnal range may be a poor diagnostic for model validation.

6.2.1.2 *Estimates of the time dependent response to changes in radiative forcing since the late 19th century using three-dimensional climate models*

Since 1990, several new studies have attempted to reproduce the instrumental temperature record from the late 19th century to present, including two using fully coupled GCMs (Hasselmann *et al.*, 1995; Mitchell *et al.*, 1995b (runs *y* and *z* respectively in Table 6.3)). As in the equilibrium simulations, the increases in greenhouse gases may be represented by an equivalent increase in CO_2, and the evolution of greenhouse gas forcing has been taken from a variety of sources including IPCC (1990).

When increases in greenhouse gases only are taken into account in simulating climate change over the last century, most GCMs (Hansen *et al.*, 1993, Mitchell *et al.* 1995b – Figure 6.3;) and energy balance models (Wigley and Raper, 1992; Schlesinger *et al.*, 1993; Murphy, 1995b) produce a greater warming than that observed to date, unless a lower climate sensitivity than that found in most GCMs is used (see Figure 8.4).

As noted above, there is growing evidence that increases in sulphate aerosols are partially counteracting the radiative forcing due to increases in greenhouse gases (Chapter 2). The temporal evolution of sulphate aerosol concentrations can be estimated from sulphur emission inventories (for example, Dignon and Hameed, 1989), assuming a constant ratio between emissions and concentrations, or through an explicit sulphur cycle model. To generate spatial distributions of sulphate aerosol concentration for use in transient GCM simulations, Hasselmann *et al.* (1995) used decadal means derived using the MOGUNTIA model (Langner and Rodhe, 1991), whereas Mitchell *et al.* (1995b) scaled the pattern given by Langner and Rodhe (1991) by the global mean emissions for each decade. Both the GCM experiments included only the direct effect of sulphate aerosols, represented by an appropriate increase in surface albedo (see Section 6.2.1.1).

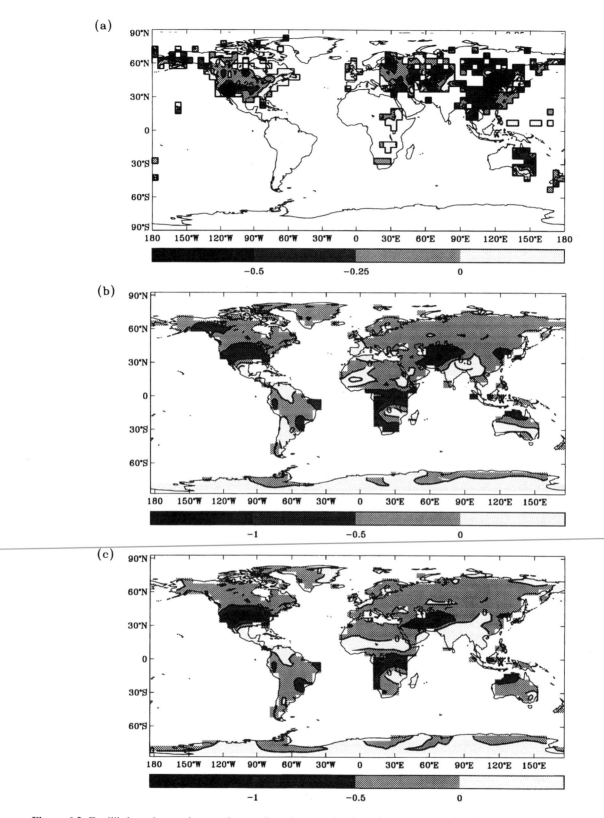

Figure 6.2: Equilibrium changes in annual mean diurnal range of surface air temperature. (a) Observations (difference in 1980–1990 and 1951–1980 means) (from Horton, 1995). Contours every 0.25°C. (b) Simulated due to a 40% increase in CO_2. Contours every 0.5°C (from Mitchell *et al.*, 1995a). (c) Simulated due to a 40% increase in CO_2 and the direct effect of current industrial sulphate aerosols. Contours every 0.5°C (from Mitchell *et al.*, 1995a).

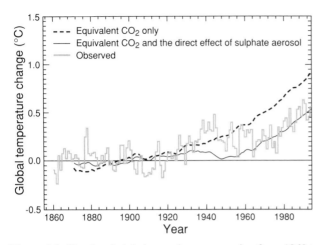

Figure 6.3: Simulated global annual mean warming from 1860 to 1990, allowing for increases in equivalent CO_2 only (dashed curve) and allowing for increases in equivalent CO_2 and the direct effects of sulphates (flecked curve) (Mitchell *et al.*, 1995a). The observed changes are from Parker *et al.* (1994). The anomalies are calculated relative to 1880–1920.

When cooling due to aerosols is included in some form, then an improved agreement with the observational record of global mean surface temperature in the last few decades is found in almost all studies (for example Figures 6.3) using models with the level of climate sensitivity found in current AOGCMs. As in the equilibrium experiments described above, the aerosol forcing considerably reduces the surface warming in the northern extratropics, and even leads to cooling in some regions. The amplification of the surface warming in northern high latitudes is also reduced – for example in *z* the warming in the Arctic from period 1880 to 1920 to the decade 1980 to 1990 is similar to the global mean warming. Mitchell *et al.* (1995b) found that the spatial correlation of simulated temperature change with that observed is also higher in the last few decades when aerosol effects are included (see also Figure 8.6). This indicates that the effect of sulphate aerosol and other forcings should be taken into account in attempts to detect or attribute climate change, as noted in Chapter 8.

Three factors need to be considered in assessing these results. First, we do not know how well a perfect model driven with the correct radiative forcing would agree with the observational record, as we do not know the level of internal variability of climate on decadal and longer time-scales. Second, it is possible in principle to get a similarly good fit to the observed global mean record if the climate sensitivity and aerosol forcing are both increased (or decreased). As there are large uncertainties in both radiative forcing (Chapter 2) and climate sensitivity, the

ability of a model to reproduce the past record does not necessarily imply that its climate sensitivity or the assumed radiative forcing is correct. Note in particular that these preliminary GCM studies include only an estimate of the direct forcing due to sulphate aerosols and neglect other potentially large sources of natural and anthropogenic forcing. Finally, there are uncertainties in the observations, particularly in remote regions and in the early part of the observational record.

6.2.2 Time-dependent Projections of Mean Climate Change Using Global Coupled Atmosphere-Ocean Models

Experiments that study the time-dependent evolution of climate are called "transient" experiments. The various coupled atmosphere-ocean climate model experiments carried out to date span a wide range of forcing scenarios, climate sensitivities (the equilibrium or long-term global mean temperature response to doubling CO_2 concentrations) and initial conditions (Table 6.3). The experiments are referenced by the letter assigned in Table 6.3, and the version of the model used by the number in Table 5.1. Fuller references to the experiments are given in Table 6.3.

Most transient experiments with AOGCMs are based on certain simplifying assumptions. First, they represent the effect of increases in all greenhouse gases by increases in CO_2 (Chapter 5). Second, apart from experiments *x,y,w* and *z*, they ignore the effects of the historical increases in forcing which leads to an underestimate of the initial rate of change (the "cold start", described in Section 6.2.4). Finally, they use simplified versions of the IPCC scenarios. A majority of the experiments assume a 1%/yr compound increase in CO_2 which gives a doubling of CO_2 after 70 years (Table 6.3). With a 1%/yr increase in equivalent CO_2, the radiative forcing is generally about 20% greater than that under Scenario IS92a which gives a doubling of equivalent CO_2 after about 95 years. For comparison, the actual increase in equivalent CO_2 (i.e., allowing for increases in CO_2 and other trace gases) over the last decade or so has been about 0.7%/yr which, if maintained, would lead to a doubling of equivalent CO_2 in 100 years.

6.2.2.1 Annual mean rates of change
Simulations with increases in greenhouse gases only
In experiments forced with a 1%/yr increase in CO_2, the rate of warming at the time of doubled CO_2 varies from about 0.17°C/decade to 0.5°C/decade (Figure 6.4). This spread is not surprising given that the estimated equilibrium warming for a doubling of CO_2 for the models in Figure 6.4 varies from 2.1 to 4.6°C (Table 6.3). The

Table 6.3: *Summary of transient coupled atmosphere–ocean GCM experiments used in this assessment. Each experiment (run) is denoted by a letter in the text, and the model is referenced by the number used in Table 5.1. The scenario gives the rate of increase of CO_2 used – most experiments use 1%/yr which gives a doubling of CO_2 after 70 years (IS92a gives a doubling of equivalent CO_2 after 95 years). The ratio of the transient response at the time of doubled CO_2 to the equilibrium (long-term) response to doubling CO_2 is given if known.*

Centre	Model No.	Experiment	Reference	Flux adjusted?	Scenario	Warming at doubling[†]	Equilibrium warming	Ratio(%)[†]
BMRC	1	a	Power et al., 1993; Colman et al., 1995	No	1%/yr	1.35	2.1	63
CCC	2	b	G. Boer (pers. comm)	Yes	1%/yr	–	3.5	
COLA	4	c	E. Schneider (pers. comm)	No	1%/yr	2.0	–	
CSIRO	5	d	Gordon and O'Farrell, 1996	Yes	1%/yr	2.0	4.3	47
GFDL	6	e	Stouffer (pers. comm.)	Yes	0.25%/yr	2.6	3.7	
..	6	f	Stouffer (pers. comm.)	Yes	0.50%/yr	2.4	3.7	
..	6	g	Manabe et al., 1991, 1992	Yes	1%/yr	2.2	3.7	59
..	6	h	Stouffer (pers. comm.)	Yes	2%/yr	1.8	3.7	
..	6	i	Stouffer (pers. comm.)	Yes	4%/yr	1.5	3.7	
..	–	j	Stouffer (pers. comm.)	Yes	1%/yr	–	–	
GISS	7	k	Russell et al., 1995; Miller and Russell, 1995	No	1%/yr	1.4	–	
IAP	9	l[A]	Keming et al., 1994	Yes	1%/yr	2.5	–	
MPI	11	m[B]	Cubasch et al., 1992, 1994b; Hasselmann et al., 1993; Santer et al., 1994	Yes	IPCC90A	1.3	2.6	50
..	11	n	Cubasch et al., 1992; Hasselmann et al., 1993; Santer et al., 1994	Yes	IPCC90D	na	2.6	
..	12	o	Hasselmann et al., 1995	Yes	IPCC90A	1.5	–	
..	–	x[C]	Hasselmann et al., 1995	Yes	IPCC90A	na	2.6	
..	–	y[D]	Hasselmann et al., 1995	Yes	Aerosols	na	2.6	
MRI	13	p	Tokioka et al., 1995	Yes	1%/yr	1.6	–	
NCAR	–	q	Washington and Meehl, 1989	No	1%/yr*	2.3	4.0	58
..	14	r[E]	Washington and Meehl, 1993, 1996; Meehl and Washington, 1996	No	1%/yr	3.8	4.6	83

Centre	Model No.	Experiment	Reference	Flux adjusted?	Scenario	Warning at doubling[†]	Equilibrium warming	Ratio(%)[†]
UKMO	–	s	Murphy 1995 a,b; Murphy and Mitchell, 1995	Yes	1%/yr	1.7	2.7	64
,,	16	t[F]	Johns et al., 1996; Keen, 1995	Yes	1%/yr	1.7	2.5	68
,,	16	w[G]	Johns et al., 1996; Tett et al., 1996;	Yes	1%/yr	na	2.5	
			Mitchell et al., 1995b; Mitchell and Johns, 1996					
,,	16	z[H]	Johns et al., 1996; Tett et al., 1996;	Yes	Aerosols	na	2.5	
			Mitchell et al., 1995b; Mitchell and Johns, 1996					

na = not available

† Numbers in italics indicate simulations with other than a 1%/yr increase in CO_2.

* 1%/yr of current CO_2 concentrations.

A Polar deep ocean quantities constrained (see Chapter 5).

B Three additional 50 year runs, each from different initial conditions.

C CO_2 from IPCC scenario 90A after forcing with greenhouse gas forcing from 1880 to 1990

D As C with a representation of aerosol forcing, with increases after 1990 based on IS92a.

E Equilibrium model excluded sea ice dynamics. Coupled model has warmer than observed tropical SSTs and a vigorous ice albedo feedback (Washington and Meehl, 1995) contributing to the high sensitivity.

F Average of three experiments from different initial conditions.

G CO_2 increased by 1%/yr from 1990. Observed greenhouse gas forcing used from 1860 to 1990.

H As G with a representation of aerosol forcing, with increases of aerosol and greenhouse gases after 1990 based on IS92a.

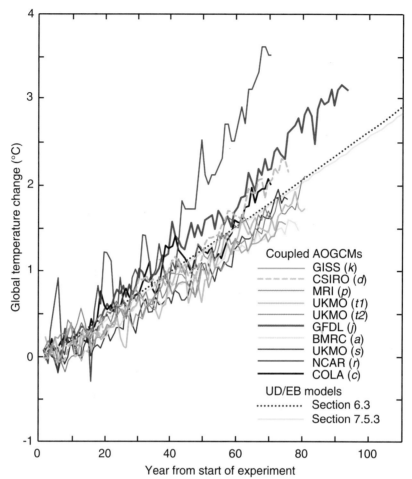

Figure 6.4: Comparison between several AOGCM simulations (climate sensitivities between 2.1 and 4.6°C), the UD/EB model of Section 6.3 (climate sensitivity 2.5°C) and the simple climate model of Section 7.5.3 (climate sensitivity of about 2.2°C). All models were forced with 1%/yr (compound) increase of atmospheric CO_2 concentration from equilibrium or near-equilibrium in 1990.

results straddle the 0.3°C/decade estimated with a simpler UD/EB model for Scenario IS92a (IPCC, 1992) when an equilibrium sensitivity of 2.5°C to doubling CO_2 is assumed. Note that the initial rate of warming in the coupled model experiments is generally small because the historical build up of greenhouse gases has been neglected (see Sections 6.2.1.2 and 6.2.4).

Two new experiments have been completed using the historical increase in greenhouse gas forcing from the end of the last century to the present, followed by CO_2 increasing by either 1%/yr (*w*), or according to the older IPCC (1990) Scenario A (*x*) which is roughly equivalent to a 1.2%/yr increase in CO_2. Hence the forcing is slightly larger than in IS92a, but climate sensitivity of both models is close to the IPCC (1990) "best estimate" value. The rate of warming in the early 21st century is close to 0.3°C/decade, in reasonable agreement with the simpler UD/EB model results noted above.

Simulations with greenhouse gas and aerosol forcing
Since IPCC (1990), attempts have been made to include the effects of increases in sulphate aerosol concentrations in models, as these may have a strong influence on patterns of climate change (see Section 2.4, Section 6.2.1, Chapter 8). The two experiments above which included the historical increase in greenhouse gas forcing (x, and w) were repeated with the direct effect of increases in sulphate aerosol represented by an increase in surface albedo (Section 6.2.1.2). The distribution of sulphate aerosols concentrations from 1990 in each case was derived using the MOGUNTIA sulphur cycle model (Langner and Rodhe, 1991) and sulphur emission scenarios under IS92a (IPCC, 1992). In z, the global mean aerosol forcing increases from 0.6 Wm^{-2} in 1990 to about 1.3 Wm^{-2} in 2050. IS92a assumes a doubling of sulphur emissions by 2050. Given the uncertainties in future sulphur emissions and in converting from sulphate concentrations to forcing,

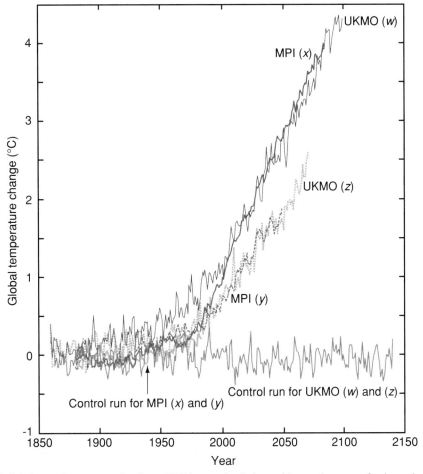

Figure 6.5: Simulated global annual mean warming from 1880 in two simulations with greenhouse gas forcing only, MPI (*x*) and UKMO(*w*), and two simulations which include both greenhouse gas and direct sulphate aerosol forcing, MPI(*y*) and UKMO(*z*) (Hasselmann *et al.*, 1995; Mitchell *et al.*, 1995a). The control runs for each model are also shown. Experiment details are given in Table 6.3.

and the neglect of other factors including indirect forcing by sulphates, these experiments should be regarded primarily as sensitivity studies which indicate the potential influence of sulphate aerosols.

The rate of warming is 0.2°C/decade, as opposed to 0.3°C/decade with increases in CO_2 alone (Mitchell *et al.* 1995b, Hasselmann *et al.* 1995) (Figure 6.5), in good agreement with earlier calculations using an energy balance model (Wigley and Raper, 1992, Figure 8.4). The addition of aerosol forcing leads to changes in the simulated patterns of change in northern mid-latitudes, as discussed below.

6.2.2.2 Patterns of annual mean temperature change
Simulations with increases in greenhouse gases only
All models produce a greater warming over land than over the sea (e.g., Figure 6.6). Although the greater thermal inertia of the oceans may contribute to this, a similar

feature is evident in equilibrium simulations with a simple ocean (for example, IPCC 1990). Simulated feedbacks are generally stronger over the land than the ocean (for example, Murphy, 1995b). One reason is that in higher latitudes, the presence of snow produces a positive temperature albedo feedback which is absent over the ocean. The temperature response will also be enhanced over dry land relative to the ocean because evaporative cooling is restricted leaving more of the increased radiative heating to raise the surface temperature.

These and other feedbacks may be enhanced or reduced by changes in cloudiness (for example, Wetherald and Manabe, 1995). The magnitude and even sign of the cloud feedback is uncertain (Chapter 4, Senior and Mitchell, 1993), so the strength of the land sea contrast will vary from model to model. Nevertheless, the enhanced warming over land remains a robust feature of all models and is physically based.

(a) BMRC AOGCM

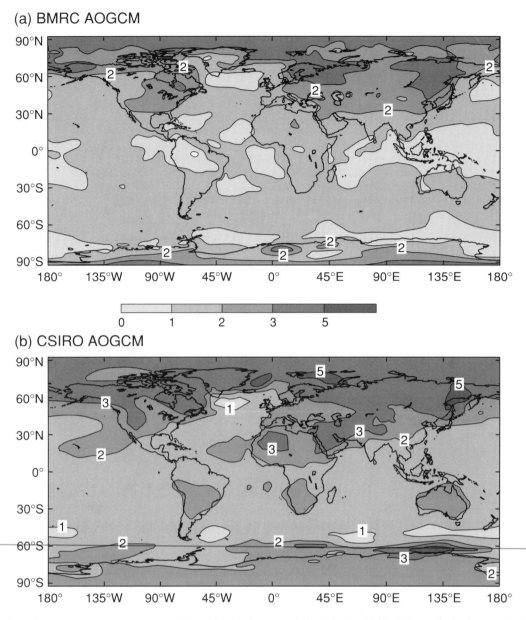

Figure 6.6: Annual mean temperature response of the BMRC and CSIRO AOGCMs (see Table 6.3) to a 1%/yr increase in CO_2, averaged around the time of doubled CO_2. (a) From model 1, experiment *a* (no flux adjustment); (b) from model 5, experiment *d* (with flux adjustment).

There is a minimum warming *(a,d,e,l,r,t,x,w)*, or even regions of cooling *(k,q,m,s)* in the high latitude southern ocean (for example, Figure 6.6). The minimum warming is associated with increases in the thermal inertia of the ocean due to deep convection and large-scale vertical motions which occur in the present climate (for example, Drijfhout *et al.*, 1996). The simulated deep mixing in this region is supported by measurements of the penetration of inert tracers on decadal time-scales (for example, Warner and Weiss, 1992). Recent studies (England, 1995; Robitaille and Weaver, 1995, also Sections 5.3.3.2 and 5.3.3.4.1)

suggest that current ocean models may produce excessive vertical mixing in this region. If this is confirmed, the degree to which the warming is retarded around the southern ocean may be exaggerated.

The position and strength of this minimum varies from model to model. It is generally around 60°S, though in *a* it is closer to Antarctica (Figure 6.6a), whereas in *d* (Figure 6.6b) and *t,w* it is further north. Note that in some models (for example, *1*) there is very little sea ice in the control. This considerably weakens the sea ice albedo feedback, but would not be expected to produce a minimum warming at

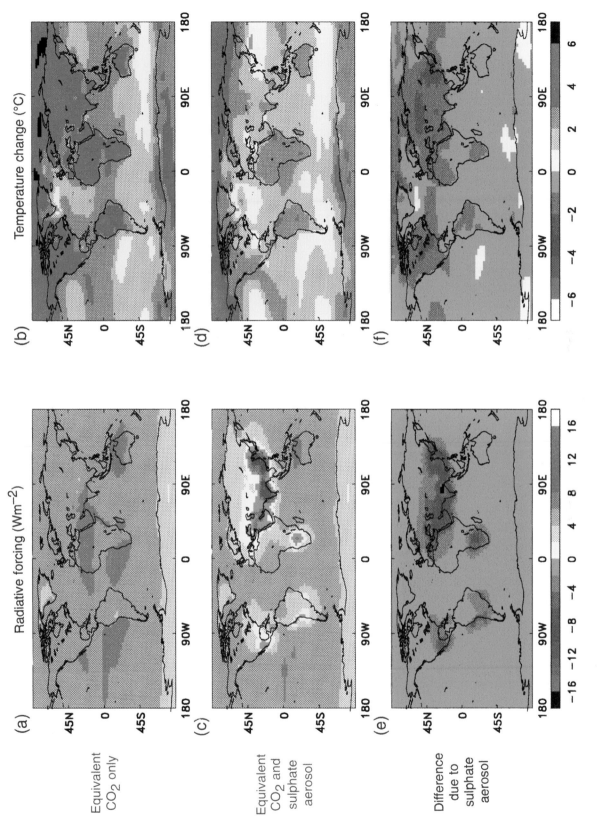

Figure 6.7: Radiative forcing (Wm^{-2}) and annual mean temperatures changes (°C) under Scenario IS92a, from 1795 to 2030–2050. (a) Equivalent CO$_2$ forcing. (b) Temperature response to equivalent CO$_2$ forcing. (c) Forcing due to aerosols and CO$_2$ combined. (d) Temperature response to combined aerosol and CO$_2$ forcing . (e) Aerosol forcing. (f) Temperature response to aerosol forcing ((d) − (b)). From Mitchell *et al.*, 1995b.

these latitudes. In *r*, which has relatively weak vertical mixing in these latitudes, the warming around Antarctica is larger than in other models.

Most models produce a region of minimum warming in the northern North Atlantic (Figure 6.6), another region of deep oceanic mixing (Weiss *et al.*, 1985; Smethie, 1993). This feature is absent in *p* which produces a very feeble North Atlantic meridional circulation in the simulation of present climate (Table 5.5), perhaps indicative that deep mixing is weak or absent, and is muted in *a*, which also has weak meridional overturning in the North Atlantic. In contrast, *k* produces a cooling which extends into the Norwegian Sea.

Many studies report a decrease in the strength of the North Atlantic thermohaline (meridional) circulation (*a,d-i,k,m,s,t,w*). Note that in two models (*a,k*), the thermohaline circulation is weak or weakening throughout the control simulation and in *p* this circulation is almost non-existent (Chapter 5). The strength of the reduction varies from model to model, from around 30% at the time of doubling (*g,s*) to less than 10% (*t*). This change has been studied in more detail in only a few experiments (*g,s,m*). In the warmer climate, the decrease is associated with an increase in the moisture supply to the surface (precipitation in high latitudes increases more than evaporation) which reduces the salinity and thus the density of surface water, and inhibits the sinking of water at high latitudes (e.g., Manabe *et al.*, 1991, Murphy and Mitchell, 1995). These authors also find a similar freshening of the upper layers in high southern latitudes. The greater penetration of heat in high latitudes may also reduce the meridional density gradient, further slowing the meridional circulation. The weakening of this circulation further inhibits warming, or may even produce local cooling (for example, Figure 6.7b), in regions of deep water formation.

In one study (*g*), the thermohaline circulation stopped altogether when the concentration of CO_2 was increased fourfold, and did not recover when the CO_2 level was maintained at this level for several centuries (Manabe and Stouffer, 1994). However, when CO_2 was gradually doubled and then held fixed in the same model, the thermohaline circulation first weakened and then returned to its original strength.

Several simulations show a differential mean warming of sea surface temperatures across the tropical Pacific with the eastern Pacific warming faster than the western Pacific (*g,k,r*). This is similar to the pattern of SST anomalies associated with ENSO events as well as the decadal time-scale pattern of SST change in the tropical Pacific (Chapter 3). In *g* (Knutson and Manabe, 1995), the difference

between the warming in the east and west is relatively small in comparison to the mean tropical sea surface temperature warming (about 1°C compared to a 4–5°C mean warming in a CO_2 quadrupling experiment), and was not clearly evident until about a century of the integration was completed with CO_2 increasing at 1%/yr. In *k* and *r*, the pattern of change becomes evident by the time of doubling (Meehl and Washington, 1996; Tett, 1995). Associated with the reduced zonal gradient in sea surface temperature in *g,q* is a mean precipitation change that resembles in some respects the precipitation anomalies associated with present day ENSO events, i.e., relatively greater increases in precipitation occur in the mean over the central equatorial Pacific, with relatively small increases or some decreases elsewhere in the Inter Tropical Convergence Zone (ITCZ), the South Pacific Convergence Zone as well as in the far western Pacific over Indonesia and northern Australia.

Finally, we consider changes in the diurnal range of temperature, since observations suggest this parameter has diminished in many regions in recent decades (Chapter 3). Some of the coupled models omit the diurnal cycle of insolation (*f–j,q,r*), and of those that do not, only a few (*a,d,k,s,t,w,z*) report changes in diurnal range. Most find small reductions in global annual mean diurnal range (e.g., Figure 6.8). The local changes vary with location and season, and from model to model, though decreases are predominant in the extratropics (*s,t,w,z*), except in *k* which reports substantial areas of increase in mid-latitudes. The mechanisms of change in diurnal range are discussed in more detail in Section 6.2.1.1.

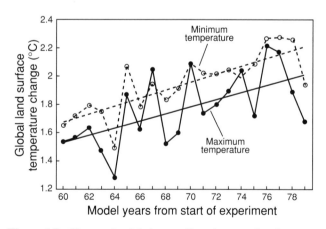

Figure 6.8: Changes in global mean diurnal range of surface temperatures during years 60 to 79 of a simulation by the BMRC AOGCM (Model 1 – see Table 6.3) in which CO_2 is increased by 1%/yr (Colman *et al.*, 1995). The linear trends in minimun and maximum temperature over the same period are also shown.

Simulations with greenhouse gas and aerosol forcing

Two simulations (x,w) were forced with the historical increase in equivalent CO_2, and then a 1%/yr increase in equivalent CO_2. The patterns of change are qualitatively similar to those in the experiments above. In Section 6.2.1.2 (see also Section 8.4.2.3), we saw that the inclusion of the direct sulphate aerosol forcing can improve the simulation of the patterns of temperature change over the last few decades. Here we consider the effect of combined greenhouse gas and direct sulphate aerosol forcing derived from IS92a on simulated patterns of temperature change to 2050 and beyond (y,z). (The details of these two experiments have been given above in Section 6.2.2.1.)

Increasing CO_2 alone leads to positive radiative forcing everywhere, with the largest radiative heating in regions of clear skies and high temperatures (experiment w shown in Figure 6.7a). The surface temperature warms everywhere except in the northern north Atlantic (Figure 6.7b). In transient simulations to 2050, the inclusion of aerosols based on IS92a (y,z) reduces the global mean radiative forcing, and leads to negative radiative forcing over southern Asia, particularly in the east (experiment z shown in Figure 6.7c). This leads to a muted warming (Figure 6.7d) or even small regions of cooling (y) in mid-latitudes. In z, China continues to warm (Figure 6.7d), albeit at a very reduced rate, even though the local net radiative forcing becomes increasingly negative (Figure 6.7c). The rate of warming over North America and western Europe, where the aerosol forcing weakens, remains below that in the simulation with greenhouse gases only (w). The cooling due to aerosols is amplified by sea ice feedbacks in the Arctic (Figure 6.7f).

In assessing these results, one should bear in mind the possible exaggeration of the sulphate aerosol concentrations under this scenario, the uncertainties in representing the radiative effects of sulphate aerosols and the neglect of other factors including the indirect effect of sulphates. Nevertheless, these experiments suggest that the direct effect of sulphate aerosols could have strong influence on future temperature changes, particularly in northern mid-latitudes.

6.2.2.3 Seasonal changes in temperature, precipitation and soil moisture

IPCC (1990) reported some broad scale changes which were evident in most of the equilibrium $2 \times CO_2$ experiments which were then available. The detailed regional changes differed from model to model. In the transient experiments reported in IPCC (1992), it was found that the large-scale patterns of response at the time

of doubling CO_2 were similar to the corresponding equilibrium experiments (IPCC 1990), except that there is a smaller warming in the vicinity of the northern North Atlantic and the southern ocean in transient experiments. Here we summarise the main features in the seasonal (December to February and June to August) patterns of change in temperature, precipitation and soil moisture in those experiments with a 1%/yr increase in CO_2 for which data were available. The changes are assessed at the time of CO_2 doubling (after 70 years). Further information on the subcontinental response in specified regions to increases in greenhouse gases only is given in Section 6.6 and Figure 6.32, and part of Figure 6.9. In experiments w–z we also contrast the continental scale response under the IS92a Scenario with and without aerosol forcing at around 2040.

Temperature

With increases in CO_2, all models produce a maximum annual mean warming in high northern latitudes (Figures 6.6 and 6.7b). The warming is largest in late autumn and winter, largely due to sea ice forming later in the warmer climate. In summer, the warming is small – if the sea ice is removed with increased CO_2, then the thermal inertia of the mixed-layer prevents substantial warming during the short summer season, otherwise melting sea ice is present in both control and anomaly simulations, and there is no change in surface temperature (see Ingram *et al.*, 1989). The details of these changes are sensitive to parametrization of sea ice, and in particular, the specification of sea ice albedo (e.g., Meehl and Washington, 1995). In one simulation (k) there is a marked cooling over the north-eastern Atlantic throughout the year which leads to a cooling over part of north-west Europe in winter. There is little seasonal variation of the warming in low latitudes or over the southern circumpolar ocean.

When aerosol effects are included $(y,z$ cf. x,w respectively), the maximum winter warming in high northern latitudes is less extensive. In mid-latitudes, there are some regions of cooling, for example, over China (Figure 6.10a, for 2040 to 2049 – the scenario is described in Section 6.2.2.1) and the mean warming in the tropics is greater than in mid-latitudes. In northern summer, there are again regions of cooling in mid-latitudes (Figure 6.10b) and the greatest warming now occurs over Antarctica. Again, including the direct forcing by sulphate aerosols has a strong effect on simulated regional temperature changes, though the reader should bear in mind the limitations of these experiments noted earlier.

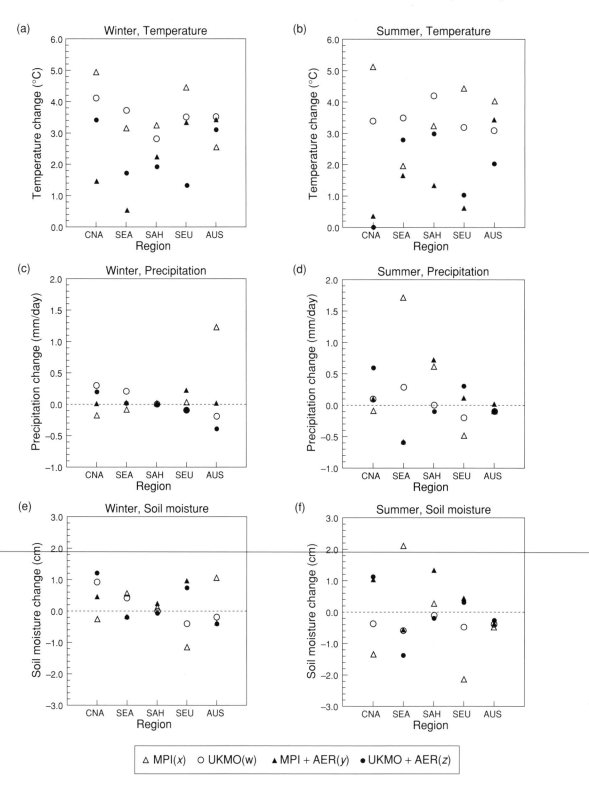

Figure 6.9: Simulated regional changes from 1880-1889 to 2040–2049 (experiments *x, y*) or from pre-industrial to 2030–2050 (experiments *w, z*). Experiments *x* and *w* include greenhouse gas forcing only, whereas *y* and *z* also include direct sulphate aerosol effects (see Table 6.3). (a) Temperature (December to February); (b) Temperature (June to August); (c) Precipitation (December to February); (d) Precipitation (June to August); (e) Soil moisture (December to February); (f) Soil moisture (June to August). CNA = Central North America; SEA = South East Asia; SAH = Sahel; SEU = Southern Europe; AUS = Australia (the regions are defined in Section 6.6.1).

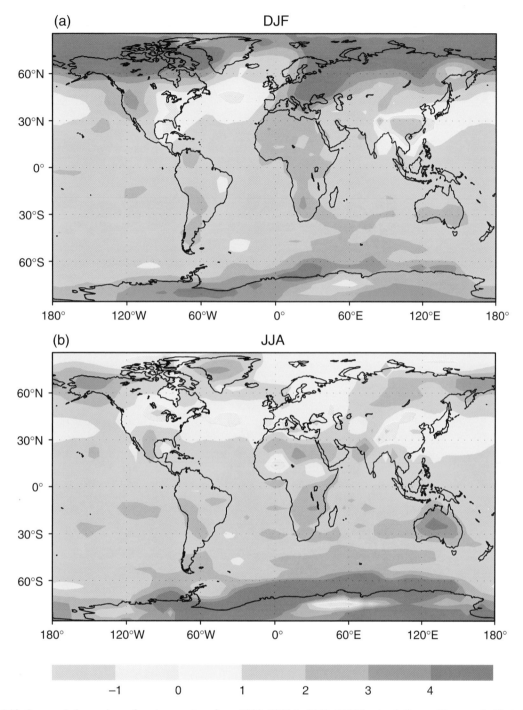

Figure 6.10: Seasonal change in surface temperature from 1880–1889 to 2040–2049 in simulations with aerosol effects included (from experiment *y* in Table 6.3, U. Cubasch, pers. comm.). Contours every 1°C . (a) December to February; (b) June to August.

Precipitation

On increasing CO_2, all models produce an increase in global mean precipitation. Precipitation increases in high latitudes in winter (except in *k* around the Norwegian Sea where there is cooling and a reduction in precipitation), and in most cases the increases extend well into mid-latitudes

(for example, Figure 6.11a). The warming of the atmosphere leads to higher atmospheric water vapour content, enhanced poleward water vapour transport into the northern high latitudes and hence enhanced water vapour convergence and precipitation (for example, Manabe and Wetherald, 1975). In the tropics, the patterns of change

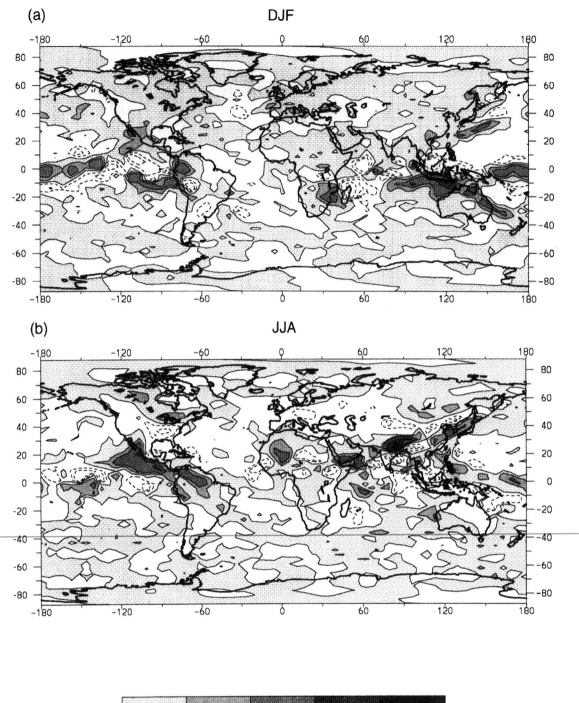

Figure 6.11: Seasonal changes in precipitation at the time of doubling CO_2 following a 1%/yr increase (from experiment *a* in Table 6.3, Colman *et al.,* 1995). Contours at ± 0.5, 1, 2 and 4 mm/day, negative contours are dashed and areas of increase areas stippled. (a) December, January and February; (b) June, July and August.

vary from model to model, with shifts or changes in intensity of the main rainfall maxima. However, many produce more rainfall over India and/or south-east Asia (*a,d,g,m,x,p,q,s,t,w*) (for example Figure 6.11b, see also Figures 6.9d and 6.32g). This is consistent with an increase in atmospheric water vapour concentration leading to enhanced low level moisture convergence associated with the strong mass convergence into the monsoon surface pressure low. All models considered apart from *p* and *q* produce a general reduction in precipitation over southern Europe (see also Figures 6.9d and 6.32g). In general, changes in the dry subtropics are small.

With the inclusion of aerosol forcing (*y,z*), there is only a small increase in global mean precipitation. The patterns of change in precipitation in northern winter are broadly similar to that in a parallel simulation with greenhouse gases only (*x,w* respectively), but less intense. In northern summer there is a net reduction in precipitation over the Asian monsoon region (Figure 6.9), because the aerosol cooling reduces the land-sea contrast and the strength of the monsoon flow. Precipitation increases on average over southern Europe (it decreases when aerosol effects are omitted, Figure 6.9), and over North America, where changes were small with increases in greenhouse gases only.

Soil moisture
Soil moisture may be a more relevant quantity for assessing the impacts of changes in the hydrological cycle on vegetation than precipitation since it incorporates the integrated effects of changes in precipitation, evaporation and run-off thorough the year. However, simulated changes in soil moisture should be viewed with caution because of the simplicity of the land-surface parametrization schemes in current models (e.g., experiments *a,e–i,l,m,n,p,q* and *r* use an unmodified "bucket" formulation – see Section 5.3.2).

Most models produce a general increase in soil moisture in the mean in high northern latitudes in winter, though in some (*a,k*) there are also substantial areas of reduction (e.g., Figure 6.12a). The increases are due mainly to increased precipitation discussed above, and the increased fraction of precipitation falling as rain in the warmer climate. At the low winter temperatures, the absolute change in potential evaporation is small, as expected from the Clausius-Clapyeron relation, so evaporation increases little even though temperature increases are a maximum in winter. Hence the increase in soil moisture in high altitudes in winter is consistent with physical reasoning and the broad scale changes are unlikely to be model dependent. However, it should be noted that in general, the models considered here do not represent the effects of freezing on ground water.

Most models produce a drier surface in summer in northern mid-latitudes (for example, Figure 6.12b). This occurs consistently over southern Europe (except *q* which produces an excessively dry surface in winter in its control climate) and North America (except *d, k* and *q*). The main factor in the drying is enhanced evaporation in summer (see Wetherald and Manabe, 1995): the absolute rate of increase in potential evaporation increases exponentially with temperature if other factors (wind, stability and relative humidity) are unchanged.

As noted in the IPCC (1990) the following factors appear to contribute to summer drying:

(i) the soil in the control simulation is close to saturation in late winter or spring: this ensures that much of the extra precipitation in winter is not stored in the soil but lost as runoff;

(ii) there is a substantial seasonal variation in soil moisture in the simulation of present climate; some of the simpler models may exaggerate the seasonal cycle of soil moisture (see Chapter 5) leading to an exaggerated response in the warmer climate;

(iii) in higher latitudes, earlier snowmelt leading to enhanced solar absorption and evaporation may contribute;

(iv) changes in soil moisture may be amplified by cloud feedbacks in regions where evaporation is being limited by low soil moisture values (e.g., Wetherald and Manabe, 1995);

(v) the drying is more pronounced in regions where precipitation is reduced in summer.

Given the varying response of different land-surface schemes to the same prescribed forcing (Chapter 5), the consistency from model to model of reductions over southern Europe in summer might be regarded as surprising. All models submitted (except *p,q*) produced a reduction in summer precipitation over southern Europe: here changes in circulation and precipitation may be more important in determining soil moisture changes than the details of the land-surface scheme. Reductions over North America are less consistent, and there is a still wider model to model variation in the response over northern Europe and northern Asia.

With aerosol forcing included (*y,z*), the patterns of soil moisture change in northern winter are similar but weaker

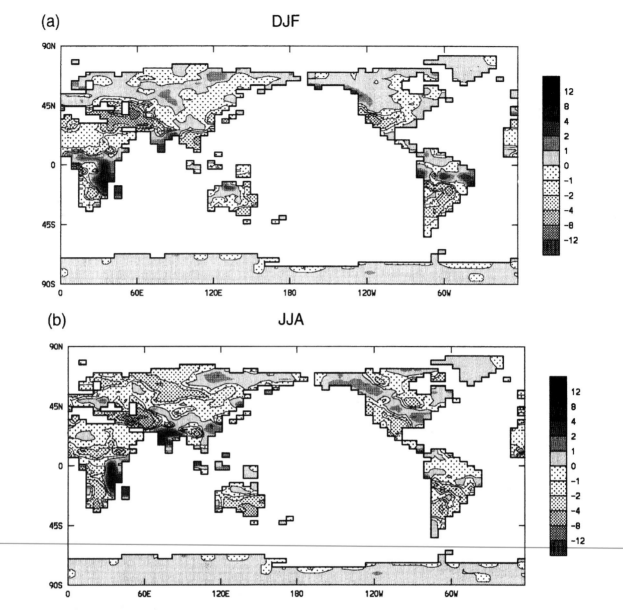

Figure 6.12: Seasonal changes in soil moisture at the time of doubling CO_2 following a 1%/yr increase. Contours at 0, ±1, 2, 4, 8 cm (from experiment *d* in Table 6.3, Gordon and O'Farrell, (1996)). (a) December to February; (b) June to August.

than with greenhouse gas forcing only (*x,w*). However, soil moisture increases over North America and southern Europe in summer when aerosol effects are included (*y,z*), presumably because of the reduced warming and its effect on evaporation, and because of increases in precipitation. The changes in the hydrological cycle are likely to be sensitive to the distribution of aerosol forcing and the coupled model used. However, it is clear that aerosol effects have a strong influence on simulated regional climate change.

6.2.3 *The influence of flux adjustments*

As noted in Section 5.2, most of the coupled models

considered in this assessment use fixed adjustments to the atmosphere-ocean surface fluxes to ensure a stable and realistic simulation of present climate. Shortcomings in model formulation lead to the need for these adjustments which are generally small averaged over the globe (for example, less than 0.5 Wm^{-2} in models 6 and 16) but reach over 200 Wm^{-2} locally in a few regions (see Chapter 5, Gates *et al.*, 1993). In a strict sense, the use of such corrections can only be defended if they are small. Although the adjustments are not small locally in current models, this is not proof that the approach is invalid: it could be that model errors in the surface fluxes change negligibly with perturbations as small as that occurring with a doubling of CO_2.

Here we address two issues:

1. Does the use of flux adjustments *per se* distort the response of a model?

2. Would correction of the deficiencies which give rise to the need for flux adjustment produce a substantially different model response?

The broad response of models with flux adjustments and those without flux adjustments are qualitatively similar. The exceptions can be traced to shortcomings in the models' simulation of features in present day climate which have been shown to be important for the simulation of climate change (for example, the region of deep mixing in the northern North Atlantic). Thus, there is no evidence from these simulations that the use of flux adjustments *per se* is substantially distorting the response to increases in greenhouse gases. However, the errors which arise without flux adjustments (for example, lack of sea ice in the Southern Hemisphere) can distort the simulated response, as seen earlier.

The flux adjustments compensate for inadequacies in the representation of physical processes in a model on present climate (Chapter 5). Running without flux adjustments (and accepting the subsequent degradation of simulation of current climate, Section 5.2) does not overcome the effect of these deficiencies on simulated climate change (Nakamura *et al.*, 1994). Flux adjustments are an engineering rather than a scientific solution.

It is more difficult to assess the likely effect of removing such deficiencies on the models' response. The largest heat flux corrections generally occur in regions of strong horizontal temperature gradient: others are associated with errors in cloud, in resolving local upwelling and in positioning the sea ice margins. As these errors vary from model to model, their effect on simulated climate change is also likely to be vary from model to model. It is not obvious that these errors have a dramatic effect on the large-scale response simulated by current models (though they could be important locally). For example, in *j*, flux adjustments were greatly reduced from the model used in *g* by enhancing model resolution and improving the simulation of cloud, but the distribution of annual mean temperature change was essentially the same as in the earlier version of the model. Many of the features noted above including reduced warming in regions of deep mixing, the maximum warming in high northern latitudes

in winter and the increased precipitation and soil moisture in high northern latitudes in winter have been linked to highly plausible physical mechanisms and hence are unlikely to be qualitatively sensitive to the removal of the errors which lead to the requirement of flux correction. Nevertheless, the reduction and removal of flux adjustments (without significantly degrading the simulation of present day climate) remains a high priority in the development of coupled models.

6.2.4 Interpretation of experiments

In this sub-section, some technical issues relevant to the interpretation of model results are discussed. Ensemble experiments will help reduce and quantify the uncertainty in single experiments arising from the natural (unforced) variability of climate. Understanding of the "cold start" is essential for interpreting the idealised experiments which ignore the effect of increases in greenhouse gases to date. If projections of equilibrium change in models with a simple ocean and those with a full representation of the ocean are similar, estimates of long-term equilibria can be made more economically.

Dependence on initial conditions and ensemble integrations

Coupled atmosphere-ocean models produce considerable variability on interanual and interdecadal time-scales (Chapter 5), as in the real world (Chapter 3). Thus, in any single realisation, unforced variations will form a substantial part of the simulated response. Hence, the evolution of the patterns of change as the forcing increases will not be smooth, and experiments repeated from slightly different initial conditions will produce responses which differ in detail. A more representative estimate of the response to the forcing may be obtained by carrying out an ensemble of experiments starting from slightly different initial conditions, and reducing the expected contribution from natural variability by averaging the results. Cubasch *et al.* (1994b) found that averaging over several simulations did lead to smoother patterns of change. It is possible to estimate the stability of features of the response by comparing different members of the ensemble. This is particularly important for prediction of regional climate change (Section 6.6), since the internal variability generally increases as the size of the averaging region is reduced (Figure 1.5). The ensemble approach also allows one to estimate the statistical distribution of possible outcomes which arise through unpredictable internal variations.

It is known that a coupled model can sustain very

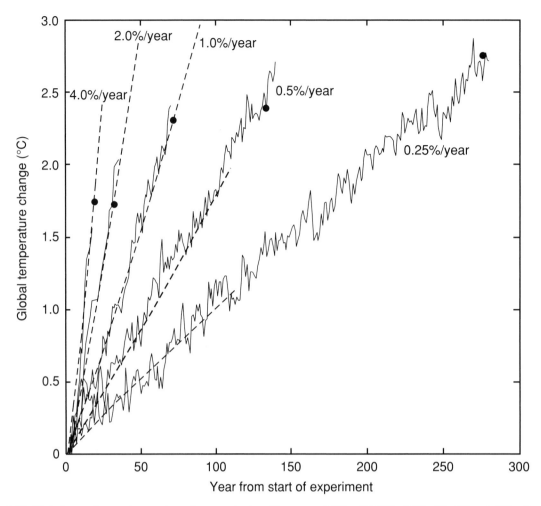

Figure 6.13: Global annual mean temperatures for annual rates of increase of CO_2 of 0.25%, 0.5%, 1%, 2% and 4%/yr from the GFDL AOGCM (solid lines) (experiments *e–i*, see Table 6.3) (R. Stouffer, pers. comm.) compared to results from an UD/EB model (dashed lines) with climate sensitivity of 3.7°C and the same rates of CO_2 increase. The time of CO_2 doubling is indicated by a dot.

different solutions with the same external forcing started from appropriate initial conditions (Manabe and Stouffer, 1988). However, evidence to date (Cubasch *et al.*, 1995b; Keen, 1995) indicates that model solutions will converge to the same solution, despite differences in the initial state which may arise from internal variability or climate drift. In other words, current simulations are not sensitive to variations in the initial state which lie within the range of the model's natural variability.

Dependence on rate of increase of CO_2

The dependence of the annual mean warming rate on the rate of change of forcing has been evaluated in a set of experiments (*e-i*) in which rate of increase in CO_2/yr is increased by factors of 2 from 0.25% to 4.0% (Figure 6.13). Increases of 0.25%, 1% and 4% /yr give a doubling of CO_2 in about 300, 70 and 18 years respectively. The

warming at the time of doubling ranges from 2.6°C in the 0.25% case to 1.5°C with a 4% increase/yr in a model with an equilibrium (long-term) response of 3.7°C to doubling CO_2. The slower the rate of increase of CO_2, the longer the time to reach a given concentration of CO_2, and hence the longer the time the ocean has to absorb the extra radiative heating (R. Stouffer, pers. comm.). At the time of doubling, the model reaches 70% of the long-term response with the slowest rate of increase (0.25%/yr), and only 40% with the fastest (4%/yr) increase.

This indicates that the response to a gradual increase in greenhouse gases depends not only on the climate sensitivity (the long-term or equilibrium response to doubling CO_2) but also on the rate of increase of the greenhouse gases (see also Hansen *et al.*, 1981). Nevertheless, a sixteenfold increase in the rate of CO_2 increase gives only a factor of two difference in warming at the time of doubling of CO_2 in these experiments. Hence the difference in model response to IPCC Scenario IS92a (equivalent to about a 0.7%/yr increase in CO_2) and the experiments using a 1%/yr increase in CO_2 are likely to be small at the time of doubling. On the other hand, the response to the much smaller rates of increase of CO_2 observed earlier this century is likely to be substantially larger than in the 1%/yr increase experiments for the observed increase in equivalent CO_2 concentration. Hence

---

The cold start

This describes the anomalously low initial rate of warming in experiments which do not take into account the effect of increases in greenhouse gases to date.

The gradual build up in gross radiative forcing due to the past increases in greenhouse gases is illustrated schematically by the curved line in Figure 6.14a. The temperature responds as in the left hand curve in Figure 6.14b. However, because of the thermal inertia of the oceans, the temperature does not equilibrate immediately. Hence the increase in radiation to space which occurs as the surface warms is smaller than the gross heating due to greenhouse gases. This gives a net heat flux into the surface (Figure 6.14c) which serves to raise the temperature of the ocean. This net heating is considerable by the 1990s (Figure 6.14c, upper curve), whereas it is zero in simulations which neglect the history of the radiative forcing.

In model simulations, it is found that in the global average, the effective heat capacity of the ocean is equivalent to that of a depth of a few hundred metres, so a heating rate of over 1 Wm^{-2} is required to maintain the 0.3 °C/decade warming typical of current transient experiments. In the experiments which commence with a 1%/yr increase in CO_2, the gross forcing starts from zero (straight line in Figure 6.14a), and it takes almost two decades for the gross forcing to reach 1 Wm^{-2}. Thus the warming starts slowly and then increases to a near steady rate (lower curve, Figure 6.14b). For comparison, the rate of warming taking into account past changes in radiative forcing, transposed to go through zero at 1990 (dashed curve, Figure 6.14b) is substantially larger in the first few decades following 1990. The difference between the response with and without the history of the forcing taken into account is defined as the "cold start".

In summary, to predict rates of climate change over the next few decades, we must allow for the net heating that has built up due to past changes in radiative forcing.

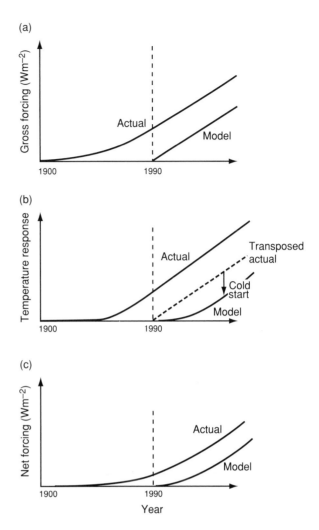

Figure 6.14: Schematic diagrams of radiative forcing and temperature response, showing the effect of neglecting the effect of past forcing (the "cold start" problem). (a) Forcing due to increases in CO_2, with a rate of increase rising gradually to 1990 as observed, and maintained at 1%/yr thereafter (left curve) and from a 1%/yr increase starting abruptly in 1990, as in idealised experiments (right curve). (b) Temperature response to the forcing in (a). The upper curve, which is the response in the case with the gradual initial increase as observed, has been transposed vertically to zero at 1990 (dashed curve) to highlight the initial slow response in the case of an abrupt increase used in idealised experiments (lower curve). The difference between the two curves is known as "the cold start" and is an artefact of the experimental design. (c) The upper curve shows the net forcing (which allows for the increased loss of radiation to space as the model warms) in the case with a gradual start to the forcing. The lower curve shows the net forcing in the idealised case. Note the net heating at 1990 which maintains the warming of the ocean in the upper curve in (b). To heat 300m depth of water by 0.3°C/decade (typical of the AOGCM experiments in Table 6.3) requires a net heating of 1.5 Wm^{-2} (cf. 4 Wm^{-2} for a doubling of CO_2) which takes several decades to build up with a 1%/yr increase in CO_2.

the 1%/yr increase experiments described above cannot be used directly to interpret the observational record. Another difficulty in using these experiments to interpret the recent temperature record is the "cold start", described below.

Dependence on past forcing on future climate change – the "cold start"

Predictions of the rate of climate change over the next few decades are of particular importance, both for the detection and attribution of climate change, and for the estimation of climate impacts. Over the last two centuries, there has been a gradual build up of the gross radiative forcing due to increases in greenhouse gases (Chapter 2). Most experiments in Table 6.3 do not account for the past build up of gases. Hence they produce a small initial rate of warming over the first one or two decades, a phenomenon known as the cold start. This complication is removed in the most recent experiments (x,y,w,z) which start from the last century when the increases in greenhouse gases were small.

Attempts to quantify the "cold start" bias have been made using simple models (Fichefet and Tricot, 1992), coupled models (Cubasch *et al.*, 1994b; Keen, 1995) or fitting coupled model results to a simple analytical model (Hasselmann *et al.*, 1993, Cubasch *et al.*, 1994b). Hasselmann *et al.* (1993) estimated a cold start of 0.4°C after 50 years with IPCC (1990) Scenario A in their run (m) started from equilibrium at 1990. Keen (1995) found a 0.2 to 0.3°C difference in long-term warming between experiments with CO_2 increasing by 1%/yr from equilibrium (t) and a parallel simulation with the historical

increase in equivalent CO_2 from 1860 first taken into account (w) (Figure 6.15).

Note that various factors including aerosols have probably reduced the heating due to increases in greenhouse gases over the last century (Chapter 2), so the net imbalance in heating and the consequent "cold start" will be smaller than that taking into account greenhouse gases alone.

In the experiments for which the information is available (g,k,m,q,s,t) the rate of increase of sea level due to thermal expansion shows a pronounced "cold start". This is because the rate of thermal expansion is more directly related to the integral of the net heating at the surface, which has time to build up from zero in experiments with a more gradual increase in CO_2, than to the instantaneous heating rate.

Climate drift is another source of error in estimating initial rates of change in transient experiments. This can occur if the ocean and atmosphere are not brought to full equilibrium before starting an experiment, and results in a systematic trend in simulated present climate as the model "drifts" towards its long-term equilibrium state. This lack of initial equilibrium may affect the evolution of a perturbed climate differently from that of present day climate, and hence distort the estimates of rates of climate change.

Long-term equilibrium of coupled models

Estimates of the equilibrium response to increases in trace gases to date have been made using atmospheric models coupled to a simple oceanic mixed-layer models. These models are much cheaper to run to equilibrium than AOGCMs, but neglect the effect of changes in ocean circulation.

In one study, (g), an AOGCM has been run to near equilibrium with doubled CO_2 and compared with results from the same experiment using a mixed-layer model (R. Stouffer, pers. comm.). The global mean warming is similar, 3.9°C compared to 3.7°C in the mixed-layer model. The warming patterns in the two models are similar, the main difference occurring over the tropics and the central Arctic, where the full model is slightly warmer, and Antarctica, especially south of Australia, where it is colder (Figure 6.16). This suggests that the oceanic heat transport may be similar in equilibrium present day and 2 × CO_2 climate, at least in this model. If this proves to be generally true, one could estimate long-term effects using mixed-layer models which neglect changes in ocean heat transports but are easily run to equilibrium. However, there are local differences in the change in horizontal temperature gradient. It is possible, for example, that the larger reduction in

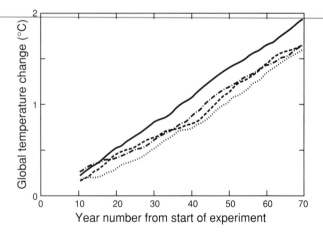

Figure 6.15: Global annual mean warming due to increasing CO_2 by 1%/yr. The solid curve is the warming when the increase in radiative forcing due to past increases in greenhouse gases is taken into account. The other curves are three estimates of warming if historical increases in greenhouse gases are neglected. The differences between the solid curve and the other curves are estimates of the "cold start" (Keen, 1995).

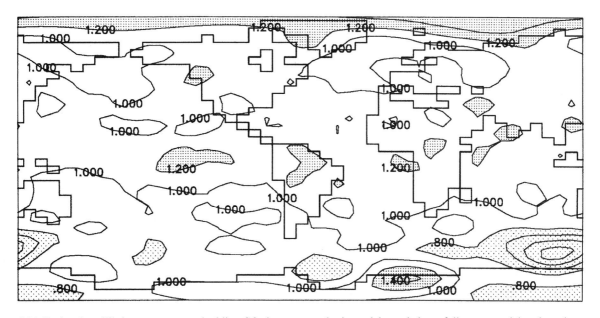

Figure 6.16: Ratio of equilibrium response to doubling CO_2 in an atmospheric model coupled to a full ocean model to that when coupled to a mixed-layer ocean. Contours every 0.2°C, light stippling where the deep ocean response is more than 20% smaller, heavy stippling where it is more than 20% greater (R. Stouffer, pers. comm.).

meridional temperature gradient associated with the greater Arctic warming in the coupled model could produce noticeable differences in the response of the North Atlantic storm track. Hence the equivalence of mixed-layer and coupled model equilibria should be assessed further.

When CO_2 concentrations were quadrupled in the AOGCM (Manabe and Stouffer, 1994), the thermohaline circulation collapsed (and failed to recover after three centuries), implying large changes in heat transport which cannot be represented in a static mixed-layer model.

6.2.5 Summary and conclusions

In 1990, preliminary results from only one transient CO_2 doubling experiment were available. 10 transient simulations of the effect of a 1%/yr(compound) increase in CO_2 have been considered in this assessment. The simulated rate of change of global mean surface air temperature ranges from just under 0.2°C/decade to 0.5°C/decade at the time of doubling. These simulations ignore the effect of any previous increase in greenhouse gases: the initial rate of warming is smaller than would be the case if earlier greenhouse gas increases are taken into account. In two simulations with the history of greenhouse gas forcing taken into account, the projected rate of global mean warming over the next decade is typically 0.3°C/decade, reduced to 0.2°C/decade in simulations allowing for the effects of sulphate aerosols, in agreement with earlier simulations using a simpler UD/EB model. In

one of these experiments, it has been shown that the pattern correlations of the patterns of the simulated and observed temperature changes in the last two decades are higher when the direct effect of sulphate aerosols is taken into account along with increased CO_2.

In experiments with a gradual increase in CO_2, all models show the following features:

- greater warming over the land than the sea;
- reduced warming, or even cooling, in the high latitude southern ocean and part of the northern North Atlantic ocean;
- a maximum warming in high northern latitudes in winter, and little warming over the Arctic in summer;
- increased precipitation and soil moisture in high latitudes in winter;
- an increase in mean precipitation in the region of the Asian summer monsoon.

All these changes are associated with identifiable physical mechanisms.

In addition, most simulations show a reduction in the strength of the North Atlantic thermohaline circulation, a widespread reduction in diurnal range of temperature (if the model includes the diurnal cycle) and reductions in summer soil moisture over southern Europe and, less consistently, North America. These features too can be explained in terms of physical mechanisms, but may be sensitive to a greater or lesser degree to the details of the parametrizations used. Nevertheless, no model has produced a stronger thermohaline circulation, or a global mean increase in the diurnal range of temperature.

Some models produce a reduction in the east to west temperature gradient in the tropical Pacific Ocean, reminiscent of the temperature pattern associated with El Niño events, though the reduction is small compared with the mean warming.

When the direct effect of sulphate aerosols is represented the simulated patterns of change to date are more like those observed. The warming in northern mid-latitudes is reduced, with regions of cooling as well as warming. The correlation with observations is also improved (see Chapter 8).

Including the effects of aerosols as well as greenhouse gases (based on IS92a) in simulations of future climate leads to much reduced warming in mid-latitudes, not only in regions where the atmosphere burden of aerosols is increasing, but also in some regions where it is decreasing. There are also marked changes in the hydrological cycle, principally in the Northern Hemisphere in summer. In particular, there is net reduction of precipitation in the Asian summer monsoon, and the summer drying over Europe and North America is weakened or even reversed. These preliminary results should be used with caution: the results are based on two studies and are likely to be quite sensitive to the choice of scenario and the assumptions in deriving the radiative forcing due to aerosols, both of which are subject to considerable uncertainty. They also neglect the effect of other anthropogenic forcing including indirect sulphate effects, soot, tropospheric ozone and aerosol from biomass burning.

The models with flux correction produce broadly similar results to those without flux correction: the main differences can be traced to shortcomings in particular features of the control simulation in the models with an anomalous response. Thus, there is no evidence from the current simulations that the use of flux adjustments is substantially distorting the response to increases in greenhouse gases.

The need for flux adjustments arises from a variety of model shortcomings and it is not obvious if or how these shortcomings would alter qualitatively the main findings summarised above. Nor is it obvious whether they would enhance or diminish simulated changes. Thus, eliminating the need for flux adjustments remains a high priority. Other uncertainties associated with cloud feedback and aerosol forcing which have been demonstrated to have a profound effect on predictions of climate change are probably of greater concern (see Section 6.7).

The models produce a high level of internal variability, as in the real world (Chapter 5), leading to a spread of possible outcomes for a given scenario, especially at the regional level. Ensemble experiments may prove useful in reducing the uncertainty in single realisations arising from internal variability.

There is some evidence that the pattern of equilibrium response of temperature to doubling CO_2 in atmospheric models coupled either to simple mixed-layer models, or to full dynamical ocean models may be similar. If this is generally true, it would allow economical estimates of the long-term equilibrium response of large-scale climate to moderate increases in greenhouse gas concentrations to be made using mixed-layer models.

6.3 Global Mean Temperature Changes for the IPCC (1992) Emission Scenarios

6.3.1 Introduction
In this section we present projections of global mean temperature change for a range of emission scenarios, specifically the IPCC (1992) scenarios of Leggett *et al.*,(1992), and consider the uncertainties in these projections. These results update the only previous comprehensive assessment of the global mean temperature and sea level implications of these scenarios (Wigley and Raper, 1992). They are complemented by an analysis that parallels the present work given in Raper *et al.* (1996).

We follow IPCC (1990) (Bretherton *et al.*, 1990) and IPCC (1992) (Mitchell and Gregory, 1992) in using a relatively simple upwelling diffusion-energy balance climate model (UD/EB) to make these projections. The specific model used (as in 1990 and 1992) is that of Wigley and Raper (1987, 1992). Although based on the original model of this type (Hoffert *et al.*, 1980), this particular model differs from others in that it distinguishes between land and ocean and between the hemispheres (by treating them as separate "boxes", with energy flows between the "boxes"). Such differentiation is important in the present context because there are large radiative forcing differences between these different parts of the globe.

Although much more sophisticated models are available (AOGCMs), which could in principle be used for producing a range of global mean temperature projections, this is not yet possible for a number of reasons. First, individual coupled atmosphere-ocean models have specific climate sensitivities, so they cannot be used to assess comprehensively uncertainties arising from uncertainties in this key climate parameter. Second, there are major uncertainties in both past and future radiative forcing, arising both from uncertainties in the forcing for given emissions (especially aerosol effects) and from uncertainties in the emissions themselves. A large number of model simulations are required to address these uncertainties, and the current computing requirements of AOGCMs precludes carrying out so many runs. By adjusting their structure and parameter values appropriately, UD/EB models can simulate well the results of AOGCMs at the global mean level.

In the UD/EB model used here, the main model parameters are: the depth of the oceanic mixed layer (taken as 90 m); the vertical diffusivity used to characterise vertical mixing processes in the ocean (taken as 1 cm^2/sec); the ratio of the temperature change for high latitude sinking water (which forms the polar branch of the model's thermohaline circulation) to the global mean temperature change (π taken as 0.2); the upwelling rate (ω), which measures the intensity of the thermohaline circulation; and the climate sensitivity ($\Delta T_{2\times}$, i.e., the equilibrium temperature change for a doubling of CO_2 concentration). The above parameter values differ from those used in IPCC (1990) for reasons that are explained and justified in Wigley and Raper (1991).

To accord with recent coupled atmosphere-ocean model results, two important changes have been made to the model's structure compared with its previous applications. First, evidence that indicates that the climate sensitivity differs between land and ocean: most models give a different equilibrium warming over these regions (see, e.g., Cubasch *et al.*, 1992; Murphy, 1995b). It is important to simulate this differential warming, not least because of its potential effect on oceanic thermal expansion, an important component of sea level rise. (Thermal expansion results produced here are used in the sea level projections given in Chapter 7.) If warming over the oceans is less than over the land, the amount of thermal expansion will be less than might be estimated on the basis of global mean temperature alone. To account for this differential sensitivity the single parameter $\Delta T_{2\times}$ is replaced by two parameters, a global mean $\Delta T_{2\times}$ and an equilibrium land/ocean warming ratio for $2 \times CO_2$. For the latter, we use the value 1.3 based on coupled atmosphere-ocean model and equilibrium GCM results. For $\Delta T_{2\times}$ we use a range of values to capture the uncertainty in this parameter.

The second major change is in the upwelling rate. Most current coupled atmosphere-ocean models show a pronounced slow-down of the thermohaline circulation as the globe warms. To account for this in the UD/EB model, ω is assumed to be a linear function of ocean mixed-layer temperature with the form

$$\omega = \omega_0(1 - \Delta T/\Delta T^*)$$

where ΔT^* is a threshold temperature change at which the thermohaline circulation shuts down. The value of ΔT^* used is 7.0°C, chosen to match results from Cubasch *et al.* (1992) and Manabe and Stouffer (1994), and ω is further constrained not to drop below 0.1 m/yr, following Manabe and Stouffer (1994), although this limit is never approached in any of the simulations presented here. The initial value of ω (ω_0 = 4 m/yr) is the standard value required to match the observed vertical ocean temperature profile under present-day conditions.

A more extensive discussion of these changes in model structure, of the model parameter values, and of the behaviour of the model is given in Raper *et al.* (1996).

To verify that the UD/EB model, when properly calibrated, is able to simulate accurately the behaviour of the AOGCMs of Chapter 5 and Section 6.2, intercomparison exercises were carried out using results from the GFDL AOGCM (R. Stouffer pers. comm.; Manabe and Stouffer, 1994; model 6 in Table 6.1a). To ensure a valid comparison, the UD/EB model climate sensitivity was set at the GFDL model's value (3.7°C equilibrium global mean warming for $2 \times CO_2$). All other UD/EB model parameters were kept at values previously chosen (see Wigley and Raper, 1991), as specified above. The value of the land/ocean sensitivity differential (1.3), chosen on the basis of other GCM results (Raper *et al.*, 1996), is similar to that for the GFDL model. The thermohaline circulation varied with time as described above. These temporal changes correspond well with those obtained with the GFDL model (Manabe and Stouffer, 1994).

Figure 6.13 shows global mean temperature results where both models were forced with a range of fixed percentage increases per year (compound) of CO_2 (the AOGCM simulations are fully discussed in Section 6.2). The UD/EB model only simulates the underlying signal in response to the external forcing, so it does not show the internally generated variability evident in the AOGCM results. Apart from this variability, both models are seen to agree well over a wide range of forcings.

Figure 6.17 shows another comparison, in this case with the $2 \times CO_2$ and $4 \times CO_2$ stabilisation simulations of Manabe and Stouffer (1994). Figure 6.17a shows global mean temperature comparisons. For the $2 \times CO_2$ case, the agreement is excellent. For the $4 \times CO_2$ case, the UD/EB model begins to show a slightly reduced warming response when the CO_2 level becomes sufficiently high. Note that none of the IS92 emission scenarios gives a forcing change as high as $4 \times CO_2$; all cases are within the range for

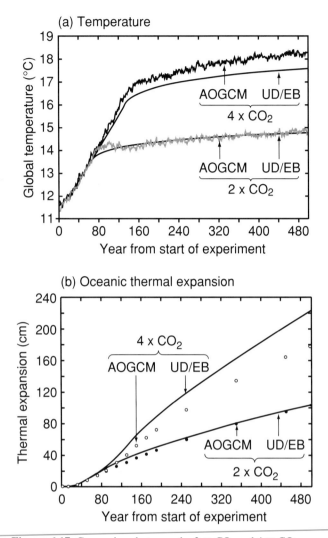

Figure. 6.17: Comparison between the $2 \times CO_2$ and $4 \times CO_2$ stabilisation simulations of Manabe and Stouffer (1994) and the UD/EB model used in Section 6.3. (a) Global mean temperatures; (b) oceanic thermal expansion.

which the UD/EB model/AOGCM agreement is good. Figure 6.17b shows a comparison of oceanic thermal expansion. Again, agreement is excellent for the $2 \times CO_2$ case, but less good for the $4 \times CO_2$ case. The fact that the UD/EB model gives higher expansion and lower warming for the $4 \times CO_2$ case implies that the model's flux of heat into the deeper layers of the ocean is greater than in the AOGCM. If this reflects a real deficiency in the UD/EB model (which is uncertain, because we do not know how well the AOGCM simulates real-world behaviour for such large forcing), then it is probably important only for situations and times outside the range to which the model is applied here.

Other UD/EB model/AOGCM comparisons have been carried out by Raper and Cubasch (1995) using the MPI AOGCM version described in Cubasch *et al.* (1992). For temperature, the authors obtain reasonable agreement between the two models. For thermal expansion, the comparison is less good. However, these comparison results are less easy to interpret than the GFDL comparisons because the MPI model shows, in the first 100–200 years of its control run, a century time-scale adjustment to small initial flux imbalances. This low-frequency variation makes it difficult to determine precisely the climate response signal (as noted by the authors), and it produces non-linear changes in the AOGCM that cannot be adequately simulated in the UD/EB model.

Figure 6.4 shows several AOGCM runs compared with the global mean warming results of the UD/EB model discussed here and those of the simple model used for alternative sea level rise calculations discussed in Section 7.5.2.2. The UD/EB model has a sensitivity of 2.5°C, while the other simple model has a sensitivity of about 2.2°C. The sensitivities of the AOGCMs range between 2.1 and 4.6. All models in Figure 6.4 were forced with a 1%/year (compound) CO_2 increase beginning in an initial equilibrium or near-equilibrium state. Differences in response occur between models, even when they are forced in an identical way and their climate sensitivities are listed as identical. These differences arise because the effective thermal inertia of their ocean components can vary between models (e.g., due to different behaviour of the thermohaline circulation during warming), and because the procedures used to estimate their climate sensitivity may differ.

6.3.2 Radiative Forcing

In order to estimate future temperature and sea level changes, it is necessary to prescribe both past and future radiative forcing changes due to the various greenhouse gases and aerosols. This section describes how radiative forcing changes were specified. Important background information is given in Shine *et al.* (1990, 1995) and these details are not repeated here. Information is given sequentially below for CO_2, methane (CH_4), tropospheric ozone, halocarbons and stratospheric ozone, and aerosols. The primary emission scenarios used are the IS92 Scenarios of IPCC (1992) (Leggett *et al.*, 1992), but with some small, but important changes. The changes are (i) to adjust (i.e., offset by a constant amount) IS92 emissions for N_2O and CH_4 in order to be consistent with the inferred budgets for 1990 based on observed atmospheric abundance and growth;[1] (ii) to ignore possible changes in carbon monoxide (CO), active nitrogen oxides (NO_x) and VOC (volatile organic compounds) emissions; and (iii) to consider only a single halocarbon emissions case, consistent with the Montreal Protocol and its later Amendments.

For CO_2, the past concentration history is the standard IPCC data set, an update of that derived for an IPCC inter-model comparison exercise by I.G. Enting and T..M.L. Wigley (Enting *et al.*, 1994). The revision only affects the most recent values, changed to give a 1980s mean atmospheric mass increase of 3.3 GtC/yr, the value given in Chapter 2 of the present report (an update from Schimel *et al.*, 1995). The precise mean is 3.28 GtC/yr.

To calculate future forcing, we use the model of Wigley (1993). Elsewhere in this report we use the model of Siegenthaler and Joos (1992), but the differences between the results of these two models are negligible in terms of implied radiative forcing. Both models use the initialisation procedure devised for the IPCC inter-model comparison exercise described by Enting *et al.* (1994), updated to use a different 1980s mean carbon budget. The current budget has a 1980s mean net deforestation flux of 1.1 GtC/yr (Schimel *et al.*, 1995) and the atmospheric mass increase stated above. The CO_2-fertilisation factor is adjusted to give a balanced 1980s mean budget, as described in Wigley (1993) and Enting *et al.* (1994).

For methane, the concentration history is that used by Osborn and Wigley (1994), consistent with information given elsewhere in this report. Future projections are made using the Osborn-Wigley model with parameter values adjusted to match closely the concentration projections in Chapter 2. In terms of 1990 to 2100 radiative forcing, the differences arising from using the Osborn-Wigley model are negligible, ranging from 0.00 Wm^{-2} to 0.03 Wm^{-2} depending on emission scenario. In calculating future concentrations we assume that the emissions of CO, NO_x and VOCs remain constant at 1990 levels. Although IPCC (1992) (Leggett *et al.*, 1992) gives scenarios for the future emissions of these gases, these are not used because their influence on methane lifetime depends on the spatial details of the emissions, details that are not specified.

In addition to direct methane forcing, we also include the indirect effects of methane-related increases in

stratospheric water vapour and tropospheric ozone. The former follows Shine *et al.* (1990), but with a scaling factor reduced from 0.30 to 0.05 following Shine *et al.* (1995). The latter assumes a global mean ozone increase of 4 ppbv since pre-industrial times due to methane increases and a radiative forcing sensitivity of 0.02 Wm^{-2}/ppbv.

A "direct" tropospheric ozone term due to emissions other than methane is also included. This is assumed to contribute a global mean forcing to 1990 of 0.32 Wm^{-2}, giving a total tropospheric ozone forcing of 0.4 Wm^{-2} (as estimated elsewhere in this report). The history of this ozone component is assumed to parallel fossil fuel emissions to 1990. For the future, the forcing is assumed to be constant. The implied total tropospheric ozone forcing changes (i.e., including those arising from methane concentration changes) are consistent with the tropospheric ozone concentration increases given in Table 2.5c. The direct tropospheric ozone forcing component is assumed to occur only in the Northern Hemisphere, where it is split 9:1 between the land and ocean areas, cf. aerosol forcing – see below.

For nitrous oxide (N_2O), the history is based on ice core data and, more recently, observed data. The initial (pre-industrial) value is taken as 270 ppbv based on Leuenberger and Siegenthaler (1992) and more recent unpublished results from D. Raynaud. Future concentration projections assume a lifetime of 120 years.

For halocarbons, the suite of gases is that used by Wigley and Raper (1992). To account for the effects of stratospheric ozone depletion, we use the chlorine loading method devised by these authors, but with a non-linear exponent (the ozone term is proportional to $(nCl)^{1.7}$ rather than $(nCl)^{1.0}$) and a higher bromine weighting (Wigley, 1994). This gives a 1990 direct halocarbon forcing of 0.28 Wm^{-2}, offset by –0.17 Wm^{-2} due to stratospheric ozone depletion; values in accord with the most recent IPCC report (Shine *et al.*, 1995). Application of a linear model with a threshold for chlorine loading gives essentially the same results.

The same halocarbon-related stratospheric ozone model is used for past and future changes. Only one emission scenario is used for halocarbons, that of IS92d and e. Emission reductions for CFC-11 and CFC-12 in this scenario are probably slower than should occur under the Protocol, but this has a negligible effect on the forcing results. Halocarbon emissions under IS92d and e are generally larger than those used to calculate the concentrations shown in Table 2.5, which more closely reflect the effects of the 1992 Copenhagen Amendments to the Montreal Protocol (Section 2.2.4).

The most complex and uncertain part of the radiative forcing is that due to aerosols from biomass burning and

---

[1] For both N_2O and CH_4, 1990 emissions specified in the IS92 scenarios do not accord with emission estimates based on observed concentration data and current lifetime estimates (see, e.g., Wigley, 1994). We therefore use these latter values to ensure a balanced contemporary budget, and assume that the scenario values are a valid representation of the changes from 1990 rather than correctly specifying absolute emissions.

Table 6.4: *Changes in radiative forcing (Wm^{-2}) between 1990 and 2100 for the IS92 emission scenarios*

Aerosol assumption	SCENARIO					
	IS92a	IS92b	IS92c	IS92d	IS92e	IS92f
No aerosols	5.76	5.55	2.55	3.45	7.90	7.07
Indirect sulphate = 0.0 Wm^{-2}	5.47	5.29	2.63	3.50	7.28	6.65
Indirect sulphate = –0.4 Wm^{-2}	5.29	5.11	2.71	3.54	6.95	6.40
Indirect sulphate = –0.8 Wm^{-2}	5.10	4.94	2.79	3.57	6.62	6.15
Wigley and Raper (1992)	5.44	5.26	2.98	3.84	6.74	6.26

from fossil fuel combustion. For the former, we assume a total forcing of –0.2 Wm^{-2} in 1990, with the history following that of gross deforestation (see Wigley, 1993, for more on the differences between gross and net deforestation). Results are insensitive to the assumed history. After 1990, this forcing value is kept constant. We also assume spatially uniform forcing for this component, although there is evidence to suggest that the forcing should be greater over land than ocean. For fossil fuel derived aerosols we assume a 1990 direct (clear-sky) effect of –0.3 Wm^{-2}, a value that accounts for the offsetting effect of soot (see Chapter 2). This term is distributed between the hemispheres in the ratio NH:SH = 4:1, and between land and ocean in the ratio 9:1, based on a range of three-dimensional modelling results.

The indirect forcing (through aerosol-induced changes in cloud albedo) is much more uncertain. Shine *et al.* (1995) give a range of possible global mean values of 0.0 to –1.5 Wm^{-2}. For the present calculations, we use –0.8 Wm^{-2} as our primary value. Because of the extreme uncertainty in this component, we also include calculations in which all future changes in aerosol forcing after 1990 are ignored (by keeping sulphur dioxide (SO$_2$) emissions at their 1990 level) to show the sensitivity of the results to this forcing component. The spatial distribution of indirect forcing is assumed to be NH:SH = 2:1, and land to ocean = 9:1.

A more detailed comparison of results for different aerosol forcings is given in Raper *et al.* (1996); these authors consider the above two cases together with two other cases where the aerosol forcings are the same as in our primary case, but with the indirect sulphate forcing in 1990 set to 0.0 Wm^{-2} and –0.4 Wm^{-2} (instead of the –0.8 Wm^{-2} used here). Raper *et al.* find that, for an indirect forcing of –0.8 Wm^{-2}, the modelled global mean warming over 1880 to 1990 is considerably less than observed (0.21–0.40°C for ΔT_{2X} = 1.5–4.5°C) and that there are noticeable differences between the warming amounts over

land and ocean and in the two hemispheres due primarily to the aerosol-induced forcing differentials. They suggest, following Wigley (1989), that these spatial differences may place a constraint on the magnitude of the aerosol forcing, a point also noted by Mitchell *et al.* (1995b).

For both past and future sulphate aerosol forcing, we use the forcing-emissions relationships employed by Wigley and Raper (1992).

The total radiative forcing change over 1765 to 1990 based on the above is 1.32 Wm^{-2}, or 2.62 Wm^{-2} if all aerosol forcing is ignored. These values should be compared with the value of 2.45 Wm^{-2} used in IPCC (1990) (Shine *et al.*, 1990). Table 6.4 shows the future radiative forcing changes for the six IPCC (1992) emissions scenarios for a range of aerosol assumptions. The Table compares these results with the earlier results of Wigley and Raper (1992). The differences are relatively small and attributable largely to how aerosol forcing is quantified. Other changes, *viz.* ignoring data on CO, NO$_x$ and VOCs emissions in calculating future CH$_4$ concentrations, the use of a single (Copenhagen-consistent) emissions scenario for halocarbons, and the adjustment of CH$_4$ and N$_2$O emissions to ensure a balanced contemporary budget, have a relatively minor effect.

Figure 6.18 shows the full forcing history to 1990 together with future projections for IS92a, modified as noted earlier, and the low (IS92c) and high (IS92e) forcing extremes out to 2100. Results are shown for both "with-aerosol" (full lines) and "aerosols constant at 1990 levels" (dashed lines). The aerosol assumption has a significant effect on both past and future forcing. It can also be seen that forcing uncertainties associated with both future emissions and aerosol effects are large. The range of uncertainty can be characterised usefully in terms of equivalent CO$_2$ concentration changes, defined by

$$C_{equiv} = 278 \exp(\Delta Q/6.3)$$

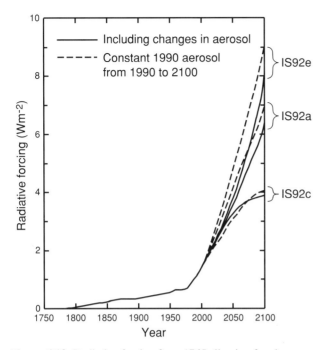

Figure 6.18: Radiative forcing from 1765 allowing for changes (full lines) and no changes (dashed lines) in aerosol concentrations after 1990. Values to 1990 are based on observed concentration changes (or emissions changes for aerosols). From 1990 to 2100 forcings are derived from the low (IS92c), mid (IS92a) and high (IS92e) emission scenarios.

where ΔQ is the radiative forcing from pre-industrial times (see Shine *et al.*, 1990). The 1990 equivalent CO_2 level is between 343 ppmv (with aerosols) and 421 ppmv (if all aerosol forcing is ignored), while the 2100 values range between 514 ppmv (IS92c) and 1201 ppmv (IS92e), with both cases using the best-guess (with-aerosol) forcing to 1990 and constant aerosol concentrations after 1990.

   Figure. 6.19 shows the forcing breakdown by component for IS92a. The values shown are from 1765, so the starting points on the left side of the graph give the 1765 to 1990 values. There are small differences in total halocarbon radiative forcing over 1990 to 2100 between Figure 6.19 and that based directly on the concentrations in Table 2.5, attributable to the effects of (modelled) stratospheric ozone changes, additional gases considered in the Figure 6.19 calculations, and to the larger emissions in IS92d and e. The "Total" shown is the value that incorporates changes of aerosol concentration beyond 1990. For IS92a, the net effect of non-CO_2 greenhouse gases is approximately cancelled out at the global mean level by aerosol forcing. At the regional level, however, for all scenarios, there is no such cancellation: aerosol effects lead to large spatial variations in forcing that will almost certainly have a major effect on the patterns of future climatic change.

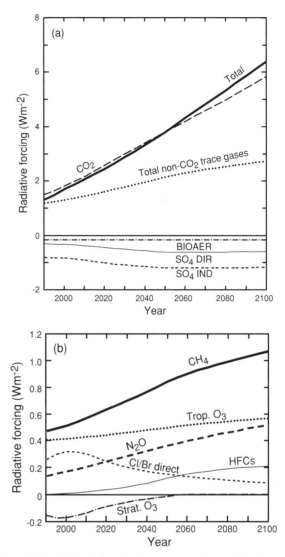

Figure 6.19: (a) Radiative forcing components resulting from the IS92a emission scenario for 1990 to 2100. The "Total non-CO_2 trace gases" curve includes the radiative forcing from methane (including methane related increases in stratospheric water vapour), nitrous oxide, tropospheric ozone and the halocarbons (including the negative forcing effect of stratospheric ozone depletion). Halocarbon emissions have been modified to take account of the Montreal Protocol and its Adjustments and Amendments. The three aerosol components are: direct sulphate (SO$_4$ DIR), indirect sulphate (SO$_4$ IND) and direct biomass burning (BIOAER). (b) Non-CO_2 trace gas radiative forcing components. "Cl/Br direct" is the direct radiative forcing resulting from the chlorine and bromine containing halocarbons; emissions are assumed to be controlled under the Montreal Protocol and its Adjustments and Amendments. The indirect forcing from these compounds (through stratospheric ozone depletion) is shown separately (Strat. O$_3$). All other emissions follow the IS92a Scenario. The tropospheric ozone forcing (Trop. O$_3$) takes account of concentration changes due only to the indirect effect of methane.

As a final point, it should be noted that all of the IPCC (1992) scenarios are "existing policies" scenarios, in that none of them assume the implementation of any strong policies to reduce emissions. The emissions (and radiative forcing) values differ markedly, however, because of different assumptions regarding, for example, population growth, economic growth, *per capita* energy use and resource availability.

6.3.3 Temperature Projections

We consider first global mean temperature projections for the central IPCC (1992) emission scenario, IS92a. Figure 6.20 shows temperature changes over 1990 to 2100 for climate sensitivities of $\Delta T_{2\times} = 1.5$, 2.5 and 4.5°C for the changing aerosol (full lines) and constant aerosol (dashed lines) cases. The central sensitivity value gives a warming of 2.0°C (changing aerosols) to 2.4°C (constant aerosols). The uncertainty due to $\Delta T_{2\times}$ uncertainties is large, and aerosol-related uncertainties are larger for larger $\Delta T_{2\times}$. These results are similar to those given by Wigley and Raper (1992) who gave a central estimate of 2.5°C warming over 1990 to 2100 for IS92a. The present values are slightly lower because of differences in the radiative forcing details (see Table 6.4), and because our present incorporation of a slow-down in the thermohaline circulation also slows down the rate of warming (although it increases the amount of oceanic thermal expansion). For a more detailed discussion of the effect of thermohaline circulation changes, see Raper *et al.* (1996).

The relative importance of CO_2, other greenhouse gases, and aerosols is shown in Figure 6.21, which gives temperature projections for the central climate sensitivity value ($\Delta T_{2\times} = 2.5$°C) for three different cases:

(1) forcing the model after 1990 with CO_2 concentration changes alone;

(2) forcing the model with greenhouse gases and aerosols;

(3) as for (2), but with aerosol amounts assumed to remain constant after 1990, i.e., no aerosol forcing changes after 1990.

Results for cases (2) and (3) are also given in Figure 6.20, as the central full (case 2) and dashed (case 3) lines. Case (3) corresponds to forcing by the full complement of greenhouse gases: the difference from case (1) represents the effect of non-CO_2 greenhouse gases. That case (1) (CO_2 alone) warms less than case (2) (all gases and aerosols) reflects the fact that in the second half of the 21st

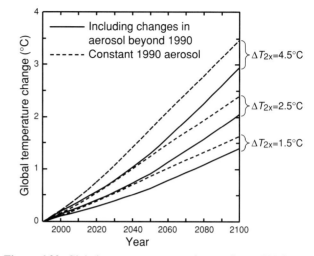

Figure 6.20: Global mean temperature changes from 1990 for IS92a for different climate sensitivities ($\Delta T_{2\times} = 1.5$, 2.5 and 4.5°C), with aerosol concentrations changing (full lines) and constant (dashed lines) beyond 1990.

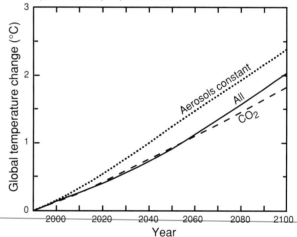

Figure 6.21: Global mean temperature changes from 1990 for IS92a for $\Delta T_{2\times} = 2.5$°C based on different assumptions regarding radiative forcing from 1990. "CO_2" shows results based on CO_2 forcing alone; "All" shows results for the full forcing complement (greenhouse gases and aerosols); "Aerosols constant" shows results where aerosol forcing changes after 1990 are ignored (i.e., as "All", but with the assumption of constant SO_2 emissions after 1990).

century the negative aerosol forcing does not fully offset the positive forcing due to non-CO_2 greenhouse gases (cf. Figure 6.19).

Figures 6.22 and 6.23 show temperature projections with changing aerosol amounts (Figure 6.22) and with constant aerosols (Figure 6.23) for the full set of IS92 emission scenarios, using the central estimate for climate sensitivity ($\Delta T_{2\times} = 2.5$°C). The effect of uncertainties in emissions remains small for many decades: for the changing aerosol case, the full range of uncertainty in 2050 is only 0.2°C

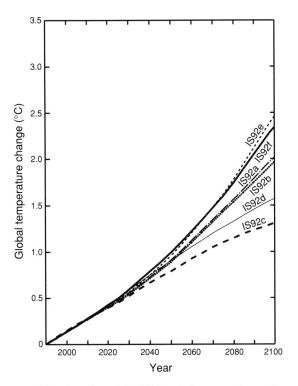

Figure 6.22: The effect of the IS92 emission scenarios on future global mean temperature changes. This is the full aerosol forcing case, with the climate sensitivity set to 2.5°C.

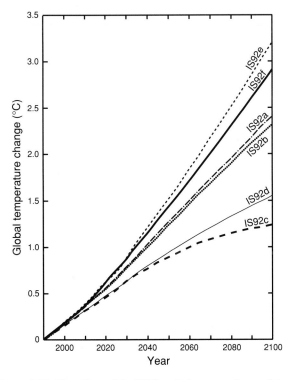

Figure 6.23: The effect of the IS92 emission scenarios on future global mean temperature changes, assuming constant aerosol concentrations beyond 1990. Results are for a climate sensitivity of 2.5°C.

(1990 to 2100 warming ranges from 0.8°C for IS92c to 1.0°C for IS92f). The different emission cases, however, give rapidly diverging temperature projections in the second half of the 21st century: for the changing aerosol case, the range of values spans 1.2°C by the year 2100. This behaviour illustrates the long-term nature of the climate change problem. There are two areas where there are appreciable lags in the system; between emission changes and concentration changes for CO_2 (and hence radiative forcing), and between radiative forcing changes and climate response. Inertia in both the carbon cycle and the climate system leads to a very long lag between CO_2 emissions changes and their eventual effect on climate.

Finally, Figure 6.24 shows an extreme range of possible warmings for the IS92 emission scenarios, together with results for the central case (IS92a) with and without changing aerosol forcing post-1990. To obtain a near-upper-limit projection, we take IS92e with a constant (1990) aerosol contribution (which maximises the forcing over 1990 to 2100) and use a climate sensitivity of $\Delta T_{2\times} = 4.5°C$; while to give a near-lower-bound projection we use IS92c, also with no post-1990 aerosol contribution (which in this

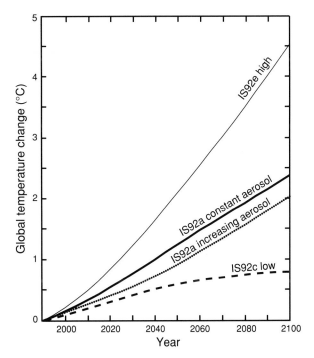

Figure 6.24: Extreme range of possible changes in global mean temperature. The topmost curve is for IS92e assuming constant aerosol concentrations beyond 1990 and a high climate sensitivity ($\Delta T_{2\times} = 4.5°C$); the lowest curve is for IS92c, also assuming constant aerosol concentrations beyond 1990, but with a low climate sensitivity ($\Delta T_{2\times} = 1.5°C$). Results for IS92a are shown for comparison, both with and without changing aerosols.

case minimises the forcing because of reductions in emissions later in the 21st century) and, in this case, $\Delta T_{2\times}$ = 1.5°C. These results quantify the full range of uncertainty associated with both the emissions uncertainties for "existing policies" and the uncertainty in the climate sensitivity.

There are, of course, additional uncertainties not accounted for in this analysis. For radiative forcing, the most important of these is that associated with the carbon cycle (see Wigley, 1994). We have used only central estimates of carbon cycle model parameters in obtaining future CO_2 concentrations: within the range of parameter uncertainties the CO_2 forcing to 2100 could be increased by a few tenths of a Wm^{-2}, corresponding to an additionalwarming of 0.1–0.2°C. There are also important uncertainties associated with the climate model. For many of these, the effects are relatively small (see Wigley and Raper, 1993). However, if the thermohaline circulation were not to slow down as assumed here, there could be an additional warming over 1990 to 2100 of more than 1°C (for constant upwelling rate, the upper limit value increases from 4.5°C to 5.7°C)[1]. (The effects of the above uncertainties on the lower bound value are small.) In assessing the significance of these extreme values, it should be noted that the probability of exceeding the bounds is likely to be small because they concatenate two possibilities that both have relatively low probability (*viz.*, emissions outside the IS92c to IS92e range, and climate sensitivity outside the range $\Delta T_{2\times}$ = 1.5–4.5°C).

The 1990–2100 warming for the central emissions scenario (IS92a) is approximately one third lower than the corresponding "best estimate" presented in 1990 (SA90): 2.0°C compared with 3.3°C for a sensitivity of 2.5°C. This arises mainly because of the following: differences in the two emission scenarios (the lower emissions for CO_2 and the halocarbons, in particular, lead to lower forcing and hence reduced warming); improvements to the carbon cycle model used to make CO_2 concentration projections (inclusion of a terrestrial sink term has led to lower concentrations); the inclusion of the cooling effect of aerosols in the current projections; the lower historical forcing to 1990 (now 1.32 Wm^{-2} compared with 2.45 Wm^{-2} previously); changes in the UD/EB model parameters; and changes in the UD/EB model structure to account for land/ocean differences in the climate sensitivity and a slow-down in the intensity of the thermohaline circulation.

_____

[1] For sea-level, in this extreme case, the constant-upwelling result is actually slightly less than for variable upwelling. This is because thr former leads to lower oceanic thermal expansion, which more than offsets the increased ice-melt contribution.

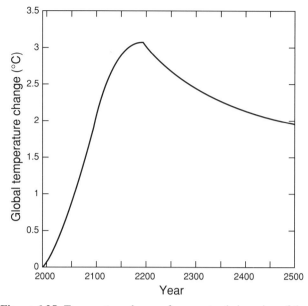

Figure 6.25: Temperature changes for an extended version of the IS92a emission scenario in which anthropogenic emissions of CO_2, CH_4, N_2O, the halocarbons and fossil fuel derived SO_2 are assumed to decline linearly to zero over 2100–2200. A climate sensitivity of $\Delta T_{2\times}$ = 2.5°C is assumed.

6.3.4 Longer Time-scale Projections

The IS92 emission scenarios give emissions only out to 2100. In almost all cases, projected values of global mean temperature are still rising rapidly at the end of the period. This raises the question of what might happen after 2100. As noted above, both the carbon cycle and the climate system have large inertia, so even if (for example) CO_2 emissions were to decline dramatically after 2100, temperature would continue to increase for some time. Figure 6.25 illustrates this. Here, anthropogenic emissions follow IS92a to 2100 and then are assumed to decline linearly to zero over 2100–2200 for CO_2, CH_4, N_2O, the halocarbons and SO_2-derived aerosols. CO_2 concentrations in this scenario peak at around 850 ppmv in 2160, and decline to around 540 ppmv in 2500. Total radiative forcing also peaks around 2160, but temperatures continue to rise to 2200, declining only slowly thereafter.

While Figure 6.25 graphically illustrates the inertia of both the carbon cycle and the climate system, a more relevant set of long-term projections arises from the CO_2 concentration stabilisation profiles given in Chapter 2. These illustrate, in an idealised way, possible future CO_2 concentration pathways and levels that might be achieved in response to Article 2 of the Framework Convention on Climate Change, which calls for eventual stabilisation of greenhouse gas concentrations at, as yet, unspecified

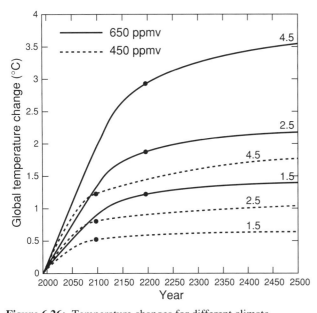

Figure 6.26: Temperature changes for different climate sensitivities ($\Delta T_{2\times}$ = 1.5, 2.5 and 4.5°C) for CO_2 concentration profiles stabilising at 450 ppmv (dashed lines) in 2100 and 650 ppmv (full lines) in 2200. Stabilisation dates are indicated by the dots. Calculations assume the "observed" history of forcing to 1990, including aerosol effects (see Figure 6.18) and then CO_2 concentration increases only beyond 1990. From Raper *et al.* (1996).

levels. Analyses of the climate and sea level implications of these stabilisation scenarios have been carried out by Raper *et al.* (1996), Wigley (1995) and Wigley *et al.* (1996).

Here we reproduce the results of Raper *et al.* for S450 and S650 (CO_2 stabilisation at 450 ppmv in 2100 and at 650 ppmv in 2200): see Figure 6.26. The changes shown are those arising from CO_2 increases alone. As noted by Raper *et al.*, the effect of other anthropogenic forcing factors might amplify the warming to 2100, based on the IS92 scenarios (compare the CO_2-alone case with the full forcing case in Figure 6.21), but the subsequent effect of these other gases is impossible to estimate. Figure 6.26 shows that global mean temperature continues to rise for centuries after the stabilisation dates (indicated by the dots on the figure), approaching an eventual asymptotic value (determined by the concentration stabilisation level) very slowly.

6.4 Simulated Changes of Variability Induced by Increased Greenhouse Gas Concentrations

Previous IPCC Reports (1990, 1992) were necessarily sparse in details concerning possible changes of variability

of the climate system with an increase of CO_2 and trace gases. The 1990 Assessment focused on changes in mean climate and relied mainly on results from atmospheric models coupled to non-dynamic slab oceans (so-called "mixed-layer models") that were not capable of simulating any variability associated with the El Niño-Southern Oscillation (ENSO). Since a major source of global interannual variability is associated with ENSO, models that cannot simulate ENSO-like oscillations are seriously constrained when used to examine possible changes in variability. The 1992 Supplement contained results from four global AOGCMs that contained some aspects of ENSO-like variability (e.g., Meehl, 1990; Lau *et al.*, 1992), but most of the reported results still involved changes in mean climate produced by those models. Even though ENSO-like variability is now represented in present global coupled models, caution must be exercised when interpreting the results since only some aspects of ENSO are simulated and the variability tends to be underestimated (see Chapter 5). Also in the earlier IPCC reports, there was a brief discussion of possible changes in variability, but most of those studies were in the early stages. Subsequently there has been additional work done to analyse how variability is represented in climate models (Chapter 5) as well as how changes in variability are simulated in climate models. More global AOGCM experiments have now been completed and there have been studies using mixed-layer and specified SST models to explore further the mechanisms of possible variability changes suggested by the global coupled models.

The definition of variability in this section is the quantification of deviations from some mean value (e.g., standard deviation or variance on monthly or longer time-scales). We will examine changes in these variability measures between simulated present-day and increased-CO_2 climates. These changes in variability are distinct from simple shifts in respective means between the two climate states. Section 6.5 will examine possible changes of extreme events on time-scales less than one month. This latter category is obviously related to changes in variability and is defined by specific meteorological occurrences that may be infrequent in time and space, and that affect society or an ecosystem at a certain location.

6.4.1 Simulated Changes in Intermonthly Variability
Model experiments with an early version of the global coarse-grid AOGCM at the National Center for Atmospheric Research (NCAR) indicated that the variability of lower tropospheric temperature in the control case compared favourably with that measured by the MSU

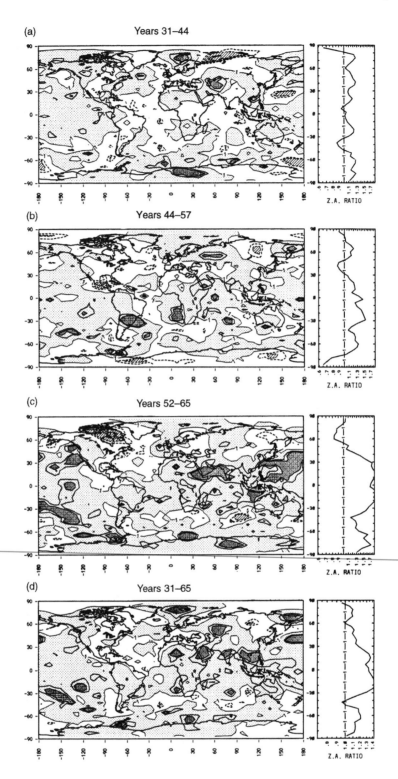

Figure 6.27: The ratio of the interannual variances (i.e., IASD squared) of the weighted lower tropospheric temperature from equivalent periods in the $2 \times CO_2$ and control of the NCAR global coupled model: (a) model years 31–44; (b) model years 44–57; (c) model years 52–65; (d) the entire period of model years 31–65. Solid contours indicate greater interannual variability in the $2 \times CO_2$ case. Light stippling indicates a ratio greater than 1 (greater interannual variability in the $2 \times CO_2$ case), dark shading indicates greater variability significant at the 5% level, and cross-hatching indicates less variability in the model significant at the 5% level. The zonal average of the ratio is shown to the right. From Meehl *et al.* (1994).

satellite data (Meehl *et al.*, 1994 – see Chapter 5). They also showed no consistent changes of zonal mean intermonthly lower tropospheric temperature variability in the tropics due to increased CO_2, but decreases near 60°N (Meehl *et al.*, 1994). In certain regions there were marginal increases in intermonthly variability (e.g., over the south Asian monsoon region). Increases of intermonthly (also termed intraseasonal) precipitation variability with increased CO_2 (as evidenced by a shift from time-scales of days to time-scales of weeks) have also been reported for several other coupled models for the Indian monsoon region (Lal *et al.*, 1994a).

The NCAR study above contrasted the results from the coupled model to an earlier version with the same atmospheric model coupled to a non-dynamic slab ocean (mixed-layer model). In that experiment there were decreases of intermonthly variability between 20 and 60°N as well as at 60°S. These mixed-layer results were consistent with earlier mixed-layer studies. The difference in the tropics compared to the coupled model had to do partly with the inclusion of ENSO-type variability in the coupled model. Comparing the NCAR coupled model to the earlier mixed-layer model, Campbell *et al.* (1995) noted that the qualitative differences in variability described by the first and second empirical orthogonal functions (EOFs) of 700 mb height were as large or larger than the differences in variability when CO_2 was doubled. These two studies clearly show the importance of including ENSO-related variability for studying such changes with increased CO_2, as well as the importance of improving the representations of ENSO in global coupled models.

In experiments with the LMD model (Parey, 1994a) a decrease in intermonthly temperature variability with increased CO_2 was noted. However in that experiment only the mean SST increases from the MPI (*m*) run were used, so there was no possibility that ENSO variability could play a role. Consistent with the decreased intermonthly temperature variability, Parey (1994a,b) also noted changes of persistent intermonthly circulation patterns, commonly referred to as "blocking" (see Section 5.4.1), in the Northern Hemisphere with increased CO_2. In northern winter in particular there were decreases of blocking frequency over the Northern Hemisphere ocean regions. Lupo *et al.* (1995) used an atmospheric model coupled to a simple mixed-layer ocean model with doubled CO_2 concentration to show that blocking events in the Northern Hemisphere were more persistent but weaker with increased CO_2, with greater blocking activity particularly over the continents. Blocking frequency had been shown to decrease generally in the Southern Hemisphere in earlier

mixed-layer model experiments (e.g., Bates and Meehl, 1986), but the centres of blocking activity shifted westward in the Northern Hemisphere. More consistent with the Parey (1994a, b) results, Meehl *et al.* (1994) in their global coupled atmosphere-ocean model study suggest that the decreases in intermonthly variability seen in the global coupled model near 60°N are associated with decreases in blocking in the Northern Hemisphere mid-latitudes.

In spite of the additional recent studies addressing changes involving persistent circulation patterns represented by blocking, there appears to be significant model-specific (and possibly blocking-definition-specific) aspects of changes in blocking activity with increased CO_2 in the various studies. For example, blocking can be defined in terms of various pattern recognition schemes, or as a certain geopotential height anomaly that persists for a certain number of days, such that different threshold values give different impressions of blocking in different models. This makes it difficult to identify consistent changes in blocking activity. However, since blocking involves stable circulation patterns and leads to periods of greater weather forecast skill, possible changes in blocking have implications for future weather forecasting as well as extreme weather events associated with blocking (see next section).

6.4.2 Possible Changes in Interannual Variability

The specified CO_2 increase and SST increase experiments with the LMD model described above also showed a general decrease of interannual temperature variability (Parey, 1994a). However, as noted in the previous section, since only mean SST increases are used, changes in variability due to ENSO are not included. Mearns (1993), Gordon and Hunt (1994) and Liang *et al.* (1995) used mixed-layer models and confirmed some of the earlier mixed-layer model results that showed some areas of increased interannual temperature variability in the tropics and subtropics, with decreases at higher latitudes. The studies with the NCAR model noted above (Meehl *et al.*, 1994; Campbell *et al.*, 1995) showed some decreases of zonal mean variability of lower tropospheric temperature between the zone 20–60°N and the zone near 60°S, but mainly increases of tropical interannual variability associated with the inclusion of ENSO-like phenomena in the coupled model (Figure 6.27).

One of the key questions to be addressed with coupled models is what changes there may be, if any, to interannual variability associated with ENSO from an increase of CO_2, and how precipitation patterns in the Pacific region may be affected.

Several model simulations with coarse-grid AOGCMs agree that ENSO-like events in those models continue to occur with an increase of CO_2 in the model atmosphere with either similar amplitude SST variability in the tropical eastern Pacific or somewhat of a decrease (Meehl *et al.*, 1993a; Knutson and Manabe, 1994; Tett, 1995; Tokioka *et al.*, 1995). However, those models show that there may be some alterations to effects associated with ENSO. Meehl *et al.* (1993a) showed that area-averaged SST variability over the eastern tropical Pacific is not appreciably changed for the relatively short time-series considered, but that area-averaged precipitation anomalies associated with ENSO are enhanced over the tropical continents. That is, in their global coupled model with ENSO-like events in the increased CO_2 climate, large-scale anomalously wet areas become wetter, and dry areas drier. Meehl *et al.* (1993a) noted that, following the earlier suggestions of Rind *et al.* (1989), an increase of mean surface temperature could result in enhanced precipitation variability due to processes associated with the non-linear relationship between surface temperature and evaporation. Meehl *et al.* (1993a) also noted that the changed mean basic state of the extratropical circulation due to increased CO_2 (i.e., changes in zonal mean flow that affects the propagation of planetary waves) is associated with an alteration of the mid-latitude teleconnections involved with ENSO-like events in the tropical Pacific. However, for definitive statistical analyses to be made, longer time-series from more model realisations would be useful.

Knutson and Manabe (1994) noted a decrease in the amplitude of ENSO-like variability over the equatorial Pacific with increased CO_2, based on three 1000-year GFDL coupled model integrations (Figure 6.28b). They also noted a slight eastward shift of the ENSO-related precipitation variability in the central equatorial Pacific. Tett (1995) and Tokioka *et al.* (1995) showed that there are no significant alterations of SST variability over the central equatorial Pacific Ocean associated with ENSO-like events, but Tett (1995) suggested that there may be small enhancements of precipitation variability. A recent study with an atmospheric GCM with SSTs specified to simulate both an ENSO event and CO_2 warming suggests that signals associated with intensification of hydrological effects may be relatively small, even though that model was not producing those changes interactively (Smith *et al.*, 1995).

Variability in another regional circulation regime that is of interest for human and economic reasons is the south Asian or Indian monsoon. Recent results from global coupled models (e.g., Meehl and Washington 1993; Singh and von Storch, 1994; Bhaskaran *et al.*, 1995; Tett, 1995) have shown an agreement with the results summarised from earlier model versions in the 1990 and 1992 IPCC reports in that mean monsoon precipitation over the south Asian or Indian monsoon regions intensifies with increased CO_2 (Figure 6.29) due to the more rapid warming of the south Asian land area compared to the Indian Ocean. This produces an enhancement of the land-sea temperature contrast in that region, and in conjunction with an increased moisture source from the warmer ocean, intensifies the monsoon rainfall. If model simulations of the monsoon include sulphate aerosols, then mean south Asian monsoon precipitation decreases owing to decreased land-sea temperature contrast (Lal *et al.*, 1995a).

Concerning variability changes, in the NCAR coarse-

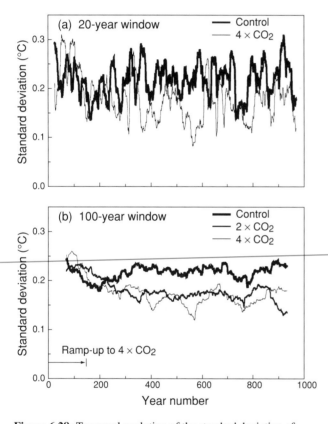

Figure 6.28: Temporal evolution of the standard deviation of ENSO-like variability in a GFDL global coupled AOGCM, and its dependence on CO_2 concentration. Shown are standard deviations of overlapping segments of bandpass (2–7 year) filtered sea surface temperature anomalies (7°N–7°S, 173°E–120°W) in °C for (a) a 20-year moving window and (b) a 100-year moving window. The different lines are for separate 1000-year experiments in which CO_2 is constant at $1 \times CO_2$ (control) or increases to $4 \times CO_2$ in the first 140 years and then remains at $4 \times CO_2$. In (b) an experiment in which CO_2 increases to $2 \times CO_2$ in the first 70 years and then remains constant at $2 \times CO_2$ is also shown.

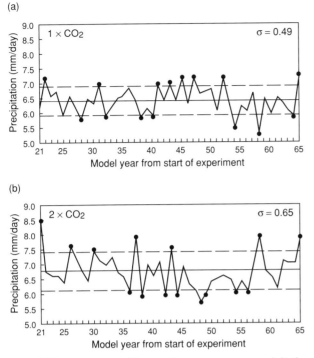

Figure 6.29: Area-averaged June to August monsoon precipitation in south Asia (land points in the area 5°N–40°N, 6 0°E–100°E) in mm/day from (a) the control case with present CO_2 concentration; and (b) the case for $2 \times CO_2$. Solid line is the 45 year mean; dashed lines indicate one standard deviation; dots indicate weak and strong monsoons (from Meehl and Washington, 1993).

grid AOGCM the interannual variability of the area-averaged summer (JJA) south Asian or Indian monsoon is enhanced (Meehl and Washington, 1993). This follows from the earlier ENSO results in that an increase of mean surface temperature could be expected to make precipitation more variable in certain tropical regions, owing to the relationship between surface temperature and evaporation. Bhaskaran *et al.* (1995) also note a significant increase in monsoon interannual variability in the Hadley Centre (UKMO)AOGCM. A marginally higher range between strong and weak Indian monsoon precipitation was documented in the MPI coupled model with increased CO_2 (Lal *et al.*, 1994b). Consistent with this result, standard deviation of Indian monsoon precipitation was somewhat higher with increased CO_2 in the CSIRO9 mixed-layer model (Chakraborty and Lal, 1994). The addition of sulphate aerosols appears to affect mean monsoon precipitation (noted above) but enhanced interannual monsoon variability still occurs as in the CO_2-only experiments (Lal *et al.*, 1995a).

6.4.3 Decadal and Longer Time-scale Variability

So-called "inherent" low-frequency variability in the climate system on decadal and longer time-scales can, for example, be induced by external sources such as volcanic activity and perhaps solar activity, and from internal sources such as alterations of the thermohaline circulation. External sources of climate variability have not been included in AOGCMs, but such models do have inherent variability on long time-scales due to interaction between the components of the coupled system. Since there are as yet few multi-century AOGCM integrations, these studies have focused less on possible changes to that variability than on the consequences of long time-scale variability on the manifestation of CO_2 climate change in the models. Clarifying these issues should be possible as more multi-century coupled model runs are completed.

Decadal variability makes it more difficult to identify the effects of increased CO_2 in the models. For example, Cubasch *et al.* (1994b) showed that a series of transient climate change experiments started from different initial states in the control integration produced as a result of decadal-scale variability somewhat different time evolutions of surface air temperature change. Tokioka *et al.* (1995) showed that decadal scale variability of sea ice volume in the MRI (Japan) global AOGCM affected the rate of warming at high latitudes in their transient CO_2 increase experiment. An increase of sea ice in the control run (and transient run) due to inherent decadal variability delays warming due to ice albedo feedback in the transient CO_2 increase experiment. However, ice albedo feedback effects can be influenced to a certain extent by the particular features of each sea ice formulation (e.g., Meehl and Washington, 1990).

As noted by Meehl *et al.* (1993b) and Karoly *et al.* (1994), climate models usually display a consistent zonal mean pattern of temperature change due to increased CO_2, with warming in the troposphere and cooling in the stratosphere. Karoly *et al.* (1995) attempted to sort out decadal variability associated with this zonal mean temperature signal from the GFDL and NCAR global AOGCMs. They conclude that the zonal mean pattern of CO_2-induced temperature change (warming in the troposphere and cooling in the stratosphere) does have inherent decadal-scale variability in the coupled models. Principal component analysis of zonal mean temperature shows some similarity of this pattern between the control runs from the models and the transient CO_2 increase experiments. Models can exhibit natural modes of variation which are similar to the signal they simulate in response to increased CO_2 (Santer *et al.*, 1994).

A set of experiments with one global coupled model

have been run long enough to address possible changes to ENSO amplitude occurring on multi-decadal to multi-century time-scales. Knutson and Manabe (1994, 1995) note that for their 1000-year integrations with the GFDL global coarse-grid AOGCM, the amplitude of ENSO-like variability can fluctuate by a factor of two or more on roughly a 50-year time-scale (Figure 6.28a). These internally generated amplitude modulations are substantially larger than the 20% decrease in mean amplitude found for a quadrupling of CO_2 (Figure 6.28b) and are reminiscent of the fluctuations in the amplitude of the real ENSO observed in this century (Section 3.4.2). Even using a 100-year sampling window (Figure 6.28b) they found that the model's internally generated fluctuations of ENSO amplitude on a multi-century time-scale can be comparable in magnitude to the long-term change due to increased CO_2. These studies point to the complications that arise due to inherent long-time-scale variability, and the difficulties that variability presents for analysing changes of the climate system on almost all time-scales due to increased CO_2, as well as for detection of a CO_2-induced climate change signal (see Chapter 8).

6.4.4 Conclusions

Analyses of climate model results (including a greater number of global coupled atmosphere-ocean models) subsequent to IPCC (1992) provide us with some indications of possible changes of variability of the climate system due to an increase of CO_2 in the context of the model limitations noted in Ch. 5:

(i)   Experiments with different model configurations indicate that zonal mean mid-latitude intermonthly temperature variability may be reduced, but there are no consistent results in regard to changes in persistent anomalies called "blocks", one of the contributors to intermonthly variability, since different definitions of blocking have been used and different models show different changes of geographical patterns of blocking.

(ii)   One study shows that intermonthly and interannual variability differences between GCMs with a simple mixed-layer and those coupled to a full ocean model are larger than changes in either type of model due to increased CO_2 alone, pointing to the importance of using a model with some representation of ENSO-like phenomena.

(iii) ENSO-like variability in several AOGCMs continues

with increased CO_2 with either little change or somewhat of a decrease of SST variability in the eastern tropical Pacific Ocean. A global AOGCM that includes some aspects of ENSO-like variability suggests that interannual variability of zonal mean lower troposphere temperature in the tropics may increase mainly due to hydrological effects on the tropical continents.

(iv)   There could be enhanced precipitation variability associated with ENSO events in the increased CO_2 climate, especially over the tropical continents, associated with the mean increase of tropical SSTs.

(v)   Enhanced interannual variability of area-averaged summer south Asian monsoon rainfall is indicated by several models.

(vi)   Decadal and longer time-scale variability affect the realisation of climate change in a number of simulations in terms of the rates of global warming and patterns of zonal mean temperature change. Multi-century variability in ENSO phenomena was in one case as large as the mean change caused by CO_2 increase. This indicates that natural climate variability on long time-scales will continue to be problematic for CO_2 climate change analysis and detection.

6.5 Changes in Extreme Events

6.5.1 Introduction

The extreme events considered here are infrequent meteorological events that surpass a certain threshold. The perceived severity of the extreme depends on the vulnerability of the natural environment and human society to the event. This implies that the definition of an extreme event can strongly depend on location.

A common property of extremes is that they refer to the deviation of meteorological elements (or the combination of several elements) from the local normals. Several mechanisms can be responsible for changes in local extremes: the appearance of large-scale meteorological phenomena (e.g., blocking or El Niño), an intensifying or weakening of either large-scale systems (e.g., stormtracks) or small-scale features such as convective storms, or a (relatively) small shift in location of weather systems possessing sharp gradients, like the inter-tropical convergence zone (ITCZ). Prediction of extremes originating from these three sources put different requirements on the quality of the GCM predictions.

Predicting changes in extremes means predicting changes in probability distributions. For example, changes in mean temperature generally have substantial impacts on exceedance probabilities because of the rather short upper tail of the temperature distribution. Changes in variability such as those described in the previous section strongly affect the occurrence of extreme events. The possibility that a change in the variance can have a larger impact on the exceedance frequencies for monthly maxima than a change in the mean is demonstrated analytically in Katz and Brown (1992). Nevertheless, enhanced greenhouse simulations indicate that the effect of changes in mean temperature are usually much larger than the effect of changes in variance (Cao *et al.*, 1992; Hennessey and Pittock, 1995).

The state-of-the-art of modelling synoptic scale extremes is briefly described in the next sub-section (6.5.2), in order to provide a general background for the subsequent discussion. Possible changes in daily variability, which might give rise to changes in the occurrence of extreme events are discussed in Section 6.5.3. Several types of the most studied extreme events (wind-related, temperature related and precipitation related), are discussed in Sections 6.5.4 to 6.5.6.

6.5.2 Background
Possible changes in extreme events were assessed by IPCC in IPCC (1990) (Mitchell *et al.*, 1990) and IPCC (1992) (Gates *et al.*, 1992).

Many scientists (e.g., Hewitson and Crane, 1992a, b; von Storch, 1993; von Storch *et al.*, 1993) consider the usefulness and realism of output from climate models as being scale-dependent. The success of the simulation of extreme events in GCMs depends on the scale of the event, and whether the events depend on regional properties (such as lake-related snow-fall in Ontario, Canada).

To simulate a reasonable global scale atmospheric or oceanic circulation, relatively coarse resolution numerical models are sufficient. Obviously, these models must resolve the scales of the forcing functions. Much finer resolution is required to model adequately the non-linear flow of energy from the large scales to the small scales. Because of the finite resolution there is a spectral cut-off point where the flux of energy to smaller scales can no longer be maintained by the non-linear dynamics, but is usually parameterized by an artificial dissipation or damping mechanism (for instance, Roeckner and von Storch, 1980).

For scales near the spatial cut-off scale, however, the dynamics are severely disrupted, and the skill of the simulation of these scales will in general be low. Further

factors which make the model output less reliable for smaller scales are the lack of sub-grid scale features within a grid box, such as a coastline, the presence of a lake or of mountains, which might be relevant for the regional circulation. Also, the parametrizations of the various processes in atmospheric GCMs describe the overall effect of a process better for some variables or regions than others.

Extreme weather events are rare and occur on the synoptic and even smaller scales (e.g., heavy showers, gusts and tornadoes). Long integrations of very high-resolution models are required to simulate them and even then there is little prospect that sub-synoptic scale events can be successfully resolved in GCMs. We present here some examples where general weather or wind patterns were analysed, from which conclusions concerning extremes may be inferred using down-scaling techniques. Hulme *et al.* (1993) assessed the control simulations (10-year) of two GCM experiments (UKHI and ECHAM1-coupled model 11) in terms of their ability to reproduce realistic "real world" weather and observed mean monthly frequencies of gales over the British Isles. The seasonality of both anticyclonic and cyclonic types of weather is much too strong in ECHAM1 and summer precipitation in this model is greatly underestimated. ECHAM1 simulates the annual cycle of temperature well, while UKHI successfully reproduces the annual cycle of precipitation. Both models underestimate the number of gales by about 50%, ECHAM1 more so than UKHI (see Figure 6.30). ECHAM1 simulates the correct form of the annual cycle with a clear winter maximum. In December and January, UKHI greatly underestimates gale frequencies, leading to a bimodal frequency distribution of simulated gale frequencies.

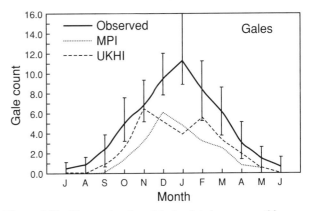

Figure 6.30: Observed and model-simulated mean monthly frequencies of gales over the British Isles. The range bars on the observed line represent the range in 10-year averages from the 110-year observed record (from Hulme *et al.*, 1993).

Beersma (1994) compared storm activity over the North Sea and the Netherlands in two climate models (CCC and ECHAM1-coupled) with observations. He observes that the models have difficulties with basic features of the large-scale circulation which affect the surface wind and concludes that the average wind speed and the frequency of gales over the North Sea are reproduced rather poorly in both models.

ENSO affects tropical cyclone behaviour and the occurrence of floods, droughts, and heat waves in many extra tropical regions. Changes in ENSO behaviour related to climate change are possible but uncertain. This is an important source of uncertainty in the following discussion.

6.5.3 Possible Changes in Daily Variability

Only a few modelling studies to date have analysed changes in daily temperature and precipitation variability (defined as variability between days) due to increased CO_2. One study used the LMD atmospheric GCM run, with specified SSTs taken from a transient climate change experiment with an AOGCM. Two series of three experiments (control, doubled and tripled CO_2 with two different model resolutions) showed a general decrease of daily temperature variability with increased CO_2 in the model (Parey, 1994a) along with an increase in mean temperatures. A similar result for daily temperature variability over a limited area was obtained from two continuous 3.5 year-long climate simulations over continental USA, one for present-day conditions and one for conditions under doubled CO_2 concentration. These were conducted with a regional climate model (RegCM), on a 60 km grid, nested in a global GCM. A moderate number of changes in daily temperature variability were statistically significant, and the majority of these were decreases. Clearest tendencies were decreases in the winter in the central and northern Great Plains, in late summer and late winter in the Great Lakes region, and in late winter in the south-east (Mearns *et al.*, 1995b). Cao *et al.* (1992) analysed global changes in daily temperature variance in the UKMO low resolution GCM for a doubling of CO_2. There is a reduction in variance in high latitudes in winter where sea ice is replaced by open water, and a generally insignificant reduction in middle latitudes over the oceans and North America in winter. Elsewhere changes in variance are small and of variable sign, though some increases over the northern continents in summer are statistically significant.

The Model Evaluation Consortium for Climate Assessment (MECCA) Phase 1 simulations have been examined to assess the possibility of changes in daily variability in surface temperature and other quantities. The largest changes in variability in surface temperature occurred at the cryospheric margin. Although there are differences between control and enhanced greenhouse values of daily standard deviation of surface temperature, differences among the five models studied were much greater than between any one model's realisation of current and future climatic variability (Henderson-Sellers *et al.*, 1995a).

Concerning changes in daily precipitation variability, analysis of the RegCM integrations showed that there were fewer areas of significant change compared to daily temperature variability change, but that most of these were increases in daily precipitation variability for doubled CO_2 conditions, particularly in winter on the north-west coast of North America.

The relationship between these changes in daily temperature and precipitation variability seems to be dependent on the geographic area considered. For example, mid-continental temperature variability could be associated with changes in clouds or soil moisture, while changes in precipitation variability could be affected by altered storm activity.

6.5.4 Extreme Wind Events

6.5.4.1 Mid-latitude storms

The main energy sources for mid-latitude depressions are the temperature contrast between the cold polar regions and the warmer sub-tropical conditions, and the release of latent heat as water vapour condenses in the warm, poleward moving, ascending air. Some GCMs show (IPCC, 1990) that in a CO_2 enhanced atmosphere, the low-level meridional temperature gradient will be reduced due to positive feedbacks at higher latitudes, although most coupled models simulate an increased temperature gradient in the Southern Hemisphere. Most models predict increased water vapour content in the warmer atmosphere. However, at upper levels the tropics warm relative to high latitudes due to the release of latent heat. It is difficult to predict the result of these competing effects.

GCMs simulate the position of the observed storm tracks well (Figure 6.31a and b) but tend to slightly underestimate storm track strength (Hall *et al.*, 1994; Murphy 1995a). GCM response to enhanced CO_2 varies somewhat with different models and resolution (e.g., Senior 1995) and different techniques are employed to study the storm tracks. Hall *et al.* (1994) found an intensification and poleward shift in both the Northern Hemisphere storm tracks in the high resolution UKMO slab model, with the most spectacular change occurring in the eastern Atlantic/

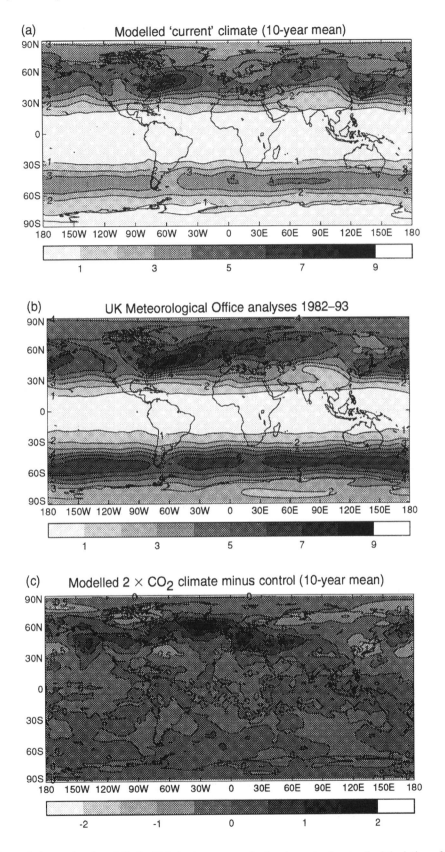

Figure 6.31: Comparison of observed and modelled mid-latitude storm tracks (as shown by the standard deviation of band-pass filtered 500 mb height) in December, January and February, from 10-year simulations with the UKMO 2.5° × 3.75° resolution slab model.

western Europe region (Figure 6.31c). Carnell *et al.* (1996) found similar results from the UKMO AOGCM. In contrast, results from the CCC slab model (Lambert, 1995) suggest no obvious shift in the Northern Hemisphere storm tracks but a slight shortening of the Atlantic tracks suggesting a reduction in storm activity over Europe. The ECHAM1 AOGCM (König *et al.,* 1993) shows a northward shift of North Atlantic cyclones and an eastward movement in the North Pacific. The most significant change is a poleward shift of Southern Hemisphere cyclones in autumn and winter. Clearly, there is little agreement between models on the changes in storminess that might occur in a warmer world. Conclusions regarding extreme storm events are obviously even more uncertain.

6.5.4.2 Tropical cyclones

The formation of tropical cyclones depends upon a number of factors, including sea surface temperatures (SSTs), the vertical lapse rate of the atmosphere, vertical wind shear, mid-tropospheric relative humidity, and the prior existence of a centre of low-level cyclonic vorticity (Gray, 1979) so predicting changes in a warming climate is not straightforward. Though climate models can simulate some of the aspects of tropical cyclone occurrence (Manabe and Broccoli, 1990; Haarsma *et al.*, 1993), the state-of-the-science remains poor because:

(i)   tropical cyclones cannot be adequately simulated in present GCMs,

(ii) some aspects of ENSO are not simulated well in GCMs,

(iii) other large-scale changes in the atmospheric general circulation which could affect tropical cyclones cannot yet be discounted, and

(iv) natural variability of tropical storms is very large, so small trends are likely to be lost in the noise.

A special study undertaken as part of the joint WMO/ICSU programme on Tropical Cyclone Disasters by Lighthill *et al.* (1994) suggests that there are no compelling reasons for expecting a major change in global tropical cyclone frequency, although substantial regional changes may occur. Models, at present, are not adequate to predict the direction of such changes. A further study by Holland (1995) indicates that there is unlikely to be more intense tropical cyclones than the worst that are experienced under current climate conditions. There may be some potential, however, for changes of cyclone intensity in regions where SST is between 26 and 29 °C.

Bengtsson *et al.* (1994, 1995) ran a 5-year experiment

for IPCC (1990)'s "Scenario A / year 2035" conditions with the T106 ECHAM model. Individual storms as well as their geographical and seasonal distribution in the control run agreed well with observations. For the enhanced greenhouse scenario, they found a decrease in the number of tropical cyclones, especially in the Southern Hemisphere while the geographical distribution appeared unchanged compared to the present climate. In this simulation the effect of warmer surface waters is counteracted by generally weaker surface winds with somewhat reduced evaporation and by changes of the vertical stability. Lal *et al.* (1995b) found no significant changes in the number and intensity of monsoon depressions in the Indian Ocean in a warmer atmosphere in the ECHAM3-T106 experiment.

In conclusion, it is not possible to say whether the frequency, area of occurrence, time of occurrence, mean intensity or maximum intensity of tropical cyclones will change.

6.5.5 Extreme Temperature Events

Zwiers (1994) made an extreme value analysis of simulated 2-metre air temperature obtained from 20-year control and doubled CO_2 simulations conducted with the Canadian Climate Centre second generation climate model (McFarlane *et al.*, 1992). Changes in both the shape and location of the 2-metre temperature distributions occur in many places. Large changes (4–6°C) in the 10-year return values of daily maximum 2-metre temperature ($T_{Max,10}$) occur over the continents (except Antarctica) as a consequence of a loss of soil moisture. Even larger changes (up to 10°C) in the 10-year return values of daily minimum 2-metre temperature ($T_{Min,10}$) over temperate northern latitudes are related to a loss of snow cover. Small changes in $T_{Min,10}$ over land masses which do not receive snow in the control simulation can be traced to losses in soil moisture and corresponding reductions in cloud. Desert areas seem to experience roughly equal changes in $T_{Min,10}$ and $T_{Max,10}$ as do tropical and temperate oceans. Formerly ice-covered parts of the polar oceans sustain relatively large changes in $T_{Max,10}$ and very large changes in $T_{Min,10}$. Parts of the polar oceans which remain ice-covered experience small changes in $T_{Max,10}$ and large changes in the $T_{Min,10}$.

Small changes in the mean climate can produce relatively large changes in the frequency of extreme events. Hennessy and Pittock (1995) performed a scenario study for the year 2030 where observed temperatures were increased according to GCM-derived warming scenarios for south-eastern Australia which vary spatially and seasonally. It was assumed that (absolute) temperature variability remains constant.

At each GCM grid-point over Victoria, Australia, the warming per degree of global warming simulated by each of five GCMs was ranked from highest (rank 1) to lowest (rank 5). The rank 2 warming was defined as the "high case" and the rank 4 warming was defined as the "low case". Multiplying these scaled regional warmings by a range of global warming estimates for a given year (Whetton *et al.*, 1993) gives an estimate of the regional warming for that year.

The low warming scenario of about 0.5°C by 2030 gives at least 25% more days over 35°C in summer and spring, and at least 25% fewer winter days below 0°C. The high warming scenario of about 1.5°C produces 50 to 100% more extremely hot summer and spring days and 50–100% fewer extremely cold winter days. Larger percentage changes occur in regions where absolute changes are smaller. Similar results were found for these and other threshold temperatures at temperate sites in the eastern Australian state of New South Wales, and at tropical sites in the Northern Territory and Western Australia.

The high warming scenario also gives a 20% increase in the probability of a run of at least 5 summer days over 35°C in the northern half of Victoria, and the risk of a run of at least 5 winter days below 0°C is reduced by up to 20% in western and coastal areas and by 25–40% in the north-east highlands.

6.5.6 Precipitation Extremes

There is now mounting evidence to suggest that a warmer climate will be one in which the hydrological cycle will in general be more intense (IPCC, 1992), leading to more heavy rain events (*ibid.*, pp.119). It should be noted, however, that, as the GCM grid sizes are much larger than convective elements in the atmosphere, daily precipitation is poorly reproduced by GCMs.

Noda and Tokioka (1989) noted in their $2 \times CO_2$ simulations an increase of cumulus type precipitation, which results in a decrease of precipitation area and an increase in precipitation intensity. Gordon *et al.* (1992) looked at changes in the frequency distribution of simulated daily rainfall. They found that return periods (between 3-month to 5-year in this study) for CSIRO4 GCM-simulated heavy rain events decreased by factors ranging from 2 to 5 under $2 \times CO_2$ conditions in the locations they examined (i.e., Australia, midwest USA, Europe and India).

Similar results were found in analyses of daily rainfall simulated by the CSIRO9 GCM (Whetton *et al.*, 1993), and

with the UKMO high resolution model (Fowler and Hennessy, 1995; Gregory and Mitchell 1995; Hennessy *et al.*, 1995). While mean rainfall intensity increased by 10–30% at most latitudes for a doubling of CO_2, increases in the frequency and intensity of 90th percentile rainfall events exceeded 50% in some regions.

Henderson-Sellers *et al.* (1995a) have compared simulations of daily precipitation from five equilibrium simulations of doubled CO_2. Overall they find shorter return periods for extreme events and increased intensity of rainfall. The largest changes occur in the tropics. This study also suggests a decrease in return period in the five regions highlighted in IPCC (1990). In addition, the MECCA models exhibit an increase in intensity in most regions except South Asia. In South Asia, however, there is an increase in the number of rain days and in the number of wet spells. The number of raindays decreases in southern Europe. In general there is an increase in the mean and variability of precipitation, a decrease in return period of extreme events and an increase in intensity and the number of wet spells (consecutive days with precipitation).

McGregor and Walsh (1994) present $2 \times CO_2$ results for a limited area model nested within the CSIRO9 GCM at 60-km resolution over Tasmania, Australia. Over 10 successive years of July conditions, the total number of rain days decreases, but the number of events exceeding 16 mm/day increases by 40%

Analysis at NCAR of a RegCM embedded in a global GCM showed areas with significant changes (both increases and decreases) of precipitation frequency and intensity under $2 \times CO_2$ conditions (Mearns *et al.*, 1995a). Precipitation variability increased significantly. In the Central Plains of the USA, small changes in mean precipitation masked large increases in intensity combined with decreases in frequency. Analysis of daily rainfall frequency over land points in the Indian monsoon region in the Hadley Centre global coupled model showed that there could be more heavy rainfall days with increased CO_2 (Bhaskaran *et al.*, 1995).

However, Parey (1994a), analysing the return period of heavy rainfall events from three 30-year simulations under $1 \times$, $2 \times$ and $3 \times CO_2$ conditions with the LMD-GCM in seven, mainly extra-tropical land regions did not find a systematic increase in heavy rainfall classes as CO_2 increases. There was no significant change in daily rainfall annual distribution between the three simulations, so the return period for heavy rainfall events was not shorter in the enhanced CO_2 simulations. In a parallel experiment with the ECHAM3 model Cubasch *et al.* (1995b) find an increase in rainfall intensity mainly over tropical land regions and the northern mid-latitudes.

Various investigators (e.g., Cubasch *et al.*, 1995b; Gregory and Mitchell, 1995; Gregory, pers. comm.) find that a shift in the distribution of daily precipitation amounts towards heavier events may in some areas be accompanied by an increase in the number of dry days. Dry days may also become more frequent where the mean precipitation decreases. For both reasons, there may be an increase in the length of dry spells (consecutive days without precipitation).

For instance, Cubasch *et al.* (1995b) found from three 30-year simulations under $1 \times$, $2 \times$ and $3 \times CO_2$ conditions, respectively, with the ECHAM3 model a generally longer dry spell length in all extratropical regions. In central North America, the probability of a 3-month dry spell is 1% in the $1 \times CO_2$, twice that in the $3 \times CO_2$ run. Gregory (pers. comm.) reports for experiment *s* that the probability of a dry spell of at least 30 days in summer in southern Europe increases by a factor of 2–5 on doubling CO_2, although the mean precipitation decreases by only 22%. This relatively small reduction also leads to a much increased incidence of low seasonal precipitation totals: the probability of a precipitation amount smaller than the driest decile of the control distribution of summer precipitation more than doubles.

Reduced precipitation may also be associated with drier soil. In Whetton *et al.* (1993) drought occurrence in Australia is assessed using an off-line soil water balance model and scenarios based upon the results of a range of GCMs. Although large changes in drought frequency were possible due to the range of uncertainty in the scenarios, the direction of change in soil moisture was uncertain at most sites examined. They demonstrate that drying was more likely in the south of Australia.

Gregory (pers. comm.) also finds higher probability of dry soil in summer in central North America and southern Europe in the Hadley Centre simulations.

6.5.7 Summary and Conclusions

Extreme weather events are important aspects of climate. They generally occur at synoptic scale and are of shorter duration than global climate change. Since possible changes in extremes may have large impacts on nature and human society, and relate to most people's direct experience, such changes will be more convincing than global changes averaged over time and space. Current climate models lack the accuracy at smaller scales and the integrations are often too short to permit analysis of local weather extremes. Except maybe for precipitation, there is little agreement between models on changes in extreme events. However, by reasoning from physical principles, or by using down-scaling techniques and looking at patterns

such as mid-latitude storm tracks from which extremes can be inferred, or by making time-slice experiments with high resolution models some tentative assessments concerning extreme events may be made:

(i)   Small changes in the mean climate or climate variability can produce relatively large changes in the frequency of extreme events; a small change in the variability has a stronger effect than a similar change in the mean.

(ii)   A general warming tends to lead to an increase in extremely high temperatures and a decrease in winter days with extremely low temperatures.

(iii) Model studies that addressed daily variability changes indicate a decrease of daily temperature variability with increased CO_2 in certain regions, with some increased daily precipitation variability over a few areas such as north-west North America in winter.

(iv) Several models suggest an increase in the precipitation intensity, suggesting a possibility for more extreme rainfall events. In some cases models also predict more frequent or severe drought periods in a warmer climate.

6.6 Simulation of Regional Climate Change

In IPCC (1990) and IPCC (1992), very low confidence was placed on the climate change scenarios produced by GCM equilibrium experiments on the sub-continental, or regional, scale (order of $10^5 - 10^7$ km^2). This was mainly attributed to coarse model resolution, limitations in model physics representations, errors in model simulation of present day regional climate features, and wide inter-model range of simulated regional change scenarios. Since then, transient runs with AOGCMs have become available which allow a similar regional analysis. In addition, different regionalisation techniques have been developed and tested in recent years to improve the simulation of regional climate change. This section examines regional change scenarios produced by new coupled GCM runs (Section 6.6.1) and progress in the application of regionalisation techniques to the simulation of regional climate change (Section 6.6.2). Following the 1990 and 1992 reports, emphasis is placed on the simulation of seasonally averaged surface air temperature and precipitation, although the importance of higher order statistics and other surface climate variables for impact assessment is recognised (Kittel *et al.*, 1995; Mearns *et al.*, 1995a, b).

6.6.1 Regional Simulations by GCMs

In IPCC (1990), five regions were identified for analysis of regional climate change simulation: Central North America (CNA, 35–50°N, 85–105°W), South East Asia (SEA, 5–30°N, 70–105°E), Sahel (Africa) (SAH, 10–20°N, 20°W–40°E), Southern Europe (SEU, 35–50°N, 10 W–45°E) and Australia (AUS, 12–45°S, 110–155°E). Output from different coupled model runs with dynamical oceans for these regions was analysed by Cubasch *et al.* (1994a), Chakraborty *et al.* (1995a, b), Whetton *et al.* (1995) and Kittel *et al.* (1996), while analysis over the Australian region from equilibrium simulations with mixed-layer ocean models was performed by Whetton *et al.* (1994). Results over two additional regions were analysed by Raisanen (1995) for Northern Europe (NEU, land areas north of 50°N and west of 60°E) and Li *et al.* (1994) for East Asia (EAS, 15–60°N, 70–140°E). To summarise the findings of these works, Figures. 6.32 (a–h) show differences between region-average values at the time of CO_2 doubling and for the control run, and differences between control run averages and observations (hereafter referred to as bias), for winter and summer surface air temperature and precipitation. The biases are presented as a reference for the interpretation of the scenarios, because it can be generally expected that the better the match between control run and observed climate, the higher the confidence in simulated change scenarios. The model runs are labelled *d, g, m, x, p, q, r, s* and *t* in Table 6.3 and the experiments do not include the effects of aerosols and human-induced changes in surface characteristics, which are likely to alter regional patterns of temperature and precipitation change (e.g., Copeland *et al.*, 1995; also see Section 6.2 of this report). The models employ different spatial resolutions and flux adjustments.

Scenarios produced by these transient experiments varied widely among models and from region to region, both for temperature and precipitation. Except for a few outliers, individual values of projected surface warming varied mostly in the range of ~1°C to ~5°C (Figure 6.32a, c), with the NCAR-*r* and MPI-*x* runs generally showing the greatest temperature sensitivity. NCAR-*q* runs showed the least temperature sensitivity because, with a 1% linear increase in CO_2/yr (Table 6.3), CO_2 had increased only by a factor of 1.7 by the end of the 70-year simulation, compared to a doubling for the other model runs. For most regions, the inter-model range of simulated temperature increase was rather pronounced, about 3–5°C. With the

exception of one or two outliers, the smallest inter-model range of simulated warming at the time of CO_2 doubling was over Australia in summer and the Sahel in winter, where the scenarios differed among models by no more than 1.3°C. It should be noted, however, that for a region such as Australia, continental-scale agreement may come from cancelling differences at the sub-continental scale.

The surface air temperature biases had positive and negative values both in winter and summer (Figures 6.32 b, d). However, biases were mostly negative in winter and positive in summer, an indication that the models tended to overestimate the seasonal temperature cycle. Most biases were in the range of –7°C to 10°C, but values as large as ~ 15°C were found. The smallest biases were found over Australia and, with the exception of one or two models, South-East Asia and Southern Europe. Over most regions, the inter-model range of temperature bias was of the order of 10°C, i.e., it was greater than the inter-model range of regional temperature increase. The surface temperature biases as well as the simulated regional warming scenarios were in the same range as those reported in IPCC (1990) for a number of equilibrium runs.

Regional precipitation biases spanned a wide range, with values as extreme as ~–90% or greater than +200% (Figure 6.32f, h). The biases were generally larger in winter than in summer, and overall, regions with the smallest biases were Southern Europe, Northern Europe, and Central North America. Regions receiving low winter precipitation (e.g., Sahel, South East Asia) tended to have large positive or negative biases because small errors in control run values appear as large biases when reported in percentage terms.

Simulated precipitation sensitivity to doubled CO_2 was mostly in the range of –20% to +20% of the control value (Figure 6.32e, g). The most salient features of simulated regional precipitation changes are summarised as follows:

(i)   All models agreed in summer precipitation increases over East Asia and, except for one model, South East Asia, reflecting an enhancement of summer monsoonal flow (see Section 6.3).

(ii)   All models agreed in winter precipitation increases over Northern Europe, East Asia and, except for one model, Southern Europe. In the other cases, agreement was not found among models even on the sign of the simulated change.

(iii)   Regions with the smallest inter-model range of simulated precipitation change were Central North

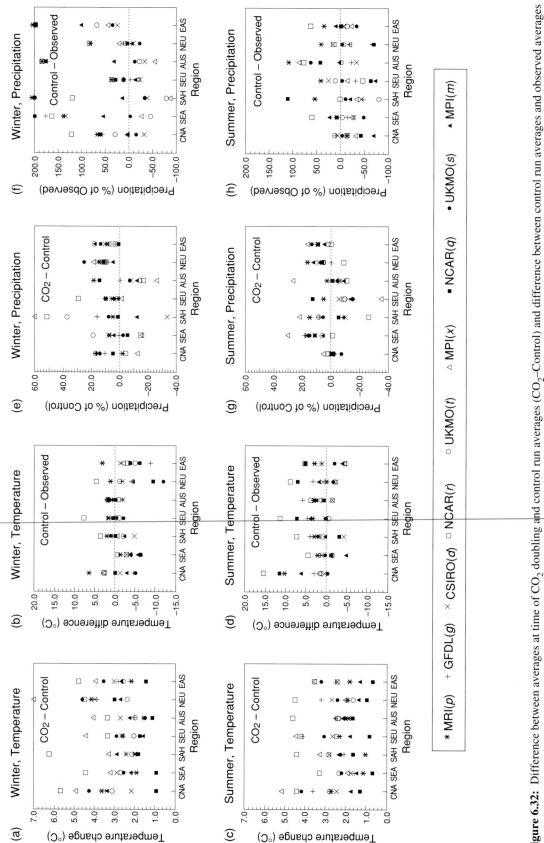

Figure 6.32: Difference between averages at time of CO_2 doubling and control run averages (CO_2–Control) and difference between control run averages and observed averages (Control–Observed) as simulated by nine AOGCM runs over seven regions (see text). (a)–(b) Temperature, winter; (c)–(d) temperature, summer; (e)–(f) precipitation, winter; (g)–(h) precipitation, summer. Units are °C for temperature and percentage of control run, or observed, averages for precipitation. In (f) and (h) values in excess of 200% have been reported at the top end of the vertical scale. In (e) values in excess of 60% have been reported at the top end of the vertical scale. CNA = Central North America, SEA = South East Asia, SAH = Sahel, SEU = Southern Europe, AUS = Australia, NEU = Northern Europe, EAS = East Asia.

America, East Asia and Northern Europe in summer and Southern Europe, Northern Europe, and East Asia in winter.

(iv) Overall, the precipitation biases were greater than the simulated changes. A rigorous statistical analysis of the model results in Figure 6.32a–h has not been carried out. However it can be expected that, due to relatively high temporal and spatial variability in precipitation, temperature changes are more likely to be statistically significant than precipitation changes.

In summary, several instances occurred in which regional scenarios produced by all models agreed, at least in sign. In fact, regardless of whether flux correction was used, the range of model sensitivity was less than the range of biases (note that the scales in Figure 6.32 a–h are different for the sensitivities and the biases). However, the range of simulated scenarios and the model regional biases were still large, so that confidence in the regional scenarios simulated by AOGCMs remains low. It should be pointed out that, while model agreement increases our confidence in the veracity of model responses, it does not necessarily guarantee their correctness because of possible systematic errors or deficiencies shared by all models. On the other hand, in spite of these errors, models are useful tools to study climate sensitivity (see Chapter 5). Even though models cannot exactly reproduce many details of today's climate, key processes that we know to exist in the real climate system are represented in these models (see Chapter 4). For example, the simulation of the seasonal cycle of winds, temperature, pressure and humidity in both the horizontal and vertical provides us with a first order qualification of the fidelity of the models' ability to capture these basic features of the Earth's climate. As another example, AOGCMs exhibit the ability to simulate essential responses of the climate system to various forcings, e.g., those involving El Niño SST anomalies and aerosols from volcanic activity. This increases our confidence in the use of AOGCM sensitivity experiments to evaluate potential changes in important climate processes.

6.6.2 Simulations Using Statistical Downscaling and Regional Climate Modelling Systems

Although computing power has substantially increased during the last years, the horizontal resolution of present coupled atmosphere-ocean models is still too coarse to capture the effects of local and regional forcings in areas of complex surface physiography and to provide information suitable for many impact assessment studies. Since IPCC

(1992), significant progress has been achieved in the development and testing of statistical downscaling and regional modelling techniques for the generation of high resolution regional climate information from coarse resolution GCM simulations.

6.6.2.1 Statistical downscaling
Statistical downscaling is a two-step process basically consisting of (i) development of statistical relationships between local climate variables (e.g., surface air temperature and precipitation) and large-scale predictors; and (ii) application of such relationships to the output of GCM experiments to simulate local climate characteristics. A range of statistical downscaling models has been developed (Karl *et al.*, 1990; Bardossy and Plate 1992; Wilks 1992, Wilson *et al.*, 1992, Von Storch *et al.*, 1993; Gyalistras *et al.*, 1994; Hughes and Guttorp 1994; Noguer 1994; Matyasowszky *et al.*, 1995; Zorita *et al.*, 1995; Pleym and Karl 1996), mostly for USA, European and Japanese locations, where better data for model calibration are available. The main progress achieved in the last few years has been the extension of many downscaling models from monthly and seasonal to daily time-scales, which allows the production of data more suitable for a broader set of impact assessment models (e.g., agriculture or hydrologic models) (Bardossy and Plate 1992; Wilks 1992; Wilson *et al.*, 1992; Matyasowszky *et al.*, 1995; Pleym and Karl 1996).

When optimally calibrated, statistical downscaling models have been quite successful in reproducing different statistics of local surface climatology (e.g., Wilks, 1992; Von Storch *et al.*, 1993; Hughes and Guttorp, 1994; Noguer, 1994; Zorita *et al.*, 1995; Pleym and Karl, 1995). Climate change scenarios have been generated through the application of statistical downscaling models (Von Storch *et al.*, 1993; Gyalistras *et al.*, 1994; Zorita *et al.*, 1995; Pleym and Karl; 1996). Confirming an early study by Wigley *et al.* (1990), these new experiments showed that, in complex physiographic settings, local temperature and precipitation change scenarios generated using downscaling methods were significantly different from, and had a finer spatial scale structure than, those directly interpolated from the driving GCMs. As an example, Pleym and Karl (1996) indicated that the application of the Climatological Prediction by Model Statistics (CPMS) to output from a doubled CO_2 GCM equilibrium run produced a decrease in simulated warming of several degrees at different Norwegian locations compared to those produced by the driving GCM.

A unique application of a statistical downscaling method

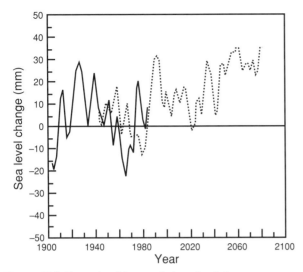

Figure 6.33: Example of downscaled sea level change at a Japanese tidal gauge site as derived from observed sea level pressure fields (solid line, from 1900 to 1988) and from simulated sea level pressure anomalies (dashed line, from "1935" to "2085"). Only the sea level component controlled by the anomalous large-scale circulation is affected by the downscaling procedure. Units are mm. From Maochang *et al.*, 1995.

has been the attempt to compute coastal sea level change at a number of Japanese coastal locations via correlations with large-scale circulation patterns (Maochang *et al.*, 1995, Figure 6.33). New techniques for the disaggregation of precipitation over mountainous terrain were proposed by Barros and Lettenmaier (1993), Miller (1994), Leung and Ghan (1995) and Goyette and Laprise (1995), in which dynamical and/or physical processes are calculated over a high resolution sub-grid of a GCM grid box without use of a full atmospheric limited area model.

6.6.2.2 Regional modelling

The (one-way) nested modelling technique has been increasingly applied to climate change studies in the last few years. This technique consists of using output from GCM simulations to provide initial and driving lateral meteorological boundary conditions for high resolution Regional Climate Model (RegCM) simulations, with no feedback from the RegCM to the driving GCM. Hence, a regional increase in resolution can be attained through the use of nested RegCMs to account for sub-GCM grid scale forcings. The most relevant advance in nested regional climate modelling activities was the production of continuous RegCM multi-year climate simulations. Previous regional climate change scenarios were mostly produced using samples of month-long simulations (Giorgi *et al.*, 1992; Marinucci and Giorgi 1992). The primary

improvement represented by continuous long-term simulations consists of equilibration of model climate with surface hydrology and simulation of the full seasonal cycle for use in impact models. In addition, the capability of producing long-term runs facilitates the coupling of RegCMs to other regional process models, such as lake models (Hostetler *et al.*, 1993, 1994), dynamical sea ice models (Lynch *et al.*, 1995) and possibly regional ocean (or coastal) and ecosystem models (Pielke *et al.*, 1992; Giorgi 1995).

Continuous month- or season-long to multi-year experiments for present day conditions with RegCMs driven either by analyses of observations or by GCMs were generated by Giorgi *et al.* (1993, 1994), Chen *et al.* (1994), Copeland *et al.* (1994, 1995), Bates *et al.* (1995) and Leung and Ghan (1995) for North American regions, Kanamitsu and Juang (1994), Liu *et al.* (1994), Hirakuchi and Giorgi (1995), Lal *et al.* (1995c) and Sasaki *et al.* (1995) for Asian regions, Cress *et al.* (1994), Machenauer *et al.* (1994), Jones *et al.* (1995), Marinucci *et al.* (1995), Podzun *et al.* (1995) and Luthi *et al.* (1996) for Europe, McGregor and Walsh (1993, 1994) and Walsh and McGregor (1995) for Australia, and Semazzi *et al.* (1994) for the Sahelian region. Fewer climate change experiments have been conducted to date. Equilibrium regional climate change scenarios due to doubled CO_2 concentration were produced by Giorgi *et al.* (1994) for the continental USA, McGregor and Walsh (1994) for Tasmania, Hirakuchi and Giorgi (1995) for Eastern Asia, and Jones *et al.* (1996) for Europe.

In the experiments mentioned above, the model horizontal grid point spacing varied in the range of 15 to 125 km and the length of runs from one month to 10 years. In addition, different nesting techniques were used, from the standard lateral boundary relaxation procedure (e.g., Giorgi 1990) to a newer spectral nesting procedure (Kida *et al.*, 1991) in which the GCM forces the low wavenumber component of the fields throughout the regional domain and the RegCM calculates the high wavenumber component.

The main results of the validation and present day climate experiments with RegCMs can be summarised in the following points:

(i)   When driven by analyses of observations, RegCMs simulated realistic structure and evolution of synoptic events. Averaged over regions of the order of 10^4 – 10^6 km^2 in size, temperature biases were mostly in the range of a few tenths of °C to a few °C and precipitation biases were mostly in the range of 10%–40% of observed values. The biases generally increased as the size of the region decreased.

(ii) The RegCM performance was critically affected by the quality of the driving large-scale fields, and tended to deteriorate when the models were driven by GCM output, mostly because of the poorer quality of the driving large-scale data compared to the analysis data (e.g., position and intensity of storm tracks).

(iii) Compared to the driving GCMs, RegCMs generally produced more realistic regional detail of surface climate as forced by topography, large lake systems, or narrow land masses. Examples of the additional detailed produced by nested RegCMs in the simulation of precipitation spatial patterns is given in Figure 6.34a,b for the USA at 60 km grid point spacing (Giorgi *et al.*, 1994) and Figure 6.35a–c for Great Britain at 50 km grid point spacing (Jones *et al.*, 1995), respectively. However, the validation experiments also showed that RegCMs can both improve and degrade aspects of regional climate compared to the driving GCM runs, especially when regionally-averaged (Table 6.5).

(iv) Overall, the models performed better at mid-latitudes than in tropical regions.

(v) The RegCM performance improved as the resolution of the driving GCM increased, mostly because the GCM simulation of large-scale circulation patterns improved with increasing resolution.

(vi) Seasonal as well as diurnal temperature ranges were simulated reasonably well.

(vii) An important problem in the validation of RegCMs has been the lack of adequately dense observational data, since RegCMs can capture the fine structure of climate patterns. This problem is especially relevant in mountainous areas, where often only a relatively small number of high elevation stations are available.

As examples of results obtained from the application of nested RegCMs to climate change simulation, Table 6.5 presents the average changes in temperature and

Table 6.5: Change in surface air temperature over 4 regions, and precipitation over 5 regions, due to doubling of carbon dioxide concentration, and difference between control run and observed values (bias), as simulated with nested RegCM and driving GCM runs Pacific North-west (Gioirgi et al., 1994) – 60-km gridpoint spacing; Great Lakes (Giorgi et al., 1994) – 60-km gridpoint spacing; Japan (Hirakuchi and Giorgi, 1995) – 50-km gridpoint spacing; Europe (Jones et al., 1996) – 50-km gridpoint spacing; Tasmania (McGregor and Walsh, 1994) – 60-km gridpoint spacing.
The precipitation change is expressed as percentage of control value, the bias in percentage of observed value.

TEMPERATURE (°C)

		Pacific North West		Great Lakes		Japan		Europe	
		winter	summer	winter	summer	winter	summer	winter	summer
RegCM	Change	4.7	3.7	5.6	3.9	6.5	4.9	4.6	3.4
	Bias	1.1	−1.6	6.2	3.9	−1.9	−2.0	−0.9	−1.1
GCM	Change	5.1	3.8	6.6	4.0	7.9	5.4	4.9	3.9
	Bias	1.6	−1.5	4.8	3.8	−3.0	−2.5	−1.7	−0.9

PRECIPITATION (%)

		Pacific North West		Great Lakes		Japan		Europe		Tasmania	
		win	sum	win	sum	win	sum	win	sum	win	sum
RegCM	Change	25	24	5	34	37	11.0	16	26	22	30
	Bias	31	21	−55	−31	−9	−34.4	17	6	26	70
GCM	Change	33	4	19	20	12	−1.3	17	6	14	−21
	Bias	82	152	14	38	34	−7.1	−8	−10	−49	−18

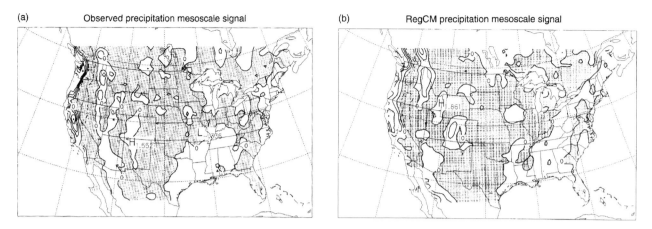

Figure 6.34 : Winter precipitation mesoscale signal over continental USA. (a) Observed and (b) Nested RegCM 3.5-year present day run. The mesoscale signal, which is a measure of the effect of sub-GCM grid scale forcings, is defined as the full field minus the large-scale component of the field, which is obtained through a spatial averaging procedure (see Giorgi *et al.*, 1994). Units are mm/day. Shading indicates negative values. From Giorgi et al., 1994.

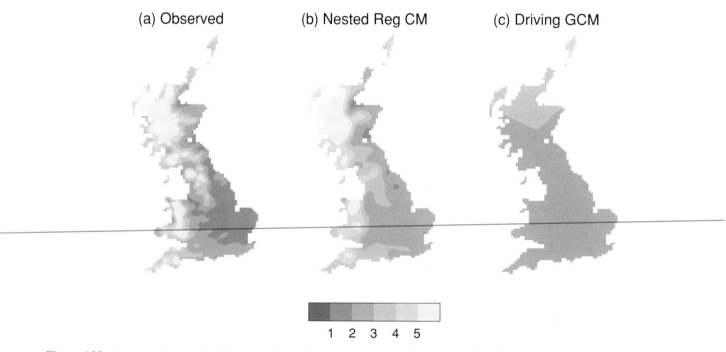

Figure 6.35: Average winter precipitation over Great Britain. (a) Observed; (b) Nested RegCM 10-year present day run and (c) driving GCM 10-year present day run. Units are mm/day. From Jones *et al.*, 1995.

precipitation induced by doubling carbon dioxide concentration over 5 regions as simulated by the nested RegCM and corresponding driving GCM equilibrium runs of Giorgi *et al.* (1994) (north-western USA and Great Lakes), McGregor and Walsh (1994) (Tasmania), Hirakuchi and Giorgi (1995) (Japan), and Jones *et al.* (1996) (western Europe). Also shown are the temperature and precipitation biases for the control runs calculated with respect to climatological observations. Each region is

characterised by complex topography and/or the presence of narrow land masses and large lake systems which were captured by the nested RegCM grids (50- to 60-km resolution) but lost at the GCM resolutions (~300- to 500-km).

It should be stressed that, since the nested modelling technique is still in its development and evaluation stages, the experiments of Table 6.5 were not intended to provide actual climate change projections, but rather to study the

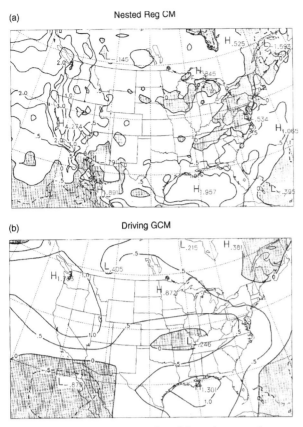

Figure 6.36: Difference between $2 \times CO_2$ and present day average winter precipitation over continental USA. (a) Nested RegCM and (b) driving GCM. Units are mm/day. Shading indicates negative values. From Giorgi *et al.*, 1994.

effects of high resolution, sub-GCM grid scale forcings as simulated by the nested RegCMs. In addition, the data of Table 6.5 cannot readily be compared with those of Figure 6.32 because they refer to equilibrium simulations under $2 \times CO_2$ forcing and not to transient simulations.

For temperature, the differences between RegCM- and GCM-simulated region-averaged change scenarios were in the range of 0.1 to 1.4 °C. The largest differences occurred in winter over the Great Lakes and Japan, because the lakes and Japanese land masses were more finely resolved by the RegCM than the GCM grids. For precipitation, the differences between RegCM and GCM region-averaged scenarios were more pronounced than for temperature, in some instances by one order of magnitude or even in sign. This was the result of the combined contributions of the different resolution of surface forcing and atmospheric circulations and, in some instances, the different behaviour of model parametrizations designed for the fine and coarse resolution models.

Examples of the regional detail included in RegCM scenarios are given in Figures 6.36a, b and 6.37a, b, for the

experiments of Giorgi *et al.* (1994) and Jones *et al.* (1996), respectively. In both cases (winter results), the large-scale patterns of precipitation change were similar in the nested and driving models, as they were mostly determined by changes in large-scale circulations simulated by the GCM, but the sub-regional detail of the scenarios in the GCM and RegCM simulations was often quite different, not only in magnitude but also in sign. For example, in the USA RegCM experiment, the simulated precipitation increase was largest in a relatively narrow region over the western USA coastal areas, whereas in the GCM the areas of pronounced precipitation increase extended farther inland. This was due to the precipitation shadowing effect of the coastal ranges which were represented by the high resolution RegCM grid but were missed by the GCM grid. Because of mountain shadowing, areas of negative change were found east of the Rockies in the RegCM over regions in which the driving GCM produced a positive change. A similar rain shadowing effect in the Alpine and Pyrenees regions was found in the experiment of Jones *et al.* (1996) (Figure 6.37a, b). Topographical enhancement of the precipitation change signal along the Italian Peninsula, the British Isles and the Carpathian region also occurred.

In summer, differences between RegCM and GCM results were generally more marked than in winter due to the greater importance of local processes (Giorgi *et al.*, 1994, Jones *et al.*, 1996). For example, Hirakuchi and Giorgi (1995) found that the representation of the Japanese Islands in the RegCM substantially affected the simulation of summer precipitation change over the region. Comparison of the biases and $2 \times CO_2$-induced temperature and precipitation changes (Table 6.5) showed that while the simulated temperature changes were generally larger than the corresponding biases, the precipitation changes were generally of the same order of, or smaller than, the precipitation biases.

Of relevance for the simulation of regional climate change is the development of a variable resolution global model recently reported by Deque and Piedelievre (1995). They compared three 10-year long runs of different resolutions: T42 (equivalent resolution of ~300 km), T106 (equivalent resolution of ~120 km), and variable resolution configuration centred over Europe (equivalent resolution of about 50 km over Europe and 500 km at the antipodes), with SSTs specified from observations for 1979 to 1988. They demonstrated that the climate simulation over the European region improved as the model resolution increased, and that the variable resolution model produced better results than the T106 model over Europe. When compared to long-term climatology, in the variable

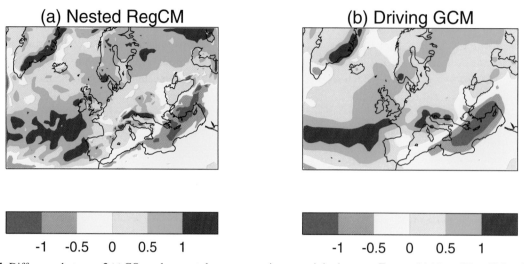

Figure 6.37: Difference between $2 \times CO_2$ and present day average winter precipitation over Europe (a) Nested RegCM and (b) driving GCM. Units are mm/day. From Jones *et al.*, 1996.

resolution experiment, seasonal temperature biases were less than a few degrees and precipitation biases were in the range of ~–45% to +20% of observations over different European sub-regions. When averaged over the entire European region, seasonal and annual precipitation biases were only a few per cent of observed values.

Although computationally very demanding, the variable resolution modelling approach can be used to produce climate change scenarios in "time slice" mode, i.e., performing snap-shot integrations several years in length with SSTs provided at times of different CO_2 levels by a coarse resolution GCM simulation of transient climate change . Time slice integrations are also under way at MPI and LMD with spectral resolutions of up to T106.

6.6.3 Conclusions

Analysis of surface air temperature and precipitation results from regional climate change experiments carried out with AOGCMs indicates that the biases in present day simulations of regional climate change and the inter-model variability in the simulated regional changes are still too large to yield a high level of confidence in simulated change scenarios. The limited number of experiments available with statistical downscaling techniques and nested regional models has shown that complex topographical features, large lake systems, and narrow land masses not resolved at the resolution of current GCMs significantly affect the simulated regional and local change scenarios, both for precipitation and (to a lesser extent) temperature. This adds a further degree of uncertainty in the use of GCM-produced scenarios for impact

assessments. In addition, most climate change experiments have not accounted for human-induced landscape changes and only recently has the effect of aerosols been investigated (see Section 6.2). Both these factors can further affect projections of regional climate change.

Compared to the global scale changes due to doubled CO_2 concentration discussed in Section 6.2 of this report, the changes at the $10^4 - 10^6$ km^2 scale derived from transient AOGCM runs and nested model runs are greater. Considering all models, at the $10^4 - 10^6$ km^2 scale, temperature changes due to CO_2 doubling varied between +0.6°C and +7°C and precipitation changes varied between –35% and + 50% of control run values, with a marked inter-regional variability. Thus the inherent predictability of climate diminishes with reduction in geographical scale. The greatest model agreement in the simulated precipitation change scenarios was found over the South East Asia (~–1% to +30%), Northern Europe (~–-9% to +16%), Central North America (~–7% to +5%) and East Asia (~+0.1 to +16%) regions in summer and Southern Europe (~–2% to +29%), Northern Europe (~+5% to +25%) and East Asia (~+0.5% to +18%) in winter. For temperature, the greatest model agreement in simulated warming occurred over Australia in summer (~+1.65°C to +2.5°C when excluding one outlier) and the Sahel in winter (~+1.8°C to +3.15°C when excluding one outlier). Regardless of whether flux correction was used, the range of model sensitivities was less than the range of biases, which suggests that models produce regional sensitivities that are more similar to each other than their biases.

The latest regional model experiments indicate that high

resolution information, of the order of a few tens of kilometres or less, may be necessary to achieve high accuracy in regional and local change scenarios in areas of complex physiography. In the last few years, substantial progress has been achieved in the development of tools for enhancing GCM information. Statistical methods were extended from the monthly/seasonal to the daily time-scale and nested model experiments were extended to the multi-year time-scale. Also, variable resolution and high resolution global models have become available for use in time-slice mode. While RegCMs allow climate sensitivity experiments to be run at a higher regional resolution, variable and high resolution global models can be used to study possible feedbacks of mesoscale forcings on the general circulation.

Regional modelling techniques, however, rely critically on the GCM performance in simulating large-scale circulation patterns at the regional scale, because these are a primary input to both empirical and physically based regional models. Although the regional performance of coarse resolution GCMs is still somewhat poor, there are indications that features such as positioning of storm track and jet stream core are better simulated as the model resolution increases (e.g., Hurrell *et al.,* 1993). The latest nested GCM/RegCM and variable resolution model experiments, which employed relatively high resolution GCMs and were run for long simulation times (up to 10 years) show an improved level of accuracy. Therefore, as a new generation of higher resolution GCM simulations become available, it is expected that the quality of simulations with regional and local downscaling models will also rapidly improve. In addition, the movement towards coupling regional atmospheric models with appropriately scaled ecological, hydrological, and mesoscale ocean models will not only improve the simulation of climatic sensitivity, but also provide assessments of the joint response of the land surface, atmosphere and/or coastal systems to altered forcings.

6.7 Reducing Uncertainties, Future Model Capabilities and Improved Climate Change Estimates

6.7.1 Recent Progress and Anticipated Climate Model Improvements

6.7.1.1 Improvements in the modelling of clouds and associated radiative processes

The single largest uncertainty in determining the climate sensitivity to either natural or anthropogenic changes are clouds and their effects on radiation and their role in the hydrological cycle. Although there are many important

unresolved issues relating to the basic physics of cloud-radiation interactions and their parametrization in climate models, even perfect parametrizations of radiation and cloud optical properties cannot produce realistic radiative fluxes and heating rates unless they are provided with a realistic distribution of cloudiness. At the present time, weaknesses in the parametrization of cloud formation and dissipation are probably the main impediment to improvements in the simulation of cloud effects on climate.

Efforts to overcome this problem have focused on the introduction of cloud microphysics into atmospheric GCMs (e.g., Sundqvist 1978; Le Treut and Li 1988; Smith, 1990; Ose 1993; Senior and Mitchell, 1993; Fowler *et al.,* 1996). There are many difficulties. The basic microphysical processes themselves are imperfectly understood. In addition, the large spatial scale of GCM grid boxes means that microphysical processes occur primarily within sub-grid regions, such as individual cloud cells, whose properties must be determined somehow. For example, according to some microphysics parametrizations, the conversion of small cloud droplets to raindrops occurs when the local small-drop concentration exceeds some threshold, but in a GCM the relevant local concentration is not known; only the generally much smaller grid-box-mean is available.

The distinction between liquid and ice phases is also of great importance for the inclusion of the commonly observed feature of supercooled water, and hence coexistence of liquid and ice particles in clouds. The difference in saturation vapour pressure then makes the ice particles grow at the expense of the liquid ones (the Bergeron-Findeisen mechanism). This mechanism enhances not only the *in situ* rate of release of precipitation, but subsequently also that in layers beneath, due to an enhanced coalescence effect as well. Consequently, whether models consider the ice phase or not has a pronounced impact on the resulting water content throughout the cloud depth (Sundqvist, 1993), and hence on the optical quality of the cloud.

Furthermore, the form (rain or snow) in which precipitation reaches the Earth's surface, is influenced by the ice and liquid microphysical processes that generate the precipitation. Hence, these processes affect the albedo of land areas. This may be an important factor in positioning and possibly moving the snow line (glacier borders). Consequently, there may be a delicate feedback from enhanced mid-latitude precipitation that is inferred to accompany a warming climate.

The distribution of cloud particle sizes is important because it affects both microphysical processes (e.g., precipitation and collection) and cloud optical properties, particularly in the solar part of the spectrum. In order to predict the particle size distribution, even crudely, it is necessary to have information about the availability of cloud condensation nuclei (CCN). Recently there have been some attempts to predict the distribution of CCN in climate models (Ghan *et al.* 1993, 1995).

Cloud-scale dynamical processes are also critically important to determine cloud formation and dissipation. The best-studied cloud dynamical processes are those associated with deep convection, for which there are several competing parametrizations. Recently there seems to be some convergence of these towards a "mass flux" formulation, although the relative merits of closures based on buoyancy and moisture convergence are still being actively debated (e.g., Tiedtke 1993). Convective clouds themselves have only modest impacts on radiation, but convective detrainment produces extensive stratiform clouds, including cirrus and anvils, which drastically affect the Earth's radiation budget (e.g., Randall 1989).

There has also been considerable work on turbulence in stratiform clouds, particularly in the boundary layer (e.g., Lilly 1968; Bretherton, 1994), with some attempts to apply these ideas to parametrizations for climate models (e.g., Suarez *et al.,* 1983).

Cloud-scale dynamics are important not only for cloud formation and dissipation, but also for the geometrical shapes of the clouds, which matter for radiative transfer (e.g., Harshvardhan and Weinman 1982).

In summary, we can anticipate improvements in the parametrization of cloud microphysics, including aerosol effects; cloud dynamics, particularly as it relates to the interactions of convective and stratiform clouds; and macroscopic cloud optical properties, particularly as they relate to sub-grid scale cloud morphology. However, these improvements will only slowly lead to a reduction in the range of climate sensitivity or need for flux adjustments. There is a great need for observations of cloud-scale dynamics and of the radiative properties of clouds, so that the parametrizations of the physical processes can improve.

6.7.1.2 *Improvements in the ocean component of climate models*

It is clear from this document that the term *climate model* has evolved from the mixed-layer ocean – atmosphere models of IPCC (1990) to fully coupled atmosphere-ocean models. In the future we should expect significant improvements in the oceanic component as the resolution increases, with increasing availability of computing resources, and as more and more sophisticated sub-grid scale parametrizations are developed and incorporated into these models. More rigorous validation of global models is becoming possible due to new sources of data such as WOCE hydrography, heat, salt and mass flux measurements over sills and through straits, tracer data, moored arrays, repeat sections and satellite-derived data such as sea level elevation. These new data sets and those which will be collected over the next few years will allow for better validation of both the mean climatic state of the ocean as well as the variability (e.g., ENSO, decadal variability) about this mean state.

Most groups are currently planning to increase the resolution of their ocean models towards a 1° latitude-longitude grid. This resolution is still not capable of resolving mesoscale eddies, but is able to better resolve straits, passages, land-sea contrasts and boundary currents which are known to be important in determining the oceans' role in climate (see Weaver and Hughes, 1992, for a review). Fully eddy-resolving climate models are still perhaps a decade or more away from being used to study climate change. Before they are ready, questions of spin-up, initialisation and climate drift must also be addressed.

Recent advances in sub-grid scale parametrizations of mixing associated with mesoscale eddies (Danabasoglu *et al.,* 1994; Gent *et al.,* 1995), of convection (Send and Marshall, 1994), of topography-eddy stress interactions (Holloway, 1992) and of flow over sills (Döscher *et al.,* 1994) are making their way into ocean GCMs. Whether they survive the test of time or whether they are superseded by more realistic parametrizations is yet to be seen. In addition, alternative model formulations (e.g., isopycnal or sigma co-ordinates) and more sophisticated numerical techniques are becoming more frequently used in the ocean modelling community. The next few years should see significant advances in ocean modelling.

The thermohaline circulation in the world oceans is an important feature which affects the coupled system. It transports large amounts of heat and water from the Southern Hemisphere to high latitudes of the North Atlantic Ocean. It also plays an important role in determining the vertical structure of the world oceans (Manabe and Stouffer, 1988) and it may respond to any climate change: many models show a weakening in the thermohaline circulation in response to increasing greenhouse gases (Section 6.2). It is very important that observational programs be developed to study this global circulation because lack of observations currently limits confidence in the simulation of its current and possible future structure.

6.7.1.3 Flux adjustment and climate models

As discussed in Section 5.2.2, when an atmospheric model is coupled to an ocean model and the implied heat and salt transports between the two models are incompatible, a drift in the climate system must occur (e.g., Weaver and Hughes, 1996). This drift adversely affects both the atmosphere and the ocean. The use of flux adjustments may stabilise this drift, but they have the undesirable effect of being physically unfounded and often of larger magnitude (in places) than the climatological mean fluxes (Manabe and Stouffer, 1988; Sausen *et al.*, 1988). As shown by ocean modelling studies (e.g., Weaver *et al.*, 1993), ocean models are capable of sustaining internal modes of oscillation under a specified freshwater flux boundary condition with thermal damping or a specified thermal flux (Greatbatch and Zhang, 1995). In coupled atmosphere-ocean simulations, any variability found when flux adjustments are used is difficult to interpret.

It has recently been suggested (e.g., Gleckler *et al.*, 1994) that the discrepancies in the implied and observed ocean heat transports lies in the parametrization of clouds (especially marine stratocumulus) within the atmospheric component of the coupled models. As these and other parametrizations are improved (see Section 6.7.1.1) the need for flux adjustments should be reduced.

6.7.1.4 Long-time integrations

With the recent increase in computer power, centennial-millennial climate integrations (like the ones in Manabe and Stouffer, 1993, Stouffer *et al.*, 1994) will become routine. It will be possible to examine transient changes in climate variability associated with climate change. One of the greatest challenges of the next decade will be for coupled atmosphere-ocean models to duplicate (statistically) the records found in the Greenland, Antarctic and other ice cores. This would be done through the statistical interpretation of an ensemble of integrations under time-dependent Milankovitch cycle solar forcing. Such simulations of past climates may provide the best validation of coupled models used to predict future changes in climate. However, this effort will need improved reconstructions and understanding of past climate forcings (CO_2, aerosols, etc.) as well as inferred temperature changes.

6.7.1.5 Sea ice model improvements

Sea ice in global coupled models has typically been approximated as a motionless thermodynamic slab. In most cases, the thermodynamic calculations include a surface energy budget and heat conduction through the ice;

however, details of parametrizations (for example, surface albedo as a function of temperature) vary substantially between models. Sea ice models need an improved understanding of leads and marginal ice zone processes. Sea ice albedo needs better treatment and greater attention should be given to the differences between processes in the two hemispheres. One-dimensional models are currently being used to investigate improvements in sea ice thermodynamics and atmosphere/ice energy exchanges (e.g., Ebert and Curry, 1993).

Inclusion of sea ice dynamics -advection and deformation – is necessary to account for the fresh water provided by melting sea ice advected from the Arctic into the Greenland and Norwegian Seas, and to reproduce dynamic/thermodynamic feedback effects. For example, experiments with stand-alone sea ice models (e.g., Lemke *et al.*, 1990; Hibler, 1984) have shown that thermodynamic-only models are more sensitive to changes in thermal forcing than those which include dynamics. Similarly, recent GCM experiments by Pollard and Thompson (1994) showed that inclusion of sea ice dynamics produced more realistic ice extent and reduced a model's CO_2-induced warming, particularly around Antarctica. Including a more realistic dynamical formulation is the principal improvement anticipated or being made in the sea ice component of global climate models. Examples include the use of the Flato and Hibler (1992) ice dynamics scheme in models being developed at NCAR (Meehl and Washington, 1995; Washington and Meehl, 1996), GFDL, UKMO and CCC, the use of Mellor and Kantha (1989) at MRI, and the use of the more sophisticated (e.g., Hibler, 1979) ice dynamics formulation in the ECHAM/OPYC and CSIRO coupled models. However, as the treatment of sea ice dynamics becomes more realistic, shortcomings in other components of the coupled model, like wind and under-ice current fields which drive ice motion, become more important. Continued improvement is therefore an iterative process.

6.7.1.6 Improvements in land-surface processes/modelling

Many improvements in modelling land-surface (including glaciers and ice sheets) processes have yet to make their way into climate models (see Chapter 9 for a more comprehensive discussion). These improvements involve advances in both *structure* and *function modelling*. In *structure modelling*, changes in land type (e.g., forest to grass) are calculated in response to changes in climate. In *function modelling*, changes in the function of a particular land type (e.g., evapotranspiration/albedo) are predicted with changes in climate. Land-surface process schemes/models are, however, most often developed and validated on scales much smaller than the grid scale of a climate model and may need further modification in order to be successfully incorporated.

Most climate modelling groups are incorporating, or will soon do so, more sophisticated land-surface representations into their GCMs in order to include processes which have the potential to influence climate sensitivity. The ultimate aim of this work is to quantify the feedbacks associated with changes in vegetation structure and function. Two recent intercomparisons have highlighted the current uncertainty in simulating evapotranspiration over land (PILPS – Henderson-Sellers *et al.*, 1995b; AMIP – Robock *et al.*, 1995). This uncertainty highlights the need for better observations and parametrizations of land-surface processes.

Land-surface schemes used in GCMs are concerned with modelling the fluxes of heat, water and momentum from a given land cover type. There have been relatively few attempts at modelling changes in vegetation type, even though such changes may produce significant climatic feedbacks (e.g., Henderson-Sellers and McGuffie, 1995). Recent work at the Hadley Centre has attempted to include these feedbacks by coupling a vegetation model to the Hadley Centre GCM. The vegetation model predicts vegetation properties (such as leaf area index, height and canopy conductance) based on climate, soil carbon and nitrogen contents, and the atmospheric CO_2 concentration. The coupling to the GCM is two-way, with the GCM providing climatic data to the vegetation model and the vegetation model predicting the vegetation properties from which the GCM land-surface parameters can be derived. Such an approach allows mutually consistent climate-vegetation equilibria to be computed and the vegetation feedbacks to be diagnosed. There remains the outstanding problem of modelling the dynamics of vegetation change, which is very dependent on little-known rates of natural and anthropogenic disturbances (e.g., fires, wind-blows, etc.).

An important land/surface process which may influence climate is changes in the areas presently covered by permafrost. Large amounts of greenhouse gases are currently trapped in permafrost so that any thawing will release these gases into the atmosphere, enhancing the greenhouse effect. This process has yet to be incorporated into a coupled climate model.

6.7.1.7 *Improvements in radiation computation*
A basic element of GCMs is the radiation scheme used to compute the short-wave and long-wave fluxes in the model. Ma *et al.* (1994) have shown that by replacing an older radiation scheme with an improved, more accurate scheme, the simulation of their coupled model greatly improved. Betts *et al.* (1993) compared the short-wave calculations of the ECMWF model with the observed fluxes obtained over Kansas during the First ISLCP Field Experiment (FIFE) from the summer and fall of 1987 for clear-sky conditions. They showed that the model systematically overestimated the incoming solar flux by 5–10% at the Earth's surface. The Japan Meteorological Agency also found a similar error in their model. They ascribe part of this error to neglect of absorption in the short-wave by either water vapour or aerosols. The improvement of the radiation schemes used in these models will continue, especially by the inclusion of vertical concentration profiles of trace substances like tropospheric aerosols and tropospheric ozone. In this regard, the comparison of the fluxes computed by the schemes to so-called line-by-line calculations will continue to be very useful.

6.7.2 *Global Carbon Cycle Models as Part of Climate Models*
The response of the carbon cycle to climate change is strongly conditional on other elements of the system, and its inclusion in comprehensive climate models is just beginning (see Chapters 9 and 10). Thus far the focus has remained on the response of various carbon reservoirs to increasing atmospheric CO_2 in a constant climate regime.

In the ocean, attempts have been made to model the response of the carbon cycle to changes in atmospheric CO_2 (e.g., Sarmiento *et al.*, 1992). These are very difficult to verify directly since the observed anomaly in stored carbon in the ocean is very small. The models are calibrated for observed uptake of carbon-14 from nuclear testing. Alternatively one can model the natural carbon cycle and perturb it with anomalous inputs. One such model is described by Maier-Reimer (1993). Such models require inclusion of a great many more processes such as biological carbon fixation and remineralisation and sedimentary interaction but also provide many more fields amenable to observational verification. Currently, these models also use highly simplified representations of biological processes, often based on forcing nutrient distributions at the surface to fit observations and empirically determined ratios of elements in tissue. Sensitivity studies such as those of Sarmiento and Orr (1991) suggest that changes in biological productivity may have a significant effect on carbon sequestration in the ocean.

In the case of terrestrial response the case is a perhaps

easier to model. Here there is no underlying ocean circulation which is not well observed. Attempts to generalise the physiological and ecological knowledge gained on individual and regional scales to the global domain are in their early stages. Perhaps the most complete attempt so far in a global study is that of Melillo *et al.* (1993).

As already mentioned, progress in global carbon cycle modelling is strongly dependent on progress in other areas of global modelling. Ocean carbon cycle models depend on the flow field of the underlying ocean general circulation model (OGCM) to describe entrainment of chemical species such as carbon into the thermocline. Terrestrial ecosystem models require specification of heat, moisture and radiative fluxes at the Earth's surface. There are, of course, areas beyond physical modelling which need improvement. In the oceans these include better experimental constraint of gas exchange coefficients in actual conditions, understanding of processes controlling isotopic fractionation, the partition of biological production between dissolved and particulate forms and the above mentioned treatment of biology.

Improvement in models which synthesise so many components awaits improvements in the process models but other developments can be expected in the more immediate term. For the oceans this means particularly coupling carbon cycle models to models of transient climate change to assess how circulation changes may affect carbon storage. For the terrestrial models , the most immediate improvements are likely to arise from more mechanistic representations of the interactions among carbon, water and nutrient fluxes and the extent to which these are modified by external factors including nitrogen deposition (Chapter 9).

Many groups are in the process of developing, or have recently developed, carbon cycle models which are to be used in conjunction with their AOGCM simulations. Since the gradients of carbon in the atmosphere are not strong, most models in the first instance will represent CO_2 as a function of time only, changes in global mean CO_2 concentration being the calculated result of modelled land/air and ocean/air interactions.

The importance of the "biological pump" is still not clear. It is clear that without marine plant and animal life the distribution of carbon in the oceans would be completely different from that of today. What is less clear is whether or not the biological pump, on the time-scales of interest here, would have any significant interaction with increasing atmospheric CO_2 (e.g., Broecker, 1991). For example, a dramatic change in marine biology would be

needed to create a change of more than 10 ppm in the atmosphere (e.g., Heinze *et al.*, 1991). Studies planned by several groups are designed to show whether the marine biological feedback is large enough to warrant its inclusion on-line in climate change integrations.

6.7.3 Climate Models Including Tropospheric Chemistry

The first AOGCM integrations to include, albeit very crudely, the effects of tropospheric aerosols were completed recently (Section 6.2.3). Only the so-called direct radiative effects of the aerosols were included in a simplistic way. This represents a first attempt to include some effects of tropospheric chemistry in an AOGCM. To date, no AOGCM experiments have studied the effects on climate of changes in tropospheric chemistry, although many institutions are planning such experiments in the near future. The main problem is how to include large, complex tropospheric chemistry models in already very large and complex atmosphere-ocean models. Also, there is a great deal of uncertainty in the estimation of future gas and aerosol concentrations . Most short term efforts will be dedicated to the inclusion of sulphates, dust and ozone into AOGCMs. With regard to dust and sulphates it is important to predict not only their distribution but also their size. For ozone, it is important to predict not only the horizontal distribution but, more importantly, the vertical distribution (Ramaswamy *et al.,* 1992).

Other atmospheric trace gases (e.g., CFCs, CH_4, N_2O) could be treated, as a first step, in a manner similar to carbon dioxide. The inclusion of methane and ozone in complex climate models will be a major challenge and will involve the coupling of GCMs to interactive tropospheric chemistry models.

References

Bardossy, A. and E.J. Plate, 1992: Space-time models for daily rainfall using atmospheric circulation patterns. *Water Resource. Res*, **28**, 1247–1259.

Barros, A. and D.P. Lettenmaier, 1993: Dynamic modeling of the spatial distribution of precipitation in remote mountainous areas. *Mon. Wea. Rev.*, **121**, 1195–1214.

Bates, G.T. and G.A. Meehl, 1986: The effect of CO_2 concentration on the frequency of blocking in a general circulation model coupled to a simple mixed layer ocean model. *Mon. Wea. Rev.*, **114**, 687–701.

Bates, G.T., S.W. Hostetler and F. Giorgi, 1995: Two-year simulation of the Great Lakes region with a coupled modeling system. *Mon. Wea. Rev.*, **123**, 1505–1522.

Beersma, J.J., 1994: Storm activity over the North Sea and the

Netherlands in two climate models compared with observations. *KNMI, WR* **94–02**.

Bengtsson, L., M. Botzet and M. Esch, 1994: Hurricane-type vortices in a general circulation model, Part II. *MPIM-Report* no. **139**.

Bengtsson, L., M. Botzet and M. Esch, 1995: Hurricane-type vortices in a general circulation model, *Tellus*, **47A**, 175–196.

Betts, A.K., J.H. Ball and A.C.M. Beljaars, 1993: Comparison between the land surface response of the ECMWF model and the FIFE- 1987 data. *Quart. J. R. Met. Soc.*, **119**, 975–1001.

Bhaskaran, B., J.F.B. Mitchell, J. Lavery and M. Lal, 1995: Climatic response of Indian subcontinent to doubled CO_2 concentration. *Intl. J. Climatol.*, **15**, 873–892.

Bretherton, C.S., 1994: A turbulence closure model of marine stratocumulus clouds. Part I: The diurnal cycle of marine stratocumulus during FIRE 1987. Submitted to *J. Atmos. Sci.*

Bretherton, F.P., K. Bryan and J.D. Woods, 1990: Time-dependent greenhouse-gas induced climate change. In: *Climate Change. The IPCC Scientific Assessment*, J.T. Houghton, G.J. Jenkins and J.J. Ephraums (eds.), Cambridge University Press, Cambridge, UK, pp. 173–193.

Broecker, W.S., 1991: Keeping global change honest. *Global Biogeochem. Cycles*, **5**, 191–193.

Campbell, G, T.G.F. Kittel, G.A. Meehl and W.M. Washington, 1995: Low-frequency variability and CO_2 transient climate change. Part 2. EOF analysis. *Global Planet. Change*, **10**, 201–216.

Cao, H-X, J.F.B. Mitchell and J.R. Lavery, 1992: Simulated diurnal range and variability of surface temperature in a global climate model for present and doubled CO_2 climates. *J.Climate*, **5**, 920–943.

Carnell, R.E., C.A. Senior, J.F.B. Mitchell, 1996: An assessment of measures of storminess: Simulated changes in Northern Hemisphere winter due to increasing CO_2. *Clim. Dyn.*, (Accepted)

Chakraborty, B. and M. Lal, 1994: Monsoon climate and its change in a doubled CO_2 atmosphere as simulated by CSIRO9 model. *Terr. Atmos. and Oceanic Sci.*, **5**(4), 515–536.

Chakraborty, B. P.H. Whetton, A.B. Pittock, and M. Lal, 1995a: Assessment of future climatic change over the Indian subcontinent as projected by General Circulation Model experiments: Part I – Comparing control climates with observations. *J. Climate* (submitted).

Chakraborty, B. P.H. Whetton, A.B. Pittock, and M. Lal, 1995b: Assessment of future climatic change over the Indian subcontinent as projected by General Circulation Model experiments: Part II – The greenhouse gas-induced scenarios. *J. Climate*, submitted.

Charlson, R.J., J. Langner, H. Rodhe, C.B. Leovy and S.G. Warren, 1991: Perturbation of the Northern Hemisphere radiative balance by backscattering from anthropogenic sulfate aerosols. *Tellus*, **43B**, 152–163.

Chen, S.-C., J. Roads, H.H.-M. Juang and M. Kanamitsu, 1994: California precipitation simulation in the NMC nested spectral

model: 1993 January event. Symposium to Share Weather Pattern Knowledge, Rocklin, CA, 25 June.

Colman, R.A., S.B. Power, B.J. MacAvaney and R.R. Dahni, 1995: A non-flux-corrected transient CO_2 experiment using the BMRC coupled atmosphere/ocean GCM. *Geophys. Res. Lett.*, **22**, 3047–3050.

Copeland, J.H., T. Chase, J. Baron, T.G.F. Kittel and R.A. Pielke, 1994: Impacts of vegetation change on regional climate and downscaling of GCM output to the regional scale. In: *Regional Impacts of Global Climate Change: Assessing Change and Response at the Scales that Matter.* Proceedings of the 32rd Hanford Symposium on Health and the Environment, Oct 18-21, 1993, Richland, WA.

Copeland, J.H., R.A. Pielke and T.G.F. Kittel, 1995: Potential climatic impacts of vegetation change: A regional modeling study. *J. Geophys. Res.*, In press.

Cress, A., H.C. Davies, C. Frei, D. Luthi and C. Schar, 1994: Regional climate simulations in the Alpine region. LAPETH-32, Atmospheric Sciences ETH, 8093 Zurich, Switzerland, 134 pp.

Cubasch, U., K. Hasselmann, H. Höck, E. Maier Reimer, U. Mikolajewicz, B.D. Santer and R. Sausen, 1992: Time-dependent greenhouse warming computations with a coupled ocean-atmosphere model. *Clim. Dyn.*, **8**, 55–69.

Cubasch, U., G. Meehl and Z.C. Zhao, 1994a: IPCC WG 1 Initiative on Evaluation of Regional Climate Change, Summary Report, 12 pp.

Cubasch, U., B.D. Santer, A. Hellach, G. Hegerl, H. Höck, E. Maier-Reimer, U. Mikolajewicz and A. Stössl, 1994b: Monte Carlo climate change forecasts with a global coupled ocean-atmosphere model. *Clim. Dyn.*, **10**,1–20.

Cubasch, U., G. Hegerl, A. Hellbach, H. Höck, U. Mikolajewicz, B.D. Santer and R. Voss, 1995a: A climate simulation starting in 1935. *Clim. Dyn.*, **11**,. 71–84.

Cubasch, U., J. Waszkewitz, G. Hegerl and J. Perlwitz, 1995b: Regional climate changes as simulated in time-slice experiments. MPI Report 153. *Clim. Change*, **31**, 273–304.

Danabasoglu, G., J.C. McWilliams and P.R. Gent, 1994: The role of mesoscale tracer transports in the global ocean circulation. *Science*, **264**, 1123–1126.

Deque, M. and J.Ph. Piedelievre, 1995: High resolution climate simulation over Europe. *Clim. Dyn.*, **11**, 321–340.

Dignon J. and S. Hameed, 1989: Global emissions of nitrogen and sulphur oxides from 1860 to 1980. *JAPCA (Journal of the Air and Waste Management Association)*, **39**, 180–186.

Döscher, R., C.W. Böning and P. Herrmann, 1994: Response of circulation and heat transport in the North Atlantic to changes in thermohaline forcing in northern latitudes. *J. Phys. Oceanogr.*, **24**, in press.

Drijfhout, S.S., C. Heinze, M. Latif and E. Maier-Reimer, 1996: Mean circulation and internal variability in an ocean primitive equation model, MPI report No. 177; *J. Phys. Oceanogr.*, (in press).

Ebert, E.E. and J.A. Curry, 1993: An intermediate one-dimensional thermodynamic sea ice model for investigating ice-atmosphere interactions. *J. Geophys. Res.*, **98**, 10085–10109.

England, M.H., 1995: Using chlorofluorocarbons to assess ocean climate models. *Geophys. Res. Lett.*, **22**, 3051–3054.

Enting, I.G., T.M.L. Wigley, and M. Heimann, 1994: Future Emissions and Concentrations of Carbon Dioxide: Key Ocean/Atmosphere/Land Analyses, *CSIRO Division of Atmospheric Physics Technical Paper No. 31*, 120 pp.

Fichefet, T. and C. Tricot, 1992: Influence of the starting date of model integration on projections of greenhouse gas induced climate change. *Geophys. Res. Lett.*, **19**, 1771–1774.

Flato, G.M. and W.D. Hibler III. 1992: Modeling pack ice as a cavitating fluid. *J. Phys. Oceanogr.*, **22**, 626–651.

Fowler, A.M. and K.J. Hennessey, 1995: Potential impacts of global warming on the frequency and magnitude of heavy precipitation. *Natural Hazards*, **11**, 283–303.

Fowler, L.D., D.A. Randall and S.A. Rutledge, 1996: Liquid and ice cloud microphysics in the CSU General Circulation Model. Part 1: model description and simulated microphysical processes. *J. Climate*, in press.

Gates, W.L., J.F.B. Mitchell, G.J. Boer, U. Cusbach and V.P. Meleshko, 1992: Climate modelling, climate prediction and model validation. In: *Climate Change 1992. The Supplementary Report to the IPCC Scientific Assessment*, J.T. Houghton, B.A. Callander and S.K. Varney (eds.), Cambridge University Press, Cambridge, UK, pp. 97–134.

Gates, W.L., U. Cusbach, G.A. Meehl, J.F.B. Mitchell and R.J. Stouffer, 1993: An intercomparison of selected features of the control climates simulated by coupled ocean-atmosphere general circulation models. *WMO/TD No 574* (Geneva).

Gent, P.R., J. Willebrand, T.J. McDougall and J.C. McWilliams, 1995: Parameterizing eddy-induced tracer transports in ocean circulation models. *J. Phys. Oceanogr.* **25**, 463–474.

Ghan, S.J., C.C. Chuang and J.E. Penner, 1993: A parameterization of cloud droplet nucleation. Part I: Single aerosol type. *Atmos. Res.*, **30(4)**, 197–221.

Ghan, S.J., C.C. Chuang, R.C. Easter and J.E. Penner, 1995: A parameterization of cloud droplet nucleation. Part II: Multiple aerosol types. *Atmos. Res.*, **36(1–2)**, 39–54.

Giorgi, F., 1990: On the simulation of regional climate using a limited area model nested in a general circulation model. *J. Climate*, **3**, 941–963.

Giorgi, F., 1995: Perspectives for regional Earth System modeling. *Global and Planetary Change*, **10**, 23–42.

Giorgi, F., G.T. Bates and S. Nieman, 1993: The multi-year surface climatology of a regional atmospheric model over the Western United States. *J. Climate*, **6**, 75–95.

Giorgi, F., M.R. Marinucci and G. Visconti, 1992: A $2 \times CO_2$ climate change scenario over Europe generated using a Limited Area Model nested in a General Circulation Model. II: Climate change scenario. *J. Geophys. Res.*, **97**, 10011–10028.

Giorgi, F., C. Shields Brodeur and G.T. Bates, 1994: Regional climate change scenarios over the United States produced with a nested regional climate model: Spatial and seasonal characteristics. *J. Climate*, **7**, 375–399.

Gleckler, P.J., D.A. Randall, G. Boer, R. Colman, M. Dix, V. Galin, M. Helfand, J. Kiehl, W. Lau, X.-Z. Liang, V. Lykossov, B. McAvaney, K. Miyakoda and S. Planton, 1994: Cloud-radiative effects on implied oceanic energy transports as simulated by atmospheric general circulation models. *PCMDI Tech. Rep. No. 15*, Univ. Cal., LLNL, Livermore, CA.

Gordon, H.B. and B.G. Hunt, 1994: Climatic variability within an equilibrium greenhouse simulation. *Clim. Dyn.*, **9**, 195–212.

Gordon, H.B. and S.P. O'Farrell, 1996: Transient climate change in the CSIRO coupled model with dynamical sea-ice. *Mon Wea. Rev.* (in press)

Gordon, H.B., P.H. Whetton, A.B. Pittock, A.M. Fowler and M.R. Haylock, 1992: Simulated changes in daily rainfall intensity due to the enhanced greenhouse effect: implications for extreme rainfall events. *Clim. Dyn.*, **8**, 83–102.

Goyette, S. and R. Laprise, 1995: Numerical investigation with a physically based regional climate interpolator: FIZR.*J. Climate.* (Submitted).

Gray, W.M., 1979: Hurricanes: their formation, structure and likely role in the tropical circulation. In: *Meteorology over the Tropical Oceans.*, Shaw O.B. (ed), Royal Meteorological Society, London, pp 155–218.

Greatbatch, R.J. and S. Zhang, 1995: An interdecadal oscillation in an idealized ocean basin forced by constant heat flux. *J.Climate*, **8**, 81–91.

Gregory, J.M. and J.F.B. Mitchell, 1995: Simulation of daily variability of surface temperature and precipitation over Europe in the current and $2 \times CO_2$ climates using the UKMO climate model. *Quart. J. R. Met. Soc.*, **121**, 1451–1476.

Gyalistras, D, H. Von Storch, A. Fischlin and M. Beniston, 1994: Linking GCM-simulated climatic changes to ecosystem models: Case studies of statistical downscaling in the Alps. *Clim.Res.*, **4**, 167–189.

Haarsma, R.J., J.F.B. Mitchell and C.A. Senior, 1993: Tropical disturbances in a GCM. *Clim. Dyn.*, **8** (5), 247–257).

Hall, N.M.J., B.J. Hoskins, P.J. Valdes and C.A. Senior, 1994: Storm tracks in a high resolution GCM with doubled CO_2. *Quart.J. R. Met. Soc.,* **120**, 1209–1230.

Hansen, J.E., D. Johnson, A. Lacis, S. Lebedeff, P. Lee, D. Rind and G. Russell 1981: Climate impact of increasing atmospheric carbon dioxide, *Science*, **213**, 957–966.

Hansen, J., A. Lacis, R. Ruedy, M. Sato and H. Wilson, 1993: How sensitive is the world's climate? *National Geographic Research and Exploration*, **9**, 142–158.

Hansen, J.E., M. Sato and R. Ruedy, 1995 : Long term changes of the diurnal temperature cycle: implications about mechanisms of global climate change. *Atmospheric Research*, **37**, 175–209.

Harshvardhan and J.A. Weinman, 1982: Infrared radiative transfer through a regular array of cuboidal clouds. *J. Atmos. Sci.*, **39**, 431–439.

Hasselmann, K, R. Sausen, E. Maier-Reimer and R. Voss, 1993:

On the cold start problem in transient simulations with coupled ocean-atmosphere models. *Clim. Dyn.*, **9**, 53–61.

Hasselmann K., L. Bengtsson, U. Cubasch, G.C. Hegerl, H. Rodhe, E. Roeckner, H. von Storch, R. Voss and J. Waszkewitz, 1995: Detection of anthropogenic climate change using a fingerprint method. In: *Proceedings of "Modern Dynamical Meteorology", Symposium in honor of Aksel Wiin-Nielsen, 1995*, P. Ditlevsen (ed.), ECMWF press, 1995.

Heinze, C., E. Maier-Reimer and K. Winn, 1991: Glacial pCO_2 reduction by the world ocean: experiments with the Hamburg carbon cycle model. *Paleoceanogr.*, **6**, 395–430.

Henderson-Sellers, A. and K. McGuffie, 1995: Global climate models and dynamic vegetation changes. *Global and Change Biology*, **1**, 63–76.

Henderson-Sellers A., J. Hoekstra, Z. Kothavala, N. Holbrook, A.-M. Hansen, O. Balachova and K. McGuffie, 1995a: Assessing simulations of daily variability by Global Climate Models for present and greenhouse climates. *Clim. Change* (submitted).

Henderson-Sellers A., A.J. Pitman, P.K. Love, P. Irannejad and T. Chen, 1995b: The project for intercomparison of land surface parameterization schemes (PILPS): Phases 2 and 3. *Bul. Am. Met. Soc.*, **76**, 489–503.

Hennessey, K.J. and A.B. Pittock, 1995: Greenhouse warming and threshold temperature events in Victoria, Australia. *Int. J. Climatol.*, **15**, 591–612.

Hewitson, B. and R.G. Crane, 1992a: Regional-scale climate prediction from the GISS GCM. *Paleogeography, Paleoclimatology, Paleoecology*, **97**, 249–267.

Hewitson, B. and R.G. Crane, 1992b: Large-scale atmospheric controls on local precipitation in tropical Mexico. *Geophys. Res. Lett.*, **19**, 1835–1838.

Hibler, W.D. III. 1979: A dynamic thermodynamic sea ice model. *J. Phys. Oceanogr.*, **9**, 815–846.

Hibler, W.D. III. 1984: The role of sea ice dynamics in modeling CO_2 increases. In: *Climate Processes and Climate Sensitivity*, Geophys. Mon. 29, AGU, 238–253.

Hirakuchi, H. and F. Giorgi, 1995: Multi-year present day and $2 \times CO_2$ simulations of monsoon climate over eastern Asia and Japan with a regional climate model nested in a general circulation model. *J. Geophys. Res.*, **100**, 21105–21126.

Hoffert, M.I., A.J. Callegari and C.-T. Hsieh, 1980: The role of deep sea heat storage in the secular response to climate forcing. *J. Geophys. Res.*, **85**, 6667–6679.

Holland, G.J., 1995: The maximum potential intensity of tropical cyclones. *J. Atmos. Sci.* (submitted.)

Holloway, G., 1992: Representing topographic stress for large-scale ocean models. *J. Phys. Oceanogr.*, **22**, 1033–1046.

Horton, B.H., 1995: The geographical distribution of changes in maximum and minimum temperatures. *Atmos. Res.*, **37**, 101–117.

Hostetler, S.W., G.T. Bates and F. Giorgi, 1993: Interactive coupling of a lake thermal model with a regional climate model. *J. Geophys. Res.*, **98**, 5045–5057.

Hostetler, S.W., F. Giorgi, G.T. Bates and P.J. Bartlein, 1994: Lake-Atmosphere feedbacks associated with paleolakes Bonneville and Lahontan. *Science*, **263**, 665–668.

Hughes, J.P. and P. Guttorp, 1994: A class of stochastic models for relating synoptic atmospheric patterns to regional hydrologic phenomena. *Water Resour. Res.*, **30**, 1535–1546.

Hulme, M., K.R. Briffa, P.D. Jones and C.A. Senior, 1993: Validation of GCM control simulations using indices of daily airflow types over the British Isles. *Clim. Dyn.*, **9**, 95–105.

Hurrell, J.W., J.J. Hack and D.P. Baumhefner, 1993: Comparison of NCAR Community Climate Model (CCM) Climates. *Technical Note NCAR/TN–395+STR*, NCAR, Boulder Colorado.

Ingram, W.J., C.A. Wilson and J.F.B. Mitchell 1989: Modelling climate change: an assessment of sea-ice and surface albedo feedbacks. *J Geophys. Res.*, **94**, 8609–8622.

IPCC (Intergovernmental Panel on Climate Change), 1990: *Climate Change: The IPCC Scientific Assessment*, J.T. Houghton, G.J. Jenkins and J.J. Ephraums (eds.). Cambridge University Press, Cambridge, UK, 365 pp.

IPCC, 1992: *Climate Change 1992: The Supplementary Report to the IPCC Scientific Assessment*, J.T. Houghton, B.A. Callander and S.K. Varney (eds.). Cambridge University Press, Cambridge, UK, 198 pp.

IPCC, 1994: *Climate Change 1994: Radiative Forcing of Climate Change and an Evaluation of the IPCC IS92 Emission Scenarios*, J.T. Houghton, L.G. Meira Filho, J. Bruce, Hoesung Lee, B.A. Callander, E. Haites, N. Harris and K. Maskell (eds.). Cambridge University Press, Cambridge, UK. 339 pp.

Johns, T.C., R.E. Carnell, J.F. Crossley, J.M. Gregory, J.F.B. Mitchell, C.A. Senior, S.F.B. Tett and R.A. Wood, 1996; The second Hadley Centre coupled ocean-atmosphere GCM: Model description, spinup and validation. (Submitted to *Clim. Dyn.*).

Jones, R.G., J.M. Murphy and M. Noguer, 1995: Simulation of climate change over Europe using a nested regional climate model. Part I. Assessment of control climate, including sensitivity to location of lateral boundaries. *Quart. J. R. Met. Soc.* **121**, 1413–1449.

Jones, R.G., J.M. Murphy, M. Noguer and A.B. Keen, 1996: Simulation of climate change over Europe using a nested regional climate model. Part II. Comparison of driving and regional model responses to a doubling of carbon dioxide concentration. (Submitted to *Quart. J. R. Met. Soc.*)

Kanamitsu, M. and H.-M. H. Juang, 1994: Simulation and analysis of an Indian Monsoon by the NMC nested regional spectral model. Extended abstract. International Conference on Monsoon Variability and Prediction. International Centre for Theoretical Physics. Trieste, Italy, 9-13 May.

Karl, T.R., W.-C. Wang, M.E. Schlesinger and R.W. Knight, 1990: A method of relating general circulation model simulated climate to the observed local climate, Part I: Seasonal Statistics. *J. Climate*, **3**, 1053–1079.

Karoly, D.J., J.A. Cohen, G.A. Meehl, J.F.B. Mitchell, A.H. Oort, R.J. Stouffer and R.T. Wetherald, 1994: An example of fingerprint detection of greenhouse climate change. *Clim. Dyn.*, **10**, 97–105.

Karoly, D.J., D. Collins, G.A. Meehl and R.J. Stouffer, 1995: An example of fingerprint detection of greenhouse climate change. *Clim. Dyn.*, **10**, 97–105.

Katz, R.W. and B.G. Brown, 1992: Extreme events in a changing climate: variability is more important than averages, *Climatic Change*, **21**, 289–302.

Keen A.B., 1995: Investigating the effects of initial conditions on the response of the Hadley Centre Coupled Model. *Hadley Centre Internal Note no. 71.*

Keming, C., J. Xiangze, L. Wuyin, Y. Yongquiang, G. Yufu and Z. Xuehong, 1994: Lecture in International Symposium on Global Change in Asia and the Pacific Region (GCAP). 8–10 Aug, 1994, Beijing.

Kida, H., T. Koide, H. Sasaki and M. Chiba, 1991: A new approach to coupling a limited area model with a GCM for regional climate simulation. *J. Met. Soc. Japan*, **69**, 723–728.

Kittel, T.G.F., N.A. Rosenbloom, T.H. Painter, D.E. Schimel and VEMAP modelling participants, 1995: The VEMAP integrated database for modeling United States ecosystem/vegetation sensitivity to climate change. *Global Ecology and Biogeography Letters* (in press).

Kittel, T.G.F., F. Giorgi and G.A. Meehl, 1996: Regional intercomparison of coupled atmosphere-ocean general circulation model climate experiments. *Geophys. Res. Lett.*, submitted.

Knutson, T.R. and S. Manabe, 1994: Impact of increased CO_2 on simulated ENSO-like phenomena. *Geophys. Res. Lett.*, **21**, 2295–2298.

Knutson, T.R. and S. Manabe, 1995: Time-mean response over the tropical Pacific to increased CO2 in a coupled ocean-atmosphere model. *J. Climate*, **8** (9), 2181–2199.

König, W., R. Sausen and F. Sielmann, 1993: Objective verification of cyclones in GCM simulations. *J. Climate,* **6**, 2217–2231.

Lal, M., B. Bhaskaran and B. Chakraborty, 1994a: Intraseasonal variability of Indian monsoon rainfall in climate models. In: *Research Activities in Atmospheric and Oceanic Modelling* , G.J.Boer (ed.), CAS/JSC Working Group Numerical Experimentation, Report **19**, February 1994, WMO/TD-No 592, pp. 7.53

Lal, M., U. Cubasch and B.D. Santer, 1994b: Effect of global warming on Indian monsoon simulated with a coupled ocean-atmosphere general circulation model. *Current Science*, **66**, 430–438.

Lal, M., U. Cubasch, R. Voss and J. Waszkewitz, 1995a: The effect of transient increase of greenhouse gases and sulphate aerosols on monsoon climate, *Current Science,* **69** (9), 752–763.

Lal, M., L. Bengtsson, U. Cubasch, M. Esch and U. Schlese, 1995b: Synoptic scale disturbances of Indian summer monsoon as simulated in a high resolution climate model. *Clim. Res.*, **5**, 243–258.

Lal, M., D. Jacob, R. Podzun and U. Cubasch, 1995c: Summer monsoon climatology simulated with a regional climate model nested in a general circulation model. In: *International Workshop on limited-area and variable resolution models, Beijing, China., 23–27 October 1995*

Lambert, S.J. 1995: The effect of enhanced greenhouse warming on winter cyclone frequencies and strengths. *J. Climate*, **8**, 1447–1452.

Langner, J. and H. Rodhe, 1991: A global three-dimensional model of the tropospheric sulfur cycle. *J. Atmos Chem.*, **13**, 225–263.

Lau, N.-C., S.G.H. Philander and M.J. Nath, 1992: Simulation of ENSO-like phenomena with a low-resolution coupled GCM of the global ocean and atmosphere. *J. Climate*, **5**, 284–307.

Le Treut, H. and Z.-X. Li, 1988: Using Meteosat data to validate a prognostic cloud generation scheme. *Atmos. Res.*, **21**, 273–292.

Le Treut, H., M. Forichon, O. Boucher and Z. X. Li, 1996: Aerosol and greenhouse gases forcing: Cloud feedbacks associated to the climate response. In: *Physical Mechanisms and their Validation*, H. Le Treut (ed.), NATO ASI Series Vol I. 34, Springer-Verlag Berlin Heidelberg, pp. 267–280.

Leggett, J.A., W.J. Pepper and R.J. Swart, 1992: Emissions scenarios for the IPCC: an update. In: *Climate Change 1992. The Supplementary Report to the IPCC Scientific Assessment*, J.T. Houghton, B.A. Callander and S.K. Varney (eds.), Cambridge University Press, Cambridge, UK, pp. 69–95.

Lemke, P., W.B. Owens and W.D. Hibler III, 1990: A coupled sea-ice mixed-layer pycnocline model for the Weddell Sea. *J. Geophys. Res.*, **95**, 9513–9525.

Leuenberger, M. and U. Siegenthaler, 1992: Ice-age atmospheric concentration of nitrous oxide from an Antarctic ice core. *Nature*, **360**, 449–451.

Leung, R.L. and S.J. Ghan, 1995: A sub-grid parameterization of orographic precipitation. *Theor. Appl. Climatology*, **52(1–2)**, 95–118.

Li, X., Z. Zongci, W. Shaowu and D. Yohui, 1994: Evaluation of regional climate change simulation: A case study. IPCC special workshop on Article 2 of the United Nations Framework Convention on Climate Change, Fortaleza, Brazil, 17–21 October 1994.

Liang, X.-L., W.-C. Wang and M.P. Dudek, 1995: Interannual climate variability and its change due to the greenhouse effect. *Global and Planet. Change*, **10**, 217–238.

Lighthill, J., G. J. Holland, W.M. Gray, C. Landsea, K. Emanuel, G. Craig, J. Evans, Y. Kunihara and C.P. Guard, 1994: Global climate change and tropical cyclones. *Bull. Am. Met. Soc.*, **75**, 2147–2157.

Lilly, D.K., 1968: Models of cloud-topped mixed layers under a strong inversion. *Quart. J. R .Met. Soc.*, **94**, 292–309.

Liu, Y.F., F. Giorgi and W.M. Washington, 1994: Simulation of summer monsoon climate over east Asia with an NCAR Regional Climate model. *Mon. Wea. Rev.*, **122**, 2331–2348.

Lupo, A.R., R.J. Oglesby and I.I. Mokhov, 1995: Climatological features of blocking anticyclones: a study of Northern Hemisphere CCM1 model blocking events in present-day and double CO_2 concentration atmospheres. *J. Climate*, (In press).

Luthi, D., A. Cress, H.C. Davies, C. Frei and C. Schar, 1996: Interannual variability and regional climate simulations. *Theor. and Applied Clim.*, (In press)

Lynch, A.H., W.L. Chapman, J.E. Walsh and G. Weller, 1995: Development of a regional climate model of the western Arctic. *J. Climate*, **8**, 1555–1570.

Ma, C.-C., C.R. Mechoso, A. Arakawa and J.D. Farrara, 1994: Sensitivity of a coupled ocean-atmosphere model to physical parameterizations. *J. Climate* **7**, 1883–1896.

Machenhauer, B., D. Jacob and M. Bozert, 1994: Using the MPI's nested limited area model for regional climate simulations. *CAS/JSC WGNE Report No. 19*, (WMO/TO–No. 592), 758–760.

Maier-Reimer, E., 1993: Geochemical cycles in an ocean general circulation model. Pre-industrial tracer distributions. *Global Biogeochem. Cycles, 7*, 645–677.

Manabe, S. and A.J. Broccoli, 1990: Can existing climate models be used to study anthropogenic changes in tropical cyclone climate? *Geophys. Res. Lett.*, **17**, 1917–1920.

Manabe, S. and R.J. Stouffer, 1988. Two stable equilibria of a coupled ocean-atmosphere model. *J Climate*, **1**, 841–866.

Manabe, S. and R.J. Stouffer, 1993: Century-scale effects of increased atmospheric CO_2 on the ocean-atmosphere system. *Nature*, **364**, 215–218.

Manabe, S. and R.J. Stouffer, 1994: Multiple century response of a coupled ocean-atmosphere model to an increase of atmospheric carbon dioxide. *J. Climate*, **7**, 5–23.

Manabe, S. and R.T. Wetherald, 1975: The effects of doubling the CO_2 concentration on the climate of a general circulation model. *J. Atmos. Sci.*, **32**, 3–15.

Manabe, S., R J. Stouffer, M.J. Spelman and K Bryan, 1991: Transient responses of a coupled-ocean atmosphere model to gradual changes of atmospheric CO_2. Part I: Annual mean response. *J. Climate, 4*, 785–818.

Manabe, S., M.J. Spelman and R.J. Stouffer, 1992: Transient responses of a coupled ocean-atmosphere model to gradual changes of atmospheric CO_2. Part II: Seasonal response. *J. Climate, 5*, 105–126.

Maochang, C., H. von Storch and E. Zorita, 1995: Coastal sea level and the large-scale climate state: A downscaling exercise for the Japanese Islands. *Tellus, 47A*, 132–144.

Marinucci, M.R. and F. Giorgi, 1992: A 2 × CO_2 climate change scenario over Europe generated using a Limited Area Model nested in a General Circulation Model. I:Present day simulation. *J. Geophys. Res.*, **97**, 9989–10009.

Marinucci, M.R., F. Giorgi, M. Beniston, M. Wild, P. Tschuck, A. Ohmura and A. Bernasconi, 1995: High resolution simulations of January and July climate over the western Alpine region with a nested regional modeling system. *Theor. Appl. Climatology*, **51**, 119–138.

Matyasovszky, I., I. Bogardi, A. Bardossy and L. Duckstein, 1995: Local temperature estimation under climate change. *Theor. Appl. Climatology*, **50**, 1–14.

McFarlane, N.A., G J. Boer, J.-P. Blanchet and M. Lazare, 1992:

The Canadian Climate Centre Second-Generation General Circulation Model and Its Equilibrium Climate. *J. Climate*, **5**, 1013–1044.

McGregor, J.L. and K.J. Walsh, 1993: Nested simulations of perpetual January climate over the Australian region. *J. Geophys. Res.*, **98**, 23283–23290.

McGregor, J.L. and K.J. Walsh, 1994: Climate change simulations of Tasmanian precipitation using multiple nestings *J. Geophys. Res.,* **99**, **D10**, 20889–20905.

Mearns, L.O., 1993: Implications of global warming for climate variability and the occurrence of extreme climate events. In: *Drought Assessment Management and Planning: Theory and Case Studies.*, D.A. Wilhite, (ed.), Kluwer, Boston, pp. 109–130.

Mearns, L.O., F. Giorgi, L. McDaniel and C. Shields, 1995a: Analysis of daily variability of precipitation in a nested regional climate model: Comparison with observations and doubled CO_2 results. *Global and Planetary Change*, **10**, 55–78.

Mearns, L.O., F. Giorgi, L. McDaniel and C. Shields, 1995b: Analysis of variability and diurnal range of daily temperature in a nested regional climate model: Comparison with observations and doubled CO_2 results. *Clim. Dyn.*, **11**, 193–209.

Meehl, G.A., 1990: Seasonal cycle forcing of El Niño-Southern Oscillation in a global coupled ocean-atmosphere GCM. *J. Climate*, **3**, 72–98.

Meehl, G.A. and W.M. Washington, 1990: CO_2 climate sensitivity and snow-sea-ice albedo parametrization in an atmospheric GCM coupled to a mixed layer ocean. *Clim. Change*, **16**, 283–306.

Meehl, G.A. and W.M. Washington, 1993: South Asian summer monsoon variability in a model with doubled atmospheric carbon dioxide concentration. *Science*, **260**, 1101–1104.

Meehl, G.A. and W.M. Washington, 1995: Cloud albedo feedback and the super greenhouse effect in a global coupled GCM. *Clim. Dyn.* **11**, 399–411.

Meehl, G.A. and W.M. Washington, 1996: El Niûo-like Pacific region climate change in a model with elevated atmosphere CO_2 concentrations. *Nature*, submitted.

Meehl, G.A., G.W. Branstator and W.M. Washington, 1993a: Tropical Pacific interannual variability and CO_2 climate change. *J. Climate*, **6**, 42–63.

Meehl, G.A., W.M. Washington and T.R. Karl, 1993b: Low-frequency variability and CO_2 transient climate change. Part 1: Time-averaged differences. *Clim. Dyn.*, **8**, 117–133.

Meehl, G.A., M. Wheeler and W.M. Washington, 1994: Low-frequency variability and CO_2 transient climate change. Part 3. Intermonthly and interannual variability. *Clim. Dyn.*, **10**, 277–303.

Melillo, J.M., A.D. McGuire, D.W. Kicklighter, B. Moore III, C.J. Vorosmarty and A.L. Schloss, 1993: Global climate change and terrestrial net primary production. *Nature*, **363**, 234–240.

Mellor, G.L. and L. Kantha, 1989: An ice-ocean coupled model. *J. Geophys. Res.*, **94**, 10937–10954.

Miller, J.R. and G.L. Russell,1995: Climate change and the Arctic hydrologic cycle as calculated by a global coupled atmosphere-ocean model. *Ann. Glac.*, **21**, 91–95.

Miller, N., 1994: *A Homogeneity Grouping Technique for Geophysical Climate Models.* Lawrence Livermore National Laboratory Report No. UCRL-JC-115996, 10 pp.

Mitchell, J.F.B. and J.M. Gregory, 1992: Climatic consequences of emissions and a comparison of IS92a and SA90. In: *Climate Change 1992: The Supplementary Report to the IPCC Scientific Assessment,* J.T. Houghton, B.A. Callander and S.K. Varney (eds.), Cambridge University Press, Cambridge, UK, pp. 171–175.

Mitchell, J.F.B. and T.J. Johns, 1996: Sensitivity of regional and seasonal climate to transient forcing by CO_2 and aerosols. (Submitted to *J. Climate*)

Mitchell, J.F.B., S. Manabe, T. Tokioka and V. Meleshko, 1990: Equilibrium Climate Change. In: *Climate Change. The IPCC Scientific Assessment ,* J.T. Houghton, G.J. Jenkins and J.J. Ephraums (eds.), Cambridge University Press, Cambridge, UK, pp. 131–172.

Mitchell, J.F.B., T.J. Johns, J.M. Gregory and S.B.F. Tett, 1995a: Climate response to increasing levels of greenhouse gases and sulphate aerosols. *Nature*, **376**, 501–504.

Mitchell, J.F.B., R.A. Davis, W.J. Ingram and C.A. Senior, 1995b: On surface temperature, greenhouse gases and aerosols: models and observations. *J. Climate,* **10**, 2364–2386.

Murphy, J.M., 1995a: Transient response of the Hadley Centre Coupled Model to increasing carbon dioxide. Part I Control climate and flux adjustment. *J. Climate*, **8**, 36–56.

Murphy, J.M., 1995b: Transient response of the Hadley Centre Coupled Model to increasing carbon dioxide. Part III Analysis of global -mean response using simple models. *J. Climate*, **8**, 496–514.

Murphy, J.M. and J.F.B. Mitchell, 1995. Transient response of the Hadley Centre Coupled Model to increasing carbon dioxide. Part II. Temporal and spatial evolution of patterns. *J. Climate*, **8**, 57–80.

Nakamura, M., P.H. Stone and J. Marotzke, 1994: Destabilisation of the thermohaline circulation by atmospheric eddy transports. *J. Climate*, **12**, 1870–1882.

Noda, A. and T. Tokioka, 1989: The effect of doubling the CO_2 concentration on convective and non-convective precipitation in a general circulation model coupled with a simple mixed layer ocean model. *J. Met. Soc. Japan*, **67**, 1057–1067.

Noguer, M., 1994: Using statistical techniques to deduce local climate distributions. An application for model validation. *Meteorological Applications*, **1**, 277–287.

Osborn, T.J. and T.M.L. Wigley, 1994: A simple model for estimating methane concentration and lifetime variations. *Clim. Dyn.*, **9**, 181–193.

Ose, T., 1993: An examination of the effects of explicit cloud water in the UCLA GCM. *J. Met. Soc. Japan*, **71**, 93–109.

Parey, S., 1994a: Simulations de trente ans $1 \times CO_2$, $2 \times CO_2$, $3 \times CO_2$ avec le modele du LMD ($64 \times 50 \times 11$) premiers resultats. *EDF, Direction des Etudes et Recherches, HE-33/94/008.*

Parey, S., 1994b: Les événements de blockage anticyclonique sur l'hemisphère nord dans des simulations réalisées avec differentes concentrations en CO_2 atmosphèrique ($1 \times CO_2$, $2 \times CO_2$, $3 \times CO_2$). *EDF, Direction des Etudes et Recherches, HE-33/94/017.*

Parker D.E., P.E Jones, C.K. Folland and A.J. Bevan, 1994: Interdecadal changes of surface temperature since the late nineteenth century. *J. Geophys..Res.*, **99**, 14373–14399.

Pielke, R.A., W.R. Cotton, R.L. Walko, C.J. Tremback, W.A. Lyons, L.D. Grasso, M.E. Nicholls, M.D. Moran, A. Wesley, T.J. Lee and J.H. Copeland, 1992: A comprehensive meteorological modeling system – RAMS. *Meteor. Atmos. Phys.*, **49**, 69–91.

Pleym, H. and T.R. Karl, 1996: Estimation of CO_2-induced local climate change in Norway using the CPMS method. *J. Climate* (Submitted).

Podzun, R, A. Cress, D. Majewski and V. Renner, 1995: Simulation of European climate with a limited area model. Part II: AGCM boundary conditions. Submitted to *Contributions to Atmos. Phys.* 68(3), 205–226.

Pollard, D. and S.L. Thompson, 1994: Sea-ice dynamics and CO_2 sensitivity in a global climate model. *Atmos.-Ocean*, (in press).

Power, S., R. Colman, B. McAvaney, R. Dahni, A. Moore and N. Smith, 1993: The BMRC coupled atmosphere/ocean/sea ice model, *The BMRC Research Report No 37.*

Raisanen, J., 1995: A comparison of the results of seven GCM experiments in Northern Europe. *Geophysica*, **30** (1–2), 3–30.

Ramaswamy, V., M.D. Schwarzkopf and K.P. Shine, 1992: Radiative forcing of climate from halocarbon-induced global stratospheric ozone loss. *Nature*, **355**, 810–812.

Randall, D.A., 1989: Cloud parameterization for climate modeling: Status and prospects. *Atmos. Res.,* **23**, 341–361.

Raper, S.C.B. and U. Cubasch, 1995: Emulation of the results from a coupled general circulation model using a simple climate model. *Geophys. Res. Lett.* (submitted)

Raper, S.C.B., T.M.L. Wigley and R.A. Warrick, 1996: Global sea level rise: Past and future. In: *Rising Sea Level and Subsiding Coastal Areas*, J.D. Milliman (ed.), Kluwer Academic Publishers, Dordrecht, The Netherlands , 384 pp.

Rind, D., R. Goldberg and R. Ruedy, 1989: Change in climate variability in the 21st century. *Clim. Change*, **14**, 537.

Robitaille, D.Y. and A.J. Weaver, 1995: Validation of sub-grid scale mixing schemes using CFCs in a global ocean model. *Geophys. Res. Lett.*, **22**, 2917–2920.

Robock, A.C., C.A. Schlosser, K.Ya. Vinnikov, S. Liu and N.A. Speranskaya, 1995: Validation of humidity, moisture fluxes and soil moisture in GCMs: report of AMIP Diagnostic Subproject 11; Part 1 – soil moisture. In *Proc. First AMIP Sci. Conf.*, WCRP-92, WMO TD-No. 732, Geneva.

Roeckner, E. and H. von Storch, 1980: On the efficiency of horizontal diffusion and numerical filtering in an Arakawa-type model. *Atmosphere-Ocean*, **18**, 239–253.

Roeckner, E., T. Siebert and J. Feichter, 1995: Climatic response to anthropogenic sulfate forcing simulated with a general circulation model. *Aerosol Forcing of Climate*, R.J. Charlson and J. Heintzenberg (eds.), John Wiley and Sons, Chichester, pp. 349–362.

Russell, G.L., J.R. Miller and D. Rind, 1995: A coupled atmosphere-ocean model for transient climate change studies. *Atmos-Ocean*, **33**, 4.

Santer, B.D., W. Bruggemann, U. Cubasch, K. Hasselmann, E. Maier-Reimer and U. Mikolajewicz, 1994: Signal to noise analysis of time dependent greenhouse warming experiments, Part 1: Pattern Analysis. *Clim. Dyn.*, **9**, 267–285.

Santer B.D., K.E. Taylor, T.M.L. Wigley, J.E. Penner, P.D. Jones and U. Cubasch, 1995: Towards the detection and attribution of an anthropogenic effect on climate. *Clim. Dyn.* **12**, 2, 77–100.

Sarmiento, J.L. and J.C. Orr, 1991: Three-dimensional simulations of the impact of Southern Ocean nutrient depletion on atmospheric CO_2 and ocean chemistry. *J. Limnol. Oceanogr.*, **36**, 1928–1950.

Sarmiento, J.L., J.C. Orr and U. Siegenthaler, 1992: A perturbation simulation of CO_2 uptake in an ocean general circulation model. *J. Geophys. Res.*, **97**, 3621–3646.

Sasaki, H., H. Kida, T. Koide and M. Chiba, 1995: The performance of long term integrations of a limited area model with the spectral boundary coupling method. *J. Met. Soc. Japan*, 73(2),165–181.

Sausen, R., K. Barthel and K. Hasselmann, 1988: Coupled ocean-atmosphere models with flux corrections. *Clim. Dyn.*, **2**, 154–163.

Schimel, D.S., I. Enting, T.M.L. Wigley, M. Heimann, D. Raynaud, D. Alves and U. Siegenthaler, 1995: The carbon cycle. In: *Radiative Forcing of Climate Change and an Evaluation of the IPCC IS92 Emissions Scenarios*, J.T. Houghton, L.G. Meira Filho, J.P. Bruce, Hoesung Lee, B.A. Callander, E.F. Haites, N. Harris and K. Maskell (eds.), Cambridge University Press, Cambridge, UK, pp. 35–71.

Schlesinger, M. E., Xinjian Jiang and R.J. Charlson, 1993: Implication of anthropogenic atmospheric sulphate for the sensitivity of the climate system. In: *"Climate change and energy policy"*, L. Rosen and R. Glasser (eds.), American Institute of Physics, pp. 75–108.

Semazzi, F.H.M., N.-H. Lin, Y.-L. Lin and F. Giorgi, 1994: A nested model study of the Sahelian climate response to sea-surface temperature anomalies. *Geophys. Res. Let.*, **20**, 2897–2900.

Send, U. and J. Marshall, 1994: Integral effects of deep convection. *J. Phys. Oceanogr*, **25**, 855–872.

Senior C.A. 1995: The dependence of climate sensitivity on the horizontal resolution of a GCM. *J Climate.*, **8**, 2860–2880.

Senior, C.A. and J.F.B. Mitchell, 1993: Carbon dioxide and climate: The impact of cloud parameterization. *J. Climate*, **6**, 393–418.

Shine, K.P., R.G. Derwent, D.J. Wuebbles and J.-J. Morcrette, 1990: Radiative forcing of climate. In: *Climate Change. The IPCC Scientific Assessment* , J.T. Houghton, G.J. Jenkins and J.J. Ephraums (eds.), Cambridge University Press, Cambridge, UK, pp. 49–68.

Shine, K.P., Y. Fouquart, V. Ramaswamy, S. Solomon and J. Srinivasan, 1995: Radiative forcing. In: *Radiative Forcing of Climate Change and an Evaluation of the IPCC IS92 Emissions Scenarios*, J.T. Houghton, L.G. Meira Filho, J.P. Bruce, Hoesung Lee, B.A. Callander, E.F. Haites, N. Harris and K. Maskell (eds.), Cambridge University Press, Cambridge, UK, pp. 163–203.

Siegenthaler, U. and F. Joos, 1992: Use of a simple model for studying oceanic tracer distributions and the global carbon cycle. *Tellus*, **44**B, 186–207.

Singh, S.V. and H. Von Storch, 1994: Downscaling Indian monsoon rainfall from the simulations of the ECHAMT21 coupled ocean-atmosphere model, *Proc. 6th. International Meeting on Statistical Climatology, 19-23 June, 1995, Galway, Ireland*, pp. 207–209.

Smethie, W.M, 1993: Tracing the thermohaline circulation in the western North Atlantic using chlorofluoromethanes, *Prog. Oceanogr*, **31**, 51–99.

Smith, I.N., M. Dix and R.J. Allan, 1995: The effect of greenhouse SSTs on ENSO simulations with an AGCM. *J. Climate*, (Submitted).

Smith, R.N.B., 1990: A scheme for predicting layer cloud and their water content in a general circulation model. *Quart. J. R. Met. Soc.*, **112**, 371–386.

Stouffer, R.J., S. Manabe and K.Va. Vinnikov, 1994: Model Assessment of the role of natural variability in recent global warming. *Nature*, **367**, 634–636.

Suarez, M., A. Arakawa and D.A. Randall, 1983: Parameterization of the planetary boundary layer in the UCLA General Circulation Model: formulation and results. *Mon. Wea. Rev.*, **111**, 2224–2243.

Sundqvist, H., 1978: A parameterization scheme for non-convective parameterization including prediction of cloud water. *Quart. J. R. Met. Soc.*, **104**, 677–690.

Sundqvist, H., 1993: Inclusion of ice phase of hydrometeors in cloud parametrisation for mesoscale and large-scale models. *Contr. Atmos. Phys.*, **66**, 137–147.

Taylor, K and J.E. Penner, 1994: Climate system response to aerosols and greenhouse gases: a model study. *Nature*, **369**, 734–737.

Tett, S., 1995: Simulation of El Niño-Southern Oscillation-like variability in a global AOGCM and its response to CO_2 increase. *J. Climate*, 8, 1473–1502.

Tett, S.F.B., T.C. Johns and J.F.B. Mitchell, 1996: Global and regional variability in a coupled AOGCM. (Submitted to *Clim. Dyn.*)

Tiedtke, M., 1993: Representation of clouds in large-scale models. *Mon. Wea. Rev.*, **121**, 3040–3061.

Tokioka, T., A. Noda, A. Kitoh, Y. Nikaidou, S. Nakagawa, T. Motoi, S. Yukimoto and K. Takata, 1995: A transient CO_2 experiment with the MRI CGCM -Quick Report-, *J. Met. Soc. Japan*, **74**(4), 817–826.

Von Storch, H., 1993: Inconsistencies at the interface of climate impact studies and global climate research. In: *Biometeorology*, A.R. Maarouf, N.N. Barthakur and W.O. Haufe (eds.), Proc. 13th International Congress of Biometeorology, Calgary, Canada, September 12-18, 1993; Part 2, Volume 1, pp. 54–87.

Von Storch, H., E. Zorita and U. Cubasch, 1993: Downscaling of climate change estimates to regional scales: An application to winter rainfall in the Iberian Peninsula. *J. Climate*, **6**, 1161–1171.

Walsh, K. and J.L. McGregor, 1995: January and July climate simulations over the Australian region using a limited area model. *J. Climate*, **8**, 2387–2403.

Wang, W.-C., M.P. Dudek and X.-Z. Liang, 1992: Inadequacy of effective CO_2 as a proxy in the regional climate change due to other radiatively active gases. *Geophys. Res. Lett.*, **19**, 1375–1378.

Warner, M.J. and R.F. Weiss, 1992: Chlorofluoromethanes in south Atlantic Antarctic Intermediate water, *Deep Sea Research*, **39**, 2053–2075.

Washington, W.M. and G.A. Meehl, 1989: Climate sensitivity due to increased CO_2: Experiments with a coupled atmosphere and ocean general circulation model. *Clim. Dyn.*, **4**, 1–38.

Washington, W.M. and G.A. Meehl, 1993: Greenhouse sensitivity experiments with penetrative cumulus convection and tropical cirrus albedo effects. *Clim. Dyn* . **8**, 211–223.

Washington, W.M., and G.A. Meehl, 1996: High latitude climate change in a global coupled ocean-atmosphere-sea ice model with increased atmospheric CO_2. *J. Geophys. Res.*, in press.

Weaver, A.J. and T.M.C. Hughes, 1992: Stability and variability of the thermohaline circulation and its link to climate. *Trends in Physical Oceanography*, Research Trends Series, Council of Scientific Research Integration, Trivandrum, India, **1**, 15–70.

Weaver, A.J. and T.M.C. Hughes, 1996: On the incompatability of ocean and atmosphere models and the need for flux adjustments. *Clim. Dyn.* **12**, 141–170.

Weaver, A.J., J. Marotzke, P.F. Cummins and E.S. Sarachik, 1993: Stability and variability of the thermohaline circulation. *J. Phys. Oceanogr.*, **23**, 39–60.

Weiss, R.F., J.L. Bullister, R.H. Gammon and M.J. Warner, 1985: Atmospheric chlorofluoromethanes in the deep equatorial Atlantic, *Nature*, **311**, 608–610.

Wetherald, R.T. and S. Manabe, 1995: The mechanisms of summer dryness induced by greenhouse warming. *J. Climate*, **8**, 3096–3108.

Whetton, P.H., A.M. Fowler, M.R. Haylock and A.B. Pittock, 1993: Implications of climate change due to the enhanced greenhouse effect on floods and droughts in Australia. *Clim. Change*, **25**, 289–317.

Whetton, P.H., P.J. Rayner, A.B. Pittock and M.R. Haylock, 1994: An assessment of possible climate change in the Australian region based on an intercomparison of General Circulation Modeling results, *J. Climate*, **7**, 441–463.

Whetton, P., M. England, S. O'Farrell, I. Waterson and B. Pittock, 1995: Global comparison of the regional rainfall results of enhanced greenhouse coupled and mixed layer ocean experiments: Implications for climate change scenario development. *Clim. Change* (Submitted).

Wigley, T.M.L., 1989: Possible climate change due to SO_2-derived cloud condensation nuclei. *Nature*, **339**, 365–367.

Wigley, T.M.L., 1993: Balancing the carbon budget. Implications for projections of future carbon dioxide concentration changes. *Tellus*, **45B**, 409–425.

Wigley, T.M.L., 1994: The contribution from emissions of different gases to the enhanced greenhouse effect. In: *Climate Change and the Agenda for Research* , T. Hanisch (ed.), Westview Press, Boulder, Colorado, pp. 193–222.

Wigley, T.M.L., 1995: Global mean temperature and sea level consequences of greenhouse gas concentration stabilization. *Geophys. Res. Lett.,*. **22**, 45–48.

Wigley, T.M.L. and S.C.B. Raper, 1987: Thermal expansion of sea water associated with global warming. *Nature* **330**, 127–131.

Wigley, T.M.L. and S.C.B. Raper, 1991: Detection of the enhanced greenhouse effect on climate. In: *Climate Change: Science, Impacts and Policy* , J. Jäger and H.L. Ferguson (eds.), Cambridge University Press, Cambridge, UK, pp. 231–242.

Wigley, T.M.L. and S.C.B. Raper, 1992: Implications for climate and sea level of revised IPCC emissions scenarios. *Nature* **357**, 293–300.

Wigley, T.M.L. and S.C.B. Raper, 1993: Future changes in global-mean temperature and sea level. In: *Climate and Sea Level Change: Observations, Projections and Implications*, R.A. Warrick, E.M. Barrow and T.M.L. Wigley (eds.), Cambridge University Press, Cambridge, UK, pp. 111–133.

Wigley, T.M.L., P.D. Jones, K.R. Briffa, and G. Smith, 1990: Obtaining sub-grid scale information from coarse-resolution general circulation model output. *J. Geophys. Res.*, **95**, 1943–1953.

Wigley, T.M.L., R. Richels and J.A. Edmonds, 1996: Alternative emissions pathways for stabilizing concentrations. *Nature*, **379**, 240–243.

Wilks, D.S., 1992: Adapting stochastic weather generation algorithms for climate change studies. *Clim.Change*, **22**, 67–84.

Wilson, L.L., D.P. Lettenmaier and E. Skyllingstad, 1992: A hierarchical stochastic model of large-scale atmospheric circulation patterns and multiple station daily precipitation. *J. Geophys. Res.*, **97**, 2791–2809.

Zorita, E, J.P. Hughes, D.P. Lettenmaier and H. von Storch, 1995: Stochastic characterization of regional circulation patterns for climate model diagnosis and estimation of local precipitation. *J. Climate*, **8**, 1023–1042.

Zwiers, F., 1994: Changes in screen temperature extremes under a doubling of CO_2, In: *Research Activities in Atmospheric and Oceanic Modelling* , G.J.Boer (ed.), pp. 7.44–7.46, CAS/JSC Working Group Numerical Experimentation, Report **19**, February 1994, WMO/TD–No 592.

7

Changes in Sea Level

R.A. WARRICK, C. LE PROVOST, M.F. MEIER, J. OERLEMANS,
P.L. WOODWORTH

Contributors:
R.B. Alley, R.A. Bindschadler, C.R. Bentley, R.J. Braithwaite, J.R. de Wolde,
B.C. Douglas, M. Dyurgerov, N.C. Flemming, C. Genthon, V. Gornitz, J. Gregory,
W. Haeberli, P. Huybrechts, T. Jóhannesson, U. Mikolajewicz, S.C.B. Raper,
D.L. Sahagian, R.S.W. van de Wal, T.M.L. Wigley

CONTENTS

SUMMARY

The purpose of this chapter is to assess the current state of knowledge regarding climate and sea level change, with special emphasis on scientific developments since IPCC (1990). The main focus is on changes that occur on the time-scale of a century. We thus look for evidence of sea level change during the last 100 years, examine the factors that could be responsible for such changes, and consider the possible changes in sea level during the next 100 years as a result of global warming.

With respect to the past, recent analyses suggest that:

- global mean sea level has risen 10–25 cm over the last 100 years. This range is slightly higher than that reported in IPCC (1990) (i.e., 10–20 cm). The higher estimate results largely from the use of geodynamic models for filtering out long-term vertical land movements, as well as from the greater reliance on the longest tide gauge records for estimating trends.

- there has been no detectable acceleration of sea level rise during this century. However, the average rise during the present century is significantly higher than the rate averaged over the last several thousand years, although century-time-scale variations of several decimetres almost certainly occurred within that longer period. The exact timing of the onset of the present, higher rate of sea level rise remains uncertain.

It is likely that the rise in sea level has been due largely to the concurrent increase in global temperature over the last 100 years. The possible climate-related factors contributing to this rise include thermal expansion of the ocean and melting of glaciers, ice caps and ice sheets. Changes in surface water and ground water storage may also have affected sea level. The assessment of the scientific evidence suggests that:

- global warming should, on average, cause the oceans to warm and expand, thus increasing sea level. The various models, from simple upwelling diffusion models to complex coupled atmosphere-ocean GCMs, all agree that oceanic thermal expansion is one consequence of global warming. The thermal expansion over the last 100 years is estimated to be 2–7 cm. Large-scale observations of changes in sub-surface ocean temperatures are beginning to support these estimates.

- global warming should, on average, increase the melt rates of glaciers and ice caps, causing sea level to rise. Observational data indicates that, globally, there has been a general retreat of glaciers during this century. Based on both observations and models, recent analyses suggest that this enhanced melting may have increased sea level by about 2–5 cm over the last 100 years.

- with respect to the Greenland ice sheet, a warmer climate should increase the melt rates at the margins. The increase in melting should dominate over any increase in accumulation rates in the interior, causing sea level to rise. However, observational evidence is insufficient to say with any certainty whether the ice sheet is currently in balance or has increased or decreased in volume over the last 100 years.

- with respect to the Antarctic ice sheet, a warmer climate should increase the accumulation rates, causing sea level to fall. Here, too, the observational evidence is insufficient to say with any certainty whether the ice sheet is currently in balance or has increased or decreased in volume over the last 100 years.

- it is unclear how changes in surface water or ground water storage have affected sea level. Estimates vary widely of the net effects of activities (largely anthropogenic) such as dam construction and reservoir filling, which lower sea level, and ground water pumping, deforestation and wetland loss, which tend to raise sea level. However, the potential future effect on sea level from such sources is probably relatively small, of the order of a few centimetres during the next century.

An exact accounting of the past sea level rise is difficult, particularly in the light of the large uncertainties associated with the mass balances of the ice sheets. However, the observed rise lies well within the combined ranges of uncertainty of the above factors.

Projections of future changes in sea level as a consequence of greenhouse-gas-induced warming were made for each of the six IPCC IS92 emission scenarios, with and without the effect of aerosol changes after 1990, for the period 1990 to 2100. In addition, high, middle and low estimates, using a range of parameter values based on key model uncertainties, were made for IS92a (the emission scenario most comparable to the IPCC (1990) Scenario A, the so-called "Business-as-usual" scenario). The results showed that:

- for Scenario IS92a, sea level is projected to be about 50 cm higher than today by the year 2100, with a range of uncertainty of 20–86 cm;

- for the range of emission scenarios IS92a–f using "best-estimate" model parameters, sea level is projected to be 38–55 cm higher than today by the year 2100;

- the extreme range of projections, taking into account both emission scenarios and model uncertainties, is 13–94 cm;

- most of the projected rise in sea level is due to thermal expansion, followed by increased melting of glaciers and ice caps. On this time-scale, the contributions made by the major ice sheets are relatively minor, but are a major source of uncertainty.

It is evident that the choice of emission scenario makes relatively little difference to the projected rise in sea level, especially for the first half of the next century. This is because much of the rise has already been determined by past changes in radiative forcing, due to lags in the response of the oceans and ice masses. For this same reason, in model simulations sea level continues to rise over many centuries even after concentrations of greenhouse gases are stabilised. In contrast, the scientific uncertainties – as reflected partly in intra-model uncertainties in the choice of individual model parameter values, and partly in inter-model uncertainties in the choice of methods for climate, glacier and ice sheet modelling – make a very large difference in the estimate of future sea level rise.

A major source of uncertainty concerns the polar ice sheets. Not only is there a lack of understanding of the current mass balance, but there is also considerable uncertainty regarding the possible dynamic responses on time-scales of centuries. Concern has been expressed that the West Antarctic Ice Sheet might "surge", causing a rapid rise in sea level. The current lack of knowledge regarding the specific circumstances under which this might occur, either in total or in part, limits the ability to quantify the risk. Nonetheless, the likelihood of a major sea level rise by the year 2100 due to the collapse of the West Antarctic Ice Sheet is considered low.

The changes in future sea level will not occur uniformly around the globe. Recent coupled atmosphere-ocean model experiments suggest that the regional responses could differ significantly, due to regional differences in heating and circulation changes. In addition, geological and geophysical processes cause vertical land movements and thus affect relative sea levels on local and regional scales. Finally, extreme sea level events – tides, waves and storm surges – could be affected by regional climate changes but are, at present, difficult to predict.

Overall, the basic understanding of climate-sea level relationships has not changed fundamentally since IPCC (1990). The estimates of global sea level rise presented here are lower than those presented in IPCC (1990), due primarily to significantly lower estimates of global *temperature* change which drive the projections of sea level rise. Thus, if global warming were to occur more rapidly than expected, the rate of sea level rise would consequently be higher.

7.1 Introduction

In terms of environmental and social consequences, sea level rise is arguably one of the most important potential impacts of global climate change. As in IPCC (1990), the primary purpose of this chapter is to assess what is known regarding how sea level has changed in the past and could change in the future, on time-scales of decades to centuries. The chapter begins by reviewing the evidence for trends in sea level over the past 100 years, based on tide gauge records. Next, the factors that could contribute to sea level change – namely, changes related to oceanic thermal expansion, glaciers and small ice caps, the large ice sheets of Greenland and Antarctica, and the possible changes in land-surface water and ground water storage – are examined. Future projections of global mean sea level rise are then made, on the basis of scenarios of future greenhouse gas emissions and projections of future global warming. The important factors that may cause such changes in sea level to vary spatially and temporally are considered, including geological, geophysical and dynamic effects.

Overall, this chapter finds that the major conclusions reached in IPCC (1990) remain qualitatively unchanged and are reinforced by the recent scientific literature. Nonetheless, improvements in observations and ocean-atmosphere modelling and in the calculation of future radiative forcing changes have led to revisions of the estimates of future sea level rise and their uncertainties.

7.2 How Has Sea Level Changed Over the Last 100 Years?

7.2.1 Sea Level Trends

Secular trends in "eustatic" global sea level (i.e., corresponding to a change in ocean volume) over the past century have been studied by a large number of authors (as summarised in Chapter 9 and Table 9.1 of the 1990 IPCC Scientific Assessment and as described in several recent reviews – see Emery and Aubrey, 1991; Woodworth, 1993; Douglas, 1995). At this time-scale, evidence for such trends is obtained from tide gauge data.

All authors of recent global sea level change reviews have used the Permanent Service for Mean Sea Level (PSMSL) data set (Spencer and Woodworth, 1993). A central problem in identifying trends in eustatic sea level from tide gauge data is how to account for vertical land movements, which also affect relative sea level change as measured by tide gauges. At the global scale, a major source of vertical land movement derives from the continuing readjustment of the Earth's crust since the

glacial retreat marking the end of the last ice age. This "post-glacial rebound" (PGR) is the only globally coherent geological contribution to long-term sea level change about which we possess detailed understanding. Since IPCC (1990), Douglas (1991) has applied post-glacial rebound corrections from the ICE-3G model of Tushingham and Peltier (1991) to the observed tide gauge data (avoiding tide gauge records in areas of converging tectonic plates, since such processes are not represented by the PGR model) and produced a highly consistent set of long-term sea level trends. The value for the mean rate of sea level rise that he obtained from a global set of 21 such stations in nine oceanic regions with an average record length of 76 years during the period 1880–1980 was 1.8 ± 0.1 mm/year. Geodynamic models of post-glacial rebound have also been employed in global analyses by Peltier and Tushingham (1989, 1991) and Trupin and Wahr (1990), who obtained similar results. Uncertainties in global sea level trends determined from PGR models may include a range of uncertainty of approximately 0.5 mm/year, depending on the Earth structure parametrization employed by the model (Mitrovica and Davis, 1995).

Other authors have used geological data directly for sites adjacent to tide gauges, a procedure which, in principle, should accommodate other geological processes in addition to post-glacial rebound (e.g., Gornitz and Lebedeff, 1987). This sort of analysis has been conducted for the North Sea region of Europe, which has an extensive tide gauge and geological sea level data set, by Shennan and Woodworth (1992). Earlier analyses (e.g., Barnett, 1984) simply averaged tide gauge records on the implicit assumption that the effects of land movements would somehow average out. It is interesting that the results of most of these studies are within the range 1–2 mm/year, with some bias towards the lower end of the range (see Gornitz, 1995a, for a review).

Gornitz (1995a) has also pointed out that estimates using the longest records obtain the highest values, closer to 2 mm/year than 1 mm/year, and that correction for post-glacial rebound by means of the Peltier models gives higher values for the trend than those analyses which employed nearby geological data values directly. This is a result of the post-glacial rebound models producing smaller extrapolated sea level trends into the present day than the linear fit to nearby geological data procedure such as Gornitz and colleagues have employed. Hence, the corrected tide gauge trends using the post-glacial rebound models will be higher than those using the nearby geological observations. In the latter case, problems could arise from the large scatter in the geological data distributed in area around the gauge locations, and from the linear extrapolation technique. Gornitz and Lebedeff (1987) found that in most cases a linear fit gives as good a description of the extrapolated trend from the geological data, as a higher order fit (see also discussion in Shennan and Woodworth, 1992). However, this ignores the additional information on the physics of the problem contained in the Peltier post-glacial rebound models.

Peltier and Tushingham (1989) also commented that the sea level trends depend strongly on the choice of minimum record length and on the particular time interval selected. Their sea level rise estimates, based on tide gauge record lengths greater than 50 years, tended to lie closer to 2 mm/yr than those based on shorter record lengths. Six of the longest tide gauge records (unadjusted for post-glacial rebound or other geological effects) for each continent are shown in Figure 7.1.

In summary, the best estimate based on recent analyses is that sea level has risen about 18 cm over the last 100 years, with a range of uncertainty of 10–25 cm – notwithstanding the fact that some authors (e.g., Gröger and Plag, 1993; Pirazzoli, 1993) maintain that a global figure for sea level rise cannot be estimated reliably, stressing the limitations of the available data set. This range is consistent with the 10–20 cm range given in IPCC (1990). As geodynamic models and measurements of vertical land movements become more reliable, confidence in such estimates will increase.

7.2.2 Has Sea Level Rise Accelerated?

Archaeological and geological data suggest that global sea levels have probably varied within a range of no more than a few tens of centimetres over the past two millennia (Flemming, 1969, 1993; Pirazzoli, 1977; Flemming and Webb, 1986; Hofstede, 1991; Tanner, 1992; Varekamp et al., 1992). The 10–25 cm rise over the past 100 years implies a comparatively recent acceleration in the rate of sea level change (Gornitz and Seeber, 1990; Shennan and Woodworth, 1992; Gornitz, 1995b). However, the exact timing of the onset of this acceleration remains uncertain. The conclusions given in Woodworth (1990), Gornitz and Solow (1991), and Douglas (1992) imply that the acceleration probably began before the 1850s. But data for the pre-instrumental period is sparse, at best. There is as yet no evidence for any acceleration of sea level rise this century (Woodworth, 1990; Gornitz and Solow, 1991; Douglas, 1992), nor would any necessarily be expected from the observed climate change to date. Small accelerations have been suggested over the past two or three centuries in European sea level data (Mörner, 1973; Ekman, 1988; Woodworth, 1990; Gornitz and Solow, 1991).

The evidence, or lack of it, for sea level accelerations over the past century depends critically on a small number of long tide gauge records which is unlikely to be supplemented significantly in the future. Nevertheless, the search for old records (or "data archaeology") can be particularly rewarding when such data are found. For example, Maul and Martin (1993) recently extended the Key West, Florida data set back to 1846 and demonstrated that no significant acceleration of sea level rise has occurred since that time.

The main difficulties in determining a more robust estimate for the sea level trends (let alone acceleration) are the unequal geographical distribution of historical tide gauge data and the considerable amount of typically decadal variability present in all records. The former is slowly being rectified with the development of a near-global tide gauge network (GLOSS) and by means of satellite radar altimetry (e.g., TOPEX/POSEIDON).

Atmospheric and oceanic dynamical processes account for a large fraction of the interannual and interdecadal variability. These fluctuations, 1–2 years to a decade in duration and coherent over long distances, reflect changes in temperature and salinity, currents, and coupled oceanic-atmospheric forcing, such as the El Niño-Southern Oscillation phenomenon in the Pacific Ocean (Komar and Enfield, 1987), or the North Atlantic Oscillation in the North Atlantic Ocean (Maul and Hanson, 1991). The limitations imposed by large interannual sea level changes in determining reliable sea level trends and accelerations have been particularly emphasised by Douglas (1992) and demonstrated by Sturges (1987) and Sturges and Hong (1995). The understanding of such interannual processes has been an aim of large international programmes such as the Tropical Ocean Global Atmosphere (TOGA) project and the World Ocean Circulation Experiment (WOCE), and will continue to be the object of research based on the results of a Global Ocean Observing System (GOOS). As ocean modelling becomes more detailed, and as surface and deep ocean observations become more routine, increased understanding of interannual processes will result, leading to an improved estimate of any underlying long-term trend and acceleration.

7.3 Factors Contributing to Sea Level Change

7.3.1 Oceanic Thermal Expansion

At constant mass the volume of ocean water, and thus sea level, varies with changes in sea water density (called "steric" changes). Density is related to temperature and salinity, hence the dependency of sea level on temperature

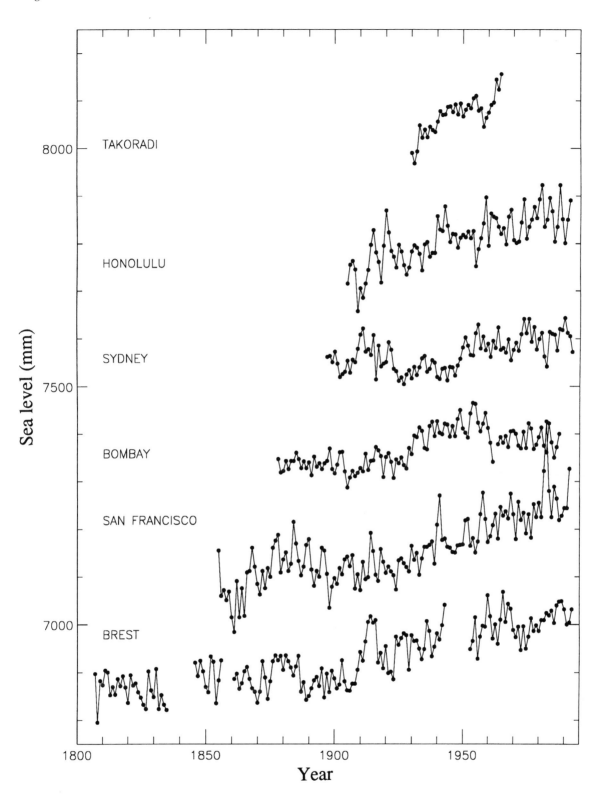

Figure 7.1: Six long sea level records from major world regions: Takoradi (Africa), Honolulu (Pacific), Sydney (Australia), Bombay (Asia), San Francisco (North America) and Brest (Europe). Each record has been offset vertically for presentational purposes. The observed trends (in mm/yr) for each record over the 20th century are, respectively, 3.1, 1.5, 0.8, 0.9, 2.0 and 1.3. The effect of post-glacial rebound (lowering relative sea level) as simulated by the Peltier ICE-3G model is less than, or of the order of, 0.5 mm/yr at each site.

and salinity variations of the oceans. The main cause of steric change at the global scale is temperature change. As density is inversely related to temperature, the volume of the ocean must expand and sea level rise if the density of the sea water decreases because of rising temperature.

Marked changes in salinity can occur at the regional scale and modify sea water density and volume. In the short term, these effects are relatively minor at the global scale. However, the main anomalous buoyancy sources are located at high latitudes where sea water finds an effective pathway from the surface to the deeper layers of the ocean. Thus, in the longer term, regional salinity changes can affect the whole ocean circulation, the distribution of heat and oceanic expansion.

More generally, because of very complex physics, there are other important regional variations and time lags involved in sea level rise by thermal expansion. The redistribution of buoyancy from source regions to other parts of the oceans involves large-scale waves and transport by currents. The adjustment processes are very slow, because they are related to the internal modes of heat and mass redistribution within the ocean. Consequently, a sea level rise due to thermal heating must be out of equilibrium, especially for rapid climate change. This implies that the ocean will rise more rapidly in some areas than in other areas.

An important feedback mechanism between thermal expansion and increased sea level caused by increased melting of ice is linked to the above processes. Increased fresh water fluxes to the ocean have an impact on density stratification and thereby on the depths of convection of the surface waters. This, in turn, affects the circulation of the interior of the ocean and thus the rates of thermal expansion for the different layers of the world ocean.

To estimate oceanic expansion from observations, temperature and salinity have to be recorded over the water column for a long period of time. In the 1990 IPCC Scientific Assessment (Subsection 9.4.1), reference was made to the works of Roemmich (1985) concerning the Panuliris series of deep hydrographic stations off Bermuda over the period 1955–1981, and to Thomson and Tabata (1987) concerning the station PAPA steric height anomalies in the north-east Pacific Ocean over a 27-year period. It was noted that the conclusions drawn from these data sets were limited geographically and that additional ocean observations are required to discern large-scale changes. Since 1990, there has been a considerable increase in the quantity of high quality hydrography and tracer sections produced from programmes like WOCE. The analyses of these data sets in relation to previous

hydrographic surveys are beginning to indicate large-scale, coherent changes in sub-surface ocean temperatures (see Chapter 3.2.4 for details). Estimates of the thermal expansion associated with some of these observed ocean temperature changes are similar in magnitude to changes in nearby sea level records (Roemmich, 1992; Salinger *et al.*, 1996). However, these estimates are still regionally specific. Such observational consistency between thermal expansion and sea level observations will improve in the near future with the availability of the new WOCE data sets and the measurement of the ocean topography through satellite altimetry (Nerem, 1995).

The problem of a steric rise in sea level due to global warming has also been examined using a variety of models. These include upwelling diffusion models, two-dimensional models, subduction models, ocean general circulation models (OGCMs), and coupled atmosphere-ocean models (AOGCMs).

The advantage of the simpler upwelling diffusion climate models is that they are computationally efficient so that the sensitivity of results to model parameters and to uncertainties in greenhouse gas emissions can easily be examined. For example, the projections of oceanic thermal expansion made by IPCC (1990) (Warrick and Oerlemans, 1990) were derived with the use of the simple upwelling diffusion-energy balance climate model of Wigley and Raper (1987, 1993). This model represents the vertical profile of area-averaged ocean temperature, which changes in response to changes in radiative forcing at the surface. Thermal expansion of sea water, and thus sea level change, is diagnosed from the temperature profiles through the use of expansion coefficients. For the period 1880 to 1990, the estimated range of sea level rise due to thermal expansion using this model is about 3.1 cm to 5.7 cm, with a best estimate of 4.3 cm (consistent with model parameters discussed in Chapter 6 and in Section 7.5.2), when the model is forced with the estimated radiative forcing changes from increases in greenhouse concentrations over the same period. A slightly more complex, 2-dimensional variant of this kind of model, using a prescribed ocean circulation and surface forcing based on observed sea surface temperature (de Wolde *et al.*, 1995), leads to similar values: 2.2 cm to 5.1 cm with a best estimate 3.5 cm. The major disadvantage of these simpler models, however, is that they do not realistically represent the processes involved in the penetration and distribution of heat from the surface to deeper layers of the ocean.

With two-dimensional, zonally averaged dynamical models coupled to an energy balance climate model (as developed by Harvey, 1992), the upwelling velocities can be related to the vertical diffusion coefficient. With such a model, Harvey (1994) has demonstrated that the 1-D upwelling diffusion models have a significantly faster surface transient response than the 2-D model, due to transient weakening of thermohaline overturning in the 2-D model which damps surface temperature warming. This difference is most pronounced if the upwelling rate in the upwelling diffusion models is constant (Raper *et al.*, 1996).

Subduction models, while still relatively simple, introduce heat into the ocean by advection along isopycnal (constant density) surfaces. Such a model has been developed by Church *et al.* (1991). In this model, the estimates of the magnitude of global thermal expansion are independent of assumptions regarding the magnitudes of eddy diffusivity. Furthermore, since this model is three-dimensional, it provides information on the regional differences in sea level change adjustments. Church *et al.* (1991) demonstrate a small dependence of the mean sea level rise on the spatial distribution of warming, for a given global mean temperature rise. With this model, Church *et al.* (1991) obtain a sea level rise due to thermal expansion of about 6.9 cm over the last one hundred years. The major disadvantage of this model is that it does not predict surface temperature changes, and it also ignores deep water formation and associated changes in convective heat fluxes.

A more realistic way to simulate ocean warming and thermal expansion is to use an OGCM. One example is the study of Mikolajewicz *et al.* (1990), who used the Large-Scale Geostrophic (LSG) OGCM of the Max-Planck-Institute (MPI). The model was forced with time-dependent surface temperature changes caused by progressively increasing atmospheric CO_2, calculated separately from an average of the equilibrium results from several GCMs for doubled CO_2.

However, in concept, the most realistic simulations involve the use of fully coupled AOGCMs. In such experiments, changes in the three-dimensional ocean and atmosphere feed back on each other as the modelled climate evolves. Although such models are very demanding on computer resources, transient experiments are increasingly being performed with coupled models (see Chapter 6). Sea level changes have been analysed from some of them. Among these are the experiments of the Geophysical Fluid Dynamics Laboratory, Princeton (see review from Gates *et al.*, 1992), of the MPI, Hamburg (Cubasch *et al.*, 1992, Cubasch *et al.*, 1994a), and of the UK Meteorological Office Hadley Centre, Bracknell (Gregory, 1993; Mitchell *et al.*, 1995). A recent simulation using the UKMO coupled model (Mitchell *et al.*, 1995)

with historical forcing changes gives a 1880 to 1990 sea level rise due to thermal expansion of 3.6 cm – in accord with simpler models.

There is no guarantee, however, that coarse resolution AOGCMs adequately represent the processes governing water mass formation and thus the penetration of surface heat into the ocean interior. Furthermore, these models also suffer from important unsolved problems to which thermal expansion calculations are particularly sensitive. First, there is the "cold start" problem. Neglecting past changes in greenhouse gas forcing before running scenarios of future forcing changes in AOGCMs leads to significant underestimation of both future warming and sea level rise due to thermal expansion of sea water. For example, Cubasch *et al.* (1994a) showed that starting the MPI coupled model from 1935 instead of 1985, but using the same future forcing scenario beyond 1985, leads to estimates of 7 cm instead of 4 cm for thermal expansion for the period 1985–2050. The GFDL model shows a much shorter time constant (about 10 years) compared to the MPI model (30 years), and different versions of the UKMO model show significant differences. A method to correct simulations for this "cold start" effect has been proposed by Hasselmann *et al.* (1993). Nonetheless, the issue has not been fully resolved, and the magnitude and relevance of this error for the surface air temperature and, especially, for sea level remains controversial.

Second, there is the problem of decadal variability and predictability. The initial conditions for AOGCMs are never fully known. This alone would be sufficient to prevent a prediction of the state of the ocean over several decades. To estimate the error that stems from this uncertainty, Cubasch *et al.* (1994a) carried out four 50-year integrations with the MPI coupled model. In all the experiments the greenhouse gas forcing started with conditions in 1986 and then followed the IPCC (1990) Scenario A. The only differences were in the initial conditions of the atmosphere and ocean. The decadal means of global warming and sea level rise showed large differences. The global mean sea level rise due to thermal expansion was 4.2 cm in the last years of integration, but in many regions of the world ocean, the local mean response had about the same size as two standard deviations of the decadal means between the runs. These problems, and others like the consequences of differences in spin-up techniques, obviously limit the reliability of the conclusions presently resulting from these coupled ocean-atmosphere model transient experiments.

Comparison of thermal expansion resulting from these various model experiments is difficult because of differences in experimental design, in model assumptions, particularly those concerning emission scenarios and climate sensitivity, and in the complexity of processes incorporated in different models. Carefully designed intermodel comparisons have not yet been carried out.

In summary, oceanic expansion is almost certainly one consequence of global warming. Although observational data are still too sparse and regionally specific to allow global estimates of thermal expansion, analyses of recent data sets in relation to previous hydrographic surveys are beginning to indicate large-scale, coherent changes in sub-surface ocean temperatures. Thermal expansion can also be predicted from a range of models, from simple to very complex. Based on model experiments, the estimated rise in sea level due to thermal expansion over the last 100 years lies within the range 2–7 cm.

7.3.2 Glaciers and Ice Caps

7.3.2.1 Processes causing change in glaciers and ice caps
The amount of land ice on Earth has fluctuated as the climate has changed. This section is concerned with the numerous mountain glaciers, ice fields and ice caps of the world, exclusive of the Greenland and Antarctic ice sheets (Figure 7.2), and their past and future contributions to changes in sea level. The potential contributions of glaciers and ice caps are not insignificant because the rates of ice accumulation and loss are more intense than those on the huge ice sheets.

The change in mass of these glaciers can be expressed as the sum of the mass balance at the surface (the difference between accumulation and loss, which can be positive or negative), plus the meltwater that is refrozen internally (internal accumulation), minus any iceberg calving flux. The change in mass of ice on land results in a change in sea level of opposite sign. The complexity stems from the fact that the surface mass balance, internal accumulation, iceberg calving flux, as well as the area and shape of the ice masses, are functions of time. The surface balance can be measured in a straightforward way, although it is a labour-intensive process (Østrem and Brugman, 1991). It can also be inferred, using hydrometeorological or climatological models, as the difference between snow accumulation (sometimes equated to winter precipitation) and snow/ice melt (often modelled using temperature or runoff as proxy); both terms are required for accurate modelling, including their seasonal differences.

Internal accumulation, on the other hand, is more difficult to measure. This term is often ignored in the modelling of glacier mass balances using climatological data, with the result that the calculated loss of water to the sea may be overestimated. Simple models to estimate the amount of internal accumulation have been developed for mountain glaciers (Trabant and Mayo, 1985) and for Arctic glaciers including Greenland (e.g., Pfeffer *et al.*, 1991; Reeh, 1991). The rate of iceberg calving can be measured (e.g., Brown *et al.*, 1982) but cannot be modelled with confidence.

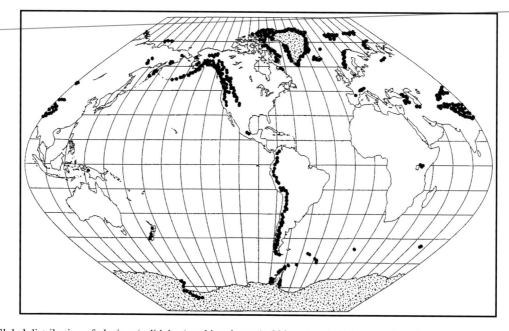

Figure 7.2: Global distribution of glaciers (solid dots) and ice sheets (pebble pattern). Note: a single dot may represent one or many individual glaciers, and many glaciers around the margin of the Antarctic ice sheet are omitted due to lack of information.

Changes in mass balance and/or iceberg calving produce changes in the area and surface profile of a glacier. Thus, models of long-term (more than a few years) changes in glaciers with respect to climate require incorporation of the flow processes – a dynamics model. Although these models are in common use and are of some numerical complexity, the basal sliding process for temperate glaciers is still very uncertain. A strong interaction between iceberg calving and glacier dynamics exists (Meier, 1994). The distribution of glacier mean thicknesses must also be considered because many small glaciers will likely disappear during the next century.

Internal accumulation and iceberg calving can be safely ignored in many regions, such as in much of Europe. In this case, conventional surface observations, or modelled relations, of snow accumulation and melt rates yield sufficient information for predicting volume changes as a result of climate change. In some areas of significant glacier cover, such as in the Arctic, all terms (and their time-dependencies) must be taken into consideration. This complexity, together with the paucity of observational data in many glacierised regions, makes a global synthesis of the present or future role of glaciers in sea level change a daunting task.

7.3.2.2 Changes in the last hundred years

Thinning of glaciers since the mid-19th century has been obvious and pervasive in many parts of the world (for example, see Figure 7.3). This long-term ice loss has considerable spatial and short-term variability. For instance, negative mass balances and rapid thinning have been observed in this century in the Alps (Haeberli and Hoelzle, 1995) and in south-central Alaska but not in the Canadian Arctic (Fisher and Koerner, 1994). Significant positive spatial correlations in annual mass balances are found between some glacierised regions, such as Kamchatka and the Northern Rockies (Trupin *et al.*, 1992), while negative correlations are found between other regions, such as Washington State and Alaska (Walters and Meier, 1989).

Table 7.1 shows improved estimates of the volume of glaciers and ice caps of the world, obtained by using data from the World Glacier Inventory (IAHS(ICSI)/UNEP/UNESCO, 1989). This publication presents the numbers of glaciers increasing by size categories from less than 0.03 to greater than 1024 km^2 for about 45 well-inventoried regions. Unfortunately, this includes only a small fraction of the world's glacier cover. For the purposes of the present report, these data were extended to a global estimate of ice cover according to size category

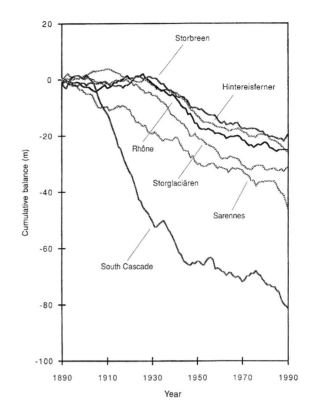

Figure 7.3: Cumulative mass balances, in metres of water equivalent, for the glaciers Hintereisferner (Austria), Rhône (Switzerland), Sarennes (France), South Cascade (United States), Storbreen (Norway), and Storglaciären (Sweden). These are among the few glaciers with long observational time series that have been extended using well calibrated hydrometeorological models. All values are relative to 1890.

by: (1) selecting 12 regions with very complete inventories that were typical of the world's glacierised regions; (2) adding an estimate for coastal Alaska and adjacent Canada using U. S. Geological Survey data on 321 large glaciers to extend the inventory, and extending these data by power-law scaling to smaller glacier sizes; (3) assembling numbers of glaciers in each size class and total area for the 13 regions; (4) assigning each of the 31 regions defined by Meier (1984) to one or more of these typical inventory regions; (5) weighting these regions according to the 31-region areas, (6) summing the above. Using a power law regression of known glacier area/volume relationships (e.g., Macheret *et al.*, 1988; Chen and Ohmura, 1990; Meier, 1993), the mean thicknesses for glaciers in each size category were estimated. On this basis, the total volume of glaciers and ice caps is estimated to be about 180,000 km^3, and the mean thicknesses range from 7 m to 655 m. Much of the glacier area and volume is comprised of glaciers in the 100 to 1000 km^2 size classes.

Table 7.1: *Some physical characteristics of ice on Earth. Accuracy is better than 10% unless indicated otherwise*

	Antarctic Ice Sheet (grounded ice only)	Greenland Ice Sheet	Glaciers and ice caps
Area (10^6 km^2)	12.1	1.71	0.68
Volume (10^6 km^3)	29[†]	2.95	0.18 ± 0.04
Volume (sea level equivalent, m)	73[†]	7.4	0.5 ± 0.10
Accumulation (10^{12} kg/yr)	1660	553	670 ± 100
Runoff (10^{12} kg/yr)	53 ± 30	237	690 ± 100*
Iceberg discharge (10^{12} kg/a)	2016 (from ice shelves)	316	50 ± 30

Data sources: Drewry, 1983; Bentley and Giovinetto, 1991; Jacobs et al., 1992; Meier, 1993; Reeh 1994; Weidick, 1995.

†   Total volume, including the amount currently below sea level (1.9×10^6 km^3, or 5 m sea level rise equivalent).

*   Includes runoff into closed basins in Central Asia (60×10^{12} kg/yr).

Table 7.2: *Estimates of global glacier mass balances during this century.*

Reference	Time period	No of observed glaciers[†] (regions)*	Global area[@] (10^3km^2),	Models: balance dynamic	Sea level change (mm/yr)
Meier (1984)	1900–61	25 (31)	540	a, f	0.46
Trupin *et al.* (1992)	1965–84	85 (31)	540	b, g	0.18
Meier (1993)	1900–61	– (31)	640	c, f	0.40
Dyurgerov (1994)	1960–93	23–76 (8)	589[§]	d, g	0.35[§]
	1985–93	"	"	"	0.60[§]

Balance Models:

a) Balance correlated with annual amplitude for measured glaciers in 13 regions, then extended globally based on measured or climatically estimated annual amplitudes.

b) Observed balance data correlated to regional precipitation, runoff data; extended globally using climatological data.

c) As in a), internal accumulation estimated.

d) Statistical analysis of all observed balance data.

Dynamics Models:

f) Balance adjusted by estimate of mean areal shrinkage (10%).

g) None (no dynamic response considered).

Notes

†   Number of observed glaciers used in correlations.

*   Number of regions into which data were aggregated.

@   Assumed total area of glaciers for calculating global contribution.

§   Does not include contribution of glaciers in drainage basins with internal runoff (52×10^3 km^2).

Internationally co-ordinated long-term monitoring of glaciers began in 1894, and now includes compilations of data on fluctuations in length, volume and mass balance by the World Glacier Monitoring Service (IASH(ICSI)/ UNESCO, 1967; IAHS(ICSI)/UNEP/UNESCO, 1973, 1977 1988, 1993).

Many attempts have been made to quantify the global rate of change of the world's glacier cover, beginning with the seminal work of Thorarinsson (1940). Recent analyses are listed in Table 7.2. The principal difficulty is the lack of observational data in many areas and some of the analyses have extended the data using climate information. Virtually no glaciers are being regularly measured in the mountains bordering the Gulf of Alaska or in Patagonia, two areas thought to contribute significantly to present-day sea level rise (Meier, 1985). The global number of glaciers on which the mass balance was directly measured was six or less through 1956, then rose to 50–60 during the period 1967 to 1989, coinciding with the new programmes of the International Hydrological Decade/Programme and the establishment of the World Glacier Monitoring Service; the number has fallen slightly since 1989 (Dyurgerov, 1994).

Clearly, insufficient data exist for a long-term global synthesis without extending the data using climate information. Simplistic models have been used to take changes in glacier dynamics into account. Table 7.2 lists examples of several recent compilations with brief sketches of the approaches used. It is obvious that further refinement is needed. Table 7.2 shows that the global contribution of glacier wastage to current sea level rise in this century is uncertain by a factor of about two. There are many reasons for this uncertainty, including: (1) different time periods used in analysis; (2) differences in the total glacier area, due to incomplete data from many of the highly glacierised regions of the world, and inconsistent consideration of small glaciers in Greenland and in Antarctica which are not part of the ice sheets; (3) incomplete climatic data from many parts of the world because of lack of weather stations near the glaciers; (4) crude approximations to dynamic feedbacks, instead of a real dynamics model; (5) general neglect of refreezing of meltwater in cold firn and of iceberg calving.

The results listed in Table 7.2 and Figure 7.4 suggest that the sea level rise contribution due to changes in glacier volumes averaged about 0.35 mm/yr between 1890 and 1990, and 0.60 mm/yr between 1985 and 1993. The uncertainty of these values is difficult to assess; the spread of the individual estimates suggests uncertainties of at least ±0.1 and ±0.05 mm/yr, respectively. None of the studies noted above used an interactive model of the balance-

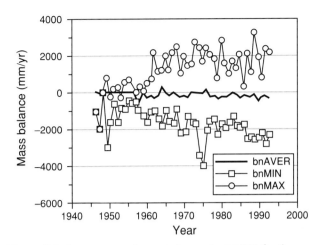

Figure 7.4: Average annual mass balances (bnAVER) for the glaciers of the world. These are weighted by area, and include virtually all of the observed glaciers (about 60). Also shown are the maximum (bnMAX) and minimum (bnMIN) mass balances for each year. To convert to equivalent sea level rise, multiply by the ratio of ice area to ocean area (1.8×10^{-3}). (From Dyurgerov and Meier, 1995).

dynamics feedback. Simple models exist (e.g., Jóhannesson *et al.*, 1989), but are difficult to apply over the vast spectrum of glaciers: large and small, temperate and cold, etc. One recent model of time-changes in glacier profiles, using typical or averaged values of ice thickness and a parameter specifying the localisation of thickness changes near the terminus, produces an estimate of sea level rise due to thinning of glaciers over the last 100 years of 0.5 mm/yr (Schwitter and Raymond, 1993).

7.3.2.3 Sensitivity to climate changes

The contribution of glacier wastage to sea level rise in the future could be calculated for any climate scenario, assuming knowledge of the dependence of the mass balance, internal accumulation and iceberg calving on the meteorological environment, and how the ensuing changes feed back to the glacier area and volume. Most of this knowledge does not currently exist. Alternatively, the contribution can be estimated using simple models developed by comparing past changes in glacier volume with observed changes in air temperature, or by calculating the present-day relation of mass balance to temperature and precipitation – the "sensitivity" (e.g., Oerlemans and Fortuin, 1992).

The term "sensitivity" can be defined in several ways. The expression often used to estimate sea level change from glacier volume change is the ratio of the rate of change of glacier volume to a small change in a climatic parameter (such as temperature). Here, this ratio is termed the *static sensitivity* of glaciers to climate change. The static sensitivity ignores changes in the configuration of the glacier. The dynamic response of the glacier and the lowering of its surface due to increased melting will change the effect of a non-zero mass balance over time-scales ranging from decades to centuries. Most glaciers in the world have not been in equilibrium with their climatic environment, at least not since the Little Ice Age, so there is a problem in computing the sensitivity value in a non-equilibrium situation with an unknown or ill-defined initial boundary condition.

Nonetheless, estimates of the static sensitivity have been obtained by different authors using different methods. The methods include: (1) hydrometeorological modelling using precipitation and temperature (e.g., Martin, 1978; Kuhn *et al.*, 1979; Tangborn, 1980; Chen and Funk, 1990); (2) degree-day modelling (e.g., Laumann and Reeh, 1993; Jóhannesson *et al.*, 1995; Wigley *et al.*, 1996); (3) mass balance or volume change observations related to an assumed linear change in global air temperature (e.g., Meier, 1993); (4) energy-balance modelling (Oerlemans and Fortuin, 1992). The resulting values range from 0.58 to 2.6 mm/yr/°C, with much of the spread related to assumptions about the initial mass balance condition (Wigley *et al.*, 1996).

7.3.3 The Greenland and Antarctic Ice Sheets

7.3.3.1 Processes causing change in the ice sheets
Most of the non-oceanic water on Earth resides in the two great ice sheets (Table 7.1), and most of their volume lies on land above sea level. Thus, loss of only a small fraction of this volume could have a significant effect on sea level.

In Antarctica, recent break-ups of the Larsen and Wordie Ice Shelves in the Antarctic Peninsula and discharges of enormous icebergs from the Filchner and Ross Ice Shelves, and the discovery of major recent changes in certain Antarctic ice streams, have focused public attention on the possibility of "collapse" of this ice reservoir within the next century, with potential impacts on sea level. Changes in floating ice shelves, of course, cannot affect sea level directly. Nevertheless, ice shelves are part of a complex, coupled ice flow system involving the inland ice and relatively fast-moving ice streams that discharge ice from land to sea, and changes in the rate of discharge can affect sea level. Whether such dynamic processes can be affected by climate changes on the time-scale of a century is a key issue.

In this sense, the Greenland ice sheet, which has no floating ice shelves of consequence, is different from the Antarctic ice sheet. In Antarctica, temperatures are so low that comparatively little surface melting occurs and the ice loss is mainly by iceberg calving, the rates of which are determined by dynamic processes involving long response times. In Greenland, ice loss from surface melting and runoff is of the same order of magnitude as loss from iceberg calving (Table 7.1). Thus, climate change in Greenland could be expected to have immediate effects on the surface mass balance of the ice sheet through melting and runoff as well as through accumulation.

On both ice sheets, the residence times for particles of ice range from the order of 10^5 yr or more for ice near an ice divide, to 10^2 yr or less for ice near the equilibrium line which separates the area of annual mass gain (accumulation) from the area of annual mass loss (melting and/or calving). These long residence times and their variations over the ice sheets further complicate the modelling of the response of ice sheets to climate change.

7.3.3.2 Current state of balance
Current changes of the ice sheets can be measured using surface mass balance observations or by geodetic (volume-change) methods. However, the ice sheets respond to processes at all time-scales, ranging from the last glacial-interglacial transition to decade-scale fluctuations in the temperature and precipitation fields. As most of our observations extend over a few decades only, this immediately poses a problem: how can we decide from observations whether a small change in ice sheet configuration is a response to a short-term climatic fluctuation or an ongoing process of slow adjustment to changes that happened a long time ago? This question requires numerical modelling studies of ice sheet dynamics.

Mass balance studies
One approach to estimating the current state of balance is to collect all available data on specific balance (net gain or loss of ice at the surface) and on iceberg production from ice shelves or outlet glaciers, and make the sum. When the interest is in the effect on sea level, the mass flux across grounding lines (the boundary between grounded and floating ice) should be considered rather than iceberg calving. With the data currently available, this procedure leads to very uncertain estimates.

The vastness of the ice sheets is the major obstacle in mass balance studies. In principle, the net mass balance of an ice sheet can be determined by summing the balance observations at the surface and the loss of ice by calving. However, measurements of the surface balance still give a poor coverage of both the Greenland and Antarctic ice

Table 7.3: *Current state of balance (in 10^{12} kg/yr) of the Antarctic ice sheet.*

Source	Accumulation (grounded ice)	Mass flux at grounding line	Balance (grounded ice)	Remarks
Budd and Smith (1985)	1800	1620	0 to +360	
Bentley and Giovinetto (1991)	1660	1260	+40 to +400	More or less reliable data for 42% of the ice sheet, extrapolated to whole grounded ice sheet
Jacobs *et al.* (1992)	1528			Estimated calving and runoff: 2613. Total accumulation (incl. ice shelves): 2144
Huybrechts (1990)			–351	Numerical ice sheet model run for last 160 kyr

Table 7.4: *Current state of balance (in 1012 kg/yr) of the Greenland ice sheet.*

Source	Accumulation	Runoff	Calving	Balance	Remarks
Bauer (1968)	500	330	280	–110	
Reeh (1985)	557	239	318	±0	
Ohmura and Reeh (1991)	535				New precipitation map
Huybrechts *et al.* (1991)	539	254			Degree-day model on grid
Reeh (1994)	553	237	316	~0	Field data
Huybrechts (1994)				+12.6	Numerical ice sheet model run for last 130 kyr

sheets, and estimating calving rates is an even more uncertain exercise. Mass balance estimates for Antarctica and Greenland are shown in Tables 7.3 and 7.4.

For the Antarctic ice sheet, two studies attempted to compare accumulation with mass flux across grounding lines. Budd and Smith (1985) used existing ice velocity measurements to estimate outflow from the grounded ice. They concluded that "...the total influx over the Antarctic ice sheet of about 2000 km³/yr (= 1800 × 10^{12} kg/yr) is probably nearly balanced by the outflow with a discrepancy most likely in the range of 0 to +20%". Bentley and Giovinetto (1991) made an assessment of the imbalance for drainage basins for which a reasonable amount of data was available. Most basins seem to have a positive balance. Extrapolating the results to the entire grounded ice sheet in three different ways yields an imbalance between +2 and +25% of the total input. These

studies thus suggest that grounded ice volume is increasing, but the error bars are very large.

For the Greenland ice sheet, early estimates (Bauer, 1968) suggested a negative balance. More recently, several mass balance studies of a more local nature have been carried out. Kostecka and Whillans (1988) compared mass balance with ice velocity measurements along two transects (International Glaciological expedition to Greenland [EGIG] traverse [71°N] and the Ohio State University traverse [65°N]). Results suggest no significant changes in ice thickness, with perhaps a very slight thickening at the Ohio State University traverse (0.06 ± 0.08 m/yr). For the Dye 3 station (65°N, 43.5°W) Reeh and Gundestrup (1986) obtain a change in ice thickness of 0.03 ± 0.06 m/yr, i.e., not significantly different from zero.

From south-west Greenland there is a large amount of information on fluctuations of outlet glaciers (Weidick,

1984, 1995). There has been a general retreat of outlet glaciers since the end of the 19th century. This retreat has slowed down in recent decades (a significant number of outlet glaciers are now advancing). It is not exactly clear how such fluctuations relate to the total volume of the ice sheet, but it appears that the Greenland ice sheet may have contributed significantly to sea level rise during the first part of this century and that its present contribution may be close to zero.

Geodetic methods

The most direct method for determining the current state of balance is to measure continuously and very accurately the surface elevation of the ice sheets. Assuming that grounding lines can be located, and changes in density and bed topography are sufficiently small or can be determined otherwise, a trend in the amount of grounded ice mass can be detected.

Ground-based levelling studies along transects on the ice sheets have been very limited. The EGIG line in Greenland, however, has been studied in some detail (Mälzer and Seckel, 1975; Seckel, 1977; Kock, 1993; Moeller, 1994). There appears to be a general thickening from 1959 to 1968 followed by a thinning to 1992, with an overall change of 0.1 m/yr or less. It is not known how representative this finding is of the whole Greenland ice sheet.

Recent attempts to measure the surface elevation include the use of satellite radar altimetry. The results are still controversial. Zwally *et al.* (1989) used satellite radar altimetry to estimate the change in surface elevation of the Greenland ice sheet south of 72°N (excluding the margins). For the period 1978–1985, they found an ice thickening rate of 0.23 ± 0.04 m/yr which implied a 25% to 45% positive balance for that period. Douglas *et al.* (1990) criticised this result on the basis of inadequate calibration of satellite orbits, but Zwally *et al.* (1990) obtained a similar result using orbits that were consistently calculated and adjusted to a common ocean reference. Lingle *et al.* (1991) used this method in the ablation zone, but due to errors in the altimetry on these sloping and undulating margins, the thickness-change results are not significantly different from zero. Ice sheet modelling by Huybrechts (1994) suggests an average thickening south of 72°N, in agreement with Zwally's results but somewhat smaller in magnitude.

Numerical modelling

A different approach simulates the evolution of the ice sheets using numerical models that include some of the relevant physical processes, attempting to constrain resulting

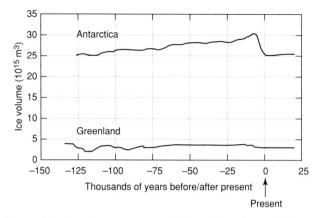

Figure 7.5: Long-term model simulations of the changes in ice volume of the Antarctic and Greenland ice sheets due to temperature and sea level changes (courtesy of Ph. Huybrechts).

ice sheet histories as much as possible by (palaeo) field evidence of both a geological and glaciological (ice cores) nature. The integrations should be performed over at least one glacial cycle to remove transient effects. One of the problems of such an approach is how to specify the forcing history (i.e., how to formulate the mass balance per unit area, which is the driving force, as a function of time and space).

Using a modelling approach, Huybrechts (1990, 1994) has carried out long integrations with a fairly comprehensive model of the Antarctic and Greenland ice sheets. This model solves the coupled mechanic and thermodynamic equations and includes ice shelves. The forcing follows a specified temperature and a sea level history. Simple parametrizations are used for the generation of the mass balance fields from a uniform temperature signal, which is taken from the Vostok ice core, and there is no direct effect of insolation variations on the mass balance. From the model, the Antarctic ice sheet shows a very strong response to the glacial-interglacial transition, as shown in Figure 7.5. This is mainly a response to the rise in sea level at that time and involves a substantial shrinkage of the West Antarctic ice sheet. The Greenland ice sheet show a much weaker response, except in the beginning of the integration (which must be considered as transient effects). The Greenland ice sheet responds solely to the temperature signal (sea level has very little effect). With respect to current mass balance, these model simulations suggest that the contribution of the Greenland ice sheet to sea level rise may be close to zero, while that of the Antarctic ice sheet may be positive.

Summary

The paucity of relevant data does not allow a meaningful judgement of the current state of balance of the Greenland

and Antarctic ice sheets. Different workers claim changes with even different sign, up to (and perhaps exceeding) 25% of the annual mass turnover (even more for south Greenland). A major problem is the use of data on the decadal time-scale to infer long-term changes. At present it can be concluded that an imbalance of up to 25% cannot be detected in a definite way by current methods/data. In terms of sea level change, a ±25% imbalance implies:

Antarctica	±1.4 mm/yr
Greenland	±0.4 mm/yr

7.3.3.3 Sensitivity to Climate Change

Tools to study the sensitivity of ice sheet mass balance comprise regression analyses, simple meteorological models (e.g., energy-balance modelling of a melting glacier surface), and GCMs. The sensitivities of Antarctica and Greenland to a 1°C warming based on these various methods are shown in Tables 7.5 and 7.6.

Ultimately, GCMs should be the best tools to study the mass balance of ice sheets, because in principle they include most relevant processes. Nevertheless, published climate change studies using GCMs rarely report or discuss changes in the water mass balance at the surface of ice sheets in spite of the crucial importance of such information for predictions of sea level change. This is an indication that confidence in such quantification is still low. Partly because ice accumulation at the surface of ice sheets is very slow, small absolute errors (and regional-scale precipitation patterns from control runs of GCMs are often significantly different from observed climate) come out as large relative errors. Even GCMs which fare well on average are not entirely reliable for polar climate studies (Genthon, 1994). Nevertheless, there have been significant improvements in recent years, mainly associated with increased resolution (Simmonds, 1990; Budd and Simmonds, 1991; Genthon, 1994). It is expected that more credible results from AOGCMs will be available in the near future.

For Greenland, a number of estimates of the sensitivity of the ice sheet to climate change have been made with simple meteorological models (degree-day models and energy balance models). The most studied quantity is the change in mean specific (surface) balance for a uniform 1°C increase in air temperature (Table 7.5). The earlier studies extrapolated calculations for a particular site to the entire ice sheet. In the later studies where calculations were made on a 20 km grid, the sensitivity values are somewhat

Table 7.5: *Sensitivity of the Greenland ice sheet to 1°C climatic warming.*

Source	Sensitivity (mm/yr in equivalent sea level change)	Remarks
Ambach and Kuhn (1989)	+0.31 [+0.24]*	Based on analysis of EGIG data; energy balance considerations at equilibrium line extrapolated to entire ice sheet.
Bindschadler (1985)	+0.57 to +0.77	Based on simple flowline model extrapolated to entire ice sheet and including estimated change in iceberg calving.
Braithwaite and Olesen (1990)	+0.36 to +0.48	Extrapolation of energy balance calculation for South-West Greenland.
Oerlemans *et al.* (1991)	+0.37 [+0.28]*	Energy balance model; ice sheet divided in four "climatic sectors".
Huybrechts *et al.* (1991)	+0.30 [+0.22]*	Degree-day model on 20 km grid.
van de Wal and Oerlemans (1994)	+0.30 [+0.21]*	Energy balance model on 20 km grid.

* Includes 5% increase in accumulation.

Table 7.6: *Sensitivity of the Antarctic ice sheet to 1°C climatic warming (mm/yr in equivalent sea level change).*

Source/Method	Sensitivity	Remarks
Muszynski and Birchfield (1985)	–0.38	Regression on 208 data points
Fortuin and Oerlemans (1990)	–0.20	Grounded ice only
		regression on 486 data points
Fortuin and Oerlemans (1992)	–0.27	2-dimensional atmospheric model: +1°C
		warming at coast and associated greenhouse
		forcing over ice sheet
Change in accumulation assumed	–0.34	20 km grid over grounded ice
proportional to saturation vapour pressure		
Temperature/sea ice/mass balance coupling	–0.7	Sensitivity of accumulation to sea ice extent
		(Giovinetto and Zwally, 1995) coupled with
		sensitivity of sea ice extent to air temperature
		(Parkinson and Bindschadler, 1984)

less. Nonetheless, the differences between the various studies are not very large. This must be due partly to the fact that empirical constants/parameters used in the different methods come mostly from the same data sets. The various approaches have some weak points in common as well (treatment of meltwater infiltration and refreezing, and of albedo). Also, accumulation rates are treated in the same way in all studies: either fixed, or changed in proportion to the saturation vapour pressure of the overlying air. However, there is little physical reason to believe that the precipitation pattern over Greenland is invariant for climate change. In fact, some studies using current information (Bromwich *et al.,* 1993; Bromwich, 1995) or results from ice cores (Kapsner *et al.,* 1995) suggest that there is little relation between temperature and precipitation changes over Greenland during the Holocene owing to the dominant effect of changes in atmospheric circulation.

Iceberg calving is a major component in the mass balances of both ice sheets. Reeh (1994) has estimated that the iceberg discharge from Greenland is 319×10^{12} kg/yr, compared with a discharge of 239×10^{12} kg/yr due to melting and meltwater runoff. The calving data are still approximate, because many major calving glaciers, especially in East Greenland, have not been measured, and because the seasonality of calving is not well defined even for the measured glaciers. Possible changes in the iceberg calving regimes due to climate change are not well understood, because a physical relation between calving rates and climate, geometry and ice flow has not yet been defined (Reeh, 1994). Much of the discharge from the Antarctic ice sheet is by calving of icebergs, but the various estimates of this flux differ. The most serious problem is that iceberg breakoff may be very episodic; in some areas the mean time between major calving events may be >100 years, so that historical data are of limited use. Also, the method of measuring the discharge using observations of current iceberg distributions is limited by uncertainty regarding mean lifetimes of icebergs of differing sizes, a critical relation needed for computing fluxes.

For the Antarctic ice sheet, possible changes in accumulation rates associated with eventual warming have been studied using regression techniques (e.g., Muszynski and Birchfield, 1985; Fortuin and Oerlemans, 1990). The problem here is the dependence within the set of predictors that are used in a multiple regression. Broadly speaking, when going from the margin of the ice sheet to the interior, the accumulation rate changes by an order of magnitude, typically from 0.4 m/yr to 0.04 m/yr (expressed as water-equivalent). This is partly due to lower air temperature, but also to other factors (e.g., distance to moisture source, topographic effects). So the sensitivity of the accumulation rate to temperature (identified as the regression coefficient for temperature) depends on the choice of the other predictors. This explains the large difference in the results of Muszynski and Birchfield (1985) and Fortuin and Oerlemans (1990); see Table 7.6. In addition, accumulation rates on the Antarctic ice sheet may be sensitive to changes in sea ice extent and concentration (Giovinetto and Zwally,

1995).

An extensive sensitivity study with a two-dimensional (vertical plane) meteorological model was conducted by Fortuin and Oerlemans (1992). In this model, the zonal mean distribution of temperature and precipitation is calculated with a four-layer dynamic model, which includes a radiation scheme. For a uniform warming, this model predicts a higher accumulation rate and a stronger evaporation in the coastal regions than in the interior. The net effect over the grounded ice is an increase in the mass balance corresponding to a –0.27 mm/yr sea level change for a 1°C warming. This value is slightly less than the –0.34 mm/yr found if the change in accumulation is set proportional to the relative change in saturation vapour pressure.

In some parts of the Antarctic Peninsula, surface temperatures are sufficiently high that melting occurs. Drewry and Morris (1992) estimated the potential contribution to sea level due to increased runoff from that part of the peninsula where melting may occur (about 20,000 km^2). They suggest a sensitivity of 0.012 mm of sea level rise/yr/°C, which is small compared to the sensitivity for the whole of Antarctica as listed in Table 7.6.

In summary, based on Tables 7.5 and 7.6, the value of the static sensitivity, in terms of equivalent sea level change, is estimated here as 0.30 mm/yr/°C for the Greenland ice sheet, and –0.30 mm /yr/°C for the Antarctic ice sheet. The spread of the individual estimates suggests uncertainties of at least ±0.15 mm/yr/°C for both Greenland and Antarctica (excluding the possibility of collapse of the West Antarctic ice sheet – see Section 7.5.5).

7.3.4 Surface Water and Ground Water Storage

Changes in the terrestrial liquid-water budget can affect sea level. Shiklomanov (1993) points out that human activity affects the hydrologic cycle through water diversions, transformations of stream networks, and changes of the surface characteristics of drainage basins, as well as causing climatic changes on regional or global scales which affect water transfers between land and sea. Both direct anthropogenic changes, as well as natural and anthropogenic effects on the climate, are involved, and it is often difficult to separate the two. Some estimates of the contributions from the major components of the liquid-water budget are as follows:

- **Ground water depletion.** Ground water pumped at a rate in excess of recharge may add to sea level. Much of this water is used for irrigation and a major fraction is transpired or evaporated to the atmosphere

or contributes to runoff, eventually reaching the sea. Estimates of the current rate of ground water depletion range from 0.07 to 0.38 mm/yr in sea level equivalent (Klige, 1982; Meier, 1983; Sahagian *et al.*, 1994).

- **Surface reservoir and lake storage.** The filling of surface water reservoirs transfers water from sea to land, tending to lower sea level. Lakes depend on the balance of precipitation and evaporation, and also respond to changes in upstream runoff which may be affected by human action. Some of the world's large lakes are currently rising; others are falling, due in part to upstream irrigation and evaporation. The current effect of these non-synchronous changes on global sea level is probably small. Estimates of the current rate of increase in water storage in artificial reservoirs range widely, from about –0.09 to –0.54 mm/yr sea level equivalent (e. g., Golubev, 1983; Chao, 1988, 1994; Shiklomanov, 1993; Gornitz *et al.*, 1994; Rodenburg, 1994; Sahagian *et al.*, 1994).

- **Deforestation.** Gornitz *et al.* (1994) estimate that the combustion and oxidation of forests transfer water to the sea at a rate of about 0.03 mm/yr in sea level equivalent, but that decreased runoff (Henderson-Sellers *et al.*, 1993) increases water on land at a rate of 0.15 mm/yr. Sahagian *et al.* (1994), on the other hand, suggest that tropical forest loss and desertification are currently contributing to sea level rise by 0.15 mm/yr, due to the loss of biomass water, soil moisture and water vapour in the atmosphere. The net effect is thus uncertain.

- **Loss of wetlands.** Sahagian *et·al.* (1994) estimate a sea level rise component of at least 0.01 mm/yr due to the loss of wetlands.

- **Other changes.** Ice-rich permafrost may contain up to twice as much water (frozen) as the same soil in a thawed state. Lachenbruch and Marshall (1986) estimate a secular thawing of permafrost in Arctic Alaska equivalent to about 10 mm of ice melt/yr. If half of the pore water runs off on thawing, and if half of the global permafrost area is similarly ice-rich and degrading (certainly an upper limit), the contribution of permafrost thawing to sea level rise could be up to 0.1 mm/yr. Gornitz *et al.* (1994) have estimated the following additional effects: the water pumped for irrigation that is returned by infiltration to soil moisture; the water added to the atmosphere as water vapour in irrigated areas; the water that seeps into the ground each year below reservoirs; and the increased water vapour content of the atmosphere in the vicinity of the reservoirs. These effects may be appreciable, causing a decrease in the overall rate of sea level rise. However, these results are conjectural, and a more accurate budget calculation will require comprehensive global hydrological modelling of the type now under development for use in general atmospheric circulation models.

The above estimates of the terrestrial liquid-water components affecting sea level vary considerably, and the spread is so large that the sum could be either positive or negative. Earlier estimates (e.g., Klige, 1982; Meier, 1983; Robin, 1986) and more recent work (Chao, 1994; Rodenburg, 1994) suggest that the sum is, in fact, close to zero. Our best estimate at this time is that the current contribution to sea level rise due to these hydrologic factors is between −0.40 and +0.75 mm/yr, with a mean estimate of about + 0.1 mm/yr. Over the last 100 years, these factors may have contributed about 0.5 cm to sea level rise, although the uncertainties are very large indeed.

Few authors have ventured an estimate for future years. Sahagian *et al.* (1994) point out that the rate of dam building has slowed markedly, but that depletion of ground water reservoirs is likely to increase with growing demands for water, apart from any effects of climate change. They estimate that the contribution of anthropogenic changes in land hydrology to sea level rise will be 2.6 cm in the next 50 years if the present rates are maintained. This estimate, largely based on the expected continued depletion of ground water reserves, is reasonable, but a future decline in dam building is not supported by the analysis of Shiklomanov (1993). Unfortunately, the contribution to sea level from changes in hydrologic practices due to climate trends has not yet been fully analysed.

7.4 Can Sea Level Changes During the Last 100 Years be Explained?

A critical issue is whether the rise in sea level observed over the last 100 years can be explained. A synthesis of the model and observational data pertaining to the factors discussed in Section 7.3 is presented in Table 7.7. In general, there is broad agreement that both thermal expansion and glaciers have contributed to the observed sea level rise, but there are very large uncertainties regarding the role of the ice sheets and other hydrologic factors.

Table 7.7: *Estimated contributions to sea level rise over the last 100 years (in cm).*

Component contributions	Low	Middle	High
Thermal expansion	2	4	7
Glaciers/small ice caps	2	3.5	5
Greenland ice sheet	−4	0	4
Antarctic ice sheet	−14	0	14
Surface water and ground water storage	−5	0.5	7
TOTAL	−19	8	37
OBSERVED	10	18	25

For thermal expansion, the range 2–7 cm reported in Table 7.7 derives from model simulations, especially those carried out for this report and other recent model results (see Section 7.5). The various studies that have addressed this issue give answers that are of the same direction and similar magnitude. Observational data related to thermal expansion are presently too sparse to make global-scale estimates. Overall, it is likely that some oceanic thermal expansion would have occurred given the global mean warming of 0.3 to 0.6 °C observed over the same time period.

It is clear that many of the world's glaciers have retreated over the last 100 years. However, continuous, long-term measurements of the mass balances of glaciers and ice caps are very limited. Based on a combination of observations and simple models (see Section 7.3.2.2), it is concluded that glaciers and ice caps may have accounted for 2–5 cm of the observed sea level rise.

With respect to the Greenland and Antarctic ice sheets, there is simply insufficient evidence, either from models or data, to say whether the average mass balances have been positive or negative. Thus, the "zero" entries in Table 7.7 should be interpreted as a reflection of the current poor state of knowledge, rather than as an estimate of the current state of balance. As mentioned previously (Section 7.3.3.2), an imbalance of up to 25% of the annual mass turnover cannot be ruled out by existing data and methods, giving the wide range of uncertainty indicated in Table 7.7. However, a large positive mass balance of both ice sheets would seem unlikely, as this would have led to a substantial sea level *lowering* and would therefore be highly inconsistent with the observed sea level rise.

The current estimates of changes in surface water and ground water storage are very uncertain and speculative.

There is no compelling recent evidence to alter the conclusion of IPCC (1990) that the most likely net contribution during the last 100 years has been near zero or perhaps slightly positive, with an uncertainty of about ±6cm.

In total, based on models and observations, the combined range of uncertainty regarding the contributions of thermal expansion, glaciers, ice sheets and land water storage to past sea level change is about –19 cm to +37 cm – a very wide band of uncertainty which easily embraces the observed sea level rise (10–25 cm). The major source of uncertainty relates to the current mass balance of the polar ice sheets. Although the apparent discrepancy between the middle values of the "total" and "observed" estimates in Table 7.7 might suggest a net positive contribution from the ice sheets, the role of the other factors cannot be ruled out within the overall uncertainties. This problem in reconciling the past change, especially in relation to climate-related factors, emphasises the uncertainties in projections of the future, as noted below.

7.5 How Might Sea Level Change in the Future?

7.5.1 Recent Projections

In 1990, the IPCC concluded that for Scenario A, or the "Business-as-Usual" scenario, sea level would rise 66 cm by the year 2100 (Warrick and Oerlemans, 1990). Since then, additional estimates of future sea level rise have been made, as shown in Table 7.8. Although these more recent estimates seem lower than that of IPCC (1990), direct comparisons cannot easily be made due to differences in assumptions regarding factors such as emission scenarios, gas concentration changes, radiative forcing changes, the climate sensitivity and initial conditions.

For example, the estimate of Church *et al.* (1991) of 35 cm by 2050 was based on a rate of global warming which was *twice* that projected by IPCC (1990). The projection of Wigley and Raper (1993) of 46 cm by 2100 was based on a lower emission scenario (IPCC (1990)-B). The Wigley and Raper (1992) estimate of 48 cm by the year 2100 was based on the revisions of gas cycle models and lower

Table 7.8: *Recent estimates of future global sea level rise (cm) (updated from Warrick and Oerlemans, 1990). It should be noted that the estimates of sea level change are not strictly comparable, as they also reflect the combination of the authors' different estimates of future emissions, radiative forcing changes and model parameters, which are difficult to disentangle.*

Source (emission scenario)	Sea level Rise Component				Total rise[a]		
	Thermal Expansion	Glaciers and ice caps	Greenland ice sheet	Antarctic ice sheet	Best Estimate	Range[b]	To (Year)
IPCC90-A (Warrick & Oerlemans, 1990)	43	18	10	-5	66	31 to 110	2100
Church *et al.* (1991)[c]	25	[.....	10 (all ice)	.....]	35	15 to 70	2050
Wigley & Raper (1992)[d]	[........	....	not specified	.......]	48	15 to 90	2100
Wigley & Raper (1993)	25	[....	21 (all ice)	.......]	46e	3 to 124[f]	2100
Titus & Narrayanan (1995)[g]	21	9	5	–1	34	5 to 77[h]	2100
IPCC projections, this report [i]	28	16	6	–1	49	20 to 86	2100
This report (Section 7.5.3.2)[j]	15	12	7	–7	27		2100

a   In most cases from 1990.

b   No confidence intervals given unless indicated otherwise.

c   Assumes rapid warming of 3°C by 2050 for best case.

d   For IPCC emission scenario IS92a.

e   For IPCC (1990) Scenario B, best estimate model parameters.

f   For IPCC (1990) Scenarios A and C, with high and low model parameters, respectively.

g   Incorporates subjective probability distributions for model parameter values based on expert opinion.

h   Represents 90% confidence interval.

i   For the IPCC IS92a forcing scenario, using a climate sensitivity of 2.5°C for the mid projection and 1.5°C and 4.5°C for the low and high projections, respectively. Also see Raper et al., 1996.

j   For IPCC IS92a forcing scenario, with a constant 2.2°C climate sensitivity (no range provided).

k   Also see Raper *et al.*, 1996.

radiative forcing changes implied by IPCC (1992). The probabilistic estimates of sea level rise provided by Titus and Narrayanan (1995, 1996) were derived from subjective estimates of probability distributions of model parameter values made by a panel of expert reviewers; the result was a median estimate of 34 cm by the year 2100. Despite the differences in methods and assumptions, all of these recent "best estimates" of future sea level rise still fall within a range of 3–6 cm /decade.

7.5.2 *Revised IPCC Projections*

In this section, we offer revised future sea level projections that are consistent with the gas concentration, radiative forcing and climate changes discussed in earlier chapters. The individual contributions from oceanic thermal expansion, glaciers and ice caps, and the ice sheets of Greenland and Antarctica are calculated separately and summed to give the total projected sea level rise to the year 2100.

7.5.2.1 *Methods and assumptions*

Projections of changes in sea level are made using a simple global climate model, a global glacier melt model, and sensitivity values relating temperature change to ice sheet mass balances. The modelling approaches are similar to those used in IPCC (1990) (Warrick and Oerlemans, 1990), but the models themselves have been substantially revised and updated as discussed below. As in Chapter 6, there are two sets of projections for each greenhouse gas emission scenario based on two alternative views of future aerosol concentrations. In the first, aerosol concentrations change in response to the changing emissions of their precursors assumed in the IS92 scenarios. In the second, future aerosol concentrations are held constant at 1990 levels for the purpose of sensitivity analyses. The latter set is included because of the large uncertainties in future aerosol forcing, related both to uncertainties in future emission changes and their consequent effects on radiative forcing change. Projections corresponding to "aerosols constant at 1990 levels" provide an estimate of the sea level response to a situation where global emissions of aerosol precursors remain similar to 1990 levels.

The projections of oceanic thermal expansion are made using an upwelling diffusion-energy balance climate model, that of Wigley and Raper (1987, 1992, 1993). Since changes in surface air temperature and oceanic thermal expansion are interactive (in general, they tend to be inversely related – the greater the thermal expansion, the less the surface warming for a positive change in radiative forcing), it is appropriate for the sake of consistency that

this is the same model used to project temperature changes in Chapter 6. The model has been recently updated and revised (Raper *et al.*, 1996) to incorporate different land/ocean climate sensitivities and temperature-dependent upwelling rates to simulate a slow-down of the thermohaline circulation with global warming, effects that are suggested by recent coupled ocean-atmosphere model experiments (Cubasch *et al.*, 1992; Manabe and Stouffer, 1994; Murphy, 1995). As demonstrated in Chapter 6, the revised model is able to emulate well the thermal expansion and temperature predictions of more complicated coupled ocean-atmosphere models (see Sections 6.3.1–6.3.3 for a description of model parameters, assumptions and results). The predicted temperature changes from this model were used to force the glacier and ice sheet models.

Changes in the volume of glaciers and ice caps were estimated with a revised version of the global glacier model used in IPCC (1990) (Wigley and Raper, 1995). In the revised model, the driving force for melting (or change in ice volume) is the difference between the ice volume and the "equilibrium" value of the ice volume, the latter itself being a function of temperature change. The model has a small number of parameters and is tuned against observationally based estimates of the glacier volume change over the period 1900 to 1961. The model incorporates a range of characteristic response times due to glacier dynamics; this is necessary because projections are needed for the next 100 years, which is of the same order of magnitude as real glacier response times. The model is regionally disaggregated to take into account the variations in the altitudinal range of the world's glaciers and their response times, so that as the more "sensitive" glaciers begin to disappear during model simulations, the average characteristic response time and other model parameters values are altered, giving a non-linear response of glacier melt to temperature change. At the initial year of the simulation (1880), it is assumed that the world's glaciers are in steady-state with respect to the prevailing climate and that they contain a volume equal to 30 cm in sea level equivalent.

For the Greenland and Antarctic ice sheets, the simplifying assumption made for the present set of model projections is that the dynamic response can be ignored on the time-scale of decades to a century (in Section 7.5.3 this assumption is relaxed). Accordingly, as for the IPCC (1990) assessment, the changes in the surface mass balance of the ice sheets are represented by static sensitivity values (in terms of sea level equivalent) as discussed above. For Greenland the sensitivity value is 0.30 ± 0.15 mm/yr/°C, and it is assumed that the Greenland ice sheet was in equilibrium for the initial year (1880) of the simulation. For Antarctica, the sensitivity value is –0.20 ± 0.25 mm/yr/°C (including a term for the possible instability of the West Antarctic ice sheet), and it is assumed that the mass balance of the Antarctic ice sheet was negative in 1880 in accordance with glacial-interglacial model simulations (Section 7.3.3.2). (An imbalance of 0.1 ± 0.5 mm/yr in sea level equivalent has been assumed as a baseline trend that is extrapolated over the simulation period).

Possible changes in surface and ground water storage are not taken into account, for three reasons: (1) the available data are insufficient for meaningful extrapolation; (2) existing studies (e.g., Sahagian *et al.*, 1994) suggest that the future contribution would, in the worst case, be rather small; (3) such changes are not, for the most part, caused directly by climate change.

For all the sea level projections, the climate model was run from 1765 to 2100. Up to 1990, the same historical radiative forcing changes (including aerosol effects) were used; thereafter, the various IS92 emission scenarios were applied. The glacier and ice sheet models, which were all run from 1880, were forced with the model-derived global mean temperature changes (except in the case of Greenland, for which the temperature changes were scaled by a factor of 1.5 in accordance with AOGCM results). Projections were made for the following sets of scenarios (see Chapter 6 for an elaboration of methods and assumptions regarding scenarios):

- IS92a–f, using "best-estimate" model parameters, both including and excluding the effects of changes in aerosol concentrations after 1990;

- IS92a: low, mid and high projections, both including and excluding the effects of changes in aerosol concentrations after 1990;

- extreme range of projections, based on the highest and lowest forcing scenarios and with model parameters chosen to maximise or minimise sea level changes.

7.5.2.2 Modelled past changes
For the period 1880 to 1990, the calculated change in sea level change ranges from –1 cm to 17 cm, with a middle estimate of 7 cm (of which more than half is due to thermal expansion, followed by glaciers). This is low in comparison to the observed range based on tide gauge

records (10–25 cm; see Section 7.2.1) because it only takes into account an estimate of the anthropogenic component of past climate forcing, giving a lower temperature change than that observed (about 0.29°C as compared to 0.45°C; see Section 6.3). Similarly, for the projections of future sea level rise only the estimated anthropogenic climate forcing is taken into account; no other assumed climate- or non-climate-related (or "unexplained") component of the observed past sea level rise is extrapolated into the future projections. For this reason, both the past and future projections are likely to be underestimated.

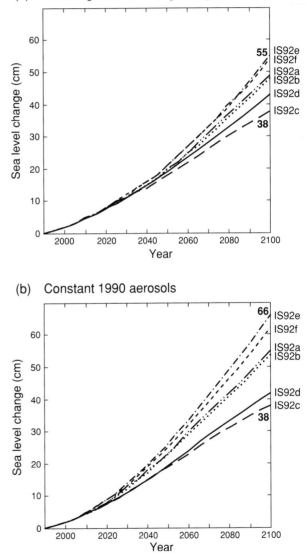

Figure 7.6: Projections of global sea level rise over the period 1990 to 2100 for Scenarios IS92a–f, using best-estimate model parameters, including the effects of changing aerosol concentrations after 1990 (a) and, to indicate the sensitivity of projections to aerosol effects, for aerosol amounts constant at 1990 levels (b).

7.5.2.3 Scenarios IS92a–f

The 1990–2100 changes in sea level for scenarios IS92a–f, with "best-estimate" model parameters, are shown in Figure 7.6 (a). The range of the projections shown is determined by the range of future emissions under the IS92 scenarios and does not include the additional uncertainties in model parameter values. For the first decades of the projection period, the choice of emission scenario has little effect on the rate of sea level rise; even by the year 2050, the range of projected sea level rise is still relatively small, 18–21 cm. This is a consequence of lags in the climate system, caused primarily by the thermal inertia of the ocean, and of the continuing response of the ocean, climate and ice masses to past changes in radiative forcing and temperature. In the short term (i.e., several decades), future sea level rise is largely determined by past emissions of greenhouse gases. During the second half of the next century, however, the curves diverge noticeably. By the year 2100, the uncertainty introduced by the emission scenarios gives a range of sea level rise of 38–55 cm. The effect of holding aerosol amounts constant at 1990 levels is to increase the range of projected sea level rise based on the six IS92 scenarios, as shown in Figure 7.6 (b).

7.5.2.4 Scenario IS92a

The 1990–2100 rise in sea level for IS92a, with high, middle and low projections based on model uncertainties, is shown in Figure 7.7 Taking into account future changes in aerosol amounts under the IS92a Scenario (Figure 7.7, solid curves), sea level is projected to rise by 20 cm by the year 2050, within a range of uncertainty of 7–39 cm. By the year 2100, sea level is estimated to rise by 49 cm, with a range of uncertainty of 20–86 cm.

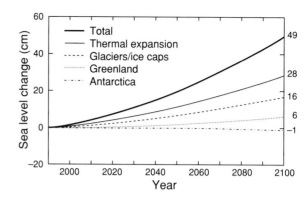

Figure 7.8: The projected individual contributions to global sea level change, 1990 to 2100, for Scenario IS92a (including the effects of changes in aerosol amounts beyond 1990).

For the middle projection under IS92a, more than half of the rise by the end of the next century is due to oceanic thermal expansion alone. This is followed by the contribution from glaciers and ice caps and from the Greenland ice sheet. The Antarctic ice sheet actually causes a slight decrease in sea level due to increased accumulation as a result of atmospheric warming (as shown in Figure 7.8).

To indicate the sensitivity of the projections to assumptions concerning aerosols, Figure 7.7 (dashed curves) shows results with aerosol concentrations held constant at 1990 levels: the IS92a projections at the year 2100 are about 10% higher than the projections that include aerosol changes. The inclusion of aerosol effects tends to lower the estimated changes in radiative forcing, compared to those due to greenhouse gases alone, both for the past and the future (see Chapters 2 and 6). This directly lowers the heating and consequent thermal expansion of the oceans, as well as reducing the surface temperature changes that drive changes in the glaciers and ice sheets.

The estimates of sea level rise presented here are lower than those given by IPCC (1990). For example, the "best estimate" value for IS92a by the year 2100 is 49 cm, compared with 66 cm for the corresponding case in IPCC (1990). This change is due primarily to the lower temperature projection (see Section 6.3.3), but it also reflects the compensating effects of a slow-down in the thermohaline circulation (which was not considered in 1990 and which leads to an increase in thermal expansion) and the changes made to the glacier model.

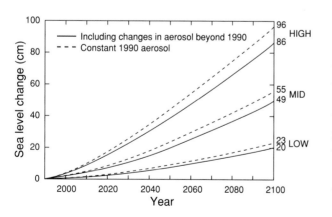

Figure 7.7: High, middle and low projections of global sea level rise over the period 1990 to 2100 for Scenario IS92a, for aerosol amounts constant at 1990 levels (dashed curves) and for changing aerosol amounts after 1990 (solid curves).

7.5.2.5 Extreme range

The extreme projections of sea level rise, taking into account uncertainties in both model parameters and future radiative forcing changes, are shown in Figure 7.9. There is

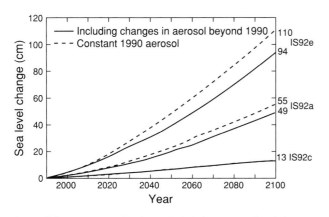

Figure 7.9: Extreme projections of global mean sea level rise from 1990 to 2100. The highest sea level rise curves assumed a climate sensitivity of 4.5°C, high ice melt parameters and the IS92e emission scenario, the lowest a climate sensitivity of 1.5°C, low ice melt parameters and the IS92c Scenario, and the middle curves a climate sensitivity of 2.5°C, mid-value ice melt parameters and the IS92a Scenario. The solid curves include the effect of changing aerosol; the dashed curves assume aerosol emissions remain constant at their 1990 levels.

an order of magnitude difference between the highest and lowest projections. The lowest projection shows sea level rising at an average rate of about 1 mm/yr over the next century, a rate comparable to that which has occurred over the last 100 years. The highest projection indicates an average rate of about 9–10 mm/yr, a rate which, on a global scale, is probably unprecedented over at least the last several thousand years. Although this range should be considered extreme, no attempt has been made to quantify the confidence interval.

7.5.2.6 Summary
The "best estimate" for IS92a is that sea level will rise by 49 cm by the year 2100, with a range of uncertainty of 20–86 cm. These projections of future sea level rise are lower than those presented in IPCC (1990). The differences are due primarily to the lower temperature projections, the inclusion of a slow-down of the thermohaline circulation, and changes to the glacier model. However, the basic understanding of climate-sea level relationships has not changed fundamentally since IPCC (1990). Thus, if future temperature change is higher than expected, sea level rise will also be higher.

Large uncertainties remain. In the particular set of models used above, these uncertainties derive mainly from uncertainties in radiative forcing and model parameters affecting temperature change, especially the value of the climate sensitivity (see Chapter 6). Relatively speaking,

uncertainties in future greenhouse gas emissions have comparatively little effect on the projected sea level rise, particularly over the first half of the next century, due largely to lags in the climate system. Nonetheless, combining the various sources of uncertainty, the extreme range of sea level projections is very large – an order of magnitude difference between the highest and lowest.

7.5.3 Possible Inter-Model Differences
In Section 7.5.2 above, a single set of integrated climate and ice-melt models was chosen to maximise consistency with the various chapters of this report. Consistency was achieved in three ways. First, the same simple climate model that was used to estimate oceanic thermal expansion was also used to estimate global-mean temperature change, thus ensuring consistency between Chapters 6 and 7. Second, this climate model was demonstrated to emulate both the temperature and thermal expansion predictions of certain AOGCMs, selected during the IPCC process as the standard for gauging the acceptability of a simple climate model as the means for examining the effects of various IPCC emission scenarios. Third, the land-ocean temperature differences, the changes in thermohaline circulation, and the magnitude of temperature changes in polar regions – as well as the global temperature and thermal expansion predictions – produced by the AOGCMS, were explicitly taken into account in using this set of models. Largely because of this consistency, these model results are promulgated as the IPCC sea level rise projections for the purposes of the present IPCC Assessment.

Using a single set of models, however, ignores the differences and uncertainties that might arise from alternative models. For this reason, in this section we present and compare the results of another set of climate, glacier and ice sheet models. This alternative set of models incorporates several recent advancements in modelling and is a credible complement to those models of Section 7.5.2. The models were forced by the identical set of IS92 radiative forcing scenarios for purposes of comparison. However, these models do not necessarily meet all the same requirements of consistency with the other chapters, and therefore were not put forward to the Summary for Policymakers as the IPCC sea-level projections. Nonetheless, the results are to be considered internally consistent, plausible and "state of the art". In this regard, the results highlight the uncertainties that could arise from various modelling approaches and emphasise the need for systematic inter-model comparisons as a post-IPCC activity in order to improve future projections, as discussed below.

7.5.3.1 Methods and assumptions

For predicting temperature changes and thermal expansion, a simple two-dimensional energy-balance climate model having latitudinal and seasonal resolution was used (Bintanja, 1995; Bintanja and Oerlemans, 1995; de Wolde *et al.*, 1995). This zonal-mean model has a climate sensitivity of approximately 2.2°C, and so would be expected to give lower sea-level changes than the best estimates described above (which use a sensitivity of 2.5°C). The model was first calibrated against the seasonal cycle of present-day observations of surface air temperature, ocean temperature (Levitus, 1982), and snow and sea-ice cover. Radiative forcing values for 1765, referenced to 1990 (see Section 6.3.2), were then used to obtain the initial state, after which the model was integrated over the period 1765–2100. Using comparable forcings, the thermal expansion results of this model have been compared to several coupled GCM results and found to be in reasonable agreement. The latitudinally and seasonally varying changes in the surface air temperature from the model were used to determine the sea level contributions from the glacier and ice sheet models.

For glaciers and ice caps, a range of sensitivity to climate change was used (from Oerlemans and Fortuin, 1992). The value of the sensitivity varies latitudinally, depending on the present-day precipitation rate, since glaciers in wetter regions are more sensitive than those in drier regions. These sensitivity values are time-independent and do not take into account the dynamic response of glaciers in a changing climate. Since the dynamic response may be very different for individual glaciers, it is uncertain how such an averaged dynamic behaviour should be included. However, it is assumed that on the time-scales considered here, the warming associated with the lowering of the ice surface and the decrease of the area due to the retreat of the glacier terminus will have a counterbalancing effect on the sensitivity values. The model calculations start in 1990. On the assumption that most glaciers are not in equilibrium with the present climate, a constant trend of 0.5 mm/yr is included (consistent with observations and the global glacier model results in Section 7.5.2).

Dynamic ice sheet models were used to estimate the sea level contributions of Greenland and Antarctica. These models take into account the effects of a changing climate on the dynamic responses of ice sheets; these flow responses were not explicitly included in the static sensitivity values used in Section 7.5.2. The two-dimensional time-dependent Greenland ice sheet model (after van de Wal and Oerlemans (1994), modified for dynamics in accordance with Mahaffy (1976) and Cadèe

(1992)) has a horizontal resolution of 20 × 20 km and is driven by atmospheric temperature changes, which change the surface mass balance. Ablation is calculated with an energy-balance model (van de Wal and Oerlemans, 1994), while the accumulation rate is kept constant at its present-day value (as described by Ohmura and Reeh, 1991). The initial state of the fully coupled model is its present-day equilibrium state.

The Antarctic ice sheet model (Huybrechts, 1990) is a three-dimensional, thermomechanic model coupled to a mass balance model that is driven by temperature changes interpolated onto a 40 km grid. Since accumulation appears to be strongly related to temperature in Antarctica, the accumulation rate is perturbed in proportion to the saturation water vapour pressure as temperature changes. The initial state of the ice sheet was obtained by integrating the ice sheet model over several glacial cycles. Although the model shows a continuing, long-term negative mass balance (refer to Figure 7.5), no trend is included for future projections because of the large uncertainties in the current state of balance due to the paucity of observations; the projections are calculated as the difference between the perturbed run and the reference run.

For both the Antarctic and Greenland ice sheet models as well as the glacier model, simulations begin in 1990, with temperature perturbations referenced to 1990. This is because of the non-linear response of the models and the tuning to present-day climate

7.5.3.2 Model results and comparisons

In general, the projections of future sea-level rise from this set of models are substantially lower than the revised IPCC projections presented in Section 7.5.2 for the identical set of IS92 forcing scenarios. To illustrate, Figure 7.10 shows the highest and lowest projections of sea level rise for the period 1990–2100, along with the "best estimates" for IS92a. For IS92a, sea level is estimated to rise by 27 cm by the year 2100 (34 cm for the sensitivity case with constant aerosols), about 45% lower than the corresponding projection in Section 7.5.2. Possible explanations for the differences are given below. It should also be noted that, unlike Section 7.5.2, the value of the climate sensitivity (2.2°C) did not vary for the high and low projections, resulting in a smaller range of estimates than would otherwise have been the case.

The individual contributions to the total projected sea level change for IS92a (including aerosols) are shown in Figure 7.11 (compare to Figure 7.8). The relative contributions shown in Figure 7.11 are similar to those presented in Section 7.5.2: most of the future rise is caused

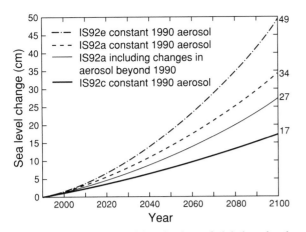

Figure 7.10: Alternative model projections of global sea level rise, 1990 to 2100, showing the range resulting from differences in IS92 scenarios and assuming a climate sensitivity of 2.2°C. The upper and lower curves show projections using the IS92e and IS92c Scenarios respectively, and assume aerosol emissions remain constant at their 1990 levels Also shown are the middle projections for IS92a, including and excluding the effects of changes in aerosol amounts beyond 1990. In contrast to Figure 7.9, these calculations do not span the full range of climate sensitivity or ice model parameters.

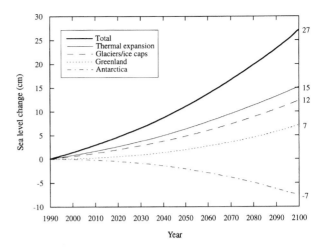

Figure 7.11: Alternative model projections of the projected individual contributions to global sea level change, 1990–2100, for Scenario IS92a (including the effects of changes in aerosol amounts beyond 1990).

by oceanic thermal expansion and increased melting of glaciers and ice caps, with a positive contribution from Greenland and a negative contribution from Antarctica (from increased accumulation). However, there are some large apparent differences regarding the absolute contributions between the two sets of model results.

For thermal expansion, the contribution to sea-level rise by 2100 is 15 cm, about half of that obtained by the simple climate model used in Section 7.5.2. This lower thermal expansion, however, can be largely explained by two factors. First, the climate sensitivity of the climate model is lower (2.2°C, as compared to 2.5°C used for the best-estimate projections in Section 7.5.2). Second, the thermohaline circulation is represented differently by the two models. For the revised IPCC projections in Section 7.5.2, the simple climate model simulates the slow-down of the thermohaline circulation with global warming, in accordance with most recent coupled GCM results (see Section 6.3), which allows greater warming at depth and larger thermal expansion. In the zonal-mean model used here, the thermohaline circulation was kept constant, with comparatively less thermal expansion. Other possible explanations for the differences in thermal explanation have yet to be examined fully.

For the Antarctic ice sheet, the dynamic model gives relatively much larger negative contributions to sea level by the year 2100 than the constant sensitivity values used

in Section 7.5.2 (–7 cm as compared to about –1 cm). For both the dynamic Antarctic and Greenland models used here, however, the portion of sea level change attributed to dynamic changes was found to be minor on this time-scale and does not appear to account for much of the inter-model differences. Rather, the inter-model differences are due largely to three other factors. First, the Antarctic temperature forcings were very different. For the revised IPCC projections, the global-mean temperature change was used (consistent with recent AOGCM results which show little enhancement of temperature changes for Antarctica). In contrast, the dynamic ice sheet model used here was forced with enhanced polar temperature changes obtained by the two-dimensional climate model. Second, the ice-sheet sensitivity value used for the revised IPCC projections was considerably lower, since it included a term for the possible instability of the West Antarctic ice sheet. Third, the revised IPCC projections included an extrapolation of an assumed 1880 negative mass balance (a relatively small effect) throughout the simulated period . Together, these factors result in a substantial difference between the two sets of model results for Antarctica.

For the Greenland ice sheet and glaciers, the sea level projections obtained by the two sets of models are in closer agreement, perhaps by coincidence, despite the different modelling approaches that were followed.

In summary, the uncertainties in sea level projections attributed to inter–model differences may be significant. Most of the differences between the model results discussed in this section can probably be attributed to assumptions about the climate sensitivity, changes in the thermohaline circulation, and the spatial patterns of temperature changes. However, no comprehensive model intercomparisons of the different component models used for sea level projections have been carried out. Such comparisons are required to identify the directions for improved models and to narrow the uncertainties in future projections.

7.5.4 Possible Longer-term (>100 years) Changes

In IPCC (1990), it was shown that on the decadal time-scale, sea level could be expected to continue to rise throughout the next century even if greenhouse forcing were stabilised by 2030 (Warrick and Oerlemans, 1990). Since IPCC (1990) several model experiments have been carried out, using both complex and simple models, that reveal the longer-term, multi-century implications of greenhouse-gas forcing on sea level.

One such transient experiment involved the use of an AOGCM in which the CO_2 concentration was increased by 1%/yr until it doubled, after 70 years (e.g., Manabe and Stouffer, 1993, 1994). Although the concentration was stabilised after doubling, sea level continued to rise from thermal expansion alone. At the end of five hundred years, sea level had risen one metre due to thermal expansion and was still rising, even though temperature changes had largely been stabilised.

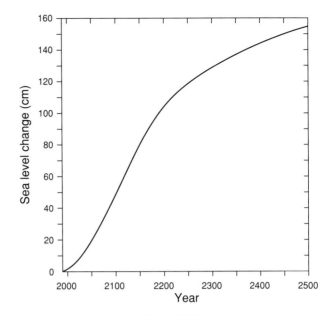

Figure 7.13: Long-term (1990 to 2500) projection of global sea level rise under an extended emission scenario comprising IS92a Scenario to 2100, with a linear decrease in greenhouse gas emissions to zero by the year 2200.

Similar results have also been derived using a simpler climate model (Raper *et al.*, 1996; Wigley, 1995). To illustrate, Figure 7.12 (reproduced from Raper *et al.*, 1996) shows the long-term effects on sea level of stabilising the atmospheric concentrations of CO_2 at 450 ppmv in 2100 and 650 ppmv in 2200 (Scenarios S450 and S650, respectively – see Chapter 2), for the high, middle and low sets of climate and ice-melt model parameters. The changes shown are those arising from CO_2 alone. In all but the lowest projections, sea level continues to rise at a scarcely unabated rate for many centuries after concentration stabilisation. Figure 7.13 provides another illustration of the long-term effects of anthropogenic forcing on sea level. In this scenario, anthropogenic emissions of CO_2, CH_4, N_2O, the halocarbons and SO_2 (an important precursor of aerosols) follow IS92a to 2100 and then are assumed to decline linearly to *zero* over 2100 to 2200. In this scenario, total radiative forcing peaks around the year 2160, but sea level is still rising by 2500, at which time it has reached 150 cm.

In both the complex coupled atmosphere-ocean models and the simpler climate models, most of the residual sea level rise is due to the thermal inertia of the oceans and continued thermal expansion. In addition, in the case where simple ice sheet models are included, the assumed large response times involving ice dynamics result in continuing effects on ice volumes after temperature changes have

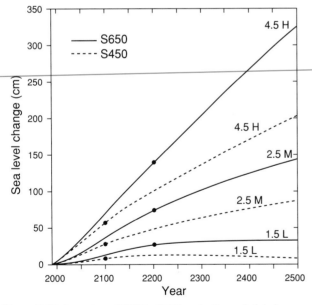

Figure 7.12: Long-term (1990 to 2500) projections of global sea level rise for stabilisation of CO_2 concentration at 450 ppmv (S450) and 650 ppmv (S650). Dots denote the dates of CO_2 stabilisation (from Raper *et al.*, 1996). Calculations assume the "observed" history of forcing to 1990, including aerosol effects (see Figure 6.18) and then CO_2 concentration increases only beyond 1990.

largely ceased. Overall, these results reinforce the conclusions of IPCC (1990) that a long-term "sea level rise commitment" must accompany greenhouse-gas-induced warming. Thus, even if greenhouse gas concentrations were stabilised, sea level would continue to rise for many centuries because of the large inertia in the ocean-ice-atmosphere climate system.

7.5.5 Possible Instabilities of the West Antarctic Ice Sheet

The West Antarctic Ice Sheet (WAIS) is a marine ice sheet – it rests on a bed well below sea level. It has long been argued (Weertman, 1974) that the WAIS may be inherently unstable because the interior, grounded ice (inland ice) cannot respond fast enough to changes in thickness of the floating portions at their junction, the grounding line. It has also been argued (Thomas, 1973, 1985) that the large abutting ice shelves create "back pressure" which prevents the collapse of the inland ice, such that ice shelf thinning or break-up could cause the grounded ice to "surge" – another critical element contributing to marine ice sheet instability.

These notions are changing. It is now known that the activity of the WAIS is dominated by fast-flowing, wet-based ice streams whose characteristics blend gradually into those of the floating ice shelves and whose response times to changes in the grounding line appear to be very rapid (Alley and Whillans, 1991). However, the effects of these dynamic ice streams on the stability of the WAIS is very much in dispute. In the view of some glaciologists, the ability of ice streams to transport ice rapidly from the interior to the ocean, on a time-scale of the order of 100 years, indicates an enhanced capability for a drastically accelerated discharge. A contrary view is that the short response time of ice streams removes the flux imbalance at the grounding line so that the purported instability may not exist.

Recent theoretical work is equivocal. Several recent treatments support the idea that the transition zone between the grounded and floating ice does not act as a source of instability (Van der Veen, 1985; Herterich, 1987; Barcillon and MacAyeal, 1993, Lestingant, 1994). On the other hand, there is also support for the idea of instability, which may include the concept of ice shelf buttressing at the grounding line (NASA, 1991). A recent theoretical development that suggests dramatic instability of a marine ice sheet is the so-called "binge-purge" cycle put forward to explain the massive outpourings of icebergs (Heinrich events) from Northern Hemisphere ice sheets during the last ice age (Alley and MacAyeal, 1994; MacAyeal, 1994). A model study of the WAIS over the last million years that incorporated ice streams and their slippery beds suggests that the ice sheet did collapse in the past but that the outflow rates were only a few times faster than at present (MacAyeal, 1992).

Recent observational work does not present a clear answer either. On the one hand, there is evidence suggesting unstable behaviour of the WAIS: Ice Stream B is currently flowing too rapidly for a steady-state; the current growth of the Crary Ice Rise is affecting the regional velocity field and perhaps reducing the discharge of Ice Stream B; and Ice Stream C has stagnated in the last 100–150 years (Retzlaff and Bentley, 1993). Furthermore, there is geologic evidence that this ice sheet has been largely or completely absent at some time after its initial formation (Scherer, 1991; Burckle, 1993), which suggests transient behaviour in this part of the Ross Ice Shelf system. On the other hand, there is evidence that does not support the notion of WAIS instability: the steady flow for the last 1500 years (except for one pulse a few hundred years ago) as suggested by flow tracers in the Ross Ice Shelf; the current growth, rather than collapse, of the glaciers feeding Pine Island Bay (which lost its ice shelf in the recent geologic past); and the lack of evidence of drastic change in the height or flow of the WAIS at Byrd Station in the last 30,000 years (Whillans, 1976).

Given our present knowledge, it is clear that while the ice sheet has had a very dynamic history, estimating the likelihood of a collapse during the next century is not yet possible. If collapse occurs, it will probably be due more to climate changes of the last 10,000 years rather than to greenhouse-induced warming. Nonetheless, such a collapse, once initiated, would be irreversible. Our ignorance of the specific circumstances under which West Antarctica might collapse limits the ability to quantify the risk of such an event occurring, either in total or in part, in the next 100 to 1000 years.

7.6 Spatial and Temporal Variability

7.6.1 Geological and Geophysical Effects

The only globally coherent geological contribution to long-term sea level change about which we possess detailed understanding due to a detailed theory of the process is post-glacial rebound (Peltier and Tushingham, 1989; Lambeck, 1990). This is the process by which the solid Earth and the ocean have continued to adjust to the effects of deglaciation throughout the Holocene period (last 10,000 years). Sea level changes due to longer time-scale geological processes (e.g., sea floor spreading) are sufficiently small to be of little interest to this report (e.g., see Harrison, 1989; Meier, 1990). Most Holocene geological sea level data have been assimilated into, or used to verify, geodynamic models of the Earth (e.g., Tushingham and Peltier, 1991; Peltier, 1994). These models attempt to achieve a consistency between the geological sea level measurements, the history of glaciation, and the physics of the solid Earth, including the resulting changes in the gravitational field of the Earth. Mantle viscosity is determined from a best fit to the data. The models result in estimates of relative sea level at any point on the Earth's surface (including the interior of continents where for "relative sea level" one infers geoid height relative to the land surface). A recent example is shown in Figure 7.14. Such models are the only ones which can be employed on a global basis to estimate the rate one would expect sea level to be changing at the present time at each location due to the continuing response of the solid Earth to deglaciation.

There is some debate at present concerning the precise form of the Earth's viscosity profile. Some authors believe that the geological sea level data constrain the viscosity to be approximately uniform, while others think either that there is a large increase in viscosity below a 660 km discontinuity, or that the present data do not constrain the profile well enough to be useful. It is not clear at the present time how the uncertainty in the viscosity profile propagates into expected rates of vertical land movement. However, it is to be expected that more sophisticated post-glacial rebound models will be developed as this debate is resolved. For example, the removal of post-glacial rebound-related land movements from tide gauge records by means of Peltier's ICE-4G model (Peltier, 1994) has not been investigated.

Superimposed upon post-glacial rebound are a variety of local and regional isostatic and tectonic effects, many of which cannot be modelled accurately (for a review, see

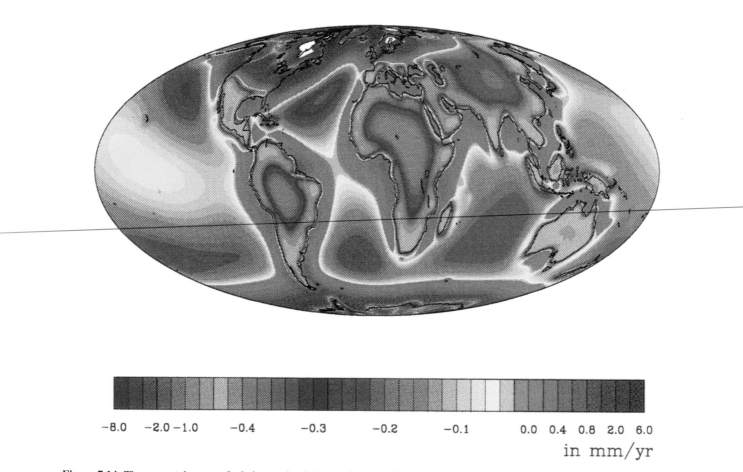

Figure 7.14: The present day rate of relative sea level change (in mm/yr) based on the topographically and gravitationally self-consistent theory of glacial isostatic adjustment of Peltier (1994). The calculation employed the ICE-4G model of the last deglaciation event of the current ice age and an internal radial viscoelastic structure comprising a 120 km thick lithosphere and an upper mantle of viscosity 2×10^{21} Pa s. This model somewhat overpredicts the ongoing rate of sea level rise due to post-glacial forebulge collapse along the east coast of the USA.

Emery and Aubrey, 1991). This is most obvious in the Mediterranean where the available archaeological sea level data primarily reflect local tectonic land movements (Flemming, 1969, 1978, 1993; Flemming and Webb, 1986) and where the Peltier post-glacial rebound models do not reproduce relative land-sea movements at all. Most large river deltas exhibit submergence associated with sedimentary isostasy which is clearly identified in tide gauge data. Very local effects, which have no relation to regional geology, can take place within the harbour of the tide gauge itself. However, differences between tide gauge sea level records can be used to provide maps of relative land movements on a regional basis, and such maps are usually consistent with previous geological knowledge of the area (Emery and Aubrey, 1991).

Studies of the vertical crustal motions on different time-scales can help elucidate patterns of tectonic deformation affecting tide gauge data. For example, along convergent plate boundaries, the uplift recorded by tide gauges or geodetic levelling measurements generally represent interseismic deformation that typically spans over 90% of the earthquake cycle (Bilham, 1991). However, the deformation during, and also shortly before or after, a major earthquake may exceed the average interseismic rate by an order of magnitude. Deformation associated with major earthquakes may temporarily cause displacements resulting in sea level changes comparable to the global sea level change (Bilham and Barrientos, 1991). These recent rates of crustal motion may differ substantially from longer-term geological trends derived from raised marine terraces (see, for example, Kelsey *et al.*, 1994; Mitchell *et al.*, 1994). Analyses of the changes in uplift rates on different time-scales may contribute to improved modelling of the earthquake cycle and extraction of the sea level signal from the tidal and geodetic data in those regions.

Anthropogenic effects (e.g., extraction of water from aquifers) can also result in considerable rates of subsidence (e.g., 10 cm/yr at Bangkok, Thailand). Ground water extraction has been an important factor in Venice sinking (Frassetto, 1991), while large recent rates of sea level rise at Manila, Philippines, has been blamed on coastal reclamation (Spencer and Woodworth, 1993).

It is important to realise that all sea level measurements, whether tide gauge, archaeological or geological, are measures of the level of the ocean relative to a land datum. The exceptions are those from space (e.g., from radar altimetry or via the Global Positioning System) for which the datum is in effect the computed satellite orbit. Such ground based measurements will always, therefore, contain some kind of inherent land-ocean level ambiguity.

However, with the advent of GPS and other forms of advanced geodetic measurement, it is now possible to measure vertical land movements independently of sea level changes (Section 7.7.1).

A good example of alternative methods of monitoring vertical land movements is provided by the recent breakthrough in development of absolute gravimeters (Carter *et al.*, 1994). Measurements are now repeatable to the 1 or 2 microgal level (1 gal = 1cm/s^2). For instance, a joint USA-Canada effort to produce the first reliable gravimetric measurement of glacial rebound shows at Churchill, Canada, that gravity has been decreasing at 1.6 microgal/yr since 1987. Theoretical models which account for viscous rebound of crust and mantle beneath Canada predict a linear relationship between gravity fall and crustal rise of about 0.15 microgal/mm. These gravity observations therefore indicate crustal uplift to be 11 mm/yr. At the same location, tide gauge records show that the Hudson Bay water level has been falling at an average rate of 11 mm/yr for the past 50 years.

Even if land levels were to be measured to good precision, there would still in principle be corresponding changes in the geoid to take into account. For example, Wagner and McAdoo (1986) presented maps of secular trends in geoid heights to be expected from post-glacial rebound. However, these are approximately an order of magnitude less than the corresponding changes in vertical land movements, and in studies of sea level changes over century time-scales or less, the geoid is considered in most applications to be time independent. Repeated space gravity missions, together with long-term laser tracking of dedicated geodetic satellites, should ideally be mounted to monitor such changes, while they can continue to be studied within geodynamic models.

7.6.2 Dynamic Effects

The variability of sea level on interannual to interdecadal time-scales is a major complication in determining reliable sea level trends and accelerations (Douglas, 1992). It is also clear that the global average sea level change is not a good indicator of local changes at any particular place.

In large part, these interannual to interdecadal fluctuations result from the fact that sea level topography, referenced to the geoid, is closely related to the dynamics and thermodynamics of the ocean. If the ocean were homogeneous and at rest, with a uniform atmospheric pressure field above it and no wind, the sea surface would correspond to the geoid (i.e., an equipotential of the gravity field). However, it does not; it differs from the geoid by ± one metre. Ocean and atmosphere are non-homogeneous and continuously moving within a variety of time and space scales, under gravitational forcing (for tides) and thermal forcing from the Sun (including variable wind stress and heat and fresh water exchanges at the sea surface).

In analysing trends, high-frequency ocean signals, swells, tides and surges are generally easily removed from sea level records by filtering techniques, as illustrated, among others, by Chelton and Enfield (1986) and Sturges (1987). But the analysis of the low-frequency residuals remains extremely difficult because these data have red spectra (the spectral energy keeps rising at low frequency), and the variability of the lowest frequencies that can generally be resolved with the longer available sea level records is of the same amplitude as (or larger than) the rising sea level signal. It is thus very important to understand the physical causes of these long-period events in order to be able to correct the data and improve the signal-to-noise ratio. Unfortunately, the forcing mechanisms are not well-known, especially at the decadal and interdecadal time-scale. The processes involve natural oscillations of the ocean-atmosphere system, which result in perturbations of the three dimensional state of the ocean, in terms of the dynamics and thermodynamics and the feedback on the state of the atmosphere.

At the short-term interannual frequencies, one major event is the El Niño-Southern Oscillation phenomenon in the Pacific Ocean. Associated sea level oscillations have been clearly observed by *in situ* tide gauges and satellite altimetry, typically with eastward propagating equatorial Kelvin waves and westward reflected or locally-forced Rossby waves (Wyrtki, 1979, 1985; Miller and Cheney, 1990). Monthly maps of sea level anomalies are now produced routinely to document and follow this interannual variability of the Pacific Equatorial Ocean.

Along single coastlines, the efficiency of wave propagation processes is a major candidate for explaining the often observed coherency of the long-period sea level variabilities, as studied by Enfield and Allen (1980) and Chelton and Davis (1982) for the Pacific coast of the USA. At basin scale, long-period baroclinic Rossby wave propagation has been demonstrated to possibly lead to coherent sea level signals; Sturges (1987) thus explained the 5–8 years period coherency in the long records available for San Francisco and Honolulu, with amplitudes of 5–15 cm at these long periods.

At the interannual to decadal periods, sea level fluctuations must often be driven primarily by atmospheric forcing, wind and pressure. Maul and Hanson (1991) have observed a significant coherency, with peaks at 4 and 13 years, between the sea level variability of the tidal records along the Atlantic coast of the USA and the long-period, basin-scale variations in North Atlantic atmospheric surface pressure known as the North Atlantic Oscillation (Rogers, 1984). Additionally, Sturges and Hong (1995) have demonstrated that, at Bermuda, sea level and thermocline variability, estimated from a simple ocean model forced by the wind, is in remarkably good agreement with observations at long periods.

However, sea surface temperature variations, and hence buoyancy forcing by the atmosphere, are coherent with changes in the wind (Kushnir, 1994). Thus, thermodynamic processes also have to be considered in interpreting long-term sea level variations. Decadal to interdecadal changes in temperature and salinity of the three-dimensional structure of the ocean have been widely reported in many places (Gordon *et al.*, 1992). Taking the North Atlantic as an example, one can refer to the Great Salinity Anomaly traced by Dickson *et al.* (1988) around the North Atlantic subpolar gyre from the mid-1960s to the late 1970s, and to the cooling at intermediate depth in the North Atlantic sub-polar gyre observed by Lazier and Gershey (1991), Read and Gould (1992) and Koltermann and Sy (1994), due to drastic changes in the production rates of Labrador Sea Water (LSW) and its property characteristics. This freshening is also present in the 24.5°N sections of hydrographic data, between 400 and 500 m (Lavin, 1993), due to some amount of cooler LSW circulating in the subtropical gyre. The complex picture emerging from these observations is related to the North Atlantic Deep Water (NADW) production, which propels the global planetary Thermohaline Conveyor Belt circulation (Broecker *et al.*, 1985). Changes in the production rate of NADW must change the northward transport of upper ocean warm water, and it could lead to an enhancement of sea rise in the mid-latitudes of the North Atlantic (Mikolajewicz *et al.*, 1990).

Large-scale ocean circulation has a direct signature on the sea surface topography, through geostrophic balance. Sea surface slope variabilities are thus thought to be good indicators of the large-scale ocean transport variabilities. Maul *et al.* (1990) have shown that it is true for the Gulf Stream for the semi-annual and annual band. However, as noticed by Sturges and Hong (1995) this has not been fully demonstrated for interannual and lower frequencies from the analysis of long coastal sea level records, until recently the only available source of data (see also Whitworth and Peterson, 1985, for the Antarctic Circumpolar Current). Hence, there is interest in long-term, deep-sea bottom pressure measurements (as used in WOCE) and satellite altimetry for studying this relation between ocean slope and ocean transport variabilities.

As noted by Gates *et al.* (1992), there are now a sufficient number of AOGCM integrations over 50–100 year and longer periods to provide preliminary information on the simulation of atmospheric and oceanic decadal variability. Over these last years, a great deal of interest has focused on the behaviour of the thermohaline circulation of the North Atlantic, the formation of NADW, the Conveyor Belt Circulation, and the possible existence of multiple equilibrium states for the global ocean circulation. The effect of these processes on the interdecadal variabilities of sea level has been noted above. Another mode of decadal climate variability, in the North Pacific Ocean, has been found in AOGCMs of the MPI, attributed to a cycle involving air-sea interactions between the sub-tropical gyre circulation and the Aleutian low pressure system (Latif and Barnett, 1994). Besides its effect on sea surface temperature, it has an effect on sea level of 3.4 cm in decadal variability.

Coupled model simulations of climate change under increasing CO_2 have also shown that the regional differences in sea level change are larger than the globally averaged change (Mikolajewicz *et al.*, 1990; Gregory, 1993; Cubasch *et al.*, 1994b). This spatial variation comes mostly from the geographical distribution of surface temperature changes, but ocean dynamics do modify it, especially in areas where temperature changes occur to considerable depth. Different climate models give different patterns of local response, although there are some common features, such as a strong relative rise off the east coast of North America, and a marked relative fall north of the Ross Sea. An example from Gregory (1993) is shown in Figure 7.15. However, the non-equilibrium response in sea level of the world ocean to warming is, as yet, very poorly understood, and it is not clear that the present generation of AOGCMs can resolve such a process correctly.

Since the launch of the TOPEX/POSEIDON altimeter satellite in 1992, it has been possible to construct similar global maps of sea level change from direct measurements (Cheney *et al.*, 1994). As in the model simulations, regional variations observed by the altimeter are quite large, with amplitudes of the order of 5 cm on monthly time-scales, as compared to the global average increase of 3 mm/yr derived from these data. The combination of altimetry with sub-surface data and winds will provide a way of interpreting these low-frequency sea level phenomena.

7.6.3 *Trends in Extremes*
The statistical treatment of data on extreme sea levels has progressed considerably over the past few years. For example, Tawn and Vassie (1991) developed a "revised joint probability method" applicable to all types of tidal regime, while Coles and Tawn (1990) employed

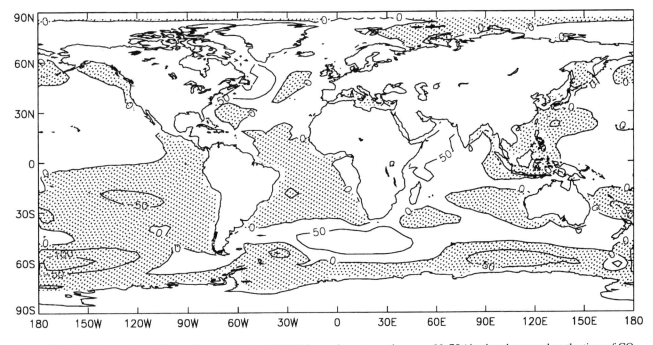

Figure 7.15: Relative sea level change from a transient AOGCM experiment over the years 66–75 (the decade centred on the time of CO_2 doubling). Relative sea level change here is calculated as the difference between the anomaly and control minus the change in global average sea level of 10.2 cm (source: Gregory, 1993).

multivariate extremes and the spatial properties of extreme processes to improve estimations at nearby sites with little or no data. Extreme levels, computed by a variety of old and new methods, are available for much of the world coastline for input to coastal impact studies (e.g., see de Mesquita and Franca, (1990) and Simon (1994), for coastal areas in Brazil and France, respectively).

The determination of a trend in extreme values is a difficult procedure since, by definition, the data become increasingly rare as the extremes are approached and estimates of statistics in the tails have increasingly wide confidence limits. Secular trends in annual maxima have tended to be studied primarily in north-west Europe where long records exist. For example, the spatial distribution of secular trends in annual maxima around the UK appears to be similar on average to those derived from tide gauge mean sea level and geological sea level data sets (Dixon and Tawn, 1992). Extreme levels, and their apparent temporal variation, have been studied at a number of "cities on water", such as Venice (Frassetto, 1991; Pirazzoli, 1991; Rusconi, 1993).

The continental shelf of north-west Europe is one example of an area where trends in storm surge activity have been studied. From analyses of storm surge frequency around the coast of the UK over the past 70 years, and at the Hook of Holland (Hoozemans and Wiersma, 1992) and the German coast over the past few centuries (Rodhe, 1980), there is no evidence of a long-term trend in storm surge activity, although Rodhe did point to a possible low-frequency (approximately 80-year) periodicity in flooding. On the other hand, evidence exists for a trend over the same period in regional winds and air pressures that would result in enhanced German Bight mean sea level (and the surge activity that it reflects) based on gridded Norwegian meteorological data for the shelf applied to a numerical model (Tsimplis *et al.*, 1994). Meanwhile, Führböter and Toppe (1991) observed tide gauge data in the German Bight directly and also indicate an increase of storm surge frequency in the area, at least since 1960 and possibly before. Local wind data for the Bight itself appear to show no obvious trend (Hoozemans and Wiersma, 1992) from 1950 to 1980, and in spite of large interdecadal variation, very little at all over century time-scales (Schmidt, 1991; Schmidt and von Storch, 1993). Clearly, findings will depend on the periods analysed and data spans. However, the difficulty in drawing conclusions relevant to climatic change studies from this relatively well-instrumented part of the world cautions against drawing conclusions for areas with considerably less data.

Numerical tidal models are frequently employed, with reductions in ocean depth, for the study of palaeotides (Scott and Greenberg, 1983) and the historical development of bedforms (Proctor and Carter, 1989; Austin, 1991). Conversely, they can be run with increased depth to predict changes in tidal amplitude that might accompany potential future sea level rise (Rijkswaterstaat, 1986). The effect of changes in depth (a surrogate for sea level rise) on storm surges can also be investigated. For example, Flather and Khandker (1993) described how tides and surges might be modified in the Bay of Bengal if sea level is modified, assuming that the character of the prevailing meteorology does not also change. In most cases, there will be little change in tide and surge (of order of a few centimetres per metre of sea level rise, unless in locally resonant situations).

7.7 Major Uncertainties and How to Reduce Them

7.7.1 Monitoring of Sea Level Change

For a future global sea level monitoring system, it is essential that information is integrated from many sources, with global and regional products (useful to scientists and non-scientists alike) derived from the blended data sets (Eden, 1990). Improvements in such a monitoring system would include five elements.

(1)   It should include a global sea level monitoring system based on a network of modern tide gauges. The Global Sea Level Observing System (GLOSS) of the Intergovernmental Oceanographic Commission (IOC) is a co-ordinated project for the monitoring of long-term global sea level change and is intended to serve the various purposes of oceanographic and climate change research into the next century (IOC, 1990; Woodworth, 1991). GLOSS consists of a network of approximately 300 tide gauges worldwide, of which over two-thirds are now operational. Technical developments in recent years have seen many of the traditional float and stilling well tide gauges replaced by modern systems based on pneumatic and acoustic principles (Spencer, 1992). Many of these have satellite or telephone data transmission equipment, enabling real-time data access and fault checking. Bottom pressure recorders (Spencer and Vassie, 1985), inverted echo sounders (Wimbush, 1990) and thermistor chain moorings for dynamic height (McPhaden, 1993) now provide quasi-sea level measurements in several areas of the deep ocean, which will provide information on the ocean circulation.

(2) An improved geodetic network is required which enables many of the sea level measurements to be placed in a stable global co-ordinate system with sub-centimetric accuracy, and which provides a decoupling of land and ocean level signals in the tide gauge records. Remarkable progress has been made in the past five years toward achieving the required network, through the development of the International Terrestrial Reference Frame (ITRF) by the International Earth Rotation Service (IERS). The ITRF is based on Very Long Baseline Interferometry (VLBI), Satellite Laser Ranging (SLR) and Global Positioning System (GPS) observations (Carter *et al.*, 1989; Bilham, 1991; Ashkenazi *et al.*, 1993; Baker, 1993; Carter, 1994). At many locations measurements of land movements near to the tide gauge sites by means of GPS will be supplemented by absolute gravity recording, with a change of gravity of 1 microgal (approximately the current accuracy of the technique) corresponding to a change of land level of 5 mm (Carter, 1994). VLBI, GPS and SLR enable precise time-series of polar motion and changes in rotation rate to be compiled, which may be pertinent to the problems addressed in this chapter.

(3) The monitoring system should include near-global observations of the ocean and ice sheets by means of satellite altimetry. Radar altimetry has become a major tool for studying sea level changes over most of the world ocean. Although a considerable amount of altimeter data have been acquired previously from, for example, the USA Navy Geosat satellite and the European ERS-1 mission, the launch of TOPEX/ POSEIDON in 1992 marked the start of a new era of precise sea level measurements from space. TOPEX/POSEIDON is providing sea level data with a single-point precision of about 5 cm (Fu *et al.*, 1994), enabling global mean sea level to be measured every 10 days with a precision of a few mm. Indeed, results from the first 2 years of the mission suggest an apparent sea level rise of approximately 3 mm/yr (Wagner *et al.*, 1994; Nerem, 1995). Given the number of potential altimeter errors at the millimetre level, it is premature to attach major significance to this result, but if measurements of this quality can be calibrated and maintained throughout a long-term program of multiple altimeter missions and integrated with the long-term GLOSS gauges, global sea level monitoring will be placed on a much firmer basis (Koblinsky *et al*, 1992). Meanwhile, ice sheet

topographies from ERS-1 and other near-polar orbiting radar altimetric satellites should result in complementary data sets of ice balance. When operational, laser altimetry over ice should be more precise than, and should complement, the radar altimeter data. However, the full potential of satellite altimetry to measure changes in ice mass will not be realised until truly global coverage is available.

Altimetry is a very important tool now available for oceanographic research, providing new insights into ocean tides, the ocean mesoscale, and basin-scale changes, in addition to global sea level change. There are corresponding advances in geophysics through the acquisition of detailed maps of the mean sea surface. In order to compute fields of absolute ocean currents for input to climate models with predictive sea level capability, precise altimetry of the mean sea surface must be complemented by detailed knowledge of the geoid by means of space gravity missions (Koblinsky *et al.*, 1992).

(4) Measurements are required of the temperature and salinity fields of the global ocean. Variations in long distance acoustic pulse travel times can provide for monitoring of average basin temperatures (Munk, 1989).

(5) For a better understanding of the ocean response to climate variability and future sea level rise, there is a need for further developments of coupled atmosphere, ocean and ice models which explicitly include sea level predictions. The continued growth in computer power will strongly support this effort, as it allows the complexity of the models to increase, thus improving their performance and realism. The scientific programmes WOCE and CLIVAR (Climate Variability programme) and the development of GOOS will provide the necessary observations to feed these models. Data assimilation techniques will thus be of primary importance in combining such data sets with models for their optimal use in understanding the climatic role of the ocean and predicting the evolution of sea level.

7.7.2 *Oceanic Thermal Expansion*
A strategy for long-term observation of the three-dimensional state of the ocean needs to be defined. This is the second goal of WOCE and will be developed under new programmes, like the oceanic component of CLIVAR and GOOS.

Observations derived from three-dimensional monitoring of the ocean will then have to be combined with models in order to make predictions. It can be anticipated that the on-going progress in ocean modelling (Semtner, 1994), and in data assimilation techniques within these models, will allow the improvement of our understanding of the interannual to interdecadal variabilities of the thermodynamic state of the ocean and the thermal contribution to changes in sea level, both temporally and spatially. Thus, continued support of both monitoring programmes and dynamic ocean modelling are necessary to enable global and regional predictions of future sea level changes.

7.7.3 Glaciers and Ice Caps

There remain large uncertainties regarding the future change in global glacier volume as a consequence of global warming. There are four major gaps in knowledge that need to be filled in order to better estimate the contributions of glaciers to sea level change.

(1) Probably the most critical need is the development of simple, yet realistic, models which include the processes linking meteorology to mass balance to dynamic response, and which include the resulting feedbacks (Meier, 1965). The importance of incorporating dynamic feedback is obvious when considering that in the next 50–100 years the estimated glacier volume change is expected to be a significant fraction of the total present volume, resulting in a major change in glacier area. Thus, the contribution of glaciers to sea level rise will reach a maximum and then decline despite continuing warming (Kuhn, 1985). The meteorological calculations can be accomplished by simple hydrometeorological (HM) models in which accumulation and energy balance are approximated by functions of seasonal precipitation and air temperature (e.g., Liestöl, 1967; Martin, 1978; Tangborn, 1980; Reeh, 1991), perhaps parameterized for mountain glaciers as functions of altitude (e.g., Kuhn, 1981; Oerlemans and Hoogendoorn, 1989; Laumann and Reeh, 1993). The dynamic processes can be modelled using conventional glacier flow models (e.g., Bindschadler, 1982; Stroeven *et al.*, 1989) or by simpler models that address the response times (Jóhannesson *et al*, 1989; Schwitter and Raymond, 1993).

(2) Model studies need to be extended to ice masses that are very different from the small mountain glaciers of the Alps and Scandinavia in order to encompass the broad spectrum of glacier behaviour. Current models are based on well-studied reference glaciers, practically all of which are small, mid-latitude glaciers (Kuhn, 1993). However, in terms of potential sea level rise, these glaciers are relatively insignificant. The glaciers that are especially critical with respect to sea level rise include the following: the large valley and piedmont glaciers of south-east Alaska; the Patagonian Ice Caps; and the monsoon-nourished glaciers of central Asia. The ice caps of the Arctic are also important and have regimes significantly different from those commonly modelled.

(3) There is a need to further quantify the process of internal accumulation (refreezing of meltwater) for incorporation into mass balance models. Internal accumulation (Trabant and Mayo, 1985; Pfeffer *et al.*, 1991; Reeh, 1991) is pervasive at high latitudes and at high altitudes, yet is frequently neglected in glacier-meteorological studies.

(4) Better understanding is required of iceberg calving flux and its interaction with ice flow dynamics. Iceberg calving is a major component of the mass balance of some large glaciers in the Arctic, Alaska, and at high southern latitudes. However, the iceberg calving flux interacts with ice flow dynamics in ways that are not well understood (Meier, 1994). This problem needs further investigation and incorporation into models.

7.7.4 Greenland and Antarctic Ice Sheets

Of all the terms that enter the sea level change equation, the largest uncertainties pertain to the Earth's major ice sheets. Relatively small changes in these ice sheets could have major effects on global sea level, yet we are not even certain of the sign of their present contribution. Obviously, better determination of the present ice sheet mass balances are needed, and additional historical (palaeoclimatic and palaeoglaciologic) studies are required in order to learn how changes have occurred in the past. The results of such studies need to be incorporated into improved modelling schemes in order to anticipate how the ice sheets will change in response to future climate perturbations.

Field (surface-based) observations of mass balances – the accumulation of snow and superimposed ice and the discharge of icebergs and meltwater – need to be continued and extended. Special attention also needs to be given to detecting change in ice dynamics, including the flow of ice streams and iceberg-calving rates. Additional observations of atmospheric water-vapour flux divergence would be useful in order to better define the processes leading to possible changes in snow accumulation. New ice cores in Greenland and high time-resolution ice cores from Antarctica will be needed to show the spatial and temporal changes in atmospheric circulation that occurred with the abrupt climate changes in the past, in order to test models of climate change in the high latitudes.

Special attention needs to be given to current changes in ice shelves, such as the break-up of a large part of the Larsen Ice Shelf on the Antarctic Peninsula in 1995, and to dynamic interactions between ice sheets, ice streams and ice shelves. While ice shelf break-up does not contribute directly to sea level rise, changes in the interconnected grounded ice will affect sea level in ways that cannot be predicted at the present. Improved understanding will require combined studies using glaciological, oceanographic and satellite observations. Because sea ice extent affects water vapour transport in the Antarctic, better definitions of variations in sea ice extent and of the distribution of ice thickness are needed.

The current mass balances of the Antarctic and Greenland Ice Sheets need to be determined and future changes monitored. Perhaps the most powerful tool for doing this in the near future is satellite altimetry. A laser altimeter is urgently needed on a polar-orbiting satellite, especially to detect changes in the low-accumulation regions and critical ice sheet margins where the existing radar altimetry is not adequate. These measurements should commence as soon as possible in order to provide a baseline for detecting greenhouse-induced changes in the future. The observed seasonal and interannual variations in surface elevation will provide information on precipitation variations, which can be used in energy-balance models and to test atmospheric general circulation model results. Accurate mapping of surface elevation changes by laser altimetry will also permit monitoring of ice sheet stability. Early warning of ice sheet collapse would thus be obtained before major ice shelf break-up or massive iceberg discharges were observed. Satellite synthetic-aperture radar interferometry may allow monitoring of ice stream velocities and grounding zone locations (Goldstein *et al.*, 1993; Herzfeld, 1994).

Improved modelling of the ice sheets is vitally needed. This should involve developing and coupling more realistic models of both atmospheric circulation and ice dynamics. Major uncertainties are associated with estimating future snowfall on the major ice sheets. Most existing models relate accumulation changes to changes in air temperature, which assumes that warmer air delivers more snow and that

atmospheric circulation will not change. Palaeoclimatic data show that storm strengths and trajectories have changed in the past and have greatly affected accumulation. This raises the possibility that future circulation changes will occur and will also affect precipitation. It is important that atmospheric and snow-surface processes in ice sheet regions be understood well enough so that model-based predictions of snow accumulation can be made directly, rather than using predictions of temperature and assumed temperature sensitivity to estimate precipitation. Palaeoclimatic data from high-resolution ice cores are needed to test these models.

Improvement is also needed in the ice dynamics models. Short-term, transient behaviour has been observed in major ice streams. The fact that these changes exist means that the physics incorporated in contemporary ice sheet models is not completely realistic. In particular, treatment of basal sliding and/or deforming basal till, together with the longitudinal stresses involved, needs to be incorporated. Another key aspect, currently not well understood and therefore impossible to model with confidence, is the rate of iceberg calving. These are the key elements in defining the interactions between ice sheets, ice streams and ice shelves. These, together with changes of snow accumulation, are the most uncertain aspects of predicting changes in the ice sheets over the next decades to centuries, and need major improvement.

References

Alley, R.B. and I.M. Whillans, 1991: Changes in the West Antarctic Ice Sheet. *Science*, **254**, 959–963.

Alley, R.B. and D.R. MacAyeal, 1994: West Antarctic ice sheet collapse: chimera or clear danger? *Antarctic Journal of the U.S.*, **28** (5), 59–60.397

Ambach, W. and M. Kuhn, 1989: Altitudinal shift of the equilibrium line in Greenland calculated from heat balance characteristics. In: *Glacier Fluctuations and Climatic Change*, J. Oerlemans (ed.), Kluwer, Dordrecht, pp 281–288.

Ashkenazi, V., R.M. Bingley, G.M. Whitmore and T.F. Baker, 1993: Monitoring changes in mean-sea-level to millimetres using GPS. *Geophys. Res. Lett.*, **20**(18), 1951–1954.

Austin, R.M., 1991: Modelling Holocene tides on the NW European continental shelf. *Terra Nova*, **3**, 276–288.

Baker, T.F., 1993: Absolute sea level measurements, climate change and vertical crustal movements. *Global Planet. Change*, **8**(3), 149–159.

Barcillon, V. and D.R. MacAyeal, 1993: Steady flow of a viscous ice stream across a no-slip/free-slip transition at the bed. *J. Glaciol.*, **39**, 167–185.

Barnett, T.P., 1984: The estimation of "global" sea level change: a problem of uniqueness. *J. Geophys. Res.*, **89** (C5), 7980–7988.

Bauer, A. 1968: Nouvelle estimation du bilan de masse de l'Inlandsis du Groenland. *Deep Sea Research* , **14**, 13–17.

Bentley, C.R. and M.B. Giovinetto, 1991: Mass balance of Antarctica and sea level change. In: *International Conference on the Role of the Polar Regions in Global Change: Proceedings of a Conference Held June 11–15, 1990 at the University of Alaska Fairbanks*, G. Weller, C.L. Wilson and B.A.B. Serverin (eds.), Geophysical Institute and Centre for Global Change and Arctic System Science, University of Alaska, Fairbanks, pp 481–488.

Bilham, R., 1991: Earthquakes and sea level: space and terrestrial metrology on a changing planet. *Rev. Geophys.*, **29**, 1–30.

Bilham, R. and S. Barrientos, 1991: Sea level rise and earthquakes. *Nature*, **350**, 386.

Bindschadler, R., 1982: A numerical model of temperate glacier flow applied to the quiescent phase of a surge-type glacier. *J. Glaciol.*, **28** (99), 239–265.

Bindschadler, R.A., 1985: Contribution of the Greenland ice cap to changing sea level: present and future. In: *Glaciers, Ice Sheets and Sea Level: Effects of a CO_2 -induced Climatic Change*. National Academy Press, Washington, pp 258–266.

Bintanja, R., 1995: The Antarctic Ice Sheet and Climate. Ph.D. Thesis, Utrecht University, 200pp.

Bintanja, R. and J. Oerlemans, 1995: The influence of the albedo-temperature feed-back on climate sensitivity. *Ann. Glaciol.*, **21**, 353–360.

Braithwaite, R.J. and O.B. Olesen, 1990: Increased ablation at the margin of the Greenland ice sheet under a greenhouse-effect climate. *Ann. Glaciol*, **14**, 20–22.

Broecker, W.S., D. Peteet and D. Rind, 1985: Does the ocean-atmosphere have more than one stable mode of operation? *Nature*, **315**, 21–25

Bromwich, D., 1995: Ice sheets and sea level. *Nature*, **373**, 18–19.

Bromwich, D.H., F.M. Robasky, R.A. Keen and J.F. Bolzan, 1993: Modeled variations of precipitation over the Greenland Ice Sheet. *J. Climate*, **6** (7),1253–1268.

Brown, C.S., M.F. Meier and A. Post, 1982: *Calving Speed of Alaska Tidewater Glaciers, With Application to Columbia Glacier*. Studies of Columbia Glacier, Alaska, Geological Survey Professional Paper 1258-C. Government Printing Office, Washington DC,13pp.

Budd, W.F. and I.N. Smith, 1985: The state of balance of the Antarctic ice sheet – an updated assessment 1984. In: *Glaciers, Ice Sheets and Sea Level: Effects of a CO_2-induced Climatic Change*. National Academy Press, Washington, pp. 172–177.

Budd, W.F. and I. Simmonds, 1991: The impact of global warming on the Antarctic mass balance and global sea level. In: *International Conference on the Role of the Polar Regions in Global Change: Proceedings of a Conference Held June 11–15, 1990 at the University of Alaska Fairbanks*, G. Weller, C.L. Wilson and B.A.B. Serverin (eds.), Geophysical Institute and Centre for Global Change and Arctic System Science, University of Alaska, Fairbanks, pp 489–494.

Burckle, L.H., 1993: Is there direct evidence for late Quaternary collapse of the West Antarctic ice sheet? *J. Glaciol.*, **39**, 491–494.

Cadèe, M., 1992: Numerieke modellering van de Groenlandse ijskap: de toepasbaarheid van een tweedimensionaal ijsstromingmodel. IMAU internal report, V92–10.

Carter, W.E. (ed.), 1994: *Report of the Surrey Workshop of the IAPSO Tide Gauge Bench Mark Fixing Committee*, 13–15 December 1993, Institute of Oceanographic Sciences Deacon Laboratory, Wormley, UK.

Carter, W.E., D.G. Aubrey, T.F. Baker, C. Boucher, C. Le Provost, D.T. Pugh, W.R. Peltier, M. Zumberge, R.H. Rapp, R.E. Schutz, K.O. Emery and D.B. Enfield, 1989: Geodetic fixing of tidegauge bench marks. *Woods Hole Oceanographic Institution Technical Report*, WHOI-89-31, 44pp.

Carter, W.E., G. Peter, G.S. Sasagawa, F.J. Klopping, K.A. Berstis, R.L. Hilt, P. Nelson, G.L. Christy, T.M. Niebauer, W.Hollander, H. Seeger, B. Richter, H. Wilmes and A. Lothammer, 1994: New gravity meter improves measurements, *EOS, Transactions,* American Geophysical Union, **75**(8), 90–92.

Chao, B. F., 1988, Excitation of the Earth's polar motion due to mass variations in major hydrological reservoirs. *J. Geophys. Res.*, **93**, B11,13811–13819.

Chao, B.F., 1994: Man-made lakes and sea level rise. *Nature*, **370**, 258.

Chelton, D.B. and R.E. Davis, 1982: Monthly mean sea level variability along the west coast of North America. *J. Phys. Oceanogr.*, **12**, 757–784.

Chelton, D.B. and D.B. Enfield, 1986: Ocean signals in tide gauge records, *J. Geophys. Res.*, **91**, 9081–9098.

Chen, J. and M. Funk, 1990: Mass balance of Rhonegletscher during 1882/83–1986/87. *J. Glaciol.*, **36**(123), 199–209.

Chen, J. and A. Ohmura, 1990: Estimation of alpine glacier water resources and their change since the 1870s. In: *Hydrology in Mountainous Regions. I – Hydrological Measurements; the Water Cycle, Proceedings of two Lausanne Symposia, August 1990*. IAHS Publication **193**, 127–135.

Cheney, R.E., L. Miller, R.W. Agreen and N.S. Doyle, 1994: TOPEX/POSEIDON: The 2-cm solution, *J. Geophys. Res.*, **99** (C12), 24555–24563.

Church, J.A., J.S. Godfrey, D.R. Jackett and T.J. McDougall, 1991: A model of sea level rise caused by ocean thermal expansion. *J. Climate*, **4**, 438–456.

Coles, S.G. and J.A. Tawn, 1990: Statistics of coastal flood prevention. *Phil. Trans. R. Soc. Lond.* A, **332**, 457–476.

Cubasch, U., K. Hasselmann, H. Höck, E. Maier-Reimer, U. Mikolajewicz, B.D. Santer, and R. Sausen, 1992: Time-dependent greenhouse warming computations with a coupled ocean-atmosphere model. *Clim. Dyn.*, **8**, 55–69.

Cubasch, U., B.D. Santer, A. Hellbach, G. Hegerl, H. Höck, E. Maier-Reimer, U. Mikolajewicz, A. Stössel and R. Voss, 1994a: Monte Carlo climate change forecasts with a global coupled ocean-atmosphere model. *Clim. Dyn.*, **10**, 1–19.

Cubasch, U., G.C. Hegerl, A. Hellbach, H. Höck, U. Mikolajewicz, B.D. Santer and R. Voss, 1994b: A climate change simulation starting from 1935. *Clim. Dyn.*, **11**, 71–84.

Dawson, A.G., D. Long and D.E. Smith, 1988: The Storegga Slides: evidence from eastern Scotland for a possible tsunami. *Mar. Geol.*, **82**, 271–276.

de Mesquita, A.R. and C.A.S. Franca, 1990: Checking the transference method for the mean and extreme sea values. *Ann. simp. Ecosis. P. Caldas, Est. Minas Geraisd.*, **1**, 30–41.

de Wolde, J.R., R. Bintanja and J. Oerlemans: 1995: On thermal expansion over the last hundred years. *J. Climate*, **8** (11), 2881–2891.

Dickson, R.R., J. Meincke, S-A. Malmberg and A.J. Lee, 1988: The "Great Salinity Anomaly" in the northern North Atlantic. *Progress in Oceanography*, **20** (2), 103–151.

Dixon, M.J. and J.A. Tawn, 1992: Trends in U.K. extreme sea levels: a spatial approach. *Geophys. J. Int.*, **111**, 607–616.

Douglas, B.C., 1991: Global sea level rise. *J. Geophys. Res.*, **96** (C4), 6981–6992.

Douglas, B.C., 1992: Global sea level acceleration. *J. Geophys. Res.*, **97**, C8, 12699–12706.

Douglas, B.C., 1995: Global sea level change: determination and interpretation. In: *Reviews of Geophysics, Supplement. U.S. National Report to International Union of Geodesy and Geophysics 1991–1994*. pp. 1425–1432.

Douglas, B.C., R.E. Cheney, L. Miller and R.W. Agreen, 1990: Greenland ice sheet: is it growing or shrinking? *Science*, **248**, 288–289.

Drewry, D. J. (ed), 1983: *Antarctic glaciological and geophysical folio*. Scott Polar Research Institute, Cambridge.

Drewry, D. and E. M. Morris, 1992: The response of large ice sheets to climatic change. *Phil. Trans. R. Soc. Lond.*, B , **338**, 235–242.

Dyurgerov, M., 1994: *Global Mass Balance Monitoring*. Report to U.S.A. Department of State, Institute of Geography of Russian Academy of Sciences.

Dyurgerov, M.B. and M.F. Meier, 1995: Year to year fluctuation in global mass balances of glaciers and their contribution to sea-level changes. *IUGG XXI Assembly Abstracts*, **B318**, American Geophysical Union.

Eden, H.F. (ed.), 1990: *Towards An Integrated System For Measuring Long Term Changes In Global Sea Level*. Report of a Workshop held at Woods Hole Oceanographic Institution, May 1990. Joint Oceanographic Institutions Inc. (JOI), Washington, DC, 178pp. & appendix.

Ekman, M., 1988: The world's longest continued series of sea level observations. *Pure and Applied Geophysics*, **127**, 73–77.

Emery, K.O. and D.G. Aubrey, 1991: *Sea Levels, Land Levels, and Tide Gauges*. Springer-Verlag, New York, 237pp.

Enfield, D.B. and J.S. Allen, 1980: On the structure and dynamics of monthly mean sea level anomalies along the Pacific coast of North and South America, *J. Phys. Oceanogr.*, **10**, 557–578.

Fisher, D.A. and R.M. Koerner, 1994: Signal and noise in four ice-core records from the Agassiz Ice Cap, Ellesmere Island, Canada: details of the last millennium for stable isotopes, melt, and solid conductivity. *Holocene*, **4** (2), 113–120.

Flather, R.A. and H. Khandker, 1993: The storm surge problem and possible effects of sea level changes on coastal flooding in the Bay of Bengal. In: *Climate and Sea Level Change: Observations, Projections and Implications*, R.A.Warrick, E.M. Barrow and T.M.L. Wigley (eds.) Cambridge University Press, Cambridge, pp 229–245.

Flemming, N.C., 1969: Archaeological evidence for eustatic change of sea level and earth movements in the western Mediterranean in the last 2,000 years. *Special Paper of the Geological Society of America*, **109**, 1–125.

Flemming, N.C., 1978: Holocene eustatic changes and coastal tectonics in the north east Mediterranean: implications for models of crustal consumption. *Phil. Trans. R. Soc. Lond.*, A, **289**, 405–458.

Flemming, N.C., 1993: Predictions of relative coastal sea level change in the Mediterranean based on archaeological, historical and tide-gauge data. In: *Climatic Change and the Mediterranean*, L. Jeftic, J.D. Milliman, G. Sestini (eds.), Edward Arnold, London, pp 247–281.

Flemming, N.C. and C.O. Webb, 1986: Regional patterns of coastal tectonics and eustatic change of sea level in the Mediterranean during the last 10,000 years derived from archaeological data. *Zeitschrift fur Geomorphologie*, December, Suppl – Bd62, 1–29.

Fortuin, J.P.G. and J. Oerlemans, 1990: Parameterization of the annual surface temperature and mass balance of Antarctica. *Ann. Glaciol.*, **14**, 78–84.

Fortuin, J.P.G. and J. Oerlemans, 1992: An axi-symmetric atmospheric model to simulate the mass balance and temperature distribution over the Antarctic ice sheet. *Z. Gletscherk. Glazialgeol.*, **26**, 31–56.

Frassetto, R. (ed.), 1991: *Impact of Sea Level Rise on Cities and Regions*. Proceedings of the First International Meeting "Cities on Water", Venice, December 11–13, 1989, Marsilio Editori, Venice, 238pp.

Fu, L.L., E.J. Christensen, C.A. Yamarone, M. Lefebvre, Y. Menard, M. Dorrer, and P. Escudier, 1994: TOPEX/ POSEIDON mission overview. *J. Geophys. Res.*, **99** (C12), 24369–24381.

Führböter, A. and A. Toppe, 1991: Duration of storm tides at high water levels. In *Storm Surges, River Flow and Combined Effects*. International workshop, held 8–12 April 1991, Hamburg, Federal Republic of Germany. A contribution to the UNESCO-IHP project H-2-2. Unesco, Paris, pp 45–54.

Gates, W.L., J.F.B. Mitchell, G.J.Boer, U. Cubash and V.P. Meleshko, 1992: Climate modelling, climate prediction and model validation. In *Climate Change 1992: The Supplementary Report to the IPCC Scientific Assessment*, J.T. Houghton, B. A. Callander and S.K. Varney (eds.), Cambridge University Press, Cambridge, UK, pp.101–134.

Genthon, C., 1994: Antarctic climate modelling with general circulation models of the atmosphere. *J. Geophys. Res.*, **99**, 12953–12961.

Giovinetto, M.B. and H.J. Zwally, 1995: Annual changes in sea ice extent and accumulation on ice sheets: implications for sea level variability. *Z. Gletscherk. Glazialgeol.*, (in press).

Goldstein, R.M., H. Engelhardt, B. Kamb and R.M. Frolich, 1993: Satellite radar interferometry for measuring ice sheet motion: application to an Antarctic ice stream. *Science*, **262**, 1525–1530.

Golubev, G.N. 1983: Economic activity, water resources and the environment: a challenge for hydrology. *Hydrol. Sci.*, **28**, 57–75.

Gordon, A.L., S. Zebiak and K. Bryan, 1992: Climate variability and the Atlantic Ocean. *EOS, Transactions*, American Geophysical Union, **73** (15).

Gornitz, V. 1995a: Sea level rise: a review of recent past and near-future trends. *Earth Surface Processes and Landforms*, **20**, 7–20.

Gornitz, V., 1995b: A comparison of differences between recent and late Holocene sea level trends from eastern North America and other selected regions. *Journal of Coastal Research, Special Issue. No.17, Holocene Cycles: Climate, Sea Levels and Sedimentation*, C.W. Finkl, Jr.(ed), pp. 287–297.

Gornitz, V. and S. Lebedeff, 1987: Global sea level changes during the past century. In: *Sea Level Change and Coastal Evolution*, D. Nummedal, O.H. Pilkey and J.D. Howard (eds.), Society for Economic Paleontologists and Mineralogists (SEPM Special Publication No.41), pp 3–16.

Gornitz, V. and L. Seeber, 1990: Vertical crustal motions along the East Coast, North America, from historic and Holocene sea level data. *Tectonophysics*, **178**, 127–150.

Gornitz, V. and A. Solow, 1991: Observations of long-term tide gauge records for indicators of accelerated sea level rise. In: *Greenhouse Gas-Induced Climatic Change: A Critical Appraisal of Simulations and Observations*, M.E.Schlesinger (ed.), Elsevier, Amsterdam, pp 347–367.

Gornitz, V., C. Rosenzweig and D. Hillel, 1994: Is sea level rising or falling? *Nature*, **37**, 481.

Gregory, J.M., 1993: Sea level changes under increasing atmospheric CO_2 in a transient coupled ocean-atmosphere GCM experiment. *J. Climate*, **6**, 2247–2262.

Gröger, M. and H.P. Plag, 1993: Estimations of a global sea level trend: limitations from the structure of the PSMSL global sea level data set. *Global Planet. Change*, **8**, 161–179.

Haeberli, W. and M. Hoelzle, 1995: Application of inventory data for estimating characteristics of and regional climate change effects on mountain glaciers – a pilot study with the European Alps. *Ann. Glaciol.*, **21**, 206–212.

Harrison, C. G. A., 1989: Rates of sea level change from ocean basin volume estimates. *EOS*, **70** (43), 1002.

Harvey, L.D.D., 1992: A two dimensional ocean model for long term climatic simulations: stability and coupling to atmosphere and sea ice models. *J. Geophys. Res.*, **97**, 9435–9453.

Harvey, L.D.D., 1994: Transient temperature and sea level response of two-dimensional ocean-climate model to greenhouse gas increases. *J. Geophys. Res.*, **99**, 18447–18466

Hasselmann, K., R Sausen, E. Maier-Reimer and R. Voss, 1993: On the cold start problem with coupled ocean-atmosphere models. *Clim. Dyn.*, **9**, 53–61.

Henderson-Sellers, A., R.E. Dickinson, T.B. Durbridge, P.J. Kennedy, K. McGuffie and A.J. Pitman, 1993: Tropical deforestation: modeling local- to regional-scale climatic change. *J. Geophys. Res.*, **98**(D4), 7289–7315.

Herterich, K., 1987, On the flow within the transition zone between ice sheet and ice shelf. In: *Dynamics of the West Antarctic Ice Sheet*, C. J. van der Veen and J. Oerlemans (eds), Riedel, Dortrecht, pp 185–202.

Herzfeld, U.C., C.S. Lingle, L-H Lee, 1994: Recent advance of the grounding line of Lambert Glacier, Antarctica, deduced from satellite radar altimetry. *Ann. Glaciol.*, **20**, 43–47.

Hofstede, J.L.A. 1991: Sea level rise in the inner German Bight (Germany) since AD 600 and its implications upon tidal flats geomorphology. In: *Von der Nordsee bis zum Indischen Ozean*, H. Bruckner and U.Radtke (eds.), Franz Steiner, Stuttgart, pp. 11–27.

Hoozemans, M.J. and J. Wiersma, 1992: Is mean wave height in the North Sea increasing? *Hydro. J.*, **63**, 13–15.

Huybrechts, P., 1990: A 3-D model for the Antarctic ice sheet: a sensitivity study on the glacial-interglacial contrast. *Clim. Dyn.*, **5**, 79–92.

Huybrechts, P. 1994: The present evolution of the Greenland ice sheet: an assessment by modelling. *Global Planet. Change* **9**, 39–51.

Huybrechts, P., A. Letreguilly and N. Reeh, 1991: The Greenland ice sheet and greenhouse warming. *Paleogeography, Paleoclimatology, Paleoecology*, **89**, 79–92.

IASH (ICSI)/UNESCO, 1967: *Fluctuations of glaciers 1959–1965.* (vol. 1, P. Kasser, ed.). UNESCO, Paris.

IAHS (ICSI)/UNESCO, 1973: *Fluctuations of glaciers 1965–1970.* (vol. 2, P. Kasser, ed.). UNESCO, Paris.

IAHS (ICSI)/UNESCO, 1977: *Fluctuations of glaciers 1970–1975.* (vol. 3, F. Muller, ed.). UNESCO, Paris.

IAHS (ICSI)/UNESCO, 1985: *Fluctuations of glaciers 1975–1980.* (vol. 4, W. Haeberli, ed.). UNESCO, Paris.

IAHS (ICSI)/UNEP/UNESCO, 1988: *Fluctuations of glaciers 1980–1985.* (vol. 5, W. Haeberli and P. Muller, ed.). UNESCO, Paris.

IAHS (ICSI)/UNEP/UNESCO, 1989: *World glacier inventory – status 1988.* (W. Haeberli, H. Bosch. K. Scherler, G. 0strem, and C. C. Wallen, ed.). UNEP, Nairobi.

IAHS (ICSI)/UNEP/UNESCO, 1993: *Fluctuations of glaciers 1985–1990.* (vol. 6, W. Haeberli and M. Hoelzle, ed.). UNESCO, Paris.

IOC(Intergovernmental Oceanographic Commission), 1990: *Global Sea Level Observing System (GLOSS) Implementation Plan.* Intergovernmental Oceanographic Commission, Technical Series, No.35, 90p.

IPCC (Intergovernmental Panel on Climate Change), 1990:

Climate Change: the IPCC Scientific Assessment, J.T. Houghton, G.J. Jenkins and J.J. Ephraums (eds.). Cambridge University Press, Cambridge, UK, 365 pp.

IPCC, 1992: *Climate Change 1992: The Supplementary Report to the IPCC Scientific Assessment*, Houghton, J.T., B.A. Callander and S.K. Varney (eds.). Cambridge University Press, Cambridge, UK, 200p.

Jacobs, S.S., H.H. Helmer, C.S.M. Doake, A Jenkins and RM Frolich, 1992: Melting of ice shelves and the mass balance of Antarctica. *J. Glaciol.*, **38**(130), 375–387.

Jóhannesson, T., C. Raymond and E. Waddington, 1989: Timescale for adjustment of glaciers to changes in mass balance. *J. Glaciol.*, **35**(121), 355–369.

Jóhannesson, T., O. Sigurdsson, T. Lauman, and M. Kennett, 1995: Degree-day glacier mass balance modelling with applications to glaciers in Iceland, Norway, and Greenland. *J. Glaciol.*, **41**, No. 138, 345–358.

Kapsner, W.R., R.B. Alley, C.A. Shuman, S. Anandakrishnan and P.M. Grootes, 1995: Dominant influence of atmospheric circulation on snow accumulation in Greenland over the past 18,000 years. *Nature*, **373**, 52–54.

Kelsey, H.M., Engebretson, D.C., Mitchell, C.E., and R.L. Ticknor, 1994: Topographic form of the Coast Ranges of the Cascadia Margin in relation to coastal uplift rates and plate subduction. *J. Geophys. Res.*, **99**, 12,245–12,255.

Klige, R.K., 1982: Oceanic level fluctuations in the history of the earth. In: *Sea and Oceanic Level Fluctuations for 15,000 years*. Moscow, USSR Academy of Sciences, pp. 11–22.

Koblinsky, C.J., P. Gaspar and G. Lagerloef (eds.), 1992: *The future of space-borne altimetry: oceans and climate change. A long-term strategy*. Washington, DC: Joint Oceanographic Institutions Inc., 75p.

Kock, H., 1993: Height determinations along the EGIG line and in the GRIP area. *GGU Open File Series 93/5*, 68–70 (Rep. 3rd WS,, Mass balance and related topics of the Greenland ice sheet, Bremerhaven 1992).

Koltermann, K.P. and A. Sy, 1994. Western North Atlantic Cools at intermediate depths. *WOCE Newsletter*, **15**, 5–6.

Komar, P.D. and D.B. Enfield, 1987: Short-term sea level changes and coastal erosion. In: *Sea level Fluctuation and Coastal Evolution*, D. Nummedal, O.H. Pilkey and J.D. Howard (eds.), SEPM Spec. Publ. No. 41, pp. 17–27.

Kostecka, J.M. and I.M Whillans, 1988: Mass balance along two transects of the west side of the Greenland Ice Sheet. *J. Glaciol.*, **34**, 31–39.

Kuhn, M., 1981: Climate and glaciers. *Int. Assoc. Hydrol. Sci.*, **131**, 3–20.

Kuhn, M., 1985: Reactions of mid-latitude glacier mass balance to predicted climatic changes. In: *Glaciers, Ice Sheets, and Sea Level: Effects of a CO_2-Induced Climatic Change*, National Academy Press, Washington DC, pp.248–254.

Kuhn, M., 1993: Possible future contributions to sea level change from small glaciers. In: *Climate and Sea Level Change: Observations, Projections and Implications*. R.A. Warrick, E.M. Barrow and T.M.L. Wigley (eds.), Cambridge University Press, Cambridge, pp. 134–143.

Kuhn, M., G. Kaser, G. Mark, H.P. Wagner and H. Schneider, 1979: *25 Jahre Massenhaushaltuntersuchen am Hintereisferner*. Universitat Innsbruk.

Kushnir, Y., 1994: Interdecadal variations in the North Atlantic sea surface temperature and associated atmospheric conditions, *J. Climate*, **7**, 141–157.

Lachenbruch, A. H. and B. V. Marshall, 1986: Changing climate: geothermal evidence from permafrost in the Alaskan Arctic. *Science*, **234**, 689–696.

Lambeck, K., 1990: Glacial rebound, sea level change and mantle viscosity. *Quart. J. R. Astron. Soc.*, **31**, 1–30.

Latif, M. and T.P. Barnett, 1994: Causes of decadal climate variability over the North Pacific and North America, *Science*, **266**, 634–637.

Laumann, T. and N. Reeh, 1993: Sensitivity to climate change of the mass balance of glaciers in southern Norway. *J. Glaciol.*, **39** (133), 656–665.

Lavin, A.M., 1993: Climatic changes in temperature and salinity in the subtropical North Atlantic, Msc Thesis, MIT, Cambridge, Mass.

Lazier, J. and R. Gershey, 1991: AR7W: Labrador sea line, *WOCE Newsletter 11*.

Lestingant, R., 1994: A two-dimensional finite-element study of flow in the transition zone between and ice sheet and an ice shelf. *Ann. Glaciol.*, **20**, 67–72.

Levitus, S., 1982: *Climatological Atlas of the World Ocean*. NOAA Professional Paper 13, U.S. Government Printing Office, Washington, DC, 177pp.

Liestöl, O., 1967: Storbreen glacier in Jotunheimen, Norway. *Norsk Polarinstitutt Skrifter* No. 141, 63p.

Lingle, C.S., A.C. Brenner and J.P. DiMarzio, 1991: Multi-year elevation changes near the west margin of the Greenland ice sheet from satellite radar altimetry. *International Conference on the Role of the Polar Regions in Global Change: Proceedings of a Conference Held June 11–15, 1990 at the University of Alaska Fairbanks*, G. Weller, C.L. Wilson and B.A.B. Serverin (eds.), Geophysical Institute and Centre for Global Change and Arctic System Science, University of Alaska, Fairbanks, pp. 35–42.

MacAyeal, D.R., 1992: Irregular oscillations of the West Antarctic ice sheet. *Nature*, **359**, 29–32.

MacAyeal, D.R., 1994: Binge/purge oscillations of the Laurentide Ice Sheet, a cause of the North Atlantic's Heinrich Events. *Paleoceanography*, **8**, 775–784.

Macharet, Y.Y., P.A. Cherkasov, and L.I. Bobrova, 1988: The thickness and volume of the Dzhungarsky Alatau glaciers from the data of airborne radio echo sounding (in Russian). *Materialy Glyatsiologischeskikh Issledovaniy*, **62**, 59–71.

Mahaffy, M.A.W., 1976: A numerical three-dimensional ice flow model. *J. Geophys. Res.*, **81**(6), 1059–1066.

Mälzer, H. and H. Seckel, 1975: Das geometrische Nivellement *f*ber das Inlandeis – Höhenänderungen zwischen 1959 and

1968 im Ost-West-Profil der EGIG. *Z. Gletscherk. Glazialgeol.* **11**, 245–252.

Manabe, S. and R.J. Stouffer, 1993: Century-scale effects of increased atmospheric CO_2 on the ocean-atmosphere system. *Nature, 364*, 215–218.

Manabe, S. and R.J. Stouffer, 1994: Multiple-Century response of a coupled ocean-atmosphere model to an increase of atmospheric carbon dioxide. *J. Climate*, **7**, 5–23.

Martin, S., 1978: Analyse et reconstitution de la série des bilans annuels du Glacier de Sarennes, sa relation avec les fluctuations du niveau de trois glaciers du Massif du Mont-Blanc (Bossons, Argentiére, Mer de Glace). *Z. Gletscherk. Glazialgeol.,* **13** (1/2), 127–153.

Maul, G.A. and K. Hanson, 1991: Interannual coherence between North Atlantic atmospheric surface pressure and composite southern U.S.A. sea level. *Geophys. Res. Lett.,* **18**, 653–656.

Maul, G.A. and D.M. Martin, 1993: Sea level rise at Key West, Florida, 1846–1992: America's longest instrument record? *Geophys. Res. Lett.,* **20**, 1955–1958.

Maul, G.A., D.A. Mayer and M. Bushnell, 1990, Statistical relationships between local sea level and weather with Florida-Bahamas Cable and Pegasus measurement of Florida Current volume transport, *J. Geophys. Res.,* **95**, C3, 3287–3296.

McPhaden, M.J., 1993: TOGA-TAO and the 1991–93 El Niño-Southern Oscillation event. *Oceanography, 6*, 36–44.

Meier, M.F., 1965: Glaciers and climate. In: *The Quaternary of the United States: A review of volume for the VIII Congress of the International Association for Quaternary Research.,* H. E. Wright Jr. and D. G. Frey (eds.), Princeton University Press. pp 795–805.

Meier, M.F., 1983: Snow and ice in a changing hydrological world. *Journal des hydrologiques,* 28(1):3–22.

Meier, M.F., 1984: Contribution of small glaciers to global sea level. *Science,* **226**, 1418–1421.

Meier, M.F., 1985: Mass balance of the glaciers and small ice caps of the world. In: *Glaciers, Ice Sheets and Sea Level: Effects of a CO_2-induced Climatic Change*, National Academy Press, Washington, pp. 139–144.

Meier, M.F., 1990: Reduced rise in sea level. *Nature,* **343**, 115–116.

Meier, M.F., 1993: Ice, climate, and sea level; do we know what is happening? In: *Ice in the Climate System,* W.R. Peltier (ed.), NATO ASI Series I, Global Environmental Change, Vol. 12, Springer-Verlag, Heidelberg, pp. 141–160.

Meier, M.F, 1994: Columbia Glacier during rapid retreat: interactions between ice flow and iceberg calving dynamics. In: *Workshop on the calving rate of West Greenland glaciers in response to climate change*, N. Reeh (ed.), Danish Polar Centre, Copenhagen, pp 63–83.

Mikolajewicz, U., B.D. Santer and E. Maier-Reimer, 1990; Ocean response to greenhouse warming. *Nature, 345*, 589–593.

Miller, L. and R. Cheney, 1990. Large-scale meridional transport in the Tropical Pacific Ocean during the 1986–1987 El Niño from Geosat, *J. Geophys. Res.,* **95**, C10, 17905–17919.

Mitchell, C.E., P. Vincent, R.J. Weldon, II and M.A. Richards, 1994: Present-day vertical deformation of the Cascadia margin, Pacific Northwest, United States. *J. Geophys. Res.,* **99**, 12257–12277.

Mitchell, J.F.B., R.A. Davis, W.J. Ingram and C.A. Senior, 1995: On surface temperature, greenhouse gases and aerosols: models and observations. *J.Climate, 10*, 2364–2386

Mitrovica, J.X. and J.L. Davis, 1995: Present-day post-glacial sea level change far from the Late Pleistocene ice sheets: implications for recent analyses of tide gauge records. *Geophys. Res. Lett. , 22*, 2529–2532.

Moeller, D., 1994: Das West-Ost-Profil der Internationalen Glaziologischen Groenland Expedition (EGIG). Rekonstruktion und Nachmessung. *Geowissenschaften* **12** (3), 80–82.

Mörner, N-A., 1973: Eustatic changes during the last 300 years. *Paleogeography, Paleoclimatology, Paleoecology,* **13**, 1–14.

Munk, W. 1989: Global ocean warming: detection by long-path acoustic travel times. *Oceanography, 2*(2), 40–41.

Murphy, J.M., 1995: Transient response of the Hadley Centre coupled model to increasing carbon dioxide. Part III, analysis of global-mean response using simple models. *J.Climate, 8*, 496–514.

Muszynski, I. and G.E. Birchfield, 1985: The dependence of antarctic accumulation rates on surface temperature and elevation. *Tellus,* **37A**, 204–208.

NASA, 1991, West Antarctic Ice Sheet Initiative, vol. 1: Science and Implementation Plan, R. A. Bindschadler (ed.), *NASA Conference Publication 3115.*

Nerem, R.S., 1995: Global mean sea level variations from TOPEX/POSEIDON altimeter data. *Science,* **268**, 708–710.

Oerlemans, J. and N. C. Hoogendoorn, 1989: Mass-balance gradients and climatic change. *J. Glaciol.,* **35**(121), 399–405.

Oerlemans, J. and J.P.F. Fortuin, 1992: Sensitivity of glaciers and small ice caps to greenhouse warming. *Science,* **258**, 115–117.

Oerlemans, J, R.S.W van de Wal and L.A. Conrads, 1991: A model for the surface balance of ice masses: Part II: application to the Greenland ice sheet. *Z. Gletscherk. Glazialgeol.* **27/28**, 85–96.

Ohmura, A. and N. Reeh, 1991: New precipitation and accumulation maps for Greenland. *J. Glaciol.,* **37** (125), 140–148.

Østrem, G. and M. Brugman, 1991: *Glacier Mass-Balance Measurements*. NHRI Science Report No 4. Saskatoon, National Hydrology Research Institute. 224 pp.

Parkinson, C.L. and R.A. Bindschadler, 1984: Response of Antarctic sea ice to uniform atmospheric temperature increases. In: *Climate Processes and Climate Sensitivity*, Geophysical Monograph 29, Maurice Ewing Vol. 5, AGU, pp. 254–264.

Peltier, W.R., 1994: Ice age paleotopography. *Science,* **265**, 195–201.

Peltier, W.R. and A.M. Tushingham, 1989: Global sea level rise and the greenhouse effect: might they be related? *Science,* **244**, 806–810.

Peltier, W.R. and A.M. Tushingham, 1991: Influence of glacial isostatic adjustment on tide gauge measurements of secular sea level change. *J. Geophys. Res.*, **96**, 6779–6796.

Pfeffer, W.T., M.F. Meier, and T.H. Illangasekare, 1991: Retention of Greenland runoff by refreezing: implications for projected future sea level change. *J. Geophys. Res.*, **96** (C12), 22117–22124.

Pirazzoli, P.A., 1977: Sea level relative variations in the world during the last 2000 years. *Zeitschrift fuer Geomorphologie*, **21** (3), 284–296.

Pirazzoli, P.A., 1991: Possible defences against a sea level rise in the Venice area, Italy. *J. Coastal Res.*, **7**(1), 231–248.

Pirazzoli, P.A., 1993. Global sea level changes and their measurement. *Global and Planetary Change*, **8**, 135–148.

Proctor, R. and L. Carter, 1989: Tidal and sedimentary response to the Late Quaternary closure and opening of Cook Strait, New Zealand: results from numerical modelling. *Paleoceanography*, **4**, 167–180.

Raper, S.C.B., T.M.L. Wigley and R.A. Warrick, 1996: Global sea level rise: past and future. In: *Rising sea level and subsiding coastal areas*, J.D. Milliman (ed.) Kluwer Academic Publishers, Dordrecht, 384 pp..

Read, J.F. and W.J. Gould, 1992: Cooling and freshening in the subpolar North Atlantic Ocean since the 1960s. *Nature*, **360**, 55–57.

Reeh, N., 1985: Greenland ice sheet mass balance and sea level change. In: *Glaciers, Ice Sheets and Sea Level: Effects of a CO_2 induced Climatic Change*. National Academy Press, Washington, pp 155–171.

Reeh, N., 1991: Parameterization of melt rate and surface temperature on the Greenland Ice Sheet. *Polarforschung*, **59**(3), 113–128.

Reeh, N., 1994: Calving from Greenland glaciers: observations, balance estimates of calving rates, calving laws. *Workshop on the calving rate of West Greenland glaciers in response to climate change*, Danish Polar Centre, Copenhagen, 171pp.

Reeh, N. and N. S. Gundestrup, 1986: Mass balance of the Greenland ice sheet at Dye 3. *J. Glaciol.*, **31**, 198–200.

Retzlaff, R. and C.R. Bentley, 1993: Timing of stagnation of Ice Stream C, West Antarctica, from short-pulse radar studies of buried surface crevasses. *Journal of Glaciology*, **39**, 553–561.

Rijkswaterstaat, 1986: Sea level rise: struggling with the rising water (In Dutch). Rijkswaterstaat, Netherlands, 148p. and figs.

Robin, G., 1986: Changing the sea level. In: *The Greenhouse Effect, Climate Change, and Ecosystems.*, B. Bolin B.R. Doos, R.A. Warrick and J. Jaeger (eds.), John Wiley & SonsInc., Chichester, pp. 323–359.

Rodenburg, E., 1994: Man-made lakes and sea level rise. *Nature*, **370**, 258p.

Roemmich, D., 1985: Sea level and thermal variability of the ocean. In: *Glacier, Ice sheets and Sea level: Effects of a CO_2 induced Climate Change*. National Academy Press, Washington, pp. 104–115.

Roemmich, D., 1992: Ocean warming and sea level rise along the southwest U.S. coast. *Science*, **257**, 273–275.

Rohde, H., 1980: Changes in sea level in the German Bight. *J. Geophys. R. Astron. Soc.*, **62**, 291–302.

Rogers, J.C., 1984: The association between the North Atlantic Oscillation and the Southern Oscillation in the Northern Hemisphere, *Mon. Weather. Rev.*, **112**, 1999.

Rusconi, A., 1993: The tidal observations in the Venice Lagoon – the variations in sea level observed in the last 120 years, In: *Sea Level Changes and Their Consequences for Hydrology and Water Management. International workshop*, 19–23 April 1993, Noordwijkerhout, Netherlands. Koblenz: Bundesanstalt fur Gewasserkunde, IHP/OHP Secretariat, pp. 115–132.

Sahagian, D.L., F.W., Schwartz, and D.K. Jacobs, 1994: Direct anthropogenic contributions to sea level rise in the twentieth century. *Nature*, **367**, 54–56.

Salinger, M.J., R. Allan, N. Bindoff, J. Hannah, B. Lavery, Z. Lin, J. Lindesay, N. Nicholls, N. Plummer and S. Torok, 1996: Observed variability and changes in climate and sea level in Oceania. In: *Greenhouse '94*, G. Pearman and M. Manning (eds.), CSIRO Press, Melbourne (in press).

Scherer, R.P., 1991: Quaternary and Tertiary microfossils from beneath Ice Stream B: evidence for a dynamic West Antarctic ice sheet history. *Paleogeography, Paleoclimatology, Paleoecology*, **90**(4), 395–412.

Schmidt, H., 1991: On the variation of the geostrophic wind from 1876 to 1989 in the German Bight and on other meteorological parameters. In: *Storm Surges, River Flow and Combined Effects*. International workshop, held 8–12 April 1991, Hamburg, Federal Republic of Germany. A contribution to the UNESCO-IHP project H-2–2, Paris: UNESCO, pp9–17.

Schmidt, H. and H. von Storch. 1993: German Bight storms analysed. *Nature*, **365**, 791p.

Schwitter, M.P. and C.F. Raymond, 1993: Changes in the longitudinal profiles of glaciers during advance and retreat. *J. Glaciol.*, **39** (133), 582–590.

Scott, D.B. and D.A. Greenberg, 1983: Relative sea level rise and tidal developments in the Fundy tidal system. *Canadian Journal of Earth Sciences*, **20**, 1554–1564.

Seckel, H, 1977: Hohenanderung im Gronlandischen Inlandeis zwischen 1959 und 1968. *Meddelelser om Gronland*, **187**(4), 58p.

Semtner, A.J., 1994: The future of global modelling within WOCE. *WOCE Note*, 6, 2, 2–3.

Shennan, I. and P.L. Woodworth, 1992: A comparison of late Holocene and twentieth-century sea level trends from the UK and North Sea region. *J. Geophys. Int.*, **109**, 96–105.

Shiklomanov, I.A., 1993: World fresh water resources. In: *Water in crisis – a guide to the world's freshwater resources*, P.H. Gleick (ed.), Oxford University Press, New York, pp.13–24.

Simmonds, I., 1990: Improvements in general circulation performance in simulating Antarctic climate. *Antarctic Science*, **2**, 287–300.

Simon, B., 1994: Statistique des niveaux marins extremes le long des cotes de France. SHOM, Rapport d'etude, 78p.

Spencer, N.E., 1992: Joint IAPSO-IOC workshop on sea level measurements and quality control, Paris, 12–13 October 1992. *Intergovernmental Oceanographic Commission Workshop Report No.81,* 167p.

Spencer, N.E. and P.L. Woodworth, 1993: *Data holdings of the Permanent Service for Mean Sea Level (November 1993).* Bidston, Birkenhead: Permanent Service for Mean Sea Level, 81p.

Spencer, R. and J.M. Vassie, 1985: Comparison of sea-level measurements obtained from deep pressure sensors, In: *Advances in Underwater Technology and Offshore Engineering, Volume 4, Evaluation, Comparison and Calibration of Oceanographic Instruments,* Graham and Trotman Ltd., for SUT, London, pp. 183–207.

Stroeven, A., R. van de Wal and H. Oerlemans, 1989: Historic front variations of the Rhone Glacier: simulation with an ice flow model. In: *Glacier Fluctuations and Climatic Change.* H. Oerlemans (eds.), Kluwer Academic Publishers, Dordrecht, pp. 391–405.

Sturges, W.E., 1987: Large-scale coherence of sea level at very low frequencies. *J. Phys. Oceanogr.,* **17**, 2084–2094.

Sturges, W. and B.G. Hong, 1995: Wind forcing of the Atlantic thermocline along 32 °N at low frequencies. *J. Phys. Oceanogr.,* **25**, 1706–1715.

Tangborn, W., 1980: Two models for estimating climate-glacier relationships in the north Cascades, Washington, U.S.A. *J. Glaciol.,* **25**(91), 3–21.

Tanner, W.F., 1992: 3000 years of sea level change. *Bull. Am. Met. Soc.,* **73**, 297–303.

Tawn, J.A. and J.M. Vassie, 1991: Recent improvements in the joint probability method for estimating extreme sea levels. In: *Tidal Hydrodynamics,* B.B.Parker (ed.), John Wiley and Sons , NY, pp. 813–828.

Thomas, R.H., 1973: The creep of ice shelves: Theory. *J. Glaciol.,* **3**, 38–42.

Thomas, R.H., 1985: Responses of the polar ice sheets to climatic warming. In: *Glaciers, Ice Sheets and Sea Level: Effects of a CO_2-induced Climatic Change,* National Academy Press, Washington, pp. 301–316.

Thomson, R.E. and S. Tabata, 1987: Steric height trends of ocean station PAPA in the northeast Pacific Ocean. *Marine Geodesy,* **11**, 103–113.

Thorarinsson, S., 1940: Present glacier shrinkage and eustatic changes of sea level. *Geografiska Annaler,* **22**, 131–159.

Titus, J.G. and V. Narrayanan, 1995: *The Probability of Sea Level Rise.* U.S. Environmental Protection Agency, Office of Policy, Planning, and Evaluation, Climate Change Division. EPA 230-R-95-008. EPA, Washington, D.C., 186p.

Titus, J.G. and V. Narrayanan, 1996: The risk of sea level rise: a Delphic Monte-Carlo analysis in which twenty researchers specify subjective probability distributions for model coefficients within their respective areas of expertise.*Clim. Change* (in press).

Trabant, D.C. and L.R. Mayo, 1985: Estimation and effects of internal accumulation on five glaciers in Alaska. *Annals of Glaciology,* **6**, 113–117.

Trupin, A. and J. Wahr, 1990: Spectroscopic analysis of global tide gauge sea level data. *Geophys. J. Intern.* **100**, 441–453.

Trupin, A.S., M.F. Meier and J.M. Wahr, 1992: Effect of melting glaciers on the Earth's rotation and gravitational field: 1965–1984. *Geophys. J. Intern.,* **108**, 1–15.

Tsimplis, M.N., R.A. Flather and J.M. Vassie, 1994: The North Sea pole tide described through a tide-surge numerical model. *Geophys. Res. Lett.,* **21**(6), 449–452.

Tushingham, A.M. and W.R. Peltier, 1991: ICE-3G: a new global model of late Pleistocene deglaciation based upon geophysical predictions of post glacial relative sea level change. *J. Geophys. Res.,* **96**, 4497–4523.

van der Veen, C.J., 1985: Response of a marine ice sheet to changes at the grounding line. *Quat. Res.,* **42**, 257–267.

van de Wal, R.S.W. and J. Oerlemans, 1994: An energy balance model for the Greenland ice sheet. *Global PlanetChange,* **9**, 115–131.

Varekamp, J.C., E. Thomas and O. Van de Plassche, 1992: Relative sea level rise and climate change over the last 1500 years. *Terra Nova,* **4**, 293–304.

Wagner, C.A., R.E. Cheney, L. Miller, C.K. Tai, N.S. Doyle and J.L. Lillibridge, 1994, Global sea level change from altimetry, *EOS,* **75** (44), 56p.

Wagner, C.A. and D.C. McAdoo, 1986: Time variations in the Earth's gravity field detectable with Geopotential Research Mission intersatellite tracking. *J. Geophys. Res,* **91**(B8), 8373–8386.

Walters, R.A. and M.F. Meier, 1989: Variability of glacier mass balances in western North America. In: *Aspects of Climate Variability in the Pacific and Western Americas, Geophysical Monograph 55,* D.H. Peterson (ed.), American Geophysical Union, pp. 365–281.

Warrick, R.A. and J. Oerlemans, 1990: Sea Level Rise, In: *Climate Change, The IPCC Assessment,* J.T. Houghton, G.J. Jenkins and J.J. Ephraums (eds.), Cambridge University Press, Cambridge, UK, pp. 257–281.

Weertman, J., 1974: Stability of the junction of an ice sheet and an ice shelf, *J. Glaciol.,* **13**, 3–11.

Weidick A., 1984: Review of glacier changes in West Greenland. *Z. Gletscherk. Glazialgeol.,* **21**, 301–309.

Weidick, A., 1995: Greenland, In: *Satellite Image Atlas of Glaciers of the World,* R.S. Williams, Jr. and J.G. Ferrigno (eds.), U.S. Geological Survey Professional Paper 1386-C, 141pp.

Whillans, I.M., 1976: Radio-echo layers and the recent stability of the West Antarctic ice sheet. *Nature,* **264**, 152–155.

Whitworth, T., III and R.G. Peterson, 1985: Volume transport of the Antarctic Circumpolar Current from bottom pressure measurements. *J. Phys.Oceanogr.,* **15**, 810–816.

Wigley, T.M.L., 1995: Global-mean temperature and sea level consequences of greenhouse gas concentration stabilization. *Geophys. Res. Lett.,* **22**, 45–48.

Wigley, T.M.L. and S.C.B. Raper, 1987: Thermal expansion of sea level associated with global warming. *Nature,* **330**, 127–131.

Wigley, T.M.L. and S.C.B. Raper, 1992: Implications for climate and sea level of revised IPCC emissions scenarios. *Nature,* **357,** 293–300.

Wigley, T.M.L. and S.C.B. Raper, 1993: Future changes in global mean temperature and sea level. In: *Climate and Sea Level: Observations, Projections and Implications.* R.A. Warrick, E.M. Barrow and T.M.L. Wigley (eds.), Cambridge University Press, Cambridge, pp. 111–133.

Wigley, T.M.L. and S.C.B. Raper, 1995: An heuristic model for sea level rise due to the melting of small glaciers. *Geophys. Res. Lett.,* **22,** 2749–2752.

Wigley, T.M.L., A. Hall, S.C.B. Raper and R.A. Warrick, 1996: Past sea level rise from glaciers based on a regional degree-day model. *J. Geophys. Res.* (submitted).

Wimbush, M. 1990: Inferring sea level variation from acoustic travel time and bottom pressure measurements. In: *Towards an Integrated System for Measuring Long Term Changes in Global Sea Level,* H.F. Eden (ed.). Report of a workshop held at Woods Hole Oceanographic Institution, May 1990, Joint Oceanographic Institutions Inc. (JOI), Washington, DC, pp. 147–153.

Woodworth, P.L., 1990: A search for accelerations in records of European mean sea level. *Int. Journal. Climatol.,* **10,** 129–143.

Woodworth, P.L., 1991: The Permanent Service for Mean Sea Level and the Global Sea Level Observing System. *Journal of Coastal Research,* **7,** 699–710.

Woodworth, P.L. 1993: A review of recent sea-level research. *Oceanography and Marine Biology. An Annual Review,* **31,** 87–109.

Wyrtki, K., 1979: The response of sea level surface topography to the 1976 El Niño. *J. Phys. Oceanogr,* **9,** 1223–1231.

Wyrtki, K., 1985: Water displacements in the Pacific and the genesis of El Niño. *J. Geophys. Res.* **90,** 7129–7132.

Zwally, H.J., A.C. Brenner, J.A. Major, R.A. Bindschadler and J.G. Marsh (1989): Growth of the Greenland Ice Sheet: Measurement. *Science,* **246,** 1587–1589.

Zwally, H.J. , A.C. Brenner, J.A. Major, R.A. Bindschadler and J.G. Marsh, 1990: Greenland Ice Sheet: is it growing or shrinking? – Response. *Science,* **248,** 288–289.

8

Detection of Climate Change and Attribution of Causes

B.D. SANTER, T.M.L. WIGLEY, T.P. BARNETT, E. ANYAMBA

Contributors:
P. Bloomfield, E.R. Cook, C. Covey, T.J. Crowley, T. Delworth, W.L. Gates,
N.E. Graham, J.M. Gregory, J.E. Hansen, K. Hasselmann, G.C. Hegerl,
T.C. Johns, P.D. Jones, T.R. Karl, D.J. Karoly, H.S. Kheshgi, M.C. MacCracken,
K. Maskell, G.A. Meehl, J.F.B. Mitchell, J.M. Murphy, N. Nicholls, G.R. North,
M. Oppenheimer, J.E. Penner, S.B. Power, A. Robock, J.L. Santer, C.A. Senior,
K.E. Taylor, S. Tett, F.W. Zwiers

GLOBAL WARMING

Industry Group Assails Climate Chapter

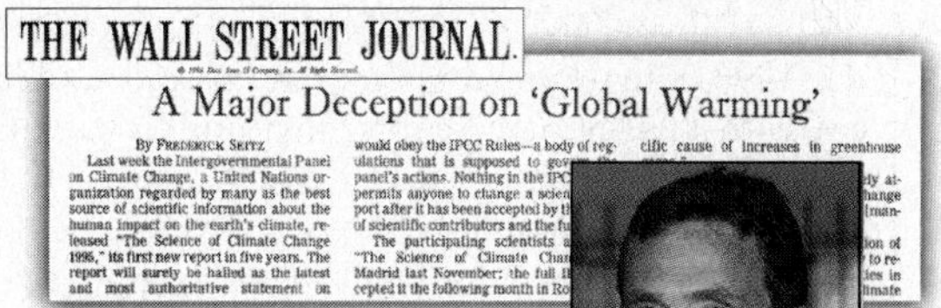

THE WALL STREET JOURNAL.

A Major Deception on 'Global Warming'

By FREDERICK SEITZ

Last week the Intergovernmental Panel on Climate Change, a United Nations organization regarded by many as the best source of scientific information about the human impact on the earth's climate, released "The Science of Climate Change 1995," its first new report in five years. The report will surely be hailed as the latest and most authoritative statement on

"My own scientific reputation and credibility [are] under attack."

—Ben Santer

The scientific debate about whether human activity is warming global climate subsided late last year when the world's leading climate researchers agreed that the answer is probably yes. But this month the political debate heated up by several degrees when an industry group charged that revisions to a crucial chapter in a United Nations (UN) report on climate change violated peer review and amount to "scientific cleansing" of doubts about human influence on climate. The charges, made 2 weeks ago, have sparked a flurry of editorials and articles repeating the charges in publications including *The Wall Street Journal*—and a spirited defense by climate researchers.

The focus of the controversy, chapter 8 in the latest report of the UN's Intergovernmental Panel on Climate Change (IPCC) Working Group I, lays out research advances since 1990 that have bolstered confidence that human activity is at least partly to blame for the gradual warming of the globe. They include better models of climate variability and a better understanding of the effects of sulfate aerosols and ozone loss, all of which tend to obscure the signal of greenhouse warming (*Science*, 8 December 1995, p. 1565). When these effects are accounted for, the warming signal seems to emerge, said the chapter. But the Washington, D.C.–based Global Climate Coalition (GCC)—a group supported by oil and coal producers and utilities—argued in a nine-page analysis and in letters to members of Congress that changes made after the draft report was issued last fall downplayed uncertainties about this conclusion.

"When the final report came out, there were sections that were not there," said John Shlaes, executive director of the coalition. "Why were they taken out when those were important elements to educate policymakers?" The answer is simple, says Ben Santer, a Lawrence Livermore National Laboratory researcher who was the lead author on the chapter and says that he made the changes himself: Reviewers requested them. He says that the coalition and other critics can't impugn the science underlying the report, so "they attack the process, the IPCC itself and the scientists."

The business coalition was particularly upset by the disappearance of the chapter's concluding summary. That section had noted that even the most telling indicator to date of human influence—so-called pattern-based computer simulations that marry the effects of aerosols and greenhouse gases to show a pattern of warming similar to the observed one—doesn't conclusively tie any change to human influence. The coalition also raised an outcry over the deletion of a phrase saying "we do not know" when scientists will be able to identify a human contribution to climate change unambiguously.

The changes were a "disturbing corruption of the peer-review process," wrote Frederick Seitz, ex-president of Rockefeller University and chair of the George C. Marshall Institute, which has also raised doubts about a human influence on climate, in a 12 June op-ed piece in *The Wall Street Journal*. Declared the GCC in its statement: "The changes quite clearly have the obvious political purpose of cleansing the underlying scientific report."

To Santer and other climatologists, it's these accusations that are politically motivated. "This is terrible what's going on, just terrible," says Santer. "I now perceive my own scientific reputation and credibility to be under attack, and that's a very hard position to be in." Backed by his three co-authors—Tom Wigley of the National Center for Atmospheric Research (NCAR), Tim Barnett of the Scripps Institution of Oceanography, and Ebby Anyamba of the NASA Goddard Space Flight Center—Santer has argued that the governments, organizations, and scientists who reviewed the draft report last October knew it would be changed to take reviewers' comments into account. He adds that the changes simply removed redundancies and fine-tuned the wording to bring the report into line with the scientific consensus.

In the case of the concluding summary, for example, Santer says he "folded [it] into other parts of the chapter" because reviewers had pointed out that, unlike any other chapter, it had summaries at both the beginning and end. As for the phrase he removed, Santer says it overstated doubts that a human effect on climate is already apparent. "The revision is now more accurate and a better reflection of prevailing scientific opinion," he says.

Kevin Trenberth, head of the climate analysis section at NCAR and lead author for chapter 1, agrees. "I think some of that redundancy was removed, but the uncertainty is clearly reflected in the chapter," he says.

Nor did the changes violate IPCC procedures, said Bert Bolin of the University of Stockholm, the IPCC chair, in a letter faxed to the GCC this week. "Your allegations are completely unfounded," he wrote. But he acknowledged that the IPCC had left an opening for such attacks by not presenting the final wording to the delegates—including the Climate Coalition—before it went to press.

–Peter Weiss

Peter Weiss covers science for the Valley Times *in Pleasanton, California.*

BIOMEDICAL RESEARCH

Salk Institute Picks a New President

Cell biologist Thomas Pollard of Johns Hopkins University is packing his bags and test tubes and heading for La Jolla, California, where he will take over as president and chief executive officer of the Salk Institute for Biological Studies. As *Science* reported last week (14 June, p. 1575), Pollard was the search committee's top choice, and on 14 June, the Salk's chairman of the board, Frederick Rentschler, announced that the deal was done. "While this has been a long search, it has ended with the right individual," wrote Rentschler to the staff.

Heading west. Cell biologist Thomas Pollard.

Pollard, 53, a member of the National Academy of Sciences (NAS) who specializes in the molecular basis of cell movement, plans to continue his research; he will take about half a dozen of the researchers in his lab with him. "There's nothing like being the boss and being up against the same challenges the faculty has," says Pollard. Those who know Pollard applaud the choice. "He's a world-class scientist. And he has an appreciation of science policy," says marine biologist John Burris, head of the Marine Biological Laboratories in Woods Hole, Massachusetts, who has worked with Pollard on NAS committees.

The Salk has been searching for a permanent president since 1994, when cancer epidemiologist Brian Henderson resigned. Nobel laureate Francis Crick and former March of Dimes executive Charles Massey ran the institute jointly before resigning last September. Pollard will take over on 1 July.

–Jon Cohen

NASA's effort to construct an attractive benefits package to lure scientists to the new institutes was also hampered by uncertainty about the true cost of the institutes. "No one really knew the economics of this," complains one Administration official involved in the debate, who said NASA estimated the transfer could cost as much as $100,000 a person. Although only about 50 to 60 senior-level scientists were thought likely to balk at making the switch in the absence of such a package, the official said, they were seen as essential for getting the new institutes off to a good start.

NASA officials are trying to put the best face on the plan's defeat. They say the agency is still committed to improving its research activities, and that the debate has highlighted the importance of closer ties with academia. "There are a lot of activities we can pursue to achieve the same goals," says NASA Chief Scientist France Cordova. "The centers already are reaching out more to universities, and in a year's time, a lot of connections have been made," she notes, citing an increased number of joint center-university research proposals for the Discovery missions and other flight projects. NASA plans to continue to encourage such collaborations. "We really want to get away from criticism that the centers are too insular, that there is too much conflict of interest, and that they are too interested in bolstering themselves," she says.

The institute idea also helped to protect NASA research programs from drastic cuts proposed in a 1995 agency memo, she said. The plan focused attention on NASA's in-house science programs and convinced senior administrators of their value to the agency. "There is no talk of getting rid of those activities now," Cordova says.

NASA still intends to create a biomedical institute in Houston, home of Johnson Space Center. That project remains alive because the life-sciences researchers there are largely contractors and not NASA employees. The center has asked potential institute operators to submit proposals by 2 August; as many as four will receive $175,000 apiece to draw up their plans. NASA intends to select a winner next March. Agency officials are also exploring other ways to bring the centers into the mainstream of the scientific community. Ames, for example, may hire more outside scientists on a temporary basis, says David Morrison, chief of the center's space science division.

But providing more opportunities for collaboration is unlikely to stave off the harmful effects of a declining budget that threatens to take large bites out of its overall work force. To do that, Goldin will need to go back to the drawing board and find another approach to protect and revitalize research that is acceptable to federal bureaucrats and legislators.

–Andrew Lawler

FEDERAL FUNDING

Appropriators Bullish on Biomedicine

When biomedical research emerged from the 1996 congressional appropriations process with a larger increase than anything else in the Department of Health and Human Services (HHS), some analysts warned that it would be a hard act for biomedicine's supporters to follow this year. But last week, a key House subcommittee put on the first act of a repeat performance. It approved a 1997 HHS appropriation bill that would give the National Institutes of Health (NIH) a budget increase of 6.9%—more than Congress allowed in 1996 (5.7%) and much more than the 4% the Administration is requesting.

Working into the wee hours of 14 June, the House appropriations subcommittee for HHS, housing, and labor plowed through more than 20 amendments before voting on the bill. In the end, Chair John Porter (R–IL) got just about everything he wanted for NIH, whose champion he has become. Not only did the panel vote to increase NIH's 1996 budget of $11.9 billion by $819 million, but it endorsed an exceptional funding plan that could make it easier for NIH to rebuild the clinical center, its aging hospital. The subcommittee members also agreed to several policy changes that could give the NIH director more administrative flexibility while curbing the authority of NIH's Office of AIDS Research (OAR). And they approved an amendment by Representative Nita Lowey (D–NY) that would partly lift a ban on human embryo research imposed earlier this year by Congress.

The vote on the clinical center could be a watershed for NIH. The agency has been trying for years to get permission to begin constructing a replacement for the Warren Grant Magnuson Clinical Center, a decaying behemoth on NIH's campus. Congress pressed for cuts, and NIH responded by shrinking the proposed center from a billion-dollar project to one that is now expected to cost $310 million. But getting approval for even this scaled-back version has been difficult.

The main problem was an accounting rule adopted by the White House Office of Management and Budget that said that any agency undertaking new construction must include the full cost of the project in the first year's appropriation. This meant that NIH could only begin building the new clinical center by holding down all other expenditures, effectively preventing growth in its budget for research and grants. NIH director Harold Varmus and HHS Secretary Donna Shalala appealed to the White House to allow NIH to spread construction costs over several years. According to House staffers, Porter also began lobbying on NIH's behalf, with hopes of

winning an exception for NIH's hospital. The result: This bill provides $90 million to start work on the new center, with a proviso that the project be paid for over 4 years. That would leave a 6.5% increase for research after construction funds are set aside.

In the policy area, one hotly contested change proposed by Porter would restrict the independent budget authority Congress gave OAR in 1993. Representative Nancy Pelosi (D–CA) and four other panel members unsuccessfully opposed Porter's proposal. However, the subcommittee did approve a compromise that one Capitol Hill staffer says aims to mollify AIDS activists and recognize the hard work of William Paul, OAR's director. The bill permits the OAR director "jointly with the director of NIH" to reallocate up to 3% of AIDS funding during the year from any institute to another program. Paul, although he would prefer greater authority, said this compromise "provides us with an enormous opportunity to do what we think should be done." The Senate is expected to seek to restore OAR's independent status.

Another provision—one that is likely to prompt intense debate later in the summer—is the change in embryo research policy. The revised version would continue to prohibit the fertilization of ova for research, but would permit studies on fertilized ova that would be discarded. And Porter may have bumped into an even nastier hornet's nest in challenging a congressional set-aside for small business. A law already in place requires NIH to devote 2.5% of research funds next year to "small business innovation research" or SBIR grants (Science, 17 May, p. 942). But biomedical groups such as the Federation of American Societies for Experimental Biology have argued that this set-aside—which cuts into funds available for basic science—is "anti-quality."

Hill staffers say that Porter first proposed capping SBIR. When the House Small Business Committee objected, he proposed that the set-aside be limited to "a pool of SBIR grants for which the median priority score is equal to or better than the median score of the pool of investigator-initiated grants." The panel approved this limit. But one Hill staffer warns: If this clause is really a spending cap, "Mr. Porter may be getting a whole lot of mail."

The Labor-HHS appropriation now goes to full committee, where it is likely to be approved this week, then to the House floor. Senate staffers say they don't expect to begin marking up the legislation until mid-July at the earliest, to be followed by a conference and final vote in late summer.

–Eliot Marshall

CONTENTS

SUMMARY

Since the 1990 IPCC Scientific Assessment considerable progress has been made in attempts to identify an anthropogenic effect on climate. The first area of significant advance is that model experiments are now starting to incorporate the possible climatic effects of human-induced changes in sulphate aerosols and stratospheric ozone. The inclusion of these factors has modified in important ways the picture of how climate might respond to human influences. Thus, the potential climate change "signal" due to human activities is now better defined, although important signal uncertainties still remain.

The second area of progress is in better defining the background natural variability of the climate system, a crucial aspect of the detection problem. "Detection of change" is the process of demonstrating that an observed change in climate is highly unusual in a statistical sense. This requires distinguishing any human effects on climate from the background "noise" of climate fluctuations that are entirely natural in origin. Such natural fluctuations can be either purely internal or externally driven, for example by changes in solar variability or the volcanic dust loading of the atmosphere. "Total" natural variability includes both internal and externally forced components. Estimating either component of natural climatic noise from observed data is a difficult problem. Recent multi-century model experiments that assume no human-induced changes in anthropogenic forcings have provided important information about the possible characteristics of the internal component of total natural variability. However, large uncertainties still apply to current estimates of the magnitude and patterns of natural climate variability, particularly on the decadal- to century-time-scales that are crucial to the detection problem.

The third area of progress is in the application of pattern-based methods in attempts to *attribute* some part of the observed changes in climate to human activities: i.e., to establish a cause-effect relationship. Most studies that have attempted to detect an anthropogenic effect on climate have used changes in global mean, annually averaged temperature. These investigations have compared observed changes over the past 10–100 years with estimates of internal or total natural variability noise derived from palaeodata, climate models, or statistical models fitted to observations. The majority of these studies show that the observed change in global mean, annually averaged temperature over the last century is unlikely to be due entirely to natural fluctuations of the climate system.

Although these global mean results suggest that there is some anthropogenic component in the observed temperature record, they cannot be considered as compelling evidence of a clear cause-and-effect link between anthropogenic forcing and changes in the Earth's surface temperature. It is difficult to achieve attribution of all or part of a climate change to a specific cause or causes using global mean changes only. The difficulties arise due to uncertainties in natural internal variability and in the histories and magnitudes of natural and human-induced climate forcings, so that many possible forcing combinations could yield the same curve of observed global mean temperature change.

To better address the attribution problem, a number of recent studies have compared observations with the spatial *patterns* of temperature-change predicted by models in response to anthropogenic forcing. The argument underlying pattern-based approaches is that different forcing mechanisms have different patterns of response or characteristic "fingerprints", particularly if the response is considered in three or even four dimensions (e.g., temperature changes as a function of latitude, longitude, height and time). Thus, a good match between observed and modelled multi-dimensional patterns of climate change increases the likelihood that the "cause" (forcing change) used in the model experiment is in fact responsible for producing the observed effect.

Several recent studies have compared observed patterns of temperature change with model patterns from simulations with simultaneous changes in carbon dioxide (CO_2) and anthropogenic sulphate aerosols. These comparisons have been made both at the Earth's surface and in vertical sections through the atmosphere. The results of these studies rest mainly on pattern similarities at the largest spatial scales, for which model predictions are most reliable: for example at the scale of temperature differences between hemispheres, land and ocean, or the troposphere and stratosphere. While there are concerns regarding the relatively simple treatment of aerosol effects in model experiments that attempt to define an anthropogenic signal, *all* such pattern comparison studies show significant correspondences between the observations and model predictions. The pattern correspondences increase with time, as one would expect as an anthropogenic signal increases in strength. Pattern correspondences using combined CO_2+aerosol signals are generally higher than those obtained if model predictions are based on changes in CO_2 alone. Furthermore, the probability is very low that these correspondences could occur by chance as a result of natural internal variability. The vertical patterns of change are also inconsistent with the response patterns expected for solar and volcanic forcing. Increasing confidence in the emerging identification of a human-induced effect on climate comes primarily from such pattern-based work.

In addition to these quantitative studies, there are areas of qualitative agreement between observations and those model predictions that either include aerosol effects or do not depend critically on their inclusion. Model and observed commonalities in which we have most confidence, but which have not been looked at in detailed detection studies, include reduction in diurnal temperature range, sea level rise, high latitude precipitation increases, and water vapour and evaporation increases over tropical oceans.

Viewed as a whole, these results indicate that the observed trend in global mean temperature over the past 100 years is unlikely to be entirely natural in origin. More importantly, there is evidence of an emerging pattern of climate response to forcings by greenhouse gases and sulphate aerosols in the observed climate record. This evidence comes from the geographical, seasonal and vertical patterns of temperature change. Taken together, these results point towards a human influence on global climate. Our ability to quantify the magnitude of this effect is presently limited by uncertainties in key factors, such as the magnitude of longer-term natural variability and the time-evolving patterns of forcing and response to changes in greenhouse gases, aerosols and other human factors.

8.1 Introduction

8.1.1 The Meaning of "Detection" and "Attribution"

The purpose of this chapter is to review work carried out over the last five years that has attempted to detect a statistically significant change in the global climate system, and attribute at least part of that change to anthropogenic factors.

The detection problem arises because any climate-change "signal" due to anthropogenic influences (such as changes in the emissions of greenhouse gases) is superimposed on the background "noise" of natural climate variability, which may partly or wholly mask the signal. Natural climate variability has both internal and external components. The internal component is due solely to interactions within the coupled atmosphere-ocean-ice-land-biosphere system. The external component is primarily caused by natural changes in the Sun's output or in the volcanic aerosol loading of the atmosphere. Because of these natural internal and external processes, climate is always changing for reasons that are unrelated to human activities. A change in climate is judged to be "significant" if it is unusual relative to the changes expected to result from natural variability.

Detection of a significant change is thus by nature a statistical problem. A standard approach is to try to disprove the statistical "null" hypothesis that an observed change in climate can be explained by natural variability. If the null hypothesis is rejected, we say that we have detected a climate change at a particular significance level.

This does not yet identify the cause of the climate change, which is the attribution problem. Cause and effect are generally investigated by carrying out a series of experiments in which the response of a system to different causes is studied systematically. Such controlled experiments cannot be carried out using the real climate system: we have only one Earth, and only one observed evolution of the climate system. Furthermore, our experimentation with the real Earth's climate is not being done in any systematic way: we are varying a number of "causes" simultaneously (e.g., by changing land surface properties, concentrations of atmospheric greenhouse gases, and anthropogenic aerosols), rather than changing an individual cause, observing a climate response, and then varying the next cause.

Experimentation therefore requires the use of numerical models. The attribution of a detected climate change to a particular causal mechanism involves tests of competing hypotheses. By carrying out model simulations, we can determine the signals for different hypothesised causes (acting either individually or in combination) and compare these with observed changes. Unique attribution of a detected "significant" climate change to human activities requires the consideration and elimination of all other plausible non-anthropogenic mechanisms.

This is clearly a very difficult task, since there are large uncertainties in defining all possible "natural" climate-change signals. Attribution, therefore, can never be "certain" in a purely statistical sense: we can state only that the available observations are consistent or inconsistent with a postulated hypothesis at a given confidence level. The claimed statistical detection of an anthropogenic signal in the observations must always be accompanied by the caveat that other explanations for the detected climate-change signal cannot be ruled out completely. There is, however, an important distinction between achieving "practically meaningful" and "statistically unambiguous" attribution. This distinction rests on the fact that scientists and policymakers have different perceptions of risk. While a scientist might require decades in order to reduce the risk of making an erroneous decision on climate change attribution to an acceptably low level (say 1–5%), a policymaker must often make decisions without the benefit of waiting decades for near-statistical certainty.

In summary, "detection of change" is the process of demonstrating that an observed change in climate is highly unusual in a statistical sense, but does not provide a reason for the change. "Attribution" is the process of establishing cause and effect, i.e., that changes in anthropogenic emissions are required in order to explain satisfactorily the observed change in climate. Unambiguous attribution is feasible only in the sense of demonstrating that the observed change is consistent or inconsistent with the climate responses to a given set of external forcing mechanisms.

Statements regarding the detection and attribution of an anthropogenic effect on climate are inherently probabilistic in nature. They do not have simple "yes-or-no" answers. The probability of successful detection of significant climate change will be given in a range rather than as a discrete value. The width of this range depends on uncertainties in model predictions of anthropogenic change (Section 8.2) and on uncertainties in estimating natural variability (Section 8.3). A probability range for successful attribution of the detected changes to human effects will be more difficult to define.

8.1.2 Types of Climate-Change Detection Studies

There is a hierarchy of settings in which researchers have attempted to identify an anthropogenic effect on climate.

This hierarchy can be conveniently organised in terms of the spatial and temporal detail considered in comparing observed climate changes with those predicted by some physically based model (e.g., in response to increasing greenhouse gases).

8.1.2.1 Stage 1 studies

At the base of the hierarchy are studies that deal with the largest spatial scales, relatively coarse temporal information, and a single variable only. Stage 1 studies exist in a variety of different types (see Section 8.4.1). All such studies to date deal with global or hemispheric mean data for a single variable only, usually annually averaged near-surface temperature.

Stage 1 studies consider whether the observed change in global mean temperature over some period of time (generally the last century or so) is unusual in a statistical sense. To define "unusual" requires some "baseline" or "yardstick" of usual behaviour. It is difficult to make direct use of the recent observed data themselves, since signal and noise are intertwined in a complex way. The yardstick most frequently applied is internally generated natural variability, either as simulated by some physically based model (e.g., Wigley *et al.*, 1989; Wigley and Raper, 1990, 1991a,b; Stouffer *et al.*, 1994), or as derived from a statistical model fitted to observations (e.g., Karl *et al.*, 1991a; Bloomfield and Nychka, 1992; Woodward and Gray, 1993, 1995). The observed global mean change is deemed to be significant when it is judged highly unlikely to have resulted from internally generated natural variability alone.

Most of the recent work in the detection field has been this type of "Stage 1" study. A number of these investigations (both pre- and post-IPCC (1990)) have claimed the detection of a highly significant change in observed global mean temperature over the last 100 years. However, none of these studies has convincingly demonstrated that this change can be uniquely attributed to anthropogenic influences (see Section 8.4.1).

8.1.2.2 Stage 2 studies

Stage 2 studies again involve a single variable, but now compare model and observed *patterns* of change, generally for temperature. The patterns of change may be defined at many locations on the Earth's surface (Barnett and Schlesinger, 1987; Santer *et al.*, 1993, 1995a; Karl *et al.*, 1995a; Hegerl *et al.*, 1996), in a vertical section through the atmosphere (Karoly *et al.*, 1994; Santer *et al.*, 1995b), or even in three spatial dimensions (Barnett and Schlesinger, 1987). Often the model signal pattern has no information on time development, and is either taken from an equilibrium response experiment or consists of a time-averaged "snapshot" from a transient response experiment. In some cases the time-dependent information is contained in a pattern of linear trends (Hasselmann *et al.*, 1995; Karl *et al.*, 1995a; Hegerl *et al.*, 1996). The observed data always have information on the time evolution of patterns.

The next step is to compute some statistical measure of the similarity between the modelled and observed patterns of change. The problem is then to determine whether changes in this statistic with time are large relative to changes obtained in the absence of any external forcing. As in Stage 1 studies, various yardsticks are used to arrive at a probability level for detection of the predicted signal in the observed data.

Detection of a significant observed change in a Stage 2 study generally implies that even hemispheric- or sub-continental-scale features of observed and model-predicted patterns of change show a level of time-increasing agreement well beyond that expected due to natural variability alone[1]. If such a pattern correspondence were observed for a particular hypothesised set of causal factors, then (because this requires simultaneous agreement in a number of spatially separated regions) it is less likely that similar agreement could be obtained for a different set of causal factors. Nevertheless, to claim attribution convincingly still requires that alternative non-anthropogenic explanations for the observed changes be tested in a rigorous way, as described in Section 8.1.1.

In a purely subjective sense, however, a scientist would probably feel more confidence in the attribution of observed changes to a specific cause or causes after "successful detection" in a Stage 2 study than in a Stage 1 study. The justification for this line of thinking is the implicit assumption that different "causes" have different climate response patterns. This need not be the case, however. For example, some evidence suggests that the patterns of surface temperature change due to solar forcing and CO_2 doubling may be similar, since both climate forcings operate via similar feedback mechanisms, and these feedbacks dictate the spatial character of the response (Wigley and Jones, 1981; Hansen *et al.*, 1984). Yet the detailed *three-dimensional* structure of the thermal response to these two different forcings may well be different (e.g., Wetherald and Manabe, 1975). In the case of combined forcing by CO_2 and anthropogenic sulphate aerosols, it is unlikely that the spatially distinctive surface temperature response could be produced by other causes.

---

[1] Except in cases where it can be shown that the pattern correspondence statistic largely provides information on global mean changes.

8.1.2.3 Stage 3 studies

Like Stage 2 studies, these investigations also involve the two- or three-dimensional signal pattern for a single climate variable. The difference is that the signal is optimised in some way before it is searched for in the observed data (Hasselmann, 1979, 1993; Bell, 1982; North *et al.*, 1995; North and Kim, 1995). Optimisation involves filtering the observed data in order to suppress components that are similar to model-estimated natural variability noise, thus making the signal more easily detectable. This filtering may be in space only, or in both space and time. The modified signal is often referred to as an "optimal fingerprint". Spatial filtering has been used in two recent detection studies (Hasselmann *et al.*, 1995; Hegerl *et al.*, 1996), while full space-time filtering still awaits a first practical application in climate-change detection studies.

8.1.2.4 Stage 4 studies

Stage 4 studies differ from the earlier categories in their use of a number of climate variables simultaneously. Thus the searched-for anthropogenic signal in a Stage 4 investigation might consist of patterns of change in surface air temperature, precipitable water, diurnal temperature range, etc. Each variable would be defined in at least two (and possibly three or four) dimensions. The individual components of this multi-variable "fingerprint" might be decided upon by preliminary analysis of the signal-to-noise properties of a wide range of climate variables in model data (see Section 8.4.3). At Stage 4, "successful detection" would mean that the multi-variate, space-time varying climate-change signal from a model experiment forced with transient increases in anthropogenic emissions was in good accord with available observations. To date, only one attempt has been made to use a full multi-dimensional and multi-variable description of signal and noise in a climate-change detection study (Barnett and Schlesinger, 1987).

8.1.3 Progress Since IPCC (1990)

Scientific understanding of the detection and attribution problem has improved markedly since IPCC (1990). We now have:

- more relevant model simulations for the definition of an anthropogenic climate change signal;
- more relevant simulations for the estimation of natural internal variability, and initial estimates from palaeoclimatic data of total natural variability on global or hemispheric scales;
- more powerful statistical methods for detection of anthropogenic change, and increased application of

pattern-based studies with greater relevance for attribution.

In 1990, most of the available information concerning the projected climate changes in response to an enhanced greenhouse effect came from experiments in which an atmospheric GCM was coupled to a relatively simple model of the uppermost 50–100 m of the ocean, and atmospheric CO_2 levels were instantaneously doubled. In the real world CO_2 is increasing gradually. The more relevant question is how the climate system – including the intermediate and deep ocean, with their characteristic time-scales of decades to centuries – will respond to slowly increasing greenhouse gas concentrations.

Before IPCC (1990), only a few experiments with time-varying CO_2 had been performed with fully coupled atmosphere-ocean general circulation models (AOGCMs) (Stouffer *et al.*, 1989; Washington and Meehl, 1989). Since 1990, at least four modelling groups have published results from such integrations (e.g., Manabe *et al.*, 1991; Cubasch *et al.*, 1992; Meehl *et al.*, 1993; Murphy and Mitchell, 1995; see Chapter 6 for further references). The climate-change signal from these experiments now has both space and time dimensions. The inclusion of full ocean dynamics has modified the climate-change projections of earlier experiments which had simpler representations of the ocean. Furthermore, at least three modelling groups have now performed both equilibrium and transient response experiments with an AOGCM (Cubasch *et al.*, 1992; Manabe and Stouffer, 1993; Murphy and Mitchell, 1995). This addresses the question of whether very different forcing histories yield fundamentally different signal patterns, and hence provides insights into the relevance of using results from equilibrium experiments in detection studies. There has also been widespread recognition of the need for ensemble experiments to obtain more reliable estimates of an anthropogenic signal (Cubasch *et al.*, 1994).

A further major advance is that modelling groups are starting to consider the effects of anthropogenic forcings other than CO_2 in climate change experiments. These include:

- anthropogenic sulphate aerosol effects, both direct (Taylor and Penner, 1994; Hasselmann *et al.*, 1995; Mitchell *et al.*, 1995a,b; Roeckner *et al.*, 1995) and indirect (Erickson *et al.*, 1995);
- effects of stratospheric ozone reduction (Hansen *et al.*, 1995a; Ramaswamy *et al.*, 1996), and of changes in other (non-CO_2) greenhouse gases (Wang *et al.*, 1991).

While some pre-IPCC (1990) studies had attempted to estimate the climate effects of anthropogenic aerosols and non-CO_2 greenhouse gases in global mean terms (e.g., Wigley, 1989; Wigley *et al.*, 1989), it is only recently that GCMs have been used to estimate their detailed spatial signatures.

Recent model simulations have also provided initial information about the background natural internal variability against which an anthropogenic signal must be detected. This is the second major area of progress since IPCC (1990). Two modelling groups have recently completed long (≥1000 years) control runs, which contain a wealth of information on the magnitude, patterns and time-scales of internal climate variability generated by AOGCMs (Delworth *et al.*, 1993; Stouffer *et al.*, 1994; Hasselmann *et al.*, 1995). These model-generated noise estimates are now being used as the "yardsticks" for judging significance in climate-change detection studies (see Section 8.4), and first attempts are underway to check the consistency of decadal to century time-scale natural variability noise estimates derived from models and palaeodata (Barnett *et al.*, 1996). Additionally, information from palaeoclimatic proxy records is now being used to estimate the natural variability of surface temperature on hemispheric or even global spatial scales.

Statistical methods are the third main area in which advances have been made since IPCC (1990). The last five years have seen an increase in the number and sophistication of statistical techniques that have been applied to the problem of identifying anthropogenic change. Two important examples are the application of Singular Spectrum Analysis (SSA) to partition signal and noise in global mean temperature data (e.g., Ghil and Vautard, 1991; Schlesinger and Ramankutty, 1994) and the definition and application of "optimal detection" techniques (Hasselmann, 1979, 1993; Bell, 1982; North *et al.*, 1995; North and Kim, 1995). Full optimal detection methods provide real promise for dealing with some of the statistical problems that will be encountered in Stage 3 and Stage 4-type studies (see Section 8.1.3).

There has also been significant progress in the application of pattern-based methods to the detection problem, and in the incorporation of refined signal and noise estimates in detection studies. Up to 1990, most detection studies were of the Stage 1 type, focusing on global mean temperature. The main exceptions were the pattern-based studies by Barnett (1986) and Barnett and Schlesinger (1987). During the last five years, a number of research groups have published results from Stage 2 and Stage 3 investigations. In addition to the enhanced

greenhouse effect (Santer *et al.*, 1993; Hegerl *et al.*, 1996), the searched-for patterns now include the signals due to anthropogenic sulphate aerosols (Karl *et al.*, 1995a) and the combined effect of greenhouse gases and sulphate aerosols (Hasselmann *et al.*, 1995; Mitchell *et al.*, 1995a; Santer *et al.*, 1995a,b). Although important uncertainties remain in these studies, they have yielded initial evidence for the existence of an anthropogenic effect on climate. Furthermore, we have now started to see pattern-based studies that directly address the attribution question, and try to rule out various non-anthropogenic forcing mechanisms as explanations for some observed pattern of climate change (e.g., Karoly *et al.*, 1994; Hansen *et al.*, 1995b).

8.2 Uncertainties in Model Projections of Anthropogenic Change

There is no direct historical or palaeoclimatic analogue for the rapid change in atmospheric CO_2 that has taken place over the last century (Crowley, 1991). This means that unless we assume that the signal pattern is independent of the spatial character of the forcing (see, e.g., Hoffert and Covey, 1992), we cannot use palaeoclimatic data or instrumental records as analogues for the regional and seasonal patterns and rate of climate change over the next century. We must therefore rely on models for this information. The aim of the following section is to consider the main uncertainties involved in such projections of anthropogenic climate change.

8.2.1 *Errors in Simulating Current Climate in Uncoupled and Coupled Models*

Model validation is one of the most important components in our efforts to predict future global climate change. Although model performance has generally improved over the last decade, both coupled and uncoupled models still show systematic errors in their representation of the mean state and variability statistics of current climate (see Chapter 5, and also Gates *et al.*, 1990, 1992). Such errors reduce our confidence in the capability of AOGCMs to predict anthropogenic change.

8.2.2 *Inadequate Representation of Feedbacks*

Realistic simulation of the present climate is probably a necessary, but not sufficient condition to ensure successful simulation of future climate. To be confident that a model has predictive skill on time-scales of decades or longer, we would also have to be sure that it incorporates correctly all of the physics and feedback mechanisms that are likely to be important as greenhouse gas concentrations or aerosol-producing emissions increase. As discussed in Chapter 4, it is unlikely that all important feedbacks have been included correctly in current AOGCMs. Feedbacks involving clouds and the surface radiation budget are poorly understood, and different schemes for parameterizing cloud processes can lead to substantially different results in greenhouse warming experiments (Cess *et al.*, 1989; Mitchell *et al.*, 1989). Other feedbacks that are either currently neglected or highly uncertain include interactions between the land biosphere and the carbon cycle, and between climate and atmospheric chemistry (see Chapters 2 and 4). Deficiencies in the treatment and incorporation of feedbacks are a source of signal uncertainty.

8.2.3 Flux Correction Problems

In present-day AOGCMs, the atmospheric and oceanic components are run independently until they achieve a steady state, and then coupled together. The coupled model is then run for some "adaptation" period prior to initiating a control or perturbation experiment. The interactive coupling of AOGCMs generally leads to a phenomenon known as climate drift – the tendency of the model's climate system to drift into a new and unrealistic mean state (Gates *et al.*, 1985; Washington and Meehl, 1989). The reasons for this drift are discussed in Chapter 5.

One strategy for dealing with this problem is to make no attempt to correct the flux mismatch. The climate response in a perturbation experiment is then assumed to be the difference between the climate change simulation and a drifting control run that started from the same initial state. This strategy may lead to erroneous results if the climate system is drifting into a state with basically different response characteristics (i.e., a climate state unlike that currently observed on Earth). An example of such a situation might be if the climate drift led to large changes in the sea ice distribution or the mean ocean circulation.

Another procedure is to reduce the drift by balancing the flux mismatch through insertion of *flux adjustment* terms in the coupled model (Sausen *et al.*, 1988; see Chapter 5). Advocates of this approach argue that flux adjustment should have no impact on the dynamical response characteristics of the system for small perturbations about the initial state. Critics, however, note that the flux adjustments in some models can in certain regions greatly exceed the forcing expected from greenhouse gases.

Such corrections are outward symptoms of underlying systematic errors in the uncoupled models. Their impact on the reliability of signals from anthropogenic change experiments and on the simulated natural variability is largely unknown. Some studies with simplified models

suggest that flux correction may seriously affect the simulated signals and natural variability noise (Nakamura *et al.*, 1994; Neelin and Dijkstra, 1995; Pierce *et al.*, 1995). Our confidence in the anthropogenic climate change signals simulated by current flux-adjusted AOGCMs, and in the results of detection studies that have used these signals, will be increased if the principal features of these signals can be reproduced by uncorrected coupled models. There are some initial indications that this may be the case (see Chapter 6).

8.2.4 Signal Estimation Problems

The anthropogenic signal from any experiment with a fully coupled AOGCM is not a "pure" signal – it is the combination of a signal and some manifestation of the model's own internally generated natural variability.

Thus, the "pure", underlying signal in response to time-evolving anthropogenic forcing cannot be estimated precisely from a single transient experiment. This problem is exacerbated by uncertainties in the initial state of the climate system, particularly the ocean. More reliable signal estimates are likely to be obtained by averaging over multiple runs, each starting from different initial conditions. Ideally, the initial states chosen should encompass the large range of uncertainty in our knowledge of the climate state around the middle of the last century.

The most relevant work in this regard is by Cubasch *et al.* (1994), who performed a suite of four 50-year "Monte Carlo" experiments with a AOGCM. The model was forced with identical increases in greenhouse gases, with each experiment starting from different "snapshots" of the climate state (taken at 30-year intervals from a long control run). Marked differences were found in the space-time structure of the surface temperature response, but these differences were no larger than those expected due to natural internal variability alone (see Chapter 6).

Clearly, it is always desirable to carry out a series of Monte Carlo experiments and determine the mean climate change response and variability statistics from this ensemble. The ensemble average is likely to be a better estimate of the pure signal than that obtained from any single realisation, and the variability statistics provide information on statistical uncertainty in the climate-change response, which should be accounted for in any detection strategy.

8.2.5 "Missing Forcing" and Uncertainties in Space-Time Evolution of Forcing

A further source of signal uncertainty relates to "missing" anthropogenic forcings that have not been included in the climate-change signals currently used in detection studies. The most important examples include the indirect effects of sulphate aerosols (Erikson *et al.*, 1995), the radiative effects of trace gases other than CO_2 (Wang *et al.*, 1991), and the forcings associated with large-scale land-use changes (Tegen and Fung, 1995) or the carbonaceous aerosols generated by biomass burning (Penner *et al.*, 1992, 1994). At present, none of these factors (other than changes in stratospheric ozone; Santer *et al.*, 1995b) have been considered in pattern-based detection studies. Some of these "missing forcings" may modify our present picture of the possible climate-change signal due to anthropogenic activities.

The current magnitude and the historical evolution of the radiative forcing is uncertain, both for these missing forcings and for anthropogenic forcings (such as sulphate aerosol direct effects and stratospheric ozone reduction) that *are* presently included in AOGCM experiments relevant to the detection issue. Uncertainties exist at both the global mean level and in terms of the forcing pattern (see Chapter 2).

For pattern-based detection studies, reliable definition of an anthropogenic signal requires, at the very least, a reliable estimate of present-day SO_2 emissions and their spatial distribution, and an accurate interpretation of these in terms of a forcing pattern (Schneider, 1994). Studies that involve a time-evolving signal additionally require information on the historical evolution of emissions and forcing. Current pattern-based detection work is now beginning to account for such forcing changes over space and time (e.g., Hasselmann *et al.*, 1995; Mitchell *et al.*, 1995a).

8.2.6 *Cold-Start Effect*

The so-called "cold start" problem results from the neglect of anthropogenic forcings (and hence some portion of the climate response) that happened before the start of the simulation. At present, only two modelling groups have performed experiments with a starting date in the latter half of the last century, thus avoiding a significant cold start error (Hasselmann *et al.*, 1995; Mitchell *et al.*, 1995a). Most climate change experiments run with AOGCMs have been initialised from a point much closer in time to the present – a time for which we have better forcing and observational data for comparison with the model signal. Such simulations are inevitably subject to a "cold start" error.

The magnitude of this effect has been estimated in a variety of ways, as discussed in Chapter 6. Errors which the cold start effect may introduce into the time evolution of an anthropogenic signal constitute a further source of uncertainty in detection studies.

8.3 Uncertainties in Estimating Natural Variability

Defining an anthropogenic climate change signal is only one part of the detection problem. The climate state of the Earth is always changing in both space and time for reasons that have nothing to do with anthropogenic forcing. The space-time structure of this natural variability must be estimated in order to decide whether the changes that have been observed in the past or that will be observed over the next 10–20 years are due primarily to human activities or natural causes. This spectrum of natural variability is a critical element in the significance-testing portion of any practical detection or attribution scheme.

Information on natural climate variability can be derived from three main sources: instrumental data, palaeoclimatic reconstructions and numerical models. Some of the major uncertainties in estimating decadal to century time-scale natural variability from these sources are summarised below.

8.3.1 *From Instrumental Data*

Surface air temperature is undoubtedly the best-observed climate variable in terms of length of the instrumental record and coverage of the Earth's surface (see Chapter 3). Even in this case, however, the length of record (ca. 130-150 years) is barely adequate to obtain a reliable estimate of the multi-decadal portion of the spectrum, and is inadequate to estimate longer time-scale variability. Furthermore, the instrumental record covers a period when both anthropogenic effects and natural variability have occurred together, making it very difficult to separate the two in an unambiguous way.

Nevertheless, several attempts have been made to estimate the background natural variability by subtracting a model-estimated anthropogenic signal from the observed data (Schlesinger and Ramankutty, 1994; Hegerl *et al.*, 1996). The latter investigation used an AOGCM to define a single pattern of temperature change in response to greenhouse gas forcing, and then employed a simple model to estimate the time evolution of this signal pattern. This estimate of the space-time evolving greenhouse signal was then subtracted from the observed data, and the residuals were used to estimate natural variability. The reliability of such variability estimates depends critically on the accuracy of the AOGCM's climate sensitivity, on the correctness of any assumed lag between radiative forcing and temperature response in the simple model, on the reliability of the signal pattern, and on the correctness of the assumed radiative forcing history. Estimating multi-decadal natural variability from the observed data in a way that accounts for such uncertainties is a difficult task.

8.3.2 *From Palaeoclimatic Records*

Changes in climate affect a wide range of biological, chemical, and geological processes. As a result, climatic information is recorded naturally in tree rings, ice cores, coral reefs, laminated sediments, etc. (e.g., Cook *et al.*, 1991; Crowley and North, 1991; Bradley and Jones, 1992; Briffa *et al.*, 1992). If we can understand the recording mechanism, for example the process by which climate imprints itself on tree growth and annual ring formation, then we have the potential to unlock a wealth of climate information stored in palaeoclimate records.

Unfortunately, unravelling the history of climatic variability contained in such records is not a simple task, as is discussed in Chapter 3. Few proxy records are capable of providing well-dated millennia-long reconstructions with at least decadal time-scale resolution. In addition, a number of general problems are common to all proxy records. Most temperature reconstructions, for example, are seasonally specific, rather than providing some integrated response to annual-mean conditions. In order to produce a reconstruction, the raw data are generally subjected to some form of statistical manipulation, through which only part of the original climate information can be retrieved (typically less than 50%). Such manipulations can also affect the frequency characteristics of the reconstruction (Briffa *et al.*, 1996). More importantly, spatial coverage is poor for high temporal resolution palaeoclimate data, and it is often difficult to compare the climate information extracted from different locations (e.g., land and ocean) or from different proxy sources.

Initial attempts are now being made to use palaeoclimate data in order to reconstruct a satisfactory, spatially comprehensive picture of climate variability over the past 1000 years (Bradley and Jones, 1993). The process of quality-controlling palaeoclimatic data, integrating information from different proxies, and improving spatial coverage should be encouraged. Better palaeoclimatic data bases for at least the past millennium are essential in order to assess the contribution of natural variability to recent observed changes and to validate coupled model noise estimates on century time-scales (Barnett *et al.*, 1996).

8.3.3 *From Numerical Models*

8.3.3 1 Energy balance models

Some of the first model-based studies of natural variability used simple energy balance models (EBMs; e.g., Hasselmann, 1976). Such models generally solve equations for the heat balance of a highly idealised representation of the Earth's atmosphere and ocean (Lemke, 1977; Hoffert and Flannery, 1985). By forcing an EBM with white noise[1], it has been possible to investigate the relationship between daily weather noise and the model's internally generated variability on time-scales of years to centuries. These early studies, together with more recent EBM studies by other groups (e.g., Wigley and Raper, 1990, 1991a; Kim and North, 1991), have demonstrated that even simple EBMs can generate useful information on decadal to century time-scale surface temperature fluctuations as an integrated response to shorter time-scale random fluctuations.

While EBMs successfully reproduce many details of observed surface temperature variability on the annual-to-decadal-time-scales (Kim and North, 1991), they may underestimate the magnitude of internal variability on longer time-scales, since they cannot explicitly simulate the changes in the horizontal and vertical transport of heat, salt and momentum necessary for an accurate representation of the ocean circulation. EBMs therefore simulate only the "passive" component of natural variability (Wigley and Raper, 1990), and fail to capture any dynamically induced variations in the ocean. Nevertheless, the spectrum of EBM-based variability is within the range of spectral estimates derived from AOGCMs that simulate both components (Santer *et al.*, 1995a).

8.3.3.2 Stochastically forced ocean models

More sophisticated estimates of natural variability may be obtained from OGCMs forced by various forms of statistical atmospheres. Virtually all such simulations show interdecadal to centennial time-scale variations no matter what the form of the ocean GCM or the character of the forcing (see Chapter 5). The key forcing variable seems to be the fresh water flux between the atmosphere and the ocean. Typical time-scales of the induced natural variability range from 10–350 years or longer. The more realistic of these simulations show a rich set of oscillatory modes specific to individual ocean basins, with the most energetic mode being global in extent with largest amplitude in the high latitudes of the southern oceans (Mikolajewicz and Maier-Reimer, 1990; Pierce *et al.*, 1995).

The problems with the class of experiments discussed above are at least threefold. First, many of the simulations are very simplified, and the realism of their simulated variability is subject to considerable doubt. Second, the OGCMs used in such experiments have almost exclusively used so-called *mixed boundary conditions* (see, e.g., Power

---

[1] e.g., by heat flux anomalies that are essentially random in time, and thus can be thought of as characteristic of the internal variability that is generated by daily weather fluctuations.

and Kleeman, 1993; Mikolajewicz and Maier-Reimer, 1994) that guarantee the model is constrained to relax from any perturbed state back towards a climatological temperature field. A number of studies have shown that the simulated oceanic variability is highly sensitive to the value of the time constant used to restore temperature to climatology (Mikolajewicz and Maier-Reimer, 1994; Power and Kleeman, 1994). Finally, the statistical atmospheres used to date do not explicitly account for feedbacks and the meridional transport of water or heat. It is possible that variations in these transports play a significant role in decadal and longer time-scale climate variations.

8.3.3.3 Coupled atmosphere-ocean general circulation models

At present there are only two long integrations with AOGCMs that might allow one to estimate the spectrum of multi-decadal to century time-scale internally generated natural variability. These are the 1000-year control integration recently performed with the GFDL (Geophysical Fluid Dynamics Laboratory) model by Stouffer *et al.* (1994), and a 1260 year integration carried out with the European Centre/Hamburg (ECHAM-1) model (Hasselmann *et al.*, 1995). This situation is likely to change in the next few years, since a number of other groups are currently performing long control integrations with coupled models (e.g., Mitchell *et al.*, 1995a).

Since the variability in the GFDL and ECHAM-1 control integrations has already been used as a "yardstick" for judging the significance of observed trends in climate (Stouffer *et al.*, 1994; Hasselmann *et al.*, 1995; Hegerl *et al.*, 1996; Karl *et al.*, 1995a; Santer *et al.*, 1995a,b), it is crucial to determine how well these models reproduce important features of observed climate variability. This is now being attempted, at least on interannual and decadal time-scales. The study by Stouffer *et al.* (1994) and work presented in Chapter 5 show that qualitative agreement exists between observed patterns of near-surface temperature variability on time-scales of one to five years and those simulated in the GFDL and ECHAM-1 control runs. The Stouffer *et al.* analysis, and more recent work by Manabe and Stouffer (1996), also indicated that some of the observed relationships between global mean and regional temperature changes are well-simulated in the GFDL control run. On longer decadal time-scales, Mehta and Delworth (1995) found considerable similarity between observed patterns of SST variability in the tropical Atlantic and those simulated by the GFDL model.

A comparison of the observed spectrum of global mean temperature variability with that obtained from the GFDL,

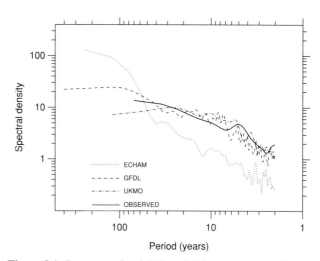

Figure 8.1: Spectrum of variability of global mean, annually averaged near-surface temperature estimated from observed data and from three AOGCM control runs. Observed data are from Jones and Briffa (1992). Model results are from the 1000-year GFDL control run (Stouffer *et al.*, 1994), the first 600 years of the 1000-year ECHAM-1/LSG control integration (Hasselmann *et al.*, 1995) and the first 310 years of the UKMO control run (Mitchell *et al.*, 1995a). The spectrum provides information on the distribution of variance on different time-scales, ranging from two years (the shortest period that can be resolved using annual-mean data) to nearly 400 years for the long GFDL control run. Spectra were estimated after first subtracting overall linear trends from the modelled and observed time-series.

ECHAM-1, and UKMO (United Kingdom Meteorological Office) (Mitchell *et al.*, 1995a) control runs shows that the GFDL and UKMO spectra are quite close to the observations over the range of periods for which they overlap, whereas ECHAM-1 has consistently less variability than the observations on time-scales of roughly 2 to 60 years (see Figure 8.1). The observations have a clear peak in the spectrum at around 3–5 years corresponding to El Niûo-Southern Oscillation (ENSO) variability. This is not well-simulated by any of the three models.

The short length of the observed record (140 years) and the UKMO control run preclude their use in estimating century time-scale variability. The GFDL and ECHAM-1 spectra differ substantially on these longer time-scales, with the GFDL spectrum flattening out at periods greater than 100 years, while variability in the ECHAM-1 control run continues to increase. A major part of the explanation for this difference in power at the low-frequency end of the spectrum is the large drift over the first 200 years of the ECHAM-1 control run. If this drift is excluded, the low-frequency temperature variability in the GFDL and ECHAM-1 control runs is much more similar, with the latter also showing a pronounced flattening.

The ubiquity of substantial power at century time-scales in palaeoclimate spectra (see, e.g., Crowley and North, 1991; Stocker and Mysak, 1992) suggests either that the palaeodata have some low-frequency forcing that the models are lacking (e.g., solar variability, volcanoes), or that both GFDL and ECHAM-1 are underestimating the magnitude of century-time-scale internally generated variability, a possible consequence of the flux-adjustment schemes used to stabilise model climate. A further possible explanation for this discrepancy is that the palaeodata are themselves erroneous[1].

One preliminary attempt to compare decadal to century time-scale natural variability estimates from coupled models[2] and from palaeoclimatic data suggests that there is disagreement between the two, both in terms of the patterns and amplitude of variability (Barnett *et al.*, 1996; see Figure 8.2). The significance of this mismatch is difficult to interpret for the reasons pointed out above. Nevertheless, this result does point towards the need for a more extensive, quality-controlled palaeoclimatic data base, and for some caution in the interpretation of results from current detection studies that use model-based noise estimates for evaluating the significance of observed changes.

8.4 Evaluation of Recent Studies to Detect and Attribute Climate Change

The following section reviews recent efforts to detect an anthropogenic climate-change signal in the observed data. It primarily covers studies of change in globally averaged variables and pattern-based approaches, and concentrates on those analyses that have used quantitative methods to compare model and observed data. The final portion of this section briefly considers studies involving model data only. Such work has attempted to identify in the model data climate variables that might be promising components of a multi-variable "fingerprint" for detecting anthropogenic change.

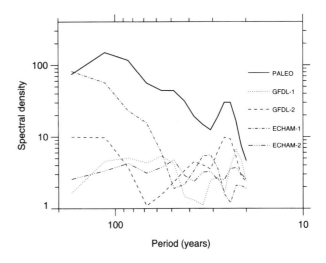

Figure 8.2: Spectra of near-surface temperature variability as estimated from two different model control runs and from palaeoclimatic temperature reconstructions. Results are from Barnett *et al.* (1996). Palaeoclimatic data were taken from Bradley and Jones (1993), and consist of temperature reconstructions from tree rings, corals, ice melt records, ice cores and historical documentary records. A total of 16 proxy records were used, each covering the common period 1600 to 1950. The space-time variability in this data set was investigated by means of an Empirical Orthogonal Function (EOF) analysis, in which the dominant modes of variability are represented by a limited number of spatial patterns (EOFs) and time-series. The bold solid line shows the spectrum of the amplitude time-series for the dominant palaeoclimatic pattern of variability. Control runs with the GFDL (Stouffer *et al.*, 1994) and ECHAM-1/LSG (Hasselmann *et al.*, 1995) AOGCMs were sampled at grid-points closest to the 16 locations of the palaeodata. Model temperature data for two separate 350-year chunks of each control run were then projected onto the dominant EOF pattern from the palaeodata. The resulting four time-series were then subjected to spectral analysis. The model data have lower variance than the palaeodata at all time-scales considered (from 2 years to > 100 years), with the largest mismatches occurring on the longer time-scales.

---

[1] As pointed out in Section 8.3.2, statistical manipulation of the paleodata may alter their low-frequency characteristics. However, this is more likely to have *decreased* low-frequency power than to have amplified it, at least for the case of temperature reconstructions from tree rings. Note also that the paleospectra are for individual sites, not for global-mean temperature, so a direct comparison between "reconstructed" variability at specific locations and model estimates of global-mean variability is not valid.

[2] As sampled at model grid-points close to the locations for which paleoclimatic temperature data were available.

8.4.1 *Studies of Changes in Global Mean Variables*

Stage 1 studies related to the detection of anthropogenic climate change may be divided into three types: detection of a statistically significant change or trend in a single climate variable; comparison of model simulations with observations; and empirical estimation of the climate sensitivity. All studies in these areas have chosen near-surface temperature as the detection variable, and have used either global mean data or some large-area subset of this (such as the average over Northern Hemisphere land areas).

8.4.1.1 Studies to detect change in observed data
Studies to detect change are concerned primarily with determining whether the observed warming is statistically significant, i.e., highly unusual relative to natural variability and unlikely to have occurred by chance alone. If the observed change is judged unlikely to have occurred due to natural processes, then an anthropogenic factor is implicated.

While this sounds straightforward, the assessment of significance is in fact quite difficult. This is primarily due to uncertainties in the magnitude of natural variability on the decadal to century time-scales relevant to the anthropogenic climate change issue. Even if a change were significant relative to internal variability, this does not mean that anthropogenic forcing is responsible: the change may be due to natural external forcing factors.

A number of studies have claimed that the observed warming is statistically significant. The credibility of these claims depends on how they have estimated the magnitude of natural variability and/or assessed the statistical significance of any change or trend. There are two basic approaches. In the first, natural variability is estimated using some form of statistical model fitted to observations, while the second approach uses an estimate derived from a physical model.

The statistical models generally attempt to represent the observed global mean, annually averaged temperature data by some model consisting of a deterministic trend (which may be either linear or non-linear) plus "residuals" about the trend, which represent natural variability. These residuals are strongly autocorrelated. Even the earliest studies (Madden and Ramanathan, 1980; Wigley and Jones, 1981) recognised that standard statistical tests, which evaluate trend significance relative to assumed "white noise" (or non-autocorrelated) residuals, were inappropriate to the climate change problem. The differences between subsequent studies arise largely from the way the residuals are modelled, i.e., how their autocorrelation structure is accounted for.

Wigley *et al.* (1989) and Allen and Smith (1994) have shown the warming trend to be significant for first-order (AR-1) autocorrelated residuals. A number of more sophisticated models of the residuals, allowing for much more complex natural variability behaviour than an AR-1 model, have also demonstrated the significance of the warming trend (see, e.g., Kuo *et al.*, 1990; Bloomfield, 1992; Bloomfield and Nychka, 1992; Galbraith and Green, 1992; Tol and de Vos, 1993; Tol, 1994). Alternately, several authors have shown that some noise models with correlation of the residuals over century time-scales are

consistent with the data but lead to a non-significant result for the observed warming trend (Kheshgi and White, 1993a; Woodward and Gray, 1993, 1995).

Claims of non-significance of the global warming trend have also been made by Ghil and Vautard (1991) using Singular Spectrum Analysis (SSA). The basic argument is that the data show a low-frequency cyclic component that is in phase with and explains most of the observed trend. While the cyclic component is a characteristic of the data, whether or not it is a valid alternative explanation of the trend depends on whether it is a basic, long-lasting, deterministic property of the climate system. It is impossible to determine whether this is correct given the shortness of the available data (see, e.g., Allen *et al.*, 1992; Cook *et al.*, 1996). Furthermore, Schlesinger and Ramankutty (1994) have shown that a better explanation of the cyclic component identified by Ghil and Vautard is to assume that it is superimposed on an anthropogenic trend. In their analysis, the 60-65 year cycle accounts for only a relatively small (<10%) fraction of the overall variation in global mean temperature.

The second approach to assessing trend significance has been to use a climate model to generate the background "unforced" variability. Using a simple upwelling diffusion-energy balance model, Wigley and Raper (1990, 1991a) have shown that the observed century time-scale trend is highly significant relative to the range of possible unforced trends. The same result has been obtained by Stouffer *et al.* (1994) using results from a 1,000-year control run simulation with the GFDL AOGCM.

Figure 8.3 shows how observed trends in global mean, annually averaged temperature over the past 10 (1984-1993) to 100 (1894-1993) years compare with the trends expected to occur through natural internal variability, based on the best estimates currently available from AOGCMs (Stouffer *et al.*, 1994; Hasselmann *et al.*, 1995; Mitchell *et al.*, 1995a). Model global mean temperature data were used to generate sampling distributions of linear trends, and the 95th percentiles are plotted for each trend length. There are model-specific differences in the magnitude of variability and its partitioning on different time-scales (see also Section 8.3.3.3 and Figure 8.1). While observed trends over the last 10 years lie within the range of 10-year "natural" trends of two of the three models, observed trends on longer time-scales are larger than (at least) 95% of the unforced trends simulated by the models. Evaluation of the true significance of the observed trends is difficult since the model noise estimates do not incorporate the effects of changes in natural external forcing factors such as the Sun's output or volcanic aerosols, and so are likely

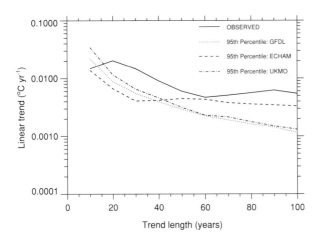

Figure 8.3: Significance of observed changes in global mean, annually averaged near-surface temperature. The solid line gives the magnitude of the observed temperature trend (in °C/year) over the recent record – i.e., over the last 10 years (1984 to 1993), 20 years (1974 to 1993), etc. to 100 years. Observed data are from Jones and Briffa (1992). Model results from three AOGCM control integrations (see Figure 8.1) were used to evaluate whether the observed linear trends are unusually large relative to the trends likely to result from internally generated natural variability of the climate system. Linear trends were fitted to overlapping "chunks" of the model temperature series, thus allowing sampling distributions of trends to be generated for the same 10 to 100 year time-scales for which observed temperature trends were estimated. The 95th percentiles of these distributions are plotted for each model control run and each trend length.

to underestimate the "total" natural variability of the real-world climate system. They may also be affected by flux correction procedures. Furthermore, the short length of the three control runs used here precludes reliable estimates of the statistical properties of internally generated variability on time-scales longer than two to three decades.

The conclusion that can be drawn from this body of work, and from earlier studies reported in Wigley and Barnett (1990), is that the warming trend to date is unlikely to have occurred entirely by chance due to natural variability of the climate system. Implicit in such studies is a weak attribution statement – i.e., some (unknown) fraction of the observed trend is being attributed to human influences. Any such attribution-related conclusions, however, rest heavily on the reliability of our estimates of both century time-scale natural variability and the magnitude of the observed global mean warming trend. At best, therefore, trend significance can only provide circumstantial support for the existence of an anthropogenic component to climate change.

8.4.1.2 Comparisons of model results and observed data

The basic strategy underlying these studies is to force some form of climate model with anthropogenic and/or natural forcing factors and then compare the resulting global mean temperature output with the observed record. The types of model used have varied from what amounts to a "null" model (i.e., assuming, unrealistically, that solar forcing and temperature response are proportional and in phase – as in the work of Friis-Christensen and Lassen, 1991), through simple regression models (as in the work of Schönwiese *et al.*, 1994, and earlier papers cited therein) that are unrealistically constrained to have a constant lag between forcing and response, to using an upwelling diffusion-energy balance model (UD/EBM), as in IPCC (1990) (Wigley and Barnett, 1990). In terms of imposed forcing, some studies have considered only a single forcing factor, such as solar variability alone in the work of Friis-Christensen and Lassen, while others have considered multiple forcing factors, both natural and anthropogenic. Since the primary area of uncertainty is in the response to anthropogenic forcing, global mean studies using only a single "natural" forcing factor and ignoring any anthropogenic component are inadequate, as pointed out by Schlesinger and Ramankutty (1992) and Kelly and Wigley (1992).

The main conclusion that can be drawn from these investigations is that the observed record of global mean temperature changes can be well simulated by a range of combinations of forcing. Best fits are obtained when anthropogenic forcing factors are included, and, when this is done, most of the observed trend is found to result from these factors. Within the range of forcing and model parameter uncertainties, there is no inconsistency between observations and the modelled global mean response to anthropogenic influences.

8.4.1.3 Empirical estimation of the sensitivity of climate to forcing

A related univariate approach attempts to use observational data to deduce $\Delta T_{2\times}$, the so-called *climate sensitivity parameter* – a measure of how sensitive the climate system is to external forcing.[1] This is done by running a series of EBM simulations with different sensitivity values and choosing the value that gives a best fit with observed global mean changes. In such experiments there are a number of "tuning knobs" other than the sensitivity, both

---

[1] The climate sensitivity is usually expressed as the equilibrium warming for a CO_2 doubling.

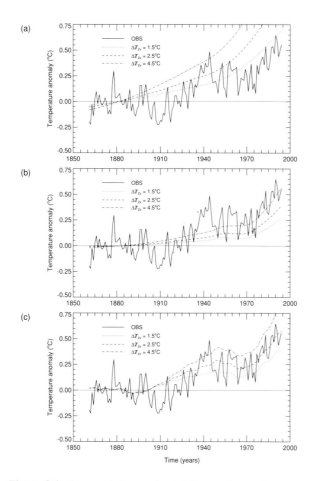

Figure 8.4: Observed changes in global mean temperature over 1861 to 1994 compared with those simulated using an upwelling diffusion-energy balance climate model. The model was run first with forcing due to greenhouse gases alone (a), then with greenhouse gases and aerosols (b), and finally with greenhouse gases, aerosols, and an estimate of solar irradiance changes (c). The radiative forcings were the best-guess values recommended in this report. Simulations were carried out with climate sensitivities ($\Delta T_{2\times}$) of 1.5, 2.5 and 4.5°C for the equilibrium CO_2-doubling temperature change.

for the input historical forcings and for the EBM parameters. A few papers have considered the joint optimisation of $\Delta T_{2\times}$ and one or more forcing factors, such as solar variability and/or sulphate aerosol forcing (e.g., Kelly and Wigley, 1992; Schlesinger *et al.*, 1992; Schlesinger and Ramankutty, 1992).

A crucial factor in studies of this type is the assumed radiative forcing. In IPCC (1990) (Wigley and Barnett, 1990), it was shown that forcing a UD-EBM with observed greenhouse gas forcing implied a $\Delta T_{2\times}$ value of around 1.5°C. Since that time, the importance of tropospheric aerosol forcing has been realised and this has a large effect

on the implied $\Delta T_{2\times}$ (Wigley, 1989; Schlesinger and Ramankutty, 1992; 1994; Wigley and Raper, 1992). This sensitivity is illustrated in Figure 8.4.

The top panel of this figure compares observations with model output for different $\Delta T_{2\times}$ for the case of greenhouse gas forcing alone. The optimum $\Delta T_{2\times}$ value is between 1°C and 2°C, consistent with IPCC (1990). There are small differences, relative to IPCC (1990), due mainly to modifications in the history of past radiative forcing. The middle panel adds a global mean aerosol forcing reaching around -1.3 Wm^{-2} in 1990 (the best-guess value from Chapter 2). As before, a reasonable fit between model and observations can be obtained, but now for a value of $\Delta T_{2\times}$ well above 4.5°C. A lower aerosol component would lead to a lower $\Delta T_{2\times}$, as in Wigley and Raper (1992) and Schlesinger et al. (1992). The bottom panel shows that an improved fit to the data and a sensitivity estimate in better agreement with our current best estimates (see Chapter 6) can be obtained if a recent estimate of solar irradiance variations (consistent with that used in Chapter 2) is added as an additional forcing factor. In this case, the implied best-fit $\Delta T_{2\times}$ lies between 3 and 4°C.

These empirical estimates of $\Delta T_{2\times}$ are subject to considerable uncertainty, as shown in a number of studies (see, e.g., Wigley and Barnett, 1990; Wigley and Raper, 1991b; Kheshgi and White; 1993b). In summary, such studies offer support for a $\Delta T_{2\times}$ value similar to that obtained from GCMs, and suggest that human activities have had a measurable impact on global climate, but they cannot establish a unique link between anthropogenic forcing changes and climate change.

8.4.2 *Studies of Modelled and Observed Spatial and Temporal Patterns of Change*

Current pattern-based studies have compared modelled and observed patterns of change at many individual points on the Earth's surface, or along a vertical section through the atmosphere. The basic approach in these studies is to compare the patterns of observed changes in climate with a single equilibrium signal pattern or a sequence of time-evolving signal patterns derived from an AOCGCM. Different comparison statistics have been used, generally classifiable as centred or uncentred statistics.[1] Since one would expect

---

[1] Uncentred statistics retain information on the spatially averaged changes in the two fields being compared. In contrast, centred statistics subtract these spatial-mean components and compare the patterns of spatial variability about the respective spatial means of the two fields. The standard product-moment pattern correlation is an example of a centred statistic.

from the history of increasing emissions of anthropogenic pollutants that the observed climate data should show a growing anthropogenic signal, the pattern similarity statistic should show an overall upward trend. Increases in the statistic with time are unlikely to be monotonic, since the behaviour of the statistic is also modulated by natural internal variability and by changes in natural external forcings. A statistically significant overall trend would constitute a form of detection.

Studies of this type have searched the observed data for the signals due to the individual effects of greenhouse gases (generally CO_2 only) or anthropogenic sulphate aerosols, or for signals due to simultaneous changes in several forcings, such as greenhouse gases and sulphate aerosols. The principal results of previous pattern-based investigations are discussed below.

8.4.2.1 Greenhouse gas signals

Most of the pioneering work on pattern-based detection studies was carried out by Barnett and collaborators (Barnett, 1986, 1991; Barnett and Schlesinger, 1987; Barnett *et al.*, 1991). These studies focused primarily on the development of statistical methods for comparing model-predicted signals with observed climatic changes. They also highlighted a number of key issues:

- the difficulty of using global mean changes to address the attribution issue;
- the need in detection strategies to account for the model-dependence of the signal and model errors in simulating present climate;
- the desirability of "pre-screening" model data to identify climate variables with high signal-to-noise ratios;
- the importance of considering a variety of statistical approaches in detection studies.

This initial pattern-based work used "CO_2-only" signals, and searched for these in recent observed records of sea surface temperature, near-surface air temperature, and the three-dimensional structure of tropospheric temperature. The various statistics used to compare model equilibrium- or transient-change patterns with observed patterns of change failed to yield any sustained, multi-decadal positive trends, beyond those trends that were related to trends in global mean quantities.

Santer *et al.* (1993) pursued some of the methodological issues raised by Barnett's work. They showed that an uncentred pattern similarity statistic could be split into two terms, one proportional to changes in the spatial average value of the variable considered (e.g., the global mean temperature), and the other proportional to the centred pattern correlation. For the CO_2-only temperature-change signals considered by Santer *et al.* (1993), the spatial mean term accounted for almost all of the variability in the uncentred statistic, i.e., the statistic contained negligible sub-global-scale pattern information. Santer *et al.* also showed that normalisation of model and observed mean-change fields by a common standard deviation field (in order to assign higher weight to grid-points with larger signal-to-noise ratios) introduced a so-called "common factor" effect which made trend interpretation difficult.

The Santer *et al.* (1993) study, like the earlier work by Barnett, failed to yield evidence for the existence of the sub-global-scale features of a model-predicted CO_2 signal in observed records of surface air temperature change. This result has three basic explanations:

- The CO_2-only signal may have been incorrect, e.g., due to the lack of a fully dynamic ocean in the models used.
- The signal may have been correct but masked by the (neglected) regional-scale effects of anthropogenic aerosols on climate (i.e., a realistic CO_2-only signal but a sub-optimal anthropogenic signal).
- The signal could have been correct but still too small to discern against the background noise of natural climatic variability at smaller than global spatial scales.

The most plausible interpretation involves some combination of these three explanations.

More recent work by Hegerl *et al.* (1996) has claimed that a model-predicted CO_2 signal pattern can be detected with a high degree of confidence in the observed surface air temperature record. This work differs from the studies described above in two important respects:

- A form of the optimum detection strategy proposed by Hasselmann (1979) was applied in order to enhance signal-to-noise ratios.
- It attempted to assess statistical significance by using natural variability noise estimates from three separate AOGCMs and from observations (from which an estimated greenhouse warming signal had been removed; see Section 8.3.1).

Hegerl *et al.* used both non-optimised and optimised forms of an uncentred pattern similarity statistic to compare observed and model-predicted patterns of

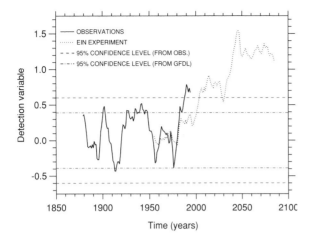

Figure 8.5: Evolution of an uncentred statistic ("detection variable") measuring the similarity between modelled and observed patterns of 20 year trends in near-surface temperature (Hegerl et al., 1996). Observed trend patterns were computed from the Jones and Briffa (1992) data set, and are overlapping in time, i.e., the first observed trend pattern spans the period 1860 to 1879, the second the period 1861 to 1880, etc. The detection variable was calculated by projecting the observed trend patterns onto a "fingerprint", here defined as the long-term pattern of temperature change (from roughly 1990 to 2080) in an "Early Industrialisation" experiment with time-increasing greenhouse gas forcing (EIN) (Cubasch et al., 1995). The fingerprint was optimally filtered in order to enhance the signal-to-noise ratio (S/N) (Hasselmann, 1979). This procedure accentuates the contribution to the total S/N of areas where the signal is large and spatially dissimilar to model-estimated natural internal variability. The detection variable was also computed using the optimised fingerprint and patterns of overlapping 20-year trends from the EIN experiment. In the latter and in the observations, the value of the detection variable increases with time, primarily reflecting the increase in global mean temperature. The behaviour of the detection variable in the absence of external forcing was determined by projecting patterns of overlapping 20-year trends from long control integrations (and from "detrended" observations; see Section 8.3.1 and Hegerl et al. for details) onto the optimised fingerprint. The most recent values of the detection variable in the observations are significantly larger than would be expected due to natural internal variability. Note that all values of the detection variable are plotted on the last year of the 20-year trend.

temperature trends. The most recent 20- to 30-year trends in this statistic were significant at the 5% level or better (relative to three out of four natural variability estimates). Significance levels were enhanced by the optimisation technique (see Figure 8.5).

This result does not mean that the regional-scale features of a model-predicted greenhouse warming pattern have been detected and convincingly attributed to increases in

atmospheric CO_2. As Hegerl *et al.* show, and as shown previously by Santer *et al.* (1993), the uncentred statistic they use is dominated by changes in observed spatial mean, annually averaged temperature. In essence, this result says that the most recent (20- to 30-year) trends in global mean temperature are significantly different from the estimated level of background noise. As noted for Stage 1 studies, this does not resolve the attribution issue.[1]

Hegerl *et al.*'s finding of significant trends in the most recent 20 and 30 years of an uncentred pattern similarity statistic is in accord with the results of Santer *et al.* (1993, 1995a). While Hegerl *et al.* had focused on annual-mean changes, Santer *et al.* (1995a) showed additionally that the most recent 20- to 50-year trends in an uncentred statistic were significant in all four seasons. However, Santer *et al.* found no significant trends in the centred statistic used to compare the sub-global-scale features of modelled and observed temperature-change patterns.

A common feature of the studies by Hegerl *et al.* (1996) and Santer *et al.* (1995a) was that both examined the significance of changes over time (trends) in model versus observed pattern similarity. A different approach was adopted in a recent study by Mitchell *et al.* (1995a), who focused instead on the significance of the absolute values of centred pattern correlations. Mitchell *et al.* compared decadally averaged patterns of near-surface temperature change from a transient greenhouse warming experiment with observed patterns of change, and assumed a one-to-one time correspondence between the experiment (with a start date in 1860) and observations. There was no evidence that sub-global-scale features of the model-predicted CO_2-only signal were becoming increasingly evident in observed records, nor that the absolute values of the centred pattern correlation were significant (see Figure 8.6).

The only other recent pattern-oriented work that has attempted to find a CO_2-only signal in observed surface air temperature data is that by Michaels *et al.* (1994). This investigation makes use of the time-dependent signal from a transient greenhouse warming experiment performed with the GFDL AOGCM (Manabe *et al.*, 1991). The premise underlying this investigation is that if the model-predicted transient signal is not found in the observed temperature record, the model is wrong. The authors fail to find this signal in the observed data, a result that is used to justify a condemnation of climate models in general.

---

[1] Optimisation does not greatly modify the pattern of the Hegerl et al. fingerprint, so the same argument applies to the optimised version of their uncentred statistic.

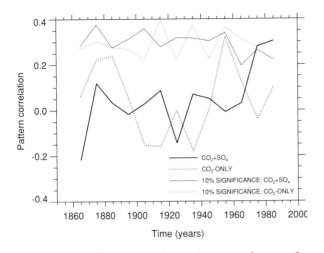

Figure 8.6: Similarity between observed patterns of near-surface temperature change and changes predicted by an AOGCM in experiments with time-dependent forcing by equivalent CO_2 only and CO_2 + anthropogenic sulphate aerosols (Mitchell et al., 1995a). All comparisons assume a one-to-one correspondence in time between the model experiments and observations. Changes were expressed relative to time averages over the first 130 years (i.e., 1860 to 1990) of the observations and the model experiment, after application of the "mask" of observational data coverage appropriate to each decade. Centred pattern correlations were then computed using decadally averaged data. The significance of these results was assessed by correlating the signal patterns from the two perturbation experiments with all possible decadal anomaly patterns from a 400-year AOGCM control integration. This procedure yields information on the degree of pattern similarity expected due to natural internal variability alone. In the most recent decades, the spatial correspondence between observed near-surface temperature changes and those predicted in the model experiment with combined CO_2 + sulphate aerosol forcing is significantly larger (at the 10% level) than the agreement expected due to natural internal climate fluctuations. Significant pattern correlations are not obtained in the CO_2-only experiment.

There are a number of serious problems with this analysis. As discussed in Section 8.2.4, a time-dependent greenhouse warming experiment performed with a fully coupled AOGCM does not have a pure signal as its output. The output consists of signal plus noise, and the early decades of such simulations, which were the focus of the Michaels *et al.* study, are often dominated by the noise. A null result on the basis of a single transient experiment such as this does not constitute "proof" that the model is erroneous, nor that the searched-for signal does not exist.

Furthermore, the Michaels *et al.* study categorically dismisses the possibility that their failure to find a time-dependent greenhouse gas signal may be due to the masking effects of anthropogenic sulphate aerosols. This dismissal is made on the following grounds. Michaels *et al.* argue that if sulphate aerosols have had an impact on climate, then the impact should be very small or close to zero in regions remote from areas where the forcing due to aerosols is large. This hypothesis is not supported by recent GCM experiments, which suggest that the atmospheric general circulation can, via dynamics, produce large remote surface temperature responses to highly regionalised forcing by sulphate aerosols (Taylor and Penner, 1994; Mitchell *et al.*, 1995b; Roeckner *et al.*, 1995).

The Michaels *et al.* results are difficult to compare with those of other Stage 2 studies that have searched for a CO_2-only signal, primarily due to differences in definition of the signal, in methodology and in the areas of the globe considered. Nevertheless, their failure to find the sub-global-scale pattern of this signal is consistent with the results of Santer *et al.* (1993, 1995a). One possible explanation for this result is that some part of the regional-scale features of a CO_2-only signal has been obscured by aerosol effects (see Section 8.4.2.3).

Changes in the vertical temperature structure of the atmosphere have also been proposed as a promising component of a multivariate, CO_2-specific fingerprint (Epstein, 1982; Karoly, 1987, 1989). Model equilibrium and transient response experiments both show a consistent picture of stratospheric cooling and tropospheric warming as a direct radiative response to the change in atmospheric CO_2. The precise details of this picture are model-dependent, e.g., the height of maximum warming in the tropical upper troposphere. Recently, Karoly *et al.* (1994) attempted to find this signal pattern in temperature measurements from radiosondes, using data from Oort and Liu (1993). The observed records spanned the period from 1963 to 1988 only. This is a relatively short record length for the detection of long-term anthropogenic climate change (Karl, 1994). Furthermore, there are known deficiencies in this data set, the most serious of which are the existence of time-varying instrumental biases and inadequate spatial coverage, particularly over the Southern Ocean (see Chapter 3; also Oort and Liu, 1993; Karoly *et al.*, 1994). However, comparisons of the Oort and Liu data with independently derived satellite estimates of vertical temperature changes indicate that the two data sets are in good global mean and hemispheric agreement for the period of overlap (see Chapter 3). This enhances confidence in the usefulness of the radiosonde temperature measurements for detection studies.

Karoly *et al.* used both centred and uncentred statistics to compare model and observed patterns of zonal mean temperature change through the atmosphere. They found

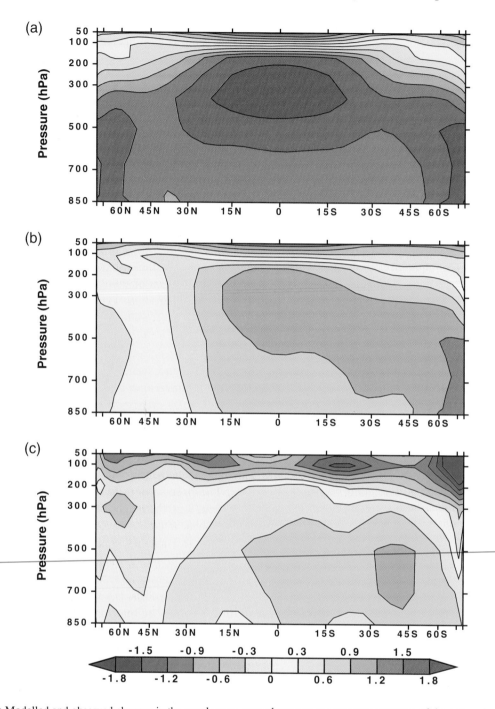

Figure 8.7: Modelled and observed changes in the zonal-mean, annual-average temperature structure of the atmosphere (°C). Model results are from equilibrium response experiments performed by Taylor and Penner (1994), in which an AGCM with a mixed-layer ocean was coupled to a tropospheric chemistry model and forced with present-day atmospheric concentrations of CO_2 (a) and by the combined effects of present-day CO_2 levels and sulphur emissions (b). Model changes are expressed relative to a control run with pre-industrial levels of CO_2 and no anthropogenic sulphur emissions. Observed changes (c) are radiosonde-based temperature measurements from the data set by Oort and Liu (1993) and are expressed as total least-squares linear trends over the 25-year period extending from May 1963 to April 1988 (i.e., °C/25 years). A common pattern of stratospheric cooling and tropospheric warming is evident in the observations and in both model experiments. In the model data, this pattern primarily reflects the direct radiative effect of changes in atmospheric CO_2. Temperature changes in the observations and in the experiment with combined CO_2 + aerosol forcing also show a common pattern of hemispherically asymmetric warming in the low- to mid-troposphere, with reduced warming in the Northern Hemisphere. This asymmetry is absent in the CO_2-only case. For further details refer to Santer et al. (1995b).

significant positive trends in both types of statistic. An innovative feature of this study was its attempt to address the attribution question, at least in a qualitative way, by comparing the characteristic CO_2 signal pattern with the patterns produced by other mechanisms, such as decreases in stratospheric ozone concentration or changes in the frequency and amplitude of El Niño/Southern Oscillation events. Karoly *et al.* concluded that these mechanisms could not mimic the CO_2-induced fingerprint of stratospheric cooling/tropospheric warming.

It is highly likely that the observed stratospheric ozone reduction over the last two decades is related to the industrial production of halocarbons, and hence represents an anthropogenic effect on climate rather than a natural forcing mechanism. One recent AGCM experiment by Hansen *et al.* (1995a) with a combined increase in CO_2 and reduction in stratospheric ozone (O_3) has shown that the inclusion of O_3 effects improves model agreement with the Oort and Liu radiosonde temperature measurements in the upper troposphere. While this improvement was demonstrated in global mean terms only, Ramaswamy *et al.* (1996) further showed that satellite-based measurements of the pattern of latitudinal and seasonal changes in lower stratospheric temperature were in good accord with a model-predicted signal pattern in response to imposed stratospheric ozone reduction.

A recent study by Santer *et al.* (1995b) confirmed the Karoly *et al.* finding of a common pattern of stratospheric cooling and tropospheric warming in the radiosonde observations and in model CO_2-only experiments (see Figure 8.7). As in the Karoly *et al.* case, the strength of the model/data pattern correspondence was found to increase in the observational record. To determine the significance of this result, Santer *et al.* used estimates of the natural internal variability of atmospheric temperature from two multi-century model control runs. They showed that the model-derived natural internal fluctuations could not plausibly explain the level of time-increasing correspondence between the model signal and observations over the period 1963 to 1988.

Some concerns remain regarding the uniqueness of this fingerprint of stratospheric cooling and tropospheric warming. Although the Santer *et al.* (1995b) study points towards important differences between natural internal variability (as simulated by two AOGCMs) and the response to changes in atmospheric CO_2, in terms of both pattern, amplitude and time-scale, other results are more ambiguous (see, e.g., Liu and Schuurmanns, 1990). A control integration and transient greenhouse warming experiment performed with a different AOGCM indicated that the dominant mode of "unforced" variability in that model was similar in pattern and amplitude to the expected greenhouse signal of stratospheric cooling and tropospheric warming, at least in the tropics (Santer *et al.*, 1994). This latter result may be suspect, however, due to the large initial drift in temperature in the control run that was examined.

There is little evidence that the observed pattern of stratospheric cooling and tropospheric warming is due to either solar variability and/or volcanic effects (Wetherald and Manabe, 1975; Hansen *et al.*, 1978). Model-predicted responses to changes in the solar constant do not show stratospheric cooling, while our best information from observations and relevant model experiments indicates that volcanically injected stratospheric aerosols tend to warm the stratosphere and cool the troposphere – a response that is the inverse of an expected greenhouse gas signal. The vertical structure of atmospheric temperature changes might therefore prove to be a fingerprint that is highly specific to anthropogenic forcing[1].

Recent studies by Karl *et al.* (1995b, 1996a), although not strictly pattern-based, have also employed a multivariate description of a greenhouse gas signal. The authors defined a "U.S. Greenhouse Climate Response Index" (GCRI) with individual components that measure selected changes in USA climate thought to be sensitive to increased emissions of greenhouse gases. The index focused on climatic events of direct relevance to the public and policymakers, for example, unequal increases in maximum and minimum temperature (10% greater for the minimum compared to the maximum), increases in cold season precipitation, severe summertime drought and the proportion of total precipitation derived from extreme one-day precipitation events, and decreases in day-to-day temperature variations. Both weighted and unweighted forms of the GCRI were computed, with weights assigned according to a subjective assessment of confidence in the GCM-predicted changes in these individual quantities. The resultant weighted and unweighted indices showed a trend towards elevated values in the last two decades, consistent with model-based projections for an enhanced greenhouse effect (Figure 8.8).

Karl *et al.* assessed the significance of these results by fitting a variety of statistical models to the index time-series, and then randomly generating hundreds of synthetic

---

[1] Note, however, that long-term trends in volcanic forcing could account for some fraction of the observed changes in the thermal structure of the atmosphere. This possibility requires further investigation.

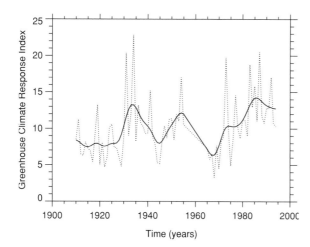

Figure 8.8: Evolution of a weighted "U.S. Greenhouse Climate Response Index" (GCRI). The index consists of five individual components that measure selected changes in USA climate thought to be sensitive to increased emissions of greenhouse gases (Karl *et al.*, 1995b, 1996a). These comprise unequal increases in maximum and minimum temperature, increases in cold season precipitation, severe summertime drought and the proportion of total precipitation derived from extreme one-day precipitation events and decreases in day-to-day temperature variations. The likelihood that increases in the GCRI since 1976 are due to purely natural causes is relatively small (between 1 and 9%).

time-series with statistical properties similar to those of the "observed" index time-series. The probability that changes in the GCRI since 1976 could be purely random in nature ranged from 1% to 20% and from 1% to 9% for unweighted and weighted versions of the GCRI (respectively), with the range dependent on the assumed statistical model. This indicated that changes in weighted and unweighted forms of the index since 1976 were unlikely to be due to natural variability alone, although this explanation could not be totally dismissed.

In summary, attempts to detect a CO_2-only signal in the climate system have given ambiguous results. Existing multi-variate detection studies have used signals from very different experiments and climate models. They differ in terms of the measures of pattern similarity and statistical significance testing procedures that were used, and in the climate variables and geographical regions that were examined. Nevertheless, it is possible to reach a few general conclusions.

As noted in this section, the searched-for greenhouse warming signal has both a global mean component and a sub-global-scale pattern component. The investigations by Hegerl *et al.* (1996) and Santer *et al.* (1995a) suggest that a global mean greenhouse warming signal is detectable with

high confidence in records of surface air temperature change over the last 20 to 50 years, in accord with Stage 1 detection studies (see Section 8.4.1). Implicit in these global mean results is a weak attribution statement – if the observed global mean changes over the last 20 to 50 years cannot be fully explained by natural climate variability, some (unknown) fraction of the changes must be due to human influences. Significant results have not been obtained for the sub-global-scale pattern component of a greenhouse warming signal, thus precluding more confident statements regarding attribution of observed changes to the specific cause of increases in greenhouse gas concentrations. The validity of conclusions on the significance of both the global mean and pattern components of a greenhouse gas signal depends on whether the estimates of natural variability noise used in these studies are realistic.

Several studies indicate that the vertical structure of atmospheric temperature changes might prove to be a fingerprint that is highly specific to combined changes in CO_2 and stratospheric ozone. The latitude-height patterns of temperature change in the available radiosonde observations over 1963 to 1988 are becoming increasingly similar to the model-predicted pattern of stratospheric cooling and tropospheric warming. Further work is necessary to clarify the uniqueness of this fingerprint and to improve the quality of the radiosonde data.

8.4.2.2 Anthropogenic sulphate aerosol signals

Until recently, investigations of the possible impact of anthropogenic sulphate aerosols on observed surface air temperature were restricted to analyses of observed changes in areas where SO_2 emissions and/or forcing are likely to have been large (Wigley *et al.*, 1992; Hunter *et al.*, 1993), or to a visual comparison of observed temperature change patterns and patterns of vertically integrated aerosol concentration predicted by a chemical-transport model (Engardt and Rodhe, 1993). These studies have shown qualitatively that some features of the recent instrumental surface temperature record are consistent with a sulphate aerosol effect on climate.

Only two studies have used quantitative techniques to search for an anthropogenic sulphate aerosol signal in observed data. The first, by Karl *et al.* (1995a), used gridded land surface air temperature data to compare linear trends from 1966 to 1980 with the mean changes in SO_2 emissions over the same period. Spatial correlations were then computed between the temperature trends and emissions changes, but along 5° latitude bands (from 20°N to 60°N) rather than over the full data fields. This yielded a series of correlation coefficients as a function of latitude and season. The significance of these results was evaluated by correlating the emissions changes with many samples of "natural" 15-year temperature trends from the 1000 year GFDL control run (Stouffer *et al.*, 1994).

In over two-thirds of the cases, the correlations were negative, indicating a decrease in temperature as emissions increased over the period 1966 to 1980. The most significant negative correlations were obtained in zones of high emissions (50°N–55°N) and at times of high insolation. The number of significant correlations was far greater than would be expected by chance alone. Karl *et al.* interpreted these results as being consistent with the anticipated direct effects of anthropogenic sulphate aerosols.

This is encouraging in terms of attribution, since the similarity between emissions changes and temperature changes is not simply at a global mean level, and since seasonal variations in the obtained significance levels are in agreement with our physical understanding of the seasonality in forcing due to direct aerosol effects. There is some scope for improving these results by using a pattern of temperature response rather than a pattern of change in emissions or forcing. Detection studies that employ some "proxy" for the temperature response, such as the pattern of observed changes in sulphur emissions used by Karl *et al.* or the pattern of sulphate aerosol distribution or aerosol forcing predicted by a sulphur chemistry model (Wigley *et al.*, 1992; Engardt and Rodhe, 1993), do not account for any modulation of the regional aerosol forcing by the general circulation of the atmosphere, and could be searching for sub-optimal signals. This may be of lesser consequence in the latitude bands (50°N–55°N) where Karl *et al.* find the strongest relationships between SO_2 emissions and temperature changes, since this is likely to be a region where there is a stronger correspondence between emissions, forcing and response.

The recent study by Santer *et al.* (1995a) differs from the work of Karl *et al.* (1995a) in that it searches directly for a temperature response pattern due to anthropogenic sulphate aerosols, rather than a forcing pattern. The response pattern was taken from experiments performed by Taylor and Penner (1994) with an AGCM coupled to a tropospheric chemistry model. As in the greenhouse gas detection studies described in Section 8.4.2.1, Santer *et al.* used both centred and uncentred statistics to compare observed temperature changes with the model predictions. These comparisons were made over the (near-global) data fields rather than at individual latitude bands. Only one result achieved significance: the most recent 50-year trend in a

centred pattern correlation for the autumn season (September to November). In this single case, the model-predicted regional scale pattern was detectable in the data.

It is difficult to compare the results of these two studies directly, because of their use of different signals and different time periods for the observations, and their consideration of pattern similarities at different spatial scales. The key point is that *both* studies show evidence supporting the idea of a sulphate aerosol cooling effect on climate, based on pattern correspondences at spatial scales smaller than the global mean.

8.4.2.3 *Combined CO_2 and sulphate aerosol signals*

Four separate multi-variate detection studies have searched for a model-predicted temperature change pattern in response to combined CO_2 and sulphate aerosol forcing. Three of these have dealt with changes in near-surface temperature (Hasselmann *et al.*, 1995; Mitchell *et al.*, 1995a; Santer *et al.*, 1995a), while one has considered modelled and observed changes in the vertical structure of atmospheric temperature (Santer *et al.*, 1995b). The latter additionally attempted to incorporate the effects of stratospheric ozone reduction in the searched-for anthropogenic signal.

The climate-change signal in the near-surface temperature study by Santer *et al.* (1995a) was derived from equilibrium response experiments by Taylor and Penner (1994), in which an AGCM coupled to a mixed-layer ocean and a model of tropospheric sulphate chemistry was forced by present-day levels of atmospheric CO_2 and anthropogenic sulphur emissions. The temperature response to the combined forcing in the Taylor and Penner experiment is very different from the response to forcing by CO_2 alone (see Figure 8.9), showing both spatially coherent cooling and warming regions – a characteristic of the way surface temperature in the real world has actually changed (see Chapter 3). This response pattern has large uncertainties. Perhaps the most important of these is that the pattern is clearly a function of the relative magnitudes of the positive forcing by CO_2 and the negative forcing by the direct effect of anthropogenic sulphate aerosols, with the latter only poorly known (see Chapter 2 and Section 8.2.5). Further signal uncertainties include those arising from the lack of a dynamic ocean, the neglect of the indirect effects of sulphate aerosols, possible model-dependence of the results, and signal estimation problems (see Section 8.2.4 and 8.2.5).

Santer *et al.* used both centred and uncentred statistics to compare the seasonally and annually averaged (equilibrium) near-surface temperature-change patterns

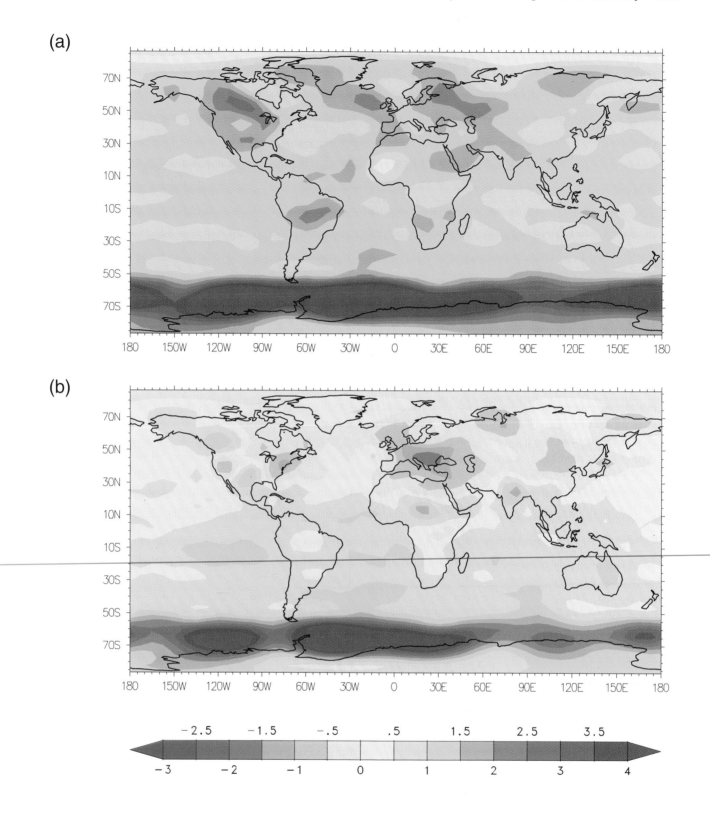

Figure 8.9: Summertime (JJA) near-surface air temperature changes (°C) in an AGCM experiment with forcing by CO_2 only (a), and by both CO_2 and anthropogenic sulphate aerosols (b). For sources of model data, refer to Figure 8.7. Temperature changes in the observations (see Chapter 3, Figure 3.14) are qualitatively more similar to the changes in the combined forcing experiment than in the CO_2 only experiment.

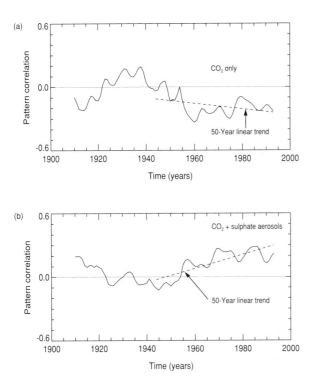

Figure 8.10: Behaviour of a centred pattern correlation statistic, *R(t)*, measuring the similarity between model-predicted and observed patterns of near-surface air temperature change. The model predictions are from equilibrium response experiments with forcing by present-day atmospheric concentrations of CO_2 (a), and by the combined effects of present-day CO_2 levels and sulphur emissions (b) (see Figure 8.7 for details). For each experiment, a single pattern characterises the temperature-change signal. This signal is then searched for in observed time-varying records of near-surface temperature change. The figure shows results for the autumn season (SON). In the case of combined CO_2 + sulphate aerosol forcing, there is a positive linear trend in the *R(t)* time-series over the last 50 years (1944 to 1993), indicating that sub-global-scale features of the observed temperature-change patterns are becoming increasingly similar to the predicted signal pattern. No such increasing similarity is found for the CO_2-only signal. The 50-year trend in *R(t)* in the experiment with combined forcing was highly significant relative to estimates of internally generated natural variability from two extended AOGCM control runs (Santer *et al.*, 1995a). The initial decrease in *R(t)* from roughly 1910 to 1945 (b) has not been fully explained, but is likely due to the fact that the observed warming in the 1930s and 1940s had some similarity to a CO_2-only signal, and was different in character from more recent changes.

from this combined forcing experiment with observed temperature changes. Their key finding was that the long-term (most recent 50-year) trends in a centred statistic for the July to August and September to November signals were statistically significant relative to two different GCM-

based estimates of internally generated variability (see Figure 8.10). In these two seasons, therefore, the sub-global-scale features of a combined CO_2 + sulphate aerosol signal pattern were increasingly evident in the observed surface air temperature data. This result is highly unlikely to be due to the noise of natural internal variability, assuming that the coupled model noise estimates were realistic.

Similar results were obtained by Mitchell *et al.* (1995a) and Hasselmann *et al.* (1995) in detection studies involving annual-mean near-surface temperature signals from AOGCM transient experiments with combined CO_2 + sulphate aerosol forcing. Direct sulphate aerosol effects in both experiments were parameterized by changes in surface albedo. As noted previously (Section 8.4.2.1), the Mitchell *et al.* study used a centred correlation statistic to compare patterns of decadally averaged temperature change from the CO_2 + sulphate aerosol experiment with observed patterns of change, under the assumption of a one-to-one time correspondence between the observations and the model signal. The absolute values of the correlation statistic increased in the most recent decades, and were deemed to be significant relative to the typical values obtained in the absence of external forcing (see Figure 8.6). Significant results were not obtained for the CO_2-only signal.

As in the investigation by Hegerl *et al.* (1996) (Section 8.4.2.1), Hasselmann *et al.* (1995) compared modelled and observed patterns of 30-year trends in near-surface temperature. A centred statistic indicated that correlations (for 30-year trends ending in 1970 to 1990) were generally higher for the combined CO_2 + aerosol signal than for a CO_2-only signal. An optimised uncentred statistic yielded the opposite result, largely reflecting the lower global mean change in the combined forcing experiment.

In work complementing the above studies, Santer *et al.* (1995b) (Section 8.4.2.1) searched for a combined CO_2 + sulphate aerosol signal in observed changes in the vertical temperature structure of the atmosphere, and additionally considered the possible effects of stratospheric ozone depletion. The CO_2 + aerosol signal was taken from the AGCM experiments by Taylor and Penner (1994), while the temperature response to stratospheric O_3 reduction was obtained from an experiment by Ramaswamy *et al.* (1996). Since experiments with simultaneous changes in CO_2, sulphate aerosols and O_3 have not yet been performed, the CO_2 + O_3 + aerosol temperature-change signals were estimated by linearly combining results from the Taylor and Penner and Ramaswamy *et al.* integrations.

A centred statistic was used to compare modelled and observed patterns of temperature change over the lower troposphere to the lower stratosphere (850 to 50 hPa) and over the low- to mid-troposphere only (850 to 500 hPa). This statistic exhibited strong trends for comparisons made over 850 to 50 hPa, both in the CO_2-only signals and in signals incorporating the added effects of sulphate aerosols and stratospheric O_3 reduction. The trends were largely due to a common pattern of stratospheric cooling and tropospheric warming in the observations and in all model experiments (see Figure 8.7).

Pattern comparisons restricted to the 850 to 500 hPa region revealed that the observations were in better statistical agreement with the temperature-change patterns due to combined forcing (CO_2 + aerosols + O_3) than with the CO_2-only pattern. This was the result of hemispheric scale temperature-change contrasts, with less warming in the Northern Hemisphere, that were common to the observations and the combined forcing signal but absent in the CO_2-only case (Figure 8.7). The closest statistical agreement between modelled and observed patterns of temperature change was achieved when stratospheric ozone effects were incorporated. The levels of model-versus-observed pattern similarity increased with time over 1963 to 1988, in all cases save for the CO_2-only signal defined over 850 to 500 hPa. These increases were found to be highly significant (at the 5% level or better) relative to model-derived estimates of natural internal variability.

In summary, all four studies show that sub-global-scale features of model-predicted CO_2 + sulphate aerosol signals are increasingly evident in the record of observed near-surface and vertical temperature changes over the last half century. "Increasingly evident" signifies that the value of the statistic used to compare modelled and observed patterns of change increases with time. This conclusion does not depend on the realism of the model-simulated natural internal variability. The four studies also show that these combined signals are detectable relative to current model estimates of natural internal climate variability. This result means that the combined CO_2+sulphate aerosol signals can be distinguished from model noise estimates with a high level of statistical confidence. The latter conclusion is dependent on the realism of model internal noise estimates in terms of pattern, amplitude and time-scale.

One of the implications of this work is that some of the sub-global-scale background noise, against which these and previous studies have attempted to detect a CO_2-only signal, is in fact part of a sulphate aerosol effect on climate. Detection studies that search for the *individual* effects of greenhouse gases and sulphate aerosols on climate therefore may be less successful in explaining observed temperature changes than studies that search for their *combined* effects.

To date, pattern-based studies have not been able to quantify the magnitude of a greenhouse gas or aerosol effect on climate. Our current inability to estimate reliably the fraction of the observed temperature changes that are due to human effects does not mean that this fraction is negligible. The very fact that pattern-based studies have been able to discern sub-global-scale features of a combined CO_2 + aerosol signal relative to the ambient noise of natural internal variability implies that there may be a non-negligible human effect on global climate.

8.4.3 Studies of Spatial and Temporal Patterns of Change in Model Data Only

A number of purely model-based studies have explored the question of whether model data can help to identify climate variables that might be promising components of a multi-variable "fingerprint" for detecting anthropogenic change. The first studies of this kind were by Barnett and Schlesinger (1987), Barnett *et al.* (1991), and Santer *et al.* (1991), and used a variety of statistical approaches to characterise the degree of similarity between a model-predicted signal and the internally generated natural variability produced by the same model in a control run. The modelled signal in this early work was invariably in response to a doubling of atmospheric CO_2, and was always generated by an atmospheric GCM coupled to a mixed-layer ocean.

More recent work has taken signal and noise data from time-dependent greenhouse warming experiments performed with AOGCMs, and explored the suitability of different statistical techniques for:

- efficiently describing the space-time evolution of greenhouse warming signals and natural variability noise (Santer *et al.*, 1994; Murphy and Mitchell, 1995), and determining at what point in time the model signal can be discriminated from the model noise (see Figure 8.11);

- identifying atmospheric variables (Santer *et al.*, 1994; Murphy and Mitchell, 1995) and oceanic variables (Mikolajewicz *et al.*, 1993; Santer *et al.*, 1995c) with high signal-to-noise ratios in the model data. Signal-to-noise has been investigated in terms of both pure pattern similarity and the relative amplitudes of an enhanced greenhouse effect signal and natural variability noise.

In general these studies have shown that for an enhanced greenhouse effect signal, changes in near-surface air temperature are likely to have a large signal-to-noise ratio,

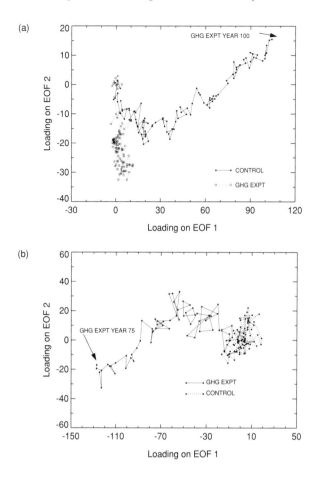

Figure 8.11: Model predictions of the time evolution of annual mean near-surface temperature changes in greenhouse warming experiments and control runs. Results are from experiments performed by Cubasch *et al.* (1992) (a), and Murphy and Mitchell (1995) (b). A statistical technique known as EOF analysis was used to describe the evolution of the modelled greenhouse warming signals and natural internal variability in terms of a small number of patterns (EOFs) and time-series (see Santer *et al.*, 1994). Temperature data from the control runs and greenhouse warming experiments were then "projected" onto the first two EOF patterns of the respective greenhouse warming experiments. Each dot represents one year of a control run or greenhouse experiment, and successive years are joined by lines. The control run results all cluster closely in a "noise cloud" that defines the limits of internally generated natural variability. If the greenhouse warming signals were similar to the (model-simulated) natural variability, both in terms of amplitude and pattern, they would be contained within these noise clouds. This is not the case, and the signals clearly separate from the noise within 20-30 years.

while mean sea level pressure and precipitation changes have less desirable signal-to-noise characteristics. Such rankings of climate variables according to their signal-to-noise characteristics may depend on the extent to which the

model noise represents *bona fide* internally generated variability rather than climate drift, and may change as anthropogenic factors other than CO_2 (such as sulphate aerosol effects) are included in model climate-change experiments. This type of work is also a useful tool for testing statistical methodology in an idealised setting with no missing data and adequate record lengths for estimating model noise properties.

8.5 Qualitative Consistency Between Model Predictions and Observations

8.5.1 Introduction

The focus of Section 8.4 has been on studies that have used quantitative approaches to compare model-predicted signals with observations. There are, however, a large number of investigations that have highlighted areas of *qualitative* agreement (or disagreement) between models and observations. Such information is necessarily less compelling than evidence obtained from quantitative detection studies. Nevertheless, claims of detection of an anthropogenic effect on climate that are based solely on complex statistical techniques are likely to be controversial unless they are supported by a broad range of evidence, some of which will be qualitative in nature.

Before such evidence is examined, it is necessary to point out that there is a fundamental difficulty in assessing "qualitative consistency". Our ideas about an expected signal in response to anthropogenic activities have changed with time, and are likely to change in the future. In the 1980s, expectations of the climate response to anthropogenic factors were largely dictated by CO_2-only results from AGCMs coupled to mixed-layer oceans. The incorporation of full ocean dynamics and the direct effect of sulphate aerosols have modified this picture in important ways (Figure 8.12).

The incorporation of sulphate aerosol effects in climate models marks another important stage in the evolutionary process of predicting and identifying future climate change, but it is clear that this process has not ended. The implication is that caution must be exercised in examining "qualitative consistency". What was qualitatively inconsistent yesterday could be qualitatively consistent today.

Chapter 3 identified observed changes in a wide range of climate parameters, and attempted to assign confidence levels to these changes based on our current understanding of data uncertainties (see Figure 3.23). A similar exercise was attempted for model projections of the climate response to CO_2 doubling in Chapter 6 (Figure 6.1). In the following section, we compare selected model predictions

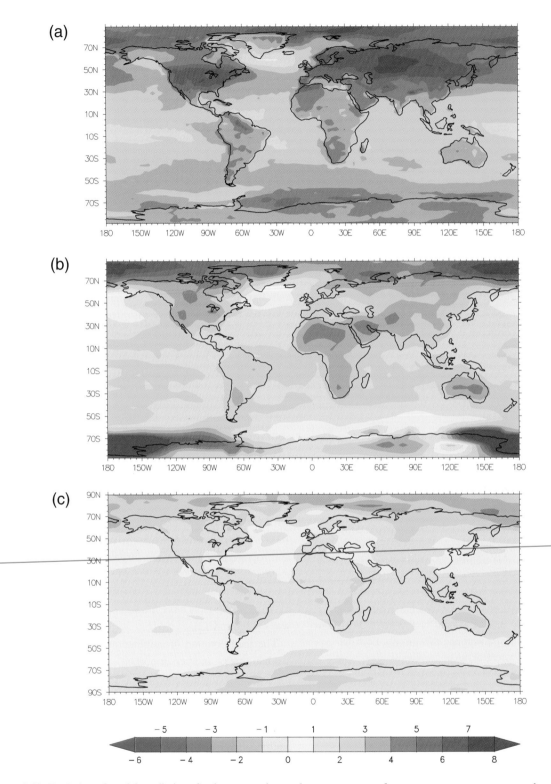

Figure 8.12: Evolution of model predictions for the expected annual average near-surface temperature response to anthropogenic activities (°C). Results are from an equilibrium CO_2 doubling experiment with an AGCM coupled to a mixed-layer model of the upper ocean (Senior, 1995) (a), from an experiment with an AOGCM driven by time-increasing atmospheric levels of equivalent CO_2 (Cubasch *et al.*, 1992) (b), and from an experiment with an AOGCM forced by time-evolving changes in equivalent CO_2 and anthropogenic sulphur emissions (Mitchell et al., 1995a) (c). All results show the temperature response at the approximate time of CO_2 doubling in the model experiment.

with observed changes in order to assess where there is and where there is not qualitative consistency. This section does not attempt to provide an exhaustive analysis of qualitative consistency for a wide range of climate variables. Instead, we focus on areas that have received recent attention in the scientific literature.

The potential problems of comparisons between a CO_2-only signal and observed changes have been discussed above. Where possible, therefore, we also make use of recent model projections based on forcing by both greenhouse gases and sulphate aerosols.

8.5.2 Global Mean Temperature and Surface Temperature Patterns

The best available evidence suggests that observed near-surface air temperature has increased by 0.3°C to 0.6°C in the last 100 years (Chapter 3). This result is in accord with both simple and more complex model predictions of the global mean temperature change in response to time-dependent changes in greenhouse gases and aerosols. As noted in Chapter 6, the time-series of global mean temperature changes simulated in two recent AOGCM experiments with combined CO_2 + sulphate aerosol forcing (Hasselmann *et al.*, 1995; Mitchell *et al.*, 1995a) are generally in better agreement with observations than the results from comparable experiments with CO_2 forcing alone.

This agreement does not, however, constitute identification of an anthropogenic effect on climate and may be serendipitous. The degree of consistency between modelled and observed global mean, annually averaged temperature changes depends on a variety of factors. These include the model's climate sensitivity, the magnitude and sign of simulated multi-decadal variability (and or climate drift), oceanic thermal inertia, and the relative strength of the positive forcing due to greenhouse gases and the negative forcing due to aerosols. It is certainly feasible that qualitative agreement could be due to compensating errors, such as a climate sensitivity that is too high being partially offset by cooling due to a residual drift, or by an overestimated aerosol effect. Nevertheless, the recent modelling work does show that it is possible to explain past changes in a global-scale property of the climate system in a plausible way.

More importantly, we have seen initial evidence of qualitative and quantitative consistency between observed *patterns* of near-surface temperature change and model predictions in response to combined greenhouse gas and aerosol forcing. Pattern correspondence is more difficult to achieve than global mean consistency, and hence is more meaningful in the context of attribution. Our best model-

based estimates of natural internal variability indicate that this correspondence is unlikely to be due to natural causes. Simultaneous model-observed agreement in terms of changes in both global means and patterns, as in the recent study by Mitchell *et al.* (1995a), is even less likely to be a chance occurrence or the result of compensating model errors.

8.5.3 Diurnal Temperature Range

Past changes in the diurnal temperature range (DTR) also show some similarity in the real world and in model predictions. Several studies in Chapter 3 reported that, since the 1950s, minimum temperatures have increased two to three times faster than maximum temperatures over large areas of land in the Northern Hemisphere (Karl *et al.*, 1991b; Karl *et al.*, 1993; Kukla and Karl, 1993; Horton, 1995). The result is a reduction in DTR, especially during summer and autumn.

Early attempts to explain these changes were unsuccessful. Model simulations with increases in greenhouse gases alone or in combination with the direct effects of sulphate aerosols could not capture the amplitude and pattern of the observed changes. New modelling and observational studies (Hansen *et al.*, 1995b; Stenchikov and Robock, 1995) have provided insight into this apparent inconsistency between model projections and the observed decrease of DTR by noting the importance of cloud effects. Thus, Hansen *et al.* (1995b) found that no plausible combination of forcings other than CO_2, sulphate aerosols and increases in cloud cover could explain the observed reduction in DTR and the increase in mean temperature in their model. Whether the required cloud changes, which are in accord with available observations, are also primarily an anthropogenic effect (i.e., due to the indirect effects of sulphate aerosols; see Chapter 2) or are due to natural climate variability is currently unknown.

8.5.4 Changes in Phase of Annual Temperature Cycle

A recent study by Thomson (1995) of the multi-century Central England time-series of monthly mean temperatures (and of other long instrumental records) employed a sophisticated spectral analysis technique to discern underlying periodicities in the data. After identifying and removing a signal due to the precession of the Earth's axis, Thomson found that the phase of the annual cycle of surface air temperature was remarkably stable until roughly 1940, at which time it changed abruptly and at an unprecedented rate (see Chapter 3). The occurrence of this phase shift at a time when atmospheric greenhouse gas concentrations and anthropogenic aerosol loadings were increasing rapidly is interesting information, but it does not prove that a causal relationship exists.

Furthermore, subsequent analysis of Central England and other time-series by Emslie and Christy (1996) and Karl *et al.* (1996b) have failed to confirm the existence of a significant post-1940 phase shift in the annual cycle (see Chapter 3). It remains to be determined whether such phase shifts are robust features of the data and are consistent with the predictions of AOGCMs in experiments with time-dependent forcing by CO_2 and aerosols.

8.5.5 *Possible Inconsistencies between Modelled and Satellite-Derived Temperature Trends*

As discussed in Chapter 3 and by Hansen *et al.* (1995a), there has been considerable confusion in the scientific literature regarding the question of whether instrumental- and model-based estimates of global temperature trends over the period 1979 to 1993 are in conflict with satellite-based observations. Here, we consider only whether *model-predicted* near-surface temperature trends are inconsistent with estimates of low to mid-tropospheric temperature trends derived from the satellite-based Microwave Sounding Unit (MSU).

Christy and McNider (1994) cited model-predicted tropospheric trend estimates of 0.3-0.4°C/decade. These were based on:

- model predictions of surface air temperature changes (in response to increases in CO_2) given in IPCC (1992);
- model predictions of more rapid warming of the global troposphere than global near-surface air.

They concluded that these values were roughly four times larger than the MSU-based trend of roughly 0.09°C/decade, which was obtained after adjusting the MSU data for the effects of El Niño and volcanic aerosols. However, as noted by Hansen *et al.* (1995b), the model predictions on which this comparison is based were taken either from equilibrium CO_2 doubling experiments that were not relevant to the problem of predicting time-dependent change, or from time-dependent "CO_2-only" simulations with rates of change of radiative forcing that were larger than observed.

More recent AOGCM experiments initialised with levels of atmospheric CO_2 corresponding to late 19th century conditions, and driven by changes in both CO_2 and sulphur emissions (Hasselmann *et al.*, 1995; Mitchell *et al.*, 1995a), show rates of near-surface temperature change for the 11- and 15-year model periods corresponding to 1979 to 1990 and

1979 to 1993 that range from 0.06–0.22°C/decade (J.F.B. Mitchell, pers. comm.). This would translate to a tropospheric trend of about 0.08–0.30°C/decade given the observed magnitude of surface versus tropospheric variability. These latter linear trend estimates are in much closer agreement with the adjusted MSU data, i.e., there is no serious inconsistency between the most recent model predictions and MSU-based trend estimates (Hansen *et al.*, 1995a).

The key point to note here is that the MSU record is short (>20 years) for the purposes of detecting a slowly evolving anthropogenic signal. This short record limits comparisons of satellite-based and model-predicted data to decadal time-scale temperature trends. These trends are strongly affected by the background noise of interannual to decadal time-scale natural variability (see, for example, the lack of significance for the most recent 10-year trends in near-surface temperature in Figure 8.3). It is therefore difficult to make a meaningful interpretation of any differences in trends on these short time-scales.

8.6 When Will an Anthropogenic Effect on Climate be Identified?

Detection of a human-induced change in the Earth's climate will be an evolutionary and not a revolutionary process. It is the gradual accumulation of evidence that will implicate anthropogenic emissions as the cause of some part of observed climate change, not the results from a single study. While there is already initial evidence for the existence of an anthropogenic climate signal, it is likely (if model predictions are correct) that this signal will emerge more and more convincingly with time. It is probable that it will be discernible at the global scale first and only later at regional scales (Briffa *et al.*, 1990; Karl *et al.*, 1991a), and that it will be clearer in some variables than in others. Convincing attribution, however, is likely to come from the analysis of full spatial patterns of change: again, as an evolutionary process.

The gradual emergence of an anthropogenic climate change signal from the background noise of natural variability guarantees that any initial pronouncement that a change in the climate has been detected and attributed to a specific cause will be questioned by some scientists. Nevertheless, if the current rate of increase of anthropogenic emissions is maintained and if the sensitivity of the climate system to anthropogenic perturbations is within the range predicted by current climate theory, it should become increasingly easy to eliminate natural variability and other natural external forcings as causes for most of the observed changes.

Finally, we come to the difficult question of when the detection and attribution of human-induced climate change is likely to occur. The answer to this question must be subjective, particularly in the light of the large signal and noise uncertainties discussed in this chapter. Some scientists maintain that these uncertainties currently preclude any answer to the question posed above. Other scientists would and have claimed, on the basis of the statistical results presented in Section 8.4, that confident detection of a significant anthropogenic climate change has already occurred.

As noted in Section 8.1, attribution involves statistical testing of alternative explanations for a detected observed change, and few would be willing to argue that *completely unambiguous* attribution of (all or part of) this change has already occurred, or was likely to happen in the next few years. However, evidence from the pattern-based studies reported on here suggests that an initial step has now been taken in the direction of attribution, since correspondences between observations and model predictions in response to combined changes in greenhouse gases and anthropogenic sulphate aerosols:

- have now been seen both at the surface and in the vertical structure of the atmosphere;
- have been found in terms of complex spatial patterns rather than changes in the global mean alone;
- show an overall increase over the last 20 to 50 years;
- are significantly different from our best model-based estimates of the correspondence expected due to natural internal climate variability.

Furthermore, although quantitative attribution studies have not explicitly considered solar and volcanic effects, our best information indicates that the observed patterns of vertical temperature change are not consistent with the responses expected for these forcings.

The body of statistical evidence in Chapter 8, when examined in the context of our physical understanding of the climate system, now points towards a discernible human influence on global climate. Our ability to quantify the magnitude of this effect is currently limited by uncertainties in key factors, including the magnitude and patterns of longer-term natural variability and the time-evolving patterns of forcing by (and response to) greenhouse gases and aerosols.

References

Allen, M.R., P.L. Read and L.A. Smith, 1992: Temperature oscillations. *Nature*, **359**, 679.

Allen, M.R. and L.A. Smith, 1994: Investigating the origins and significance of low-frequency modes of climate variability. *Geophys. Res. Lett.*, **21**, 883–886.

Barnett, T.P., 1986: Detection of changes in global tropospheric temperature field induced by greenhouse gases. *J. Geophys. Res.*, **91**, 6659–6667.

Barnett, T.P., 1991: An attempt to detect the greenhouse-gas signal in a transient GCM simulation. In: *Greenhouse-Gas-Induced Climatic Change: A Critical Appraisal of Simulations and Observations*, M.E. Schlesinger (ed.), Elsevier, Amsterdam, pp. 559–568.

Barnett, T.P. and M.E. Schlesinger, 1987: Detecting changes in global climate induced by greenhouse gases. *J. Geophys. Res.*, **92**, 14772–14780.

Barnett, T.P., M.E. Schlesinger and X. Jiang, 1991: On greenhouse-gas detection strategies. In: *Greenhouse-Gas-Induced Climatic Change: A Critical Appraisal of Simulations and Observations*, M.E. Schlesinger (ed.), Elsevier, Amsterdam, pp. 537–558.

Barnett, T.P., B.D. Santer, P.D. Jones, R.S. Bradley and K.R. Briffa, 1996: Estimates of low-frequency natural variability in near-surface air temperature. Holocene (in press).

Bell, T.L., 1982: Optimal weighting of data to detect climatic change: Application to the carbon dioxide problem. *J. Geophys. Res.*, **87**, 11161–11170.

Bell, T.L., 1986: Theory of optimal weighting of data to detect climatic change. *J. Atmos. Sci.*, **43**, 1694–1710.

Bloomfield, P., 1992: Trends in global temperature. *Clim. Change*, **21**, 1–16.

Bloomfield, P. and D. Nychka, 1992: Climate spectra and detecting climate change. *Clim. Change*, **21**, 275–287.

Bradley, R.S. and P.D. Jones, 1992: *Climate Since A.D. 1500*. Routledge, London, 679 pp.

Bradley, R.S. and P.D. Jones, 1993: "Little Ice Age" summer temperature variations: their nature and relevance to recent warming trends. *Holocene*, **3**, 367–376.

Briffa, K.R., T.S. Bartholin, D. Eckstein, P.D. Jones, W. Karlén, F.H. Schweingruber and P. Zetterberg, 1990: A 1,400-year tree-ring record of summer temperatures in Scandinavia. *Nature*, **346**, 434–439.

Briffa, K.R., P.D. Jones, T.S. Bartholin, D. Eckstein, F.H. Schweingruber, W. Karlén, P. Zetterberg, and M. Eronen, 1992: Fennoscandian summers from A.D. 500: Temperature changes on short and long time-scales. *Clim. Dyn.*, **7**, 111–119.

Briffa, K.R., P.D. Jones, F.H. Schweingruber, W. Karlén and S.G. Shiyatov, 1996: Tree-ring variables as proxy climate indicators: Problems with low-frequency signals. In: *Climatic Variations and Forcing Mechanisms of the Last 2000 years*, P.D. Jones, R.S. Bradley and J. Jouzel (eds.), Springer Verlag, Berlin, pp. 9–41.

Cess, R.D., G.L. Potter, J.P. Blanchet, G.J. Boer, S.J. Ghan, J.T. Kiehl, H. Le Treut, Z.-X. Li, X.-Z. Liang, J.F.B. Mitchell, J.-J. Morcrette, D.A. Randall, M.R. Riches, E. Roeckner, U. Schlese, A. Slingo, K.E. Taylor, W.M. Washington, R.T. Wetherald and I. Yagai, 1989: Interpretation of cloud-climate feedback as produced by 14 atmospheric general circulation models. *Science*, **245**, 513–516.

Christy, J.R. and R.T. McNider, 1994: Satellite greenhouse signal. *Nature*, **367**, 325.

Cook, E.R., T. Bird, M. Peterson, M. Barbetti, B. Buckley, R. D'Arrigo, R. Francey and P. Tans, 1991: Climatic change in Tasmania inferred from a 1089-year tree-ring chronology of Huon pine. *Science*, **253**, 1266–1268.

Cook, E.R., B. Buckley and R. D'Arrigo, 1996: Inter-decadal climate variability in the Tasmanian sector of the Southern Hemisphere: Evidence from tree rings over the past three millennia. In: *Climatic Variations and Forcing Mechanisms of the Last 2000 years*, P.D. Jones, R.S. Bradley and J. Jouzel (eds.), Springer Verlag, Berlin, pp.141–160.

Crowley, T.J., 1991: Utilization of paleoclimate results to validate projections of a future greenhouse warming. In: *Greenhouse-Gas-Induced Climatic Change: A Critical Appraisal of Simulations and Observations*, M.E. Schlesinger (ed.), Elsevier, Amsterdam, pp. 35–45.

Crowley, T.J. and G.R. North, 1991: *Paleoclimatology*. Oxford University Press, New York, 339 pp.

Cubasch, U., K. Hasselmann, H. Höck, E. Maier-Reimer, U. Mikolajewicz, B.D. Santer and R. Sausen, 1992: Time-dependent greenhouse warming computations with a global coupled ocean-atmosphere model. *Clim. Dyn.*, **8**, 55–69.

Cubasch, U., B.D. Santer, A. Hellbach, G. Hegerl, H. Höck, E. Maier-Reimer, U. Mikolajewicz, A. Stössel and R. Voss, 1994: Monte Carlo climate change forecasts with a global coupled ocean-atmosphere model. *Clim. Dyn.*, **10**, 1–19.

Cubasch, U., G. Hegerl, A. Hellbach, H. Höck, U. Mikolajewicz, B.D. Santer and R. Voss, 1995: A climate change simulation starting at an early time of industrialization. *Clim. Dyn.*, **11**, 71–84.

Delworth, T., S. Manabe and R.J. Stouffer, 1993: Interdecadal variability of the thermohaline circulation in a coupled ocean-atmosphere model. *J. Climate*, **6**, 1993–2011.

Emslie, A.G. and J.R. Christy, 1996: Astronomical influences and long-term trends in terrestrial temperature. *Geophys. Res. Lett.* (submitted).

Engardt, M. and H. Rodhe, 1993: A comparison between patterns of temperature trends and sulphate aerosol pollution. *Geophys. Res. Lett.*, **20**, 117–120.

Epstein, E.S., 1982: Detecting climate change. *J. Appl. Met.*, **21**, 1172–1182.

Erickson, D.J., R.J. Oglesby and S. Marshall, 1995: Climate response to indirect anthropogenic sulfate forcing. *Geophys. Res. Lett.*, **22**, 2017–2020.

Friis-Christensen, E. and K. Lassen, 1991: Length of the solar cycle: An indicator of solar activity closely associated with climate. *Science*, **254**, 698–700.

Galbraith, J.W. and C. Green, 1992: Inference about trends in global temperature data. *Clim. Change*, **22**, 209–221.

Gates, W.L., Y.J. Han, and M.E. Schlesinger, 1985: The global climate simulated by a coupled atmosphere-ocean general circulation model: preliminary results. In: *Coupled Ocean-Atmosphere Models*, J.C.J. Nihoul (ed.), Elsevier, Amsterdam, pp. 131–151.

Gates, W.L., P.R. Rowntree and Q.C. Zeng, 1990: Validation of climate models. In: *Climate Change. The IPCC Scientific Assessment.*, J.T. Houghton, G.J. Jenkins and J.J. Ephraums (eds.), Cambridge University Press, Cambridge, UK, pp. 93–130.

Gates, W.L., J.F.B. Mitchell, G.J. Boer, U. Cubasch, and V.P. Meleshko, 1992: Climate modeling, climate prediction and model validation. In: *Climate Change 1992, The Supplementary Report to the IPCC Scientific Assessment*, J.T. Houghton, B.A. Callander and S.K. Varney (eds.), Cambridge University Press, Cambridge, UK, pp. 97–134.

Ghil, M. and R. Vautard, 1991: Interdecadal oscillations and the warming trend in global temperature time series. *Nature*, **350**, 324–327.

Hansen, J.E., W.-C. Wang and A.A. Lacis, 1978: Mount Agung provides a test of a global climatic perturbation. *Science*, **199**, 1065–1068.

Hansen, J.E., A. Lacis, D. Rind and G. Russell, 1984: Climate sensitivity: Analysis of feedback mechanisms. In: *Climate Processes and Climate Sensitivity*, J.E. Hansen and T. Takahasi (eds.), Maurice Ewing Series 5, American Geophysical Union, pp 130–163.

Hansen, J.E., H. Wilson, M. Sato, R. Ruedy, K.P. Shah and E. Hansen, 1995a: Satellite and surface temperature data at odds? *Clim. Change*, **30**, 103–117.

Hansen, J.E., M. Sato and R. Ruedy, 1995b: Long-term changes of the diurnal temperature cycle: Implications about mechanisms of global climate change. *Atmos. Res.*, **37**, 195–209.

Hasselmann, K., 1976: Stochastic climate models. Part I: Theory. *Tellus*, **28**, 473–485.

Hasselmann, K., 1979: On the signal-to-noise problem in atmospheric response studies. In: *Meteorology of Tropical Oceans*, D.B. Shaw (ed.), Royal Meteorological Society London, pp. 251–259.

Hasselmann, K., 1993: Optimal fingerprints for the detection of time dependent climate change. *J. Climate*, **6**, 1957–1971.

Hasselmann, K., L. Bengtsson, U. Cubasch, G.C. Hegerl, H. Rodhe, E. Roeckner, H. v. Storch and R. Voss, 1995: Detection of anthropogenic climate change using a fingerprint method. *Max-Planck Institut für Meteorologie Report No. **168**, 20 pp.

Hegerl, G., H. v. Storch, K. Hasselmann, B.D. Santer, U. Cubasch and P.D. Jones, 1996: Detecting anthropogenic climate change with a fingerprint method. *J. Climate* (in press).

Hoffert, M.I. and B.P. Flannery, 1985: Model projections of the time-dependent response to increasing carbon dioxide. In: *Projecting the Climatic Effects of Increasing Carbon Dioxide*,

M.C. MacCracken and F.M. Luther (eds.), U.S. Department of Energy, Carbon Dioxide Research Division, Washington, DC, pp. 149–190.

Hoffert, M.I. and C. Covey, 1992: Deriving global climate sensitivity from paleoclimate reconstructions. *Nature*, **360**, 573–576.

Horton, E.B., 1995: Geographical distribution of changes in maximum and minimum temperature. *Atmos. Res.*, 37, 102–117.

Hunter, D.E., S.E. Schwartz, R. Wagoner and C.M. Benkovitz, 1993: Seasonal, latitudinal, and secular variations in temperature trend: Evidence for influence of anthropogenic sulphate. *Geophys. Res. Lett.*, **20**, 2455–2458.

IPCC (Intergovernmental Panel on Climate Change), 1990: *Climate Change: The IPCC Scientific Assessment*, J.T. Houghton, G.J. Jenkins and J.J. Ephraums (eds). Cambridge University Press, Cambridge, UK, 365 pp.

IPCC, 1992: *Climate Change 1992: The Supplementary Report to the IPCC Scientific Assessment*, J.T. Houghton, B.A. Callander and S.K. Varney (eds). Cambridge University Press, Cambridge, UK, 198 pp.

Jones, P.D. and K.R. Briffa, 1992: Global surface air temperature variations during the Twentieth century: Part1. Spatial, temporal and seasonal details. *Holocene*, **3**, 77–88.

Karl, T.R., 1994: Smudging the fingerprints. *Nature*, **371**, 380–381.

Karl, T.R., R.R. Heim Jr. and R.G. Quayle, 1991a: The greenhouse effect in central North America: If not now, when? *Science*, **251**, 1058–1061.

Karl, T.R., G. Kukla, V.N. Razuvayev, M.J. Changery, R.G. Quayle, R.R. Heim Jr., D.R. Easterling and C.B. Fu, 1991b: Global warming: Evidence for asymmetric diurnal temperature change. *Geophys. Res. Lett.*, **18**, 2253–2256.

Karl, T.R., P.D. Jones, R.W. Knight, G. Kukla, N. Plummer, V. Razuvayev, K.P. Gallo, J. Lindseay, R.J. Charlson and T.C. Peterson, 1993: Asymmetric trends of daily maximum and minimum temperature. *Bull. Am. Met. Soc.*, **74**, 1007–1023.

Karl, T.R., R.W. Knight, G. Kukla and J. Gavin, 1995a: Evidence for radiative effects of anthropogenic sulphate aerosols in the observed climate record. In: *Aerosol Forcing of Climate*, R. Charlson and J. Heintzenberg (eds.), John Wiley and Sons, Chichester, pp. 363–382.

Karl, T.R., R.W. Knight, D.R. Easterling and R.G. Quayle, 1995b: Trends in U.S. climate during the Twentieth Century. *Consequences*, **1**, 3–12.

Karl, T.R., R.W. Knight, D.R. Easterling and R.G. Quayle, 1996a: Indices of climate change for the United States. *Bull. Am. Met. Soc.* (in press).

Karl, T.R., P.D. Jones and R.W. Knight, 1996b: A note on David J. Thomson's article: "The Seasons, Global Temperature, and Precession". *Science* (in press).

Karoly, D.J, 1987: Southern Hemisphere temperature trends: A possible greenhouse gas effect? *Geophys. Res. Lett.*, **14**, 1139–1141.

Karoly, D.J, 1989: Northern Hemisphere temperature trends: A possible greenhouse gas effect? *Geophys. Res. Lett.*, **16**, 465–468.

Karoly, D.J., J.A. Cohen, G.A. Meehl, J.F.B. Mitchell, A.H. Oort, R.J. Stouffer and R.T. Wetherald, 1994: An example of fingerprint detection of greenhouse climate change. *Clim. Dyn.*, **10**, 97–105.

Kelly, P.M. and T.M.L. Wigley, 1992: Solar cycle length, greenhouse forcing and global climate. *Nature*, **360**, 328–330.

Kheshgi, H.S. and B.S. White, 1993a: Does recent global warming suggest an enhanced greenhouse effect? *Clim. Change*, **23**, 121–139.

Kheshgi, H.S. and B.S. White, 1993b: Effect of climate variability on estimation of greenhouse parameters: Usefulness of a pre-instrumental temperature record. *Quat. Sci. Rev.*, **12**, 475–481.

Kim, K.-Y. and G.R. North, 1991: Surface temperature fluctuations in a stochastic climate model. *J. Geophys. Res.*, **96**, 18573–18580.

Kukla, G. and T.R. Karl, 1993: Nighttime warming and the greenhouse effect. *Environ. Sci. and Tech.*, **27**, 1468–1474.

Kuo, C., C. Lindberg and D.J. Thomson, 1990: Coherence established between atmospheric carbon dioxide and global temperature. *Nature*, **343**, 709–713.

Lemke, P., 1977: Stochastic climate models. Part 3: Application to zonally averaged energy models. *Tellus*, **29**, 385–392.

Liu, Q. and C.J.E. Schuurmanns, 1990: The correlation of tropospheric and stratospheric temperatures and its effect on the detection of climate changes. *Geophys. Res. Lett.*, **17**, 1085–1088.

Madden, R.A. and V. Ramanathan, 1980: Detecting climate change due to increasing carbon dioxide. *Science*, **209**, 763–768.

Manabe, S. and R.J. Stouffer, 1993: Century-scale effects of increased atmospheric CO_2 on the ocean-atmosphere system. *Nature*, **364**, 215–218.

Manabe, S. and R.J. Stouffer, 1996: Low-frequency variability of surface air temperature in a 1,000-year integration of a coupled ocean-atmosphere model. *J. Climate*, **9**, 376–393.

Manabe, S., R. J. Stouffer, M. J. Spelman and K. Bryan, 1991: Transient responses of a coupled ocean-atmosphere model to gradual changes of atmospheric CO_2. Part I: Annual mean response. *J. Climate*, **4**, 785–818.

Meehl, G.A., W.M. Washington and T.R. Karl, 1993: Low-frequency variability and CO_2 transient climate change. *Clim. Dyn.*, **8**, 117–133.

Mehta, V.M. and T. Delworth, 1995: Decadal variability of the tropical Atlantic Ocean surface temperature in shipboard measurements and in a global ocean-atmosphere model. *J. Climate*, **8**, 172–190.

Michaels, P.J., P.C. Knappenberger and D.A. Gay, 1994: General circulation models: Testing the forecast. *Journal of the Franklin Institute*, **331A**, 123–133.

Mikolajewicz, U. and E. Maier-Reimer, 1990: Internal secular variability in an ocean general circulation model. *Clim. Dyn.*, **4**, 145–156.

Mikolajewicz, U. and E. Maier-Reimer, 1994: Mixed boundary conditions in ocean general circulation models and their influence on the stability of the model's conveyor belt. *J. Geophys. Res.*, **99**, 22633–22644.

Mikolajewicz, U., E. Maier-Reimer and T.P. Barnett, 1993: Acoustic detection of greenhouse-induced climate changes in the presence of slow fluctuations of the thermohaline circulation. *J. Phys. Oceanogr.*, **23**, 1099–1109.

Mitchell, J.F.B., C.A. Senior and W.J. Ingram, 1989: CO_2 and climate: a missing feedback? *Nature*, **341**, 132–134.

Mitchell, J.F.B., T.C. Johns, J.M. Gregory and S.F.B. Tett, 1995a: Transient climate response to increasing sulphate aerosols and greenhouse gases. *Nature*, **376**, 501–504.

Mitchell, J.F.B., R.A. Davis, W.J. Ingram and C.A. Senior, 1995b: On surface temperature, greenhouse gases and aerosols: Models and observations. *J. Climate*, **8**, 2364–2386.

Murphy, J.M. and J.F.B. Mitchell, 1995: Transient response of the Hadley Centre coupled ocean-atmosphere model to increasing carbon dioxide. Part II. Spatial and temporal structure of response. *J. Climate*, **8**, 57–80.

Nakamura, M., P.H. Stone, and J. Marotzke, 1994: Destabilization of the thermohaline circulation by atmospheric eddy transports. *J. Climate*, **7**, 1870–1882.

Neelin, J.D. and H.A. Dijkstra, 1995: Ocean-atmosphere interaction and the tropical climatology. Part I: The dangers of flux correction. *J. Climate*, **8**, 1325–1342.

North, G.R. and K.-Y. Kim, 1995: Detection of forced climate signals. Part II: Simulation results. *J. Climate*, **8**, 409–417.

North, G.R., K.-Y. Kim, S.P. Shen and J.W. Hardin, 1995: Detection of forced climate signals. Part I: Filter theory. *J. Climate*, **8**, 401–408.

Oort, A.H. and H. Liu, 1993: Upper-air temperature trends over the globe, 1958-1989. *J. Climate*, **6**, 292–307.

Penner, J.E., R.E. Dickinson, and C.A. O'Neill, 1992: Effects of aerosol from biomass burning on the global radiation budget. *Science*, **256**, 1432–1434.

Penner, J.E., R.J. Charlson, J.M. Hales, N.S. Laulainen, R. Leifer, T. Novakov, J. Ogren, L.F. Radke, S.E. Schwartz and L. Travis, 1994: Quantifying and minimizing uncertainty of climate forcing by anthropogenic aerosols. *Bull. Am. Met. Soc.*, **75**, 375–400.

Pierce, D.W., T.P. Barnett and U. Mikolajewicz, 1995: On the competing roles of heat and fresh water flux in forcing thermohaline oscillations. *J. Phys. Oceanog.*, **25(9)**, 2046–2064.

Power, S.B. and R. Kleeman, 1993: Multiple equilibria in a global ocean general circulation model. *J. Phys. Oceanog.*, **23**, 1670–1681.

Power, S.B. and R. Kleeman, 1994: Surface heat flux parameterization and the response of ocean general circulation models to high-latitude freshening. *Tellus*, **46A**, 86–95.

Ramaswamy, V., M.D. Schwarzkopf and W.J. Randel, 1996: An unanticipated climate change in the global lower stratosphere due to ozone depletion. *Nature* (submitted).

Roeckner, E., T. Siebert and J. Feichter, 1995: Climatic response to anthropogenic sulphate forcing simulated with a general circulation model. In: *Aerosol Forcing of Climate*, R. Charlson and J. Heintzenberg (eds.), John Wiley and Sons, Chichester, pp. 349–362.

Santer, B.D., T.M.L Wigley, P.D. Jones and M.E. Schlesinger, 1991: Multivariate methods for the detection of greenhouse-gas-induced climate change. In: *Greenhouse-Gas-Induced Climatic Change: A Critical Appraisal of Simulations and Observations*, M.E. Schlesinger (ed.), Elsevier, Amsterdam, pp. 511–536.

Santer, B.D., T.M.L. Wigley and P.D. Jones, 1993: Correlation methods in fingerprint detection studies. *Clim. Dyn.*, **8**, 265–276.

Santer, B.D., W. Brüggemann, U. Cubasch, K. Hasselmann, H. Höck, E. Maier-Reimer and U. Mikolajewicz, 1994: Signal-to-noise analysis of time-dependent greenhouse warming experiments. Part 1: Pattern analysis. *Clim. Dyn.*, **9**, 267–285.

Santer, B.D., K.E. Taylor, T.M.L. Wigley, J.E. Penner, P.D. Jones and U. Cubasch, 1995a: Towards the detection and attribution of an anthropogenic effect on climate. *Clim. Dyn.*, **12**, 79–100.

Santer, B.D., K.E. Taylor, T.M.L. Wigley, P.D. Jones, D.J. Karoly, J.F.B. Mitchell, A.H. Oort, J.E. Penner, V. Ramaswamy, M.D. Schwarzkopf, R.J. Stouffer and S. Tett, 1995b: A search for human influences on the thermal structure of the atmosphere. *PCMDI Report no. 27*, Lawrence Livermore National Laboratory, Livermore, 26pp.

Santer, B.D., U. Mikolajewicz, W. Brüggemann, U. Cubasch, K. Hasselmann, H. Höck, E. Maier-Reimer and T.M.L. Wigley, 1995c: Ocean variability and its influence on the detectability of greenhouse warming signals. *J. Geophys. Res.*, **100**, 10693–10725.

Sausen, R., K. Barthel and K. Hasselmann, 1988: Coupled ocean-atmosphere models with flux corrections. *Clim. Dyn.*, **2**, 154–163.

Schlesinger, M.E. and N. Ramankutty, 1992: Implications for global warming of intercycle solar irradiance variations. *Nature*, **360**, 330–333.

Schlesinger, M.E. and N. Ramankutty, 1994: An oscillation in the global climate system of period 65-70 years. *Nature*, **367**, 723–726.

Schlesinger, M.E., K. Jiang and R.J. Charlson, 1992: Implication of anthropogenic atmospheric sulphate for the sensitivity of the climate system. In: *Climate Change and Energy Policy*, L. Rosen and R. Glasser (eds.), American Institute of Physics, pp. 75–108.

Schneider, S.H., 1994: Detecting climatic change signals: Are there any "fingerprints"? *Science*, **263**, 341–347.

Schönwiese, C.-D., R. Ulrich, F. Beck and J. Rapp, 1994: Solar signals in global climatic change. *Clim. Change*, **27**, 259–281.

Senior, C.A., 1995: The dependence of climate sensitivity on the horizontal resolution of a GCM. *J. Climate*, **8**, 2860–2880.

Stenchikov, G.L. and A. Robock, 1995: Diurnal asymmetry of climatic response to increased CO_2 and aerosols: Forcings and feedbacks. *J. Geophys. Res.,* **100**, 26211–26227.

Stocker, T.F. and L.A. Mysak, 1992: Climatic fluctuations on the century time scale: A review of high-resolution proxy data and possible mechanisms. *Clim. Change,* **20**, 227–250.

Stouffer, R.J., S. Manabe and K. Bryan, 1989: Interhemispheric asymmetry in climate response to a gradual increase of atmospheric CO_2. *Nature,* **342**, 660–662.

Stouffer, R.J., S. Manabe and K. Ya. Vinnikov, 1994: Model assessment of the role of natural variability in recent global warming. *Nature,* **367**, 634–636.

Taylor, K.E. and J.E. Penner, 1994: Anthropogenic aerosols and climate change. *Nature,* **369**, 734–736.

Tegen, I. and I. Fung, 1995: Contribution to the atmospheric mineral aerosol load from land surface modification. *J. Geophys. Res.,* **100**, 18707–18726.

Thomson, D.J., 1995: The seasons, global temperature, and precession. *Science,* **268**, 59–68.

Tol, R.S.J., 1994: Greenhouse statistics – time series analysis: Part II. *Theor. Appl. Climatol.* **49**, 91–102.

Tol, R.S.J. and A.F. de Vos, 1993: Greenhouse statistics – time series analysis. *Theor. Appl. Climatol.* **48**, 63–74.

Wang, W.-C., M.P. Dudek, X.-Z. Liang and J.T. Kiehl, 1991: Inadequacy of effective CO_2 as a proxy in simulating the greenhouse effect of other radiatively active gases. *Nature,* **350**, 573–577.

Washington, W.M. and G.A. Meehl, 1989: Seasonal cycle experiments on the climate sensitivity due to a doubling of CO_2 with an atmospheric general circulation model coupled to a simple mixed layer ocean model. *J. Geophys. Res.,* **89**, 9475–9503.

Wetherald, R.T, and S. Manabe, 1975: The effects of changing the solar constant on the climate of a general circulation model. *J. Atmos. Sci.,* **32**, 2044–2059.

Wigley, T.M.L., 1989: Possible climate change due to SO_2-derived cloud condensation nuclei. *Nature,* **339**, 365–367.

Wigley, T.M.L and P.D. Jones, 1981: Detecting CO_2-induced climatic change. *Nature,* **292**, 205–208.

Wigley, T.M.L. and T.P. Barnett, 1990: Detection of the greenhouse effect in the observations. In: *Climate Change. The IPCC Scientific Assessment,* J.T. Houghton, G.J. Jenkins and J.J. Ephraums (eds.), Cambridge University Press, Cambridge, UK, pp. 239–256.

Wigley, T.M.L. and S.C.B. Raper, 1990: Natural variability of the climate system and detection of the greenhouse effect. *Nature,* **344**, 324–327.

Wigley, T.M.L and S.C.B. Raper, 1991a: Internally-generated natural variability of global-mean temperatures. In: *Greenhouse-Gas-Induced Climatic Change: A Critical Appraisal of Simulations and Observations,* M.E. Schlesinger (ed.), Elsevier, Amsterdam, pp. 471–482.

Wigley, T.M.L. and S.C.B. Raper, 1991b: Detection of the enhanced greenhouse effect on climate. *Proceedings of the Second World Climate Conference,* World Meteorological Organization, Geneva.

Wigley, T.M.L. and S.C.B. Raper, 1992: Implications for climate and sea-level of revised IPCC emissions scenarios. *Nature,* **357**, 293–300.

Wigley, T.M.L., P.D. Jones, P.M. Kelly and S.C.B. Raper, 1989: Statistical significance of global warming. *Proceedings 13th Annual Climate Diagnostics Workshop,* pp. A1-A8.

Wigley, T.M.L., P.D. Jones, P.M. Kelly, and M. Hulme, 1992: Recent global temperature changes: Ozone and aerosol influences. *Proceedings 16th Annual Climate Diagnostics Workshop,* pp. 194–202.

Woodward, W.A. and H.L. Gray, 1993: Global warming and the problem of testing for trend in time series data. *J. Climate,* **6**, 953–962.

Woodward, W.A. and H.L. Gray, 1995: Selecting a model for detecting the presence of a trend. *J. Climate,* **8**, 1929–1937.

9

Terrestrial Biotic Responses to Environmental Change and Feedbacks to Climate

J.M. MELILLO, I.C. PRENTICE, G.D. FARQUHAR, E.-D. SCHULZE, O.E. SALA

Contributors:
P.J. Bartlein, F.A. Bazzaz, R.H.W. Bradshaw, J.S. Clark, M. Claussen,
G.J. Collatz, M.B. Coughenhour, C.B. Field, J.A. Foley, A.D. Friend,
B. Huntley, C.H. KÖrner, W. Kurz, J. Lloyd, R. Leemans, P.H. Martin,
A.D. McGuire, K.G. McNaughton, R.P. Neilson, W.C. Oechel,
J.T. Overpeck, W.A. Parton, L.F. Pitelka, D. Rind, S.W. Running,
D.S. Schimel, T.M. Smith, T. Webb III, C. Whitlock

CONTENTS

SUMMARY

Terrestrial ecosystems and climate are closely coupled. Changes in climate and the carbon dioxide concentration of the atmosphere cause changes in the structure and function of terrestrial ecosystems. In turn, changes in the structure and function of terrestrial ecosystems influence the climatic system through biogeochemical processes that involve the land-atmosphere exchanges of radiatively active gases such as carbon dioxide (CO_2), methane (CH_4) and nitrous oxide (N_2O), and changes in biogeophysical processes that involve water and energy exchanges. The combined consequences of these effects and feedbacks must be taken into account when evaluating the future state of the atmosphere or of terrestrial ecosystems.

The global carbon budget and net CO_2 exchange between the land and the atmosphere

Analyses based on atmospheric CO_2 and $^{13}CO_2$ measurements suggest that the terrestrial biosphere is currently a net carbon sink. Such analyses quantify the strength of this sink as 0.5–1.9 GtC/yr during the 1980s, and as high as 2.6 GtC/yr during 1992–3; they also suggest that the tropics have been a net carbon source, implying even greater rates of carbon storage in mid- to high latitudes. Direct observations to establish the processes responsible for this carbon storage are, however, lacking. Possiblities include post-harvest regrowth of mid- and high latitude forests (0.5–0.9 GtC/yr according to IPCC WGII (1995), Chapter 24), enhanced vegetation growth due to physiological effects of increasing CO_2 (0.5–2.0 GtC/yr) and nitrogen deposition (0.5–1.0 GtC/yr) (figures from model calculations), and, probably, a substantial range (0–2.0 GtC/yr) of interannual variation due to climatic anomalies. These processes are not additive: for example, the rate of carbon sequestration in mid-latitude forests may include effects of all of the other processes.

The future role of the terrestrial biosphere in controlling atmospheric CO_2 concentrations is difficult to predict because we do not know which of these processes will dominate. As long as CO_2 increases, the CO_2 fertilisation effect is expected to play a role in enhancing terrestrial carbon storage. However, this role will disappear if atmospheric CO_2 is stabilised; also, plant growth may not continue to increase if the atmosphere's CO_2 concentration rises above some level, perhaps around 1000 ppmv, because the CO_2 fertilisation response is then saturated. Nitrogen fertilisation may continue to promote carbon storage in the forests of the Northern Hemisphere, but this effect would cease if the cumulative nitrogen inputs reach levels, yet to be defined, beyond which the impacts on plant growth become deleterious. Deforestation, which is already causing a net release of carbon from tropical lands to the atmosphere of 1.6 ± 1.0 GtC/yr, may increase to meet the food needs of an expanding human population, and extensive deforestation could adversely affect the biosphere's continued capacity to act as a carbon sink.

Further uncertainties arise because changes in climate and atmospheric CO_2 over the next decades to a century are likely to produce changes in the structure of natural and managed ecosystems. Structural changes include changes in the local abundance of species and genetic sub-groups (genotypes), and in the global geographic distributions of assemblages of species (biomes). There will be transient effects, varying according to the rate of climate change. With slow change, shifts in competitive balance among species might occur subtly with minor effects on terrestrial carbon storage. With rapid change, direct impacts on the growth and survival of particular types of plants could cause die back and carbon loss before better adapted types become established. This possible asymmetry of terrestrial carbon loss and accumulation under rapid climate change has led to the concern that climate-induced transient vegetation changes could release CO_2 into the atmosphere, counteracting the biosphere's capacity to take up CO_2. The magnitude of this feedback is highly uncertain: it could be near zero or, with low probability, as much as 200 GtC over the next one to two centuries. The more rapid the climate change, the greater the probability of a large transient carbon release. The probability also depends on the extent, intensity and impact of mid-continental drought, which is a major discrepancy both among general circulation models[1] (GCMs) and among ecosystem models. More reliable projections of drought and its impacts on ecosystems will require the incorporation of ecosystem dynamics in global carbon cycle models and the coupling of ecosystem models with GCMs.

Methane (or CO₂) release from wetlands

Methane is produced in flooded organic soils as a result of the metabolic activities of micro-organisms in the absence of oxygen. Methane emissions from natural wetlands contribute about 20% to the global emissions of this gas to the atmosphere. Methane flux from wetlands could either increase due to rising temperatures and CO_2, or decrease due to drying of the soils. If high latitude wetlands dry out, however, there will be a release of stored carbon as CO_2. The rate of CO_2 release with drying is uncertain, but potentially large since as much as 450 GtC may be stored in high latitude wetlands.

Nitrous oxide release from soils

The global nitrous oxide budget is dominated by microbial processes in soils, especially those in the moist tropics. The major N_2O-producing process is denitrification. Denitrification is promoted by high nitrate supply and low soil oxygen concentration. Warmer soils promote more rapid nitrogen cycling and often more nitrate, while wetter soils lead to low soil oxygen levels. Where soils become warmer and wetter, the production of N_2O will increase, but the global magnitude of this increase has not been estimated.

Effects of land-surface changes on climate

Vegetation mediates the exchange of water and energy between the land surface and the atmosphere, and thereby affects climate. As biomes shift, the climate will be affected. For example, high latitude warming is expected to cause forests to spread into tundra. This change would be expected to increase the warming in northern mid- to high latitudes by more than 50% over 50–150 years because of the lower albedo of forests during the snow season. Such feedbacks will, however, be modified by land-use changes such as deforestation.

As atmospheric CO_2 increases, stomatal conductance declines, so that the effectiveness of water conservation by plants is increased and the effects of drought on plant growth ameliorated. But declining stomatal conductance will also have feedback effects on climate. It has been estimated that a global halving of stomatal conductance, with no change in leaf area, would lead to an additional surface air warming of about 0.5°C averaged over the land.

---

[1] Throughout this chapter a GCM refers to an atmospheric general circulation model coupled to a mixed-layer ocean model.

9.1 Introduction

Terrestrial ecosystems can influence the climate system through biogeochemical and biogeophysical processes. Key biogeochemical effects and feedbacks of ecosystems on climate involve the land–atmosphere exchange of radiatively active gases, such as CO_2, CH_4 and N_2O. These gases exert a major control on the climate system. Their atmospheric content can be increased or decreased by changes in the structure and function of terrestrial ecosystems as these systems respond to environmental changes including changes in climate. Structural and functional changes in land ecosystems can also cause biogeophysical effects and feedbacks by altering the exchange of water and energy, leading to changes in surface temperatures, atmospheric circulation and precipitation patterns.

In this chapter we will consider major feedbacks between terrestrial ecosystems and the climatic system for three time intervals: (1) the present; (2) a future period of "climate transition" that is likely to range from decades to centuries; and (3) a hypothesised "new equilibrium condition" that is probably centuries into the future. We deal with feedbacks in the global carbon, methane and nitrous oxide budgets. We also discuss how changes in ecosystem structure and function affect climate through biogeophysical feedbacks.

The range of topics covered in this chapter is more restricted than the range covered in the first IPCC Scientific Assessment (Melillo *et al.*, 1990). This is because an extensive assessment of terrestrial impacts is given in IPCC WGII (1995).

9.2 Land–atmosphere CO_2 Exchange and the Global Carbon Balance: the Present

Land ecosystems of the Earth contain about 2,200 GtC; an estimated 600 GtC in vegetation and 1600 GtC in soils. These land carbon stocks are changing now and are likely to continue to change in the future in response to changes in any or all of the following factors; area of agricultural land, age structure of forests, climate, and chemistry of the atmosphere and precipitation.

9.2.1 Changes in the Area of Agricultural Land

Agricultural land occupies almost one fifth of the Earth's terrestrial surface (Olson, 1983). A substantial portion of this land was once forested and so contained relatively large carbon stocks in both trees and soils. The conversion of forests to agricultural lands releases carbon, mostly from

trees, to the atmosphere through burning and decay. Conversely, the regrowth of forests on abandoned lands withdraws carbon from the atmosphere and stores it again in trees and soil. The net flux of carbon from the land to the atmosphere primarily associated with agricultural expansion for 1980 has been estimated at between 0.6 and 2.5 GtC/yr (Houghton *et al.*, 1987; Hall and Uhlig, 1991; Houghton, 1991; Houghton, 1995). IPCC (1994) indicates that the net emission from changes in tropical land-use was 1.6 ± 1.0 GtC/yr for the period 1980 through 1989 (Schimel *et al.*, 1995). Houghton (1995) estimated that in 1990 the net flux to the atmosphere, essentially all from the tropics, was 1.7 GtC/yr with an uncertainty of $\pm 30\%$.

9.2.2 Changes in the Age Structure of Forests

Young and middle-aged forests accumulate carbon, while old-growth forests accumulate little if any carbon. Forests of the Northern Hemisphere's mid-latitudes that were harvested in the early and middle parts of the 20th century are still regrowing and accumulating carbon (e.g., Kauppi *et al.*, 1992; Wofsy *et al.*, 1993). Estimates of rates of carbon accumulation related to forest regrowth in these regions range between 0.7 and 0.8 GtC/yr for the 1980s (Melillo *et al.*, 1988; Sedjo, 1992; Dixon *et al.*, 1994). Melillo *et al.* (1988) have argued that the fate of the cut wood must also be taken into account when evaluating the net effect of forest harvest and regrowth in the global carbon budget; that is, the rate at which the carbon in the cut wood is returned to the atmosphere as a result of burning and decay must be considered. When Melillo *et al.* did this, they concluded that the net effect of forest harvest and regrowth for the middle and high latitudes of the Northern Hemisphere on terrestrial net carbon storage was approximately zero in the 1980s. It must be recognised that there are considerable uncertainties associated with estimating the fate of cut wood, including wood left at the harvest sites, fire wood, and wood products such as paper and lumber (Melillo *et al.*, 1988; Harmon *et al.*, 1990). Because of these uncertainties, IPCC (1994) estimated that forest harvest and regrowth resulted in an accumulation on land of 0.5 ± 0.5 GtC/yr during the 1980s (Schimel *et al.*, 1995).

Changes in the frequency of fires, insect outbreaks and other disturbances can also alter the age structure of forests and affect their capacity to store carbon. Disturbance regimes are affected by climatic conditions such as warming and drought. Kurz and Apps (1995) have reported that since 1970, the boreal forest regions of Canada have experienced increased rates of disturbance, especially spruce budworm outbreaks and fire. As a consequence, these forests have switched from being a sink for atmospheric CO_2 (0.15 GtC/yr) to being a carbon source to the atmosphere (0.05 GtC/yr), albeit a small one. Kurz *et al.* (1995) consider warming to be one of several factors possibly responsible for the recent increase in fire frequency in Canada. While climate-related increases in disturbance frequencies may be occurring in other regions (Woodwell, 1995), we do not have a good global accounting of them or of their consequences for terrestrial carbon storage.

9.2.3 Changes in Ecosystem Metabolism

Each year, about 5% of the land's total carbon stock is exchanged with the atmosphere as a result of plant and soil metabolic activities. Through the process of photosynthesis, land plants take up on the order of 120 GtC/yr in the form of CO_2 from the atmosphere. This carbon uptake is approximately balanced by plant and soil respiration, which release carbon as CO_2 to the atmosphere. A change in the balance between photosynthesis and respiration will change the carbon stock on land and also has the potential to alter the CO_2 content of the atmosphere.

The relationships among the various metabolic processes in terrestrial ecosystems can be defined by three equations:

$$NEP = GPP - R_a - R_h \quad (9.1)$$
$$NPP = GPP - R_a \quad (9.2)$$
$$NEP = NPP - R_h \quad (9.3)$$

where, at an annual time step:

ΣNEP is net ecosystem production, the yearly rate of change in carbon storage in an ecosystem. It can be either positive or negative. A positive *NEP* indicates that the ecosystem has accumulated carbon during the year, while a negative *NEP* indicates that it has lost carbon during the year.

ΣGPP is gross primary production, the amount of carbon fixed through the process of photosynthesis by the ecosystem's green plants in a year.

ΣR_a is autotrophic respiration, the amount of carbon released to the atmosphere as CO_2 by the ecosystem's green plants through respiration in a year.

ΣR_h is heterotrophic respiration, the amount of carbon released to the atmosphere as CO_2 by the ecosystem's animals and micro-organisms through respiration in a year.

ΣNPP is net primary production, the difference between gross primary production (*GPP*) and plant respiration (R_h) in a year.

As we face the prospect of climate change, an important question to ask ourselves is: how will the relationship between *NPP* and R_h change as the Earth warms in response to the accumulation of carbon dioxide and other heat-trapping gases in the atmosphere? An answer to this question is complicated by the fact that both positive and negative metabolic feedbacks may be simultaneously involved.

The major positive metabolic feedback is associated with warming. The suggestion is that an increase in temperature will increase rates of both plant and microbial respiration and speed the release of carbon into the atmosphere from plants and soils (Woodwell, 1983, 1989, 1995). Negative feedbacks have been proposed that are associated with an increase in atmospheric CO_2 and with warming. The increase in atmospheric CO_2 may cause a stimulation in rates of photosynthesis and an increase in carbon storage (e.g., Oechel and Strain, 1985). In nutrient-limited forests, warming may increase terrestrial carbon storage by relieving plant nutrient limitations (McGuire *et al.*, 1992; Shaver *et al.*, 1992).

In addition to the possibility of simultaneous feedbacks of opposite sign, there are other aspects of global change that may affect terrestrial ecosystem metabolism and carbon storage. Some of the major effects are associated with the burning of fossil fuels, which causes changes in the chemistry of the atmosphere and precipitation. Fossil fuel burning can lead to the production of air pollutants such as sulphur dioxide and ozone that are toxic to plants. These air pollutants can decrease *NPP* and carbon storage (Allen and Amthor, 1995). The burning of fossil fuels can also lead to increases in nitrogen in precipitation. Up to some cumulative level, increased nitrogen inputs to nitrogen-limited ecosystems, such as many temperate and boreal forests of the northern hemisphere, can cause increases in *NPP* and carbon storage. Nitrogen inputs beyond the threshold level can lead to decreases in *NPP* and carbon storage (Aber *et al.*, 1989; Schulze *et al.*, 1989).

9.2.3.1 Climate change

Climate changes may affect *NPP* in a variety of ways. Elevated temperature may increase *NPP* by enhancing photosynthesis (Larcher, 1983) or through increased nutrient availability if decomposition and nutrient mineralisation are enhanced (Bonan and Van Cleve, 1992; McGuire *et al.*, 1992; Melillo *et al.*, 1993, 1995a). It may decrease *NPP* by decreasing soil moisture which may reduce photosynthesis through decreased stomatal conductance (Hsaio, 1973; Gifford, 1994) or decreased decomposition and mineralisation (McGuire *et al.*, 1992, 1993; Pastor and Post, 1988; Parton *et al.*, 1995). Elevated temperature may also increase plant respiration and so reduce *NPP* (McGuire *et al.*, 1992, 1993; Running and Nemani, 1991), although this effect may have been overestimated (Gifford, 1993). The effects of precipitation and cloudiness on *NPP* can also be positive or negative in different situations. In dry regions, lower precipitation or lower cloudiness may decrease *NPP* by lowering soil moisture. In moist regions, increased cloudiness may decrease *NPP* by reducing the availability of photosynthetically active radiation (PAR). Climate changes may further influence *NPP* by affecting leaf phenology in deciduous vegetation (Long and Hutchin, 1991).

Climate affects carbon storage in terrestrial ecosystems because temperature, moisture, and radiation influence both carbon gain through photosynthesis and carbon loss through respiration. Soil respiration is generally accelerated by higher temperatures, producing an increase in the release of CO_2 from terrestrial ecosystems (Houghton and Woodwell, 1989; Melillo *et al.*, 1990; Shaver *et al.*, 1992; Townsend *et al.*, 1992; Oechel *et al.*, 1993; Peterjohn *et al.*, 1993, 1994; Lloyd and Taylor, 1994; Schimel *et al.*, 1994; Kirschbaum, 1995). It is thought that in many ecosystems the increase of soil respiration with temperature is steeper than any increase of *NPP* with temperature, so that the net effect of warming is to reduce carbon storage. Lloyd and Taylor (1994), however, caution that valid conclusions regarding the effects of possible changes in temperatures on soil carbon pools cannot yet be made because we lack a detailed knowledge of the temperature sensitivity of carbon input into the soil via *NPP*.

Oechel *et al.* (1993) report that the recent general pattern of warming of the north slope of Alaska and the Canadian Arctic may have led to a reduction of carbon storage in tundra ecosystems of the region. At a set of sites in the tundra of the north slope of Alaska, they measured whole ecosystem CO_2 flux over five summers between 1983 and 1990, and found a net flux of carbon from the land to the atmosphere. Extrapolating their results from the Alaskan tundra to the circumpolar Arctic, Oechel *et al.* (1993) estimate that regional warming could have caused a net flux from the land to the atmosphere of about 0.2 GtC/yr during the 1980s.

Warming may not always lead to a reduction of carbon storage in terrestrial ecosystems. It has also been suggested

---

The Link Between Carbon Storage and Nitrogen Distribution in Terrestrial Ecosystems

Changes in the distribution of nitrogen between plants and soil can change the amount of carbon stored in an ecosystem. Plant and detrital tissues in terrestrial ecosystems vary widely in their stoichiometric ratios of C and N (Melillo and Gosz, 1983; Schindler and Bayley, 1993; Gifford, 1994; Kinzig and Socolow, 1994; Schulze *et al.*, 1994). If warming of the Earth accelerates mineralisation of N from soil organic matter, where the C:N ratio usually ranges from 10 to 25, and if the N is taken up by plants and used to produce more woody biomass with a C:N ratio between 40 and 400, then the ecosystem would experience a net increase in carbon storage (Gifford, 1994; Peterjohn *et al.*, 1994; Melillo, 1995; Melillo *et al.*, 1995a). On the other hand, if the processes of microbial mineralisation of soil organic matter and plant uptake of mineral N are decoupled in space or time, then N might be lost from the ecosystem via leaching or denitrification, resulting in a net loss of both C and N from that ecosystem.

---

that in nutrient-limited forests, particularly in cold climates, warming may increase carbon storage through a two-step mechanism (McGuire *et al.*, 1992; Shaver *et al.*, 1992; Melillo *et al.*, 1993; see Box on The Link Between Carbon Storage and Nitrogen Distribution). First, warming may increase the decay rate of low carbon-to-nutrient ratio soil organic matter which may release nutrients to soil solution and some CO_2 to the atmosphere (Peterjohn *et al.*, 1994; Melillo *et al.*, 1995a). Second, there may be increased nutrient uptake by trees which store nutrients in woody tissues which have a high carbon-to-nutrient ratio.

By differentially affecting *NPP* and R_h, climate variations in temperature, precipitation and cloudiness on the annual to decadal time-scale may have affected terrestrial carbon storage. Dai and Fung (1993) and Keeling *et al.* (1995) have suggested that these variations could have resulted in a substantial terrestrial carbon sink in recent years. Ciais *et al.* (1995a) suggested that cooling arising from the effects of Mt. Pinatubo's eruption in June, 1991, may have increased terrestrial carbon storage and contributed to the observed reduction in atmospheric growth rate during the 1991 to 1992 period. The magnitude of the effect of these climatic variations on global terrestrial carbon storage is highly uncertain, but may be between 0 and 2.0 GtC/yr. Terrestrial carbon storage caused by climate anomalies may be spatially complex due to the potential for a spatial mosaic structure of the anomalies. In the near term, some regions may experience increases in carbon storage and others decreases.

9.2.3.2 *Carbon dioxide fertilisation*
Results of laboratory and field studies
A short-term increase in CO_2 concentration causes an increase in photosynthesis at the level of individual leaves (e.g., Idso and Kimball, 1993). Leaves may acclimate to increased CO_2 such that long-term increases in CO_2 result in smaller increases in leaf-level photosynthesis (Wong, 1979; Tissue and Oechel, 1987; Fetcher *et al.,* 1988; Sage *et al.,* 1989); but this does not always occur (e.g., Masle *et al.,* 1993; Gunderson and Wullschleger, 1994; Luo *et al.,* 1994; Sage, 1994). At the level of the whole plant, feedbacks from non-leaf organs can influence leaf photosynthetic responses to CO_2. If photosynthesis increases to a greater extent than the rate at which carbon assimilated during photosynthesis can be used in additional growth, photosynthesis can be slowed by negative metabolic feedbacks in plants (Arp, 1991; Bowes, 1991; Luo *et al.,* 1994; Sage, 1994). On the other hand, an increased carbohydrate supply which is allocated below ground may help to alleviate nutrient limitations on photosynthesis by enhancing biological nitrogen fixation and/or increasing mychorrizal activity leading to greater uptake of phosphorus (Norby, 1987; O'Neill *et al.,* 1987; Arnone and Gordon, 1990; Lewis *et al.,* 1994; O'Neill, 1994).

Studies of both crop and non-crop plants grown in doubled CO_2 have shown increased growth responses often in the range of 15 to 71%. The total range is from a negative response of 43% to a positive response of up to 375% (Kimball, 1975; Kimball and Idso, 1983; Eamus and Jarvis, 1989; Poorter, 1993; Ceulemans and Mousseau, 1994, Idso and Idso, 1994; McGuire *et al.,* 1995a; Wullschleger *et al.,* 1995). Across the tissue, plant and ecosystem levels the experimental evidence shows that the proportional response of *NPP* to elevated CO_2 is greater when soil moisture is depleted. Elevated CO_2 generally decreases stomatal conductance and hence transpiration, enhancing *NPP* by promoting greater water use efficiency (Gifford, 1979; Wong, 1979; Morrison and Gifford, 1984; Polley *et al.,* 1993). There is some evidence that nitrogen-limited plants are less responsive to increases in CO_2 level (McGuire *et al.,* 1995a), but a recent review suggests that this is not always true (Lloyd and Farquhar, 1995).

Increased CO_2 may lead to increased carbon storage, but the magnitude of this effect is highly uncertain. Carbon sequestration at the ecosystem level depends not only on plant photosynthesis, respiration and growth, but also on the fluxes of carbon out of litter and soil carbon pools. At equilibrium, plant *NPP* would be balanced by these microbially mediated soil fluxes (R_h). However, if there is a progressive change in *NPP* such as a progressive increase in *NPP* due to the effect of rising CO_2, the system will not be in equilibrium. In general the soil CO_2 flux (R_h) is proportional to the size of the decomposing carbon pools, whereas *NPP* is determined by the activity of the leaves. If CO_2 increases, *NPP* may therefore increase immediately while the size of the decomposing pools takes time to build up. If *NPP* were to stop increasing (e.g., due to CO_2 stabilisation), the sink would weaken, with a decay time determined by the residence times of the pools to which carbon has been added. This is the theoretical explanation for why rising CO_2 is thought to be producing a carbon sink, and why the sink may be a transient one (Taylor and Lloyd, 1992).

A number of CO_2-enrichment field experiments have been conducted on intact ecosystems (Drake, 1992a; Nie *et al.,* 1992; Owensby *et al.,* 1993; Oechel *et al.,* 1994; Jackson *et al.,* 1994; Körner and Diemer, 1994). When the *NEP* responses to CO_2 increases were measured, the results ranged from relatively large increases in a temperate wetland (Drake, 1992a,b), to a small increase in a tundra (Oechel *et al.,* 1994). There have been no comparable experiments on forests and this has resulted in an important gap in our understanding of how terrestrial ecosystems respond to increased atmospheric CO_2. Carbon dioxide fertilisation studies in forests are a high research priority.

Modelling studies
The first model-based analyses of the current effects of CO_2 increase on terrestrial carbon storage were performed with simple models in which the terrestrial biosphere was represented by one or a few boxes. In one such analysis, Gifford (1993) worked within a plausible range of values (10, 25 and 40%) for the increase in *NPP* at doubled CO_2. Assuming that the terrestrial biosphere was in equilibrium in pre-industrial times, he forced his model with the historical record to give an idea of the extent to which the CO_2 increase would produce an imbalance of carbon uptake and release. The resulting terrestrial carbon storage rates due to this mechanism were 0.5, 2.0 and 4.0 GtC/yr during the 1980s. Rotmans and den Elzen (1993) showed, in an analogous model experiment, that CO_2 fertilisation was stimulating carbon storage at a rate of 1.2 GtC/yr. A new version of a global biogeochemistry model, the Terrestrial Ecosystem Model (TEM), (McGuire *et al.,*

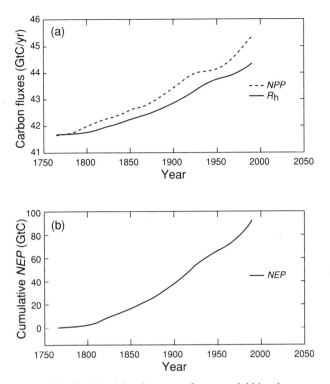

Figure 9.1: Simulated development of a terrestrial biosphere carbon sink due to CO_2-induced increases in net primary production (*NPP*), as simulated with the Terrestrial Ecosystem Model (Melillo *et al.*, 1995b). The biosphere was assumed to be in equilibrium before 1750. Climate was kept constant (as present). (a) Simulated time course of net primary production (*NPP*) and heterotrophic respiration (R_h). (b) Cumulative net ecosystem production (*NEP*), i.e., sink strength.

1995b) was used by Melillo *et al.* (1995b) in the transient mode to explore the effects of CO_2 enrichment, taking into account the variety of responses by different ecosystems. The model indicated that the mean CO_2-stimulated carbon storage (*NPP* minus R_h) during the 1980s was 0.9 GtC/yr, and that it reached almost 1.0 GtC/yr in 1990 (Figure 9.1). Based on our current knowledge from laboratory, field and modelling studies, we consider a plausible range of the CO_2 fertilisation effect to be 0.5–2.0 GtC/yr for the 1980s.

9.2.3.3 Air pollution
The world's two major phytotoxic air pollutants are ozone (O_3) and sulphur dioxide (SO_2). They are spread throughout the industrial and agricultural regions (Heck, 1984). Experimental studies have documented significant effects of ozone on a variety of plant processes. It can reduce photosynthesis, increase respiration and promote early leaf senescence. These effects, either singly or in combination, can lead to reduced plant growth (Allen and

Amthor, 1995). Chameides *et al.* (1994) have presented evidence that the growth of urban and industrialised areas and intensive agriculture, through their production of tropospheric ozone, are reducing agricultural yields and, presumably, plant growth in adjacent non-agricultural ecosystems. Sulphur dioxide's main effect appears to be to reduce photosynthesis, which leads to reduced plant growth (Allen and Amthor, 1995). In combination, ozone and sulphur dioxide can interact to magnify the negative effects of either pollutant alone on plant growth (Ormrod, 1982). While it is not disputed that ozone and sulphur dioxide pollution have the potential to reduce both *NPP* and carbon storage in terrestrial ecosystems, there is little basis for making global estimates of these effects (Chameides *et al.*, 1994).

9.2.3.4 Nitrogen fertilisation associated with acid rain
Those mid-latitude ecosystems of the Northern Hemisphere that are near or downwind of industrial and agricultural areas are receiving large inputs of nitrogen, mostly from fossil fuel burning. Forests in western and central Europe and eastern North America receive up to 50 kg N/ha/yr (Melillo *et al.*, 1989; Schulze *et al.*, 1989) or more. These nitrogen inputs may be enhancing *NPP* and terrestrial carbon storage. For example, Kauppi *et al.* (1992) suggested that increases in the growth rates of European forests observed between 1950 and 1980 were due to nitrogen fertilisation, although other mechanisms such as CO_2 fertilisation, climate variability and management changes may also be involved.

Experimental evidence supports the nitrogen-stimulation mechanism for increasing carbon storage in mid-latitude forests. In a field study in which 50 kg N/ha/yr has been added to large plots of healthy evergreen and deciduous forests for several years, carbon storage in aboveground woody tissue has increased by 10–20%/yr (Aber *et al.*, 1993). However, this increase in carbon storage is expected to reach a threshold level beyond which further additions of nitrogen are likely to result in *NPP* reduction, tree death, and carbon loss (Aber *et al.*, 1989; Schulze *et al.*, 1989).

Many terrestrial ecosystems, especially those in middle and high latitudes, are nitrogen limited; that is, added nitrogen will produce a growth response and additional carbon storage (e.g., Melillo and Gosz, 1983; Vitousek and Howarth, 1991; Schimel *et al.*, 1994; Melillo, 1995). Nitrogen deposition from fertilisers and oxides of nitrogen released from the burning of fossil fuel during the 1980s is estimated to amount to a global total of 0.05–0.08 GtN/yr (Peterson and Melillo, 1985; Duce *et al.*, 1991; Galloway *et al.*, 1995; Melillo, 1995), spatially concentrated in the

Table 9.1: *Recent estimates of nitrogen-stimulated carbon storage in terrestrial ecosystems.*

	Region(s) considered	N input TgN/yr	C store GtC/yr
Kohlmaier *et al.*, 1988	30 – 60°N	21 ± 7	up to 0.7
Schindler and Bayley, 1993	globe	13	~ 0.7 – 2.0
Hudson *et al.*, 1994	Northern Hemisphere	30	~ 0.7
Townsend *et al.*, 1994	global (excluding cultivated areas)	7[†]	0.2 – 1.2[*]
Melillo, 1995	temperate and boreal forests of North America and Europe	18	0.6 – 0.9[+]

[†] NO_3 only.

[*] Best estimate 0.3 – 0.6.

[+] Lower end of range more likely.

northern mid-latitudes. The carbon sequestration which results from this added nitrogen is estimated to be 0.2–2.0 GtC/yr (see Table 9.1), depending on assumptions about: (1) the proportion of nitrogen that remains in ecosystems; (2) the relative distribution of nitrogen between vegetation and soil; and (3) the C:N ratios of the vegetation and soil. Estimates higher than 1 GtC/yr seem unrealistic because they assume that all of the N would be stored, and in forms with high C:N ratios, which is improbable. Recent studies of the fate of nitrogen added to forest ecosystems suggests that between 70 and 90% ends up in the soil, which has a relatively low C:N ratio (Aber *et al.*, 1993; Nadelhoffer *et al.*, 1995). In addition, much atmospheric nitrogen is in reality deposited on grasslands and agricultural lands where storage occurs in soils with low average C:N ratios. A more plausible range for this effect is therefore 0.5–1.0 GtC/yr.

Model calculations have not yet considered the fact that CO_2 and nitrogen fertilisation are likely to be interacting. The additional nitrogen may enhance photosynthesis and plant growth (Bazzaz and Fajer, 1992; Diaz *et al.*, 1993; McGuire *et al.*, 1995a – see Section 9.2.3.1), and possibly lead to increased carbon storage.

9.2.4 Estimates of Current Terrestrial Carbon Sinks from Atmospheric Measurements

Observations at stations monitoring background concentrations of CO_2 indicate that only a portion of the fossil fuel-derived CO_2 remains airborne, while the remainder is absorbed by the oceans and the terrestrial biosphere. A first order estimate of the global distribution of sinks by latitude can be made using CO_2 concentration data from networks of monitoring stations (Denning *et al.*, 1995), but the distinction between land and ocean sinks and

sources within a particular latitudinal band requires additional information. The relative roles of terrestrial versus ocean ecosystems in these bands can be estimated with several approaches (Tans *et al.*, 1995), including ones that use the $^{13}C/^{12}C$ ratio of the CO_2. The carbon isotope ratio of CO_2 is useful because terrestrial uptake discriminates strongly against ^{13}C while the ocean uptake shows little discrimination (Keeling *et al.*, 1989, 1995; Francey *et al.*, 1995). Francey *et al.* (1995), using a terrestrial discrimination value of 18 parts per thousand, estimated a net terrestrial storage over the last decade of 1 to 3 GtC/yr.

Lloyd and Farquhar (1994) pointed out that the magnitude of ^{13}C discrimination among land plants depends on the biochemical pathway they use in photosynthesis and that this affects the analysis of atmospheric observations. The C_3 pathway results in a discrimination of 18 parts per thousand, while the C_4 pathway results in very little discrimination.[1] Considering the distribution of C_3 and C_4 plants and their share in global *NPP*, Lloyd and Farquhar (1994) estimated that the global average terrestrial discrimination factor is about 15 parts per thousand. This means that Francey *et al.* (1995) may have underestimated the net terrestrial uptake by up to 25%. Ciais *et al.* (1995b) included the C_4 effect in an

---

[1] C_3 photosynthesis is the normal biochemical pathway of carbon fixation in plants. C_4 photosynthesis is a more complex pathway, found for example in tropical grasses, which includes a mechanism for concentrating CO_2 at the sites of carbon fixation. The C_4 pathway accounts for a substantial fraction of total *NPP* in tropical grasslands and savannahs. Some tropical crops, including sugar cane, also use this pathway

analysis of data from a greatly expanded global network. They found that the net terrestrial carbon sink averaged 2.6 GtC/yr during the years 1992 and 1993.

The analyses of Ciais *et al.* (1995b) indicate that for the land ecosystems outside of the equatorial band from 30°S to 30°N, net carbon storage (i.e. *NEP*) averaged 4.3 Gt/yr for the years 1992 and 1993. Most of this storage, an average of 3.5 GtC/yr, was between 30° and 60°N. These are large numbers in the global carbon cycle when compared to the annual fossil fuel emissions for the same two years, which was about 6.1 GtC/yr. It follows that the terrestrial biosphere is currently playing a major role in the carbon cycle, in some years more than compensating for terrestrial ecosystem sources of carbon, including those associated with deforestation (see Section 9.2.1) and enhanced respiration in response to regional warming in the circumpolar tundra (see Section 9.2.3.1). The future maintenance of this role is crucially important to scenarios of CO_2 stabilisation.

A major complication in developing future scenarios of CO_2 stabilisation that accurately include the role of terrestrial ecosystems in the global carbon cycle is that the magnitude of the extra-tropical sink appears to vary substantially through time (see Table 9.2, Keeling *et al.*, 1989; Tans *et al.*, 1990; Ciais et *al.*, 1995b; Francey *et al.*, 1995) and may be responding to climatic variations on the scale of years to decades (Dai and Fung, 1993; Keeling *et al.*, 1995). The magnitude of the response to climatic anomalies is highly uncertain but may be in the range of ±2.0 GtC/yr.

Longer time-series with information on carbon isotope composition of CO_2 will provide us with a better idea of how much the extra-tropical carbon sink changes through time. Refined estimates of the role of terrestrial ecosystems in the global carbon balance will likely soon come from data on the oxygen isotope composition of CO_2 (Francey and Tans, 1987; Farquhar *et al.*, 1993) which can also be obtained from the global network. Further information will come from measurements of the atmospheric concentration of oxygen and its spatial variation (Keeling and Shertz, 1992).

While various pieces of evidence support a substantial terrestrial carbon sink in the Northern Hemisphere, direct observations to confirm the hypothesis and to establish the processes for this carbon storage are lacking. Expanding our understanding of the processes responsible for net annual storage of carbon by the terrestrial biosphere is among our greatest challenges.

Table 9.2: Estimated global carbon budgets based on atmospheric observations, obtained using a range of methods for various years during the 1980s and 1990s (modified from Tans et al., 1995).

Reference	Year(s)	Global net flux	Tropical net flux[§]	Extra-tropical net flux	Primary constraints
		GtC/yr	GtC/yr	GtC/yr	
Keeling *et al.*,1989	1984	−0.5[$]	+0.3[$]	0.7[†]	history of atmospheric [CO_2], and its $^{13}C/^{12}C$ ratio
Tans *et al.*, 1990	1981–1987	−1.9	+0.5	−2.4[†]	observed latitudinal gradient of atmospheric [CO_2], and its global annual rate of increase
Ciais *et al.*, 1995a	1992–1993	−2.6	+1.7	−4.3[*]	observed latitudinal gradient of atmospheric [CO_2], and its $^{13}C/^{12}C$ ratio

§   The net tropical flux term represents both the release of carbon from the land to the atmosphere due to deforestation and any storage occurring in undisturbed forests associated with processes such as CO_2 fertilisation.

$ A negative sign represents uptake of carbon by terrestrial ecosystems from the atmosphere, a positive sign represents a release of carbon by terrestrial systems to the atmosphere.

†   Extra-tropical region includes all land outside of a band between 15°N and 15°S.

*   Extra-tropical region includes all land outside of a band between 30°N and 30°S.

9.3 Possible Effects of Climate Change and Atmospheric Carbon Dioxide Increases on Ecosystem Structure

Changes in climate and atmospheric carbon dioxide concentration affect the species composition and structure of ecosystems because the environment limits both the types of organisms that can thrive and the amounts of plant tissues that can be sustained. Compositional and structural changes, in turn, affect ecosystem function (Schulze, 1982; Schulze and Chapin, 1987; Schulze, 1994). Compositional and structural changes will occur over a longer period than functional changes and these may not keep pace with rapid environmental change, so complex transient effects may result.

Experimental approaches have shown strong effects of climate change on vegetation composition and structure (Harte and Shaw, 1995). However, longer-term changes cannot be investigated experimentally, so much of what we understand about these changes is based on modelling and palaeodata.

9.3.1 Environmental Controls on Vegetation Structure: Ecophysiological Constraints on Plant Types

The ecosystems of the world are usually classified into 15–20 "biomes", each characterised by the dominance of one or more structural/functional types of plant. The global distributions of biomes are determined by ecophysiological constraints on the dominant plant types of each biome (Schulze, 1982; Woodward, 1987). The main constraints are related to temperature and water.

Cold tolerance in woody plants ranges from tropical trees, some damaged by temperatures of less than 10°C, to boreal deciduous trees which in their leafless, cold-hardened state have apparently unlimited tolerance (Sakai and Weiser, 1973; Larcher, 1983; Woodward, 1987). Warmer winters would cause an expansion of more diverse woody plant types towards the poles and continental interiors. Warmer summers, and/or longer growing seasons, would also allow some cold-adapted trees to spread poleward (Emanuel et al., 1985; Pastor and Post, 1988) especially at the polar tree-line where growing-season warmth is limiting. Cold-tolerant woody plants have, however, evolved mechanisms to delay spring budburst until day length is adequate, and/or a long enough chilling period has occurred. These mechanisms insure against the possibility of premature budburst in a mild period during winter. Premature budburst would be damaging because the leaves could be killed by a subsequent frost. Chilling requirements keep continental

woody plants out of maritime climates. Warmer winters will force such species to retreat from their low latitude and maritime limits (Dahl, 1990; Overpeck et al., 1991; Davis and Zabinski, 1992).

Plants vary enormously in their ability to tolerate drought, ranging from rain forest trees that can tolerate only a few weeks without rain unless they have access to deep water, to succulents that store water, and other hot desert species that can persist in a dry state. Some evergreen trees survive drought by reducing transpiration to near zero, while drought deciduous trees lose their leaves. As water availability declines further, trees give way to fire-adapted grasses and/or shrubs, then ultimately to the low shrubs, forbs and other drought-adapted life forms characteristic of steppes and deserts. Changes in water availability, whether positive or negative, will lead to shifts along this continuum.

9.3.2 Environmental Controls on Vegetation Structure: Resource Availability

Vegetation structure is determined not only by the types of plants present but also by the height and foliage cover they attain. Foliage cover, often expressed as leaf area index (LAI), is constrained by resource availability (water, carbon, nitrogen). LAI decreases as water availability declines. Along moisture gradients, vegetation composition and structure change due partly to replacement of drought-sensitive by drought-tolerant (or more deep-rooted) plant types, and partly to reductions in the LAI of each type (Walter, 1979). Values for LAI are typically low enough to prevent drought damage in most years (Specht, 1972; Woodward, 1987; Neilson, 1995), so maintaining annual *NPP* near maximal for the environment (Eagleson, 1978; Haxeltine et al., 1996). Changes in water availability will therefore affect LAI (Smith et al., 1993), but the response may be modified by changes in CO_2. In particular, increasing CO_2 (by reducing transpiration) would be expected to compensate for the effect of reduced water availability on LAI (Jarvis, 1989; Woodward, 1992; VEMAP Members, 1995).

Vegetation structure also changes with light availability, summer temperature and growing season length. From boreal forest to high-arctic tundra, trees are gradually replaced by shrubs and grasses as *NPP* declines. Comparable vegetation gradients occur at high elevations in all latitudes (Walter, 1979). In these cold environments warming should generally increase *NPP* and therefore LAI. Rising CO_2 may also increase LAI in so far as CO_2 fertilisation can increase *NPP* (see Section 9.2.3.2). *NPP* in cold climates today may be further limited by the slow rate of nitrogen mineralisation (see Section 9.2.3.1). The combination of warming with CO_2 increase could therefore increase LAI especially in high latitudes.

The ecophysiological and resource-availability constraints that determine the natural distribution of biomes also closely determine potential (non-agricultural) land-use, for example, suitability for forestry or grazing, the types of tree which can be exploited for forestry, sustainable density of grazing animals. Changes in these constraints strongly affect both natural and managed ecosystems.

9.3.3 Environmental Mediation of the Competitive Balance

The physical environment also contributes to determining the natural competitive balance among those plant types that can co-exist in a given environment. For example:

(1) Wet tropical climates support evergreen rain forest; longer dry seasons favour drought-deciduous trees in seasonal and dry forests (Chabot and Hicks, 1982; Haxeltine *et al.*, 1996). Any change in monsoon duration would shift the natural boundaries among these forest types.

(2) Savannahs represent an equilibrium of trees and grasses. In seasonal tropical climates, grasslands are favoured by summer rainfall and slow-percolating clay soils, woodlands by winter rainfall and sandy soils (Walter, 1979; Walker and Noy-Meir, 1982; Lauenroth *et al.*, 1993). Natural savannahs occur in intermediate situations, where the combination is favoured (Eagleson and Segarra, 1985; Haxeltine *et al.*, 1996). Changes in rainfall seasonality, as well as changes in land management, would affect the tree–grass balance.

(3) Seasonal shifts between C_3 and C_4 grass dominance (Groves and Williams, 1981) occur because C_4 plants have a higher light use efficiency than C_3 plants wherever temperatures exceed a threshold of about 22°C, at current CO_2 concentration (Ehleringer and Monson, 1993). An increase in atmospheric CO_2 raises this temperature threshold (Drake, 1992b; Arp *et al.*, 1993; Johnson *et al.*, 1993). Warming should therefore favour C_4, but increasing CO_2 may favour C_3. Increasing CO_2 might have caused the recent invasion of some C_4 grasslands by C_3 woody plants (e.g., Johnson *et al.*, 1993), although other factors such as land-use changes may be involved as well (Archer *et al.*, 1995).

(4) In temperate and cold climates, evergreen woody plants are favoured over winter-deciduous woody plants wherever carbon and/or nutrient costs of replacing leaves annually exceed the costs of maintaining leaves during winter (Chabot and Hicks, 1982; Reich *et al.*, 1992; Chapin, 1991; Mooney *et al.*, 1991; Shaver and Chapin, 1991; Arris and Eagleson, 1994; Chapin *et al.*, 1995). Changes in the seasonality of either temperature or precipitation could alter the balance between evergreen and deciduous trees.

9.3.4 Global Biome Model Projections

Global models for ecosystem dynamics, that can simulate the transient response of vegetation structure to changes in climate and atmospheric CO_2, are being developed by several groups but none has yet been published. Simple biome models project equilibrium distributions of broad vegetation types ("potential natural vegetation") from ecophysiological constraints, but without consideration of CO_2 effects on plant physiology. The BIOME model (Prentice *et al.*, 1992) reproduces well the present vegetation as mapped by Olson *et al.* (1983), except where intensive agriculture predominates. BIOME has been applied to various $2 \times CO_2$ equilibrium climate scenarios developed using outputs from atmospheric general circulation models (GCMs) (Solomon *et al.*, 1993; Claussen, 1994; Claussen and Esch, 1994; Prentice and Sykes, 1995). Consistent results include poleward shifts of the northern-hemisphere taiga, temperate deciduous, and warm-temperate evergreen/warm mixed forest belts, a northward shift of the Eurasian taiga, and a slight expansion of tropical seasonal and rain forests into areas of warm-temperate evergreen forests (IPCC WG II, 1995). Differences among GCM scenarios in the extent of these changes are consistent with the GCMs' different climate sensitivities. Some scenarios (not all) show reductions of temperate forests in the continental interiors.

These results differ from analyses based on earlier, empirical models, e.g., in the tropics where the Holdridge (Prentice and Fung, 1990; Smith *et al.*, 1992, 1993) and Budyko (Tchebakova *et al.*, 1992) models showed greater expansions of tropical rain forest. Such expansions are probably an artefact, due to ignoring rainfall seasonality. The Holdridge model also showed a smaller northward and eastward shift of the Eurasian taiga than other models (including Tchebakova *et al.*, 1992; Monserud *et al.*, 1993), because temperature seasonality is ignored.

Comparisons with past biome distributions are instructive. At 6000 yr bp (before present) there was more insolation than today in the Northern Hemisphere summer, and more total annual insolation (leading to warmer year-round temperatures) at high latitudes. 6000 yr bp is not an analogue for the future, because the Earth's orbital configuration was different from present, while the CO$_2$ concentration was similar to pre-industrial (Mitchell *et al.,* 1990; Monserud *et al.,* 1993). Palaeoecological data for 6000 yr bp show that climate change alone does indeed cause major biome shifts. GCM simulations (e.g., Kutzbach and Guetter, 1986) driven by the insolation changes ("orbital forcing") capture the general direction of these shifts (COHMAP Members, 1988; Wright *et al.,* 1993; Foley, 1994). Their magnitude tends to be underestimated, however, possibly because the GCM modelling procedure disregards biogeophysical feedback (Henderson-Sellers and McGuffie, 1995; Section 9.7).

9.3.5 Regional Biome Model Projections

"Second-generation" biome models include resource limitation and competitive balance effects. Examples are MAPSS (Neilson, 1995; Neilson and Marks, 1995) and BIOME2 (Haxeltine *et al.,* 1996). In an application to the continental USA both models produced good simulations of potential natural vegetation (VEMAP Members, 1995; Figure 9.2). Temperature-controlled biome boundaries responded to 2 × CO$_2$ climates in a similar way to the earlier BIOME results. Moisture-controlled boundaries (e.g., the forest-grassland boundary) responded differently, due to different evapotranspiration parametrizations, in the two models. In the central and eastern USA, some climate scenarios without corresponding CO$_2$ doubling caused partial replacement of forests by grasslands. This effect was more extensive in MAPSS, but the effect was

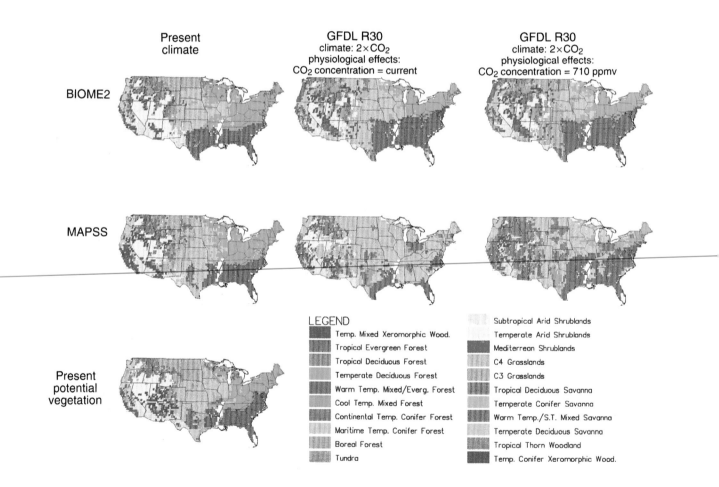

Figure 9.2: Changes in potential natural vegetation of the conterminous USA, simulated with the BIOME2 and MAPPS vegetation models. Left: simulations based on present climate. The actual distribution of potential natural vegetation types is shown for comparison. Centre: simulations based on "2 × CO$_2$" climate scenario. This scenario was derived by using an equilibrium simulation from the GFDL R30 atmospheric/mixed-layer ocean model to modify the climate data. Right: simulations based on the same climate scenario, but including the physiological effects of a CO$_2$ concentration of 710 ppmv. After VEMAP Members (1995).

mitigated in both models by the physiological effects of CO_2. This mitigation was more effective in MAPSS. Note that the mitigation is an effect of CO_2 only; it does not apply to the other greenhouse gases.

BIOME2 also simulates the differential effects of CO_2 on photosynthesis in C_3 and C_4 plants. Warming alone pushes the area of potential C_4 grasslands northwards, but CO_2 doubling overwhelms this effect, favouring C_3 grasses down to the subtropics. The competition between C_4 grasses and woody plants is determined mainly by the climate change and only slightly by CO_2, according to this model. MAPSS simulates only the differential effects of CO_2 on water use, which in combination with greater simulated drought produces the opposite effect favouring C_4 plants up to the boreal zone.

BIOME2 and MAPSS probably bracket the plausible range for the sensitivity of moisture-related biome boundaries to temperature and CO_2, and for the response of C_3 versus C_4 plants to changes in CO_2. More realistic treatment of the coupling between carbon and water fluxes should help to resolve the discrepancies. These discrepancies represent an important uncertainty about the ecosystem impacts of climate change, comparable in magnitude with the uncertainty due to differences in the regional climate anomalies predicted by GCMs.

9.3.6 Changes in Biome Distribution Since the Last Glacial Maximum

Climatically controlled changes in biome distributions are well documented for the recent glacial–interglacial cycle, especially from the last glacial maximum (LGM, 21,000 yr bp) to the present (Wright *et al.*, 1993). Atmospheric CO_2 increased by about 90 ppmv to the pre-industrial value of 280 ppmv after the LGM (Chapter 2). Large changes occurred in global climate and vegetation patterns (Overpeck and Bartlein, 1989; Webb, 1992).

Climate simulations of the LGM are based mainly on the ice sheet distributions, sea level, sea surface temperatures and CO_2 (Kutzbach and Guetter, 1986; Broccoli and Manabe, 1987); the Earth's orbital configuration was also slightly different from present. BIOME results derived from the LGM climate simulation by Lautenschlager and Herterich (1990) agree qualitatively with palaeoclimatic and palaeoecological data (COHMAP Members, 1988; Street-Perrott *et al.*, 1989; Prentice *et al.*, 1993a). Thus, major changes in biome distribution after the LGM can be explained as a result of changes in climate. Direct effects of low CO_2 may also have been involved, due to reduced carbon assimilation or low water-use efficiency of C_3 plants at the LGM (Beerling and Woodward, 1993; Giresse *et al.*, 1994).

9.3.7 Dynamics of the Vegetational Response to Climate Change

The geographic distribution of biomes, and the composition of natural and managed vegetation types, will not remain in equilibrium with the changing climate during the next 100 years. There will be transient effects, varying according to the rate of climate change, which may depart considerably from the pattern indicated by biome models.

These effects depend on the particular mechanisms involved. Shifts in competitive balance might occur subtly over time. Direct impacts on the growth or survival of particular types of plant could cause die back (and carbon loss) before better-adapted types become established. This possible asymmetry of carbon loss and accumulation has led to the concern that transient vegetation changes could produce a CO_2 "spike" (King and Neilson, 1992; Smith *et al.*, 1992; Smith and Shugart, 1993). We return to this issue in Section 9.4.1.

Some aspects of vegetation's dynamic response to environmental changes, on time-scales from 10 to 10^4 years, have been established from palaeoecology (Prentice, 1986; Bennett, 1986; Huntley, 1988; Davis, 1990; Webb, 1992). For perennial plants, the evolution of new adaptations is too slow to be an effective response. Instead, changes occur both in the local abundances of species and genotypes (Type A response, Webb, 1986) and in the geographic distributions of species (Type B response) (Figure 9.3). In this process biomes do not change or move *en bloc*; instead, species react to climate change individually according to their biology. When climate changes to produce novel regional climates, novel associations of plants arise (Webb, 1992). Successive interglacial periods have therefore been characterised by different vegetation assemblages (Watts, 1988). The general directions of change are predictable from a knowledge of the climate change and the present-day climatic tolerances (realised niches) of the species.

Type A vegetation responses are caused primarily by the differential effects of climate on the growth and regeneration of different taxa and plant types. These responses can occur at different rates depending on the rate of climate change. In periods of rapid climate changes, Type A vegetation responses have been reported to occur within 150 yr (MacDonald *et al.*, 1993), 50–100 yr (Mayle and Cwynar, 1995) and even within 30 yr (Gear and Huntley, 1991; Zackrisson *et al.*, 1995). Type A vegetation responses can be modelled with vegetation-dynamics models such as forest succession models (Botkin *et al.*, 1972; Shugart, 1984). These represent regeneration, growth and mortality of statistical populations of individual plants

Observed pollen (radiocarbon dated)

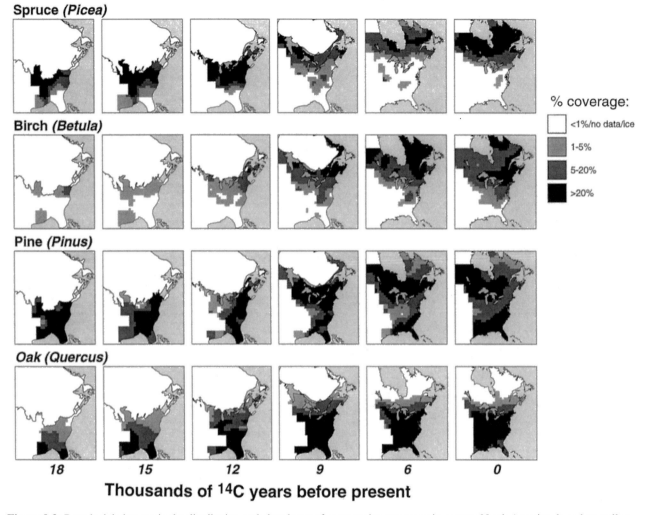

Figure 9.3: Postglacial changes in the distribution and abundance of some major tree types in eastern North America, based on pollen analysis data. After Webb *et al.* (1993).

on a plot whose size reflects the scale of competition. With no environmental change, such models simulate the natural cyclical behaviour of ecosystems (Shugart, 1984; Prentice and Leemans, 1990). Climate effects modify the growth and regeneration functions. Applied to the past, such models have shown responses of forest composition to climate change that agree well with changes shown in pollen records (Solomon and Bartlein, 1992; Campbell and McAndrews, 1993).

When driven by climate-change scenarios, forest succession models can describe the interaction of climate and natural cycles (Solomon, 1986; Prentice, *et al.*, 1991a). Various transient effects have appeared, including transient die backs of species whose growth becomes reduced (Solomon and Bartlein, 1992). However, many succession

models assume stylised climatic responses (e.g., dome-shaped responses of total annual growth to annual growing degree days) which have no physiological basis and may give misleading results (Bonan and Sirois, 1992). The growth equation used in many such models is also physiologically unrealistic (Moore, 1989), producing too sudden mortality in large trees. Because of these and other uncertainties, including the potential significance of short-lived climatic extremes, the structure of vegetation dynamics models is undergoing re-examination (Bugmann, 1993; Prentice *et al.*, 1993b).

Climate change may further alter ecosystem structure and composition by affecting the disturbance regime (Grimm, 1983; Torn and Fried, 1992; Clark, 1993). Natural disturbances generally facilitate plant migration and the attainment of compositional equilibrium with climate (Davis and Botkin, 1985; Bradshaw and Hannon, 1992). But climatically induced transitions between biomes may also show hysteresis, due to positive feedbacks that tend to maintain vegetation structure (Grimm, 1983). Such effects can cause delays of up to 100 yr (Grimm, 1984). The total response time of vegetational composition (combined Type A and B responses) has been estimated as between 300 and 1500 yr; fast enough to track the "envelope" of climate changes since the LGM (Overpeck and Bartlein, 1989; Prentice *et al.,* 1991b), but too slow to track interdecadal climate variability without considerable lag (Davis and Botkin, 1985; Campbell and McAndrews, 1993).

Faced with climate change, species presumably can (in the right conditions) spread at least as fast as they have done before. Species spread (migration) involves a number of factors including dispersal, regeneration on a suitable site, growth to maturity and seed production. Dispersal seems not to have been a major limitation in the past, at least to the major species detected over the long time-periods represented in the palaeo-record (see Biogeographic Dynamics Box). The mechanisms of rapid spread as observed in the Holocene may involve coalescence of many populations, each starting from an "infection centre" which could either be a pre-existing small population in a favourable microhabitat, or a new population founded by long-distance seed dispersal (birds, mammals and tornadoes are possible vectors; see references in Box). This way, species distribution ranges could expand orders of magnitude faster than they would if they advanced along an orderly front.

The palaeo-record tells us mainly about migration in a natural matrix. Future migrations may be very different. For example, the modern landscape provides fewer regeneration sites. Migrating species in the past exploited recently disturbed sites, whereas now natural vegetation is often fragmented (Peters, 1992) and confined to undisturbed sites. On the other hand, humans are already spreading many species beyond their natural ranges both deliberately and accidentally. Outlier populations spread by humans could be future infection centres even if they are not regenerating now (Davis and Zabinski, 1992). The net effect of human activities on plant species' ability to migrate is unclear.

Climate changes implied by the IS92 emission scenarios (see Chapter 6) call for migration rates up to ten times faster than historically observed for many taxa, so vegetation composition may be out of equilibrium with the changing climate (Davis, 1989, 1990; Davis and Zabinski, 1992). Non-equilibrium vegetation types could take many

forms. At one extreme, an existing vegetation type might persist until gradually invaded by other species, with minimal implications for the atmosphere. At the other extreme, large areas of forests might die (e.g., due to heat stress, drought and fire; Auclair and Carter, 1993) and be temporarily replaced by shrublands. This would result in different structural and functional properties, the transition being accompanied by changes in albedo, canopy roughness and rooting depth, reductions in *NPP*, and losses of plant and soil carbon to the atmosphere (Neilson, 1993; Section 9.4.1). The larger and more rapid the climate change, the greater the chance that effects of the latter type will occur.

**Biogeographic Dynamics:
The Issue of Dispersal Rate**

Rates of species spread in response to Holocene climate changes have been reconstructed by mapping pollen data from networks of ^{14}C-dated sediment cores. They range from 50–2000 m/yr for most woody species in Europe and North America (Davis, 1976; Huntley and Birks, 1983; Huntley, 1988). The puzzle is how species could spread so fast. Several explanations have been offered. For example, the rapid spread of *Picea glauca* into interior Canada immediately after deglaciation has been attributed to long-distance dispersal aided by strong anticyclonic winds (Ritchie and MacDonald, 1986). Migrations into already-vegetated regions that appear instantaneous over hundreds of kilometres, such as the spread of *Tsuga* into the Great Lakes region in response to precipitation and winter temperature increases ~ 7000 yr bp (Prentice *et al.,* 1991b), may be better described as infilling by the expansion of scattered "advance populations" (Clark, 1993; Davis, 1983; MacDonald *et al.,* 1993).

Rates of spread of different species within genera (e.g., *Picea*, *Fagus*) were similar regardless of whether the migration route lay in mountain regions or lowlands (Davis, 1983; Bennett, 1986). Species spread at comparable rates whether their seeds are adapted to wind- or animal transport (Davis, 1983). Migrating species were not stopped by water bodies such as Lake Michigan or the Baltic (Davis *et al.,* 1986; Huntley and Webb, 1989). The lesson for the future is that dispersal as such will not necessarily limit species migration rates. Other factors, such as the availability of sites for regeneration, may well provide greater barriers.

9.4 Effects of Climate Change and Carbon Dioxide Increases on Regional and Global Carbon Storage: Transient and Equilibrium Analyses

9.4.1 Possible Transient Effects of Climate Change on Global Carbon Storage

While biome shifts may have a major effect on terrestrial carbon storage during the climate transient, many other things may be happening at the same time that have the potential to affect terrestrial carbon storage. Continued forest clearing for agriculture to meet the food needs of a growing human population combined with an expansion and intensification of air-pollution stress on terrestrial ecosystems could further reduce carbon storage on land. Alternatively, mechanisms such as CO_2 fertilisation may continue to result in enhanced carbon storage in land areas not involved in biome shifts. Improved quantification of these transient processes is a research priority.

9.4.2 Equilibrium Analyses: Regional Ecosystem Model Projections

In a rapidly changing environment, there may be complex transient effects including possible releases of carbon due to climate-change-induced forest die back as mentioned in Section 9.3.7. At the present time there is no consensus on the likely magnitude of such releases. In a modelling study, Smith and Shugart (1993) began to explore how large they might be. In Smith and Shugart's model, climate-induced vegetation redistribution initially led to a large transient release of about 200 GtC during 100–200 years due to die back of forests. While an important conceptual advance, this transient analysis contains many crude assumptions such as very long lags due to slow migration. This estimate probably represents an upper bound to the effect of biome shifts alone in causing a transient loss of carbon from the terrestrial biosphere. Such losses would be expected to depend strongly on the rate of climate change; that is, the faster the rate of climate change, the greater the likelihood of large carbon losses from terrestrial ecosystems due to biome shifts during the transient (Woodwell, 1995).

A number of terrestrial biogeochemistry models are now capable of evaluating how terrestrial carbon storage might change with shifts in climate and atmospheric CO_2 concentrations from one equilibrium condition to another. Examples are BIOME–BGC (Hunt and Running, 1992; Running and Hunt, 1993), CENTURY (Parton *et al.*, 1987, 1988, 1993) and TEM (Raich *et al.*, 1991; McGuire *et al.*, 1992; Melillo *et al.*, 1993). In an application to the continental United States, these models were run for a range of future equilibrium climates at doubled CO_2. The

BIOME–BGC model projected terrestrial carbon losses up to 33% relative to the current condition, and CENTURY and TEM projected carbon gains of between 6 and 16% (VEMAP Members, 1995).

As part of the VEMAP activity, these three biogeochemistry models were coupled with three biogeography models[1] (see Section 9.3.5); BIOME2 (Haxeltine *et al.*, 1996), DOLY (Woodward and Smith, 1994; Woodward *et al.*, 1995), and MAPSS (Neilson, 1995; Neilson and Marks, 1995). Each of the model pairs was then run with three GCM-generated climate scenarios for doubled CO_2. The terrestrial carbon storage response ranged from a loss of 39% for the BIOME–BGC/MAPSS pair with a UKMO–GCM climate, to a 32% gain for the TEM/MAPSS pair with an OSU–GCM climate (Figure 9.4). The BIOME–BGC/MAPSS response for the UKMO climate scenario was primarily caused by decreases in forested area and temperature-induced water stress. The TEM/MAPSS response for the OSU climate was largely attributable to forest expansion and temperature-enhanced nitrogen cycling. This range of responses represents the pooled uncertainty for structural and functional ecosystem responses and for the regional climate anomalies inferred from three GCMs.

9.4.3 Equilibrium Analyses: Global Ecosystem Model Projections

Several process-based ecological models including TEM and IMAGE2 (Alcamo *et al.*, 1994) have been used to evaluate the effects of equilibrium climate and atmospheric carbon dioxide shifts on global terrestrial carbon storage. Melillo *et al.* (1995b) used a new version of TEM (McGuire *et al.*, 1995b) to assess the equilibrium response of global ecosystem carbon storage to a CO_2 doubling alone, with no climate change and no change in vegetation distribution. Under these conditions, terrestrial carbon storage was projected to increase by 360 GtC (Figure 9.5).

The TEM model was also used to simulate the response of terrestrial carbon storage with constant CO_2, but with CO_2-induced climate changes inferred from equilibrium simulations with the GFDL–GCM. The result was a projected net loss of terrestrial carbon of 130 Gt (Figure 9.5). This result was obtained under the assumption of no vegetation redistribution. When TEM was coupled to a modified version of BIOME (Prentice *et al.*, 1992), the

---

[1] Terrestrial biogeochemistry models simulate carbon, water and nutrient fluxes through ecosystems, assuming a prescribed vegetation structure. Thus they contrast with biome (or biogeography) models, which simulate the geographic distributions of vegetation structural types (biomes).

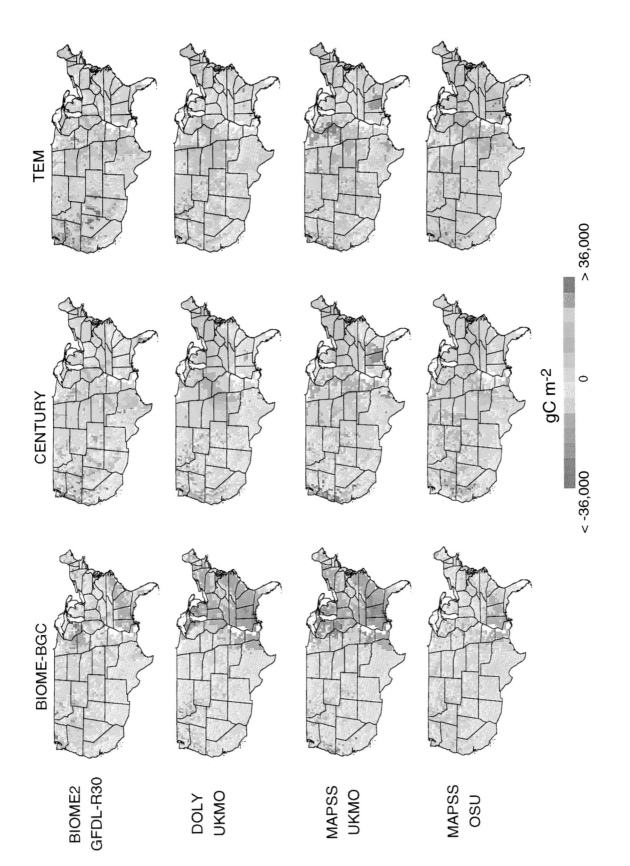

Figure 9.4: Simulated changes in equilibrium terrestrial carbon storage for the conterminous USA (VEMAP Members, 1995). Biogeochemistry models (BIOME-BGC, CENTURY and TEM) are run with the vegetation distributions of biogeography models (BIOME2, DOLY Y and MAPSS) for particular climate scenarios (GFDL-R30, UKMO and OSU).

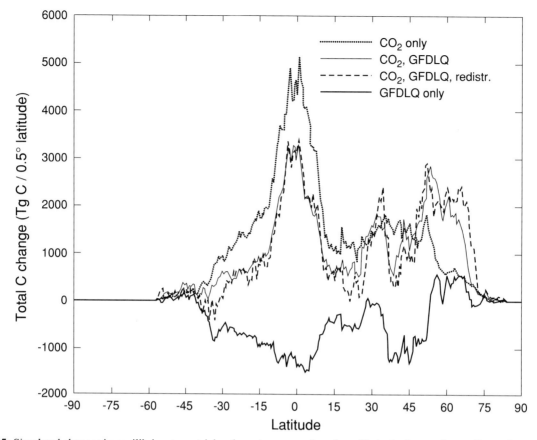

Figure 9.5: Simulated changes in equilibrium terrestrial carbon storage as a function of latitude due to: direct effects of a doubling of CO_2 (dotted line); effects of climate change, from the GFDL $2 \times CO_2$ scenario (thick solid line); combined effects of CO_2 and climate without vegetation redistribution (thin solid line); combined effects of CO_2 and climate with vegetation redistribution (dashed line). After Melillo *et al.* (1995b).

projected carbon loss was of a similar magnitude (Figure 9.5). Although some other modelling studies have projected long-term increases in carbon storage of up to about 100 GtC over several hundred years (Cramer and Solomon 1993; Smith and Shugart 1993; Prentice and Sykes 1995), due to climate change alone, these results are probably unrealistic. The TEM–BIOME result (Figure 9.5) includes the effects of both biogeochemistry and biogeography; the other results included only biogeographical changes, assuming fixed carbon densities in each biome.

Ecosystem models have also been used to evaluate the combined response of terrestrial carbon storage to CO_2 fertilisation and climate change. TEM (Melillo *et al.*, 1995b) projected a net increase of 290 GtC (Figure 9.5). The gain due to CO_2 fertilisation outweighed the loss due to warming. Again, allowing vegetation redistribution did not appreciably change this result (Figure 9.5). Alcamo *et al.* (1994) used the IMAGE2 model to assess the combined effects of CO_2 fertilisation and climate change plus various scenarios of land-use change. Their study projected an

increase in terrestrial carbon storage, in the range 200–250 Gt depending upon future land-use assumptions (Figure 9.6). These equilibrium model results suggest that the effect of CO_2 in increasing carbon storage could dominate over any warming-induced reduction in carbon storage, but this result must be considered as preliminary.

9.4.4 *The Palaeo-record and Implications for a Future Climate Equilibrium: Increasing Terrestrial Carbon Storage with Increases in Atmospheric Carbon Dioxide and Global Mean Temperature after the Last Glacial Maximum*

Between the last glacial maximum and the start of the present (Holocene) interglacial there were large increases both in atmospheric greenhouse gas concentrations and global mean temperature. CO_2 increased by about 90 ppmv. There may have been several causes of the increase in atmospheric CO_2, but they probably all included an ocean source (Siegenthaler, 1989; Heinze *et al.*, 1991; Peltier *et al.*, 1993; Archer and Maier-Reimer 1994;

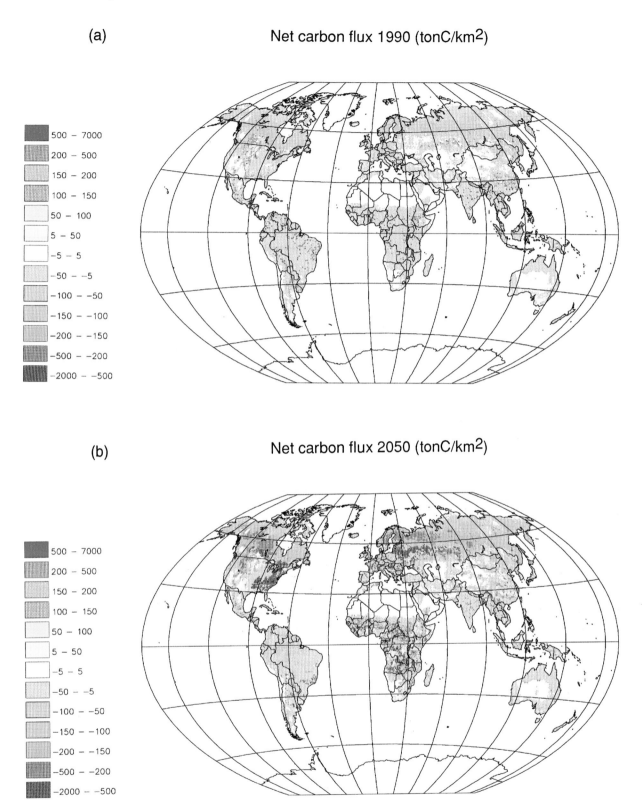

(a) Net carbon flux 1990 (tonC/km^2)

	500 – 7000
	200 – 500
	150 – 200
	100 – 150
	50 – 100
	5 – 50
	–5 – 5
	–50 – –5
	–100 – –50
	–150 – –100
	–200 – –150
	–500 – –200
	–2000 – –500

(b) Net carbon flux 2050 (tonC/km^2)

	500 – 7000
	200 – 500
	150 – 200
	100 – 150
	50 – 100
	5 – 50
	–5 – 5
	–50 – –5
	–100 – –50
	–150 – –100
	–200 – –150
	–500 – –200
	–2000 – –500

Figure 9.6: Net carbon flux from the terrestrial biosphere to the atmosphere simulated by the IMAGE2 integrated assessment model for (a) 1990 and (b) 2050 under a "conventional wisdom" scenario. The net flux includes deforestation and forest regrowth as well as climate and CO_2 effects. The large sinks shown in the northern temperate regions for 2050 are due to abandonment of agricultural land. After Alcamo *et al.* (1994).

Ganeshram *et al.*, 1995; Chapter 10). Along with the atmospheric CO_2 increase, global mean temperature increased by 4–7°C according to climate model estimates (Kutzbach and Guetter, 1986; Broccoli and Manabe, 1987; Kutzbach *et al.*, 1995). Palaeodata indicate local temperature increases on the order of 5°C or more in both temperate and tropical regions (Wright *et al.*, 1993; Stute *et al.*, 1995; Thompson *et al.*, 1995). Over the same period, global terrestrial carbon storage increased by 310–550 Gt (Duplessy *et al.*, 1988; Bird *et al.*, 1994).

For several reasons this is not a strict analogy for future CO_2 and climate changes. The detailed time courses of change in global mean temperature, CO_2 and especially carbon storage are not known. The increase in temperature was partly due to the disappearance of the ice sheets, and was about twice as large as would be expected due to the radiative-forcing changes alone (Lorius *et al.*, 1990). Deglaciation increased the area available for vegetation; however, a similar area of the exposed continental shelf was flooded. The increase in unglaciated land area from the LGM to the present was, at most, 2% (Prentice *et al.*, 1993b). The lesson is that a combined CO_2 increase and global warming can lead to a substantial (in this case, about 25%) increase in the amount of carbon stored in terrestrial vegetation and soils.

9.5 Methane: Effects of Climate Change and an Increase in Atmospheric CO_2 on Methane Flux and Carbon Balance in Wetlands

Methane (CH_4) is produced in flooded organic soils as a result of anaerobic respiration (methanogenesis), and CH_4 emissions from natural wetlands are estimated to contribute about 20% to the global emissions of this gas to the atmosphere (Chapter 2; Prather *et al.*, 1995). Climate change could either increase or decrease CH_4 flux from wetlands. Factors increasing CH_4 flux would include northward spread of peat-forming areas into the high latitudes (enhanced by increased precipitation) and faster carbon turnover due to warmer temperatures (Crill *et al.*, 1988; Christensen and Cox, 1995) and/or higher CO_2 (Hutchin *et al.*, 1995). Factors decreasing CH_4 flux would include drier conditions (lower water table) in extant peatlands (Whalen and Reeburgh, 1990; Roulet *et al.*, 1992), and drying-out and/or permafrost melting leading to loss of peat-forming areas in the continental interiors (Lachenbruch and Marshall, 1986; Oechel *et al.*, 1993; Oechel and Vourlitis, 1994; Gorham, 1995).

However, if CH_4 flux declines in some regions due to drying, this would imply an additional flux of CO_2 to the atmosphere due to enhanced aerobic respiration and,

perhaps, large-scale oxidation of the peat by erosion and fire (Hogg *et al.*, 1992). This scenario is of concern because as much as 450 GtC may be stored in high latitude peats (Gorham, 1991; Botch *et al.*, 1995). It has been suggested that the large warming that climate models predict for the high latitudes could thus threaten the integrity of this carbon store (Gorham, 1991, 1995). Like the possible carbon "spike" due to transient vegetation changes, this possible source of carbon to the atmosphere represents a potential positive feedback that has not been adequately quantified (Nisbet and Ingham, 1995).

Analysis of the palaeo-record of CH_4 support the idea that the net CH_4 flux from wetlands could either increase due to warming and CO_2 increase, or decline due to drying. Atmospheric CH_4 concentration increased (by 300 ppbv) after the LGM, and closely tracked global climate changes through the deglaciation (Stauffer *et al.*, 1988; Chapellaz *et al.*, 1990; Chapellaz *et al.*, 1993a). The increase in CH_4 concentration contributed to the global warming that followed the LGM (Lorius *et al.*, 1990; Lorius and Oeschger, 1994). In contrast to CO_2, the increase in CH_4 is thought to represent a positive feedback involving the terrestrial biosphere (Schimel *et al.*, 1995).

Model calculations suggest that the low CH_4 concentration at the LGM was primarily due to low CH_4 production by terrestrial ecosystems (Chapellaz *et al.*, 1993b), rather than to high OH concentrations in the atmosphere (Pinto, 1991; Thompson *et al.*, 1993; Crutzen and Bruhl, 1993; Martinerie *et al.*, 1995). Low CH_4 production would be expected at the LGM, because tropical and high latitude wetlands were less extensive (implying a reduced source area: Petit-Maire *et al.*, 1991) and temperatures and CO_2 were lower, both factors implying slower carbon turnover in the source area.

Atmospheric CH_4 concentration stood at 700–750 ppbv in the early Holocene, had fallen to < 600 ppbv by the mid-Holocene, then rose gradually in the late Holocene to once again reach the "pre-industrial" level of 700–750 ppbv. These variations may reflect a trade-off between the extent of tropical and high latitude wetlands (Blunier *et al.*, 1995). Tropical wetlands were most extensive in the early Holocene due to increased monsoonal precipitation (Petit-Maire *et al.*, 1991; Street-Perrott, 1992). High latitude wetlands have increased in extent during the late Holocene due to decreasing temperatures and evaporation (Ovenden, 1990; Zoltai and Vitt, 1990; Gorham, 1991; Botch *et al.*, 1995). The lesson is that warming in high latitudes can either increase or decrease natural CH_4 production, depending on the extent to which increased temperatures are matched by increased precipitation and/or CO_2.

9.6 Nitrous Oxide

Our current understanding of the global budget of nitrous oxide (N_2O) is reviewed in Chapter 2 and by Prather *et al.* (1995). The budget is largely controlled by microbial processes in soils. Today, the warm, moist soils of the tropical forests are probably the single most important source of N_2O. Land-use and the intensification of agriculture in the tropics appear to be increasing the size of the N_2O source from this region.

The microbial process responsible for the production of most of the N_2O is denitrification; the dissimilatory reduction of oxides of nitrogen that produces N_2 as well as N_2O. The rate of denitrification is controlled by oxygen (O_2), nitrate (NO_3) and carbon. Moisture has an indirect effect on denitrification by influencing O_2 content of soil. If other conditions are appropriate, then temperature becomes an important controller of denitrification.

Denitrifiers in natural environments are capable of producing either N_2O or N_2 as end products. Numerous factors have been reported to affect the proportion of N_2O produced relative to N_2. Perhaps most important are the relative supplies of nitrate and carbon (Firestone and Davidson, 1989). The dominant product of denitrification may be N_2O in systems where, at least for a time, nitrate supply is high and carbon supply is low, but not excessively so.

Probably due to a combination of more extensive wetland areas and increased rates of nitrogen cycling in terrestrial ecosystems under warmer and/or moister climates after the LGM, atmospheric N_2O concentration increased from ~200 ppbv to ~270 ppbv (Leuenburger and Siegenthaler, 1992). In the future, a wetter climate may lead to increased N_2O production, although this potential positive feedback has not been adequately quantified.

9.7 Global-Scale Biogeophysical Feedbacks: Changes in Ecosystem Structure and Function Affect Climate

9.7.1 Effects of Vegetation Structure on Land-surface Characteristics

Vegetation mediates the exchange of water and energy between the land surface and the atmosphere (Rind, 1984; Sud *et al.*, 1990; Hostetler *et al.*, 1994). Land–atmosphere interactions are physically represented by land-surface models designed for implementation in current GCMs (see Chapter 5, Section 5.3.2). These models require a global vegetation type (biome) distribution to be prescribed. Most climate change analyses with GCMs have assumed no change in vegetation distribution. However, both land-use change (e.g., deforestation and human-induced deforestation) and climatically induced natural vegetation change can significantly alter global vegetation distribution over decades to centuries.

The main land-surface parameters influenced by vegetation structure are surface albedo (normal and snow-covered), roughness length (affecting boundary-layer conductance), canopy conductance and rooting depth. Snow-free surface albedo for total short-wave radiation ranges from ~0.15 in closed forests to 0.4–0.5 in hot deserts (Henderson-Sellers and McGuffie, 1987). The largest effect of snow cover is on low vegetation types such as grasslands and tundra, where the snow-covered albedo can be up to 0.8. Roughness length increases with vegetation height: tall forests therefore present a boundary-layer conductance that is much larger than short grasslands. Canopy conductance is influenced by foliage density, plant nitrogen content, atmospheric CO_2 content and drought stress (Schulze *et al.*, 1994). At present ambient CO_2, most natural vegetation types have a maximum stomatal conductance of 3–6 mm/s while field crops have a higher conductance, up to 12 mm/s (Kelliher *et al.*, 1993). Canopy conductance increases asymptotically with leaf area index, towards a value of about 3–4 times stomatal conductance for a closed canopy (Schulze *et al.*, 1994). Maximum stomatal conductance is lowered under increased ambient CO_2. Stomatal closure under midday conditions of high evaporative demand acts to restrict canopy conductance. This closure occurs sooner as soil moisture supply is reduced and vapour pressure deficit increases.

9.7.2 Effects of Land-surface Changes on Climate

The sensitivity of climate to changes in these different land-surface properties varies regionally. For example, albedo effects are important in controlling precipitation in climatic regimes where precipitation is controlled by large-scale dynamics or convection; canopy conductance and rooting depth are important in regimes where a large proportion of precipitation arises by recycling of evapotranspiration from the land surface (Rind, 1984).

Sensitivity analyses with GCMs have been used to estimate the response of climate to major changes in the land surface, such as tropical deforestation (Dickinson, 1989; Shukla *et al.*, 1990; Nobre *et al.*, 1991; Henderson-Sellers *et al.*, 1993; Polcher and Laval, 1994). Through raising albedo and/or lowering evapotranspiration, large-scale deforestation tends to reduce moisture convergence and precipitation. The potential area of tropical rain forests and seasonal forests is therefore reduced. More generally, albedo exerts a strong control over evapotranspiration and precipitation in the tropics and subtropics (Charney, 1975; Charney *et al.*, 1977; Mylne and Rowntree, 1992).

Albedo changes in the high latitudes can also have major effects. Bonan *et al.* (1992) and Chalita and Le Treut (1994) examined the sensitivity of global climate to boreal deforestation (replacement of the boreal forests by tundra). The large increase in snow-covered albedo resulted in colder winters and a longer snow season. In Bonan *et al.*'s study, which included a mixed-layer ocean model, the thickness and duration of Arctic sea ice was increased and the cooling thereby extended from winter into summer. The total effect was felt across the Northern Hemisphere and summer temperatures became too cold for the persistence of boreal forests in their present range. Such sensitivity studies suggest that large biogeophysical effects of vegetation structure on climate could be brought into play by land-use change.

Climate-induced changes in biome distribution would presumably also cause feedbacks. Palaeoclimate simulations suggest that such feedbacks are important. Street-Perrott *et al.* (1990) showed that a GCM with surface boundary conditions kept as today could simulate the phenomenon, but not the extent and magnitude, of African monsoon expansion at 9000 yr bp. When a realistic expansion of vegetation across the present-day Sahara was included (*via* the albedo effect), precipitation increased by a further ~1 mm/day over large areas, including northern and western regions that were scarcely affected in the original simulation. Foley *et al.* (1994) showed that the warming-induced poleward expansion of boreal forests into what is now tundra at 6000 yr bp could have doubled the initial effects of orbital forcing (Figure 9.7). In both examples, biogeophysical feedbacks amplified the initial radiative forcing.

More detailed investigation of biogeophysical feedbacks requires coupling vegetation models with GCMs (Henderson-Sellers, 1993; Claussen, 1994). Claussen (1993) coupled the BIOME model asynchronously to an GCM, using an iterative procedure: 4–6 year ensembles of simulated climate were used to modify the biome distribution, which in turn modified the climate. The coupled system proved stable with respect to small perturbations of the biome distribution, as also found by Henderson-Sellers (1993). However a very large perturbation triggered a complex response in the atmospheric circulation and caused the coupled system to return to an alternative state in which the south-western Sahara remained wooded. With 6000 yr bp orbital forcing, the coupled system approached an even more extreme state in which most of the Sahara turned to savannah (Claussen

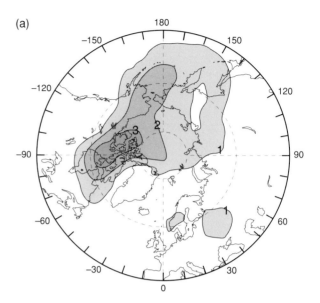

Δt (°C) – Annual average

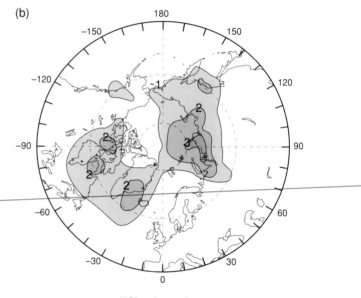

Δt (°C) – Annual average

Figure 9.7: Amplification of 6,000 yr bp radiative forcing due to biogeophysical feedback, as simulated with the GENESIS atmospheric/mixed-layer ocean model. (a) Increase in annual mean temperature due to radiative forcing alone. (b) *Additional* increase due to northward extension of the boreal forest. After Foley *et al.* (1994).

and Gayler, 1995). This state is similar to the 6000 yr bp situation shown by palaeoecological data.

CO_2-induced changes in canopy conductance are expected to have further feedback effects on climate

(Martin *et al.,* 1989; Field *et al.,* 1995). As ambient CO_2 increases, stomatal conductance declines. GCM sensitivity studies indicate that a global halving of surface conductance, with no change in leaf area, would lead to reduced evapotranspiration rates, increased surface air warming about 0.5°C averaged over terrestrial areas, compared with 1.1 to 2.5°C which is predicted for the combined effects of radiative forcing and aerosols (Chapter 6), and in some regions increase soil moisture storage (Henderson-Sellers *et al.,* 1995; Pollard and Thompson 1995). This latter effect of CO_2 could mitigate or even reverse the mid-latitude drying seen in some $2 \times CO_2$ simulations with fixed surface conductance (Chapter 6). Compensatory increases in leaf area would work against the effect of CO_2-induced reductions in stomatal conductance. However, a study with a coupled (single column) model of the soil-ecosystem-atmosphere system, including interactive adjustment of leaf area in response to CO_2, suggested that the stomatal conductance response would dominate over the compensatory leaf area response (Friend and Cox, 1995). Thus, the net "physiological" effect of CO_2 on climate would be to reduce evapotranspiration and increase soil moisture, relative to the scenarios based on radiative forcing alone.

Such studies underline the sensitivity of the simulated hydrological cycle to land-surface properties that are determined by ecosystem functional and structural responses to climate and CO_2. In particular, GCM simulations of regional changes in soil moisture in a high-CO_2 world are questionable because such simulations have not, as yet, taken into account changes in ecosystem structure and function that would have major feedback effects on the hydrological cycle. The uncertainty is compounded because most current GCMs do not resolve the vertical structure of the planetary boundary layer (PBL), which may limit the effect of changes in stomatal conductance on evapotranspiration (Jarvis and McNaughton, 1986; Martin, 1989; McNaughton and Jarvis, 1991; Monteith, 1995).

Proper representation of these feedbacks in models requires a physiologically based representation of the processes controlling canopy conductance (e.g., Collatz *et al.,* 1991; Friend and Cox, 1995), a sufficiently resolved representation of the PBL (e.g., Troen and Mahrt, 1986; MacNaughton and Jarvis, 1991; Jacobs and de Bruin, 1992), and these elements to be fully interactive in the GCM. Such a coupling of atmospheric and ecosystem processes appears to be a high priority because the potential for future drought is: (a) a key issue in assessing the biological and societal impacts of global change; and

(b) one of the main hypotheses underlying predictions of positive biogeochemical feedbacks leading to CO_2 release and further warming, as discussed in Section 9.2.3.1 and in Woodwell and Mackenzie (1995).

In conclusion, biogeophysical feedbacks involve interactions between the atmosphere and biosphere which can only be assessed quantitatively through the further development of coupled models including the terrestrial biosphere as an integral component of the climate system. Model sensitivity studies (including palaeoclimate simulation) suggest that these feedbacks are potentially of similar magnitude to the direct effects of changes in radiative forcing.

References

Aber, J.D., J.K. Nadelhoffer, P.A. Steudler and J.M. Melillo, 1989: Nitrogen saturation in northern forest ecosystems – hypotheses and implications. *BioSci.,* **39**, 378–386.

Aber, J.D., A. Magill, R. Boone, J.M. Melillo, P. Steudler and R. Bowden, 1993: Plant and soil responses to chronic nitrogen additions at the Harvard Forest, Massachusetts. *Ecol. Appl.,* **3** (1), 156–166.

Alcamo, J., G.J. Van den Born, A.F. Bouwman, B. de Haan, K. Klein-Goldewijk, O. Klepper, R. Leemans, J.A. Oliver, B. de Vries, H. van der Woerd and R.F. van den Wijngaard, 1994: Modeling the global society-biosphere-climate system. Part 2: Computed scenarios. *Water Air Soil Pollut.,* **76**, 37–78.

Allen, L.H. Jr., and J.S. Amthor, 1995: Plant physiological responses to elevated CO_2, temperature, air pollution, and UV-B Radiation. In: *Biotic Feedbacks in the Global Climatic System,* G.M. Woodwell and F.T. Mackenzie (eds), Oxford University Press, New York, pp. 51–84.

Archer, D. and E. Maier-Reimer, 1994: Effect of deep-sea sedimentary calcite preservation on atmospheric CO_2 concentration. *Nature,* **367**, 260–263.

Archer, S., D.S. Schimel and E.A. Holland, 1995: Mechanisms of shrubland expansion: land use, climate or CO_2? *Clim. Change* **29**, 91–99.

Arnone, J.A., III and J.C. Gordon, 1990: Effect of nodulation, nitrogen fixation, and CO_2 enrichment on the physiology, growth and dry mass allocation of seedlings of *Alnus rubra* Bong. *New Phytol.,* **116**, 55–66.

Arp, W.J., 1991: Effects of source–sink relations on photosynthetic acclimation to elevated CO_2. *Plant, Cell and Environ.,* **14**, 869–75.

Arp, W.J., B.G. Drake, W.T. Pockman, P.S. Curtis and D.F. Whigman, 1993: Interactions between C_3 and C_4 salt marsh plant species during four years of exposure to elevated atmospheric CO_2. *Vegetatio,* **104/105**, 133–143.

Arris, L.L., and P.S. Eagleson, 1994: A water use model for locating the boreal/deciduous forest ecotone in eastern North America. *Water Resour. Res.,* **30**, 1–9.

Auclair, A.N.D. and T.B. Carter, 1993: Forest wildfires as a recent source of CO_2 at northern latitudes. *Can. J. For. Res.*, **23**, 1528–1536.

Bazzaz, F.A. and E.D. Fajer, 1992: Plant life in a CO_2-rich world. *Sci. Am.*, **266**, 68–74.

Beerling, D.J. and F.I. Woodward, 1993: Ecophysiological responses of plants to global environmental change since the last glacial maximum. *New Phytol.*, **125**, 641–648.

Bennett, K.D., 1986: The rate of spread and population increase of forest trees during the postglacial. *Phil. Trans. R. Soc. London, B*, **314**, 523–31.

Bird, M.I., J. Lloyd and G.D. Farquhar, 1994: Terrestrial carbon storage at the LGM. *Nature*, **371**, 585.

Blunier, T., J. Chapellaz, J. Schwander, B. Stauffer and D. Raynaud, 1995: Variation in atmospheric methane concentration during the Holocene epoch. *Nature*, **374**, 46–49.

Bonan, G.B. and L. Sirois, 1992: Air temperature, tree growth, and the northern and southern range limits to *Picea mariana*. *J. Veg. Sci.*, **3**, 495–506.

Bonan, G.B.and K. Van Cleve, 1992: Soil temperature, nitrogen mineralization, and carbon source–sink relationships in boreal forests. *Can. J. For. Res.*, **22**, 629–639.

Bonan, G.B., D. Pollard and S.L. Thompson, 1992: Effects of boreal forest vegetation on global climate. *Nature*, **359**, 716–718.

Botch, M.S., K.I. Kobak, T.S. Vinson and T.P. Kolchugina, 1995: Carbon pools and accumulation in peatlands of the former Soviet Union. *Glob. Biogeochem. Cycles*, **9**, 37–46.

Botkin, D.B., J.F.Janak and J.R. Wallis, 1972: Some ecological consequences of a computer model of forest growth. *J. Ecol.*, **60**, 948–972.

Bowes, G., 1991: Growth at elevated CO_2: photosynthetic responses mediated through Rubisco. *Plant, Cell and Environ.*, **14**, 795–806.

Bradshaw, R.H.W. and G.E. Hannon, 1992: Climatic change, human influence and disturbance regime in the control of vegetation dynamics within Fiby forest, Sweden. *J. Ecol.*, **80**, 625–632.

Broccoli, A.J. and S. Manabe, 1987: The influence of continental ice, atmospheric CO_2 and land albedo on the climate of the last glacial maximum. *Clim. Dyn.*, **1**, 87–99.

Bugmann, H.K.M., 1993: On the Ecology of Mountainous Forests in a Changing Climate: A Simulation Study, Ph. D. thesis, Swiss Federal Institute of Technology, Zürich.

Campbell, I.D. and J.H. McAndrews, 1993: Forest disequilibrium caused by rapid Little Ice Age cooling. *Nature*, **366**, 336–338.

Ceulemans, R. and M. Mousseau, 1994: Effects of elevated atmospheric CO_2 on wood plants. *New Phytol.*, **127**, 425–426.

Chabot, B.F. and D.J. Hicks, 1982: The ecology of leaf life spans. *Ann. Rev. Ecol. Syst.*, **13**, 229–259.

Chalita, S. and H. Le Treut, 1994: The albedo of temperate and boreal forest and the northern hemisphere climate: A sensitivity experiment using the LMD GCM. *Clim. Dyn.*, **10**, 231–240.

Chameides, W.L., P.S. Kasibhatla, J. Yienger and H. Levy II, 1994: Growth of continental-scale metro-agro-plexes, regional ozone production, and world food production. *Science*, **264**, 74–77.

Chapellaz, J., J.M. Barnola, D. Raynaud, Y.S. Korotkevitch and C. Lorius, 1990: Ice core record of atmospheric methane over the past 160,000 years. *Nature*, **345**, 127–131.

Chapellaz, J., T. Blunier, D. Raynaud, J.M. Barnola, J. Schwander and B. Stauffer, 1993a: Synchronous changes in atmospheric CH_4 and Greenland climate between 40 and 8 kyr bp. *Nature*, **366**, 443–445.

Chapellaz, J.A., I.Y. Fung and A.M. Thompson, 1993b: The atmosphere CH_4 increase since the last glacial maximum:1. Source estimates. *Tellus*, **45B**, 228–241.

Chapin, F.S. III, 1991: Effects of multiple environmental stresses on nutrient availability and use. In: *Response of Plants to Multiple Stresses*, H.A. Mooney *et al.*, (eds.), Academic Press, San Diego, CA, pp. 67–88.

Chapin, F.S. III, G.R. Shaver, A.E. Giblin, K.J. Nadelhoffer and J.A. Laundre, 1995: Reponses of arctic tundra to experimental and observed changes in climate. *Ecology*, **76**, 694–711.

Charney, J.G., 1975: Dynamics of deserts and drought in the Sahel. *Quart. J. R. Met. Soc.*, **101**, 193–202.

Charney, J.G, W. Quirk, J. Chow and J. Kornfield, 1977: A comparative study of the effects of albedo change on drought in semi-arid regions. *J. Atmos. Sci.*, **34**, 1366–1385.

Christensen, T.R. and P. Cox, 1995: Response of methane emission from arctic tundra to climatic change: results from a model simulation. *Tellus*, **47B**, 301–310.

Ciais, P., P.P. Tans, J.W.C. White, M. Trolier, R.J. Francey, J.A. Berry, D.R. Randall, P.J. Sellers, J.G. Collatz and D.S. Schimel, 1995a: Partitioning of ocean and land uptake of CO_2 as inferred by $\delta^{13}C$ measurements from the NOAA Climate Moinitoring and Diagnostics Laboratory Global Air Sampling Network, *J. Geophys. Res.*, **100**, 5051–5057.

Ciais, P., P.P. Tans, M. Trolier, J.W.C. White and R.J. Francey, 1995b: A large northern hemisphere terrestrial CO_2 sink indicated by $^{13}C/^{12}C$ of atmospheric CO_2. *Science*, **269**, 1098–1102.

Clark, J.S., 1993: Paleoecological perspectives on modeling broad scale responses to global change. In: *Biotic Interactions and Global Change*, P.M. Kareiva, J.G. Kingsolver, R.B. Huey (eds.), Sinauer, Sunderland, MA, pp. 315–332.

Claussen, M., 1993: Shift of biome patterns due to simulated climate variability and climate change. *Max-Planck-Institut für Meteorologie, Report No. 115*, Hamburg , 30 pp.

Claussen, M., 1994: On coupling global biome models with climate models. *Clim. Res.*, in press.

Claussen, M. and M. Esch, 1994: Biomes computed from simulated climatologies. *Clim. Dyn.*, **9**, 235–243.

Claussen, M and V. Gayler. 1995: Modelling paleo and present-day vegetation patterns. *Annales Geophysicae*, Suppl. II to Volume 13, C357.

COHMAP Members, 1988: Climate changes of the last 18,000 years: Observations and model simulations. *Science,* **241,** 1043–1052.

Collatz, G.J., J.T. Ball, C. Grivet and J.A. Berry, 1991: Physiological and environmental regulation of stomatal conductance, photosynthesis and transpiration: a model that includes a laminar boundary layer. *Agric. For. Meteorol.,* **54,** 107–136.

Cramer, W.P. and A.M. Solomon, 1993: Climate classification and future global redistribution of agricultural land. *Clim. Res.,* **3,** 97–110.

Crill, P.M., K.B. Bartlett, R.C. Harriss, E. Gorham, E.S. Verry, D.I. Sebacher, L. Madzar and W. Sanner, 1988: Methane flux from Minnesota Peatlands. *Glob. Biogeochem. Cycles,* **2,** 371–384.

Crutzen, P.J. and C. Br*f*hl, 1993: A model study of atmospheric temperatures and concentrations of ozone, hydroxyl, and some other photochemically active gases during the glacial, the preindustrial Holocene, and the present. *Geophys. Res. Lett.,* **20,** 1047–1050.

Dahl, E., 1990: Probable effects of climatic change due to the greenhouse effect on plant productivity and survival in North Europe. In: *Effects of Climate Change on Terrestrial Ecosystems,* J.I. Holten (ed.), Nina Notat 4, 7–17. Norwegian Institute for Nature Research, Trondheim.

Dai, A. and I.Y. Fung, 1993: Can climate variability contribute to the "missing" CO_2 sink? *Glob. Biogeochem. Cycles,* **7,** 599–609.

Davis, M.B., 1976: Pleistocene biogeography of temperate deciduous forests. *Geoscience and Man,* **13,** 13–26.

Davis, M.B., 1983: Quatenary history of deciduous forests of eastern North America and Europe. *Annals of the Missouri Botanical Garden,* **70,** 550–563.

Davis, M.B., 1989: Lags in vegetation response to greenhouse warming. *Clim. Change,* **15,** 75–82.

Davis, M.B., 1990: Climatic change and the survival of forest species. In: *The Earth in Transition: Patterns and Processes of Biotic Impoverishment,* G.M. Woodwell (ed.), Cambridge University Press, Cambridge, UK, pp. 99–110.

Davis, M.B. and D.B. Botkin, 1985: Sensitivity of cool-temperate forests and their fossil pollen record to rapid temperature change. *Quat. Res.,* **23,** 327–40.

Davis, M.B. and C. Zabinski, 1992: Changes in geographical range resulting from greenhouse warming: effects on biodiversity in forests. In: *Global Warming and Biological Diversity,* R.L. Peters and T.E. Lovejoy (eds.), Yale University Press, New Haven, pp. 297–308.

Davis, M.B., K.D. Woods, S.L. Webb and R.P. Futyma, 1986: Dispersal versus climate: expansion of Fagus and Tsuga into the Upper Great Lakes region. *Vegetatio,* **67,** 93–103.

Denning, A.S., I.Y. Fung and D. Randall, 1995: Latitudinal gradient of atmospheric CO_2 due to seasonal exchange with land biota. *Nature,* **376,** 240–243.

Diaz, S., J.P. Grime, J. Harris and E. McPherson, 1993: Evidence of a feedback mechanism limiting plant response to elevated carbon dioxide. *Nature,* **364,** 616–617.

Dickinson, R.E., 1989: Modeling the effects of Amazonian deforestation on regional surface climate: a review. *Agric. For. Meteorol.,* **47,** 339–347.

Dixon, R.K., S.A. Brown, R.A. Houghton, A.M. Solomon, M.C. Trexler and J. Wisniewski, 1994: Carbon pools and flux of global forest ecosystems. *Science,* **263,** 185–190.

Drake, B.G., 1992a: A field study of the effects of elevated CO_2 on ecosystem processes in a Chesapeake Bay wetland. *Austr. J. Bot.,* **40,** 579–595.

Drake, B.G., 1992b: The impact of rising CO_2 on ecosystem production. *Water Air Soil Pollut.,* **64,** 25–44.

Duce, R.A., P.S. Liss, J.T. Merrill, E.L. Atlas, P. Buat-Menard, B.B. Hicks, J.M. Miller, J.M. Prospero, R. Arimoto, T.M. Church, W. Ellis, J.N. Galloway, L. Hansen, T.D. Jickells, A.H. Knap, K.H. Reinhardt, B. Schneider, A. Soudine, J.J. Tokos, S. Tsunogai, R. Wollast and M. Zhou, 1991: The atmospheric input of trace species to the world ocean. *Glob. Biogeochem. Cycles,* **5,** 193–259.

Duplessy, J.C., N.J. Shackleton, R.G. Fairbanks, L. Labeyrie, D. Oppo and N. Kallel, 1988: Deepwater source variations during the last climatic cycle and their impact on the global deepwater circulation. *Paleoceanography,* **3,** 343–360.

Eagleson, P.S., 1978: Climate, soil and vegetation 6. Dynamics of the annual water balance. *Water Resour. Res.,* **14,** 749–764.

Eagleson, P.S. and R.I. Segarra, 1985: Water-limited equilibrium of savanna vegetation systems. *Water Resour. Res.,* **21,** 1483–1493.

Eamus, D. and P.G. Jarvis, 1989: The direct effects of increase in the global atmsopheric CO_2 concentration on natural and commercial temperate trees and forests. *Adv. Ecol. Res.,* **19,** 1–55.

Ehleringer, J.R. and R.K. Monson, 1993: Evolutionary and ecological aspects of photosynthetic pathway variation. *Ann. Rev. Ecol. Syst.,* **24,** 411–439.

Emanuel, W.R., H.H. Shugart and M.P. Stevenson, 1985: Climatic change and the broad-scale distribution of terrestrial ecosystem complexes. *Clim. Change,* **7,** 29–43.

Farquhar, G.D., J. Lloyd, J.A. Taylor, L.B. Flanagan, J.P. Syvertsen, K.T. Hubick, S.C. Wong and J.R. Ehleringer, 1993: Vegetation effects on the isotopic composition of oxygen in atmospheric CO_2. *Nature,* **363,** 439–443.

Fetcher, N., C.H. Jaeger, B.R. Strain and N. Sionit, 1988: Long-term elevation of atmospheric CO_2 concentration and the carbon exchange rates of saplings of *Pinus* and *Liquidamber styraciflua* (L.). *Tree Physiol.,* **4,** 255–262.

Field, C.B., R.B. Jackson and H.A. Mooney, 1995: Stomatal responses to increased CO_2: Implications from the plant to the global scale. *Plant, Cell Environ.,* in press.

Firestone, M.K. and E.A. Davidson, 1989: Microbial basis for NO and N_2O production and consumption in soil. In: *Exchange of Trace Gases between Terrestrial Ecosystems and the Atmosphere,* M.O. Andreae and D.S. Schimel (eds.), John Wiley & Sons, Ltd., Chichester, pp. 7–22.

Foley, J.A., 1994: Sensitivity of the terrestrial biosphere to climatic change: A simulation of the middle holocene. *Glob. Biogeochem. Cycles,* **8,** 405–425.

Foley, J.A., J.E. Kutzbach, M.T. Coe and S. Levis, 1994: Feedbacks between climate and boreal forests during the mid-Holocene. *Nature,* **371,** 52–54.

Francey, R.J. and P.P. Tans, 1987: Latitudinal variation in oxygen-18 in atmospheric CO_2. *Nature,* **327,** 495–497.

Francey, R.J., P.P. Tans, C.E. Allison, I.G. Enting, J.W.C. White and M. Trolier, 1995: Changes in oceanic and terrestrial carbon uptake since 1982. *Nature,* **373,** 326–330.

Friend, A.D. and P.M. Cox, 1995: Modeling the effects of atmospheric CO_2 on vegetation–atmosphere interactions. *Agric. For. Meteorol.,* **73,** 285–295.

Galloway, J.N., W.H. Schlesinger, H. Levy II, A. Michaels and J.L. Schnoor, 1995: Nitrogen fixation: Anthropogenic enhancement–environmental response. *Glob. Biogeochem. Cycles,* **9,** 235–252.

Ganeshram, R.S., T.F. Pedersen, S.F. Calvert and J.W. Murray, 1995: Large changes in oceanic nutrient inventories from glacial to interglacial periods. *Nature,* **376,** 755–758.

Gear, A.J. and B. Huntley, 1991: Rapid changes in the range limits of Scots pine 4000 years ago. *Science,* **251,** 544–547.

Gifford, R.M., 1979: Growth and yield of CO_2-enriched wheat under water-limited conditions. *Aust. J. Plant Physiol.,* **6,** 367–378.

Gifford, R.M., 1993: Implications of CO_2 effects on vegetation for the global carbon budget. In: *The Global Carbon Cycle,* M. Heimann (ed.), Proceedings of the NATO Advanced Study Institute, Il Ciocco, Italy, September 8–20, 1991, pp. 165–205.

Gifford, R.M., 1994: The global carbon cycle: A viewpoint on the missing sink. *Austr. J. Plant Physiol.,* **21,** 1–15.

Giresse, P., J. Maley and P. Brenac, 1994: Late Quaternary paleoenvironments in the Lake Barombi Mbo (West Cameroon) deduced from pollen and carbon isotopes of organic matter. *Paleogeogr., Paleoclim., Paleoecol.,* **107,** 65–78.

Gorham, E., 1991: Northern peatlands: role in the carbon cycle and probable responses to climatic warming. *Ecol. Appl.,* **2,** 182–195.

Gorham, E., 1995: The biogeochemistry of northern peatlands and its possible response to global warming. In: *Biotic Feedback in the Global Climatic System,* G.M. Woodwell and F.T. Mackenzie (eds.), Oxford University Press, New York, pp. 169–187.

Grimm, E.C., 1983: Chronology and dynamics of vegetation change in the prairie-woodland region of southern Minnesota, U.S.A. *New Phytol.,* **93,** 311–350.

Grimm, E.C., 1984: Fire and other factors controlling the Big Woods vegetation of Minnesota in the mid-nineteenth century. *Ecol. Monogr.,* **54,** 291–311.

Groves, R.H. and O.B. Williams, 1981: Natural grasslands. In: *Australian Vegetation,* R.H. Groves (ed), Cambridge University Press, Cambridge, UK, pp. 293–316.

Gunderson, C.A. and S.D. Wullschleger, 1994: Photosynthetic acclimation in trees to rising atmospheric CO_2: A broader perspective. *Photosynth. Res.,* **39,** 369–388.

Hall, C.A.S. and J. Uhlig, 1991: Refining estimates of carbon released from tropical land-use change. *Can. J. For. Res.,* **21,** 118–131.

Harmon, M.E., W.K. Ferrel and J.F. Franklin, 1990: Effects on carbon storage of the conversion of old-growth forests to young forests. *Science,* **247,** 699–702.

Harte, J. and R. Shaw, 1995: Shifting dominence in a montane vegetation community: results of a climate-warming experiment. *Science,* **267,** 876–880.

Haxeltine, A., I.C. Prentice and I.D. Cresswell, 1996: A coupled carbon and water flux model to predict vegetation structure. *J. Veg. Sci.,* in press.

Heck, W.W., 1984: Defining gaseous pollution problems in North America. In: *Gaseous Air Pollutants and Plant Metabolism,* M.J. Koziol and F.R. Whatley (eds.), Butterworths, London, pp. 35–48.

Heinze, C., E. Maier-Reimer and K. Winn, 1991: Glacial pCO_2 reduction by the world ocean: experiments with the Hamburg ocean model. *Paleoceanography ,* **6,** 395–430.

Henderson-Sellers, A., 1993: Continental vegetation as a dynamic component of global climate models: a preliminary assessment. *Clim. Change,* **23,** 337–378.

Henderson-Sellers, A. and K. McGuffie, 1987: *A Climate Modeling Primer.* J. Wiley & Sons, New York, 217pp.

Henderson-Sellers, A. and K. McGuffie, 1995: Global climate models and 'dynamic' vegetation changes. *Global Change Biology,* **1,** 63–76.

Henderson-Sellers, A., R.E. Dickinson, T.B. Durbidge, P.J. Kennedy, K. McGuffie and A.J. Pitman, 1993: Tropical deforestation: modeling local- to regional-scale climate change. *J. Geophys. Res.,* **98,** 7289–7315.

Henderson-Sellers, A., K. McGuffie and C. Gross, 1995: Sensitivity of global climate model simulations to increased stomatal resistance and CO_2 increases. *J. Climate,* in press.

Hogg, E.H., V.J. Lieffers and R.W. Wein, 1992: Potential carbon losses from peat profiles. *Ecol. Appl.,* **2,** 298–306.

Hostetler, S.W., F. Giorgi, G.T. Bates and P.J. Bartlein, 1994: Lake-atmosphere feedbacks associated with Paleolakes Bonneville and Lahontan. *Science,* **263,** 665–668.

Houghton, R.A., 1991: Tropical deforestation and atmospheric carbon dioxide. *Clim. Change,* **19,** 99–118.

Houghton, R.A., 1995: Effects of land-use change, surface temperature, and CO_2 concentration on terrestrial stores of carbon. In: *Biotic Feedbacks in the Global Climatic System,* G.M. Woodwell and F.T. Mackenzie (eds), Oxford University Press, New York, pp. 333–366.

Houghton, R.A. and G.M. Woodwell, 1989: Global climatic change. *Sci. Am.,* **260,** 36–47.

Houghton, R.A., R.D. Boone, J.R. Fruci, J.E. Hobbie, J.M. Melillo, C.A. Palm, B.J. Peterson, G.R. Shaver, G.M. Woodwell, B. Moore, D.L. Skole and N. Myers, 1987: The flux of carbon from terrestrial ecosystems to the atmosphere in 1980 due to changes in land use: Geographic distribution of the global flux. *Tellus,* **39B,** 122–139.

Hsaio, T., 1973: Plant responses to water stress. *Annu. Rev. Plant Physiol.*, **24**, 519–570.

Hudson, R.J.M., S.A. Gherini and R.A. Goldstein, 1994: Modeling the global carbon cycle: Nitrogen fertilization of the terrestrial biosphere and the "missing" CO_2 sink. *Glob. Biogeochem. Cycles,* **8**, 307–333.

Hunt, E.R. Jr. and S.W. Running, 1992: Simulated dry matter yields for aspen and spruce stands in the North American Boreal Forest. *Can. J. Rem. Sens.*, **18**, 126–133.

Huntley, B., 1988: Europe. In: *Vegetation History*, B. Huntley and T. Webb III (eds.), Kluwer Academic Publishers, Dordrecht, pp. 341–383.

Huntley, B. and H.J.B. Birks, 1983: *An Atlas of Past and Present Pollen Maps for Europe: 0–13000 B.P.* Cambridge University Press, Cambridge, UK.

Huntley, B. and T. Webb III, 1989: Migration: species' response to climatic variations caused by changes in the earth's orbit. *J. Biogeogr.*, **16**, 5–19.

Hutchin, P.A., M.C. Press, J.A. Lee and T.W. Ashenden, 1995: Elevated concentrations of CO_2 may double methane emissions from mires. *Global Change Biology*, **1**, 125–128.

Idso, S.B. and B.A. Kimball, 1993: Tree growth in carbon dioxide enriched air and its implications for global carbon cycling and maximum levels of atmospheric CO_2. *Glob. Biogeochem. Cycles,* **7**, 537–555.

Idso, K.E. and S.B. Idso, 1994: Plant responses to atmospheric CO_2 enrichment in the face of environmental constraints: a review of the last 10 years' research. *Agric. For. Meteorol.*, **69**, 153–203.

IPCC, 1994: *Climate Change 1994: Radiative Forcing of Climate Change and an Evaluation of the IPCC IS92 Emission Scenarios*, J.T. Houghton, L.G. Meira Filho, J. Bruce, Hoesung Lee, B.A. Callander, E. Haites, N. Harris and K. Maskell (eds.). Cambridge University Press, Cambridge, UK.

IPCC WGII, 1995: *Climate Change 1995–Impacts, Adaptations and Mitigations of Climate Change: Scientific-Technical Analyses:The Second Assessment Report of the Inter-Governmental Panel on Climate Change.* R.T. Watson, N.C. Zinyowera and R.H. Moss (eds.). Cambridge University Press, New York, USA.

Jackson, R.B., O.E. Sala, C.B. Field and H.A. Mooney, 1994: CO_2 alters water use, carbon gain, and yield in a natural grassland. *Oecologia*, **98**, 257–262.

Jacobs, C.M.J. and H.A.R. de Bruin, 1992: The sensitivity of regional transpiration to land-surface characteristics: significance of feedback. *J. Climate*, **5**, 683–698.

Jarvis, P.G., 1989: Atmospheric carbon dioxide and forests. *Phil. Trans. R. Soc. London B*, **324**, 369–392.

Jarvis, P.G. and K.G. McNaughton, 1986: Stomatal control of transpiration: scaling up from leaf to region. *Adv. Ecol. Res.*, **15**, 1–49.

Johnson, H.B., W. Polley and H.S. Mayeux, 1993: Increasing CO_2 and plant–plant interactions: effects on natural vegetation. *Vegetatio*, **104/105**, 157–170.

Kauppi, P.E., K. Mielkäinen and K. Kuusela, 1992: Biomass and carbon budget of European forests. *Science,* **256**, 70–74.

Keeling, C.D., R.B. Bacastow, A.F. Carter, S.C. Piper, T.P. Whorf, M. Heimann, W.G. Mook and H. Roeloffzen, 1989: A three dimensional analysis of atmospheric CO_2 transport based on observed winds. I. Analysis of observational data. In: *Aspects of Climatic Variability in the Pacific and Western Americas*, J.H. Peterson (ed.), Geophysical Monograph 55, American Geophysical Union, Washington, DC, pp 165–236.

Keeling, C.D., T.P. Whorf, M. Wahlen and J. van der Plicht, 1995: Interannual extremes in the rate of rise of atmospheric carbon dioxide since 1980. *Nature*, **375**, 666–670.

Keeling, R.F. and S.R. Shertz, 1992: Seasonal and interannual variations in atmsopheric oxygen and implications for the global carbon cycle. *Nature*, **358**, 723–727.

Kelliher, F.M., R. Leuning, R. and E.-D. Schulze, 1993: Evaporation and canopy characteristics of coniferous forests and grasslands. *Oecologia*, **95**, 153–163.

Kimball, B.A., 1975: Carbon dioxide and agricultural yield: An assemblage and analysis of 430 prior observations. *Agronomy Journal*, **75**, 779–788.

Kimball, B.A. and S.B. Idso, 1983: Increasing atmospheric CO_2: Effects on crop yield, water use and climate. *Agric. Water Manage.*, **7**, 55–72.

King, G.A. and R.P. Neilson, 1992: The transient response of vegetation to climate change: a potential source of CO_2 to the atmosphere. *Water Air Soil Pollut.*, **64**, 365–383.

Kinzig, A.P. and R.H. Socolow, 1994: Human impacts on the Nitrogen Cycle. *Physics Today*, **47**(11), 24–31.

Kirschbaum, M.V.F., 1995: The temperature dependence of soil organic matter decomposition and the effect of global warming on soil organic carbon storage. *Soil Biol. Biochem.*, **27**, 753–760.

Kohlmaier, G.H., A. Janecek and M. Plöchl, 1988: Modeling response of vegetation to both excess CO_2 and airborne nitrogen compounds within a global carbon cycle model. In: *Advances in Environmental Modeling*, A. Marani (ed.), Elsevier, New York, pp. 207–234.

Körner, C. and M. Diemer, 1994: Evidence that plants from high altitudes retain their greater photosynthetic efficiency under elevated CO_2. *Funct. Ecol.*, **8**, 58–68.

Kurz, W.A. and M.J. Apps, 1995: Retrospective assessment of carbon flows in Canadian boreal forests. In: *Forest Ecosystems, Forest management and the Global Carbon Cycle*, M.J. Apps and D.T. Price (eds.), Springer-Verlag, NATO Advanced Science Institute Series, in press.

Kurz, W.A., M.J. Apps, B.J. Stocks and W.J.A. Volney, 1995: Global climate change: disturbance regions and tropospheric feedbacks of temperate and boreal forests. In: *Biotic Feedbacks in the Global Climatic System*, G. M. Woodwell, G.M. and F.T. Mackenzie (eds.), Oxford University Press, New York.

Kutzbach, J.E. and P. Guetter, 1986: The influence of changing orbital parameters and surface boundary conditions on climate simulations for the past 18,000 years. *J. Atmos. Sci.*, **43**, 1726–1759.

Kutzbach, J.E., R. Gallimore, S.P. Harrison, P. Behling, R. Selin and F. Laurif, 1995: Climate and biome simulations for the past 21,000 years. *Quant. Sci. Rev.*, in press.

Lachenbruch, A.H. and B.V. Marshall, 1986: Changing climate: Geothermal evidence from permafrost in the Alaskan Arctic. *Science*, **234**, 689–696.

Larcher, W., 1983: *Physiological Plant Ecology*, 2nd edn. Springer-Verlag, Berlin.

Lauenroth, W.K., D.L. Urban, D.P. Coffin, W.J. Parton, H.H. Shugart, T.B. Kirchner and T.M. Smith, 1993: Modeling vegetation structure–ecosystem process interactions across sites and ecosystems. *Ecol. Model.*, **67**, 49–80.

Lautenschlager, M. and K. Herterich, 1990: Atmospheric response to ice age conditions – climatology near the earth's surface. *J. Geophys. Res.*, **95**, 22547–22557.

Leuenburger, M. and U. Siegenthaler, 1992: Ice-age atmospheric concentration of nitrous oxide from an Antarctic ice core. *Nature*, **360**, 449–451.

Lewis, J.D., R.B. Thomas and B.R. Strain, 1994: Effect of elevated CO_2 on mycorrhizal coloniation of loblolly pine (*Pinus taeda* L.) seedlings. *Plant and Soil*, **165**, 81–88.

Lloyd, J. and G.D. Farquhar, 1994: ^{13}C discrimination during CO_2 assimilation by the terrestrial biosphere. *Oecologia*, **99**, 201–215.

Lloyd, J. and G.D. Farquhar, 1995: The CO_2 dependence of photosynthesis, plant growth responses to elevated atmospheric CO_2 concentrations, and their interaction with soil nutrient status. I. General principles and forest ecosystems. *Funct. Ecol.*, in press.

Lloyd, J. and J.A. Taylor, 1994: On the temperature dependence of soil respiration. *Funct. Ecol.*, **8**, 315–323.

Long, S.P. and P.R. Hutchin, 1991: Primary production in grasslands and coniferous forests with climate change: An overview. *Ecol. Appl.*, **1**, 139–156.

Lorius, C. and H. Oeschger, 1994: Paleo-perspectives: reducing uncertainties in global change? *Ambio*, **23**, 3036.

Lorius, C., J. Jouzel, D. Raynaud, J. Hansen and H. Le Treut, 1990: The ice-core record: Climate sensitivity and future greenhouse warming. *Nature*, **347**, 139–145.

Luo, Y., C.B. Field and H.A. Mooney, 1994: Predicting responses of phyotosynthesis and root fraction to elevated [CO_2]: interactions among carbon, nitrogen, and growth. *Plant, Cell and Environ.*, **17**, 1195–1204.

MacDonald, G.M., T.W.D. Edwards, K.A. Moser, R. Pienitz and J.P. Smol, 1993: Rapid response of treeline vegetation and lakes to past climatic warming. *Nature*, **361**, 243–46.

Martin, P.H., 1989. The significance of radiative coupling between vegetation and atmosphere. *Agric. For. Meteorol.*, **49**, 45–53.

Martin, P., N.J. Rosenberg and M.S. McKenney, 1989: Sensitiviy of evapotranspiration in a wheat field, a forest, and a grassland to changes in climate and the direct effects of carbon dioxide. *Clim. Change*, **14**, 117–151.

Martinerie, P., G.P. Brasseur and C. Granier, 1995: The chemical composition of ancient atmospheres: a model study constrained by ice core data. *J. Geophys. Res.*, **100**, 14291–12304.

Masle, J., G.S. Hudson and M.R. Badger, 1993: Effects of ambient CO_2 concentration on growth and nitrogen use in tobacco (*Nicotiana tabacum*) plants transformed with an antisense gene to the small subunit of ribulose-1, 5-bisphosphate carboxylase/oxygenase. *Plant Physiology*, **103**, 1075–1088.

Mayle, F.E. and L.C. Cwyner, 1995: Impact of the Younger Dryas cooling event upon lowland vegetation of Maritime Canada. *Ecol. Monogr.*, **65**, 129–154.

McGuire, A.D., J.M. Melillo, L.A. Joyce, D.W. Kicklighter, A.L. Grace, B. Moore III and C.J. Vörösmarty, 1992: Interactions between carbon and nitrogen dynamics in estimating net primary productivity for potential vegetation in North America. *Glob. Biogeochem. Cycles*, **6**, 101–124.

McGuire, A.D., L.A. Joyce, D.W. Kicklighter, J.M. Melillo, G. Esser and C.J. Vorosmarty, 1993: Productivity response of climax temperate florests to elevated temperature and carbon dioxide: A North American comparison between two global models. *Clim. Change*, **24**, 287–310.

McGuire, A.D., J.M. Melillo and L.A. Joyce, 1995a: The role of nitrogen in the response of forest net primary production to elevated atmospheric carbon dioxide. *Annu. Rev. Ecol. Syst.*, **26**, 473–503.

McGuire, A.D., J.M. Melillo, D.W. Kicklighter and L.A. Joyce, 1995b: Equilibrium responses of soil carbon to climate change: Empirical and process-based estimates. *Glob. Ecol. Biogeogr. Lett.*, in press.

McNaughton, K.G. and P.G. Jarvis, 1991: Effects of spatial scale on stomatal control of transpiration. *Agric. For. Meteorol.*, **54**, 279–301.

Melillo, J.M., 1995: Human influences on the global nitrogen budget and their implications for the global carbon budget. In: *Toward Global Planning of Sustainable Use of the Earth: Development of Global Eco-Engineering*, S. Murai and M. Kimura (eds.), Elsevier, Amsterdam, pp. 117–134.

Melillo, J.M., and J.R. Gosz, 1983: Interactions of biogeochemical cycles in forest ecosystems. In: *The Major Biogeochemical Cycles and Their Interactions*, B. Bolin and R.B. Cook (eds.), John Wiley and Sons, New York, pp. 177–222.

Melillo, J.M, J.R. Fruce, R.A. Houghton, B. Moore and D.L. Skole, 1988: Land-use change in the Soviet Union between 1850 and 1980: Causes of a net relase of CO_2 to the atmosphere. *Tellus*, **40B**, 116–128.

Melillo, J.M., P.A. Steudler, J.D. Aber and R.D. Bowden, 1989: Atmospheric deposition and nutrient cycling. In: *Exchange of Trace Gases between Terrestrial Ecosystems and the Atmosphere*, M.O. Andreae and D.S. Schimel (eds), Dahlem Conference Proceedings, John Wiley and Sons, New York and Chichester, pp. 263–280.

Melillo, J.M., T.V. Callaghan, F.I. Woodward, E. Salati and S.K.

Sinha, 1990: Effects on ecosystems. In: *Climate Change: The IPCC Scientific Assessment*, J.T. Houghton, G.J. Jenkins and J.J. Ephraums, (eds), Cambridge University Press, Cambridge, UK, pp. 283–310.

Melillo, J.M., A.D. McGuire, D.W. Kicklighter, B. Moore III, C.J. VörÖsmarty and A.L. Schloss, 1993: Global climate change and terrestrial net primary production. *Nature*, **363**, 234–240.

Melillo, J.M., D.W. Kicklighter, A.D. McGuire, W.T. Peterjohn and K. Newkirk, 1995a: Global change and its effects on soil organic carbon stocks. In: *Dahlem Conference Proceedings*, John Wiley and Sons, New York, John Wiley & Sons, Ltd., Chichester, pp. 175–189.

Melillo, J.M., D.W. Kicklighter, A.D. McGuire, B. Moore III, C.J. Vörösmarty and A.L. Schloss, 1995b: The effect of CO_2 fertilization on the storage of carbon in terrestrial ecosystems: a global modeling study. *Glob. Biogeochem. Cycles*, (submitted)

Mitchell, J.F.B., S. Manabe, V. Melesko and T. Tokioka, 1990: Equilibrium climate change and its implications for the future. In: *Climate Change: The IPCC Scientific Assessment* , J.T. Houghton, G.J. Jenkins and J.J. Ephraums, (eds.), Cambridge University Press, Cambridge, UK, pp. 131–172.

Monserud, R.A., N.M. Tchebakova and R. Leemans, 1993: Global vegetation change predicted by the modified Budyko model. *Clim. Change, **25**, 59–83.

Monteith, J.L., 1995: Accommodation between transpiring vegetation and the convective boundary layer. *J. Hydrol.*, **166**, 251–263.

Mooney, H.A., B.G. Drake, R.J. Luxmoore, W.C. Oechel and L.F. Pitelka, 1991: Predicting ecosystem responses to elevated CO_2 concentrations. *BioScience*, **41**, 96–104.

Moore, A.D., 1989: On the maximum growth equation used in forest gap simulation models. *Ecol. Model.*, **45**, 63–67.

Morrison, J.I.L. and R.M. Gifford, 1984: Plant growth and water use with limited water supply in high CO_2 concentration: II Plant dry weight, partitioning and water use efficiency. *Austr. J. Plant Physiol.*, **11**, 374–384.

Mylne, M.F. and P.R. Rowntree, 1992: Modeling the effects of albedo change associated with tropical deforestation. *Clim. Change*, **21**, 317–343.

Nadelhoffer, K.J., M.R. Downs, B. Fry, J.D. Aber, A.H. Magill and J.M. Melillo, 1995: The fate of ^{15}N labeled nitrate additions to a northern hardwood forest in eastern Maine, USA. *Oecologia* (in press).

Neilson, R.P., 1993: Vegetation redistribution: a posssible biosphere source of CO_2 during climatic change. *Water, Air and Soil Pollut.*, **70**, 659–673.

Neilson, R.P., 1995: A model for predicting continental scale vegetation distribution and water balance. *Ecol. Appl.*, **5**, 362–385.

Neilson, R.P. and D. Marks, 1995: A global perspective of regional vegetation and hydrologic sensitivites from climate change. *J. Veg. Sci.*, **5**, 715–730.

Nie, D., M.B. Kirkham, L.K. Ballou, D.J. Lawlor and E.T. Kanemasu, 1992: Changes in prairie vegetation under elevated carbon dioxide levels and two soil moisture regimes. *J. Veg. Sci.*, **3**, 673–678.

Nisbet, E.G. and B. Ingham, 1995: Methane output from natural and quasinatural sources: a review of the potential for change and for biotic and abiotic feedbacks. In: *Biotic Feedbacks in the Global Climatic System*, G.M. Woodwell and F.T. Mackenzie (eds.), Oxford University Press, New York, pp. 190–218.

Nobre, C.A., P.J. Sellers and J. Shukla, 1991: Amazonian deforestation and regional climatic change. *J. Climate*, **4**, 957–988.

Norby, R.J., 1987: Nodulation and nitrogenase activity in nitrogen-fixing woody plants stimulated by CO_2 enrichment of the atmosphere. *Physiol. Plant.*, **71**, 77–82.

Oechel, W.C. and B.R. Strain, 1985: Native species responses to increased atmospheric carbon dioxide concentrations. In: *Direct Effects of Increasing Carbon Dioxide on Vegetation.*, B.R. Strain and J.D. Cure (eds.), DOE/ER–0238, U.S. Department of Energy, Carbon Dioxide Research Division, Washington, D.C., pp. 117–154.

Oechel, W.C. and G.L. Vourlitis, 1994: The effects of climate change on land-atmosphere feedbacks in arctic tundra regions. *TREE*, **9**, 324–329.

Oechel, W.C., S.J. Hastings, G. Vourlitis, M. Jenkins, G. Riechers and N. Grulke, 1993: Recent change of Arctic tundra ecosystems from a net carbon dioxide sink to a source. *Nature*, **361,** 520–523.

Oechel, W.C., S. Cowles, N. Grulke, S.J. Hastings, B. Lawrence, T. Prudhomme, G. Riechers, B. Strain, D. Tissue and G. Vourlitis, 1994: Transient nature of CO_2 fertilization in Arctic tundra. *Nature*, **371**, 500–503.

Olson, J.S., A. Watts and L.J. Allison, 1983: *Carbon in Live Vegetation of Major World Ecosystems*. ORNLO–5861, Oak Ridge National Laboratory, Oak Ridge, Tennessee.

O'Neill, E.G., 1994: Responses of soil biota to elevated atmospheric carbon dioxide. *Plant and Soil*, **165**, 55–65.

O'Neill, E.G., R.J. Luxmoore and R.J. Norby, 1987: Increases in mycorrhizal colonization and seedling growth in *Pinus echinata* and *Quercus alba* in an enriched CO_2 atmosphere. *Can. J. For. Res.*, **17**, 878–883.

Ormrod, D.P., 1982: Air pollutant interactions in mixtures. In: *Effects of Air Pollution in Agriculture and Horticulture*, M.H. Unsworth and D.P. Ormrod (eds.), Butterworths, London, pp. 307–331.

Ovenden, L., 1990: Peat accumulation in northern wetlands. *Quat. Res.*, **33**, 377–386.

Overpeck, J.T. and P.J. Bartlein, 1989: Assessing the response of vegetation to future climate change: ecological response surfaces and paleoecological model validation. In: *The potential effects of global climate change on the United States. Appendix D. Forests*, United States Environmental Protection Agency Report PM–221.

Overpeck, J.T., P.J. Bartlein and T. Webb III, 1991: Potential magnitude of future vegetation change in eastern North America: comparisons with the past. *Science*, **254**, 692–695.

Owensby, C.E., P.I. Coyne, J.M. Ham, L.M. Auen and A.K. Knapp, 1993: Biomass production in a tallgrass prairie ecosystem exposed to ambient and elevated CO_2. *Ecol. Appl.*, **3**, 644–653.

Parton, W.J., D.S. Schimel, C.V. Cole and D.S. Ojima, 1987: Analysis of factors controlling soil organic matter levels in Great Plains grasslands. *Soil Sci. Soc. Am. J.*, **51**, 1173–1179.

Parton, W.J., J.W.B. Stewart and C.V. Cole, 1988: Dynamics of C, N, P and S in grassland soils: A model. *Biogeochem.*, **5**, 109–131.

Parton, W.J., J.M.O. Scurlock, D.S. Ojima, T.G. Gilmanov, R.J. Scholes, D.S. Schimel, T.B. Kirchner, J.-C. Menaut, T. Seastedt, E. Garcia Moya, Apinan Kamnalrut and J.I. Kinyamario, 1993: Observations and modeling of biomass and soil organic matter dynamics for the grassland biome worldwide. *Glob. Biogeochem. Cycles*, **7**, 785–809.

Parton, W.J., J.M.O. Scurlock, D.S. Ojima, D.S. Schimel, D.O. Hall and SCOPEGRAM Group Members, 1995: Impact of climate change on grassland production and soil carbon worldwide. *Global Change Biology*, **1**, 13–22.

Pastor, J. and W.M. Post, 1988: Response of northern forests to CO_2-induced climate change. *Nature*, **334**, 55–58.

Peltier, W.R., C.A. Burga, J.-C. Duplessy, K. Herterick, I. Levin, E. Maier-Reimer, M. McElroy, J.T. Overpeck, D. Raynaud and U. Siegenthaler, 1993: Group report: how can we use paleodata to evaluate the internal variability and feedbacks in the climate system? In: *Global Changes in the Perspective of the Past*, J.A. Eddy and H. Oeschger (eds.), John Wiley & Sons, Ltd., Chichester, pp. 239–263.

Peterjohn, W.T., J.M. Melillo, F.P. Bowles and P.A. Steudler, 1993: Soil warming and trace gas fluxes: Experimental design and preliminary flux results. *Oecologia*, **93**, 18–24.

Peterjohn, W.T., J.M. Melillo, P.A. Steudler, K.M. Newkirk, F.P. Bowles and J.D. Aber, 1994: Responses of trace gas fluxes and N availability to experimentally elevated soil temperatures. *Ecol. Appl.*, **4**, 617–625.

Peters, R.L., 1992: Conservation of biological diversity in the face of climate change. In: *Global Warming and Biological Diversity*, R.L. Peters and T.E. Lovejoy (eds), Yale University Press, New Haven, pp. 15–30.

Peterson, B.J. and J.M. Melillo, 1985: The potential storage of carbon caused by eutrophication of the biosphere. *Tellus*, **37B**, 117–127.

Petit-Marie, N., M. Fontugne and C. Rouland, 1991: Atmospheric methane ratio and environmental changes in the Sahar and Sahel during the last 130 kyrs. *Paleogeogr., Paleoclimatol., Paleoecol.*, **86**, 197–204.

Pinto, J.P., 1991: The stability of tropospheric OH during ice ages, interglacial epochs and modern times. *Tellus*, **43B**, 347–352.

Polcher, J. and K. Laval, 1994: A statistical study of the regional impact of deforestation on climate in the LMD GCM. *Clim. Dyn.* **10**, 205–219.

Pollard, D. and S.L. Thompson, 1995: The effect of doubling stomatal resistance in a global climate model. *Global and Planetary Change*, in press.

Polley, H.W., H.B. Johnson, B.D. Marino and H.S. Mayeux, 1993: Increase in C_3 plant water-use efficiency and biomass over Glacial to present CO_2 concentrations. *Nature*, **361**, 61–63.

Poorter, H., 1993: Interspecific variation in the growth response of plants to an elevated ambient CO_2 concentration. *Vegetatio*, **104/105**, 77–97.

Prather, M., R. Derwent, D. Ehhalt, P. Fraser, E. Sanhueza and X. Zhou, 1995: Other trace gases and atmospheric chemistry. In: *Climate Change 1994*. J.T. Houghton, L.G. Meira Filho, J. Bruce, H. Lee, B.A. Callander, E. Haites, N. Harris and K. Maskell (eds.), Cambridge University Press, Cambridge, UK, pp. 72–126.

Prentice, I.C., 1986: Vegetation responses to past climatic variation mechanisms and rates. *Vegetatio*, **67**, 131–141.

Prentice, I.C. and R. Leemans, 1990: Pattern and process and the dynamics of forest structure: a simulation approach. *J. Ecol.*, **78**, 340–55.

Prentice, I.C. and M.T. Sykes, 1995: Vegetation geography and global carbon storage changes. In: *Biotic Feedbacks in the Global Climate System*, G.M. Woodwell and F.T. Mackenzie (eds.), Oxford University Press, New York, pp. 304–312.

Prentice, I.C., M.T. Sykes and W. Cramer, 1991a: The possible dynamic response of northern forests to global warming. *Glob. Ecol. Biogeogr. Let.*, **1**, 129–135.

Prentice, I.C., P.J. Bartlein and T. Webb III, 1991b: Vegetation change in eastern North America since the last glacial maximum: a response to continuous climatic forcing. *Ecology*, **72**, 2038–2056.

Prentice, I.C., W. Cramer, S.P. Harrison, R. Leemans, R.A. Monserud and A.M. Solomon, 1992: A global biome model based on plant physiology and dominance, soil properties and climate. *J. Biogeogr.*, **19**, 117–134.

Prentice, I.C., M.T. Sykes, M. Lautenschlager, S.P. Harrison, O. Denissenko and P.J. Bartlein, 1993a: Modeling global vegetation patterns and terrestrial carbon storage at the last glacial maximum. *Glob. Ecol. Biogeogr. Lett.*, **3**, 67–76.

Prentice, I.C., M.T. Sykes and W. Cramer, 1993b: A simulation model for the transient effects of climate change on forest landscapes. *Ecol. Model.*, **65**, 51–70.

Prentice, K. and I.Y. Fung, 1990: The sensitivity of terrestrial carbon storage to climate change. *Nature*, **346**, 48–50.

Raich, J.W., E.B. Rastetter, J.M. Melillo, D.W. Kicklighter, P.A. Steudler, B.J. Peterson, A.L. Grace, B. Moore III and C.J. Vörösmarty, 1991: Potential net primary productivity in South America: Application of a global model. *Ecol. Appl.*, **1**, 399–429.

Reich, P.B., M.B. Walter and D.S. Ellsworth, 1992: Leaf life-span in relation to leaf, plant, and stand characteristics among diverse ecosystems. *Ecol. Monogr.*, **62**, 365–392.

Rind, D., 1984: The influence of vegetation on the hydrologic cycle in a global climate model. In: *Climate Processes and Climate Sensitivity*, J.E. Hansen and T. Takahashi (eds.), Geophysical Monographs 29, Volume 5. American Geophysical Union, Washington, DC, pp. 73–91.

Ritchie, J.C. and G.M. MacDonald, 1986: The patterns of post-glacial spread of white spruce. *J. Biogeogr.*, **13**, 527–540.

Rotmans, J. and M.G.J. den Elzen, 1993: Modeling feedback mechanisms in the carbon cycle: balancing the carbon budget. *Tellus*, **45B**, 301–320.

Roulet, N.T., T.R. Moore, J. Bubier and P. Lafleur, 1992: Northern fens: methane flux and climatic change. *Tellus.*, **44B**, 100–105.

Running, S.W. and R.R. Nemani, 1991: Regional hydrologic and carbon balance responses of forests resulting from potential climate change. *Clim. Change*, **19**, 349–368.

Running, S.W. and E.R. Hunt Jr., 1993: Generalization of a forest ecosystem process model for other biomes, BIOME–BGC, and an application for global-scale models. In: *Scaling Processes Between Leaf and Landscape Levels*, J.R. Ehleringer and C. Field (eds.), Academic Press, Orlando, pp. 141–158.

Sage, R.F., T.D. Sharkey and J.R. Seeman, 1989: Acclimation of photosynthesis to elevated CO_2 in five C_3 species. *Plant Physiol.*, **89**, 590–596.

Sage, R.F., 1994: Acclimation of photosynthesis to increasing atmospheric CO_2: The gas exchange perspective. *Photosynth. Res.* **39**, 351–68

Sakai, A. and C.J. Weiser, 1973: Freezing resistance of trees in N. America with reference to tree regions. *Ecology*, **54**, 118–126.

Schimel, D.S., B.H. Braswell, Jr., E.A. Holland, R. McKeown, D.S. Ojima, T.H. Painter, W.J. Parton and A.R. Townsend, 1994: Climatic, edaphic and biotic controls over storage and turnover of carbon in soils. *Glob. Biogeochem. Cycles*, **8**, 279–293.

Schimel, D.S., I.G. Enting, M. Heimann, T.M.L. Wigley, D. Raynaud, D. Alves and U. Siegenthaler, 1995: CO_2 and the carbon cycle. In: *Climate Change 1994: Radiative Forcing of Climate Change and an Evaluation of the IPCC IS92 Emission Scenarios*, J.T. Houghton, L.G. Meira Filho, J. Bruce, H. Lee, B.A. Callander, E. Haites, N. Harris and K. Maskell (eds.), Cambridge University Press, Cambridge, UK, pp. 35–71.

Schindler, D.W. and S.E. Bayley, 1993: The biosphere as an increasing sink for atmospheric carbon: estimates from increased nitrogen deposition. *Glob. Biogeochem. Cyc.les* **7**, 717–734.

Schulze, E.-D., 1982: Plant life forms and their carbon, water and nutrient relations. *Encyclopedia of Plant Physiology*, **12B**, 615–676, Springer-Verlag, Berlin.

Schulze, E.-D., 1994: Flux control at the ecosystem level. *Trends Ecol. Evol.*, **10**, 40–43.

Schulze, E.-D. and F.S. Chapin III, 1987: Plant specialization to environments of different resource availability. In: *Potentials and Limitations in Ecosystem Analysis*, E.-D. Schulze and H. Zwolfer (eds), Springer-Verlag, Berlin, pp. 120–148.

Schulze, E.-D., W. DeVries, M. Hauhs, K. Rosén, L. Rasmussen, O.-C. Tann and J. Nilsson, 1989: Critical loads for nitrogen deposition in forest ecosystems. *Water, Air, Soil Pollut.*, **48**, 451–456.

Schulze, E.-D., F.M. Kelliher, C. Körner, J. Lloyd and R. Leuning, 1994: Relationships between maximum stomatal conductance, ecosystem surface conductance, carbon assimilation rate and plant nitrogen nutrition. *Ann. Rev. Ecol. Syst.*, **25**, 629–660.

Sedjo, R.A., 1992: Temperate forest ecosystems in the global carbon cycle. *Ambio*, **21**, 274–277.

Shaver, G.R. and F.S. Chapin III, 1991: Production: Biomass relationships and element cycling in contrasting arctic vegetation types. *Ecol. Monogr.*, **61**, 1–31.

Shaver, G.R., W.D. Billings, F.S. Chapin III, A.E. Giblin, K.J. Nadelhoffer, W.C. Oechel and E.B. Rastetter, 1992: Global change and the carbon balance of Arctic ecosystems. *BioSci.*, **42**, 433–441.

Shugart, H.H., 1984: *A Theory of Forest Dynamics*. Springer-Verlag, New York.

Shukla, J., C. Nobre and P. Sellers, 1990: Amazon deforestation and climate change. *Science*, **247**, 1322–1325.

Siegenthaler, U., 1989: Glacial-interglacial atmospheric CO_2 variations. In: *Global Changes of the Past*, R.S. Bradley (ed.), UCAR/OIES, pp. 245–260.

Smith, T.M. and H.H. Shugart, 1993: The transient response of terrestrial carbon storage to a perturbed climate. *Nature*, **361**, 523–526.

Smith, T.M., H.H. Shugart, G.B. Bonan and J.B. Smith, 1992: Modeling the potential response of vegetation to global climate change. *Adv. Ecol. Res.*, **22**, 93–116.

Smith, T.M., W.P. Cramer, R.K. Dixon, R. Leemans, R.P. Nielsen and A.M. Solomon, 1993: The global terrestrial carbon cycle. *Water ,Air ,Soil Pollut.*, **70**, 19–37.

Solomon, A.M., 1986: Transient response of forests to CO_2-induced climate change: simulation modeling experiments in eastern North America. *Oecologia*, **68**, 567–579.

Solomon, A.M. and P.J. Bartlein, 1992: Past and future climate change: response by mixed deciduous-coniferous forest ecosystems in northern Michigan. *Can. J. For. Res.*, **22**, 1727–1738.

Solomon, A.M., K.C. Prentice, R. Leemans and W.P. Cramer, 1993: The interaction of climate and land use in future terrestrial carbon storage and release. *Water, Air, Soil Pollut.* **70**, 595–614.

Specht, R.L., 1972: Water use by perennial evergreen plant communities in Australia and Papua, New Guinea. *Austr. J. Bot.*, **20**, 273–299.

Stauffer, B., E. Lochbronner, H. Oeschger and J. Scwander, 1988: Methane concentration in the glacial atmosphere was only half that of the preindustrial Holocene. *Nature*, **332**, 812–814.

Street-Perrott, F.A., 1992: Atmospheric methane: tropical wetland sources. *Nature*, **355**, 23–24.

Street-Perrott, F.A., D.S. Marchand, N. Roberts and S.P. Harrison, 1989: *Global lake-level variations from 18,000 to 0 years ago: a paleoclimatic analysis.* U.S. Department of Energy Report ER/6–304-H1.

Street-Perrott, F.A., J.F.B. Mitchell, D.S. Marchand and J.S. Brunner, 1990: Milankovitch and albedo forcing of the tropical monsoons: a comparison of geological evidence and numerical simulations for 9000 yr bp. *Trans. R. Soc. Edinburgh: Earth Sci.,* **81**, 407–427.

Stute, M., M.Foster, H. Frischkom, A. Serejo, J.F. Clark, P. Schlosser, W.S. Broecker and G. Bonani, 1995: Cooling of tropical Brazil (5°C) during the last glacial maximum. *Science,* **179**, 371–383.

Sud, Y.C., P.J. Sellers, Y. Mintz, M.D. Chou, G.K. Walker and W.E. Smith, 1990: Influence of the biosphere on the global circulation and hydrologic cycle – a GCM simulation experiment. *Agric. For. Meteorol.,* **52**, 133–180.

Tans, P.P., I.Y.Fung and T. Takahashi, 1990: Observational constraints on the global atmospheric carbon dioxide budget. *Science,* **247**, 1431–1438.

Tans, P.P., I.Y. Fung and I.G. Enting, 1995: Storage versus flux budgets: The terrestrial uptake of CO_2 during the 1980s. In: *Biotic Feedbacks in the Global Climatic System,* G.M. Woodwell and F.T. Mackenzie (eds.), Oxford University Press, New York, pp. 351–374.

Taylor, J.A., and J. Lloyd, 1992: Sources and sinks of CO_2. *Austr. J. Bot.,* **40**, 407–418.

Tchebakova, N.M., R.A. Monserud, R.A. Leemans and D.I. Nazimova, 1992: Possible vegetation shifts in Siberia under climatic change. In: *Impacts of Climate Change on Ecosystems and Species,* R. Leemans, O. Elder and J. Pernetta (eds.), Chapman & Hall, New York.

Thompson, A.M. J.A. Chapellaz, I.Y. Fung and T.L. Kuscera, 1993: The atmospheric CH_4 increase since the last glacial maximum: 2. Interaction with oxidents. *Tellus,* **45B**, 242–257.

Thompson, L.G., E. Mosley-Thompson, M.E. Davis, P.-N. Lin, K.A. Henderson, J.Cole-Dau, J.F. Bolzan and L.-B. Liu, 1995: Late glacial stage and Holocene tropical ice core records from Huascarçn, Peru. *Science,* **269**, 46–50.

Tissue, D.T. and W.C. Oechel, 1987: Response of *Eriophorum vaginatum* to elevated CO_2 and temperature in the Alaskan arctic tundra. *Ecology,* **68**(2), 401–410.

Torn, M.S. and J.S. Fried, 1992: Predicting the impacts of global warming on wildland fire. *Clim. Change,* **21**, 257–274.

Townsend, A.R., P.M. Vitousek and E.A. Holland, 1992: Tropical soils could dominate the short-term carbon cycle feedbacks to increased global temperatures. *Clim. Change,* **22**, 293–303.

Townsend, A.R., B.H. Braswell, E.A. Holland and J.E. Penner, 1994: Spatial and temporal patterns in potential terrestrial carbon storage resulting from deposition of fossil fuel derived nitrogen. *Ecological Applications,* (in press).

Troen, I.B. and L. Mahrt, 1986: A simple model of the atmospheric boundary layers: sensitivity to surface evaporation. *Boundary-Layer Meteorology,* **37**, 129–148.

VEMAP Members, 1995: VEMAP: A comparison of biogeography and biogeochemistry models in the context of global climate change. *Glob. Biogeochem.* Cycles, **9**, 407–437.

Vitousek, P.M. and R.W. Howarth, 1991: Nitrogen limitation on land and in the sea: how can it occur? *Biogeochem.* **13**, 87–115.

Walker, B.H. and I. Noy-Meir, 1982: Aspects of stability and resilience of savanna ecosystems. In: *Ecology of Tropical Savannas,* B. Huntley (ed.), Springer-Verlag, Berlin, pp. 577–590.

Walter, H., 1979: *Vegetation of the Earth and Ecological Systems of the Geo-biosphere,* 2nd edn. Springer-Verlag, New York.

Watts, W.A., 1988: Europe. In: *Vegetation History,* B. Huntley and T. Webb III (eds.), Kluwer Academic Publishers, Dordrecht, pp. 155–192.

Webb, T. III, 1986: Is vegetation in equilibrium with climate? How to interpret late-Quaternary pollen data. *Vegetatio,* **67**, 75–91.

Webb, T. III 1992: Past changes in vegetation and climate: lessons for the future. In: *Global Warming and Biological Diversity,* R.L. Peters and T.E. Lovejoy (eds.), Yale University Press, New Haven, pp. 59–75.

Webb, T. III and P.J. Bartlein, 1992: Global changes during the last 3 million years: climatic controls and biotic responses. *Ann. Rev. Ecol. Syst.,* **23**, 141–173.

Webb, T, III, P. J. Bartlein, S. P. Harrison and K. H. Anderson, 1993: Vegetation, lake levels, and climate in eastern North America for the past 18,000 years. In: *Global Climates Since the Last Glacial Maximum.* H. E. Wright, J. E. Kutzbach, T. Webb III, W. F. Ruddiman, F. A. Street-Perrott and P. J. Bartlein (eds.), University of Minnesota Press, Minneapolis, pp. 415–467.

Whalen, S.C. and W.S. Reeburgh, 1990: Consumption of atmospheric methane by tundra soils. *Nature,* **346**, 160–162.

Wofsy, S.C., J.E. Munger, P.S. Bakwin, B.C. Daube and T.R. Moore, 1993: Net CO_2 uptake by northern woodlands. *Science,* **260**, 1314–1317.

Wong, S.C., 1979: Elevated atmospheric partial pressures of CO_2 and plant growth: I. Interactions of nitrogen nutrition and photosynthetic capacity in in C_3 and C_4. *Oecologia,* **44**, 68–74.

Woodward, F.I., 1987: *Climate and Plant Distribution.* Cambridge University Press, Cambridge, UK.

Woodward, F.I., 1992: A review of the effects of climate on vegetation: ranges, competition, and composition. In: *Global Warming and Biological Diversity,* R.L. Peters and T.E. Lovejoy (eds.), Yale University Press, New Haven, pp. 105–123.

Woodward, F.I. and T.M. Smith, 1994: Predictions and measurements of the maximum photosynthetic rate at the global scale. In: *Ecophysiology of Photosynthesis,* E.-D. Schulze and M.M. Caldwell (eds.), *Ecological Studies,* vol. 100, Springer-Verlag, Berlin, pp. 491–509.

Woodward, F.I., T.M. Smith and W.R. Emanuel, 1995: A global primary productivity and phytogeography model. *Glob. Biogeochem. Cycles* , **9**, 471–490.

Woodwell, G.M., 1983: Biotic effects on the concentration of atmospheric carbon dioxide: A review and projection. In: *Changing Climate*, National Academy of Science Press, Washington, DC, pp. 216–241.

Woodwell, G.M., 1989: The warming of the industrialized middle latitudes 1985–2050: Causes and consequences. *Clim. Change,* **15**, 31–50.

Woodwell, G.M., 1995: Will the warming speed the warning? In: *Biotic Feedbacks in the Global Climatic System*, G.M. Woodwell and F.T. Mackenzie (eds), Oxford University Press, New York.

Woodwell, G.M. and F.T. Mackenzie, 1995: *Biotic Feedbacks in the Global Climatic System*, Oxford University Press, New York, 416 pp.

Wright, H.E. Jr., J.E. Kutzbach, T. Webb III, W.F. Ruddiman, F.A. Street-Perrott and P.J. Bartlein, 1993: *Global Climates since the Last Glacial Maximum*. University of Minnesota Press, Minneapolis.

Wullschleger, S.D., W.M. Post and A.W. King, 1995: On the potential for a CO_2 fertilization effect in forest trees – an assessment of 58 controlled-exposure studies and estimates of the biotic growth factor. In: *Biotic Feedbacks in the Global Climatic System*, G.M. Woodwell and F.T. Mackenzie (eds.), Oxford University Press, New York, pp. 85–107.

Zackrisson, O., M.-C. Nilsson, I. Steijlen and G. Hörnberg, 1995: Regeneration pulses and climate–vegetation interactions in non-pyrogenic boreal Scots pine stands. *J. Ecol.,* in press.

Zoltai, S.C. and D.H. Vitt, 1990: Holocene climatic change and the distribution of peatlands in western interior Canada. *Quat. Res.,* **33**, 231–240.

10

Marine Biotic Responses to Environmental Change and Feedbacks to Climate

K. DENMAN, E. HOFMANN, H. MARCHANT

Contributors:
*M.R. Abbott, T.S. Bates, S.E. Calvert, M.J. Fasham, R. Jahnke,
S. Kempe, R.J. Lara, C.S. Law, P.S. Liss, A.F. Michaels, T.F. Pedersen, M.A. Peña,
T. Platt, J. Sharp, D.N. Thomas, K.A. Van Scoy, J.J. Walsh, A.J. Watson*

CONTENTS

SUMMARY

Marine biogeochemical processes both respond to and influence climate. Atmospheric carbon dioxide (CO_2) is the most important greenhouse gas increasing rapidly due to human activities. The oceans contain about 40,000 GtC in dissolved, particulate, and living forms. By contrast, land biota, soils and detritus total about 2200 GtC. Living and dead biogenic matter in the ocean contains at least 700 GtC, almost equal to the amount of CO_2 in the atmosphere (about 750 GtC). The increase in atmospheric carbon since pre-industrial times (150 GtC) is equivalent to that cycled through the marine planktonic ecosystem in less than 5 years. Simulation models calibrated with oceanic observations indicate that instead of the pre-industrial atmospheric CO_2 concentration of 280 ppmv, the atmospheric concentration of CO_2 would have been 450 ppmv in the absence of marine biota. Clearly, it is imperative that we understand the contribution of biogeochemical processes in maintaining the steady state functioning of the ocean carbon cycle. In addition to their importance in the global cycling of CO_2, marine organisms are significant sources of climatically active trace gases, especially dimethyl sulphide (DMS).

Because of the complexity of biological systems, we cannot yet say whether some likely feedbacks from the marine biota in response to climate related changes will be positive or negative. However improved quantitative understanding is likely within the next decade. The principal result of this chapter is the assessment of our scientific knowledge on the following topics:

- New nutrients (including iron) coming from outside the ocean (both as a result of increased atmospheric deposition or coastal runoff from human activities) would increase organic carbon C_{org} production, its export to the deep ocean (the "biological pump"), and the drawdown of atmospheric CO_2 (probably by less than 1 GtC/yr). Increased export carbon production would also result in increased remineralisation of organic matter at depth with an accompanying increase in production of nitrous oxide (N_2O), a potent greenhouse gas. The oceans contribute about 20% of the total input to the atmosphere, but it is unlikely that any possible increases could even double the ocean's contribution.

- Carbon is considered to be taken up by phytoplankton during primary production and released during remineralisation of organic matter in constant proportion to the major nutrients C_{org}:N:P, referred to as the Redfield ratios. If these ratios (or their vertical gradients) change in time in response to changes in ocean circulation or other properties, model simulations indicate that there is a large potential for the biological pump to influence atmospheric CO_2 concentrations.

- Calcium carbonate ($CaCO_3$) is fixed during photosynthesis by those marine algae and animals that have "hard" parts of $CaCO_3$: this $CaCO_3$ also sinks out of the surface layer with the exported organic carbon, but each molecule removed is accompanied by creation of a molecule of CO_2 in the surface ocean, counteracting the removal of the organic carbon. Globally, the ratio C_{org}:$CaCO_3$ exported from the surface ocean is about 4:1, but a shift in phytoplankton species causing a shift to a ratio of 1:1 would neutralise the effect of the "biological pump" on surface pCO_2. At present, we cannot quantify the probability or extent of such a shift occurring as a response to climate change.

- Biogeochemical processes occur principally in the top few hundred metres of the sea. The continental margins play a major, but poorly assessed, role in oceanic biogeochemical cycling and are the burial sites of a substantial amount of organic carbon derived both from these oceanic processes as well as terrestrial sources. A potential feedback may result from anthropogenic eutrophication of continental shelf areas. There, increased nutrient availability may promote denitrification and thus the release of elemental nitrogen (N_2) and methanogenesis releasing a minor amount of methane (CH_4).

- Iron in atmospheric dust may have contributed to the transition from glacial to interglacial periods. Lack of iron appears to limit phytoplankton growth in oceanic regions where the macronutrients, particularly nitrogen, phosphorous and silica, are abundant yet phytoplankton concentrations are generally low. Intentional iron fertilisation of the Southern Ocean to promote phytoplankton growth and hence the drawdown of atmospheric CO_2 appears not to be viable for two main reasons: first, it would require fertilisation of 25% of the world ocean continuously and indefinitely, and second, if it worked perfectly, it would only reduce the increase in atmospheric CO_2 concentration over the next century by about 50 ppmv.

- DMS is the principal volatile sulphur containing compound in the sea and the major natural source of sulphur to the atmosphere. In the atmosphere DMS is oxidised to produce aerosols which promote the development of clouds and thus influence climate. The production of DMS is a function of the composition and concentration of the plankton. DMS is removed from sea water by physical, chemical and biological processes. In temperate and tropical waters the rate of biological consumption can be more than ten times greater than atmospheric ventilation. A much better understanding of sulphur cycling in the surface ocean is required for quantitative assessment.

- Solar ultraviolet (UV) radiation has increased as a consequence of stratospheric ozone depletion. UV-B (280-320 nm) exposure depresses primary production and growth of phytoplankton. As species differ in their tolerance to UV-B, a shift in community structure favouring the more tolerant would be expected. UV-B radiation inhibits marine bacterial activity and has differing effects on dissolved organic carbon (DOC) derived from various sources. The effects of UV-B on marine biogeochemical processes must be taken into account for a realistic assessment of the role of these processes in environmental change.

Coupled time-dependent general circulation models (GCMs) are principal tools in probing the possible responses of the land-atmosphere-ocean system to climate change. For these models to be more effective they must include feedback terms to account for the role of biogeochemical processes in modulating the production and degradation of radiatively active gases in the upper ocean. At present, models constructed to consider biogeochemical processes are severely limited by the inability to parameterize important biological activities and to specify the temporal and spatial variation in these parametrizations over scales longer than a few weeks and more than tens of kilometres.

10.1 Introduction

In IPCC (1990), the role of marine biota in climate change was treated in a cursory manner. Since then, there has been increasing awareness of the likelihood of a variety of responses of marine biota to climate/environmental change and feedbacks to the climate system. The purpose of this chapter is to assess the manner in which these responses and feedbacks might occur, in preparation for the inclusion over the next five to ten years of ocean biogeochemical processes in the models used for climate change projections. IPCC WG II (1995) deals in detail with possible impacts of climate change on marine ecosystems, but does not consider feedbacks to the climate system.

Several critical issues require explicit attention because of their demonstrated or perceived importance in terms of feedback to climate. They include: possible responses of the ocean biological pump and any resulting change in the rate of CO_2 cycling; iron fertilisation which might speed up the biological pump and the draw down of atmospheric CO_2; changes in the production by phytoplankton of dimethyl sulphide (DMS) which in the atmosphere may contribute to the production of sulphate aerosols; and potential consequences of increased ultraviolet radiation in the wavelength range 280-320 nm (UV-B) on plankton production and community composition. Climatic variation has been documented on temporal scales from sub-decadal (for example, the El Niño-Southern Oscillation (ENSO) phenomenon) to millennia. Most projections in the IPCC scientific assessments have been for the years 2050 and 2100, so we consider ENSO to century time-scales and regional to global spatial scales. Physical and biogeochemical processes, in the very heterogeneous continental margins especially, vary strongly on regional scales and must be understood on those scales in order to determine possible global effects.

10.2 Ocean Processes – Biogeochemical Responses

10.2.1 Oceanic Space and Time-scales in the Context of Climate Change Effects

The physico-chemical habitat of the ocean and its interaction with marine biota is fundamental in determining biogeochemical responses and feedbacks that may arise in response to climate change. The range of possible space and time-scales over which these oceanographic processes operate covers many orders of magnitude (Figure 10.1a). For physical processes, these range from dissipative processes, which operate at short space and time-scales, to scales associated with ocean gyres and the thermohaline circulation. However, the wide ranges of space and time-scales associated with physical oceanographic processes are not independent. Rather, these should be regarded as a cascade of scales through which information can be transferred from large to small scales and vice versa. Consequently, perturbations introduced by climate change can potentially affect oceanic circulation processes at a large number of scales. It is at the largest scales that the effects of environmental perturbations are often most dramatic. These take the form of the approximate decadal scale of El Niño-Southern Oscillation events to the longer millennial scale of modifications to the thermohaline circulation. However, shorter term variations associated with changing frequency of occurrence of oceanic eddies or upwelling as well as changes in the mixing environment of the upper ocean can be manifestations of changes to the environment.

Similarly, the range of space and time-scales over which individual organisms and populations of organisms exist is large (Figure 10.1b). The population scales especially overlap with a wide variety of physical scales and consequently with physical processes. The range of sizes of marine biota allow for connections between physical and biological processes. Therefore, modifications that affect circulation dynamics can potentially influence a wide range of biological processes, and modifications that affect individual biological components have the potential of cascading throughout the marine food web. Most marine biotic responses to environmental and climate change that might have feedbacks to climate involve zooplankton and smaller organisms. From Figure 10.1b we would expect that their responses might occur on time-scales of less than decades. If, for example, the marine biota respond as a pulse of elevated primary production injecting (mostly as sinking particles) more organic carbon into the deep ocean, physical transport processes might then redistribute the anomalous deep carbon to complete the response, on time-scales shown schematically in Figure 10.2. These processes might take considerably longer, especially if the largest scale mode, the thermohaline circulation, is involved.

However, perturbations in temperature, fresh water discharge and wind strength that may arise from climate change can significantly alter the intensity and the depth of penetration of the thermohaline circulation. In fact, evidence from seafloor sediments indicates that the thermohaline circulation has varied with time-scales of 20,000 years down to decades (e.g., Dansgaard *et al.*, 1993; GRIP Project Members, 1993), variations that are now being incorporated into computer models of large scale thermohaline circulation patterns (e.g., Weaver *et al.*, 1993; Fichefet *et al.*, 1994; Rahmstorf, 1994; Weaver and

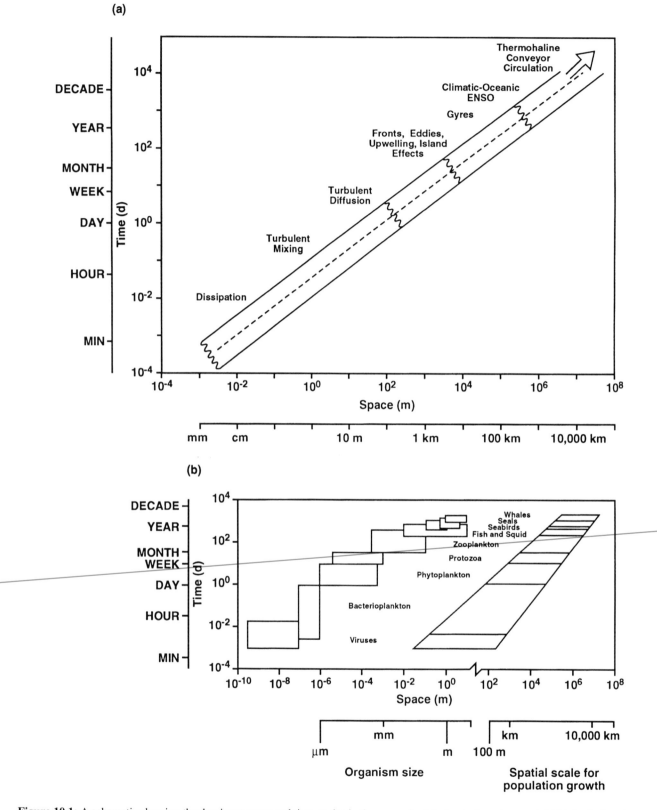

Figure 10.1: A schematic showing the dominant space and time-scales in the ocean for (a) physical motions and (b) biological scales. In (b) the left-hand overlapping boxes represent typical size ranges (on the x-axis) and typical times for population doubling (on the y-axis) for each type of organism; the right-hand boxes represent typical spatial ranges of each organism during their lifetime. Adapted from Murphy *et al.* (1988).

Hughes, 1994). Variations in the thermohaline circulation have been implicated in glacial-interglacial changes in the carbon cycle of the ocean and the atmosphere (e.g., Broecker, 1991; Oppo and Lehman, 1993).

The transfer of information between biological and physical processes is frequently thought of as being in one direction: physical to biological. However, in some regions of the ocean biological processes can provide a feedback to physical processes. For example, the feedback between light absorption by marine phytoplankton and heating in the upper water column can modify mixing dynamics (e.g., Sathyendranath *et al.*, 1991), which can alter vertical structure and circulation in surface waters.

10.2.2 Marine Biological Production and Climate

Marine primary production, through which phytoplankton convert carbon to organic and structural molecules, requires light, CO_2 and dissolved nutrients. The rate of primary production is not controlled by the supply of CO_2, considered to be almost always available in excess of requirements for marine primary production, but rather by the supply of new nutrients to the sunlit layer. Total primary production can be partitioned (over long enough time-scales) into two fractions: recycled production supported by nutrients cycled within the sunlit layer of the ocean, and export production supported by nutrients entering the sunlit layer of the ocean from land, the atmosphere, and primarily the deep ocean, that is available to be harvested or transported vertically to the ocean interior (see Eppley and Peterson, 1979).

The "biological carbon pump" refers to the process by which export production removes carbon from the ocean surface layer to the ocean interior, as sinking organic particles and as dissolved organic matter transported by ocean currents. Eventually this carbon is "remineralised" by bacteria at depth into dissolved inorganic form, except for a small fraction that is buried in the sediments. Thus, the biological pump reduces the total carbon dioxide in the ocean surface layer (and the atmosphere) and increases the total carbon content in the deep ocean. For pre-industrial equilibrium conditions (atmospheric concentration of CO_2: ~280 ppmv), model simulations indicate that a fully efficient biological pump (using all available surface nutrients) would result in an atmospheric CO_2 level of ~160 ppmv, whereas extinction of all marine production would lead to a level of ~450 ppmv (e.g., Sarmiento and Toggweiler, 1984, Shaffer, 1993). However, the biological pump is considered to have been roughly in steady state over the last century, unaffected by the increase in CO_2 availability.

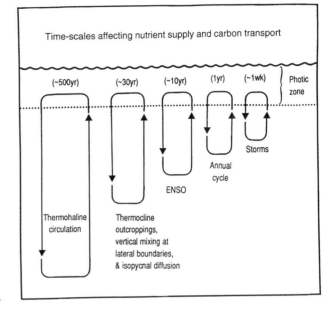

Figure 10.2: Schematic showing characteristic time-scales for various processes of vertical exchange between the euphotic zone and the ocean interior. Redrawn from Denman (1993).

The global pattern of marine primary production (Figure 10.3, after Berger and Wefer, 1991) shows that much of the highest production is associated with continental margins. The regions of enhanced primary production away from continental margins are generally associated with open ocean upwelling, such as occurs in equatorial regions, and with deep winter mixing that entrains more nutrients into the upper ocean. Open-ocean primary production outside of upwelling regions tends to be lower, but extends over much larger areas. The nutrients coming from the ocean interior must be supplied by vertical exchange processes that operate on shorter space and time-scales than the thermohaline circulation (Figure 10.2). Berger and Wefer (1991) took the pattern of vertically exported primary production to be identical with (but not linearly proportional to) that of total primary production, with the largest values found along the continental margins; their pattern is remarkably similar to the annual average pattern of Ekman upwelling in Xie and Hsieh (1995), suggesting a strong dependence on wind-driven upwelling.

The high primary production in continental margins is usually the direct result of coastal upwelling and mixing processes. This production is supported by nutrients that are brought into the euphotic (sunlit) zone from depths of 200 m or less. The upwelling that supplies these nutrients results from wind-driven circulation processes or from interactions of western boundary currents with shelf break

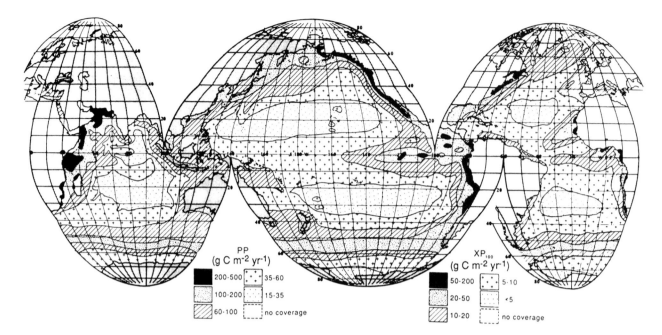

Figure 10.3: World map of primary production (PP) and export production at 100 m depth (XP$_{100}$). Productivity estimated from a literature compilation of radiocarbon-based measurements complemented by estimates based on phosphate at 100 m depth, distance from land, and latitude, according to an algorithm fitted to radiocarbon measurements for areas with poor coverage. XP$_{100}$ is rounded off from the expression: XP$_{100}$ = (PP)$^{3/2}$/50. From Berger and Wefer (1991).

regions. Thus, the time-scale for nutrient supply and for returning carbon to the upper portion of the water column is shorter in these regions than in the open ocean. However, changes that might affect the intensity of upwelling (Bakun, 1990) or the location of a western boundary current relative to the continental margin (Atkinson, 1977) can alter these time-scales. For example, long-term variability in primary production as determined from the palaeorecord in the Arabian Sea has been related to changes in the strength and location of the monsoonal wind system (Prell and Van Campo, 1986; Luther and O'Brien, 1990; Shimmield *et al.*, 1990).

If much of the global oceanic primary production results from upwelling of waters shallower than 200 m, there are two ramifications. First, if climate-induced changes in ocean circulation patterns change the supply of nutrients to the ocean surface layer, the time-scale involved for this and the biological response might be decades rather than centuries. This time-scale is comparable with the 2050 and 2100 projections of interest to policymakers, and hence the possible response of the marine biota and potential for feedback to climate merit attention. Second, the contribution of continental margins to global estimates of total and export oceanic primary production, and eventual burial in the sediments, may not have been properly evaluated.

Carbon is considered to be taken up by phytoplankton during primary production and released during remineralisation of organic matter in constant proportion to the major nutrients, referred to as the Redfield ratios (Redfield *et al.*, 1963). The Redfield ratios between the elements C$_{org}$, N, P, and O are taken here as 103:16:1:-72 after Takahashi *et al.* (1985). If climate change causes changes in ocean circulation, or in inputs from land and the atmosphere, that affect the supply of new nutrients to the sunlit upper ocean (and the production of export carbon), then the rate of sequestering of carbon by the biological pump could change with possible feedbacks to the build-up of greenhouse gases in the atmosphere. Similarly, if these changes cause changes in the Redfield ratios, then more or less carbon would be transferred by the biological pump for each unit of nutrient, again with possible implications for oceanic carbon sequestration (see Section 10.3.2.2).

10.2.3 Direct Effects of Climate Change on Marine Biota

Climate change can affect marine biota directly (e.g., increased temperature or changed solar radiation and their effects on the physiology or behaviour of marine organisms), or indirectly (e.g., changes to ocean circulation patterns that result in changes in distributions or functioning of marine communities).

Temperature influences biological production and can have a profound effect on growth and metabolic processes. Observations show that biological rates double or halve with a 10°C increase or decrease in temperature, respectively, i.e., a Q_{10} of 2. General relationships have been derived for predicting the effect of temperature variation on large phytoplankton (Eppley, 1972) and zooplankton (Huntley and Boyd, 1984) maximum growth rates. These relationships suggest that over the range of temperatures encountered in the ocean, –2 to 30 °C, maximum growth rates vary by about a factor of ten. At high latitudes where the temperature range is smaller, –2 to 3 °C, temperature changes may have a relatively larger effect (Smith and Sakshaug, 1990). Hence, even small changes in temperature could have pronounced effects on biological rates such as growth and development. Q_{10} values are usually obtained from laboratory experiments conducted over short (days) time-scales. Over longer time-scales, physiological adaptation may occur, and the effects may be more subtle, consisting of changes in community composition rather than in altered rates of the original species. For this reason a direct correlation between temperature and community production cannot be drawn. Changes consistent with climate warming have been reported by Barry *et al.* (1995) in the invertebrate fauna of an intertidal community, and by Roemmich and McGowan (1995) in zooplankton biomass over the continental shelf. Temperature-induced change in biological rates or ecosystem structure may impact carbon cycling in the ocean by modifying the biological carbon pump. Furthermore, as shown by Figure 10.1, impacts that occur at one size class of organism can transfer to those above and below, thereby potentially affecting all components of the marine food web in a given region.

The quality and quantity of radiation can also be altered by climate change, directly affecting marine organisms. Between 1930 and 1981 the total cloud amount over Northern Hemisphere oceans increased by 6% (Parungo *et al.*, 1994), which must affect total incoming solar radiation, the spectral distribution of the incoming radiation, and the amount of outgoing radiation. Primary production by marine phytoplankton depends both on the amount and the spectral composition of the incoming solar radiation. Given that primary production might be linearly dependent on light levels much of the time when there are sufficient nutrients, we would expect a 6% increase in cloudiness to result in a decrease in primary production but somewhat less than that amount, perhaps a few per cent. However, the distinction between short term physiological response, and biogeochemically-significant community response remains unresolved.

UV-B radiation, which has been shown to affect biological production and a wide variety of marine organisms, and which is increasing especially at high latitudes, will be treated in detail in Section 10.3.6.

10.2.4 Responses to Changes in Circulation Patterns and Other Oceanographic Processes

Marine ecosystem structure and productivity is largely a function of physical processes controlling the supply of nutrients to the sunlit surface layer of the ocean (e.g., Denman and Gargett, 1995). Alteration of oceanographic patterns from climate or environmental change may alter the structure of oceanic ecosystems on large scales with changes in the marine biotic production and/or fixation of radiatively active gases, primarily CO_2, DMS and N_2O, with possible feedback to climate. Especially productive marine ecosystems are those associated with wind-driven upwelling in eastern boundary currents and retreating sea ice regions. It has been argued by Bakun (1990) that global warming may enhance the upwelling. Projections of greater warming over continents (especially the already observed higher minimum night-time temperatures) than over the adjacent ocean would strengthen the onshore temperature gradients that drive the equatorward winds along eastern boundaries of oceanic gyres. Bakun presents evidence for the last several decades from four major upwelling areas that alongshore windstress has already been increasing, but other possible feedbacks might explain these trends. On longer scales, Altabet *et al.* (1995) have recently found climate-related variations in nitrogen isotope ratios in the Arabian Sea, suggesting changes in denitrification within the sediments that may have resulted in changes in release of N_2, and N_2O to the atmosphere.

The largest climate anomalies are associated with ENSO, and effects of the ENSO phenomenon on marine ecosystems continue to be documented. The presence of an El Niño reduces the surface nitrate concentration and the ΔpCO_2 between the atmosphere and the surface ocean to at most half the normal values (Wong *et al.*, 1993; Murray *et al.*, 1994). Harris *et al.* (1991) documented significant changes in the marine ecosystem in the subtropical convergence zone off Tasmania associated with the La Niña "Cold Event" of 1988. Karl *et al.* (1995) have observed at a time-series station in the subtropical gyre near Hawaii a complete change in the planktonic ecosystem during the 1991-92 ENSO event. There was a significant change in the phytoplankton community structure – an increase in nitrogen-fixing micro-organisms *Trichodesmium*. This change caused a shift from a primarily nitrogen-limited to a primarily phosphorus-limited habitat with accompanying changes in total production by phytoplankton and in the amount of that production lost to the deep sea. Most importantly, the primary production supported by the nitrogen fixed by the *Trichodesmium* organisms removes CO_2 from the surface waters without requiring upwelled nitrate which would be accompanied by dissolved inorganic carbon (DIC).[1]

A number of climate change scenarios suggest that temperature changes would be greatest at high latitudes (see Section 6.2.2.3). Jacka and Budd (1991), for example, have documented the sensitivity of Antarctic sea ice to warming. The seasonal development and regression of sea ice is of substantial biological importance in addition to its profound role in global ocean-atmosphere exchange of heat, water and gases. Blooms of phytoplankton at the ice edge, induced by upwelling of nutrient-rich water and/or the stabilising effect of melting ice-water, are significant at high latitudes in both hemispheres. Changes in the extent and thickness of the ice affect the timing, magnitude and duration of the seasonal pulse of primary production in polar regions, while the mode of sea ice formation influences the composition of the biota associated with it. This in turn affects the quantity and quality of food available to grazers and the timing and the magnitude of vertical flux of carbon and other elements (Eicken, 1992). Again, recent findings by Gloersen (1995) indicate that sea ice extent in both hemispheres correlates with ENSO cycles.

In summary, the marine biota and their environment interact on a wide variety of space and time-scales. Especially productive areas are on continental margins and along the equator because of the input of new nutrients from land, from the atmosphere, and mostly from the ocean interior via upwelling. Climate change can affect marine biota by changing temperatures altering physiological rates and/or community composition, by changing radiation affecting biological production or damaging cells, and by changes to circulation patterns and other oceanographic processes altering ecosystem structure and function. Such changes offer a variety of possible feedbacks to the climate system.

10.2.5 Evidence from the Palaeorecord

The glacial/interglacial record suggests a decrease of ~80 ppmv in atmospheric CO_2 occurred during the last glacial period. The ocean contains more than 50 times the carbon contained in the atmosphere. During glaciation, the carbon stored in land biota, soil and detritus was significantly reduced by ice cover and a colder, drier climate. Therefore, the glacial-interglacial change in atmospheric CO_2 content must have been driven by changes in ocean chemistry and/or biology. Ignoring changes in solubility due to changing temperatures (too small and largely offset by an opposite change in salinity) and effects due to sea level change (compare the ~150 m rise during deglaciation with the projection of less than 1 m over the next century), there still exist several hypotheses for oceanic chemical and/or biotic processes causing or triggering glacial/interglacial transitions (Broecker, 1982). Changes in deep ocean $CaCO_3$ dissolution or burial caused by changes in concentrations of deep ocean organic carbon may have accounted for part of the recent glacial/interglacial increase in atmospheric CO_2, but again the time-scale would be millennia (Boyle, 1988; Archer and Maier-Reimer, 1994), too long for the 2050 to 2100 IPCC interests.

Other hypotheses to explain the glacial drawdown of atmospheric CO_2 with the potential for change on decade to century time-scales have been simulated in an ocean general circulation model (GCM) by Heinze *et al.* (1991). They include increased Redfield ratios of C:N or C:P, increased ocean nutrient inventory causing increased production of particulate organic carbon (POC), increased high latitude primary production, changes in the "ventilation" of deep waters (i.e., a shallower thermocline circulation), and changes in the proportion of production by phytoplankton of CaCO3 relative to organic carbon. Heinze *et al.* found that their model was most sensitive to changes in Redfield ratios, ventilation, and changes in the inventories of nutrients or POC. Qualitative consistency with palaeoceanographic data was greatest for changes in Redfield ratios and in the oceanic inventory of POC. In the next sections, we will consider the potential for marine biotic responses in these processes, and in processes where there is potential for feedbacks to climate through related biogeochemical processes. Two examples of the latter are: (i) Increases in POC on continental shelves could lead to increased anoxia and loss of nitrogen to the atmosphere by denitrification, and reduced production of POC (e.g., Christensen, 1994). (ii) A shift to $CaCO_3$-fixing phytoplankton (*Coccolithophores*) would increase both surface pCO_2, the marine production of DMS (with a possible increase in marine cloudiness), and the near-surface reflectance of visible light with possible increases in sea surface reflectance or near surface temperatures (e.g., Holligan, 1992).

---

[1] Dissolved inorganic carbon (DIC) consists of aqueous CO_2, bicarbonate and carbonate ions, i.e. DIC = $CO_2 + HCO_3 + CO_3^{2-}$.

10.3 Feedbacks: Influence of Marine Biota on Climate Change

10.3.1 Critical Issues

We are concerned here with potential feedbacks to climate by environmentally induced changes to marine biota. In particular, how might changes in the marine biota in response to changing climate affect the atmospheric distributions of radiatively active gases? In addition to maintaining a reduced concentration of CO_2 in the surface ocean (relative to the ocean interior), marine biota are a significant source of N_2O, another greenhouse gas, and of dimethyl sulphide (DMS), which when ventilated to the atmosphere can form cloud condensation nuclei. Several critical issues (CO_2 cycling via the biological pump, the role of iron in sequestering carbon to the deep ocean, DMS production, and UV-B radiation damage) have received considerable attention in the scientific and public media and will be addressed individually.

10.3.2 Response of the Biological Pump to Climate-Induced Change and Possible Feedbacks

10.3.2.1 The open ocean biological pump

The role played by the biological pump in the ocean carbon cycle, discussed only briefly in IPCC, 1994 (Sections 1.3.3.3 and 1.4.3), is expanded and updated here. According to our understanding of the current operation of the global carbon cycle (Figure 10.4), on average the ocean takes up about half of the anthropogenic input of CO_2 to the atmosphere that does not remain there. This amount (e.g., Siegenthaler and Sarmiento, 1993) is in the order of 2 GtC/yr (1 Gt = 1 Pg, where P = Peta = 10^{15}), relative to a gross flux of CO_2 in each direction between the ocean and the atmosphere in the order of 90 GtC/yr. The surface ocean mediates the fluxes of carbon (dioxide) between the atmosphere and the ocean interior and between the land and the ocean sediments or ocean interior. The gross flux in each direction between the surface ocean and the ocean

CARBON CYCLE 1980 - 89

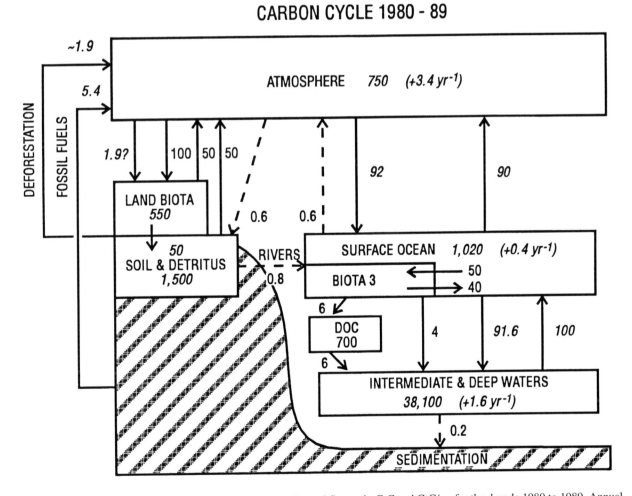

Figure 10.4: Box diagram of the global carbon cycle reservoirs and fluxes, in GtC and GtC/yr, for the decade 1980 to 1989. Annual changes in reservoirs are in parentheses, and italic numbers represent those believed to have changed since pre-industrial times due to human activities. Adapted from Siegenthaler and Sarmiento (1993).

interior is in the order of 100 GtC/yr (a number that can vary considerably depending on what depth is taken as the base of the surface layer), and the net flux is 1-2 GtC/yr (estimated to be less than the flux from atmosphere to the surface layer to account for observations that show the surface-ocean total carbon concentration to be increasing by about 0.4 GtC/yr (Siegenthaler and Sarmiento, 1993)).

Globally, marine plants are responsible for more than a third of the total gross photosynthetic production that incorporates carbon from CO_2 into organic molecules; the other photosynthesis is accomplished by terrestrial plants. On an annual basis, the gross rate of carbon fixation by marine algae (phytoplankton) may be as high as 50 GtC/yr, of which more than half is supported by nutrients recycled within the euphotic layer, and the rest supported by new nutrients introduced into the euphotic zone, primarily from the ocean interior, but also from the atmosphere and the land via runoff. From the point of view of the balance of CO_2 between the atmosphere and ocean, the quantity of interest is the export production – the fraction of the primary production that eventually passes from the euphotic layer to the ocean interior, either as DOC[1] through vertical transport by water motions, or as sinking POC. This export production is estimated to be 10 GtC/yr (range 5-20 GtC/yr) (Siegenthaler and Sarmiento, 1993; Murray *et al.*, 1994; Ducklow, 1995).

The approximately 50 GtC cycled annually by the ocean biota between living and dead material represents more than ten times their standing stock of 3-4 GtC (i.e., the standing stock is cycled approximately monthly), and roughly 5 times the (poorly known) amount of biogenic carbon that is transported downward to the ocean interior. To address how this biological carbon pump might respond to climate-related or other environmental change, and with what potential feedbacks to the climate system, requires a brief description of its workings (see Longhurst, 1991 and Sarmiento and Siegenthaler, 1992 for recent reviews).

The organic carbon export from the surface ocean to the ocean interior (i.e., the biological pump) is equal to photosynthetic production minus remineralisation in the surface ocean. Some phytoplankton die, some are grazed by protozoa, zooplankton or fish, which in turn die or are eaten by larger zooplankton or fish, and so on up to seabirds and mammals (seals, whales and humans). These processes generate organic matter in dissolved and

particulate phases, from dying plants or animals, from waste products, and from messy feeding. Much of this material is reusable by phytoplankton, either immediately or after processing by bacteria. The export carbon production consists of POC which sinks out of the surface layer, and DOC which can be transported downwards by subduction or convective sinking of surface waters. Organic matter that reaches great depths is returned mostly as DIC after remineralisation or as long-lived DOC, on time-scales of centuries to millennia. However, most of the organic material is remineralised in the upper 500 m of the water column and is reintroduced into the euphotic zone on much shorter time-scales through a variety of processes – coastal and equatorial upwelling, deep winter convective mixing, surface outcropping of density surfaces at high latitudes, etc. Only a small proportion (order 1%) of the POC that sinks out of the surface layer reaches the ocean bottom where it may eventually be buried in the sediments. The operation of the biological pump is depicted schematically in Figure 10.5.

Related to the biological pump (or part of it depending on the definition) is the uptake of carbon by those marine algae and animals that have "hard" parts constructed of calcium carbonate. This $CaCO_3$ sinks out of the euphotic zone with the export production. Globally, 1 carbon atom is incorporated by calcareous phytoplankton into $CaCO_3$ for about every 4 carbon atoms into organic molecules (Broecker and Peng, 1982; Tsunogai and Noriki, 1991), but the ratio can be 1:1 in areas with blooms of certain coccolithophores (Robertson *et al.*, 1994). The $CaCO_3$ particles do not remineralise during sinking; rather they mostly redissolve at depths of several thousands of metres (e.g., Tsunogai and Noriki, 1991; Anderson and Sarmiento, 1994), although some must obviously settle on the sea floor, such as ocean ridges where the bottom is above the dissolution horizon, i.e., the depth (differing in different oceans) below which sinking $CaCO_3$ particles would tend to redissolve (Broecker and Peng, 1982). Factors controlling the $CaCO_3$ component of photosynthesis have not been well studied, but efforts are under way to develop a quantitative basis for assessment (e.g., Holligan *et al.*, 1993; Westbroek *et al.*, 1993).

Conventional wisdom states that carbon availability does not limit marine photosynthesis. Therefore the increasing amount of CO_2 in the surface layer of the ocean is thought not to cause an increase in either the gross or net rate of operation of the biological pump, which to first order is thought to have been in steady state through the last century (e.g., Sarmiento and Siegenthaler, 1992; Siegenthaler and Sarmiento, 1993). Nitrogen is sufficiently

---

[1] Dissolved organic carbon (DOC) is operationally that fraction of particulate organic carbon (POC) that passes through a filter of pore size in the order of 0.5 μm.

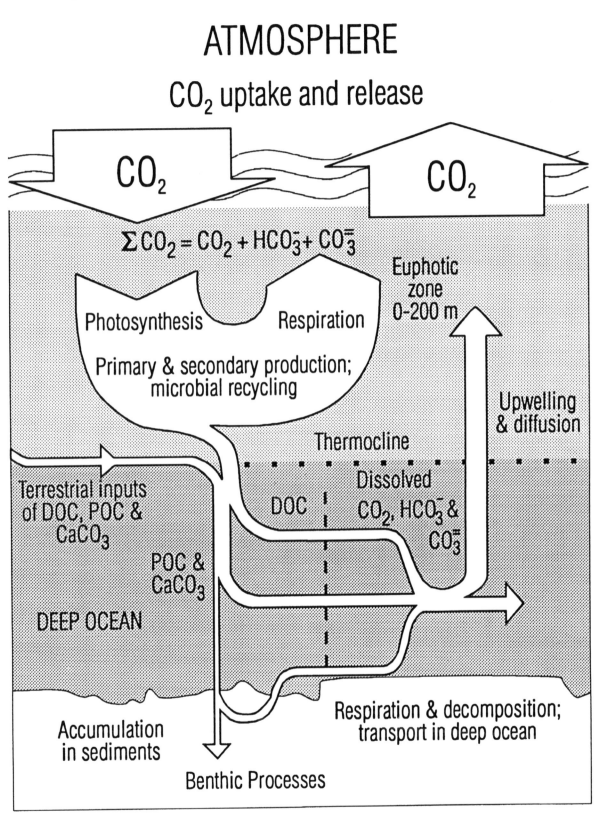

Figure 10.5: Schematic showing the dominant biogenic fluxes of carbon in the ocean and the biological processes transforming the carbon into different forms. Widths of the arrows depict semi-qualitatively the relative sizes of the fluxes. Vertical transports of DIC are not shown (see Figure 10.4). Adapted from SCOR/JGOFS brochure "Oceans, Carbon and Climate Change", 1990.

scarce that it is thought to be the nutrient limiting the amount of carbon fixed by photosynthesis on geologically short time-scales. The loss rate of organic carbon (and nitrogen) from the surface layer to the ocean interior, the export production, is considered to be balanced by the input of new nitrogen to the surface layer – from freshwater runoff, from the atmosphere, and primarily by advective and diffusive transport of dissolved nitrate from the ocean interior back up into the euphotic layer. Thus geochemists conclude that the biological pump cannot react directly to the increase of CO_2 in the atmosphere and in the upper ocean, but only indirectly in response *to changes in the supply* of new nitrogen caused by climate or environmental change. This indirect response could be a change in the structure of the biological pump, i.e., a change in ecosystem structure altering (in the context of carbon cycling) the proportion of recycling versus loss to the ocean interior. Such a change might be different in different regions (biogeographic "provinces"); whether they would tend to cancel out or add to a change significant at the global scale is also important.

Riebesell *et al.* (1993) have recently challenged this conventional wisdom, arguing that diatom growth can be limited by the supply of CO_2 and that the approximate doubling in surface water pCO_2 since the last glaciation from 180 to 355+ ppmv could have stimulated marine primary productivity, in such a way that the biological pump may already be participating in sequestering to the ocean interior some fraction of the atmospheric CO_2 produced by human activity. Turpin (1993), on the other hand, argued that CO_2-limited photosynthesis does not necessarily imply CO_2-limited growth rates in the sea. Raven (1993) pointed out that cell-membrane-mediated inorganic carbon-limitation concerns the rate rather than the extent of production, i.e., a spring bloom might take longer but the annual global export production would be unchanged. Obviously, more thinking, if not more research, is required.

The estimate in Figure 10.4 for total oceanic primary production of 50 GtC/yr is roughly double that of 27 GtC/yr made by Berger (1989), with similar data and formulae to those used to generate the original map on which Figure 10.3 is based (Berger and Wefer, 1991). Longhurst *et al.* (1995) have examined primary productivity data from the North Atlantic to partition the ocean basins into four domains and four seasons according to the nature and intensity of physical forcing likely to affect phytoplankton dynamics. This classification, with the global satellite ocean colour data base, has been used to estimate the global annual oceanic primary production,

giving a value of about 50 GtC/yr for total (recycled + export) production. Recent estimates of annual total or export production obtained from tracer studies, *in situ* ^{14}C incubations with "clean" techniques, and from sediment traps, when compared with their positions on the Berger and Wefer map, also support the higher global estimates (e.g., Jenkins and Goldman, 1985; Platt and Harrison, 1985; Knauer *et al.*, 1990; Lohrenz *et al.*, 1992; Welschmeyer *et al.*, 1993; Karl *et al.*, 1995). There is some evidence, especially in the North Pacific, that primary production and standing stocks of phytoplankton (Venrick *et al.*, 1987; Parsons and Lalli, 1988) and zooplankton (Brodeur and Ware, 1992) may have been increasing over the last two decades, although Roemmich and McGowan (1995) documented decreasing zooplankton biomass and increasing temperatures over the last 45 years in the California Current area. Methodological changes, e.g., in filters and in "clean" techniques, may account for most if not all of the apparent increases in open ocean phytoplankton primary production and standing stocks.

Studies using the most recent methodology for DOC (e.g., Sharp *et al.*, 1995) show significant changes in DOC concentrations over short temporal and spatial scales (Carlson *et al.*, 1994; Carlson and Ducklow, 1995; Peltzer and Hayward, 1995). Surface values of 60 to 80 micromolar carbon can drop to 50-60 micromolar over a depth range of 150-300 m below the surface; horizontal variations over scales of tens of kilometres and temporal variations from day-to-day can be seen on the order of 20 micromolar carbon. With these observations, it is possible to recognise different components of DOC ranging from the traditional "refractory" pool (not readily usable by biological organisms), with a turnover time on the order of ocean mixing cycles, to the "very labile" pool, with a turnover time of hours to a day (Carlson *et al.*, 1994).

It is important to be able to assess whether there are changes in the inventory of carbon above the permanent thermocline when trying to understand the ocean-atmosphere linkage. A series of model simulations (Toggweiler, 1989, 1990; Bacastow and Maier-Reimer, 1991; Najjar *et al.*, 1992; and Siegenthaler and Sarmiento, 1993) have been able to achieve closer congruence with observations if vertical transports of DOC are comparable with vertical fluxes of sinking organic particles (POC). Recent careful measurements of DOC concentrations made in conjunction with measurements of primary production and sinking particles (export production) support the model simulations. Along the equatorial Pacific, Murray *et al.* (1994) concluded that 60 to 90% of the new production was transported away from the equator by currents, possibly as DOC. In the Sargasso Sea, Carlson *et al.* (1994) found the DOC advected downwards from the euphotic zone via deep mixing each winter to be greater than the POC captured by sediment traps at a depth of 150 m. In the next few years, careful measurements of DOC with a common methodology should provide estimates of the global pool of DOC, its spatial distribution and its temporal dynamics within the marine foodweb.

The Joint Global Ocean Flux Study (JGOFS) is currently undertaking research to increase our understanding of the global ocean carbon cycle, so that we might address the above mechanisms on a more quantitative basis. It should be pointed out that at the present time the global numbers mentioned in this discussion with regard to the ocean carbon cycle have uncertainties in the order of 100%. It is within this context of large uncertainties and tightly coupled physical, biological and chemical processes that we shall describe possible responses and feedbacks of the biological pump to climate change.

10.3.2.2 The Redfield ratios – are they constant?

Recent model simulations of ocean carbon cycle changes demonstrate a high sensitivity to changes in the Redfield ratios between carbon and the limiting nutrient (Heinze *et al.*, 1991; Shaffer, 1993). These ratios are often assumed to be constant in coupled analyses of the cycles of these elements in the ocean and have great importance in global biogeochemistry. If there is significant departure from these ratios, either in their uptake during primary production or in their remineralisation back to inorganic forms in the water column, then our understanding of how the oceanic carbon, nitrogen and phosphorus cycles are coupled needs to be revised.

Recent observations by Sambrotto *et al.* (1993) suggest elevated levels in the uptake of carbon relative to nitrogen, especially during the spring bloom. Banse (1994) also found a decoupling of the uptake by marine phytoplankton of dissolved inorganic carbon and nitrate, but with C:N ratios lower than those of Redfield. Utilisation of nitrate is often converted to carbon fixation by use of a constant Redfield ratio. The elevated ratios of C:N fixation reported by Sambrotto *et al.* would lessen the discrepancy between estimated O_2 utilisation rates and nitrate upwelling by allowing more carbon to be fixed for the same amount of nitrogen, and would increase the global estimate of carbon sequestering by export production, usually calculated from the ratio of new nitrate uptake to total nitrate uptake by marine phytoplankton.

For remineralisation, simple models (e.g., Evans and Fasham, 1993; Shaffer, 1993) demonstrated that more rapid remineralisation of nitrogen relative to carbon in sinking organic particles leads to lowered surface CO_2 concentrations, and Shaffer's results also demonstrated a significant reduction of atmospheric CO_2 concentration in response to doubling of the remineralisation depth scale of carbon relative to that of the limiting nutrient. Lohrenz *et al.* (1992), Honjo and Manganini (1993), and Karl *et al.* (1996) all found elevated C:N ratios in sediment trap samples that increased markedly with depth, indicating that the remineralisation depth scale for nitrogen was less than that for carbon, i.e., suggesting that the remineralised nitrogen might reach the surface layer more rapidly than the remineralised carbon. On the other hand, Anderson and Sarmiento (1994) found regenerated ratios to be in rough agreement with Redfield values, but with considerable variability.

The widely observed departures from Redfield ratios indicate that past conclusions concerning operation of the biological pump based on adherence to Redfield ratios need to be reviewed. Furthermore, if the Redfield ratios (or their vertical gradients) change in time in response to changes in ocean circulation or other properties, the model simulations indicate that there is a larger potential for the biological pump to influence atmospheric CO_2 concentrations.

10.3.2.3 Carbonate production

Changes in ecosystem structure can have an additional, seldom appreciated, effect on the oceanic carbon cycle. A shift from organic- or silica-walled organisms to carbonate (i.e., $CaCO_3$) secreting organisms (e.g., coccolithophorids), possibly accompanying temperature changes or coastal eutrophication, would cause an increase in surface water pCO_2 because of the overall calcification reaction:

$$Ca^{2+} + 2HCO_3^- \rightarrow CaCo_3 + CO_2 + H_2O$$

Thus, although carbonate production reduces total DIC in the surface layer, the reaction tends to increase the surface ocean pCO_2 by decreasing the surface alkalinity (e.g., Broecker and Peng, 1982) thus driving CO_2 from the ocean to the atmosphere (Frankignoulle and Gattuso, 1993). The released CO_2 constitutes roughly 60% of that fixed as carbonate and this fraction increases with an increase in sea water pCO_2, thereby increasing this feedback (Frankignoulle and Canon, 1994). Although this fraction may be reduced if the CO_2 released intracellularly is fixed photosynthetically, thereby decreasing its release to sea water (Sikes and Fabry, 1994), observations in the North Atlantic during an extensive coccolithophorid bloom

(Holligan *et al.*, 1993; Robertson *et al.*, 1994) clearly showed that surface water pCO$_2$ decreased less rapidly than would be expected on the basis of photosynthesis alone, which also resulted in the undersaturation of surface waters being less pronounced. Palaeo-oceanographic evidence suggests changes in the carbonate cycle influenced atmospheric CO$_2$ changes that accompanied glacial-interglacial periods. Boyle (1988) argued that during the last glacial maximum a shift in labile nutrients and metabolic CO$_2$ to greater depths increased the dissolved CO$_2$ raising the depth horizon below which CaCO$_3$ would dissolve. The increased Ca^{2+} slowly raised the alkalinity of the whole ocean, lowering the surface pCO$_2$ and reducing atmospheric CO$_2$ concentrations. During the last deglaciation, the change in the locus of carbonate production in the ocean from the deep sea (pelagic carbonates) to the continental shelves (coral reefs) following the rise in sea level (the coral reef hypothesis) has been suggested as a cause of the increase in the atmospheric pCO$_2$ level (Berger, 1982). Total atmospheric pCO$_2$ change due to this effect has been shown from modelling to approximate that observed in the Vostok ice core (Opdyke and Walker, 1992). However, another model, based on the hypothesis that a change in the relative rates at which organic carbon and calcium carbonate are deposited on the sea floor should drive a compensating change in ocean pH, also offers an explanation of the last deglaciation (Archer and Maier-Reimer, 1994).

Thus, although the effect of carbonate production is to remove carbon from the sea surface in the form of settling carbonate skeletons, the effect on the CO$_2$ flux is the reverse of that occurring during the production of non carbonate-producing plankton. Moreover, carbonate production is not governed by the Redfield ratios (or their variation). However, the existence of competing models to explain the recent palaeorecord (e.g., Heinze *et al.*, 1991; Westbroek *et al.*, 1993) suggests that a clear understanding of the factors controlling carbonate versus organic export production (Tsunogai and Noriki, 1991) is yet to emerge.

10.3.2.4 Role of continental margins

The continental margins play a major role in oceanic biogeochemical cycles (Walsh, 1991) and are burial sites of large masses of organic carbon, derived both from terrestrial sources via rivers or from local production in the coastal ocean (Berner, 1982). The reservoir of buried organic carbon in these areas is approximately six times greater than that in abyssal sediments (Holser *et al.*, 1988) in spite of a much smaller global area of deposition. The most recent estimate suggests that approximately 70 TgC/yr are buried

off deltas, and a similar amount accumulates on continental shelves and the upper parts of the continental slopes (Hedges and Keil, 1995). These two regimes account for more than 90% of the total amount of organic carbon buried on the sea floor; shallow-water carbonate platforms (4%), abyssal sediments (3%) and anoxic (oxygen-free) basins (0.5%) account for the remainder. Furthermore, the rate of accumulation of sediment along continental margins is higher by a factor of at least 100 than that in the deep ocean (Calvert and Pedersen, 1992), so that the deposited carbon is rapidly buried below the principal source of oxidants and the abundance maximum of bacteria at the sediment/water interface. In concert with adsorption of organic molecules to the surfaces of fine-grained sediment particles, this contributes to the preferential preservation of the less reactive organic materials (Ittekkot, 1993; Hedges and Keil, 1995).

Much of the material deposited on the margins is recycled in the benthic boundary layer, both by microbial activities and by the feeding of benthic animals. Dissolved carbon may be transported offshore into deeper waters or returned to the shallow water column, from where it can be recycled to the sea surface in these dynamic regimes on relatively short time-scales (Smith and Hollibaugh, 1993). A significant fraction is apparently permanently buried, as attested by the high organic carbon contents of many outer shelf and upper slope deposits of the world ocean (Premuzic *et al.*, 1982), although a recent intensive study of the Middle Atlantic Bight of eastern North America found that less than 5% of new biogenic particulate material is exported to the adjacent slope region (Biscaye *et al.*, 1994). Globally, Smith and Hollibaugh (1993) estimate an offshore export in the order of 0.2 GtC/yr.

Most authors caution that a global estimate of (especially) net primary fixation of carbon over continental shelves is subject to large (ca100%) errors. A recent numerical analysis of the seasonal cycles of CO$_2$, DOC and POC in relation to pristine nutrient conditions on Alaskan shelves and eutrophication in the northern Gulf of Mexico emphasises the complex heterogeneous nature of the continental margins (Walsh and Dieterle, 1994). They suggest that even pristine shelf areas such as the Bering and Chukchi Seas may have switched from being a source to a sink for CO$_2$ over the last 250 years in response to rising atmospheric CO$_2$ levels, with ~50% of the present invasion of CO$_2$ stored as DOC. Globally, they estimate that polar and temperate continental shelves currently sequester 1.0-1.2 GtC/yr, but that increased metabolism on tropical continental shelves may cause evasion of CO$_2$ to the atmosphere, resulting in continental shelves being a net world sink for CO_2 of 0.6-0.8 GtC/yr. Studies suggest that anthropogenic effects have already taken place to a much greater extent than in the open ocean, making the determination of an "unperturbed" carbon budget for continental margins even more difficult. Kempe (1993) estimates that the total transport of nitrate + nitrite to the continental shelf by the Danube River increased between the 1950s and 1989 from 12,500 t/yr to 273,000 t/yr, a 22-fold increase. Globally, Walsh (1989) suggested a possible tenfold increase of anthropogenic nitrogen loading to the coastal zone had already occurred by 1980, accounting for roughly half (0.4-0.5 GtC/yr) of the export to the continental slope regions, and Paerl (1993) estimated that atmospheric depositon of anthropogenic nitrogen contributes 10-50% of the total external nitrogen load to coastal areas. Smith and Hollibaugh (1993) argued that anthropogenic organic carbon loading to the coastal zone may have nearly doubled. Much of this organic carbon is oxidised, "net organic metabolism" in their terminology, contributing not to atmospheric carbon removal but rather "biotically mediated gas evasion" back to the atmosphere. Until we improve our understanding of the carbon and nitrogen dynamics in the coastal zone with sufficient spatial and temporal resolution (e.g., Kempe and Pegler, 1991) to allow reasonable global estimates, we cannot expect such estimates to converge on consensus values.

Christensen (1994) reported on a box-model study of the global carbon cycle including continental shelves, calibrated with atmospheric and oceanic $\delta^{14}CO_2$ observations. Atmospheric CO_2 levels were relatively insensitive to most model runs except that if most of the global ocean denitrification (due to remineralisation of organic carbon in low oxygen environments) was shifted from open ocean intermediate depths to the waters and sediments of continental shelves (to simulate increasing eutrophication of coastal regions), the atmospheric pCO_2 increased by ~10 ppmv, because shelf denitrification immediately removed nitrogen from the continental shelves, thereby reducing shelf primary production and utilisation of CO_2. On the other hand, Orr and Sarmiento (1992) concluded from simulated enhanced CO_2 fixation in large areas of an ocean general circulation model that drawdown of atmospheric CO_2 would be minor on a century time-scale, but they did not include a feedback that might lead to shelf denitrification.

10.3.2.5 Biological pump summary

In summary, while the rate of sequestering of organic carbon from the surface ocean is primarily limited by the supply of new nitrogen, the biological pump is sufficiently

complex that there are a number of possibilities for feedback to climate for which we possess inadequate understanding for quantitative assessment. Chief among them are: (1) factors controlling the Redfield ratios and their gradients with depth: increasing the C:N (or C:P) ratios will fix more carbon for each molecule of nutrient used; (2) the potential for changes in the proportion of $CaCO_3$ to organic carbon (C_{org}) formed by phytoplankton: an increase in the ratio $CaCO_3$:C_{org} from the current 1:4 to a possible 1:1 would eliminate the role of the biological pump in maintaining a reduced upper ocean pCO_2; (3) the role of the continental margins on a global scale in sequestering carbon and how they are changing as a result of anthropogenic inputs.

10.3.3 Fertilisation of Ocean Productivity – Natural and Anthropogenic

10.3.3.1 Inputs of nitrogen and iron from the atmosphere and rivers

The increased deposition over the oceans of atmospheric nitrogen from anthropogenic sources (fertilisers) over the last century may have resulted in increased export carbon production. Duce *et al.* (1991) estimated the total atmospheric deposition of nutrient nitrogen to the oceans to be about 30 TgN/yr. Schindler and Bayley (1993) estimated another 21 TgN/yr from riverine input. Using a C:N atomic ratio of 7:1 gives an equivalent increased production of export carbon in the oceans of 0.31 GtC/yr, most of which they assumed was due to human activities. Galloway *et al.* (1995) divided these inputs into present and pre-industrial, estimating a nitrogen input from human activities to the oceans of 18 TgN/yr via atmospheric deposition and 6 TgN/yr via rivers. A C:N atomic ratio of 7 gives (for the total 24 TgN/yr) an equivalent fertilised carbon production of 0.14 GtC/yr, only about 1% of estimated global marine export production. However a C:N ratio of ~10 as found by Sambrotto *et al.* (1993) would yield an equivalent fertilised production of 0.21 GtC/yr. In the Sargasso Sea, which is often nitrate-depleted, nitrogen deposition is an important source of export production during short episodes (Owens *et al.*, 1992), but for only 2% of the time is it capable of supporting the measured export production (Michaels *et al.*, 1993). Episodic deposition of anthropogenically produced nitrogen may then alter the functioning of regional ecosystems, without apparently having a significant effect on global export production. However, we point out that the highest estimate of human-induced fertilised export production, 0.21 GtC/yr, represents about 10% of 2 GtC/yr, believed to be the net flux from the atmosphere to the ocean. The key point about export production driven by deposition of atmospheric nitrogen compared with that driven by input of new nitrogen from upwelling is that it is not accompanied by upwelled deep DIC at Redfield ratios: it potentially converts to a net flux of DIC from the surface to the ocean interior.

Elemental nitrogen comprises over three-quarters of the atmosphere. The cyanobacterium *Oscillatoria* (*Trichodesmium*) can fix nitrogen for biological utilisation in the sea: recent estimates of total global nitrogen fixation of 10 TgN/yr, by Carpenter and Romans (1991), and of 40-200 TgN/yr, by Galloway *et al.* (1995) (including larger areas), correspond (for a Redfield C:N ratio in the order of 7) to a rate of carbon fixation by photosynthesis in the range of about 0.06-1.2 GtC/yr. Shifts induced by climate change

in the phytoplankton community to a greater proportion of *Oscillatoria* (as observed by Karl *et al.*, (1995) during the 1991 to 1992 warming event) could potentially increase the sequestering of anthropogenic CO_2 by a sizeable fraction in less than a decade.

Since the input of iron from the atmosphere may have been a factor in transitions between glacial and interglacial periods, estimates of current natural and anthropogenic inputs to the atmosphere and to the oceans are critical to improved understanding of the control of iron on marine primary production in high nutrient – low chlorophyll (HNLC) regions. Duce *et al.* (1991) and Duce and Tindale (1991) estimate the global input to the oceans of iron in atmospheric dust to be 32 Tg/yr of which about 10% is

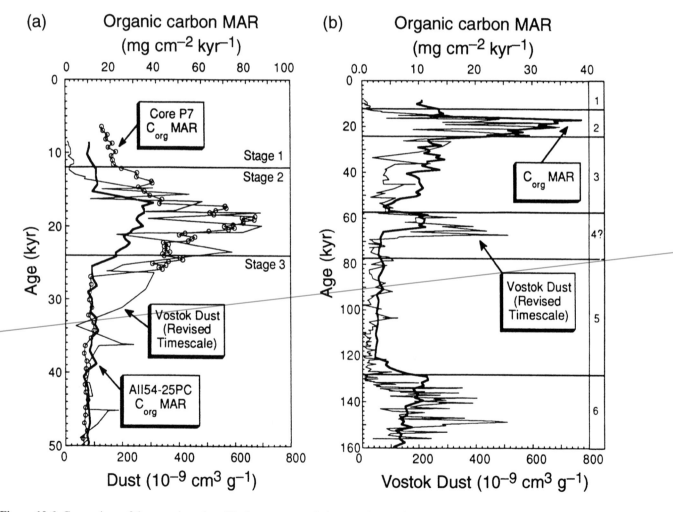

Figure 10.6: Comparison of the organic carbon (C_{org}) mass accumulation rate (MAR) from cores P7 (open circles) and AII54-25PC (bold line) and the Vostok dust record (thin line, plotted as volume of dust per gram of ice) as a function of time for the last 50 kyr (a) and the last 160 kyr (b). The correlation of the dust content in the Vostok core with the organic carbon accumulation rate in a deep ocean core does not necessarily indicate causation, but it does suggest the possibility of iron fertilisation in high nutrient - low chlorophyll (HNLC) regions during the last two periods of glaciation. The two cores were obtained from bottom depths of 3085 and 3225 m in the Panama Basin north-east of the Galapagos Islands, an HNLC region. From Pedersen *et al.* (1991).

readily soluble and hence biologically usable. Riverine input of dissolved iron is estimated to be about a third that from atmospheric dust. Most of the atmospheric input occurs in the Northern Hemisphere, much of it east of China. The inputs are also highly episodic.

Another intriguing process is the correlation of deep organic carbon fluxes with total mass accumulation rate, interpreted by Ittekkot (1993) as resulting from deposition of atmospheric dust acting as ballast sequestering more organic carbon deeper in the water column. On a time-scale less than decades, increased dust would by this process reduce atmospheric CO_2 levels. Increased remineralisation of organic carbon at depth would, according to Boyle (1988), redissolve more $CaCO_3$, leading to increased ocean alkalinity, further reducing the atmospheric CO_2 but on the millennium time-scale required for the alkalinity increase to reach the surface ocean.

10.3.3.2 *Fertilisation of phytoplankton productivity by iron*

Early results from the Vostok ice core showed that iron concentrations (as inferred from aluminium concentrations) were high during glacial periods, coincident with minima in atmospheric CO_2 concentrations (Martin, 1990). Comparison of such information with records of burial of marine organic matter (e.g., Lyle *et al.*, 1988; Pedersen *et al.*, 1991) shows a striking positive correlation: organic carbon burial was higher in the past when the atmosphere was more turbid (Figure 10.6). Although this correlation may be spurious, it suggests there may have been an effect from enhanced inputs of dust-borne iron to surface waters, at least in the tropical Pacific region.

Under the present climate, Southern Ocean phytoplankton productivity is assumed to be limited by lack of iron, since macro-nutrients in the surface layer there are apparently only rarely depleted (Lizotte and Sullivan, 1992), a characteristic in common with other HNLC regions – the subarctic North Pacific Ocean and the eastern Equatorial Pacific. Thus, Martin (1990) hypothesised that iron could be added to the Southern Ocean to stimulate primary production there to use up the large supply of nutrients and at the same time reduce the future rate of increase of atmospheric CO_2 concentration. There are several unknowns in this strategy: firstly, would the foodweb respond as hypothesised; secondly, are the magnitudes sufficient and would the coupled atmosphere-ocean system respond in such a straightforward way; and thirdly, would there be other unexpected responses with potential feedbacks to climate?

The first question – whether the primary production by phytoplankton in HNLC regions would respond to iron fertilisation – has recently received much attention. Although experiments conducted in small laboratory containers suggest that addition of iron stimulates phytoplankton growth, it is difficult to apply these results to the ocean because the very nature of container experiments precludes heterotrophic grazers and mixing, both factors which limit productivity in the ocean (Banse, 1991). To test this hypothesis in the ocean, acidified iron sulphate tagged with an inert chemical tracer was added to an area of 64 km^2 in the equatorial Pacific Ocean in October 1993 (Kolber *et al.*, 1994; Martin *et al.*, 1994; Watson *et al.*, 1994). Early results suggest that while the addition of iron had an easily measurable effect on the marine ecosystem (with an approximate threefold increase in both productivity and chlorophyll concentrations), it had only a small effect on the carbon dioxide levels. After a brief disequilibrium the ecosystem apparently responded by recycling carbon rapidly back to the inorganic form, although there was evidence that the whole patch sunk out of the surface layer after several days A second fertilisation experiment was conducted in the same area in May 1995 with more positive results, but the Equatorial Pacific is not a likely site for a purposeful widespread iron fertilisation because the surface layer is isolated from the deep ocean by the main thermocline (Sarmiento and Orr, 1991). Recently, de Baar *et al.* (1995) found spring blooms in the Polar Front of the Southern Ocean with biomass levels ten times those of adjacent areas. The Polar Front was characterised by high iron concentrations in the top 150 m of the ocean, as compared with lower dissolved iron in the remainder of the Circumpolar current where no blooms were found.

The second of these questions has been addressed recently with 3-dimensional ocean general circulation models (Sarmiento and Orr, 1991; Kurz and Maier-Reimer, 1993). The models assumed that iron fertilisation of the Southern Ocean would occur every year for 50 or 100 years (a single addition would basically achieve nothing), the phytoplankton would utilise all nutrients in the surface layer (although de Baar *et al.* (1995) observed that nutrients were not exhausted in regions of elevated iron concentration), and the area of coverage (10 to 15% of the global ocean) could actually be fertilised. The rate and extent of fertilisation depended primarily on the rate and extent of vertical mixing in the Southern Ocean, whereby the nutrient in the surface ocean can be replenished from below, and on the rate of exchange of CO_2 between the atmosphere and the ocean. The simulations gave similar results: that the atmospheric reduction after 100 years of iron fertilisation would be less than 10% of total atmospheric concentration in 2100 for the "Business as usual" and less than 20% for the "Constant emission" scenarios, as outlined in IPCC (1990).

With regard to the third question, all model simulations indicated significant reduction in oxygen concentrations below the surface layers of the Southern Ocean due to the remineralisation (oxidation by bacteria) of the increased rain of organic matter out of the euphotic layer. Fuhrman and Capone (1991) reviewed potential biogeochemical consequences of oceanic iron fertilisation, including changes in DMS production and possible consequences of widespread subsurface decrease in oxygen concentrations. In fact, in the 1993 iron enrichment study, Martin *et al.* (1994) observed a significant increase within the enriched patch in particulate DMSP (dimethyl sulphoniopropionate), the precursor to DMS (see Section 10.3.4). In addition, if oxygen concentrations were to become severely depleted, anoxic bacterial processes such as denitrification, sulphate reduction and methanogenesis could increase in importance. Denitrification would result in loss of nitrate, counteracting the fertilisation, and methanogenesis could lead to additional release from the oceans of another important greenhouse gas, methane (CH_4) (see Section 10.3.5.1). The formation and release of N_2O to the atmosphere could be increased by the increased remineralisation of nitrogen compounds caused by the increased primary production resulting from the iron fertilisation. Using the fact that the per molecule Global Warming Potential for N_2O is about 300 times that of CO_2 on the 20 to 100 year time horizon (IPCC, 1994), and the results of the model simulations by Joos *et al.* (1991), Sarmiento and Orr (1991) and Fuhrman and Capone (1991) concluded that increased release of N_2O could offset the effect of a significant fraction of the removal of atmospheric CO_2 by iron fertilisation.

In summary, there is mounting evidence, from both the palaeorecord and contemporary studies that iron can enhance phytoplankton production where macro-nutrients are present in excess. However, it is clear from all the studies conducted to date, both experimental and model simulations, that iron fertilisation is not a feasible mitigation tool, given our current knowledge of the potential ramifications of such a procedure. Fertilisation of the ocean with iron or nitrates through inputs from increased land runoff and atmospheric deposition can lead to eutrophication or increased biological production. Increased remineralisation of organic materials can increase the production of N_2O, and in low oxygen environments (especially on continental shelves) can promote denitrification and thus the release of N_2 and CH_4

to the atmosphere. Removing nitrogen from the shelves can also reduce phytoplankton production and the sequestering of carbon. All three processes represent potential positive feedbacks to the climate system.

10.3.4 DMS Production and Possible Changes to Cloud Condensation Nuclei

DMS is the dominant volatile sulphur-containing compound in the oceans and the major natural source of sulphur to the atmosphere (Bates *et al.*, 1992; Liss and Galloway, 1993). Rough global estimates suggest that DMS emissions to the atmosphere are approximately 25 TgS/yr, compared with estimates of sulphate (from sulphur dioxide) of 10 TgS/yr from volcanoes and 75 TgS/yr from fossil fuels. Sulphur emissions in the Southern Hemisphere are clearly dominated by oceanic DMS. When ventilated to the atmosphere, DMS is oxidised to produce aerosol particles which affect the radiative properties of marine stratus clouds and thereby affect climate (Charlson *et al.*, 1987). Although the magnitude of this effect is highly uncertain, Charlson *et al.* calculated that a 30% increase in the number of particles serving as cloud condensation nuclei (CCN) in marine stratiform clouds would decrease the global-average surface temperature of the Earth by 1.3°C. Correlations between DMS and CCN have been observed (Ayers and Gras, 1991; Hegg *et al.*, 1991; Andreae *et al.*, 1995), although the processes controlling the formation and growth of particles in the marine boundary layer are still controversial. Furthermore, Boers *et al.* (1994) observed significant in-phase coherence in the seasonal cycles between 7-year records of marine boundary layer CCN concentrations and satellite-derived cloud optical depth in the vicinity of an open ocean Southern Hemisphere observing station. Quantifying these processes is a major focus of the Aerosol Characterisation Experiments (ACE) of the International Global Atmospheric Chemistry (IGAC) Program.

A key factor in this natural sulphur/aerosol/climate system is the air-sea exchange of DMS which is a function of the surface sea water DMS concentration. Unfortunately, the factors controlling oceanic DMS concentrations and the parameters needed to model these concentrations are not well characterised. Ice core measurements of the atmospheric DMS oxidation product, methane sulphonate (MSA), suggest that DMS emissions (and presumably oceanic DMS concentrations) may have changed by a factor of 6 between glacial and interglacial times (Legrand *et al.*, 1991). Seasonal studies of oceanic DMS concentrations have shown that average surface sea water DMS concentrations can vary by as much as a factor of 50 between summer and winter in the mid- and high latitudes. On large regional and temporal scales, DMS concentrations have been correlated with sea water chlorophyll concentrations, but in general, oceanic DMS distributions are poorly correlated with phytoplankton production or biomass (Holligan *et al.*, 1987; Andreae, 1990; Leck *et al.*, 1990). Part of the difficulty in establishing these correlations is that the production of the DMS precursor, dimethyl sulphoniopropionate (DMSP), is highly species specific. Prymnesiophytes, including coccolithophorids and Phaeocystis, and some dinoflagellates produce orders of magnitude more DMSP per cell than other algae such as diatoms (Keller, 1991; Liss *et al.*, 1994). Thus, blooms of coccolithophorids in the relatively nutrient-poor open ocean can produce extremely high concentrations of DMSP (Malin *et al.*, 1993) as does *Phaeocystis* in nutrient-rich polar and some coastal waters (Turner *et al.*, 1988). Another complicating factor in modelling DMS concentrations is that the conversion of DMSP to DMS is often associated with the decline of a phytoplankton bloom during the senescence phase or during active zooplankton grazing as opposed to the active growth phase (Dacey and Wakeham, 1986; Turner *et al.*, 1988; Leck *et al.*, 1990). Furthermore, DMS

is only one product of DMSP and the percentage conversion to DMS is dependent on the plankton cell density (Wolfe *et al.*, 1994). The production of DMS in surface sea water is therefore a strong function of the species composition, the total biomass and the local trophic structure. Processes which affect these biological parameters such as eutrophication, enhanced UV-B radiation or changing temperature will potentially affect the concentration of DMS and its subsequent flux to the atmosphere.

DMS is removed from sea water by various physical, chemical and biological pathways (Figure 10.7). Kiene and Bates (1990) and Bates *et al.* (1994) have shown that in tropical and temperate waters of the Pacific, DMS removal is principally by biological processes with microbial consumption removing DMS more than ten times faster than atmospheric ventilation. Thus it appears that DMS concentrations are intimately linked to both the structure and level of activity of the microbial food web (Belviso *et al.*, 1990). If bacterial activities are being substantially inhibited by present levels of UV radiation (see Section 10.3.6), attention needs to be given to the impact of UV-B not only on the organisms that produce DMS and its precursor, but also on those other organisms involved,

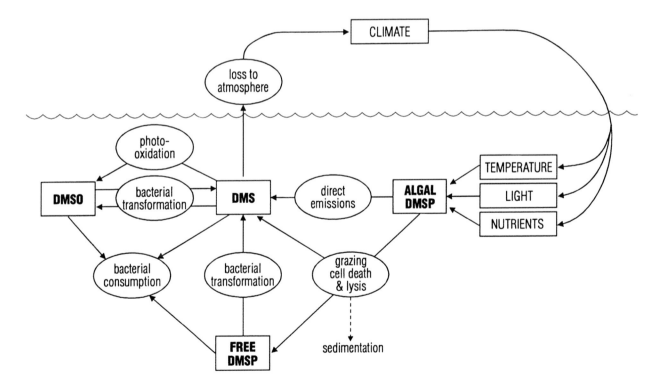

Figure 10.7: The marine biogeochemical cycle of DMS (dimethyl sulphide): production of DMSP (dimethyl sulphonioproprionate) by phytoplankton, transformation by bacteria to DMS and by bacteria or photo-oxidation to DMSO (dimethyl sulphoxide), and alternative utilisation pathways that may influence the quantity of DMS lost to the atmosphere. From Malin *et al.* (1992).

directly and indirectly in its utilisation. In addition, another fate of DMS in sea water is photochemical oxidation (Brimblecombe and Shooter, 1986). The extent to which an increase in UV-B flux changes the rate of photochemical degradation of DMS is yet to be ascertained. From these data it is apparent that reliable parametrizations of surface oceanic DMS concentrations will require a better understanding of the processes involved in the cycling of sulphur in the upper water column. It remains to be resolved how changes in climate will affect this cycle and the exchange of DMS with the atmosphere. Furthermore, in the Northern Hemisphere the much larger change in atmospheric aerosols from human activities will make climate-related changes in natural DMS outgassing difficult to detect.

In summary, the link between DMS production by phytoplankton and increased cloud condensation nuclei is reasonably well established. However, the biological and chemical processes in the ocean and the chemical processes in the atmosphere are complex, and are only sufficiently well understood to allow their parametrization for incorporation into preliminary models (Gabric *et al.* 1993).

10.3.5 Marine Biota and Other Greenhouse Gases

10.3.5.1 Methane

Although the concentration of methane in the atmosphere is well documented, the magnitudes of its sources and sinks and the processes involved are poorly understood. The annual input from the oceans to the atmosphere is of order 10 Tg/yr, representing about 2% of all sources (IPCC, 1994). Although Bange *et al.* (1994) estimated that ~75% of the oceanic outgassing of methane comes from the continental shelves where environmental change is greatest, it is unlikely that changes in marine biota could alter the oceanic (net) contribution by more than a few Tg/yr, or less than 1% of the total amount emitted to the atmosphere annually.

10.3.5.2 Nitrous oxide

The atmospheric concentration of nitrous oxide (N_2O) in 1992 was 311 ppbv, increasing by around 0.25%/yr. Sources of atmospheric N_2O and estimates of annual emissions are very uncertain. It is thought that the oceans contribute about 3 Tg/yr to a total annual input of around 15 Tg/yr (IPCC, 1994). Production of N_2O in the sea occurs during remineralisation/oxidation of organic matter in intermediate layers and at sediment interfaces (e.g., Elkins *et al.*, 1978; Cohen and Gordon, 1979; Codispoti and Christensen, 1985; Law and Owens, 1990; Naqvi and Noronha, 1991; Christensen, 1994).

10.3.5.3 Other trace gases

Carbon monoxide, organic carbonyl sulphide (OCS) and non-methane hydrocarbons (NMHCs) appear indirectly linked to marine productivity, but the main mechanism of their production is photochemical breakdown of dissolved organic matter (DOM) in surface waters (e.g., Ratte *et al.*, 1993; Bates *et al.*, 1995; Weiss *et al.*, 1995). Photochemical production of trace gases in the surface ocean influences their ocean to atmosphere flux. Carbon monoxide is an important atmospheric constituent because its reaction with the hydroxyl radical (OH) may control OH concentration and thus influence global biogeochemical processes. Oxidation of NMHCs may also influence OH concentrations in the atmosphere. OCS is a precursor of stratospheric sulphate aerosols and is thus likely to participate in heterogeneous reactions related to ozone depletion. The influence of enhanced UV-B exposure on the concentrations of these trace chemical species has received very little attention.

Organo-halogen gases, such as bromoform and methyl halides, formed by biological and/or photochemical processes in the oceans may have a significant effect on the oxidising capacity of the atmosphere, following their emission from sea water. For example, the decreases in tropospheric ozone concentrations observed at Arctic polar sunrise have been attributed to destruction of ozone by bromine species arising from marine-produced organo-bromine compounds (Barrie *et al.*, 1988). How such sea-to-air fluxes might change under an altered climate is hard to assess, particularly in view of our lack of understanding of the processes by which these gases are formed in the oceans and broken down in the atmosphere.

10.3.6 Effect of UV-B on Marine Plankton with Potential Feedback to Climate

Over the millennia, equilibrium has been reached between incident sunlight, its natural variability and the components of the marine ecosystem. Altering this balance by suddenly (on evolutionary time-scales) depleting stratospheric ozone can have serious consequences for the organisms and their activities. Concern has been expressed that any diminution in marine primary production may lead to a positive feedback with respect to atmospheric CO_2 that would exacerbate the greenhouse effect. That exposure to UV-B decreases primary productivity and inhibits growth of marine microorganisms has been well documented (e.g., Cullen and Lesser, 1991; Cullen *et al.*, 1992). Also it has been found that there are substantial differences between species in their response to UV-B exposure indicating the possibility of shifts in community structure which has consequences for both foodweb dynamics and biogeochemical cycles. Marine organisms employ a range of responses and adaptations to minimise their UV-B exposure. These include the synthesis of photoprotective pigments, DNA repair mechanisms, and avoidance mechanisms (Karentz, 1994). The linkages between, and consequences of, increased UV-B exposure and biogeochemical reactions are at present only poorly known and in obvious need of investigation (Marchant, 1994).

Inhibition of phytoplanktonic photosynthesis results from exposure to UV-A (wavelengths 320-400 nm) and PAR (photosynthetically available radiation: 400-700 nm) as well as UV-B. Helbling *et al.* (1992) reported that UV-A was responsible for over 50% of the photo-inhibition, with wavelengths shorter than 305 nm accounting for 15-20% of the inhibition. The action spectrum, or biological weighting function of UV inhibition of photosynthesis, varies with both the absolute amount and the ratio of UV-B:UV-A + PAR but not UV:PAR (Neale *et al.*, 1994). Tropical phytoplankton exhibit substantially higher resistance to UV exposure than Antarctic organisms. However, phytoplankton from below the mixed layer in Antarctic waters appear to be more tolerant of solar radiation than organisms from below the mixed layer in tropical waters. This is likely to be due to the relative instability and deeper mixing in the Southern Ocean and the photo-adaptive processes of these organisms from surface waters. Vertical mixing plays a crucial role in mitigating photo-inhibition and UV-B exposure.

UV-B radiation penetrates to ecologically significant depths of the ocean, i.e., 10% of surface UV radiation can penetrate to depths of 5 to 25 m (Smith and Baker, 1989). Incubations of phytoplankton under both natural and laboratory sources of UV-B radiation result in species-specific reduced productivity, growth and reproduction, nitrogen metabolism and motility. As a consequence, changes in community structure favouring those organisms that are more tolerant of UV would be expected. A better understanding of the nature and magnitude of these changes requires further work to elucidate the photobiology of key species. In addition, the eggs and larval stages of invertebrates and fish have been shown to be sensitive to UV-B exposure (Hunter *et al.*, 1981; Damkaer and Dey, 1983).

Over the last decade our understanding of the importance of bacteria in marine processes has developed dramatically. It is now recognised that these highly abundant organisms play a pivotal role in biogeochemical pathways, in foodweb dynamics, and as well in such physical processes as light scattering and combining with trace metals (Cho and Azam,

1990; Simon *et al.*, 1992). However, only a few studies have been conducted on the effects of UV on bacteria, and much of the impact of solar UV exposure has been ascribed previously to UV-A. But recent investigations indicate a high sensitivity to UV over a broader wavelength band (Herndl *et al.*, 1993): a 30-minute exposure to solar UV-B resulted in a 40% reduction in bacterial activity in the top 5 m of coastal waters. Bacterial extracellular enzymic activity was inhibited by around 70%. Production of extracellular enzymes was reduced by UV-B exposure, but photolysis of these enzymes in solution accounted for the largest diminution of activity.

Most of the work on the impacts of UV-B on natural systems has concentrated on living organisms with only relatively few studies on the effect of UV photolysis on dissolved organic and inorganic material. Oceanic DOC is one of the largest reservoirs of carbon on Earth; yet the processes regulating the cycling of this material are not well understood. It is estimated that some 15% of marine DOC exists as humic substances, assumed to be slowly degradable macromolecules. However, the bulk of oceanic DOC appears to be comprised of small molecules (molecular weight less than 1000 daltons), relatively unavailable to microorganisms (Amon and Benner, 1994). Aquatic refractory macromolecules can be photochemically degraded by sunlight into biologically labile compounds (Kieber *et al.*, 1989). Mopper *et al.* (1991) proposed that such photochemical degradation is the rate limiting step for the removal of a large fraction of oceanic DOC, and it is assumed that this will increase under increased solar UV-B flux. Recently, this has lead to heightened interest in this phenomenon, although the implications of such photochemical degradation of organic matter in sea water had been discussed in the late 1970s (Duursma and Dawson, 1981, and papers therein). One of the difficulties in extrapolating trends from the work done to date in the world ocean is the different nature of the DOC pools investigated. Several studies have concentrated on coastal waters where terrestrially derived organic matter may be more susceptible to photodegradation than purely marine DOC (Moran and Hodson, 1994). Others have worked on samples from deep or oligotrophic waters, and Mopper *et al.* (1991) proposed that the photodegradable fraction of the DOC might originate from sediment porewater. Data from productive surface waters are lacking, which is surprising considering the greatest amount of oceanic DOC is produced by planktonic organisms in this UV-influenced layer. Thomas and Lara (1995) have shown that algal-derived DOC is resistant to naturally occurring levels of UV radiation. This finding cautions against the extrapolation of photodegradation of poorly characterised, mixed DOC pools to estimates of oceanic carbon budgets. Bacteria are the principal consumers of labile DOC. Increased substrate concentrations could plausibly sustain higher bacterial activity, leading to accelerated nutrient remineralisation. However, as previously mentioned, elevated UV-B levels may also inhibit bacterial activity, resulting in a net accumulation of labile dissolved organic matter in surface waters. It is also possible that certain photolytic products have a toxic or adverse ecophysiological effect on organisms (McKnight *et al.*, 1990). Although photochemical degradation of DOC dominates present thinking, photoreactions may even have an opposite effect, playing a role in the production of refractory compounds (Lee and Henrichs, 1993). It is clearly premature to adopt the scenario of increased UV-B flux causing higher concentrations of labile substrates, with subsequent increased bacterial respiration leading to enhanced production of CO_2.

Thus we know that bacteria and phytoplankton can be damaged by UV-B exposure and that because of the differential response of various species to UV, community structure may be altered. We do not know yet if these changes will significantly alter exchanges of radiatively active gases with the atmosphere. Furthermore, if there are changes, it is not clear whether they will be of global consequence. At present it is unclear what the effects of a change in global temperature would be on ozone depletion and increased UV-B radiation, but Austin *et al.* (1992) predicted the regular formation of Arctic ozone holes as a consequence of stratospheric cooling.

In summary, solar UV-B has increased due to stratospheric ozone depletion. Species-specific sensitivity to UV-B by phytoplankton could lead to widespread shifts in community structure with possible changes in the ability for high latitude ecosystems to sequester carbon. UV-B inhibits marine bacterial activity and has differing effects on DOC derived from different sources, both factors that could lead to significant alteration of ecosystem structure.

10.4 The State of Biogeochemical Ocean Modelling

If we are to perform realistic projections of the possible responses (and feedbacks) of the coupled land-atmosphere-ocean system to climate variation and environmental change over the next 50 to 100 years, then it is clear that coupled time-dependent GCMs of these systems will play a key role in formulation of these projections. At present, published results exist for model simulations with coupled atmosphere-ocean models forced by an atmospheric CO_2

concentration of 2 or 4 times the pre-industrial value (see Chapters 5 and 6).

These coupled ocean-atmosphere models include only the physical climate system. The chemistry and biology of CO_2, $CaCO_3$, N_2O, DMS and other radiatively-active gases that may alter the efficacy of the greenhouse effect or the concentration of atmospheric aerosols are not included explicitly in these coupled models. CO_2 is included only in terms of how its atmospheric concentration alters the strength of the greenhouse effect. The ocean carbon cycle and how it affects the CO_2 partial pressure difference across the air-sea boundary, and hence the air-sea transfer, is not included. Feedback effects, such as how increasing temperatures will affect the partial pressures of CO_2 in the atmosphere and the ocean, and hence the air-sea fluxes, are also yet to be included. Initial box models were successful at demonstrating the probable role of increased high latitude ocean primary production in reducing the atmospheric CO_2 concentrations during glacial periods (Knox and McElroy, 1984; Sarmiento and Toggweiler, 1984; and Siegenthaler and Wenk, 1984). At present geochemical cycles have been simulated in 2- and 3-dimensional ocean GCMs (e.g., Maier-Reimer, 1993; Drange, 1994; Stocker *et al.*, 1994), and no doubt initial simulations with coupled models that include active geochemical cycling are now being undertaken. It is likely that contemporary limitations on computing power will limit sensitivity studies. This will constrain the rate of progress in developing coupled models that will reproduce an "equilibrium" climate not just of the physical cycles but also of the biogeochemical cycles. Most present climate GCMs do not include continental margins or their processes. These models predict that the largest physical changes will occur around the model boundaries in some basins (e.g., the North Atlantic). The effect of these changes on biological processes related to global greenhouse gas fluxes and aerosol concentrations needs to be elucidated. We need to acquire sufficient understanding of the processes that control the vertical fluxes of climate-related gases and biogenic particulates (e.g., Erickson, 1993), and to develop improved models of the upper ocean and its upper and lower boundaries for inclusion in the future development of multi-disciplinary coupled ocean-atmosphere models. At present we do not know enough about the processes governing the production, degradation and sources of light-sensitive compounds in the upper ocean, in particular DOC and DMS, to parameterize their behaviour in these models.

The availability of models for assessment of climate change effects on marine biotic systems is limited to box models of various levels of sophistication (e.g., Shaffer, 1993; Christensen, 1994; Shaffer and Sarmiento, 1995). Models for investigating marine ecosystems have focused on understanding or identifying important processes, that is they are process-oriented models. The majority of the existing models consider interactions between circulation and a truncated form of the marine ecosystem. The circulation component of these models ranges from simple prescribed conditions (e.g., Fasham *et al.*, 1990) to large-scale primitive equation general circulation models (e.g., Sarmiento *et al.*, 1993; Fasham *et al.*, 1993). The marine ecosystem representations are generally in terms of aggregate dynamics, e.g., single terms representing all phytoplankton or all zooplankton (Wroblewski, 1977; Wroblewski *et al.*, 1988). However, some attempts have been made to include realistic size- or stage-structured phytoplankton and zooplankton population dynamics (Hofmann, 1988; Moloney and Field, 1991).

The current generation of models that have been constructed to consider biogeochemical cycling in the ocean are limited in application by the inability to prescribe adequately the temporal and spatial variation in the parametrizations of biological processes known to exist over scales of more than a few days and more than tens of kilometres. Consequently, most of the models consider regional processes and have been focused on understanding short-term episodic events such as coastal upwelling. The few attempts at larger scale applications (North Atlantic basin) have shown that the solutions degrade over time due to inadequate representation of the marine ecosystem and deficiencies in the simulated circulation and mixing patterns (Sarmiento *et al.*, 1993: Slater *et al.*, 1993). Efforts are now being focused on developing algorithms and parametrizations for biological processes such that they are valid over larger regions and over longer time-scales (Evans and Fasham, 1993). In addition, circulation models are usually constructed to aid in understanding ocean circulation dynamics: thus, the horizontal and vertical resolution of these models is usually not adequate for biogeochemical modelling needs. Also, circulation models often lack dynamics, such as mixed-layer circulation, that are important in regulating the flux of biogeochemical properties. Future advances in modelling of climate effects on marine systems will require that the circulation and biogeochemical models be developed together.

Perhaps the strongest limitation at present is the lack of data that are adequate to formulate, calibrate and evaluate marine climate effects models. This limitation has been recognised for marine modelling in general and marine ecosystem modelling in particular (GLOBEC, 1993a, b).

The need for multi-disciplinary data sets has led to proposals to develop integrated multi-platform data acquisition systems (Dickey, 1991). This in turn has prompted the development of marine ecosystem and circulation models that are capable of assimilation of a wide range of data sets. Unified modelling and observational efforts are critical for efficient model use and advancement of interdisciplinary research, such as that required for addressing questions of climate change responses and feedbacks.

As a final note of caution, in the ocean many algae, bacteria, viruses, protozoa and metazoa interact with the physico-chemical environment and with each other. There are interspecific differences both in responses to environmental conditions and in specific biogeochemical effects. It is from this almost infinite multiplicity of connections that the climatic role of the marine biota emerges. A further complication is that these interactions may span a range of time-scales from millisecond to millennia and beyond, as climate change affects the molecular biological constitution of the marine biota, and this in turn feeds back on climate. For an adequate understanding of the climate system these fundamental aspects cannot be ignored. The fact that the present models oversimplify the issue by emphasising bulk effects and responses of the marine biota further illustrates our limited understanding of the climate system. However, new generations of biogeochemical climate models are being developed, providing a more satisfactory representation of the diversity and operational range of the marine biota (e.g., Westbroek *et al.*, 1993).

In summary, coupled ocean-atmosphere models used for climate change projections do not presently contain any marine biogeochemical processes. However, component biogeochemical models are rapidly improving, and they will be incorporated into coupled ocean-atmosphere models over the next few years, as dictated by available computing power. The extent to which they can adequately simulate the complexities of the marine ecosystem will ultimately be determined by the availability of observations required for parameter estimation.

References

Altabet, M.A., R. Francois, D.W. Murray and W.L. Prell, 1995: Climate-related variations in denitrification in the Arabian Sea from sediment $^{15}N/^{14}N$ ratios. *Nature*, **373**, 506–509.

Amon, R.M.W. and R. Benner, 1994: Rapid cycling of high molecular weight dissolved organic matter in the ocean. *Nature*, **369**, 549–552.

Anderson, L.A. and J.L. Sarmiento, 1994: Redfield ratios of remineralization determined by nutrient data analysis. *Global Biogeochem. Cycles*, **8**, 65–80.

Andreae, M.O., 1990: Ocean-atmosphere interactions in the global biogeochemical sulfur cycle. *Mar. Chem.*, **30**, 1–29.

Andreae, M.O., W. Elbert and S.J. deMora, 1995: Biogenic sulfur emissions and aerosols over the tropical south Atlantic, 3: atmospheric dimethylsulfide, aerosols and cloud condensation nuclei. *J. Geophys. Res.*, **100**, 1133–11356.

Archer, D. and E. Maier-Reimer, 1994: Effect of deep-sea sedimentary calcite preservation on atmospheric CO_2 concentration. *Nature*, **367**, 260–263.

Atkinson, L.P., 1977: Modes of Gulf Stream intrusion into the South Atlantic Bight shelf waters. *Geophys. Res. Lett.*, **4**, 583–586.

Austin, J., N. Butchart and K.P. Shine, 1992: Possibility of an Arctic ozone hole in a doubled-CO_2 climate. *Nature*, **360**, 221–225.

Ayers, G.P. and J.L. Gras, 1991: Seasonal relationships between cloud condensation nuclei and aerosol methanesulphonate in marine air. *Nature*, **353**, 834–835.

Bacastow, R. and E. Maier-Reimer, 1991: Dissolved organic carbon in modeling oceanic new production. *Global Biogeochem. Cycles*, **5**, 71–85.

Bakun, A., 1990: Global climate change and intensification of coastal ocean upwelling. *Science*, **247**, 198–201.

Bange, H.W., U.H. Bartell, S. Rapsomanikis and M.O. Andreae, 1994: Methane in the Baltic and North Seas and a reassessment of the marine emissions of methane. *Global Biogeochem. Cycles*, **8**, 465–480.

Banse, K., 1991: Rates of phytoplankton cell division in the field and in iron enrichment experiments. *Limnol. Oceanogr.*, **36**, 1886–1898.

Banse, K., 1994: Uptake of inorganic carbon and nitrate by marine plankton and the Redfield ratio. *Global Biogeochem. Cycles*, **8**, 81–84.

Barrie, L.A., J.W. Bottenheim, R.C. Schnell, P.J. Crutzen and R.A. Rasmussen, 1988: Ozone destruction and photochemical reactions at polar sunrise in the lower Arctic atmosphere. *Nature*, **334**, 138–141.

Barry, J.P., C.H. Baxter, R.D. Sagarin and S.E. Gilman, 1995: Climate-related, long-term faunal changes in a California rocky intertidal community. *Science*, **267**, 672-675.

Bates, T.S., B. K. Lamb, A.B. Guenther, J. Dignon and R.E. Stoiber, 1992: Sulfur emissions to the atmosphere from natural sources. *J. Atmos. Chem.*, **14**, 315–337.

Bates, T.S., R.P. Kiene, G.V. Wolfe, P.A. Matrai, F.P. Chavez, K.R. Buck, B.W. Blomquist and R.L. Cuhel, 1994: The cycling of sulfur in surface seawater of the northeast Pacific. *J. Geophys. Res.*, **99**, 7835–7843.

Bates, T. S., K.C. Kelly, J.E. Johnson and R.H. Gammon, 1995: Regional and seasonal variations in the flux of oceanic carbon monoxide to the atmosphere. *J. Geophys. Res.*, **100**, 23093–23101.

Belviso, S., S.K. Kim, F. Rassoulzadegan, B. Krajka, B.C. Nguyen, N. Mihalopoulos and P. Buat-Menard, 1990: Production of dimethylsulfonium propionate (DMSP) and dimethylsulfide (DMS) by a microbial food web. *Limnol. Oceanogr.*, **35**, 1810–1821.

Berger, W.H., 1982: Increase of carbon dioxide in the atmosphere during deglaciation: The coral reef hypothesis. *Naturwissenschaften*, **69**, 87–88.

Berger, W.H., 1989: Global maps of ocean productivity. In: *Productivity of the Ocean: Present and Past*, W.H. Berger, V.S. Smetacek, and G. Wefer (eds.), John Wiley and Sons, Chichester, pp. 429–455.

Berger, W.H. and G. Wefer, 1991: Productivity of the glacial ocean: Discussion of the iron hypothesis. *Limnol. Oceanogr.*, **36**, 1899–1918.

Berner, R.A., 1982: Burial of organic carbon and pyrite sulphur in the modern ocean: its geochemical and environmental significance. *Am. J. Sci.*, **282**, 451–473.

Biscaye, P.E., C.N. Flagg and P.G. Falkowski, 1994: The Shelf-Edge Exchange Processes experiment, SEEP-II: an introduction to hypotheses, results and conclusions. *Deep-Sea Res. II*, **41**, 231–252.

Boers, R., G.P. Ayers and J.L. Gras, 1994: Coherence between seasonal variation in satellite-derived cloud optical depth and boundary layer CCN concentrations at a Mid-latitude Southern Hemisphere station. *Tellus*, **46B**, 123–131.

Boyle, E.A., 1988: The role of vertical chemical fractionation in controlling late Quaternary atmospheric carbon dioxide. *J. Geophys. Res.*, **93**, 15701–15714.

Brimblecombe, P. and D. Shooter, 1986: Photo-oxidation of dimethylsulfide in aqueous solution. *Mar. Chem.*, **19**, 343–353.

Brodeur, R.D. and D.M. Ware, 1992: Long-term variability in zooplankton biomass in the subarctic Pacific Ocean. *Fish. Oceanogr.*, **1**, 32–38.

Broecker, W.S., 1982: Ocean chemistry during glacial time. *Geochimica et Cosmochimica Acta*, **46**, 1689–1705.

Broecker, W.S., 1991: The Great Ocean Conveyor. *Oceanography*, **4**, 79–89.

Broecker, W.S. and T-H. Peng, 1982: *Tracers in the Sea*. Lamont-Doherty Geological Observatory, Columbia University.

Calvert, S.E. and T.F. Pedersen, 1992: Organic carbon accumulation and preservation in marine sediments: How important is anoxia? In: *Productivity, Accumulation and Preservation of Organic Matter in Recent and Ancient Sediments*, J.K. Whelan and J.W. Farrington (eds.), Columbia University Press, New York, pp. 231–263.

Carlson, C.A. and H.W. Ducklow, 1995: Dissolved organic carbon in the upper ocean of the central Equatorial Pacific, 1992: Daily and finescale vertical variations. *Deep-Sea Res. II*, **42**, 639–656.

Carlson, C.A., H.W. Ducklow and A.F. Michaels, 1994: Annual flux of dissolved organic carbon from the euphotic zone in the northwestern Sargasso Sea. *Nature*, **371**, 405–408.

Carpenter, E.J. and K. Romans, 1991: Major role of the *Cyanobacterium Trichodesmium* in nutrient cycling in the North Atlantic Ocean. *Science*, **254**, 1356–1358.

Charlson, R.J., J.E. Lovelock, M.O. Andreae and S.G. Warren, 1987: Oceanic phytoplankton, atmospheric sulfur, cloud albedo and climate. *Nature*, **326**, 655–661.

Cho, B.C. and F. Azam, 1990: Biogeochemical significance of bacterial biomass in the ocean's euphotic zone. *Marine Ecology Progress Series*, **63**, 253–259.

Christensen, J.P., 1994: Carbon export from continental shelves, denitrification and atmospheric carbon dioxide. *Continental Shelf Research*, **14**, 547–576.

Codispoti, L.A. and J.P. Christensen, 1985: Nitrification, denitrification and nitrous oxide cycling in the eastern tropical South Pacific Ocean. *Mar. Chem.*, **16**, 277–300.

Cohen, Y. and L.I. Gordon, 1979: Nitrous oxide production in the ocean. *J. Geophys. Res.*, **84**, 347–353.

Cullen, J.J. and M.P. Lesser, 1991: Inhibition of photosynthesis by ultraviolet radiation as a function of dose and dosage rate: results for a marine diatiom. *Mar. Biol.*, **111**, 183–190.

Cullen, J.J., P.J. Neale and M.P. Lesser, 1992: Biological weighting function for the inhibition of phytoplankton photosynthesis by ultraviolet radiation. *Science*, **258**, 646–650.

Dacey, J.W.H. and S.G. Wakeham, 1986: Oceanic dimethylsulfide: production during zooplankton grazing on phytoplankton. *Science*, **233**, 1314–1316.

Damkaer, D.M. and D.B. Dey, 1983: UV damage and photoreactivation potentials of larval shrimp *Pandalus platyceros* and adult euphausiids *Thysanoessa raschii*. *Oecologia*, **60**, 169–175.

Dansgaard, W., S.J. Johnsen, H.B. Clausen, D. Dahl-Jensen, N.S. Gundestrup, C.U. Hammer, C.S. Hvidberg, J.P. Steffensen, A.E. Sveinbjörnsdottir, J. Jouzel and G. Bond, 1993: Evidence for general instability of past climate from a 250-kyr ice-core record. *Nature*, **364**, 218–220.

de Baar, H.J., J.T. de Jong, D.C. Bakker, B.M. Loscher, C. Veth, U. Bathmann and V. Smetacek, 1995: Importance of iron for plankton blooms and carbon dioxide drawdown in the Southern Ocean. *Nature*, **373**, 412–415.

Denman, K.L., 1993: The ocean carbon cycle and climate change: an analysis of interconnected scales. In: *Patch Dynamics*, S.A. Levin, T.M. Rowell and J.H. Steele (eds.), Springer Verlag, Berlin, pp. 213, 223.

Denman, K.L. and A.E. Gargett, 1995: Biological-physical interactions in the upper ocean: the role of vertical and small scale transport processes. *Annual Reviews of Fluid Mechanics*, **27**, 225–255.

Dickey, T.D., 1991: The emergence of concurrent high-resolution physical and bio-optical measurements in the upper ocean and their applications. *Reviews of Geophysics*, **29**, 383–413.

Drange, H., 1994: An isopycnic coordinate carbon cycle model for the North Atlantic; and the possibility of disposing of fossil fuel CO_2 in the ocean. Dr. Scient. Thesis, Department of Mathematics and Nansen Environmental and Remote Sensing Center, University of Bergen, Norway, 286 p.

Duce, R.A. and N.W. Tindale, 1991: Atmospheric transport of iron and its deposition in the ocean. *Limnol. Oceanogr.*, **36**, 1715–1726.

Duce, R.A., P.S. Liss, J.T. Merril, E.L. Atlas, P. Buat-Menard, B.B. Hicks, J.M. Miller, J.M. Prospero, R. Arimoto, T.M. Church, W. Ellis, J.N. Galloway, L. Hansen, T.D. Jickells, A.H. Knap, K.H. Reinhardt, B. Schneider, A. Soudine, J.J. Tokos, S. Tsunogai, R. Wollast and M. Zhou, 1991: The atmospheric input of trace species to the world ocean. *Global Biogeochem. Cycles*, **5**, 193–259.

Ducklow, H.W, 1995: Ocean biogeochemical fluxes: new production and export of organic matter from the upper ocean. *Reviews of Geophysics*, **33**, Supplement, 1271–1276.

Duursma, E.K. and R. Dawson (eds.), 1981: Marine organic chemistry. Evolution, composition, interactions and chemistry of organic matter in seawater. *Elsevier Oceanography Series*, **31**, 521

Eicken, H. 1992: The role of sea ice in structuring Antarctic ecosystems. *Polar Biology*, **12**, 3–13.

Elkins, J.W., S.C. Wofsy, M.B. McElroy, C.E. Kolb and W.A. Kaplan, 1978: Aquatic sources and sinks for nitrous oxide. *Nature*, **275**, 602–606.

Eppley, R.W., 1972: Temperature and phytoplankton growth in the sea. *Fishery Bulletin*, **70**, 1063–1085.

Eppley, R.W. and B.J. Peterson, 1979: Particulate organic matter flux and planktonic new production in the deep ocean. *Nature*, **282**, 677–680.

Erickson, D.J., 1993: A stability dependent theory for air-sea gas exchange. *J. Geophys. Res.*, **98**, 8471–8488.

Evans, G.T. and M.J. Fasham, 1993: Themes in modelling ocean biogeochemical processes. In: *Towards a Model of Ocean Biogeochemical Processes*, G.T. Evans and M.J. Fasham (eds.), Springer-Verlag, Berlin, pp. 1–19.

Fasham, M.J.R., H.W. Ducklow and S.M. McKelvie, 1990: A nitrogen-based model of plankton dynamics in the oceanic mixed layer. *J. Mar. Res.*, **35**, 357–394.

Fasham, M.J.R., J.L. Sarmiento, R.D. Slater, H.W. Ducklow and R. Williams, 1993: Ecosystem behavior at Bermuda Station "S" and Ocean Weather Station "India": a general circulation model and observational analysis. *Global Biogeochem. Cycles*, **7**, 379–415.

Fichefet, T., S. Hovine and J.-C. Duplessy, 1994: A model study of the Atlantic thermohaline circulation during the last glacial maximum. *Nature*, **372**, 252–255.

Frankignoulle, M. and J.-P. Gattuso, 1993: Air-sea CO_2 exchanges in coastal ecosystems. In: *Interactions of C, N, P and S Biogeochemical Cycles and Global Change*, R. Wollast, F.T. Mackenzie and L. Chou (eds.), Springer-Verlag, Berlin, pp. 233–248.

Frankignoulle, M. and C. Canon, 1994: Marine calcification as a source of carbon dioxide: Positive feedback of increasing atmospheric CO_2. *Limnol. Oceanogr.*, **39**, 458–462.

Fuhrman, J.A. and D.G. Capone, 1991: Possible biogeochemical consequences of ocean fertilization, *Limnol. Oceanogr.*, **36**, 1951–1959.

Gabric, A., N. Murray, L. Stone and M. Kohl, 1993: Modelling the production of dimethylsulfide during a phytoplankton bloom. *J. Geophys. Res.*, **98**, 22805–22816.

Galloway, J.N., W.H. Schlesinger, H. Levy II, A.F. Michaels, and J.L. Schnoor, 1995: Nitrogen fixation: Anthropogenic enhancement-environmental response. *Global Biogeochem. Cycles*, **9**, 235-252.

GLOBEC, 1993a: *Report of the first meeting of the international GLOBEC working group on Sampling and Observation Systems.* GLOBEC International Report No. 3.

GLOBEC, 1993b: *Report of the first meeting of the international GLOBEC working group on Numerical Modeling.* GLOBEC International Report No. 6, 60 pp.

Gloersen, P., 1995: Modulation of hemispheric sea-ice cover by ENSO events. *Nature*, **373**, 503–506.

GRIP Project Members, 1993: Climatic instability during the last interglacial period revealed in the Greenland summit ice-core. *Nature*, **364**, 203–207

Harris, G.P., F.B. Griffiths, L.A. Clementson, V. Lyne and H. Van der Doe, 1991: Seasonal and interannual variability in physical processes, nutrient cycling, and the structure of the food chain in Tasmanian shelf waters. *Journal of Plankton Research*, **13**, Suppl., 109–131.

Hedges, J.I. and R.G. Keil, 1995: Sedimentary organic matter preservation: an assessment and speculative synthesis. *Mar. Chem.*, **49**, 81–115.

Hegg, D.A., R.J. Ferek, P.V. Hobbs and L.F. Radke, 1991: Dimethyl sulfide and cloud condensation nucleus correlations in the Northeast Pacific Ocean. *J. Geophys. Res.* **96**, 13 189–13 191.

Heinze, C., E. Maier-Reimer and K. Winn, 1991: Glacial pCO_2 reduction by the world ocean: experiments with the Hamburg carbon cycle model. *Paleoceanography*, **6**, 395–430.

Helbling, E.W., V. Villifane, M. Ferrario and O. Holm-Hansen, 1992: Impact of natural ultraviolet radiation on rates of photosynthesis and on specific marine phytoplankton species. *Marine Ecology Progress Series*, **80**, 89–100.

Herndl, G.J., G. Muller-Niklas and J. Frick, 1993: Major role of ultraviolet-B in controlling bacterioplankton growth in the surface layer of the ocean. *Nature*, **361**, 717–719.

Hofmann, E.E., 1988: Plankton dynamics on the southeastern U.S. continental shelf. Part III. A coupled physical-biological model. *J. Mar. Res.*, **46**, 919–946.

Holligan, P.M., 1992: Do marine phytoplankton influence global climate? In: *Primary Productivity and Biogeochemical Cycles in the Sea*, P.G. Falkowski and A.D. Woodhead (eds.), Plenum, New York, pp. 487–501.

Holligan, P.M., S.M. Turner and P.S. Liss, 1987: Measurements of dimethyl sulphide in frontal regions, *Continental Shelf Research*, **7**, 213–224.

Holligan, P.M., E. Fernández, J. Aiken, W.M. Balch, P. Boyd, P.H. Burkill, M. Finch, S.B. Groom, G. Malin, K. Muller, D.A. Purdie, C. Robinson, C.C. Trees, S.M. Turner and P. van der Wal, 1993: A biogeochemical study of the coccolithophore, *Emiliania huxleyi*, in the North Atlantic. *Global Biogeochem. Cycles*, **7**, 879–900.

Holser, W.T., M. Schindlowski, F.T. MacKenzie and J.B. Maynard, 1988: Biogeochemical cycles of carbon and sulfur. In: *Chemical Cycles in the Evolution of the Earth*, C.B. Gregor, R.M. Garrels, F.T. MacKenzie and J.B. Maynard (eds.), John Wiley and Sons, New York, pp. 105–174.

Honjo, S. and S.J. Manganini, 1993: Annual biogenic particle fluxes to the interior of the North Atlantic Ocean; studied at 34°N 21°W and 48°N 21°W. *Deep-Sea Res. II*, **40**, 587–607.

Hunter, J.R., S.E.Kaup and J.H. Taylor, 1981: Effects of solar and artificial ultraviolet-B radiation on larval Northern Anchovy, *Engraulis mordax*. *Photochemistry and Photobiology*, **34**, 477–486.

Huntley, M. E. and C. Boyd, 1984: Food-limited growth of marine zooplankton. *American Naturalist*, **124**, 455–478.

IPCC, 1990: *Climate Change: The IPCC Scientific Assessment*, J.T. Houghton, G.J. Jenkins and J.J. Ephraums (eds.), Cambridge University Press, Cambridge, UK, 365 pp.

IPCC, 1994: *Climate Change 1994: Radiative Forcing of Climate Change and an Evaluation of the IPCC 1S92 Emission Scenarios*, J.T. Houghton, L.G. Meira Filho, J. Bruce, Hoesung Lee, B. A. Callander, E.F. Haites, N. Harris and K. Maskell (eds.), Cambridge University Press, Cambridge, UK.

IPCC WGII, 1995: *Climate Change 1995-Impacts, Adaptations and Mitigations of Climate Change: Scientific-Technical Analyses: The Second Assessment Report of the Inter-Governmental Panel on Climate Change*, R.T. Watson, M.C. Zinyowera and R.H. Moss (eds.), Cambridge University Press, New York, USA.

Ittekkot, V., 1993: The abiotically driven biological pump in the ocean and short-term fluctuations in atmospheric CO_2 contents. *Global and Planetary Change*, **8**, 17–25.

Jacka, T.H. and W.F. Budd, 1991: Detection of temperature and sea ice extent changes in the Antarctic and Southern Ocean. In: *Proceedings of the International Conference on the Role of Polar Regions in Global Change*, G. Weller, C. L. Wilson, and B.A.B Severin (eds.), Geophysical Institute, Fairbanks, pp. 63–70.

Jenkins, W.J. and J.C. Goldman, 1985: Seasonal oxygen cycling and primary production in the Sargasso Sea. *J. Mar. Res.*, **43**, 465–491.

Joos, F., J.L. Sarmiento and U. Siegenthaler, 1991: Estimates of the effect of Southern Ocean iron fertilization on atmospheric CO_2 concentrations. *Nature*, **349**, 772–775.

Karentz, D., 1994: Ultraviolet tolerance mechanisms in Antarctic marine organisms. In: *Ultraviolet Radiation in Antarctica: Measurements and Biological Effects*, C.S. Weiler and P. Penhale, (eds.), American Geophysical Union Antarctic Research Series, 62, pp. 93–110.

Karl, D.M., R. Leteller, D. Hebel, L. Tupas, J. Dore , J. Christian and C. Winn, 1995: Ecosystem changes in the North Pacific subtropical gyre attributed to the 1991-92 El Niño. *Nature*, **373**, 230–234.

Karl, D.M., J.R. Christian, J.E. Dore, D.V. Hebel, R.M. Letalier, L.M. Tupas and C.D. Winn, 1996, Seasonal and interannual variability in primary production and particle flux at station ALOHA. *Deep-Sea Res. I*, in press.

Keller, M.D., 1991: Dimethylsulfide production and marine phytoplankton: the importance of species composition and cell size. *Biol. Ocean.*, **6**, 375–382.

Kempe, S., 1993: Are coastal systems a source of CO_2 to the atmosphere? In: *Proceedings of the First IGBP/LOICZ Core Project Meeting*, T.S. Hopkins and C.A. Kinder (eds.), Department of Marine, Earth and Atmospheric Sciences, North Carolina State University, pp. 27–31.

Kempe, S. and K. Pegler, 1991: Sinks and sources of CO_2 in coastal waters. *Tellus*, **43B**, 224–235.

Kieber, D.J., J. McDaniel and K. Mopper, 1989: Photochemical source of biological substrates in seawater: implications for carbon cycling. *Nature*, **341**, 637–639.

Kiene, R.P. and T.S. Bates, 1990: Biological removal of dimethyl sulphide from sea water. *Nature*, **345**, 702–705.

Knauer, G.A., D.G. Redalje, W.G. Harrison and D.M. Karl, 1990: New production at the VERTEX time-series site. *Deep-Sea Research*, **37**, 1121–1134.

Knox, F. and M.B. McElroy, 1984: Changes in atmospheric CO_2: influence of the marine biota at high latitude. *J. Geophys. Res.*, **89**, 4629–4637.

Kolber, Z.S., R.T. Barber, K.H. Coale, S.E. Fitzwater, R.M. Greene, K.S. Johnson, S. Lindley and P.G. Falkowski, 1994: Iron limitation of phytoplankton photosynthesis in the equatorial Pacific Ocean. *Nature*, **371**, 145–149.

Kurz, K.D. and E. Maier-Reimer, 1993: Iron fertilization of the austral ocean – the Hamburg model assessment. *Global Biogeochem. Cycles*, **7**, 229–244.

Law, C.S. and N.J.P. Owens, 1990: Significant flux of atmospheric nitrous oxide from the northwest Indian Ocean. *Nature*, **346**, 826–828.

Leck, C., U. Larsson, L.E. Bagander, S. Johansson and S. Hajdu, 1990: Dimethylsulfide in the Baltic Sea: annual variability in relation to biological activity. *J. Geophys. Res.*, **95**, 3353–3364.

Lee, C. and S.M. Henrichs, 1993: How the nature of dissolved organic matter might affect the analysis of dissolved organic carbon. *Marine Chemistry*, **41**, 105–120.

Legrand, M., C. Feniet-Saigne, E.S. Saltzman, C. Germain, N.I. Barkov and V.N. Petrov, 1991: Ice-core record of oceanic emissions of dimethylsulphide during the last climate cycle. *Nature*, **350**, 144–146.

Liss, P.S. and J.N. Galloway, 1993: Air-sea exchange of sulphur and nitrogen and their interaction in the marine atmosphere. In: *Interactions of C, N, P and S Biogeochemical Cycles and Global Change*, R. Wollast, F.T. Mackenzie and L. Chou (eds.), Springer-Verlag, Berlin, pp. 259–281.

Liss, P.S., G. Malin, S.M. Turner and P.M. Holligan, 1994: Dimethyl sulphide and *Phaeocystis*: A review. *J. Marine Systems*, **5**, 41–53.

Lizotte, M.P. and C.W. Sullivan, 1992: Biochemical composition and photosynthate distribution in sea ice microalgae of McMurdo Sound, Antarctica: evidence for nutrient stress during the spring bloom. *Antarctic Science*, **4**, 23–30.

Lohrenz, S.E., G.A. Knauer, V.L. Asper, M. Tuel, A.F. Michaels and A.H. Knap, 1992: Seasonal and interannual variability in primary production and particle flux in the northwestern Sargasso Sea: U.S. JGOFS Bermuda Atlantic Time Series. *Deep-Sea Research*, **39**, 1373–1391.

Longhurst, A.R., 1991: Role of the marine biosphere in the global carbon cycle. *Limnol. and Oceanogr.*, **36**, 1507–1526.

Longhurst, A., S. Sathyendranath, T. Platt and C. Caverhill, 1995: An estimate of global primary production in the ocean from satellite radiometer data. *J. Plankton Res.*, **17**, 1245–1271.

Luther, M. E. and J.J. O'Brien, 1990: Variability in upwelling fields in the northwestern Indian Ocean 1. Model experiments for the past 18,000 years. *Paleoceanography*, **5**, 433–445.

Lyle, M., D.W. Murray, B.P. Finney, J. Dymond, J.M. Robbins and K. Brookforce, 1988: The record of late Pleistocene biogenic sedimentation in the eastern tropical Pacific Ocean. *Palaeoceanography*, **3**, 39–60.

Maier-Reimer, E., 1993: Geochemical cycles in an ocean general circulation model. Preindustrial tracer distributions. *Global Biogeochem. Cycles*, **7**, 645–677.

Malin, G., S.M. Turner and P.S. Liss, 1992: Sulfur: the plankton/climate connection. *J. Phycol.*, **28**, 590–597.

Malin, G., S. Turner, P. Liss, P. Holligan and D. Harbour, 1993: Production of dimethylsulfide and dimethylsulphonio-propionate in the north east Atlantic during the summer coccolithophore bloom, *Deep Sea Res. I*, **40**, 1487–1508.

Marchant, H.J., 1994: Biological impacts of seasonal ozone depletion. In: *Antarctic Science – Global Concerns*, Hempel, G. (ed.), Springer-Verlag, Berlin. pp. 95–109.

Martin, J., 1990: Glacial–interglacial CO_2 change: the iron hypothesis. *Paleoceanography*, **5**, 1–13.

Martin, J.H., K.H. Coale, K.S. Johnson, S.E. Fitzwater, R.M. Gordon, S.J. Tanner, C.N. Hunter, V. Elrod, J. Nowicki, T. Coley, R. Barber, S. Lindley, A. Watson, K. Van Scoy, C. Law, M. Liddicoat, R. Ling, T. Stanton, J. Stockel, C. Collins, A. Anderson, R. Bidigare, M. Ondrusek, M.M. Latasa, F. Millero, K. Lee, W. Yao, J. Zhang, G. Friederich, C. Sakamoto, F. Chavez, K. Buck, Z. Kolber, R. Greene, P. Falkowski, S. Chisholm, F. Hoge, B. Swift, S. Turner, P. Nightingale, P. Liss and N. Tindale, 1994: The iron hypothesis: Ecosystem tests in Equatorial Pacific waters. *Nature*, **371**, 123–129.

McKnight, D.M., P. Behmel, D.A. Franko, E.T.Gjessing, U. Münster, R.C. Petersen Jr., O.M. Skulberg, C.E.W. Steinberg, E. Tipping, S.A. Visser, P.W. Werner and R.G. Wetzel, 1990: Group report, how do organic acids interact with solutes, surfaces and organisms? In: *Organic Acids In Aquatic Ecosystems*, E.M. Perdue and E.T. Gjessing (eds.), Dahlem Workshop Reports – Life Sciences Research Report 48, John Wiley and Sons, pp. 223–243.

Michaels, A.F., D.A. Siegel, R.J. Johnson, A.H. Knap and J.N. Galloway, 1993: Episodic inputs of atmospheric nitrogen to the

Sargasso Sea: Contributions to new production and phytoplankton blooms. *Global Biogeochem. Cycles*, **7**, 339–351.

Moloney, C.L. and J.G. Field, 1991: The size-based dynamics of plankton food webs. I. A simulation model of carbon and nitrogen flows. *J. Plankton Res.*, **13**, 1003–1038.

Mopper, K., X. Zhou, R.J. Kieber, D.J. Kieber, R.J. Sikorski and R.D. Jones, 1991: Photochemical degradation of dissolved organic carbon and its impact on the oceanic carbon cycle. *Nature*, **353**, 60–62.

Moran, M.A. and R.E. Hodson, 1994: Support of bacterioplankton production by dissolved humic substances from three marine environments. *Marine Ecology Progress Series*, **100**, 241–247.

Murphy, E.J., D.J. Morris, J.L. Watkins and J. Priddle, 1988: Scales of interaction between Antarctic krill and the environment. In: *Antarctic Ocean and Resources Variability*, D. Sahrhage (ed.), Springer-Verlag, Berlin, pp. 120–303.

Murray, J.W., R.T. Barber, M.R. Roman, M.P. Bacon and R.A. Feely, 1994: Physical and biological controls on carbon cycling in the equatorial Pacific. *Science*, **266**, 58–65.

Najjar, R.G., J.L. Sarmiento and J.R. Toggweiler, 1992: Downward transport and fate of organic matter in the ocean: simulations with a general circulation model. *Global Biogeochem. Cycles*, **6**, 45–76.

Naqvi, S.W. and R.J. Noronha, 1991: Nitrous oxide in the Arabian Sea. *Deep-Sea Res.*, **38**, 871–890.

Neale, P.J., M.P. Lesser and J.J. Cullen, 1994: Effects of ultraviolet radiation on photosynthesis of phytoplankton in the vicinity of McMurdo Station, Antarctica. In: *Ultraviolet Radiation in Antarctica: Measurements and biological effects*, C.S. Weiler and P. Penhale (eds.), American Geophysical Union Antarctic Research Series, 62, pp. 125–142.

Opdyke, B.N. and J.C.G. Walker, 1992: Return of the coral reef hypothesis: Basin to shelf partitioning of $CaCO_3$ and its effect on atmospheric CO_2. *Geology*, **20**, 733–736.

Oppo, D.W., and S.J. Lehman, 1993: Mid-depth circulation of the subpolar North Atlantic during the last glacial maximum. *Science*, **259**, 1148–1152.

Orr, J.C. and J.L. Sarmiento, 1992: Potential of marine macroalgae as a sink for CO_2: constraints from a 3-D general circulation model of the global ocean. *Water, Air, and Soil Pollution*, **64**, 405–421.

Owens, N.P., J.N. Galloway and R.A. Duce, 1992: Episodic atmospheric nitrogen deposition to oligotrophic oceans. *Nature*, **357**, 397–399.

Paerl, H., 1993: Emerging role of atmospheric nitrogen deposition in coastal eutrophication: biogeochemical and trophic perspectives. *Can. J. Fish. Aquatic Sci.*, **50**, 2254–69.

Parsons, T.R. and C.M. Lalli, 1988: Comparative oceanic ecology of the plankton communities of the subarctic Atlantic and Pacific Oceans. *Oceanogr. Mar. Biol. Annu. Rev.*, **26**, 51–68.

Parungo, F., J.F. Boatman, H. Sievering, S.W. Wilkison and B.B. Hicks, 1994: Trends in global marine cloudiness and anthropogenic sulfur. *J. Climate*, **7**, 434–440.

Pedersen, T.F., B. Nielsen and M. Pickering, 1991: The timing of Late Quaternary productivity pulses in the Panama Basin and implications for atmospheric CO_2. *Paleoceanography*, **6**, 657–677.

Peltzer, E.T. and N.A. Hayward, 1995: Spatial distribution and temporal veriability of dissolved organic carbon along 140°W in the Equatorial Pacific Ocean in 1992. *Deep-Sea Res.*, in review.

Platt, T. and W.G. Harrison, 1985: Biogenic fluxes of carbon and oxygen in the ocean. *Nature*, **318**, 55–58.

Prell, W.L. and E. Van Campo, 1986: Coherent response of Arabian Sea upwelling and pollen transport to late Quaternary monsoonal winds. *Nature*, **323**, 526–528.

Premuzic, E.T., C.M. Benkovitz, J.S. Gaffney and J.J. Walsh, 1982: The nature and distribution of organic matter in the surface sediments of world oceans and seas. *Org. Geochem.*, **4**, 63–77.

Rahmstorf, S., 1994: Rapid climate transitions in a coupled ocean-atmosphere model. *Nature*, **372**, 82–85.

Ratte, M., C. Plass-Dulmer, R. Koppmann, J. Rudolph and J. Denga, 1993: Production mechanisms of C2-C4 hydrocarbons in seawater: field measurements and experiments. *Global Biogeochem. Cycles*, **7**, 369–378.

Raven, J.A., 1993: Limits on growth rates. *Nature*, **361**, 209-210.

Redfield, A.C., B.H. Ketchum and F.A. Richards, 1963: The influence of organisms on the composition of sea water. In: *The Sea, Vol. 2*, M.N. Hill (ed.), Interscience, New York, pp. 26–77.

Riebesell, U., D.A. Wolf-Gladrow and V. Smetacek, 1993: Carbon dioxide limitation of marine phytoplankton growth rates. *Nature*, **361**, 249–251.

Robertson, J.E. C. Robinson, D.R. Turner, P. Holligan, A.J. Watson, P.Boyd, E. Fernandez and M. Finch, 1994: The impact of a coccolithophore bloom on oceanic carbon uptake in the northeast Atlantic during summer 1991. *Deep-Sea Res. I*, **41**, 297–314.

Roemmich, D. and J. McGowan, 1995: Climatic warming and the decline of zooplankton in the California Current. *Science*, **267**, 1324–1326.

Sambrotto, R.N., G. Savidge, C. Robinson, P. Boyd, T. Takahashi, D.M. Karl, C. Langdon, D. Chipman, J. Marra and L. Codespoti, 1993: Elevated consumption of carbon relative to nitrogen in the surface ocean. *Nature*, **363**, 248–250.

Sarmiento, J.L. and J.R. Toggweiler, 1984: A new model for the role of the oceans in determining atmospheric $p\mathrm{CO_2}$. *Nature*, **308**, 621–624.

Sarmiento, J.L. and J.C. Orr, 1991: Three-dimensional simulations of the impact of Southern Ocean nutrient depletion on atmospheric CO_2 and ocean chemistry. *Limnol. Oceanogr.*, **36**, 1928–1950.

Sarmiento, J.L. and U. Siegenthaler, 1992: New production and the global carbon cycle. In: *Primary Productivity and Biogeochemical Cycles in the Sea*, P.G. Falkowski and A.D. Woodhead (eds.), Plenum, New York, pp. 317–332.

Sarmiento, J.L., R.D. Slater, M.J.R. Fasham, H.W. Ducklow, J.R. Toggweiler and G.T. Evans, 1993: A seasonal three-dimensional ecosystem model of nitrogen cycling in the North Atlantic euphotic zone. *Global Biogeochem. Cycles, 7*, 417–450.

Sathyendranath, S., A.D. Gouveia, S.R. Shetye, P. Ravindran and T. Platt, 1991: Biological control of surface temperature in the Arabian Sea. *Nature*, **349**, 54–56.

Schindler, D.W. and S.E. Bayley, 1993: The biosphere as an increasing sink for atmospheric carbon: estimates from increased nitrogen deposition. *Global Biogeochem.Cycles*, **7**, 717–733.

Shaffer, G., 1993: Effects of the marine biota on global carbon cycling. In: *The Global Carbon Cycle*, M. Heimann (ed.), Springer-Verlag, Berlin, pp. 431–455.

Shaffer, G. and J.L. Sarmiento, 1995: Biogeochemical cycling in the global ocean 1. A new, analytical model with continuous vertical resolution and high-latitude dynamics. *J. Geophys. Res.*, **100**, 2659–2672.

Sharp, J.H., R. Benner, L. Bennett, C.A. Carlson, S.E. Fitzwater, E.T. Peltzer and L.M. Tupas, 1995: Analyses of dissolved organic carbon in seawater: the JGOFS EqPac methods comparison. *Mar. Chem.*, **48**, 91–108.

Shimmield, G.B., S.R. Mowbray and G.P. Weedon, 1990: A 350 ky history of the Indian Southwest monsoon – evidence from deep-sea cores, northwest Arabian Sea. *Transactions of the Royal Society of Edinburgh*, **81**, 289–299.

Siegenthaler, U. and Th. Wenk, 1984: Rapid atmospheric CO_2 variations and ocean circulation. *Nature*, **308**, 624–626.

Siegenthaler, U. and J.L. Sarmiento, 1993: Atmospheric carbon dioxide and the ocean. *Nature*, **365**, 119–125.

Sikes, C.S. and V.J. Fabry, 1994: Photosynthesis, $CaCO_3$ deposition, coccolithophorids and the global carbon cycle. In: *Regulation of Atmospheric CO_2 and O_2 by Photosynthetic Carbon Metabolism*, N.E. Tolbert and J. Preiss (eds.), Oxford University Press, New York, pp. 217–233.

Simon, M., B.C. Cho and F. Azam, 1992: Significance of bacterial biomass in lakes and the ocean: comparison to phytoplankton biomass and biogeochemical implications. *Marine Ecology Progress Series*, **86**, 103–110.

Slater, R.D., J.L. Sarmiento and M.J.R. Fasham, 1993: Some parametric and structural simulations with a three-dimensional ecosystem model of nitrogen cycling in the North Atlantic euphotic zone. In: *Towards a Model of Ocean Biogeochemical Processes*, G.T. Evans and M.J.R. Fasham (eds.), Springer-Verlag, Berlin, pp. 261–294.

Smith, R.C. and K.S. Baker, 1989: Stratospheric ozone, middle ultraviolet radiation and phytoplankton productivity. *Oceanography*, **2**, 4–10.

Smith, S.V. and J.T. Hollibaugh, 1993: Coastal metabolism and the oceanic organic carbon balance. *Rev. Geophys.*, **31**, 75–89.

Smith, W.O. Jr. and E. Sakshaug, 1990: Polar phytoplankton. In: *Polar Oceanography, Part B Chemistry, Biology and Geology*, W.O. Smith, Jr. (ed.), Academic Press, pp. 477–525.

Stocker, T.F., W.S. Broecker and D.G. Wright, 1994: Carbon uptake experiments with a zonally-averaged global ocean circulation model. *Tellus*, **46B**, 103–122.

Takahashi, T., W.S. Broecker and S. Langer, 1985: Redfield ratio based on chemical data from isopycnal surfaces. *J. Geophys. Res.*, **90**, 6907–6924.

Thomas, D.N. and R.J. Lara, 1995: Photodegradation of algal derived dissolved organic carbon. *Marine Ecology Progress Series*, **116**, 309–310.

Toggweiler, J.R., 1989: Is the downward dissolved organic matter (DOM) flux important in carbon transport? In: *Productivity of the Ocean: Present and Past*, W.H. Berger, V.S. Smetacek and G. Wefer (eds.), J. Wiley and Sons, New York, pp. 65–85.

Toggweiler, J.R., 1990: Bombs and ocean carbon cycles. *Nature*, **347**, 122–123.

Tsunogai, S.,and S. Noriki, 1991: Particulate fluxes of carbonate and organic carbon in the ocean. Is the marine biological activity working as a sink of the atmospheric carbon? *Tellus*, **43B**, 256–266.

Turner, S.M., G. Malin, P.S. Liss, D.S. Harbour and P.M. Holligan, 1988: The seasonal variation of dimethyl sulfide and dimethylsulfoniopropionate concentrations in nearshore waters. *Limnol. Oceanogr.*, **33**, 364–375.

Turpin, D.H., 1993: Phytoplankton growth and CO_2. *Nature*, **363**, 678–679.

Venrick, E.L., J.A. McGowan, D.R, Cayan and T.L. Hayward, 1987: Climate and chlorophyll, a: Longterm trends in the central North Pacific Ocean. *Science*, **238**, 70–72.

Walsh, J.J., 1989: How much shelf production reaches the deep sea? In: *Productivity of the Ocean: Present and Past*, W.H. Berger, V.S. Smetacek and G. Wefer (eds.), John Wiley and Sons, Chichester, pp. 175–191.

Walsh, J.J., 1991: Importance of continental margins in the marine biogeochemical cycling of carbon and nitrogen. *Nature*, **350**, 53–55.

Walsh, J.J. and D.A. Dieterle, 1994: CO_2 cycling in the coastal ocean. I. A numerical analysis of the southeastern Bering Sea with applications to the Chukchi Sea and the northern Gulf of Mexico. *Progress in Oceanography*, **34**, 335–392.

Watson, A.J., C.S. Law, K.A. Van Scoy, F.J. Millero, W. Yao, G. Friederich, M.I. Liddicoat, R.H. Wanninkhof, R.T. Barber and K. Coale, 1994: Implications of the "Ironex" iron fertilisation experiment for atmospheric carbon dioxide concentrations. *Nature*, **371**, 143–145.

Weaver, A.J. and T.M.C. Hughes, 1994: Rapid interglacial climate fluctuations driven by North Atlantic ocean circulation. *Nature*, **367**, 447–450.

Weaver, A.J., J. Marotzke, P.F. Cummins and E.S. Sarachik, 1993: Stability and variability of the thermohaline circulation. *J. Phys. Oceanogr.* **23**, 39–60.

Weiss, P.S., J.E. Johnson, R.H. Gammon and T.S. Bates, 1995: A reevaluation of the open ocean source of carbonyl sulfide to the atmosphere. *J. Geophys. Res.*, **100**, 23083–23092.

Welschmeyer, N.A., S. Strom, F. Goericke, G. DiTullio, M. Belvin and W. Petersen, 1993: Primary production in the subarctic Pacific Ocean: Project SUPER. *Prog. Oceanogr.*, **32**, 101–135.

Westbroek, P., C.W. Brown, J. van Bleijswijk, C. Brownlee, G.J. Brummer, M. Conte, J. Egge, E. Fernández, R. Jordan, M. Knappersbusch, J. Stefels, M. Veldhuis, P. van der Wal and J. Young, 1993: A model system approach to biological climate forcing: The example of *Emiliania huxleyi*. *Global Planetary Change*, **8**, 27–46.

Wolfe, G.V., E.B. Sherr and B.F. Sherr. 1994: Release and consumption of DMSP from *Emiliania huxleyi* during grazing by *Oxyrrhis marina*. *Marine Ecology Progress Series*, **111**, 111–119.

Wong, C.S., Y.-H. Chan, J.S. Page, G.E. Smith and R.D. Bellegay, 1993: Changes in equatorial CO_2 flux and new production estimated from CO_2 and nutrient levels in Pacific surface waters during the 1986/87 El Niño. *Tellus*, **45B**, 64–79.

Wroblewski, J. S., 1977: A model of phytoplankton plume formation during variable Oregon upwelling. *J. Mar. Res.*, **35**, 357–394.

Wroblewski, J.S., J.L. Sarmiento and G.L. Flierl, 1988: An ocean basin scale model of plankton dynamics in the North Atlantic. 1. Solutions for the climatological oceanographic conditions in May. *Global Biogeochem. Cycles*, **2**, 199–218.

Xie, L., and W.W. Hsieh, 1995: The global distribution of wind-induced upwelling. *Fisheries Oceanography*, **4**, 52–67.

11

Advancing our Understanding

G.A. MCBEAN, P.S. LISS, S.H. SCHNEIDER

CONTENTS

SUMMARY

- Since the 1990 Scientific Assessment significant progress has been made in furthering our understanding of climate change. The inclusion of radiative forcing due to aerosols, in particular their spatial patterns, in climate models has provided the means to diagnose recent climate warming and given growing confidence that a significant part of the observed climate warming can be attributed to human activities. The global role of aerosols in climate change to date has begun to be quantified and provides further confidence in simulations of future climate change.

- Our understanding of the climate system typically advances on time-scales of 5-10 years and a co-ordinated spectrum of approaches, ranging from individual research projects through to global-scale experiments and observational systems and infrastructure, is necessary. Lack of support for these activities will slow advances. An international frame-work for climate studies (the international Climate Agenda, including WCRP, IGBP, GCOS) exists.

- Complex systems, such as the climate system, can respond in non-linear ways and produce surprises. The nature of the system forces us to take a flexible approach in research planning and societal response and makes exact predictions of research advances impossible. Although, by definition, surprises cannot be anticipated, non-traditional, multi-scale, multi-disciplinary and multi-institutional research efforts with supporting organisational infrastructure are the best means of addressing them. There is a need for continual re-assessment of climate change and variability and for means of communicating this information to policymakers.

Priorities to address climate research requirements have been grouped below, although many of them are interconnected:

(i)   related to all topics
- systematic and sustained global observations of key variables
- capacity building in all nations

(ii)   the magnitude of global and continental scale climate change and of sea level rise
- the factors controlling the distribution of clouds and their radiative characteristics
- the hydrological cycle, including precipitation, evaporation and runoff
- the distribution of aerosols and their radiative characteristics
- the response of terrestrial and marine systems to climate change and their positive and negative feedbacks
- monitoring and modelling of land ice sheets

(iii)   the rate of climate change
- human activities influencing emissions (research priority of Working Group III)
- the coupling between the atmosphere and ocean, and ocean circulation
- the factors controlling the atmospheric concentrations of carbon dioxide and other greenhouse gases

(iv)   the detection and attribution of climate change
- systematic observations of key variables and data and model diagnostics
- relevant proxy data to construct and test palaeo-climatic time series to describe internal variability of the climate system

(v)   regional patterns of climate change
- land surface processes and their atmospheric linkages
- coupling between scales represented in global climate models and those in regional and smaller-scale models
- simulations with higher resolution climate models

11.1 Introduction

Global climate change moved to the forefront of the international agenda even before the recent growing confidence that a significant part of the observed global climate change can be attributed to human activities. Governments have agreed on the Climate Convention owing to the plausibility of the scientific arguments and measurements showing significant increases in the atmospheric concentrations of greenhouse gases. Global climate models have been used to project how the climate might change, based on plausible scenarios of anthropogenic emissions of greenhouse gases through the next century.

Scientific knowledge from a wide range of disciplines needs to be gathered and integrated to fashion scenarios of potential anthropogenic climatic changes over the next century. The process begins with projecting population growth, living standards and the technologies that will be used to achieve them. This information is used to produce scenarios of greenhouse gas and other chemical emissions such as sulphur dioxide, as well as projected land-use changes, which can affect the flows of energy and constituents between the atmosphere and the surface. This effort necessitates input from the natural, social and engineering and other applied sciences and, within the IPCC, is the domain of Working Group III. It provides a basis for model predictions of the atmospheric concentrations of radiatively active gases and aerosols as a function of time and location. Then meteorology, oceanography, hydrology, ecology, glaciology and other natural science disciplines are called upon to calculate potential climatic responses to concentrations derived from a wide range of plausible emission and land-use change scenarios. Review of this work is the domain of IPCC Working Group I and hence the subject of this report. The climate projections are then used to produce estimates of a range of plausible environmental and societal impacts (the domain of IPCC Working Groups II and III). In so far as those impacts on marine or terrestrial ecosystems may have feedback effects on climate, they are treated by Working Group I.

The UN Framework Convention on Climate Change has as its ultimate objective:

> ...stabilization of greenhouse gas concentrations in the atmosphere at a level that would prevent dangerous anthropogenic interference with the climate system. Such a level should be achieved within a time-frame sufficient to allow ecosystems to adapt naturally to climate change, to ensure that food production is not

threatened and to enable economic development to proceed in a sustainable manner.

To achieve this objective, it is necessary to understand and be able to model adequately: the greenhouse gas cycles and how atmospheric concentrations relate to anthropogenic emissions; how the climate system responds, both temporally and spatially, to the changing greenhouse gas concentration; and how ecosystems and human activities respond to this changing climate.

The IPCC function is to assess the state of our understanding and to judge the confidence with which we can make projections of climate change and its impacts. These tentative projections will aid policymakers in deciding on actions to mitigate or adapt to anthropogenic climatic change, which will need to be re-assessed on a regular basis. It is recognised that many remaining uncertainties need to be reduced in each of the above-named disciplines, which is why IPCC projections and scenarios are often expressed with upper and lower limits. These ranges are based on the collective judgement of the IPCC authors and the reviewers of each chapter, but it may be appropriate in the future to draw on formal methods from the discipline of decision analysis to achieve more consistency in setting criteria for high and low range limits.

In this chapter, we stress the need to narrow the wide range of outcomes and informal probability estimates in many of the subdisciplines which collectively comprise the assessment of human influence on the climate system. In addition to the requirements for each subdiscipline to reduce uncertainties in its speciality, there is the need to improve techniques for conducting the integration process itself, which is an overall IPCC goal.

Chapters 1 to 10 of this report examine the state of knowledge and identify the uncertainties in the topic area of each chapter. The objective of the present chapter is to outline the individual research activities and national and internationally co-ordinated programmes needed to advance our understanding. It should be noted that each increment of better understanding may not necessarily reduce uncertainties, at least initially. Eventually, however, such increased understanding should help to narrow uncertainties, i.e., to move the high and low range limits closer together. Further, we pay attention to areas where currently imaginable but highly uncertain "surprise" outcomes may exist. One "surprise" could be that elements of the climate system are inherently unpredictable; however, evaluation exercises indicate that many phenomena can be predicted even if the current uncertainties are large.

11.2 Framework for Analysis

As a framework for the analysis of uncertainties, it is appropriate to consider a simplified model of the process of simulating future climate changes in response to increasing atmospheric concentrations of greenhouse gases and other climatically-important substances and changes in land-use practices. Although presented in a linear fashion for illustrative purposes, it must be recognised that the problem of global climate change is complex and often non-linear and that the phenomena represented below as distinct questions are often interactive. Consider the model in terms of the following questions:

(i)   What will be the future emissions of greenhouse gases and other climatically-important substances and changes in land-use practices?

(ii) What will be the future atmospheric concentrations of greenhouse gases and other radiatively important substances?

(iii) What will be the resulting additional radiative forcing?

(iv) How will the climate system respond (globally and regionally) to this altered energy input? The tools to be used for climate projections are climate models. How good are they and how can they be improved?

(v)   Natural climate variations will be occurring simultaneously with these human-induced changes. How can we ascertain the causes of observed variability and distinguish between natural and human-induced changes?

(vi) What will be the impacts on sea level, natural and managed ecosystems, and socio-economic systems? How will these impacts feed back on the concentrations of greenhouse gases and other radiatively important substances?

11.3 Anthropogenic Emissions

First, what will be the future concentrations of greenhouse gases and other climatically-important substances due to human activities including land-use change? Future emissions of greenhouse gases and other relevant substances are primarily being addressed by IPCC Working Group III. Actual emissions and land-use practices will depend on policy decisions and the response of humans to potential climate change. Humans' response will depend, at least in part, on their confidence that scientists are right in their projections. As scientific knowledge increases, consensus of support for any necessary remedial action should become stronger.

11.4 Atmospheric Concentrations

What will be the future atmospheric concentration of greenhouse gases and other radiatively important substances? For some greenhouse gases, such as carbon dioxide, anthropogenic emissions are small but climatologically significant perturbations superimposed on large natural cycles of the gases; for others, such as halocarbons and perhaps methane, anthropogenic emissions dominate. These natural cycles were in relative steady state for several thousand years before the marked increase in anthropogenic emissions began with the industrial revolution. Chapter 2 analyses the global cycles of the important radiatively active gases and the role of aerosols, while Chapters 9 and 10 deal with the terrestrial and marine biotic responses to environmental change and identify possible feedbacks to climate.

For the global carbon cycle, the key question is whether anthropogenic emissions will remain in the atmosphere or be taken up by oceanic or terrestrial ecosystems. Is the uptake limited absolutely or will it be some continuing constant fraction of anthropogenic emissions? How will natural variability change the uptake? Will the effective lifetime of carbon dioxide added to the atmosphere change as the climate warms? These are the types of question that must be answered to reduce uncertainties in the computation of future atmospheric carbon dioxide concentrations for a given emission scenario. For methane, it is also important to understand the complex chemical feedbacks in order to determine the response time of a methane perturbation.

Continued monitoring of the atmospheric concentrations of the main greenhouse gases is needed in order to identify long-term trends and establish annual cycles and year-to-year variability. It is clear that changes in both tropospheric and stratospheric ozone play important roles in the radiative balance. Measurements of ozone, nitrogen oxides, carbon monoxide and hydrocarbons are needed to estimate global distributions of the short-lived hydroxyl radical, OH, since direct measurements are unlikely to be possible for some time to come. The OH radical is important for the destruction of methane and the conversion of sulphur dioxide to sulphate aerosol.

A combination of focused process studies and model development, validation and application, linked to the continued global monitoring of gases and relevant isotopes is needed to better quantify the cycles of carbon dioxide and other greenhouse gases. Reconstruction of the regional terrestrial carbon balance from the historical period through to present and the future, including the annual cycle, requires ecosystem models of the transient terrestrial ecosystem (Chapter 9), that include both biogeochemistry and vegetation redistribution. These will permit quantitative estimates of land-atmosphere carbon dioxide exchange and biophysical feedbacks to the global climate. The impacts of forest regrowth on the carbon cycle needs to be better quantified. In order to complete these reconstructions, it will be necessary to have an expanded soils' data base and maps of land cover/land-use and vegetative carbon stocks for historical, contemporary and future periods. Because nitrogen plays a major role in vegetation dynamics, estimates of nitrogen (both oxidised and reduced) depositions to regional ecosystems are needed for the periods of consideration. Recognising that it is impossible to monitor in detail the global ecosystem, it is important to define key indices and to establish permanent monitoring plots. Process studies are needed to improve representation of processes, both those resolved and especially those on sub-grid scales, through better parametrizations in models of the important processes of whole ecosystem CO_2 fertilisation (alone and with simultaneous warming and nitrate inputs) and to understand how ecosystems respond to changing carbon and nitrogen inputs.

Traditional global maps of oceanic productivity (Chapter 10) show that the most intense primary productivity occurs along the continental margins. Two important scientific questions arise. First, the fixation of carbon from CO_2 depends on the supply of nutrients, a significant fraction of which comes from upwelling from depths less than 500 m. Changes in ocean circulation that would alter the nutrient supply can occur on time-scales of weeks to decades. Thus, there is potential for significant change in the biological carbon pump in response to climate-induced change on time-scales relevant over the next century. Second, assessments of global scale removal of carbon from the surface via the biological pump concentrate on the open ocean, and the difficult assessment of the potentially significant contribution along the continental margins (integrated up to the global scale) had been largely ignored until recently. New estimates of the biogenic flux and the effect of a 10% variation (in response to oceanic circulation variations) suggest that the functioning of the biological pump requires careful quantitative assessment.

Atmospheric aerosols (Chapter 2) also play an important role in the Earth's radiative budget. There are fairly reliable estimates of the amount of sulphur burned but these do not translate directly into number density of aerosols, for which the size, hygroscopic and optical properties, as well as their vertical, horizontal and temporal distributions, have not been well observed. Emissions of anthropogenic sulphate emissions are estimated to be larger than natural emissions in the Northern Hemisphere and smaller in the Southern Hemisphere. Biomass and industrial burning are also important sources of non-sulphate aerosols, such as those containing black carbon, which can lead to local heating. Over the oceans, where aerosol influences on clouds may be particularly important, natural emissions play a larger role. Quantifying the atmospheric sulphur budget requires accounting for dimethyl sulphide (DMS) generated by phytoplankton, primarily at high latitudes and in coastal areas. Even the direction of possible feedback processes are currently speculative.

11.5 Radiative Forcing

For a given atmospheric concentration of greenhouse gases and aerosols, what will be the resulting additional radiative forcing? Computation of direct radiative forcing requires information on the distribution and radiative properties of the additional greenhouse gases and aerosols (Chapter 2). There is also indirect forcing due to the production or destruction of other greenhouse gases by chemical reaction or, in the case of aerosols, through effects on the radiative properties of clouds. Clouds themselves play a critical role in radiative forcing (Chapter 4); cloud feedback will be discussed in Section 11.6. Generally, the confidence level in the computation of the direct radiative forcing due to greenhouse gases is high. The indirect effects of greenhouse gases, mainly due to their influences on both stratospheric and tropospheric ozone, is generally small, but the confidence in these computations is low (more than 50% uncertainty). Research is needed into the radiative properties of newly produced chemical compounds (e.g., hydrochloro-fluorocarbons (HCFCs) and hydrofluoro-carbons (HFCs)) and on the chemical and dynamical processes that define the indirect effects of various greenhouse gases. The influence of solar variability is also thought to be small but the confidence level is very low.

The direct radiative cooling effect of tropospheric aerosols is estimated to be about 20% of the direct radiative heating effect of greenhouse gases, with a factor of two uncertainty; the indirect effect, via the effect of aerosol on cloud properties, is much more uncertain. Thus, at present the uncertainty in aerosol radiative forcing is the largest source of uncertainty in the total radiative forcing of climate over the past industrial period. Since aerosols are very patchy in their distribution, they could create significant regional climate changes regardless of their effect on globally averaged forcing. There is an array of scientific issues that needs to be resolved, including the horizontal, vertical and temporal distributions of aerosols and their optical properties. Because of the short lifetimes of most aerosols (days relative to decades for the main greenhouse gases), aerosols may become a decreasingly important global problem in the long term if countries adopt policies of low sulphur emission. It is important that the role of aerosols, including those of volcanic origin, be better understood in order to determine their current and future role in masking climate changes due to increasing greenhouse gases and to understand potential regional effects. Finally, the effect of land-use changes on surface albedo and vegetation needs to be considered in the context of changes in the regional radiation budget and hydrological cycle.

For policy considerations, the concept of a global warming potential (GWP) is useful in comparing the effects of various emission scenarios for different gases with lifetimes sufficiently long (2 yrs approximately) to become well-mixed in the atmosphere. In addition to the comparative radiative properties of constituents, it is important to know their lifetime in the atmosphere. GWPs have an uncertainty of about $\pm 35\%$. For aerosols, short-lived gases and the indirect effects through chemical reactions, there is need for an approach beyond the GWP.

11.6 Response of the Climate System

Given altered radiative forcing, how will the climate system respond to this changed energy input? Although feedbacks within the climate system (Chapter 1) appear to amplify the initial greenhouse radiative forcing, mainly due to interactions of radiation and the water cycle, our understanding of these feedbacks is relatively poor. Because interactions within the climate system are non-linear, important questions surround the predictability of the responses and the best means of separating the signal (the response due to altered radiative forcing) from the natural variability of the system. Uncertainties (Chapters 4, 5, 6, 9 and 10) in predictions of the magnitude of climate warming arise owing to the complex nature of the interactions of water vapour, clouds and aerosols with radiation, as well as from the neglect of feedbacks from biogeochemical systems.

To document how global and regional climates are changing and to evaluate and improve models and the processes incorporated in models, global data sets are required. These should include the history of the external forcing of the climate system for model sensitivity and validation studies. Furthermore, better coverage, the use of model assimilation to identify key data requirements, improved data archaeology and rescue, and the removal of inhomogeneities in historical and future data are necessary. Finally, it is vital for long time-series data records to be maintained when they are relevant to the substantiation of global change.

Estimates of climate sensitivity (Chapters 4, 5 and 6), have remained unaltered since the 1990 IPCC Assessment, at between 1.5 and 4.5°C. Uncertainties in modelling cloud-radiation interactions are the largest factor in determining this range. It needs to be noted that this estimate of warming is based on models of the physical climate system and does not include biogeochemical responses. To narrow this range will require improved understanding and improved models, incorporating improved parametrizations and numerical procedures. It will probably require at least a decade more of monitoring key variables, such as atmospheric and oceanic temperatures, precipitation, and surface solar radiation, both to document actual changes and to validate the models and the processes incorporated therein.

Whereas the feedback processes involving the atmosphere and its interactions with the surface are the chief factors in determining the equilibrium sensitivity of the climate system, the ocean plays the major role in determining the response time of the climate system and is important for regional impacts of climate change. Although ocean modelling has advanced considerably in the past few years, there are still difficulties in modelling phenomena such as temperature and salinity in deep waters, narrow boundary currents, oceanic eddies and details of the thermohaline circulation. Because global heat transport and storage are essential for regional assessment of climate change, it is important to determine the extent to which these omissions affect the simulation skill. Recent comparisons of model simulations with new tracer data sets gathered as part of the World Ocean Circulation Experiment are exposing these deficiencies and allowing new parametrizations to be tested. Another concern is that many coupled atmosphere-ocean models use energy, salt and water flux adjustments to keep the models' control simulations closer to the present climate. Until model sub-components are improved to the extent that flux adjustments are not needed, it is important to investigate the effects of these adjustments on simulations of climate under changing radiative forcing.

For coupled ocean-atmosphere-land surface modelling, narrowing uncertainties is a process going far beyond direct model improvements. Dedicated climate process studies are needed in order to improve parametrizations. Models with better physical, biological and biogeochemical processes and increased resolution both in the horizontal and vertical are needed in order to resolve land-surface heterogeneity, synoptic-scale disturbances in the ocean and to include mesoscale feedbacks, especially of the tropical atmosphere. The following are priority research areas:

(i)   cloud/aerosol/radiation interaction
(ii) convection and precipitation
(iii) sea ice formation, drift and melting
(iv) land-surface
(v)   ocean processes

In the long run, we need to be able to use given emission scenarios to calculate greenhouse gas concentrations within Earth system (atmosphere-ocean-ice-land surface-biota) models and obtain time-evolving scenarios of changes. It should be anticipated that the behaviour of such coupled non-linear systems may well produce unexpected responses (i.e., "surprises") that might not have been uncovered by studying each sub-system in isolation. For example, coupled atmosphere-ocean models have already revealed highly non-linear behaviour during the approach to equilibrium. Furthermore, the palaeo-climatic record indicates a number of very rapid climatic and trace gas changes, and it is important to know whether similar events could be triggered by anthropogenic forcing.

11.7 Natural Climate Variations and Detection and Attribution of Climate Change

It is important to understand how natural climate variations interact with human-induced changes (Chapter 8). There are several issues here. First, as already noted, there needs to be a continuing high quality global climate monitoring system to better establish the changing state of the climate. Then, we must know how the climate system varies in the absence of anthropogenic forcing. For example, a better understanding of El Niño should include the factors that determine its variable intensity and frequency. El Niño may be affected by anthropogenic climate change. Through better observations, palaeo-reconstructions and improved knowledge and understanding of natural variability, it will be possible to detect the anthropogenic climate signal with greater confidence.

To detect and attribute human-induced climatic change, it is necessary to account for the regional components of all significant climatic forcings (e.g., sulphate or biomass burning aerosols, land-use changes and tropospheric ozone) since such forcings can produce unique regional and altitudinal climatic response patterns (i.e., "fingerprints"). When such fingerprints occur in both modelled and actual climates, increased confidence in cause and effect linkages is justified. It is also necessary to determine the likelihood that such patterns may have occurred by chance.

The scientific community is now starting to use low-frequency (decadal-to-century time-scale) internal variability generated by fully coupled global climate models to determine the probability that a trend of a given magnitude (such as the ~0.5°C in observed global mean surface temperature increase over the past 100 years; Chapter 3) could have been generated by natural internal variability alone. The issue of detection must also include attribution of observed variability to other known forcings, such as volcanoes and solar variability. Noise information from long control runs of coupled-models has the advantage of being spatially complete and available for virtually any climate parameter of interest. However, the effect of artificial corrections, such as flux adjustments, on the conclusions, is generally unknown.

To build confidence in the decade-to-century time-scale natural variability simulated by models, there is a need to compare model attempts to mimic the climate of the last 1000 years with variability estimates from palaeo-climate data with comparable time resolution. Reconstructions of the near surface temperature of the past 1000 years based on such proxies as tree rings, ice cores and corals are valuable but have their own inherent deficiencies. Further complications are added by the need to identify the forcings, for example solar, volcanoes, deforestation, that may have taken place over this time period. Nevertheless, model control runs make specific "predictions" concerning the variation of surface temperature on a wide range of space and time-scales. They imply phase lags and leads at very large spatial scales which are potentially testable given appropriate palaeo-data.

11.8 Impacts of Climate Change

What will be the impacts of climate change on: sea level (Chapter 7); natural and managed ecosystems (Chapters 9, 10); and socio-economic systems? Climate change affects sea level through oceanic thermal expansion and changes in glaciers and ice sheets. To determine better the role of ocean thermal expansion requires continued support of both monitoring programmes and ocean modelling. There remain large uncertainties regarding future changes in global glacier volume and a combination of modelling, process studies and better data is needed to reduce these uncertainties. Possible instabilities of the major ice sheets need to be clarified. Chapters 9 and 10 deal with the role of changes in the terrestrial and marine biotic systems that feed back into modifying climate change. One area for research is the direct effects of atmospheric CO_2 increases on plant physiology and how these effects will interact at ecosystem scales with climatic impacts on the vegetation. Large- and small-scale field experiments combined with micro and boundary layer scale atmospheric knowledge will be needed to address this research issue which is at the interface of ecological and atmospheric sciences. Societal and biological impacts, not dealt with here, are primarily the responsibility of Working Groups II and III.

11.9 Cross-Cutting Issues

It is important to develop and maintain the capability to diagnose the system as it evolves. There is a common need for observations of the climate system. The Global Climate Observing System, built on the base of the World Weather Watch and other programmes, should provide the basis, but the research programmes also provide indispensable observations. For global issues such as climate change, satellites can play an important role in providing globally consistent data sets of long duration, along with *in-situ* measurements. There is justifiable concern over the consistency of some data sets, which may need reanalysis.

Another key strategy in understanding the climate system is a hierarchy of climate and Earth system models. These are tools for integrating information, identifying interfacial issues and interpreting results. A basic question is the extent to which climate is predictable. Predictability problems fall into at least two categories. First is the predictability of external forcings like solar variability, explosive volcanic eruptions, human land-use changes and energy system technologies. Second, complex systems often allow deterministic predictability of some characteristics (e.g., temperature response to volcanic dust veils or ENSO events given a history of atmospheric forcings) yet do not permit skilful forecasts of other phenomena (e.g., the evolution of specific weather systems for more than a few weeks time). The Earth system undoubtedly possesses mixed deterministic, stochastic and chaotic sub-systems, which implies that there are varying

degrees of predictability for different aspects at different scales. Researchers need to be cognisant of these issues, especially since such complex systems may exhibit behaviour not expected by most analysts (more commonly called "surprises"). One strategy to help identify such potential surprises (or to narrow the probability estimates for imaginable but poorly understood possibilities) is to foster co-operative interdisciplinary research teams that address problems across many scales and disciplines. Such problem areas possess the highest potential for unexpected outcomes, and call for non-traditional, multi-scale, multi-disciplinary and multi-institutional research efforts and supporting organisational infrastructure.

Climatic forecasts, for example, often are given at the global climate model grid box scales (i.e., 200 km × 200 km) whereas ecological field research is typically conducted on 20 m × 20 m plots, requiring a 10,000 fold interpolation. Research on techniques to "downscale" climatic information as well as more large-scale ecological studies are needed to help bridge the "scale gap" across these two disciplines, whose interaction is essential for addressing climatic changes and their biological consequences.

11.10 International Programmes

Scientific research that will contribute to advancing our understanding of climate change is being performed in a wide variety of institutions and laboratories around the world. Much of the research is unco-ordinated and independently motivated and funded. This research is essential and provides important input to the overall advancement of climate science. At the international level, the Climate Agenda (see Box), the framework for advancement in understanding the climate system and its interactions with the global biogeochemical cycles, is based on knowledge derived from both the World Climate Research Programme (WCRP, a component of the comprehensive World Climate Programme), sponsored by the World Meteorological Organisation, the International Council of Scientific Unions (ICSU) and the Intergovernmental Oceanographic Commission of UNESCO, and from the International Geosphere-Biosphere Programme (IGBP) of ICSU. The Human Dimensions of Global Environmental Change Programme (HDP) of the International Social Sciences Council (ISSC) and recently ICSU develop activities in the realm of the social sciences, which primarily contribute to Working Groups II and III. Projects on the impacts of climate variability and change are included within the IGBP as well as the World Climate

The Climate Agenda – A Framework for the International Co-ordination of Climate Research

The components of particular relevance to Working Group I fall within the World Climate Research Programme, the International Geosphere-Biosphere Programme and the Global Climate Observing System. Other components include the World Climate Monitoring and Data Programme, the World Climate Applications and Services Programme, the World Climate Impacts and Response-Strategies Programme, the Human Dimensions of Global Environmental Change Programme and the climate-related programmes of FAO and UNESCO.

The components of the World Climate Research Programme are:

- Global Energy and Water Cycle Experiment (GEWEX)

- World Ocean Circulation Experiment (WOCE)

- Climate Variability and Predictability Programme (CLIVAR)

- Working Group on Numerical Experimentation (WGNE)

- Arctic Climate System Study (ACSyS)

- Stratospheric Processes and their Role in Climate (SPARC)

The components of the International Geosphere-Biosphere Programme are:

- Biospheric Aspects of the Hydrological Cycle (BAHC)

- International Global Atmospheric Chemistry (IGAC) Project

- Global Change and Terrestrial Ecosystems (GCTE)

- Joint Global Ocean Flux Study (JGOFS)

- Land-Ocean Interactions in the Coastal Zone (LOICZ)

- Past Global Changes (PAGES)

- Data and Information System (DIS)

- Global Analysis, Interpretation and Modelling (GAIM)

- Land use/cover change (LUCC-jointly with HDP)

The Global Climate Observing System (GCOS) is built upon the World Weather Watch and the Global Atmosphere Watch and includes the climate portions of the Global Ocean Observing System and the Global Terrestrial Observing System.

The System for Analysis, Research and Training (START) is co-sponsored by IGBP, WCRP and HDP.

Programme; these activities are primarily of concern to IPCC Working Group II. For areas within the mandate of Working Group I and this report, basically the natural sciences and their role in understanding climate change, the IGBP and the WCRP are the principal international scientific research initiatives, while the Global Climate Observing System (GCOS) should provide the systematic observational basis. They provide for international planning and co-ordination of mainly large-scale endeavours, which are then funded directly by national sources. Working closely together, the WCRP places its emphasis on the physical aspects of the climate system while the IGBP concentrates on biogeochemical cycles.

The IGBP and WCRP do not attempt to organise all climate research, but instead focus on those activities which require international co-operation, either because of their global or large regional scale or because of their large human, technical and financial resource requirements. Co-operation across appropriate disciplines is fostered, and standardised methods and intercomparisons contribute to enhancing the Programmes' output. As noted, much essential climate research is being done, and is best done, by individuals or in small research groups at universities or governmental institutes, around the world. The WCRP, IGBP and GCOS function through international co-ordination of related parts of national programmes, based on the premise that the co-ordinated whole will be worth more than the sum of the individual elements. Through these and related programmes, considerable progress has been made in understanding the climate system, through a balanced set of monitoring, process studies and modelling activities. These are not independent activities but strongly interact, with results or deficiencies identified in one often leading to design of new research studies. Through the System for Analysis, Research and Training (START), co-sponsored by IGBP, WCRP and HDP, the capability of developing countries to address their regional climate change concerns and to contribute to global initiatives, is being addressed.

Climate science requires effort on a range of scales, from individual investigators to co-ordinated international programmes. All are supported through science councils within countries. National participation in large internationally co-ordinated projects, such as the WCRP, IGBP and GCOS is essential to the overall co-ordinated research strategy. It needs to be stressed that this participation is through contributing national activities, not through centrally funded science. There is need for significant increase in resources for capacity building in developing countries and international co-ordination of the scientific programmes.

Climate science requires multi-disciplinary and cross-institutional research. Relevant organisations need to adapt to facilitate the functioning of interdisciplinary teams addressing climate system problems. This will require the recruitment and training of bright young people with diverse backgrounds, and provision of the right environment to allow them to flourish.

11.11 Research Priorities

Since the completion of the 1990 Scientific Assessment, substantial progress has been made and we are confident that it will continue. The rate of progress will depend on the scale of effort and the degree of co-ordination, particularly for large-scale projects. Because there are always the elements of good fortune and scientific breakthroughs which are unpredictable, it is difficult to place with confidence a time-scale on the advancement of science and technology. However, the timing of major international programmes provides an indication. Three illustrative examples of the time scale for major scientific undertakings are: first, the completion of the 7-year field phase of the World Ocean Circulation Experiment in 1997 will give oceanographers the most comprehensive data set for understanding ocean circulation, model development and validation. The subsequent scientific analyses of the data sets generated can be realistically expected to take place over 5-10 years, at least. Second, clarification of the important questions of the role of clouds in the energetics of the atmosphere and testing of models on global cloud data sets may require global measurements that will only be possible by a space borne cloud radar coupled with extensive ground based measurements. The earliest that such a radar may be launched is after the turn of the century and scientific analysis will again span a 5-10 year period. Third, adequate observations to develop and test refined models of the oceanic carbon cycle will also take several years (with the successful completion of the JGOFS observation phase). All of these time lines indicate that advances in our knowledge of the climate system will take place gradually but on time-scales of 5-10 years, after which, depending on progress, more refined studies can be defined and implemented, also on similar time-scales.

Research priorities need to be focused on the determination of the likely magnitude and rate of human-induced climate change and its regional variations, the detection and attribution of climate change, the likely magnitude of sea level rise and the impacts of climate change on ecosystems (for areas within the purview of IPCC Working Group I). It needs be stressed that research studies to address these objectives are not independent but very much interconnected and their grouping below should be considered as demonstrative, not exclusive.

Priorities to address climate research requirements are listed below, although many of them are interconnected:

(i) related to all topics
- systematic and sustained global observations of key variables
- capacity building in all nations

(ii) the magnitude of global and continental-scale climate change and of sea level rise
- the factors controlling the distribution of clouds and their radiative characteristics
- the hydrological cycle, including precipitation, evaporation and runoff
- the distribution of ozone and aerosols and their radiative characteristics
- the response of terrestrial and marine systems to climate change and their positive and negative feedbacks
- monitoring and modelling of land ice sheets

(iii) the rate of climate change
- human activities influencing emissions (research priority of Working Group III)
- the coupling between the atmosphere and ocean, and ocean circulation

- the factors controlling the atmospheric concentrations of carbon dioxide and other greenhouse gases

(iv) detection and attribution of climate change
- systematic observations of key variables and data and model diagnostics
- relevant proxy data to construct and test palaeo-climatic time series to describe internal variability of the climate system

(v) regional patterns of climate change
- land surface processes and their atmospheric linkages
- coupling between scales represented in global climate models and those in regional and smaller-scale models
- simulations with higher resolution climate models

The international framework of integrated global climate programmes of WCRP, IGBP, GCOS, WCP, HDP and related programmes provides for international co-ordination but nations must undertake to support both their own activities and their contributions to these international activities. In this they should ensure that to the maximum extent possible, these activities and contributions are mutually supportive, so that progress can be made in responding to the needs of the Framework Convention on Climate Change and the needs of nations to provide climate services within their countries.

Appendix 1

Organisation of the IPCC

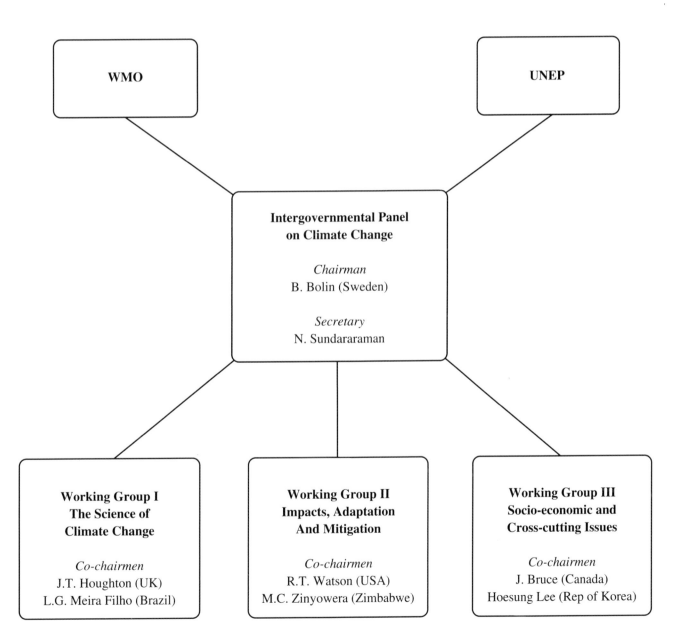

WMO

UNEP

**Intergovernmental Panel
on Climate Change**

Chairman
B. Bolin (Sweden)

Secretary
N. Sundararaman

**Working Group I
The Science of
Climate Change**

Co-chairmen
J.T. Houghton (UK)
L.G. Meira Filho (Brazil)

**Working Group II
Impacts, Adaptation
And Mitigation**

Co-chairmen
R.T. Watson (USA)
M.C. Zinyowera (Zimbabwe)

**Working Group III
Socio-economic and
Cross-cutting Issues**

Co-chairmen
J. Bruce (Canada)
Hoesung Lee (Rep of Korea)

Appendix 2

List of Major IPCC Reports (in English unless otherwise stated)

Climate Change – The IPCC Scientific Assessment. The 1990 report of the IPCC Scientific Assessment Working Group (*also in Chinese, French, Russian and Spanish*).

Climate Change – The IPCC Impacts Assessment. The 1990 report of the IPCC Impacts Assessment Working Group (*also in Chinese, French, Russian and Spanish*)

Climate Change – The IPCC Response Strategies. The 1990 report of the IPCC Response Strategies Working Group (*also in Chinese, French, Russian and Spanish*)

Emissions Scenarios. Prepared for the IPCC Response Strategies Working Group, 1990.

Assessment of the Vulnerability of Coastal Areas to Sea Level Rise – a Common Methodology, 1991 (*also in Arabic and French*).

Climate Change 1992 – The Supplementary Report to the IPCC Scientific Assessment. The 1992 report of the IPCC Scientific Assessment Working Group.

Climate Change 1992 – The Supplementary Report to the IPCC Impacts Assessment. The 1992 report of the IPCC Impacts Assessment Working Group.

Climate Change: The IPCC 1990 and 1992 Assessments – IPCC First Assessment Report Overview and Policymaker Summaries, and 1992 IPCC Supplement.

Global Climate Change and the Rising Challenge of the Sea. Coastal Zone Management Subgroup of the IPCC Response Strategies Working Group, 1992.

Report of the IPCC Country Studies Workshop, 1992.

Preliminary Guidelines for Assessing Impacts of Climate Change, 1992.

IPCC Guidelines for National Greenhouse Gas Inventories (3 volumes), 1994 (*also in French, Russian and Spanish*).

IPCC Technical Guidelines for Assessing Climate Change Impacts and Adaptations, 1995 (*also in Arabic, Chinese, French, Russian and Spanish*).

Climate Change 1994 – Radiative Forcing of Climate Change and an Evaluation of the IPCC IS92 Emission Scenarios.

Climate Change 1995 – Impacts, Adaptations and Mitigation of Climate Change: Scientific-Technical Analyses. Contribution of Working Group II to the IPCC Second Assessment Report.

Climate Change 1995 – Economic and Social Dimensions of Climate Change. Contribution of Working Group III to the IPCC Second Assessment Report.

IPCC Second Assessment Synthesis of Scientific-Technical Information Relevant to Interpreting Article 2 of the UN Framework Convention on Climate Change.

Enquiries: IPCC Secretariat, c/o World Meteorological Organisation, P O Box 2300, CH1211 Geneva 2, Switzerland.

Appendix 3

Contributors to IPCC WGI Report

Technical Summary

D. Albritton	NOAA Aeronomy Laboratory, USA
B. Bolin	IPCC Chairman, Sweden
B. Callander	IPCC WGI Technical Support Unit, UK
K. Denman	Institute of Ocean Sciences, Canada
R. Dickinson	University of Arizona, USA
L. Gates	Lawrence Livermore National Laboratory, USA
H. Grassl	World Meteorological Organisation, Switzerland
M. Grubb	Royal Institute of International Affairs, UK
N. Harris	European Ozone Research Co-ordinating Unit, UK
J. Houghton	IPCC WGI Co-Chairman, UK
P. Jonas	UMIST, UK
A. Kattenberg	Royal Netherlands Meteorological Institute (KNMI), Netherlands
K. Maskell	IPCC WGI Technical Support Unit, UK
G. McBean	Atmospheric Environment Canada, Canada
M. Mcfarland	United Nations Environment Programme (UNEP), Kenya
G. Meira	Agência Espacial Brasileira, Brazil
J. Melillo	Woods Hole Oceanographic Institution, USA
N. Nicholls	Bureau of Meteorology Research Centre, Australia
L. Ogallo	University of Nairobi, Kenya
M Oppenheimer	Environmental Defense Fund, USA
M. Prather	University of California @ Irvine, USA
B. Santer	Lawrence Livermore National Laboratory, USA
D. Schimel	National Center for Atmospheric Research, USA
K. Shine	University of Reading, UK
K. Trenberth	National Center for Atmospheric Research, USA
R. Warrick	The University of Waikato, New Zealand
R. Watson	Office of Science and Technology Policy, USA
J. Zillman	Bureau of Meteorology, Australia

Chapter 1: The Climate System; An overview

Convening Lead Author
K. Trenberth   National Center for Atmospheric Research, USA

Lead Authors
J. Houghton   IPCC WGI Co-Chairman, UK
G. Meira   Agência Espacial Brasileira, Brazil

Chapter 2: Radioactive Forcing of Climate Change

Convening Lead Author

D. Albritton	NOAA Aeronomy Laboratory, USA
P. Jonas	UMIST, UK
M. Prather	University of California @ Irvine, USA
D. Schimel	National Centre for Atmospheric Research, USA
K. Shine	University of Reading, UK

Lead Authors

D. Alves	Instituto Nacional de Pesquisas Espaciais (INPE), Brazil
R. Charlson	University of Washington, USA
R. Derwent	Meteorological Office, UK
D. Ehhalt	Institut für Chemie der KFA Jülich GmbH, Germany
I. Enting	CSIRO Division of Atmospheric Research, Australia
Y. Fouquart	LOA/Université des Science & Technologie de Lille, France
P. Fraser	CSIRO Division of Atmospheric Research, Australia
M. Heimann	Max-Plank Institut für Meteorologie, Germany
I. Isaksen	University of Oslo (Geophysics), Norway
F. Joos	University of Bern, Switzerland
M. Lal	Centre for Atmospheric Sciences, India
V. Ramaswamy	Geophysical Fluid Dynamics Laboratory, USA
D. Raynaud	CNRS Laboratoire de Glaciologie, France
H. Rodhe	University of Stockholm, Sweden
S. Sadasivan	Bhabha Atomic Research Centre, India
E. Sanhueza	Instituto Venezolano de Investigaciones Cientificas, Venezuela
S. Solomon	NOAA Aeronomy Laboratory, USA
J. Srinivasan	Langley Research Centre, USA
T. Wigley	Office for Interdisciplinary Earth Studies @ UCAR, USA
D. Wuebbles	Lawrence Livermore National Laboratory, USA
X. Zhou	Academy of Meteorological Sciences, China

Contributors

F. Alyea	Georgia Institute of Technology, USA
T. Anderson	University of Washington, USA
M. Andreae	Max-Planck Institut für Chemie, Germany
D. Blake	University of California @ Irvine, USA
O. Boucher	Laboratoire de Météorologie Dynamique du CNRS, France
C. Brühl	Max-Planck Institut für Chemie, Germany
J. Butler	NOAA, Climate Monitoring & Diagnostics Lab, USA
D. Cunnold	Georgia Institute of Technology, USA
J. Dignon	Lawrence Livermore National Laboratory, USA

E. Dlugokenchy	NOAA ERL @ Boulder, USA
J. Elkins	NOAA ERL @ Boulder, USA
I. Fung	University of Victoria, Canada
M. Geller	New York State University, USA
D. Hauglustaine	Service d'Aeronomie du CNRS, France
J. Haywood	Geophysical Fluid Dynamics Laboratory, USA
J. Heintzenberg	Institut für Troposphärenforschung, Germany
D. Jacob	Harvard University, USA
A. Jain	University of Illinois, USA
C. Keeling	Scripps Institute of Oceanography, USA
S. Khmelevtsov	Institute of Experimental Meteorology, Russian Federation
J. Lelieveld	University of Wageningen, Netherlands
H. Le Treut	Laboratoire de Météorologie Dynamique du CNRS, France
I. Levin	Universität Heidelberg, Germany
M. Maiss	Universität Heidelberg, Germany
G. Marland	Oak Ridge National Laboratory, USA
S. Marshall	University of Washington, USA
P. Midgley	M+D Consulting, Germany
B. Miller	Scripps Institute of Oceanography, USA
J. Mitchell	Meteorological Office, UK
S. Montzka	NOAA/CMDL, USA
H. Nakane	National Institute for Environmental Studies, Japan
P. Novelli	NOAA, Climate Monitoring & Diagnostics Lab, USA
B. O'Neill	Environmental Defense Fund, USA
D. Oram	University of East Anglia, UK
S. Penkett	University of East Anglia, UK
J. Penner	Lawrence Livermore National Laboratory, USA
S. Prinn	Massachusetts Institute of Technology, USA
P. Quay	University of Washington, USA
A. Robock	University of Maryland, USA
S. Schwartz	Brookhaven National Laboratory, USA
P. Simmonds	Bristol University, UK
S. Singh	Indian Institute of Tropical Meteorology
A. Slingo	Meteorological Office, UK
F. Stordal	Norwegian Institute for Air Research, Norway
E. Sulzman	National Center for Atmospheric Research, USA
P. Tans	NOAA,Climate Monitoring & Diagnostics Lab, USA
R. Weiss	Scripps Institute of Oceanography, USA
A. Wharner	Institut für Chemie der KFA Jülich GmbH, Germany
T. Whorf	Scripps Institute of Oceanography, USA

Chapter 3: Observed Climate Variability and Change

Convening Lead Author

N. Nicholls	Bureau of Meteorology Research Centre, Australia

Lead Authors

G. Gruza	Institute for Global Climate and Ecology, Russian Federation
J. Jouzel	LMCE/DSM, France
T. Karl	NOAA National Climate Data Center, USA

| L. Ogallo | University of Nairobi, Kenya |
| D. Parker | Meteorological Office, UK |

Contributors

J. Angell	NOAA ERL @ Silver Springs, USA
S. Anjian	National Meteorological Centre, China
P. Arkin	NWS W/NMC, USA
R. Balling Jr	Arizona State University, USA
M. Bardin	Institute for Global Climate and Ecology, Russian Federation
R. Barry	University of Colorado, USA
W. Bomin	National Meteorological Centre, China
R. Bradley	University of Massachusetts, USA
K. Briffa	University of East Anglia, UK
A. Carleton	Indiana University, USA
D. Cayan	Scripps Institute of Oceanography, USA
F. Chiew	University of Melbourne, Australia
J. Christy	University of Alabama @ Huntsville, USA
J. Churc,	CSIRO Division of Oceanography, Australia
E. Cook	Lamont-Doherty Earth Observatory, USA
T. Crowley	Texas A&M University, USA
N. Datsenko	Hydrometeorological Research Centre of Russia, Russian Federation
R. Davis	University of Virginia, USA
B. Dey	Howard University, USA
H. Diaz	NOAA ERL @ Boulder, USA
W. Drosdowsky	Bureau of Meteorology Research Centre, Australia
M. Duarte	Ciudad Universitaria, Argentina
J. Duplessy	Centre des Faibles Radioactivités, France
D. Easterling	National Climatic Data Center, USA
J. Eischeid	University of Colorado, USA
W. Elliott	NOAA ERL @ Silver Springs, USA
B. Findlay	Environment Canada, Canada
H. Flohn	Universitat Bonn, Germany
C. Folland	Meteorological Office, UK
R. Franke	Deutscher Wetterdienst Seewetteramt, Germany
P. Frich	Danish Meteorological Institute, Denmark
D. Gaffen	NOAA ERL @ Silver Springs, USA
V. Georgievsky	State Hydrological Institute, Russian Federation
T. Ginsburg	Hydrometeorological Research Centre of Russia, Russian Federation
W. Gould	Institute of Oceanographic Sciences, UK
P. Groisman	State Hydrological Institute, Russian Federation
D. Gullet	Environment Canada, Canada
W. Haeberli	Verschanstalt für Wasserbau Institut, Switzerland
S. Hastenrath	University of Wisconsin @ Madison, USA
A. Henderson-Sellers	Macquarie University, Australia
M. Hoelzle	Verschanstalt für Wasserbau Institut, Switzerland
W. Hogg	Environment Canada, Canada
G. Holland	Bureau of Meteorology Research Centre, Australia
L. Hopkins	Bureau of Meteorology Research Centre, Australia
M. Hulme	Climatic Research Unit (UEA), UK
N. Ivachtchenko	Hydrometeorological Research Centre of Russia, Russian Federation

P. Jones	Climatic Research Unit (UEA), UK
R. Katz	National Center for Atmospheric Research, USA
B. Kininmonth	Bureau of Meteorology, Australia
R. Knight	NOAA National Climate Data Center, USA
N. Kononova	Russian Agricultural Academy, Russian Federation
L. Korovkina	Institute for Global Climate & Ecology, Russian Federation
G. Kukla	Lamont-Doherty Geological Laboratory, USA
K. Kumar	Indian Institute of Tropical Meteorology, India
P. Lamb	University of Oklahoma, USA
C. Landsea	Colorado State University, USA
S. Levitus	NOAA/NODC (E/OC23), USA
T. Lewis	Pacific Geoscience Centre, Canada
H. Lins	US Geological Survey (WGII Liaison Grp), USA
J. Lough	Australian Institute of Marine Science, Australia
L. Malone	Environment Canada, Canada
J. Marengo	CPTEC-INPE, Brazil
T. McMahon	University of Melbourne, Australia
E. Mekis	Environment Canada, Canada
A. Meshcherskya	Main Geophysical Observatory, Russian Federation
P. Michaels	University of Virginia, USA
S. Nicholson	Florida State University, USA
J. Oerlemans	University of Utrecht, Netherlands
G. Ohring	NOAA/NESDIS, USA
G. Pant	Indian Institute of Tropical Meteorology, India
N. Plummer	National Climate Centre, Australia
F. Quinn	Great Lakes Environmental Research Lab, USA
E. Ran'kova	Institute for Global Climate and Ecology, Russian Federation
E.V. Rocheva	Institute for Global Climate and Ecology, Russian Federation
C. Ropelewski	Climate Analysis Center, USA
B. Santer	Lawrence Livermore National Laboratory, USA
H. Schmidt	Deutscher Wetterdienst Seewetteramt, Germany
E. Semenyuk	Institute for Global Climate and Ecology, Russia
I. Shiklomanov	State Hydrological Institute, Russian Federation
M. Shinoda	University of Tokyo, Japan
N. Sidorenkov	Hydrometeorological Research Centre of Russia, Russian Federation
I. Soldatova	Hydrometeorological Research Centre of Russia, Russian Federation
D. Sonechkin	Hydrometeorological Research Centre of Russia, Russian Federation
R. Spencer	Marshall Space Flight Center, USA
N. Speranskaya	State Hydrological Institute, Russian Federation
K. Trenberth	National Center for Atmospheric Research, USA
C. Tsay	Central Weather Bureau, Taiwan
J. Walsh	University of Illinois @ Urbana-Champaign, USA
K. Wang	Pacific Geoscience Centre, Canada
N. Ward	IMGA-CNR, Italy
S. Warren	University of Washington, USA
T. Yasunari	University of Tsukuba, Japan
Q. Zu	Jiangsu Meteorological Institute, China

Chapter 4: Climate Processes

Convening Lead Author

R. Dickinson   University of Arizona, USA

Lead Authors

V. Meleshko   Voeikov Main Geophysical Observatory, Russian Federation
D. Randall   Colorado State University, USA
E. Sarachik   University of Washington, USA
P. Silva-Dias   University of São Paulo, Brazil
A. Slingo   Meteorological Office, UK

Contributors

A. Barros   Pennsylvania State University, USA
O. Boucher   Laboratoire de Météorologie Dynamique du CNRS, France
R. Cess   University of New York State, USA
A. Del Genio   GISS, USA
L. Dumenil   Max-Planck Institut für Meteorologie, Germany
R. Fu   University of Arizona, USA
P. Gleckler   Lawrence Livermore National Laboratory, USA
J. Hansen   Goddard Institute for Space Studies, USA
R. Lindzen   Massachusetts Institute of Technology, USA
E. Maier-Reimer   Max-Planck Institut für Meteorologie, Germany
K. McNaughton   Hort Research, New Zealand
J..McWilliams   UCLA, USA
G. Meehl   National Centre for Atmospheric Research, USA
M. Miller   ECMWF, UK
D. Neelin   University of California, USA
E. Olaguer   Dow Chemical Co., USA
T. Palmer   ECMWF, UK
C. Penland   University of Colorado, USA
R. Pinker   University of Maryland, USA
V. Ramaswamy   Geophysical Fluid Dynamics Laboratory, USA
D. Rind   Goddard Institute of Space Studies, USA
A. Robock   University of Maryland, USA
M. Salby   University of Colorado, USA
M. Schlessinger   University of Illinois at Urbana-Champaign, USA
H. Schmid   Swiss Federal Institute of Technology, Switzerland
C. Senior   Meteorological Office, UK
Q. Shao   University of Arizona, USA
K. Shine   University of Reading, UK
H. Sundquist   University of Stockholm, Sweden
A. Vogelmann   University of Arizona, USA
A. Weaver   University of Victoria, Canada

Chapter 5: Climate Models – Evaluation

Convening Lead Author

W. Gates   Lawrence Livermore National Laboratory, USA

Lead Authors

G. Boer   Canadian Centre for Climate Modelling & Analysis, Canada
A. Henderson-Sellers   Macquarie University, Australia
C. Folland   Meteorological Office, UK
A. Kitoh   Japan Meteorological Agency, Japan
B. McAvaney   Bureau of Meteorology Research Centre, Australia
F. Semazzi   North Carolina State University, USA
N. Smith   Bureau of Meterology Research Centre, Australia
A. Weaver   University of Victoria, Canada
Q. Zeng   Institute of Atmospheric Physics, China

Contributors

J. Boyle   Lawrence Livermore National Laboratory, USA
R. Cess   University of New York State, USA
T. Chen   Macquarie University, Australia
J. Christy   University of Alabama @ Huntsville, USA
C. Covey   Lawrence Livermore National Laboratory, USA
T. Crowley   Texas A&M University, USA
U. Cubasch   Deutsches Klimarechenzentrum, Germany
J. Davies   Meteorological Office, UK
M. Fiorino   Lawrence Livermore National Laboratory, USA
G. Flato   Canadian Centre for Climate Modelling & Analysis, Canada
C. Fredericksen   BMRC, Australia
F. Giorgi   National Center for Atmospheric Research, USA
P. Gleckler   Lawrence Livermore National Laboratory, USA
J. Hack   NCAR, USA
J. Hansen   Goddard Institute for Space Studies, USA
G. Hegerl   Max-Planck Institut für Meteorologie, Germany
R. Huang   Woods Hole Oceanographic Institution, USA
P. Irannejad   Macquarie University, Australia
T. Johns   Meteorological Office, UK
J. Kiehl   National Center for Atmospheric Research, USA
H. Koide   Meteorological Research Institute, Japan
R. Koster   University of Maryland, USA
J. Kutzbach   University of Wisconsin @ Madison, USA
S. Lambert   University of Victoria, Canada
R. Latif   Max-Planck Institut für Meteorologie, Germany
N. Lau   NOAA, USA
P. Lemke   Alfred-Wegener Institute for Polar & Marine Research, Germany
R. Livezey   NOAA, USA
P. Love   Macquarie University, Australia
N. McFarlane   Canadian Climate Centre, Canada
K. McGuffie   University of Technology, USA
G. Meehl   National Centre for Atmospheric Research, USA
I. Mokhov   Institute of Atmospheric Physics, Russia

A. Noda	Japan Meteorological Agency,Japan
B. Otto-Bliesner	University of Texas, USA
T. Palmer	ECMWF, UK
T. Phillips	Lawrence Livermore National Laboratory, USA
A. Pitman	Macquarie University, Australia
J. Polcher	Laboratoire de Météorologie Dynamique du CNRS, France
G. Potter	Lawrence Livermore National Laboratory, USA
S.B. Power	Bureau of Meteorology Research Centre, Australia
D. Randall	Colorado State University, USA
P. Rasch	National Center for Atmospheric Research, USA
A. Robock	University of Maryland, USA
B. Santer	Lawrence Livermore National Laboratory, USA
E. Sarachik	University of Washington, USA
N. Sato	Japan Meteorological Agency, Japan
A. Semtner Jr	Naval Postgraduate School, USA
J. Slingo	University of Reading, UK
I. Smith	IEA Coal Research, London, UK
K. Sperber	Lawrence Livermore National Laboratory, USA
R. Stouffer	Geophysical Fluid Dynamics Laboratory, USA
M. Sugi	National Research Institute for Earth Science & Disaster Prevention, Japan
J. Syktus	CSIRO, Australia
K. Taylor	Lawrence Livermore National Laboratory , USA
S. Tett	Meteorological Office, UK
S. Tibaldi	University of Bologna, Italy
W. Wang	State University of New York @ Albany, USA
W. Washington	National Center for Atmospheric Research, USA
B. Weare	University of California, USA
D. Williamson	National Center for Atmospheric Research, USA
T. Yamagata	University of Tokyo, Japan
Z. Yang	University of Arizona, USA
R. Zhang	Institute of Atmospheric Physics, China
M. Zhang	State University of New York @ Stony Brook, USA
F. Zwiers	Canadian Climate Centre, Canada

Chapter 6: Climate Models – Projections of Future Climate

Convening Lead Author

| A. Kattenberg | Royal Netherlands Meteorological Institute (KNMI), Netherlands |

Lead Authors

F. Giorgi	National Centre for Atmospheric Research, USA
H. Grassl	World Meteorological Organisation, Switzerland
G. Meehl	National Centre for Atmospheric Research, USA
J. Mitchell	Meteorological Office, UK
R. Stoufer	Geophysical Fluid Dynamics Laboratory, USA
T. Tokioka	Japan Meteorological Agency, Japan
A. Weaver	University of Victoria, Canada
T. Wigley	Office for Interdisciplinary Earth Studies @ UCAR, USA

Contributors

A. Barros	Pennsylvania State University, USA
M. Beniston	ETH Institute of Geography, Switzerland
G. Boer	Canadian Centre for Climate Modelling & Analysis, Canada
T. Buishand	Royal Netherlands Meteorological Institute (KNMI), Netherlands
J. Christensen	Danish Meteorological Institute, Denmark
R. Colman	Bureau of Meteorology Research Centre, Australia
J. Copeland	Colorado State University, USA
P. Cox	Meteorological Office, UK
A. Cress	Deutscher Wetterdienst Zentralamt, Germany
U. Cubasch	Deutsches Klimarechenzentrum, Germany
M. Deque	Centre National de Recherches Météorologiques, France
G. Flato	Canadian Centre for Climate Modelling & Analysis, Canada
C. Fu	Institute of Atmospheric Physics, China
I. Fung	University of Victoria, Canada
J. Garratt	CSIRO, Australia
S. Ghan	Battelle Pacific Northwest Laboratories, USA
H. Gordon	CSIRO Division of Atmospheric Research, Australia
J. Gregory	Meteorological Office, UK
P. Guttorp	University of Washington, USA
A. Henderson-Sellers	Macquarie University, Australia
K. Hennessy	CSIRO Division of Atmospheric Research, Australia
H. Hirakuchi	CRIEPI, Japan
G. Holland	Bureau of Meteorology Research Centre, Australia
B. Horton	Meteorological Office, UK
T. Johns	Meteorological Office, UK
R. Jones	Meteorological Office, UK
M. Kanamitsu	W/NMC2, USA
T. Karl	NOAA National Climate Data Center, USA
D. Karoly	Monash University, Australia
A. Keen	Meteorological Office, UK
T. Kittel	National Center for Atmospheric Research, USA
T. Knutson	Geophysical Fluid Dynamics Laboratory, USA
T. Koide	Meteorological Research Institute, Japan
G. Können	Royal Netherlands Meteorological Institute (KNMI), Netherlands
M. Lal	Centre for Atmospheric Sciences, India
R. Laprise	University of Quebec at Montreal, Canada
R. Leung	Battelle Pacific Northwest Laboratory, USA
A. Lupo	Purdue University, USA
A. Lync	University of Tasmania, Australia
C. Ma	University of California, USA
B. Machenhauer	Max-Planck Institut für Meteorologie, Germany
E. Maier-Reimer	Max-Planck Institut für Meteorologie, Germany
M. Marinucci	NCAR, USA
B. McAvaney	Bureau of Meteorology Research Centre, Australia
J. McGregor	CSIRO Division of Atmospheric Research, Australia
L. Mearns	National Center for Atmospheric Research, USA
N. Miller	Lawrence Livermore National Laboratory, USA
J. Murphy	Meteorological Office, UK
A. Noda	Japan Meteorological Agency, Japan

M. Noguer	Meteorological Office, UK
J. Oberhuber	Deutsches Klimarechenzentrum, Germany
S. Parey	Electricité de France, France
H. Pleym	Telemark College of Engineering, Norway
J. Raisanen	University of Helsinki, Finland
D. Randall	Colorado State University, USA
S. Raper	Climatic Research Unit (UEA), UK
P. Rayner	Princeton University, USA
J. Roads	Scripps Institution of Oceanography, USA
E. Roeckner	Max-Planck Institut für Meteorologie, Germany
G. Russell	Goddard Institute for Space Studies, USA
H. Sasaki	Meteorological Research Institute, Japan
F. Semazzi	North Carolina State University, USA
C. Senior	Meteorological Office, UK
C. Skelly	James Cook University, Australia
K. Sperber	Lawrence Livermore National Laboratory, USA
K.Taylor	Lawrence Livermore National Laboratory, USA
S. Tett	Meteorological Office, UK
H. von Storch	Max- Planck Institut für Meteorologie, Germany
K. Walsh	CSIRO, Australia
P. Whetton	CSIRO Division of Atmospheric Research, Australia
D. Wilks	Cornell University, USA
I. Woodward	University of Sheffield, UK
F. Zwiers	Canadian Climate Centre, Canada

Chapter 7: Changes in Sea Level

Convening Lead Author

D. Warrick	The University of Waikato, New Zealand

Lead Authors

C. Le Provost	Institute de Mécanique de Grenoble, France
M. Meier	Institute of Artic & Alpine Research, USA
J. Oerlemans	University of Utrecht, Netherlands
P. Woodworth	Permanent Service for Mean Sea Level (PSMSL), UK

Contributors

R. Alley	Pennsylvania State University, USA
C. Bentley	University of Wisconsin @ Madison, USA
R. Bindschadler	Goddard Space Flight Center, USA
R. Braithwaite	University of Manchester, UK
B. Douglas	NOAA/NODC, USA
M. Dyurgerov	Institute of Geography, Russia
N. Flemming	Institute of Oceanographic Sciences Deacon Labotatory, UK
C. Genthon	Laboratoire de Glaciologie, CNRS, France
V. Gornitz	Goddard Institute for Space Studies, USA
J. Gregory	Meteorological Office, UK
W. Haeberli	Verschanstalt für Wasserbau Institut, Switzerland
P. Huybrechts	Alfred-Wegener Institute for Polar & Marine Research, Germany
T. Jóhannesson	Orkustofnun (National Energy Authority), Iceland

U. Mikolajewicz	Max-Planck Institut für Meteorologie, Germany
S. Raper	Climatic Research Unit (UEA), UK
D. Sahagian	University of New Hampshire, USA
T. Wigley	OIES @ UCAR, USA
J. de Wolde	University of Utrecht, Netherlands

Chapter 8: Detection of Climate Change and Attribution of Causes

Convening Lead Author

| B. Santer | Lawrence Livermore National Laboratory, USA |

Lead Authors

E. Anyamba	NASA Goddard Space Flight Center, USA
T. Barnett	Scripps Institute Of Oceanography, USA
T. Wigley	OIES @ UCAR, USA

Contributors

P. Bloomfield	Merrill Lynch Derivative Products, USA
E. Cook	Lamont-Doherty Earth Observatory, USA
C. Covey	Lawrence Livermore National Laboratory, USA
T. Crowley	Texas A&M University, USA
T. Delworth	Geophysical Fluid Dynamics Laboratory, USA
L. Gates	Lawrence Livermore National Laboratory, USA
N. Graham	Scripps Institute of Oceanography, USA
J. Gregory	Meteorological Office, UK
J. Hansen	Goddard Institute for Space Studies, USA
K. Hasselmann	Max-Planck Institut für Meteorologie, Germany
G. Hegerl	Max- Planck Institut für Meteorologie, Germany
T. Johns	Meteorological Office, UK
P. Jones	Climatic Research Unit (UEA), UK
T. Karl	NOAA National Climate Data Center, USA
D. Karoly	Monash University, Australia
H. Kheshgi	EXXON, USA
M. MacCracken	Office of the USGCRP, USA
K. Maskell	IPCC WGI Technical Support Unit, UK
G. Meehl	National Centre for Atmospheric Research, USA
J. Mitchell	Meteorological Office, UK
J. Murphy	Meteorological Office, UK
N. Nicholls	Bureau of Meteorology Research Centre, Australia
G. North	Texas A & M University, USA
M. Oppenheimer	Environmental Defense Fund, USA
J. Penner	Lawrence Livermore National Laboratory, USA
S. Power	Bureau of Meteorology Research Centre, Australia
A. Robock	University of Maryland, USA
C. Senior	Meteorological Office, UK
K. Taylor	Lawrence Livermore National Laboratory, USA
S. Tett	Meteorological Office, UK
F. Zwiers	Canadian Climate Centre, Canada

Chapter 9: Terrestial Biotic Responses to Environmental Change and Feedbacks to Climate

Convening Lead Author

J. Melillo   Woods Hole Oceanographic Institution, USA

Lead Authors

G. Farquhar   Australian National University, Australia
C. Prentice   Lund University, Sweden
O. Sala   University of Buenos Aires, Argentina
E. Schulze   Bayreuth University, Germany

Contributors

P. Bartlein   University of Oregon, USA
F. Bazzaz   Harvard University, USA
R. Bradshaw   Swedish University of Agricultural Sciences, Sweden
J. Clark   Duke University, USA
M. Claussen   Max Planck Institut für Meteorologie, Germany
G. Collatz   NASA Goddard Space Flight Center, USA
M. Coughenour   Colorado State University, USA
C. Field   Carnegie Institute of Washington, USA
J. Foley   James Cook University, Australia
A Friend   Institute of Terrestrial Ecology, UK
B. Huntley   University of Durham, UK
C. Körner   University of Basel, Switzerland
W. Kurz   ESSA Ltd, Canada
R. Leemans   RIVM, Netherlands
J. Lloyd   Australian National University, Australia
P. Martin   European Commision, Italy
K. McNaughton   Hort Research, New Zealand
A. McGuire   University of Alaska-Fairbanks, USA
R. Neilson   US Dept of Agriculture, USA
W. Oechel   San Diego State University, USA
J. Overpeck   NOAA/National Geophysical Data Center, USA
W. Parton   Colorado State University, USA
L. Pitelka   Electric Power Research Institute (EPRI), USA
D. Rind   Goddard Institute of Space Studies, USA
S. Running   University of Montana. USA
D. Schimel   National Center for Atmospheric Research, USA
T. Smith   University of Virginia, USA
T. Webb   Brown University, USA
C. Whitlock   University of Oregon, USA

Chapter 10: Marine Biotic Responses to Environmental Change and Feedbacks to Climate

Convening Lead Author

K. Denman   Institute of Ocean Sciences, Canada

Lead Authors

E. Hofmann   Crittenton Hall Old Dominion University, USA
H. Marchant   DEST, Australia

Contributors

M. Abbott   Oregon State University, USA
T. Bates   NOAA, USA
S. Calvert   University of British Columbia, Canada
M. Fasham   James Rennell Centre (NERC), UK
R. Jahnke   Skidaway Institution of Oceanography, USA
S. Kempe   University of Hamburg, Germany
R. Lara   Alfred-Wegener Institute for Polar & Marine Research, Germany
C. Law   Plymouth Marine Laboratory, UK
P. Liss   University of East Anglia, UK
A. Michaels   Bermuda Biological Research Station, Bermuda
T. Pederson   University of British Columbia, Canada
M. Peña   Institute of Ocean Sciences, Canada
T. Platt   Bedford Institute of Oceanography, Canada
K. Van Scoy   Plymouth Marine Laboratory, UK
J. Sharp   University of Delaware, USA
D. Thomas   The Interuniversity Institute for Marine Sciences, Israel
J. Walsh   University of South Florida, USA
A. Watson   Plymouth Marine Laboratory, UK

Chapter 11: Advancing our Understanding

Convening Lead Author

G. McBean   Atmospheric Environment Canada, Canada

Lead Authors

P. Liss   University of East Anglia, UK
S. Schneider   Stanford University, USA

Appendix 4

Reviewers of the IPCC WGI Report[1]

Albania

E. Demiraj   Hydrometeorological Institute

Australia

R. Allan	CSIRO Division of Atmospheric Research
I. Allison	University of Tasmania
P. Baines	CSIRO Division of Atmospheric Research
S. Barrell	Bureau of Meteorology
N. Bindoff	University of Tasmania
W. Bouma	CSIRO Division of Atmospheric Research
H. Bridgman	University of Newcastle
J. Bye	The Flinders University of South Australia
R. Byron-Scott	The Flinders University of South Australia
F. Chiew	University of Melbourne
J. Church	CSIRO Division of Oceanography
K. Colls	Bureau of Meteorology
R. Colman	Bureau of Meteorology Research Centre
B. Curran	Dept of Primary Industries & Energy
M. Dix	CSIRO Division of Atmospheric Research
B. Dixon	Bureau of Meteorology
M. England	University of New South Wales
I. Enting	CSIRO Division of Atmospheric Research
D. Etheridge	CSIRO Division of Atmospheric Research
G. Farquhar	Australian National University
P. Fraser	CSIRO Division of Atmospheric Research
C. Fredericksen	Bureau of Meteorology Research Centre
I. Galbally	CSIRO Division of Atmospheric Research
J. Garratt	CSIRO Division of Atmospheric Research
T. Gibson	University of Tasmania
R. Gifford	CSIRO Division of Plant Industry

---

1 The list may not include some experts who contributed to the peer review only through their governments.

L. Glover	Department of the Environment, Sports & Territories
A. Gordon	Flinders University
G. Harris	CSIRO Division of Atmospheric Research
A. Henderson-Sellers	Macquarie University
K. Hennessy	CSIRO Division of Atmospheric Research
J. Jakka	University of Tasmania
J. Jensen	CSIRO Division of Atmospheric Research
B. Kininmonth	Bureau of Meteorology
M. Kirschbaum	CSIRO Division of Atmospheric Research
R. Kleeman	Bureau of Meteorology Research Centre
J. Lloyd	Australian National University
J. Lough	Australian Institute of Marine Science
M. Manton	Bureau of Meteorology Research Centre
H. Marchant	Department of the Environment, Sports & Territories
B. McAvaney	Bureau of Meteorology Research Centre
T. McDougall	CSIRO Division of Atmospheric Research
J. McGregor	CSIRO Division of Atmospheric Research
C. Mitchell	CSIRO Division of Atmospheric Research
N. Nicholls	Bureau of Meteorology Research Centre
S. O'Farrell	CSIRO Division of Atmospheric Research
G. Paltridge	University of Tasmania
J. Parslow	CSIRO Division of Atmospheric Research
G. Pearman	CSIRO Division of Atmospheric Research
A. Pitman	Macquarie University
N. Plummer	National Climate Centre
S. Power	Bureau of Meteorology Research Centre
P. Rhines	CSIRO Division of Atmospheric Research
L. Rikus	Bureau of Meteorology Research Centre
S. Rintoul	CSIRO Division of Atmospheric Research
G. Rumantir	Macquarie University
B. Ryan	CSIRO Division of Atmospheric Research
P. Schwerdtfeger	The Flinders University of South Australia
C. Skelly	James Cook University
N. Smith	Bureau of Meteorology Research Centre
I. Smith	CSIRO Division of Atmospheric Research
G. Stephens	CRC For Southern Hemisphere Meteorology
B. Tucker	CSIRO Division of Atmospheric Research
Y. Wang	CSIRO Division of Atmospheric Research
I. Watterson	CSIRO Division of Atmospheric Research
T. Weir	Department of Primary Industries & Energy
P. Whetton	CSIRO Division of Atmospheric Research
M. Williams	University Of Adelaide
B. Wright	National Climate Centre
C. Zammit	Department of Foreign Affairs & Trade
J. Zillman	Bureau of Meteorology

Austria

H. Hojesky	Federal Ministry for Environment

Barbados

L. Nurse   Coastal Conservation Project Unit

Belguim

A. Berger   Université Catholique de Louvain

Benin

E. Ahlonsou   Service Météorologique

Brazil

P. Fearnside   National Institute of Research on the Amazon

Canada

G. Boer   Canadian Centre for Climate Modelling & Analysis
H. Boyd   Environment Canada
G. Brunet   Environment Canada
R. Allyn Clarke   Bedford Institute of Oceanography
R. Daley   Environment Canada
K. Denman   Institute of Ocean Sciences
M. Pe   Institute of Ocean Sciences
L. Dupigny-Giroux   McGill University
A. Dyke   Geological Survey of Canada
P. Egginton   Geological Survey of Canada
G. Evans   DFO Science Branch
B. Findlay   Environment Canada
G. Flato   Canadian Centre for Climate Modelling & Analysis
I. Fung   University of Victoria
R. Grant   University of Alberta
D. Gullett   Environment Canada
H. Hengeveld   Canadian Climate Centre
K. Higuchi   Atmospheric Environment Service
W. Hogg   Environment Canada
J. Jerome   McGill University
H. Jette   Geological Survey of Canada
B. Johnson   Dalhousie University
M. Johnston   Lakehead University
A. Judge   Geological Survey of Canada
R. Koerner   Geological Survey of Canada
S. Lambert   University of Victoria
L. Malone   Environment Canada
C. McElroy   Atmospheric Environment Service (ARQX)
P. Merilees   Environment Canada
P. Mudie   Bedford Institute of Oceanography
L. Mysak   McGill University
I. Perry   Fisheries and Oceans Canada

N. Roulet	York University
J. Shaw	Bedford Institute of Oceanography
J. Stone	Canadian Climate Centre
A. Weaver	University of Victoria
D. Whelpdale	Environment Canada
F. Zwiers	Canadian Climate Centre

Chile

H. Antolini	Servicio Hidrográfico y Oceanográfico de la Armada
H. Fuenzalida	Universidad de Chile
I. González	Dirección General de Territorio Marítimo y Marina
J. Gutiérrez	Universidad de la Serena
O. Jara	Instituto de Fomento Pesquero
M. Manzur	CODEFF
J. Searle	Comision Nacional del Medio Ambiente
R. Serra	Instituto de Fomento Pesquero

China

S. Anjian	National Meteorological Centre
D. Yihui	Academy of Meteorological Sciences
M. Dong	National Climate Centre
W. Futang	Academy of Meteorological Sciences
G. Shi	Institute of Atmospheric Physics
S. Wang	Beijing University
L. Yong	National Climate Centre
Z. Zhao	National Climate Centre

Colombia

P. Leyva	Instituto de Hidrologia
E. Rangel	Instituto de Hidrologia
N. Sabogal	Institute of Colombia

Denmark

J. Christensen	Danish Meteorological Institute
E. Friis-Christensen	Danish Meteorological Institute
J. Gundermann	Danish Energy Agency
A. Jørgensen	Danish Meteorological Institute
E. Kaas	Danish Meteorological Institute
P. Laut	Engineering Academy of Denmark
N. Reeh	Danish Polar Centre
G. Shaffer	University of Copenhagen

Finland

R. Heino	Finnish Meteorological Institute
E. Holopainen	University of Helsinki

| I. Savolainen | VTT-Energy |
| J. Sinisalo | VTT-Energy |

France

M. Deque	Centre National de Recherches Météorologiques
J. Guiot	Labo Botanique Historique
J. Jouzel	LMCE/DS
H. Le Treut	Laboratoire de Météorologie Dynamique du CNRS
P. Monfray	Centre des Faibles Radioactivités
V. Moron	Université de Bourgogne
C. Nadine	CNRS
E. Nesme-Ribes	Observatoire de Paris
C. Waelbroeck	LMCE CEA Saclay

Germany

K. Arpe	Max-Planck Institut für Meteorologie
E. Augstein	Alfred Wegener Institut
U. Böhn	Potsdam Institut
M. Claussen	Max Planck Institut für Meteorologie
R. Conrad	Universität Marburg
W. Cramer	Potsdam Institut
P. Crutzen	Max-Planck Institut für Chemie
U. Cubasch	Deutsches Klimarechenzentrum
K. Dehne	Meteorologisches Observatorium Potsdam
L. Dumenil	Max-Planck Institut für Meteorologie
H. Flohn	Universität Bonn
A. Ganopolski	Potsdam Institut
J. Heintzenberg	Institut für Troposphärenforschung
K. Koltermann	Federal Maritime and Hydrographic Agency
K. Lange	Federal Ministry for the Environment
R. Lara	Alfred Wegener Institut
B. Machenhauer	Max-Planck Institut für Meteorologie
E. Raschke	GKSS-Forschungszentrum
P. Sachs	GEOMAR-Forschungszentrum
H. Schellnhuber	Potsdam Institut
C. Schönwiese	Universität J W Goethe
V. Smetacek	Alfred Wegener Institut
J. Thiede	GEOMAR
V. Vent-Schmidt	Deutscher Wetterdienst Zentralamt
G. Wefer	Universität Bremen

Hungary

P. Ambrózy	Hungarian Meteorological Society
E. Antal	University for Agricultural Sciences
J. Bartholy	Euötvös Loránd University
T. Faragó	Ministry for Environment & Regional Policy
Z. Iványi	Eötvös Loránd University

G. Koppány	József Attila University
J. Mika	Hungarian Meteorological Service
T. Pálvölgyi	Ministry for Environment & Regional Policy

Iceland

| T. Johannesson | Orkustofnun (National Energy Authority) |

India

| M. Lal | Centre for Atmospheric Sciences |
| S. Sadasivan | Bhabha Atomic Research Centre |

Iran

| M. Abduli | University of Tehran |

Israel

| D. Thomas | The Inter-university Institute for Marine Sciences |

Italy

| G. Visconti | Università Degli Studi dell'Aquila |

Jamaica

| S. McGill | Meteorological Service |

Japan

S. Asano	Japan Meteorological Agency
M. Hirota	Meteorological Research Institute JMA
H. Inoue	Japan Meteorological Agency
T. Ito	Japan Meteorological Agency
M. Kimoto	Japan Meteorological Agency
A. Kitoh	Japan Meteorological Agency
I. Koike	University of Tokyo
S. Kusunoki	Environment Agency
K. Mabuchi	Meteorological Research Institute JMA
T. Matsuno	University of Hokkaido
S. Nakagawa	Japan Meteorological Agency
H. Nakane	National Institute for Environmental Studies
T. Nitta	University of Tokyo
A. Noda	Japan Meteorological Agency
T. Okita	Ohobirin University
N. Ono	National Institute of Polar Research
M. Shinoda	University of Tokyo
K. Shuto	Meteorological Research Institute JMA
M. Sugi	National Research Institute for Earth Science & Disaster Prevention

A. Sumi	University of Tokyo
T. Takeda	Nagoya University
T. Tokioka	Japan Meteorological Agency
S. Tsunogai	University of Hokkaido
R. Yamamoto	Kyoto University
T. Yasunari	University of Tsukuba

Kenya

M. Indeje	Kenya Meteorological Department
W. Kimani	Kenya Meteorolgical Department
M. Kinyanjui	Permanent Mission of Kenya, Geneva
G. Mailu	Ministry of Research, Technical Training & Technology
F. Mutua	University of Nairobi
H. Muturi	Ministry of Research, Technical Training & Technology
R. Mwangi	University of Nairobi
J. Ng'ang'a	University of Nairobi
N. Njau	Kenya Meteorological Department
J. Njihia	Kenya Meteorological Department
W. Nyakwada	Kenya Meteorological Department
R. Odingo	University of Nairobi
L. Ogallo	University of Nairobi
A. Owino	Kenya Meteorological Department

Morocco

| M. Abderrahmane | Direction de Météorologie Nationale |
| S. Larbi | Direction de Météorologie Nationale |

Netherlands

H. de Baar	Netherlands Institute for Sea Research
A. Baede	Royal Netherlands Meteorological Institute (KNMI)
J. Beck	Air Research Laboratory
J. Beersma	Royal Netherlands Meteorological Institute (KNMI)
P. Builtjes	University of Utrecht
T. Buishand	Royal Netherlands Meteorological Institute (KNMI)
G. Burgers	Royal Netherlands Meteorological Institute (KNMI)
H. de Bruin	Wageningen Agricultural University
S. Drijfhout	Royal Netherlands Meteorological Institute (KNMI)
R. Guicherit	TNO/IMW
L. Janssen	RIVM, Netherlands Institute for Public Health & Environment
W. Kieskamp	Netherlands Energy Research Foundation ECN
G. Komen	Royal Netherlands Meteorological Institute (KNMI)
G. Können	Royal Netherlands Meteorological Institute (KNMI)
R. Leemans	RIVM, Netherlands Institute for Public Health & Environment
G. Mohren	IBN-DLO
J. Oerlemans	University of Utrecht
J. Opsteegh	Royal Netherlands Meteorological Institute (KNMI)
M. Roemer	TNO/IMW

C. Schuurmans	University of Utrecht
A. Sterl	Royal Netherlands Meteorological Institute (KNMI)
H. ten Brink	Netherland Energy Research Foundation
H. Tennekes	Royal Netherlands Meteorological Institute (KNMI)
H. The	RIVM, Netherlands Institute for Public Health & Environment
A. van Ulden	Royal Netherlands Meteorological Institute (KNMI)
A. van Amstel	RIVM, Netherlands Institute for Public Health & Environment
R. van Dorland	Royal Netherlands Meteorological Institute (KNMI)
G. Velders	RIVM, Netherlands Institute for Public Health & Environment
H. Visser	KEMA
C. Vreugdenhil	University of Utrecht
H. Vugts	Institute for Planetary Sciences
P. Westbroek	University of Leiden

New Zealand

B. Campbell	NZ Pastoral Agriculture Research Institute
A. Carran	NZ Pastoral Agricultural Research Institute
P. Clinton	New Zealand Forest Research Institute
C. de Freitas	The University of Auckland
J. Grieve	NIWA
J. Hall	NIWA
J. Hannah	Otago University
M. Harvey	Nat Institute of Water & Atmospheric Research Ltd
J. Kidson	NIWA
B. Kirk	University of Canterbury
B. Liley	NIWA
D. Lowe	NIWA
M. Manning	National Institute of Water & Atmospheric Research
M. McGlone	Landcare Research
R. McKenzie	Nat Institute of Water & Atmospheric Research Ltd
K. McNaughton	Hort Research Institute
B. Mullan	New Zealand Meteorological Service
L. Phillips	University of Canterbury
H. Plume	Ministry for the Environment
M. Salinger	NIWA – CLIMATE
A. Sturman	University of Canterbury
K. Tate	Landcare Research NZ Ltd
D. Whitehead	Landcare Research NZ Ltd
D. Wratt	NIWA
P. Whiteford	Institute of Geological & Nuclear Sciences

Norway

| T. Martinsen | State Pollution Control Authority |
| F. Stordal | Norwegian Institute for Air Research |

Poland

W. Andrzej	Institute of Oceanology, PAS
E. Bulewicz	Institute of Chemistry and Technology
B. Jakubiak	Institute of Meteorology & Water Management
M. Mietus	Institute of Meteorolgy & Water Management
M. Sadowski	Institute of Environmental Protection
P. Sowinski	Plant Breeding & Acclimatization Institute

Russian Federation

G. Golitsyn	Institute of Atmospheric Physics
P. Groisman	State Hydrological Institute
G. Gruza	Institute for Global Climate and Ecology
I. Karol	Main Geophysical Observatory
V. Meleshko	Main Geophysical Observatory
D. Sonechkin	Hydrometeorological Research Centre of Russia

Spain

L. Balairon	National Meteorological Institute
A. Labajo	National Meteorological Institute

Sri Lanka

J. Ratnasiri	Ministry of Environment and Parliamentary Affairs

Sweden

B. Bolin	University of Stockholm, IPCC Chairman
S. Craig	University of Stockholm
T. Hedlund	Swedish Commission on Climate Change
A. Johansson	Swedish Meteorological & Hydrological Institute
W. Josefsson	Swedish Meteorological & Hydrological Institute
C. Prentice	Lund University
H. Rodhe	University of Stockholm

Switzerland

A. Niederberger	Office Fédéral de l'Environnement
J. Beer	EAWAG
W. Haeberli	Verschanstalt für Wasserbau Institut
M. Hoelzle	Verschanstalt für Wasserbau Institut
D. Imboden	EAWAG
J. Innes	Swiss Federal Institute for Forest, Snow & Landscape Research
F. Joos	University of Bern
P. Francis	Office Fédéral de l'Environnement
H. Oeschger	University of Bern
B. Sevruk	ETH Institute of Geography
J. Staehelin	ETH Atmospheric Physics Laboratory

Tanzania

D. Mussa	Directorate of Meteorology
B. Nyenzi	Directorate of Meteorology

Thailand

J. Boonjawat	Chulalongkorn University
M. Brikshavana	Meteorolgical Department
S. Piamphongsant	Office of Environmental Policy and Planning

Trinidad & Tobago

E. Henry	Meteorological Services Unit
V. Mendez-Charles	Town & Country Planning Division
P. Atherley-Rowe	Ministry of Health
E. Caesar	Tobago House of Assembly
F. Campayne	University of the West Indies
B. Chatoor	University of the West Indies
C. O'Brian	Delpesh Institute of Marine Affairs
R. Maysingh	Ministry of Works & Transport
R. Ramdin	Water Resources Agency
P. Samuel	Ministry of Planning & Development
A. Wharton	University of the West Indies
H. Wilson	Ministry of Agriculture, Land & Marine Resources

United Kingdom

T. Anderson	James Rennell Centre (NERC)
H. Bryden	James Rennell Centre (NERC)
D. Cushing	Retired (MAFF)
C. Folland	Meteorological Office, Hadley Centre
W. Gould	Institute of Oceanographic Sciences
J. Grove	University of Cambridge
T. Guymer	James Rennell Centre (NERC)
G. Jenkins	Meteorological Office, Hadley Centre
C. Johnson	Meteorological Office, Hadley Centre
P. Jones	Climatic Research Unit (UEA)
M. Kelly	University of East Anglia
K. Law	University of Cambridge
P. Liss	University of East Anglia
P. Mallaburn	Department of the Environment
J. Mitchell	Meteorological Office, Hadley Centre
N. Owens	Plymouth Marine Laboratory
T. Palmer	ECMWF
D. Parker	Meteorological Office, Hadley Centre
D. Pugh	Southampton Oceanography Centre
S. Raper	Climatic Research Unit (UEA)
D. Roberts	Meteorological Office, Hadley Centre
P. Rowntree	Meteorological Office, Hadley Centre

K. Shine	University of Reading
P. Smithson	University of Sheffield
S. Tett	Meteorological Office, Hadley Centre
S. Tudhope	University of Edinburgh
A. Watson	Plymouth Marine Laboratory
D. Webb	Institute of Oceanographic Sciences
T. Wigley	OIES @ UCAR, USA

United States of America

J. Aber	University of New Hampshire
R. Alley	Pennsylvania State University
A. Ameko	Los Alamos National Laboratory
T. Anderson	University of Washington
J. Angell	NOAA ERL @ Silver Springs
E. Anyamba	NASA Goddard Space Flight Center
J. Arnold	USDA/ARS
T. Barnett	Scripps Institute of Oceanography
A. Barros	Pennsylvania State University
T. Bates	NOAA
J. Bates	NOAA Department of Commerce
R. Beardsley	Woods Hole Oceanographic Institution
C. Bentley	University of Wisconsin @ Madison
R. Birdsey	US Forest Service/USDA
G. Branstator	NCAR
K. Bryan	Geophysical Fluid Dynamics Laboratory
T. Charlock	NASA
R. Charlson	University of Washington
J. Christy	University of Alabama @ Huntsville
T. Crowley	Texas A&M University
R. Dahlman	US Department of Energy
E. Davidson	Woods Hole Research Center
R. Davis	University of Virginia
A. Del Genio	GISS
R. Dickinson	University of Arizona
P. Dirmeyer	IGES
H. Ducklow	Virginia Institute of Marine Sciences
J. Dutton	Pennsylvania State University
W. Elliott	NOAA ERL @ Silver Springs
H. Ellsaesser	Lawrence Livermore National Laboratory
J. Evans	Pennsylvania State University
G. Evans	US Dept of Agriculture
C. Field	Carnegie Institute of Washington
D. Gaffen	NOAA ERL @ Silver Springs
J. Gavin	Lamont-Doherty Geological Laboratory
E. Gorham	University of Minnesota
J. Hansen	Goddard Institute for Space Studies
J. Hanson	USDA/ARS
L. Harper	USDA/ARS
S. Hastenrath	University of Wisconsin @ Madison

T. Hayward	Scripps Institution of Oceanography
W. Heilman	US Forest Service/USDA
J. Hom	US Forest Service/USDA
R. Houghton	Woods Hole Research Center
T. Hughes	University of Maine
J. Hunter	NOAA/ National Marine Fisheries Service
C. Jackman	NASA Goddard Space Flight Center
D. Jacob	Harvard University
S. Jacobs	Columbia University
R. Jahnke	Skidaway Institution of Oceanography
A. Jenkins	Columbia University
G. Johnson	USDA/ARS
T. Joyce	Woods Hole Oceanographic Institution
C. Jim Kao	Los Alamos National Laboratory
R. Katz	National Center for Atmospheric Research
C. Keeling	Scripps Institute of Oceanography
J. Kiehl	National Center for Atmospheric Research
B. Kimball	US Water Conservation Laboratory
T. Knutson	Geophysical Fluid Dynamics Laboratory
R. Koster	University of Maryland
G. Kukla	Lamont-Doherty Geological Laboratory
Y. Kushnir	Columbia University
A. Lacis	Goddard Institute for Space Studies
M. Landry	University of Hawaii
C. Landsea	Colorado State University
D. Latham	US Forest Service/USDA
R. Lindzen	Massachusetts Institute of Technology
C. Loehle	Argonne National Laboratory
J. Logan	Harvard University
M. Lupo	Purdue University
D. Macayeal	University of Chicago
M. MacCracken	Office of the USGCRP
R. Madden	National Center for Atmospheric Research
J. Mahlman	Geophysical Fluid Dynamics Laboratory
J. Mak	State University of New York @ Stony Brook
T. Malone	North Carolina State University
S. Manabe	Geophysical Fluid Dynamics Laboratory
P. Mayewski	University of New Hampshire
J..McWilliams	UCLA
C. Mechoso	UCLA
G. Meehl	National Centre for Atmospheric Research
P. Michaels	University of Virginia
C. Miller	Oregon State University
C. Milly	US Geological Survey
R. Neilson	US Department of Agriculture
A. Nicks	USDA/ARS/SPA
W. Nierenberg	Scripps Institute of Oceanography
R. Norby	Oak Ridge National Laboratory
J. Penner	Lawrence Livermore National Laboratory
R. Pinkel	University of California @ San Diego

W. Porch	Los Alamos National Laboratory
L. Porter	USDA/ARS
C. Potter	Ames Research Center
M. Prather	University of California @ Irvine
R. Prinn	Massachusetts Institute of Technology
D. Randall	Colorado State University
A. Ravishankara	NOAA Aeronomy Laboratory
R. Reck	Argonne National Laboratory
C. Richardson	IUSDA/ARS
B. Ridley	National Center for Atmospheric Research
M. Rienecker	Goddard Space Flight Center
J. Roads	Scripps Institution of Oceanography
D. Robinson	Rutgers University
A. Robock	University of Maryland
G. Russell	Goddard Institute for Space Studies
P. Russell	NASA Ames Research Center
B. Santer	Lawrence Livermore National Laboratory
J. Sarmiento	Geophysical Fluid Dynamics Laboratory
S. Schwartz	Brookhaven National Laboratory
T. Smith	University of Virginia
W. Smith	Yale University
A. Solomon	Oregon State University
A. Solow	Woods Hole Oceanographic Institution
P. Stone	Massachusetts Institute of Technology
R. Stouffer	Geophysical Fluid Dynamics Laboratory
Y. Sud	NASA Goddard Space Flight Center
E. Sundquist	US Geological Survey
K. Taylor	Lawrence Livermore National Laboratory
A. Thompson	NASA Goddard Space Flight Center
K. Trenberth	National Center for Atmospheric Research
C. van der Veen	Ohio State University
J..Wahr	Department of Physics, University of Colorado
M. Wallace	University of Washington
J. Walsh	University of Illinois @ Urbana Champaign
W. Wang	State University of New York @ Albany
W. Washington	National Center for Atmospheric Research
B. Weare	University of California
T. Webb	Brown University
J. White	University of Colorado
D.Winstanley	US Department of Energy
G. Woodwell	Woods Hole Research Center
Z. Yang	University of Arizona
X. Zeng	University of Arizona
M. Zhang	State University of New York @ Stony Brook
H. Zwally	NASA

Venezuela

L. Hidalgo	University of Venezuela

United Nations Organisations and Specialised Agencies

J. van de Vate	International Atomic Energy Agency, Vienna
M. Cheatle	United Nations Environment Programme, Nairobi
R. Christ	United Nations Environment Programme, Nairobi
M. Mcfarland	United Nations Environment Programme, Nairobi
H. Grassl	World Meteorological Organisation, Geneva

Non-Governmental Organisations

B. Kuemmel	2morrows Climate and Environment
A. McCulloch	AFEAS/ICI C&P, UK
S. Nishininomiya	CRIEPI
K. Gregory	Centre for Business and the Environment
R. Whitney	Coal Research Association of New Zealand
W. Hennesey	Coal Research Association of New Zealand
V. Grey	Coal Research Association of New Zealand
E. Olaguer	Dow Chemical Co.
J. Kinsman	Edison Electtric Institute, USA
W. Guyker	Edison Electric Institute, USA
J. Kennedy	Edison Electric Institute, USA
L. Pitelka	Electric Power Research Institute(EPRI)
S. Parey	Electricité de France
J. Hales	Envair
M. Oppenheimer	Environmental Defense Fund, USA
S. Japar	Ford Motor Company
B. Gardner	Global Climate Coalition
W. Hare	Greenpeace International, Amsterdam
E. Jackson	Greenpeace International, Australia
B. Flannery	IPIECA & Exxon, USA
T. Murray	MAF Fisheries
L. Paul	MAF Fisheries
D. Lashof	Natural Resources Defense Council, USA
J. LeCornu	Shell
V. Narayanan	Technical Resources International Inc.
M. Jefferson	World Energy Council, London

Appendix 5

Acronyms

ACE	Aerosol Characterisation Experiments
ACSyS	Arctic Climate System Study
AFEAS	Alternative Fluorocarbons Environmental Acceptability Study
AGAGE	Atmospheric Lifetime Experiment/Global Atmospheric Gases Experiment
AMIP	Atmospheric Model Intercomparison Project
AOGCM	Atmosphere-Ocean General Circulation Model
BAHC	Biospheric Aspects of the Hydrological Cycle
BATS	Biosphere-Atmosphere Transfer Scheme
BMRC	Bureau of Meteorology Research Centre (Australia)
CCC	Canadian Centre for Climate
CCM	Community Climate Model
CCN	Cloud Condensation Nuclei
CGCMs	Coupled General Circulation Models
CLIVAR	Climate Variability and Predictability Programme
CMDL	Climate Monitoring and Diagnostics Laboratory (NOAA)
CNRM	Centre National de Recherches Météorologiques (France)
COADS	Comprehensive Ocean Air Data Set
COHMAP	Co-operative Holocene Mapping Programme
COLA	Center for Ocean, Land and Atmosphere
CRF	Cloud Radiative Forcing
CSIRO	Commonwealth Scientific and Industrial Research Organization (Australia)
CSU	Colorado State University (USA)
DIC	Dissolved Inorganic Carbon
DIS	Data and Information System
DOC	Dissolved Organic Carbon
DOLY	Dynamic Global Photogeography Model
DOM	Dissolved Organic Matter
DTR	Diurnal Temperature Range
EBM	Energy Balance Models
ECHAM	European Centre/Hamburg Model (ECMWF/MPI)
ECMWF	European Centre for Medium-Range Weather Forecasts
EGIG	International Glaciological expedition to Greenland
EOF	Empirical Orthogonal Function
ERBE	Earth Radiation Budget Experiment

ET	Evapotranspiration
FAO	Food and Agriculture Organization (of the UN)
FIFE	First ISLCP Field Experiment
FIRE	First ISCCP Regional Experiment
FRAM	Fine Resolution Antarctic Model
GAIM	Global Analysis, Interpretation and Modelling
GCEM	Goddard Cumulus Ensemble Model
GCM	General Circulation Model
GCOS	Global Climate Observing System
GCRI	Greenhouse Climate Response Index
GCTE	Global Change and Terrestrial Ecosystems
GEWEX	Global Energy and Water Cycle Experiment
GFDL	Geophysical Fluid Dynamics Laboratory
GHCN	Global Historical Climate Network
GISP	Greenland Ice Sheet Project
GISS	Goddard Institute for Space Studies
GLOSS	Global Sea Level Observing System
GOOS	Global Ocean Observing System
GPS	Global Positioning System
GRIP	Greenland Icecore Project
GST	Ground Surface Temperature
GWP	Global Warming Potential
HDP	Human Dimensions of Global Environmental Change Programme
HNLC	High Nutrient-Low Chlorophyll
HRC	Highly Reflective Clouds
IAHS	International Association of Hydrological Science
IASH	International Association of Scientific Hydrology
ICE	International Cirrus Experiment
ICSI	International Commission on Snow and Ice
ICSU	International Council of Scientific Unions
IERS	International Earth Rotation Service
IGAC	International Global Atmospheric Chemistry Project
IGBP	International Geosphere-Biosphere Programme
IMAGE	Integrated Model for Assessment of the Greenhouse Effect
INPE	Instituto Nacional de Pesquisas Espaciais, Brazil
IOC	Intergovernmental Oceanographic Commission
ISCCP	International Satellite Cloud Climatology Project
ISLCP	International Satellite Land-surface Climatology Project
ISSC	International Social Sciences Council
ITRF	International Terrestrial Reference Frame
JGOFS	Joint Global Ocean Flux Study
LAI	Leaf Area Index
LANL	Los Alamos National Laboratory (USA)
LGM	Last Glacial Maximum
LLNL	Lawrence Livermore National Laboratory, USA
LMD	Laboratoire de Météorologie Dynamique du CNRS (France)
LOICZ	Land-Ocean Interactions in the Coastal Zone
LSW	Labrador Seawater
LSX	Land Surface Transfer
MAST	Monterey Area Ship Track experiment

MECCA	Model Evaluation Consortium for Climate Assessment
MGO	Main Geophysical Observatory (Russia)
MJO	Madden-Julian Oscillation
MOGUNTIA	Model of the General Universal Tracer transport In the Atmosphere
MPI	Max-Planck Institute for Meteorology (Germany)
MRI	Meteorological Research Institute (Japan)
MSLP	Mean Sea Level Pressure
MSU	Microwave Sounding Unit
NADW	North Atlantic Deep Water
NASA	National Aeronautics and Space Administration (USA)
NCAR	National Center for Atmospheric Research (USA)
NMAT	Night Marine Air Temperature
NOAA	National Oceanic and Atmospheric Administration (USA)
NRC	National Research Council (USA)
NWP	Numerical Weather Prediction
OCCAM	Ocean Circulation and Climate Advanced Modelling (UK)
OLR	Outgoing Long-wave Radiation
OSU/IAP	Oregon State University /Institute of Atmospheric Physics (USA)
PAGES	Past Global Changes
PAR	Photosynthetically Available Radiation
PBL	Planetary Boundary Layer
PGR	Post-glacial Rebound
PILPS	Project for Intercomparison of Land-surface Parametrization Schemes
PMIP	Palaeoclimate Modelling Intercomparison Project
POC	Particulate Organic Carbon
PSMSL	Permanent Service for Mean Sea Level
SLR	Satellite Laser Ranging
SOI	Southern Oscillation Index
SPARC	Stratospheric Processes and their Role in Climate
SSA	Singular Spectrum Analysis
SSTs	Sea Surface Temperatures
START	System for Analysis, Research and Training
SVATs	Soil Vegetation Atmosphere Transfer Schemes
TEM	Terrestrial Ecosystem Model
TOGA	Tropical Ocean Global Atmosphere
TOPEX/POSEIDON	US/French Ocean Topography Satellite Altimeter Experiment
UD-EBM	Upwelling Diffusion-Energy Balance Model
UGAMP	University Global Atmospheric Modelling Project (Reading, UK)
UKMO	United Kingdom Meteorological Office
UNEP	United Nations Environment Programme
UNESCO	United Nations Education, Cultural and Scientific Organisation
VEMAP	Vegetation/Ecosystem Modelling and Analysis Project
VLBI	Very Long Baseline Interferometry
WAIS	West Antarctic Ice Sheet
WCRP	World Climate Research Programme
WGNE	Working Group on Numerical Experimentation
WMO	World Meteorological Organisation
WOCE	World Ocean Circulation Experiment

Appendix 6

Units

SI (Système Internationale) Units:

Physical Quantity	Name of Unit	Symbol
length	metre	m
mass	kilogram	kg
time	second	s
thermodynamic temperature	kelvin	K
amount of substance	mole	mol

Fraction	Prefix	Symbol	Multiple	Prefix	Symbol
10^{-1}	deci	d	10	deca	da
10^{-2}	cent	c	10^2	hecto	h
10^{-3}	milli	m	10^3	kilo	k
10^{-6}	micro	μ	10^6	mega	M
10^{-9}	nano	n	10^9	giga	G
10^{-12}	pico	p	10^{12}	tera	T
10^{-15}	femto	f	10^{15}	peta	P
10^{-18}	atto	a			

Special Names and Symbols for Certain SI-Derived Units:

Physical Quantity	Name of SI Unit	Symbol for SI Unit	Definition of Unit
force	newton	N	$kg\ m\ s^{-2}$
pressure	pascal	Pa	$kg\ m^{-1}s^{-2}(=N\ m^{-2})$
energy	joule	J	$kg\ m^2s^{-2}$
power	watt	W	$kg\ m^2s^{-3}(=J\ s^{-1})$
frequency	hertz	Hz	s^{-1}(cycles per second)

Decimal Fractions and Multiples of SI Units Having Special Names:

Physical Quantity	Name of Unit	Symbol for Unit	Definition of Unit
length	ångstrom	Å	$10^{-10}\ m = 10^{-8}\ cm$
length	micron	μm	$10^{-6}\ m$
area	hectare	ha	$10^4\ m^2$
force	dyne	dyn	$10^{-5}\ N$
pressure	bar	bar	$10^5\ N\ m^{-2} = 10^5\ Pa$
pressure	millibar	mb	$10^2\ N\ m^{-2} = 1\ hPa$
weight	tonne	t	$10^3\ kg$

Non-SI Units:

°C	degrees Celsius (0 °C = 273 K approximately)
	Temperature differences are also given in °C (=K)
	rather than the more correct form of "Celsius degrees".
ppmv	parts per million (10^6) by volume
ppbv	parts per billion (10^9) by volume
pptv	parts per trillion (10^{12}) by volume
bp	(years) before present
kpb	thousands of years before present
mbp	millions of years before present

The units of mass adopted in this report are generally those which have come into common usage, and have deliberately not been harmonised, e.g.,

kt	kilotonnes
GtC	gigatonnes of carbon (1 GtC = 3.7 Gt carbon dioxide)
PgC	petagrams of carbon (1PgC = 1 GtC)
MtN	megatonnes of nitrogen
TgC	teragrams of carbon (1TgC = 1 MtC)
TgN	teragrams of nitrogen
TgS	teragrams of sulphur